Theory of Transformation Groups I

Sophus Lie

Theory of Transformation Groups I

General Properties of Continuous Transformation Groups. A Contemporary Approach and Translation

With the collaboration of Friedrich Engel

Editor and Translator: Joël Merker

Springer

Author
Sophus Lie (1842–1899)

Collaborator
Friedrich Engel (1861–1941)

Editor and Translator
Joël Merker
Laboratoire de Mathématiques, Faculté des Sciences d'Orsay
Université Paris-Sud 11
Orsay
France

First edition published in German language by B.G. Teubner, Leipzig, with the title: Theorie der Transformationsgruppen in 1888, 1930

ISBN 978-3-662-46210-2 ISBN 978-3-662-46211-9 (eBook)
DOI 10.1007/978-3-662-46211-9

Library of Congress Control Number: 2015931246

Mathematics Subject Classification: 22E05, 22E10, 22E60, 22-03, 01A05, 01A55, 17B30, 17B40, 17B45, 17B56, 17B66, 17B70, 22F30, 12H05, 14P05, 14P15

Springer Heidelberg New York Dordrecht London

Cover picture taken from "Ny illustreret Tidende", 1886

Printed on acid-free paper

Springer-Verlag GmbH Berlin Heidelberg is part of Springer Science+Business Media
(www.springer.com)

THEORIE

DER

TRANSFORMATIONSGRUPPEN

ERSTER ABSCHNITT

UNTER MITWIRKUNG VON Prof. Dr. FRIEDRICH ENGEL

BEARBEITET VON

SOPHUS LIE,

WEIL. PROFESSOR DER GEOMETRIE AND DER UNIVERSITÄT LEIPZIG
UND PROFESSOR I TRANSFORMASJONSGRUPPENES TEORI AN DER
KÖNIGLICHEN FREDERIKS UNIVERSITÄT ZU OSLO

UNVERÄNDERTER NEUDRUCK
MIT UNTERSTÜTZUNG DER KÖNIGLICHEN
FREDERIKS UNIVERSITÄT ZU OSLO

1930

**VERLAG UND DRUCK VON B.G. TEUBNER IN LEIPZIG
UND BERLIN**

Foreword

This modernized English translation grew out of my old simultaneous interest in the mathematics itself and in the metaphysical thoughts governing its continued development. I owe to the books of Robert Hermann, Peter Olver, Thomas Hawkins, and Olle Stormark my introduction to Lie's original vast field.

Up to the end of the 18th Century, the universal language of Science was Latin, until its centre of gravity shifted to German during the 19th Century, while nowadays — needless to say — English is widespread. Being intuitively convinced that Lie's original works contain much more than what has been modernized up to now, three years ago I started to learn German *from scratch* just in order to read Lie, with two main goals in mind:

□ to complete and modernize the Lie-Amaldi classification of finite-dimensional local Lie group holomorphic actions on spaces of complex dimensions 1, 2 and 3 for various applications in complex and Cauchy-Riemann geometry;

□ to better understand the roots of Élie Cartan's achievements.

Then it gradually appeared to me that *Lie's mathematical thought is universal and transhistorical*, hence it deserves *per se* to be translated. The present adapted English translation follows an earlier monograph[1] written in French and specially devoted to Engel and Lie's treatment of the so-called *Riemann-Helmholtz problem* in Volume III of the *Theorie der Transformationsgruppen*.

A few observations are in order concerning the chosen format. For several reasons, it was essentially impossible to directly translate the first few chapters in which Lie's intention was to set up the beginnings of the theory in the highest possible generality, especially in order to eliminate the axiom of inverse, an aspect never dealt with in modern treatises. As a result, I decided in the first four chapters to reorganize the material and to reprove the relevant statements, nevertheless retaining all of the embraced mathematical content. But starting with Chap. 5, Engel and Lie's exposition is so smooth, so rigorous, so understandable, so systematic, so astonishingly well organized — *so beautiful for thought* — that a pure translation is essential.

[1] Merker, J.: Sophus Lie, Friedrich Engel et le problème de Riemann-Helmholtz, Hermann Éditeur des Sciences et des Arts, Paris, xxiii+325 pp, 2010.

Lastly, the author is grateful to Gautam Bharali, Philip Boalch, Egmont Porten, and Masoud Sabzevari for a few fine suggestions concerning the language and for misprint chasing, but is of course solely responsible for the lack of idiomatic English.

Paris, École Normale Supérieure, *Joël Merker*
16 March 2010

Contents

Part I Modern Presentation

Part I
Modern Presentation

Chapter 1
Three Principles of Thought Governing the Theory of Lie

Let $x = (x_1, \dots, x_n)$ be coordinates on an n-dimensional real or complex euclidean space $\mathbb{C}^n$ or $\mathbb{R}^n$, considered as a source domain. The archetypal objects of Lie's Theory of Continuous Transformation Groups are *point transformation equations*:

$$x_i' = f_i(x_1, \dots x_n; a_1, \dots, a_r) \qquad {\scriptstyle (i=1 \cdots n)},$$

parametrized by a finite number r of real or complex parameters $(a_1, \dots, a_r)$, namely each map $x' = f(x; a) =: f_a(x)$ is assumed to constitute a diffeomorphism from some domain[1] (Translator's note: By a *domain*, we will always mean a *connected*, nonempty open set.) in the source space into some domain in a target space of the same dimension equipped with coordinates $(x_1', \dots, x_n')$. Thus, the functional determinant:

$$\det \operatorname{Jac}(f) = \begin{vmatrix} \frac{\partial f_1}{\partial x_1} & \cdots\cdots & \frac{\partial f_1}{\partial x_n} \\ \vdots & \ddots & \vdots \\ \frac{\partial f_n}{\partial x_1} & \cdots\cdots & \frac{\partial f_n}{\partial x_n} \end{vmatrix} = \sum_{\sigma \in \mathsf{Perm}_n} \operatorname{sign}(\sigma) \frac{\partial f_1}{\partial x_{\sigma(1)}} \frac{\partial f_2}{\partial x_{\sigma(2)}} \cdots \frac{\partial f_n}{\partial x_{\sigma(n)}}$$

does not vanish at any point of the source domain.

§ **15**. ([1], pp. 25–26) [The concepts of transformation $x' = f(x)$ and of transformation equations $x' = f(x; a)$ are of a purely analytic nature.] However, these concepts are given a graphical interpretation [ANSCHAULICH AUFFASUNG] when the concept of an n-times extended space [RAUM] is introduced.

If we interpret $x_1, \dots, x_n$ as the coordinates of the points [PUNKTCOORDINATEN] of such a space, then a transformation $x_i' = f_i(x_1, \dots, x_n)$ appears as a point transformation [PUNKTTRANSFORMATION]; consequently, this transformation can be interpreted as an *operation* such that every point x_i is transferred at the same time into the new position x_i'. One expresses this as follows: the

S. Lie, *Theory of Transformation Groups I*,
DOI 10.1007/978-3-662-46211-9_1

transformation in question is an operation via which the points of the space $x_1,\dots,x_n$ are permuted with each other.

Before introducing the continuous group axioms in Chap. 3 below, the very first question to be settled is: how many different transformations $x_i' = f_i(x;a)$ correspond to the ∞^r different systems of values $(a_1,\dots,a_r)$? Some parameters might indeed be superfluous, hence they should be removed from the outset, as will be achieved in Chap. 2. For this purpose, it is crucial to formulate explicitly, once and for all, *three principles of thought concerning the admission of hypotheses that hold throughout the theory of continuous groups developed by Lie*.

General Assumption of Analyticity

Curves, surfaces, manifolds, groups, subgroups, coefficients of infinitesimal transformations, etc., all mathematical objects of the theory will be assumed to be *analytic*, i.e. their representing functions will be assumed to be locally expandable in convergent, univalent power series defined in a certain domain of an appropriate $\mathbb{R}^m$.

Principle of Free Generic Relocalization

Consider a local mathematical object which is represented by functions that are analytic in some domain U_1, and suppose that a certain "generic" nice behavior holds on $U_1\backslash \mathsf{D}_1$ outside a certain proper closed analytic subset $\mathsf{D}_1 \subset U_1$; for instance: the invertibility of a square matrix composed of analytic functions holds outside the zero-locus of its determinant. Then relocalize the considerations in some subdomain $U_2 \subset U_1\backslash \mathsf{D}_1$.

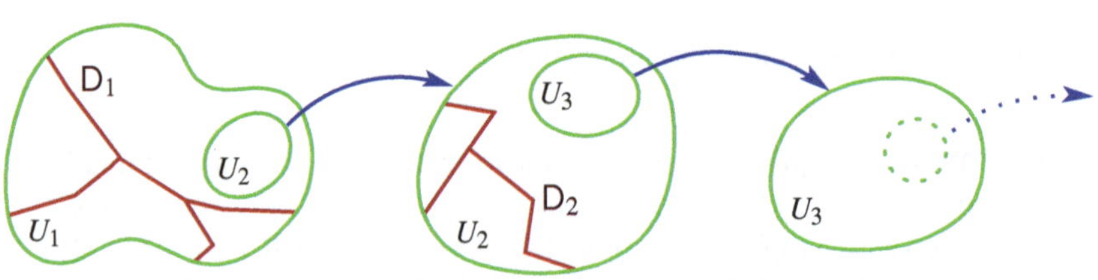

Fig. 1.1 Relocalizing finitely many times in neighborhoods of generic points

Then, in U_2, further reasoning may necessitate avoiding another proper closed analytic subset D_2, hence to relocalize the considerations in some subdomain $U_3 \subset U_2\backslash \mathsf{D}_2$, and so on. Most proofs of the *Theorie der Transformationsgruppen*, and especially the classification theorems, allow such relocalizations a great number of times, often without mention, such an *act of thought* being considered as implicitly clear, and free relocalization being justified by the necessity of *initially studying generic objects.*

Giving no Name to Domains or Neighborhoods

Without providing a systematic notation, Lie and Engel commonly wrote *the* neighborhood [*der* UMGEBUNG] (of a point), similarly as one speaks of *the* neighborhood of a house, or of *the* surroundings of a town, whereas contemporary topology conceptualizes *a* (say, sufficiently small) given neighborhood amongst an infinity. Contrary to what the formalistic, twentieth-century mythology sometimes says, Lie and Engel did emphasize the local nature of the concept of transformation group in terms of narrowing down neighborhoods; we shall illustrate this especially when presenting Lie's attempt to economize the axiom of inverse. Certainly, it is true that most of Lie's results are stated without specifying domains of existence, but in fact, it is quite plausible that Lie soon realized that giving no name to neighborhoods, and avoiding superfluous denotation, is efficient and expeditious in order to perform far-reaching classification theorems.

Therefore, adopting the economical style of thought in Engel-Lie's treatise, our "modernization-translation" of the theory will, while nevertheless providing frequent reminders, presuppose that:

- mathematical objects are analytic;
- relocalization is freely allowed;
- open sets are often small, usually unnamed, and always *connected.*

Introduction ([1], pp. 1–8)

If the variables $x_1',\dots,x_n'$ are determined as functions of $x_1,\dots,x_n$ by n equations, solvable with respect to $x_1,\dots,x_n$:

$$x_i' = f_i(x_1,\dots,x_n) \qquad (i=1\cdots n),$$

then one says that these equations represent a transformation [TRANSFORMATION] between the variables x and x'. In the sequel, we will have to deal with such transformations; unless the contrary is expressly mentioned, we will restrict ourselves to the case where the f_i are *analytic* [ANALYTISCH] *functions* of their arguments. However, because a not negligible portion of our results are independent of this assumption, we will occasionally indicate how various developments take shape by taking into consideration functions of this sort.

When the functions $f_i(x_1,\dots,x_n)$ are analytic and are defined inside a common region [BEREICH], then according to the known studies of CAUCHY,

WEIERSTRASS, BRIOT and BOUQUET, one can always delimit, in the manifold of all real and complex systems of values $x_1,\dots,x_n$, a region (x) such that all functions f_i are univalent in the complete extension [AUSDEHNUNG] of this region, and so that, in the neighborhood [UMGEBUNG] of every system of values $x_1^0,\dots,x_n^0$ belonging to the region (x), the functions behave regularly [REGULÄR VERHALTEN], that is to say, they can be expanded in ordinary power series with respect to $x_1 - x_1^0, \dots, x_n - x_n^0$ with only whole positive powers.

For the solvability of the equations $x_i' = f_i(x)$, a unique condition is necessary and sufficient, namely the condition that the functional determinant:

$$\sum \pm \frac{\partial f_1}{\partial x_1} \cdots \frac{\partial f_n}{\partial x_n}$$

should not vanish identically. If this condition is satisfied, then the region (x) defined above can specially be defined so that the functional determinant does not take the value zero for any system of values in (x). Under this assumption, if one lets the x gradually take all systems of values in the region (x), then the equations $x_i' = f_i(x)$ determine, in the domain [GEBIETE] of the x', a region of such a nature that $x_1,\dots,x_n$, in the neighborhood of every system of values ${x_1'}^0,\dots,{x_n'}^0$ in this new region, behave regularly as functions of $x_1',\dots,x_n'$, and hence can be expanded as ordinary power series of $x_1' - {x_1'}^0, \dots, x_n' - {x_n'}^0$. It is well known that from this, it does not follow that the x_i are univalent functions of $x_1',\dots,x_n'$ in the complete extension of the new region; but when necessary, it is possible to narrow down the region (x) defined above so that two different systems of values $x_1,\dots,x_n$ of the region (x) always produce two, also different, systems of values $x_1' = f_1(x), \dots, x_n' = f_n(x)$.

Thus, the equations $x_i' = f_i(x)$ establish a univalent invertible relationship [BEZIEHUNG] between regions in the domain of the x and regions in the domain of the x'; to every system of values in one region, they associate one and only one system of values in the other region, and conversely.

If the equations $x_i' = f_i(x)$ are solved with respect to the x, then in turn, the resulting equations:

$$x_k = F_k(x_1',\dots,x_n') \qquad (k=1\cdots n)$$

again represent a transformation. The relationship between this transformation and the initial one is evidently a reciprocal relationship; accordingly, one says: the two transformations are *inverse* to one another. From this definition, it visibly follows:

If one first executes the transformation:

$$x_i' = f_i(x_1,\dots,x_n) \qquad (i=1\cdots n)$$

and then the transformation inverse to it:

$$x_i'' = F_i(x_1', \dots, x_n') \qquad (i=1\cdots n),$$

then one obtains the identity transformation:

$$x_i'' = x_i \qquad (i=1\cdots n).$$

Here lies the real definition of the concept [BEGRIFF] of two transformations inverse to each other.

In general, if one executes two arbitrary transformations:

$$x_i' = f_i(x_1, \dots, x_n), \qquad x_i'' = g_i(x_1', \dots, x_n') \qquad (i=1\cdots n)$$

one after the other, then one obtains a new transformation, namely the following one:

$$x_i'' = g_i\big(f_1(x), \dots, f_n(x)\big) \qquad (i=1\cdots n).$$

In general, this new transformation naturally changes when one changes the order [REIHENFOLGE] of the two transformations; however, it can also happen that the order of the two transformations is indifferent. This case occurs when one has identically:

$$g_i\big(f_1(x), \dots, f_n(x)\big) \equiv f_i\big(g_1(x), \dots, g_n(x)\big) \qquad (i=1\cdots n);$$

we then say, as in the process of the Theory of Substitutions [VORGANG DER SUBSTITUTIONENTHEORIE]: *the two transformations:*

$$x_i' = f_i(x_1, \dots, x_n) \qquad (i=1\cdots n)$$

and:

$$x_i' = g_i(x_1, \dots, x_n) \qquad (i=1\cdots n)$$

are interchangeable [VERTAUSCHBAR] *with one another.* —

A finite or infinite family [SCHAAR] *of transformations between the x and the x' is called a group of transformations or a transformation group when any two transformations of the family executed one after the other give a transformation which again belongs to the family.*[1]

A transformation group is called *discontinuous* when it consists of a discrete number of transformations, and this number can be finite or infinite. Two transformations of such a group are finitely different from each other. The discontinuous groups belong to the domain of the *Theory of Substitutions*, so in the sequel, they will remain out of consideration.

[1] Sophus LIE, Gesellschaft der Wissenschaften zu Christiania 1871, p. 243. KLEIN, Vergleichende Betrachtungen über neuere geometrische Forschungen, Erlangen 1872. LIE, Göttinger Nachrichten 1873, 3. Decemb.

The discontinuous groups stand in opposition to the *continuous* transformation groups, which always contain infinitely many transformations. A transformation group is called *continuous* when it is possible, for every transformation belonging to the group, to indicate certain other transformations which differ only infinitely little from the transformation in question, and when by contrast, it is not possible to reduce the complete totality [INBEGRIFF] of transformations contained in the group to a single discrete family.

Now, amongst the continuous transformation groups, we again consider two separate categories [KATEGORIEN] which, in the nomenclature [BENENNUNG], are distinguished as *finite continuous* groups and as *infinite continuous* groups. To begin with, we can only give provisional definitions of the two categories, and these definitions will be apprehended precisely later.

A *finite continuous transformation group* will be represented by *one* system of n equations:

$$x_i' = f_i(x_1, \dots, x_n, a_1, \dots, a_r) \qquad (i=1\cdots n),$$

where the f_i denote analytic functions of the variables $x_1, \dots, x_n$ and of the arbitrary parameters $a_1, \dots, a_r$. Since we have to deal with a group, two transformations:

$$x_i' = f_i(x_1, \dots, x_n, a_1, \dots, a_r)$$
$$x_i'' = f_i(x_1', \dots, x_n', a_1, \dots, a_r),$$

when executed one after the other, must produce a transformation which belongs to the group, hence which has the form:

$$x_i'' = f_i\big(f_1(x,a), \dots, f_n(x,a), b_1, \dots, b_r\big) = f_i(x_1, \dots, x_n, c_1, \dots, c_r).$$

Here, the c_k are naturally independent of the x and so, are functions of only the a and the b.

Example. A known group of this sort is the following:

$$x' = \frac{x + a_1}{a_2 x + a_3},$$

which contains the three parameters a_1, a_2, a_3. If one executes the two transformations:

$$x' = \frac{x + a_1}{a_2 x + a_3}, \qquad x'' = \frac{x' + b_1}{b_2 x' + b_3}$$

one after the other, then one obtains:

$$x'' = \frac{x + c_1}{c_2 x + c_3},$$

where c_1, c_2, c_3 are defined as functions of the a and the b by the relations:

$$c_1 = \frac{a_1 + b_1 a_3}{1 + b_1 a_2}, \qquad c_2 = \frac{b_2 + a_2 b_3}{1 + b_1 a_2}, \qquad c_3 = \frac{b_2 a_1 + b_3 a_3}{1 + b_1 a_2}.$$

The following group with n^2 parameters a_{ik} is also well known:

$$x_i' = \sum_{k=1}^{n} a_{ik} x_k \qquad (i=1\cdots n).$$

Here, if one sets:

$$x_\nu'' = \sum_{i=1}^{n} b_{\nu i} x_i' \qquad (i=1\cdots n),$$

then we have:

$$x_\nu'' = \sum_{i,\,k}^{1\cdots n} b_{\nu i} a_{ik} x_k = \sum_{k=1}^{n} c_{\nu k} x_k,$$

where the $c_{\nu k}$ are determined by the equations:

$$c_{\nu k} = \sum_{i=1}^{n} b_{\nu i} a_{ik} \qquad (\nu, k=1\cdots n). -$$

In order to arrive at a usable definition of a finite continuous group, we first want to somehow reshape the definition of finite continuous groups. On the occasion, we shall use a proposition from the theory of differential equations which we will return to later in a more comprehensive way (cf. Chap. 10).

Let the equations:

$$x_i' = f_i(x_1, \dots, x_n, a_1, \dots, a_r) \qquad (i=1\cdots n)$$

represent an arbitrary continuous group. According to the proposition in question, it is then possible to define the functions f_i through a system of differential equations, insofar as they depend upon the x. To this end, one only has to differentiate the equations $x_i' = f_i(x,a)$ with respect to $x_1, \dots, x_n$ sufficiently often and then to set up all equations that may be obtained by elimination of $a_1, \dots, a_r$. If one has gone sufficiently far by differentiation, then by elimination of the a, one obtains a system of differential equations for $x_1', \dots, x_n'$, whose most general system of solutions is represented by the initial equations $x_i' = f_i(x,a)$ with the r arbitrary parameters. Now, since by assumption the equations $x_i' = f_i(x,a)$ define a group, it follows that the concerned system of differential equations possesses the following remarkable property: if $x_i' = f_i(x_1, \dots, x_n, b_1, \dots, b_r)$ is a system of solutions of it, and if $x_i' = f_i(x,a)$ is a second system of solutions, then:

$$x_i' = f_i\big(f_1(x,a), \dots, f_n(x,a), b_1, \dots, b_r\big) \qquad (i=1\cdots n)$$

is also a system of solutions.

From this, we see that the equations of an arbitrary finite continuous transformation group can be defined by a system of differential equations which possesses certain specific properties. Firstly, from two systems of solutions of the concerned differential equations one can always derive, in the way indicated above, a third system of solutions: it is precisely in this that we have to deal with a group. Secondly, the most general system of solutions of the concerned differential equations depends only upon a finite number of arbitrary constants: this circumstance expresses that our group is finite.

Now, we assume that there is a family of transformations $x_i' = f_i(x_1,\dots,x_n)$ which is defined by a system of differential equations of the form:

$$W_k\left(x_1',\dots,x_n',\frac{\partial x_1'}{\partial x_1},\dots,\frac{\partial^2 x_1'}{\partial x_1^2},\dots\right)=0 \qquad (k=1,2\cdots).$$

Moreover, we assume that this system of differential equations possesses the first of the two mentioned properties, but not the second; therefore, if $x_i' = f_i(x_1,\dots,x_n)$ and $x_i' = g_i(x_1,\dots,x_n)$, then $x_i' = g_i\big(f_1(x),\dots,f_n(x)\big)$ is also a system of solutions of these differential equations, and the most general system of solutions of them does not only depend upon a finite number of arbitrary constants, but also upon higher sorts of elements, as for example, upon arbitrary functions. Then the totality of all transformations which satisfy the concerned differential equations evidently again forms a group, and in general, a continuous group, though no more a finite one, but one which we call *infinite continuous*.

Straightaway, we give a few simple examples of infinite continuous transformation groups.

When the differential equations which define the concerned infinite group reduce to the identity equation $0 = 0$, then the transformations of the group read:

$$x_i' = \Pi_i(x_1,\dots,x_n) \qquad (i=1\cdots n)$$

where the Π_i denote arbitrary analytic functions of their arguments.

The equations:

$$\frac{\partial x_i'}{\partial x_k}=0 \qquad (i\neq k;\ i,k=1\cdots n)$$

also define an infinite group, namely the following one:

$$x_i' = \Pi_i(x_i) \qquad (i=1\cdots n),$$

where again the Π_i are absolutely arbitrary.

Furthermore, it is to be observed that the concept of an infinite continuous group can be understood more generally than what has been said here. Actually, one could call infinite continuous any continuous group which is not finite. However, this definition does not coincide with the one given above.

For instance, the equations:

$$x' = F(x), \qquad y' = F(y)$$

in which F, for the two cases, denotes the same function of its arguments, represent a group. This group is continuous, since all its transformations are represented by a single system of equations; in addition, it is obviously not finite. Consequently, it would be an infinite continuous group if we interpreted this concept in the more general sense indicated above. But on the other hand, it is not possible to define the family of the transformations:

$$x' = F(x), \qquad y' = F(y)$$

by differential equations that are free of arbitrary elements. Consequently, the definition stated first for an infinite continuous group does not apply to this case. Nevertheless, we find it suitable to consider only the infinite continuous groups which can be defined by differential equations and hence, we always set as fundamental our first, tight definition.

We wish to emphasize that the concept of "transformation group" is still not exhausted by the difference between discontinuous and continuous groups. Rather, there are transformation groups which are subordinate to neither of these two classes but which have something in common with each of them. In the sequel, we must at least occasionally treat this sort of group. Provisionally, two examples will suffice.

The totality of all coordinate transformations of a plane by which one transfers an ordinary right-angled system of coordinates to another forms a group which is neither continuous nor discontinuous. Indeed, the group in question contains two separate categories of transformations between which a continuous transition is not possible: firstly, the transformations by which the old and the new systems of coordinates are congruent, and secondly, the transformations by which these two systems are not congruent.

The first transformations have the form:

$$x' - a = x \cos\alpha - y \sin\alpha, \qquad y' - b = x \sin\alpha + y \cos\alpha,$$

while the analytic expression of the second transformations reads:

$$x' - a = x \cos\alpha + y \sin\alpha, \qquad y' - b = x \sin\alpha - y \cos\alpha.$$

Each of these systems of equations represents a continuous family of transformations, hence the group is not discontinuous; but it is also not continuous, because both systems of equations taken together provide all transformations of the group; thus, the transformations of the group decompose into two discrete families. If one imagines the x, y plane in ordinary space and if one adds

z as a third right-angled coordinate, then one can imagine the totality of coordinate transformations of the plane $z = 0$ is a totality of certain movements [BEWEGUNGEN] of the space, namely the movements for which the plane $z = 0$ keeps its position. Correspondingly, these movements separate into two classes, namely the class which only translates the plane into itself and the class which rotates the plane.

As a second example of such a group, one can take the totality of all projective and dualistic transformations of the plane.

According to these general remarks on the concept of a transformation group, we turn ourselves to the consideration of the finite continuous transformation groups which constitute the object of the following studies. These studies are divided into three volumes [ABSCHNITTE]. The *first* volume treats finite continuous groups in general. The *second* volume treats the finite continuous groups whose transformations are so-called *contact transformations* [BERÜHRUNGSTRANSFORMATIONEN]. Lastly, in the *third* volume, certain general problems of group theory will be carried out in great detail for a small number of variables.

References

1. Engel, F., Lie, S.: Theorie der Transformationsgruppen. Erster Abschnitt. Unter Mitwirkung von Prof. Dr. Friedrich Engel, bearbeitet von Sophus Lie, Verlag und Druck von B.G. Teubner, Leipzig und Berlin, xii+638 pp. (1888). Reprinted by Chelsea Publishing Co., New York, N.Y. (1970)

Chapter 2
Local Transformation Equations and Essential Parameters

Abstract Let $\mathbb{K} = \mathbb{R}$ or $\mathbb{C}$, throughout. As said in Chap. 1, transformation equations $x_i' = f_i(x; a_1, \dots, a_r)$, $i = 1, \dots, n$, which are local, analytic diffeomorphisms of $\mathbb{K}^n$ parametrized by a finite number r of real or complex numbers $a_1, \dots, a_r$, constitute the archetypal objects of Lie's theory. The preliminary question is to decide whether the f_i really depend upon *all* parameters, and also, to get rid of superfluous parameters, if there are any.

Locally in a neighborhood of a fixed x_0, one expands $f_i(x;a) = \sum_{\alpha \in \mathbb{N}^n} \mathscr{U}_\alpha^i(a)\,(x - x_0)^\alpha$ in power series and one looks at the *infinite coefficient mapping* $\mathsf{U}_\infty : a \longmapsto \big(\mathscr{U}_\alpha^i(a)\big)_{\alpha \in \mathbb{N}^n}^{1 \leqslant i \leqslant n}$ from $\mathbb{K}^r$ to $\mathbb{K}^\infty$, which is expected to faithfully describe the dependence with respect to a in question. If ρ_∞ denotes the maximal, generic and locally constant rank of this map, with of course $0 \leqslant \rho_\infty \leqslant r$, then the answer says that locally in a neighborhood of a generic a_0, there exist both a local change of parameters $a \mapsto \big(\mathsf{u}_1(a), \dots, \mathsf{u}_{\rho_\infty}(a)\big) =: \mathsf{u}$ decreasing the number of parameters from r down to ρ_∞, and new transformation equations:

$$x_i' = g_i\big(x; \mathsf{u}_1, \dots, \mathsf{u}_{\rho_\infty}\big) \qquad (i=1 \cdots n)$$

depending *only* upon ρ_∞ parameters which give again the old ones:

$$g_i\big(x; \mathsf{u}(a)\big) \equiv f_i(x;a) \qquad (i=1 \cdots n).$$

At the end of this brief chapter, before giving a precise introduction to the local Lie group axioms, we present an example due to Engel which shows that the axiom of inverse cannot be deduced from the axiom of composition, contrary to one of Lie's *Idées fixes*.

2.1 Generic Rank of the Infinite Coefficient Mapping

Thus, we consider local transformation equations:

S. Lie, *Theory of Transformation Groups I*,
DOI 10.1007/978-3-662-46211-9_2

$$x'_i = f_i(x_1, \ldots, x_n; a_1, \ldots, a_r) \qquad (i=1\cdots n).$$

We want to illustrate how the principle of free generic relocalization described above on p. 4 helps to get rid of superfluous parameters a_k. We assume that the f_i are defined and analytic for x belonging to a certain (unnamed, connected) domain of $\mathbb{K}^n$ and for a belonging to some domain of $\mathbb{K}^r$.

Expanding the f_i of $x'_i = f_i(x;a)$ in power series with respect to $x - x_0$ in some neighborhood of a point x_0:

$$f_i(x;a) = \sum_{\alpha \in \mathbb{N}^n} \mathcal{U}^i_\alpha(a)\,(x-x_0)^\alpha,$$

we get an infinite number of analytic functions $\mathcal{U}^i_\alpha = \mathcal{U}^i_\alpha(a)$ of the parameters that are defined in some uniform domain of $\mathbb{K}^r$. Intuitively, this infinite collection of coefficient functions $\mathcal{U}^i_\alpha(a)$ should show how $f(x;a)$ depends on a.

To make this claim precise, we thus consider the map:

$$\mathsf{U}_\infty : \quad \mathbb{K}^r \ni a \longmapsto \big(\mathcal{U}^i_\alpha(a)\big)^{1 \leqslant i \leqslant n}_{\alpha \in \mathbb{N}^n} \in \mathbb{K}^\infty.$$

For the convenience of applying standard differential calculus in finite dimensions, we simultaneously consider all of its κ-th truncations:

$$\mathsf{U}_\kappa : \quad \mathbb{K}^r \ni a \longmapsto \big(\mathcal{U}^i_\alpha(a)\big)^{1 \leqslant i \leqslant n}_{|\alpha| \leqslant \kappa} \in \mathbb{K}^{n\frac{(n+\kappa)!}{n!\,\kappa!}},$$

where $\frac{(n+\kappa)!}{n!\,\kappa!}$ is the number of multiindices $\alpha \in \mathbb{N}^n$ whose length $|\alpha| := \alpha_1 + \cdots + \alpha_n$ satisfies the upper bound $|\alpha| \leqslant \kappa$. We call U_κ, U_∞ the *(in)finite coefficient mapping(s)* of $x'_i = f_i(x; a)$.

The *Jacobian matrix* of U_κ is the $r \times \big(n\frac{(n+\kappa)!}{n!\,\kappa!}\big)$ matrix:

$$\Big(\frac{\partial \mathcal{U}^i_\alpha}{\partial a_j}(a)\Big)^{|\alpha| \leqslant \kappa,\, 1 \leqslant i \leqslant n}_{1 \leqslant j \leqslant r},$$

its r rows being indexed by the partial derivatives. The *generic rank* of U_κ is the largest integer $\rho_\kappa \leqslant r$ such that there is a $\rho_\kappa \times \rho_\kappa$ minor of $\operatorname{Jac} \mathsf{U}_\kappa$ which does not vanish identically, but all $(\rho_\kappa + 1) \times (\rho_\kappa + 1)$ minors do vanish identically. The uniqueness principle for analytic functions then insures that the common zero-set of all $\rho_\kappa \times \rho_\kappa$ minors is a *proper* closed analytic subset D_κ (of the unnamed domain where the $\mathcal{U}^i_\alpha$ are defined), so it is stratified by a finite number of submanifolds of codimension $\geqslant 1$ ([8, 2, 3, 5]), and in particular, it has empty interior, hence it is intuitively “thin”.

So the set of parameters a at which there is a least one $\rho_\kappa \times \rho_\kappa$ minor of $\operatorname{Jac} \mathsf{U}_\kappa$ which does not vanish is open and *dense*. Consequently, “for a generic point a”, the map U_κ is of rank $\geqslant \rho_\kappa$ at every point a' sufficiently close to a (since the corresponding $\rho_\kappa \times \rho_\kappa$ minor does not vanish in a neighborhood of a), and because all $(\rho_\kappa + 1) \times (\rho_\kappa + 1)$ minors of $\operatorname{Jac} \mathsf{U}_\kappa$ were assumed to vanish identically, the map

U_κ happens to be in fact of *constant* rank U_κ in a (small) neighborhood of every such generic a.

Insuring constancy of a rank is one important instance of why free relocalization is useful: a majority of theorems of the differential calculus and of the classical theory of (partial) differential equations hold under specific local constancy assumptions.

As κ increases, the number of columns of $\mathrm{Jac}\, U_\kappa$ increases, hence $\rho_{\kappa_1} \leqslant \rho_{\kappa_2}$ for $\kappa_1 \leqslant \kappa_2$. Since $\rho_\kappa \leqslant r$ is bounded, the generic rank of U_κ becomes constant for all $\kappa \geqslant \kappa_0$ bigger than some sufficiently large κ_0. Thus, let $\rho_\infty \leqslant r$ denote this maximal possible generic rank.

Definition 2.1. The parameters $(a_1, \dots, a_r)$ of given point transformation equations $x_i' = f_i(x; a)$ are called *essential* if, after expanding $f_i(x;a) = \sum_{\alpha \in \mathbb{N}^n} \mathscr{U}_\alpha^i(a)\,(x - x_0)^\alpha$ in power series at some x_0, the generic rank ρ_∞ of the coefficient mapping $a \longmapsto \big(\mathscr{U}_\alpha^i(a)\big)_{\alpha \in \mathbb{N}^n}^{1 \leqslant i \leqslant n}$ is maximal, equal to the number r of parameters: $\rho_\infty = r$.

Without entering into technical details, we make a remark. It is a consequence of the principle of analytic continuation and of some reasonings with power series that the *same* maximal rank ρ_∞ is enjoyed by the coefficient mapping $a \mapsto \big({\mathscr{U}'}_\alpha^i(a)\big)_{\alpha \in \mathbb{N}^n}^{1 \leqslant i \leqslant n}$ for the expansion of $f_i(x;a) = \sum_{\alpha \in \mathbb{N}^n} {\mathscr{U}'}_\alpha^i(a)\,(x - x_0')^\alpha$ at another, arbitrary point x_0'. Also, one can prove that ρ_∞ is independent of the choice of coordinates x_i and of parameters a_k. These two facts will not be needed, and the interested reader is referred to [9] for proofs of quite similar statements holding true in the context of *Cauchy-Riemann geometry*.

2.2 Quantitative Criterion for the Number of Superfluous Parameters

It is not very practical to compute the generic rank of the infinite Jacobian matrix $\mathrm{Jac}\,\mathsf{U}_\infty$. To check essentiality of parameters in concrete situations, a helpful criterion due to Lie is **(iii)** below.

Theorem 2.1. *The following three conditions are equivalent:*

(i) *In the transformation equations*

$$x_i' = f_i(x_1, \dots, x_n; a_1, \dots, a_r) = \sum_{\alpha \in \mathbb{N}^n} \mathscr{U}_\alpha^i(a)\,(x - x_0)^\alpha \qquad (i=1\cdots n),$$

the parameters $a_1, \dots, a_r$ are not *essential.*

(ii) *(By definition) The generic rank ρ_∞ of the infinite Jacobian matrix:*

$$\mathrm{Jac}\,\mathsf{U}_\infty(a) = \Big(\frac{\partial \mathscr{U}_\alpha^i}{\partial a_j}(a)\Big)_{1 \leqslant j \leqslant r}^{\alpha \in \mathbb{N}^n,\, 1 \leqslant i \leqslant n}$$

is strictly less than r.

(iii) *Locally in a neighborhood of* every (x_0, a_0), *there exists a not identically zero analytic vector field on the parameter space:*

$$\mathscr{T} = \sum_{k=1}^{n} \tau_k(a) \frac{\partial}{\partial a_k}$$

which annihilates all the $f_i(x;a)$*:*

$$0 \equiv \mathscr{T} f_i = \sum_{k=1}^{n} \tau_k \frac{\partial f_i}{\partial a_k} = \sum_{\alpha \in \mathbb{N}^n} \sum_{k=1}^{r} \tau_k(a) \frac{\partial \mathscr{U}_\alpha^i}{\partial a_k}(a) \, (x - x_0)^\alpha \qquad (i = 1 \cdots n).$$

More generally, if ρ_∞ *denotes the generic rank of the infinite coefficient mapping:*

$$\mathsf{U}_\infty : \qquad a \longmapsto \left(\mathscr{U}_\alpha^i(a)\right)_{\alpha \in \mathbb{N}^n}^{1 \leqslant i \leqslant n},$$

then locally in a neighborhood of every (x_0, a_0), *there exist exactly* $r - \rho_\infty$, *and no more, analytic vector fields:*

$$\mathscr{T}_\mu = \sum_{k=1}^{n} \tau_{\mu k}(a) \frac{\partial}{\partial a_k} \qquad (\mu = 1 \cdots r - \rho_\infty),$$

with the property that the dimension of $\mathrm{Span}\big(\mathscr{T}_1\big|_a, \dots, \mathscr{T}_{r-\rho_\infty}\big|_a\big)$ *is equal to* $r - \rho_\infty$ *at every parameter a at which the rank of* U_∞ *is maximal, equal to* ρ_∞, *such that the derivations* $\mathscr{T}_\mu$ *all annihilate the* $f_i(x;a)$*:*

$$0 \equiv \mathscr{T}_\mu f_i = \sum_{k=1}^{r} \tau_{\mu k}(a) \frac{\partial f_i}{\partial a_k}(x;a) \qquad (i = 1 \cdots n;\ \mu = 1 \cdots r - \rho_\infty).$$

Proof. Just by the chosen definition, we have **(i)** $\Longleftrightarrow$ **(ii)**. Next, suppose that condition **(iii)** holds, in which the coefficients $\tau_k(a)$ of the concerned nonzero derivation $\mathscr{T}$ are locally defined. Recalling that the Jacobian matrix $\mathrm{Jac}\,\mathsf{U}_\infty$ has r rows and an infinite number of columns, we then see that the n annihilation equations $0 \equiv \mathscr{T} f_i$, when rewritten in matrix form as:

$$0 \equiv \big(\tau_1(a), \dots, \tau_r(a)\big) \left(\frac{\partial \mathscr{U}_\alpha^i}{\partial a_j}(a)\right)_{1 \leqslant j \leqslant r}^{\alpha \in \mathbb{N}^n, 1 \leqslant i \leqslant n}$$

just say that the transpose of $\mathrm{Jac}\,\mathsf{U}_\infty(a)$ has nonzero kernel at each a where the vector $\mathscr{T}\big|_a = \big(\tau_1(a), \dots, \tau_r(a)\big)$ is nonzero. Consequently, $\mathrm{Jac}\,\mathsf{U}_\infty$ has rank strictly less than r locally in a neighborhood of every a_0, hence in the whole a-domain. So **(iii)** $\Rightarrow$ **(ii)**.

Conversely, assume that the generic rank ρ_∞ of $\mathrm{Jac}\,\mathsf{U}_\infty$ is $< r$. Then there exist $\rho_\infty < r$ "basic" coefficient functions $\mathscr{U}_{\alpha(1)}^{i(1)}, \dots, \mathscr{U}_{\alpha(\rho_\infty)}^{i(\rho_\infty)}$ (there can be several choices) such that the generic rank of the extracted map $a \mapsto \big(\mathscr{U}_{\alpha(l)}^{i(l)}\big)_{1 \leqslant l \leqslant \rho_\infty}$ equals ρ_∞ already. We abbreviate:

$$\mathsf{u}_l(a) := \mathscr{U}_{\alpha(l)}^{i(l)}(a) \qquad {\scriptstyle (l=1\cdots\rho_\infty)}.$$

The goal is to find vectorial local analytic solutions $\big(\tau_1(a),\dots,\tau_r(a)\big)$ to the infinite number of linear equations:

$$0 \equiv \tau_1(a)\,\frac{\partial \mathscr{U}_\alpha^i(a)}{\partial a_1}(a) + \cdots + \tau_r(a)\,\frac{\partial \mathscr{U}_\alpha^i(a)}{\partial a_r}(a) \qquad {\scriptstyle (i=1\cdots n;\ \alpha\in\mathbb{N}^n)}.$$

To begin with, we look for solutions of the finite, extracted linear system of ρ_∞ equations with the r unknowns $\tau_k(a)$:

$$\begin{cases} 0 \equiv \tau_1(a)\dfrac{\partial \mathsf{u}_1}{\partial a_1}(a) + \cdots + \tau_{\rho_\infty}(a)\dfrac{\partial \mathsf{u}_1}{\partial a_{\rho_\infty}}(a) + \cdots + \tau_r(a)\dfrac{\partial \mathsf{u}_1}{\partial a_r}(a) \\ \cdots\cdots\cdots\cdots\cdots\cdots\cdots\cdots\cdots\cdots\cdots\cdots\cdots\cdots\cdots\cdots \\ 0 \equiv \tau_1(a)\,\dfrac{\partial \mathsf{u}_{\rho_\infty}}{\partial a_1}(a) + \cdots + \tau_{\rho_\infty}(a)\,\dfrac{\partial \mathsf{u}_{\rho_\infty}}{\partial a_{\rho_\infty}}(a) + \cdots + \tau_r(a)\,\dfrac{\partial \mathsf{u}_{\rho_\infty}}{\partial a_r}(a). \end{cases}$$

After possibly renumbering the variables $(a_1,\dots,a_r)$, we can assume that the left $\rho_\infty \times \rho_\infty$ minor of this system:

$$\Delta(a) := \det\left(\frac{\partial \mathsf{u}_l}{\partial a_m}(a)\right)_{1\leqslant m\leqslant\rho_\infty}^{1\leqslant l\leqslant\rho_\infty}$$

does not vanish identically. However, it can vanish at some points, and while endeavoring to solve the above linear system by an application of the classical Cramer rule, the necessary division by the determinant $\Delta(a)$ introduces poles that are undesirable, for we want the $\tau_k(a)$ to be analytic. So, for any μ with $1 \leqslant \mu \leqslant r-\rho_\infty$, we look for a solution (rewritten as a derivation) in the specific form:

$$\mathscr{T}_\mu := -\Delta(a)\,\frac{\partial}{\partial a_{\rho_\infty+\mu}} + \sum_{1\leqslant k\leqslant\rho_\infty} \tau_{\mu k}(a)\,\frac{\partial}{\partial a_k} \qquad {\scriptstyle (\mu=1\cdots r-\rho_\infty)},$$

in which we introduce in advance a factor $\Delta(a)$ designed to compensate the unavoidable division by $\Delta(a)$. Indeed, such a $\mathscr{T}_\mu$ will annihilate the u_l:

$$0 \equiv \mathscr{T}_\mu \mathsf{u}_1 \equiv \cdots \equiv \mathscr{T}_\mu \mathsf{u}_{\rho_\infty}$$

if and only its coefficients are solutions of the linear system:

$$\begin{cases} \Delta(a)\,\dfrac{\partial \mathsf{u}_1}{\partial a_{\rho_\infty+\mu}}(a) \equiv \tau_{\mu 1}(a)\,\dfrac{\partial \mathsf{u}_1}{\partial a_1}(a) + \cdots + \tau_{\mu\rho_\infty}(a)\,\dfrac{\partial \mathsf{u}_1}{\partial a_{\rho_\infty}}(a) \\ \cdots\cdots\cdots\cdots\cdots\cdots\cdots\cdots\cdots\cdots\cdots\cdots \\ \Delta(a)\,\dfrac{\partial \mathsf{u}_{\rho_\infty}}{\partial a_{\rho_\infty+\mu}}(a) \equiv \tau_{\mu 1}(a)\,\dfrac{\partial \mathsf{u}_{\rho_\infty}}{\partial a_1}(a) + \cdots + \tau_{\mu\rho_\infty}(a)\,\dfrac{\partial \mathsf{u}_{\rho_\infty}}{\partial a_{\rho_\infty}}(a). \end{cases}$$

Cramer's rule then yields the unique solution:

$$\tau_{\mu k}(a) = \frac{1}{\Delta(a)} \begin{vmatrix} \frac{\partial \mathsf{u}_1}{\partial a_1}(a) & \cdots & \Delta(a)\,\frac{\partial \mathsf{u}_1}{\partial a_{\rho_\infty+\mu}}(a) & \cdots & \frac{\partial \mathsf{u}_1}{\partial a_{\rho_\infty}}(a) \\ \cdots\cdots & \cdots & \cdots\cdots\cdots & \cdots & \cdots\cdots \\ \frac{\partial \mathsf{u}_{\rho_\infty}}{\partial a_1}(a) & \cdots & \Delta(a)\,\frac{\partial \mathsf{u}_{\rho_\infty}}{\partial a_{\rho_\infty+\mu}}(a) & \cdots & \frac{\partial \mathsf{u}_{\rho_\infty}}{\partial a_{\rho_\infty}}(a) \end{vmatrix} \qquad (k=1\cdots r),$$

where, as predicted, the overall factor $\Delta(a)$ of the k-th column compensates the division by the determinant $\Delta(a)$ of this system, so that the $\mathscr{T}_\mu$ indeed all have *analytic* coefficients.

Clearly, $\mathscr{T}_1, \dots, \mathscr{T}_{r-\rho_\infty}$ are linearly independent at all (generic) points a where $\Delta(a) \neq 0$. It remains to show that the $\mathscr{T}_\mu$ also annihilate all other coefficient functions $\mathscr{U}_\alpha^i$.

Thus, let $(i,\alpha) \neq (i(1),\alpha(1)), \dots, (i(\rho_\infty),\alpha(\rho_\infty))$. By the very definition of ρ_∞, the generic rank of any $r \times (1+\rho_\infty)$ extracted subJacobian matrix:

$$\begin{pmatrix} \frac{\partial \mathscr{U}_\alpha^i}{\partial a_1}(a) & \frac{\partial \mathsf{u}_1}{\partial a_1}(a) & \cdots & \frac{\partial \mathsf{u}_{\rho_\infty}}{\partial a_1}(a) \\ \cdots & \cdots & \cdots & \cdots \\ \frac{\partial \mathscr{U}_\alpha^i}{\partial a_r}(a) & \frac{\partial \mathsf{u}_1}{\partial a_r}(a) & \cdots & \frac{\partial \mathsf{u}_{\rho_\infty}}{\partial a_r}(a) \end{pmatrix}$$

must always be equal to ρ_∞, the generic rank of its last ρ_∞ columns. Consequently, in a neighborhood of every point a at which its top right $\rho_\infty \times \rho_\infty$ minor $\Delta(a)$ does not vanish, the first column is a certain linear combination:

$$\frac{\partial \mathscr{U}_i^\alpha}{\partial a_k}(a) = \lambda_1(a)\,\frac{\partial \mathsf{u}_1}{\partial a_k}(a) + \cdots + \lambda_{\rho_\infty}(a)\,\frac{\partial \mathsf{u}_{\rho_\infty}}{\partial a_k}(a) \qquad (i=1\cdots n;\ \alpha\in\mathbb{N}^n;\ k=1\cdots r)$$

of the last ρ_∞ columns in question, where, again thanks to an appropriate application of Cramer's rule, the coefficients $\lambda_l(a)$ are *analytic* in the concerned generic neighborhood, for their denominator $\Delta(a)$ is $\neq 0$ there. It then follows immediately by appropriate scalar multiplication and summation that:

$$\begin{aligned} \sum_{k=1}^r \tau_{\mu k}(a)\,\frac{\partial \mathscr{U}_\alpha^i}{\partial a_k}(a) &\equiv \lambda_1(a) \sum_{k=1}^r \tau_{\mu k}(a)\,\frac{\partial \mathsf{u}_1}{\partial a_k}(a) + \cdots + \lambda_{\rho_\infty}(a) \sum_{k=1}^r \tau_{\mu k}(a)\,\frac{\partial \mathsf{u}_{\rho_\infty}}{\partial a_k}(a) \\ &\equiv \lambda_1(a)\,\mathscr{T}_\mu\,\mathsf{u}_1 + \cdots + \lambda_{\rho_\infty}(a)\,\mathscr{T}_\mu\,\mathsf{u}_{\rho_\infty} \\ &\equiv 0 \qquad\qquad (i=1\cdots n;\ \alpha\in\mathbb{N}^n). \end{aligned}$$

But since these analytic equations hold on the dense open set where $\Delta(a) \neq 0$, we deduce by continuity that the equations:

$$0 \equiv \mathcal{T}_1 \mathcal{U}_\alpha^i \equiv \cdots \equiv \mathcal{T}_{r-\rho_\infty} \mathcal{U}_\alpha^i \qquad {\scriptstyle (i=1\cdots n;\ \alpha\in\mathbb{N}^n)}$$

do hold *everywhere*, as desired. In conclusion, we have shown the implication **(ii)** $\Rightarrow$ **(iii)**, and simultaneously, we have established the last part of the theorem. □

Corollary 2.1. *Locally in a neighborhood of every generic point a_0 at which the infinite coefficient mapping $a \mapsto \mathsf{U}_\infty(a)$ has maximal, locally constant rank equal to its generic rank ρ_∞, there exist both a local change of parameters $a \mapsto (\mathsf{u}_1(a), \dots, \mathsf{u}_{\rho_\infty}(a)) =: \mathsf{u}$ decreasing the number of parameters from r down to ρ_∞, and new transformation equations:*

$$x_i' = g_i\big(x; \mathsf{u}_1, \dots, \mathsf{u}_{\rho_\infty}\big) \qquad {\scriptstyle (i=1\cdots n)}$$

depending only *upon ρ_∞ parameters which give again the old ones:*

$$g_i\big(x; \mathsf{u}(a)\big) \equiv f_i(x;a) \qquad {\scriptstyle (i=1\cdots n)}.$$

Proof. Choose ρ_∞ coefficients $\mathcal{U}_{\alpha(l)}^{i(l)}(a) =: \mathsf{u}_l(a)$, $1 \leqslant l \leqslant \rho_\infty$, with $\Delta(a) := \det\big(\frac{\partial \mathsf{u}_l(a)}{\partial a_m}(a)\big)_{1\leqslant m\leqslant\rho_\infty}^{1\leqslant l\leqslant\rho_\infty} \not\equiv 0$ as in the proof of the theorem. Locally in some small neighborhood of any a^0 with $\Delta(a_0) \neq 0$, the infinite coefficient map U_∞ has constant rank ρ_∞, hence the constant rank theorem provides, for every (i,α), a certain function $\mathcal{V}_\alpha^i$ of ρ_∞ variables such that:

$$\mathcal{U}_\alpha^i(a) \equiv \mathcal{V}_\alpha^i\big(\mathsf{u}_1(a), \dots, \mathsf{u}_{\rho_\infty}(a)\big).$$

Thus, we can work out the power series expansion:

$$\begin{aligned} f_i(x;a) &= \textstyle\sum_{\alpha\in\mathbb{N}^n} \mathcal{U}_\alpha^i(a)\,(x-x_0)^\alpha \\ &= \textstyle\sum_{\alpha\in\mathbb{N}^n} \mathcal{V}_\alpha^i(\mathsf{u}_1(a), \dots, \mathsf{u}_{\rho_\infty}(a))\,(x-x_0)^\alpha \\ &=: g_i(x, \mathsf{u}_1(a), \dots, \mathsf{u}_{\rho_\infty}(a)) \end{aligned}$$

which yields the natural candidate for $g_i(x;\mathsf{u})$. Lastly, one may verify that any Cauchy estimate for the growth decrease of $\mathcal{U}_\alpha^i(a)$ as $|\alpha| \to \infty$ insures a similar Cauchy estimate for the growth decrease of $b \mapsto \mathcal{V}_\alpha^i(\mathsf{u})$, whence each g_i is analytic, and in fact, termwise substitution was legitimate. □

Definition 2.2. The transformation equations $x_i' = f_i(x_1, \dots, x_n; a_1, \dots, a_r)$, $i = 1, \dots, n$, are called *r-term* if all the parameters $(a_1, \dots, a_r)$ are essential.

2.3 The Axiom of Inverse and Engel's Counterexample

Every analytic diffeomorphism of an n-times extended space permutes all the points in a certain differentiable, invertible way. Although they act on a set of infinite cardinality, diffeomorphisms can thus be regarded as a kind of analog of the *substitutions* on a finite set. In fact, in the years 1873–80, Lie's *Idée fixe* was to build, in the geometric realm of n-dimensional continua, a counterpart of the Galois theory of substitutions of roots of algebraic equations ([6]).

As above, let $x' = f(x; a_1, \dots, a_r) =: f_a(x)$ be a family of (local) analytic diffeomorphisms parametrized by a finite number r of parameters. For Lie, the basic, single group axiom should just require that such a family be *closed under composition*, namely that one always has $f_a\big(f_b(x)\big) \equiv f_c(x)$ for some c depending on a and on b. More details on this definition will given in the next chapter, but at present, we ask whether one can really economize the other two group axioms: existence of an identity element and existence of inverses.

Lemma 2.1. *If H is any subset of some abstract group G with* $\mathrm{Card}\, H < \infty$ *which is closed under group multiplication:*

$$h_1 h_2 \in H \qquad \text{whenever} \qquad h_1, h_2 \in H,$$

then H contains the identity element e of G and every $h \in H$ has an inverse in H, so that H itself is a true subgroup of G.

Proof. Indeed, picking arbitrary $h \in H$, the infinite sequence $h, h^2, h^3, \dots, h^k, \dots$ of elements of the finite set H must become eventually periodic: $h^a = h^{a+n}$ for some $a \geqslant 1$ and for some $n \geqslant 1$, whence $e = h^n$, so $e \in H$ and h^{n-1} is the inverse of h. □

For more than thirteen years, Lie was convinced that a purely similar property should also hold with $G = \mathsf{Diff}_n$ being the (infinite continuous pseudo)group of analytic diffeomorphisms and with $H \subset \mathsf{Diff}_n$ being any continuous family closed under composition. We quote a characteristic excerpt of [7], pp. 444–445.

As is known, one shows in the theory of substitutions that the permutations of a group can be ordered into pairwise inverse couples of permutations. Now, since the distinction between a permutation group and a transformation group only lies in the fact that the former contains a finite and the latter an infinite number of operations, it is natural to presume that the transformations of a transformation group can also be ordered into pairs of inverse transformations. In previous works, I came to the conclusion that this should actually be the case. But because in the course of my investigations in question, certain *implicit* hypotheses have been made about the nature of the functions appearing, I think that it is necessary to *expressly add the requirement that the transformations of the group can be ordered into pairs of inverse transformations.* In any case, I conjecture that this is a necessary consequence of my original definition of the concept [BEGRIFF] of transformation group. However, it has been impossible for me to prove this in general.

In his first year working with Lie (1884), Engel proposed the following counterexample. Consider the family of transformation equations:

$$x' = \zeta x,$$

where $x, x' \in \mathbb{C}$ and the parameter $\zeta \in \mathbb{C}$ is restricted to $|\zeta| < 1$. Of course, this family is closed under any composition, say: $x' = \zeta_1 x$ and $x'' = \zeta_2 x' = \zeta_1 \zeta_2 x$, with indeed $|\zeta_2 \zeta_1| < 1$ when $|\zeta_1|, |\zeta_2| < 1$, but neither the identity element nor any inverse transformation belongs to the family. However, the requirement $|\zeta| < 1$ here is too artificial: in fact the family trivially extends as the complete group $(x' = \zeta x)_{\zeta \in \mathbb{C}}$ of dilations of the line. Engel's idea was to appeal to a Riemann map ω having $\{|\zeta| = 1\}$ as a boundary of nonextendability. The map used by Engel is the following.[1] (Translator's note: In the treatise [4], this example is presented at the end of Chap. 9, *see* below p. 179.) Let od_k denote the number of odd divisors (including 1) of any integer $k \geqslant 1$. The theory of holomorphic functions in one complex variables yields the following.

Lemma 2.2. *The infinite series:*

$$\omega(a) := \sum_{\nu \geqslant 1} \frac{a^\nu}{1 - a^{2\nu}} = \sum_{\nu \geqslant 1} \left(a^\nu + a^{3\nu} + a^{5\nu} + a^{7\nu} + \cdots\right) = \sum_{k \geqslant 1} \mathrm{od}_k \, a^k$$

converges absolutely in every open disc $\Delta_\rho = \{z \in \mathbb{C} : |z| < \rho\}$ of radius $\rho < 1$ and defines a univalent holomorphic function $\Delta \to \mathbb{C}$ from the unit disc $\Delta := \{|z| < 1\}$ to $\mathbb{C}$ which does not *extend holomorphically across any point of the unit circle $\partial\Delta := \{|z| = 1\}$.*

In fact, any other similar Riemann biholomorphic map $\zeta \longmapsto \omega(\zeta) =: \lambda$ from the unit disc Δ onto some simply connected domain $\Lambda := \omega(\Delta)$ having fractal boundary which is not a Jordan curve, e.g. the Von Koch Snowflake Island, would do the job.[2] (Translator's note: A concise presentation of Carathéodory's theory may be found in Chap. 17 of [10].) Denote then by $\lambda \longmapsto \chi(\lambda) =: \zeta$ the inverse of such a map and consider the family of transformation equations:

$$\left(x' = \chi(\lambda)\, x\right)_{\lambda \in \Lambda}.$$

By construction, $|\chi(\lambda)| < 1$ for every $\lambda \in \Lambda$. Any composition of $x' = \chi(\lambda_1)x$ and of $x'' = \chi(\lambda_2)x'$ is of the form $x'' = \chi(\lambda)x$, with the uniquely defined parameter $\lambda := \omega\big(\chi(\lambda_1)\chi(\lambda_2)\big)$, hence the group composition axiom is satisfied. However, there is again no identity element, and again, no transformation has an inverse. Furthermore, crucially (and lastly), there does not exist any extension of the family to a larger domain $\widetilde{\Lambda} \supset \Lambda$ together with a holomorphic extension $\widetilde{\chi}$ of χ to $\widetilde{\Lambda}$ so that $\widetilde{\chi}(\widetilde{\Lambda})$ contains a neighborhood of $\{1\}$ (in order to include the identity) or *a fortiori* a neighborhood of $\overline{\Delta}$ (in order to include inverses of transformations $x' = \chi(\lambda)x$ with $\lambda \in \Lambda$ close to $\partial\Lambda$).

Observation

In Vol. I of the *Theorie der Transformationsgruppen*, this example appears only in Chap. 9, on pp. 163–165, and it is written in small characters. In fact, Lie still believed that a deep analogy with substitution groups should come out as a theorem. Hence *the structure of the first nine chapters insist on setting aside*, whenever possible, *the two axioms of existence of identity element and of existence of inverses*. To do justice to this great treatise, we shall translate in Chap. 9 how Master Lie managed to produce Theorem 26 on p. 177, which he considered to provide the sought analogy with finite group theory, after taking Engel's counterexample into account.

References

1. Ackerman, M., Hermann, R.: Sophus Lie's 1880 Transformation Group paper. Math. Sci. Press, Brookline, Mass. (1975)
2. Bochnak, J., Coste, M., Roy, M.-F.: Géométrie algébrique réelle. Ergenisse der Mathematik und ihrer Grenzgebiete (3), 12. Springer-Verlag, Berlin, x+373 pp. (1987)
3. Chirka, E.M.: Complex analytic sets. Mathematics and its applications (Soviet Series), 46. Kluwer Academic Publishers Group, Dordrecht, xx+372 pp. (1989)
4. Engel, F., Lie, S.: Theorie der Transformationsgruppen. Erster Abschnitt. Unter Mitwirkung von Prof. Dr. Friedrich Engel, bearbeitet von Sophus Lie, Verlag und Druck von B.G. Teubner, Leipzig und Berlin, xii+638 pp. (1888). Reprinted by Chelsea Publishing Co., New York, N.Y. (1970)
5. Gunning, R.: Introduction to Holomorphic Functions of Several Variables, 3 vols. Wadsworth & Brooks/Cole. I: Function theory, xx+203 pp. II: Local theory, +218 pp. III: Homological theory, +194 pp. (1990)
6. Hawkins, T.: Emergence of the theory of Lie groups. An essay in the history of mathematics 1869–1926. Sources and studies in the history of mathematics and physical sciences, Springer-Verlag, Berlin, xiii+564 pp. (2001)
7. Lie, S.: Theorie der Transformationsgruppen. Math. Ann. **16**, 441–528 (1880). Translated in English and commented in: [1].
8. Malgrange, B.: Ideals of Differentiable Functions. Tata Institute of Fundamental Research Studies in Mathematics, No. 3, Bombay. Oxford University Press, London, vii+106 pp. (1967)
9. Merker, J.: On the local geometry of generic submanifolds of $\mathbb{C}^n$ and the analytic reflection principle. Journal of Mathematical Sciences (N. Y.) **125**, 751–824 (2005)
10. Milnor, J.: Dynamics in one complex variable. Annals of Mathematics Studies, 160. Princeton University Press, Princeton, Third Edition, viii+304 pp. (2006)

Chapter 3
Fundamental Differential Equations for Finite Continuous Transformation Groups

Abstract A *finite continuous local transformation group* in the sense of Lie is a family of local analytic diffeomorphisms $x_i' = f_i(x;a)$, $i = 1\ldots,n$, parametrized by a finite number r of parameters $a_1,\ldots,a_r$ that is closed under composition and under taking inverses:

$$f_i\big(f(x;a);b)\big) = f_i\big(x;\mathbf{m}(a,b)\big) \qquad \text{and} \qquad x_i = f_i\big(x';\mathbf{i}(a)\big),$$

for some *group multiplication map* $\mathbf{m}$ and for some *group inverse map* $\mathbf{i}$, both local and analytic. Also, it is assumed that there exists an $e = (e_1,\ldots,e_r)$ yielding the *identity* transformation $f_i(x;e) \equiv x_i$.

Crucially, these requirements imply the existence of *fundamental partial differential equations*:

$$\boxed{\frac{\partial f_i}{\partial a_k}(x;a) = -\sum_{j=1}^{r} \psi_{kj}(a)\,\frac{\partial f_i}{\partial a_j}(x;e)} \qquad (i=1\cdots n,\ k=1\cdots r)$$

which, technically speaking, are cornerstones of the basic theory. What matters here is that the group axioms guarantee that the $r\times r$ matrix (ψ_{kj}) depends only on a and it is locally invertible near the identity. Geometrically speaking, these equations mean that the r infinitesimal transformations:

$$X_k^a\big|_x = \frac{\partial f_1}{\partial a_k}(x;a)\,\frac{\partial}{\partial x_1} + \cdots + \frac{\partial f_n}{\partial a_k}(x;a)\,\frac{\partial}{\partial x_n} \qquad (k=1\cdots r)$$

corresponding to an infinitesimal increment of the k-th parameter computed at a:

$$f(x;a_1,\ldots,a_k+\varepsilon,\ldots,a_r) - f(x;a_1,\ldots,a_k,\ldots,a_r) \approx \varepsilon X_k^a\big|_x$$

are *linear combinations*, with certain coefficients $-\psi_{kj}(a)$ *depending only on the parameters*, of the same infinitesimal transformations computed at the identity:

S. Lie, *Theory of Transformation Groups I*,
DOI 10.1007/978-3-662-46211-9_3

$$X_k^e\big|_x = \frac{\partial f_1}{\partial a_k}(x;\, e)\,\frac{\partial}{\partial x_1} + \cdots + \frac{\partial f_n}{\partial a_k}(x;\, e)\,\frac{\partial}{\partial x_n} \qquad {\scriptstyle (k=1\cdots r)}.$$

Remarkably, the process of removing superfluous parameters introduced in the previous chapter applies to local Lie groups without the necessity of relocalizing around a generic a_0, so that everything can be achieved around the identity e itself, without losing it.

3.1 The Concept of a Local Transformation Group

Let $\mathbb{K} = \mathbb{R}$ or $\mathbb{C}$, let $n \geqslant 1$ be an integer and let $x = (x_1, \dots, x_n) \in \mathbb{K}^n$ denote variables of an n-times extended space. We shall constantly employ the sup-norm:

$$|x| := \max_{1 \leqslant i \leqslant n} |x_i|,$$

where $|\cdot|$ denotes the absolute value on $\mathbb{R}$, or the modulus on $\mathbb{C}$. For various "radii" $\rho > 0$, we shall consider the precise open sets centered at the origin that are defined by:

$$\Delta_\rho^n := \big\{x \in \mathbb{K}^n : |x| < \rho\big\};$$

in case $\mathbb{K} = \mathbb{C}$, these are of course standard open polydiscs, while in case $\mathbb{K} = \mathbb{R}$, these are open cubes.

On the other hand, again[1] (Translator's note: When $x \in \mathbb{R}^n$ is real, while studying local analytic Lie group actions $x_i' = f_i(x;a)$ below, we will naturally require that $a \in \mathbb{R}^r$ also be real (unless we complexify both x and a). When $x \in \mathbb{C}^n$, we can assume a is either real or complex.) with $\mathbb{K} = \mathbb{R}$ or $\mathbb{C}$, let $r \geqslant 1$ be another integer and introduce parameters $a = (a_1, \dots, a_r)$ in $\mathbb{K}^r$, again equipped with the sup-norm:

$$|a| := \max_{1 \leqslant k \leqslant r} |a_k|.$$

For various $\sigma > 0$, we similarly introduce the precise open sets:

$$\square_\sigma^r := \big\{a \in \mathbb{K}^r : |a| < \sigma\big\}.$$

3.1.1 Transformation Group Axioms

Let

$$x_i' = f_i(x_1, \dots, x_n; a_1, \dots, a_r) \qquad {\scriptstyle (i=1\cdots n)}$$

be local transformation equations, as presented in Chapter 1. To fix the local character, the $f_i(x;a)$ will be assumed to be defined when $|x| < \rho_1$ and when $|a| < \sigma_1$, for some $\rho_1 > 0$ and for some $\sigma_1 > 0$. We shall assume that for the parameter $a := e$, equal to the origin $0 \in \mathbb{K}^r$, the transformation corresponds to the identity, so that:

$$f_i(x_1, \ldots, x_n : 0, \ldots, 0) \equiv x_i \qquad (i = 1 \cdots n).$$

Consequently, for the composition of two successive such transformations $x' = f(x;a)$ and $x'' = f(x';b)$ to be well defined, it suffices to shrink ρ_1 to ρ_2 with $0 < \rho_2 < \rho_1$ and σ_1 to σ_2 with $0 < \sigma_2 < \sigma_1$ in order to insure that:

$$|f(x;a)| < \rho_1 \qquad \text{for all} \qquad |x| < \rho_2 \quad \text{and all} \quad |a| < \sigma_2.$$

This is clearly possible thanks to $f(x;0) = x$. Now, we can present the *local* transformation group axioms, somehow with a rigorous control of existence domains.

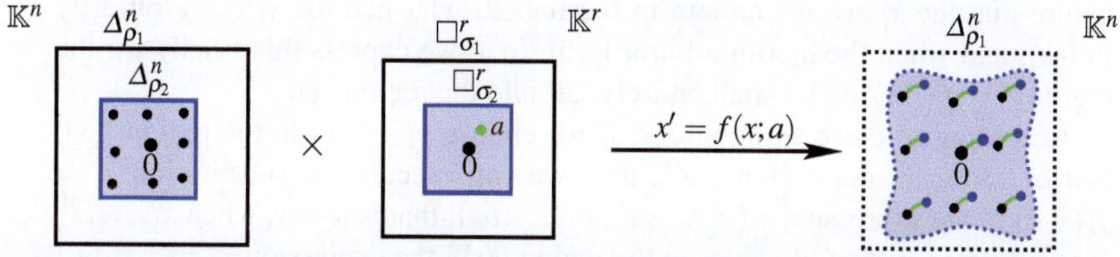

Fig. 3.1 Local transformation group in terms of cubes

Group composition axiom. For every $x \in \Delta^n_{\rho_2}$, and $a, b \in \square^r_{\sigma_2}$, an arbitrary composition:

(1)
$$x'' = f(f(x;a);b) = f(x;c) = f(x;\mathbf{m}(a,b))$$

always identifies to an element $f(x;c)$ of the same family, for a unique parameter $c = \mathbf{m}(a,b)$ given by an auxiliary *group-multiplication* local analytic map:

$$\mathbf{m} : \quad \square^n_{\sigma_1} \times \square^n_{\sigma_1} \longrightarrow \mathbb{K}^r$$

which satisfies $\mathbf{m}\big(\square^r_{\sigma_2} \times \square^r_{\sigma_2}\big) \subset \square^r_{\sigma_1}$ and $\mathbf{m}(a,e) \equiv \mathbf{m}(e,a) \equiv a$.

For $a, b, c \in \square^n_{\sigma_3}$ with $0 < \sigma_3 < \sigma_2 < \sigma_1$ small enough so that three successive compositions are well defined, the associativity of diffeomorphism composition yields:

$$f\big(x;\mathbf{m}(\mathbf{m}(a,b),c)\big) = f\big(f(f(x;a);b);c\big) = f\big(x;\mathbf{m}(a,\mathbf{m}(b,c))\big),$$

whence, thanks to the assumed uniqueness of $c = \mathbf{m}(a,b)$, we obtain the group associativity: $\mathbf{m}\big(\mathbf{m}(a,b),c\big) = \mathbf{m}\big(a,\mathbf{m}(b,c)\big)$ for such restricted values of a, b, c.

Contrary to what his opponents sometimes claimed, e.g. Study, Slocum and others, Lie was conscious of the necessity of emphasizing the local character of transformation groups that is often required in applications. Amongst the first 50 pages of the *Theorie der Transformationsgruppen*, at least 15 pages (written in small characters) are devoted to rigorously discussing when and why domains of definition should be shrunk. We translate for instance a relevant excerpt ([1], pp. 15–16) in which the symbol (x), due to Weierstrass and introduced by Engel and Lie just beforehand, denotes a region of the coordinate space and (a) a region of the parameter space, with the f_i analytic there.

Here one has to observe that we have fixed the behavior of the functions $f_i(x;a)$ only inside the regions (x) and (a).

Consequently, we have permission to substitute the expression $x'_\nu = f_\nu(x,a)$ in the equations $x''_i = f_i(x',b)$ only when the system of values $x'_1,\dots,x'_n$ lies in the region (x). That is why we are compelled to add the following assumption to those already imposed on the regions (x) and (a): it shall be possible to indicate, inside the regions (x) and (a), respective subregions $((x))$ and $((a))$ of such a nature that the x'_i always remain in the region (x) when the x_i run arbitrarily in $((x))$ and when the a_k run arbitrarily in $((a))$; we express this briefly as: the region $x' = f\big(((x))\,((a))\big)$ shall entirely fall into the region (x).

According to these assumptions, if we choose $x_1,\dots,x_n$ in the region $((x))$ and $a_1,\dots,a_r$ in the region $((a))$, then we can execute the substitution $x'_k = f_k(x,a)$ in the expression $f_i(x'_1,\dots,x'_n,\,b_1,\dots,b_n)$; that is to say, when $x^0_1,\dots,x^0_n$ is an arbitrary system of values in the region $((x))$, the expression:

$$f_i\big(f_1(x,a),\dots,f_n(x,a),\,b_1,\dots,b_n\big)$$

can be expanded, in the neighborhood of the system of values x^0_k, as an ordinary power series in $x_1-x^0_1,\dots,x_n-x^0_n$; the coefficients of this power series are functions of $a_1,\dots,a_r,\,b_1,\dots,b_r$ and behave regularly, when the a_k are arbitrary in $((a))$ and the b_k are arbitrary in (a).

Existence of an inverse-element map. There exists a local analytic map:

$$\mathbf{i}\colon\ \square^r_{\sigma_1} \longrightarrow \mathbb{K}^r$$

with $\mathbf{i}(e)=e$ (namely $\mathbf{i}(0)=0$) such that for every $a\in\square^r_{\sigma_2}$:

$$\begin{aligned} &e=\mathbf{m}(a,\mathbf{i}(a))=\mathbf{m}\big(\mathbf{i}(a),a\big)\\ &\text{whence in addition:}\quad x=f\big(f(x;a);\mathbf{i}(a)\big)=f\big(f(x;\mathbf{i}(a));a\big),\end{aligned}$$

for every $x\in\Delta^n_{\rho_2}$.

3.1.2 Some Conventions

In the sequel, the diffeomorphism $x\mapsto f(x;a)$ will occasionally be written $x\mapsto f_a(x)$. Also, we shall sometimes abbreviate $\mathbf{m}(a,b)$ by $a\cdot b$ and $\mathbf{i}(a)$ by a^{-1}. Also, a finite number of times, it will be necessary to further shrink ρ_2 and σ_2. This will be done automatically, without emphasizing it.

3.2 Changes of Coordinates and of Parameters

In the variables $x_1, \dots, x_n$, let the equations:

$$x_i' = f_i(x_1, \dots, x_n, a_1, \dots, a_r) \qquad (i=1\cdots n)$$

of an r-term group be presented. Then there are various means to derive from these equations other equations which again represent an r-term group.

On one hand, in place of the a, we can introduce r arbitrary independent functions of them:

$$\overline{a}_k = \beta_k(a_1, \dots, a_r) \qquad (k=1\cdots r)$$

as new parameters. By resolution with respect to $a_1, \dots, a_r$, one can obtain:

$$a_k = \gamma_k(\overline{a}_1, \dots, \overline{a}_r) \qquad (k=1\cdots r)$$

and by substitution of these values, one may set:

$$f_i(x_1, \dots, x_n, a_1, \dots, a_r) = \mathfrak{f}_i(x_1, \dots, x_n, \overline{a}_1, \dots, \overline{a}_r).$$

Then if we set:

$$\beta_k(b_1, \dots, b_r) = \overline{b}_k, \qquad \beta_k(c_1, \dots, c_r) = \overline{c}_k \qquad (k=1\cdots r),$$

the composition equations:

$$f_i\big(f_1(x,a), \dots, f_n(x,a), b_1, \dots, b_r\big) = f_i(x_1, \dots, x_n, c_1, \dots, c_r)$$

take, without effort, the form:

$$\mathfrak{f}_i\big(\mathfrak{f}_1(x,\overline{a}), \dots, \mathfrak{f}_n(x,\overline{a}), \overline{b}_1, \dots, \overline{b}_r\big) = \mathfrak{f}_i(x_1, \dots, x_n, \overline{c}_1, \dots, \overline{c}_r),$$

from which it results that the equations:

$$x_i' = \mathfrak{f}_i(x_1, \dots, x_n, \overline{a}_1, \dots, \overline{a}_r) \qquad (i=1\cdots n)$$

with the r essential parameters $\overline{a}_1, \dots, \overline{a}_r$ represent in the same way an r-term group.

Certainly, the equations of this new group are different from those of the original group, but these equations obviously represent exactly the same transformations as the original equations $x_i' = f_i(x,a)$. Consequently, the new group is fundamentally identical to the old one.

On the other hand, we can also introduce new independent variables $y_1,\dots,y_n$ in place of the x:

$$y_i = \omega_i(x_1,\dots,x_n) \qquad (i=1\cdots n),$$

or, if resolved:

$$x_i = w_i(y_1,\dots,y_n) \qquad (i=1\cdots n).$$

Then, we have to set:

$$x_i' = w_i(y_1',\dots,y_n') = w_i', \qquad x_i'' = w_i(y_1'',\dots,y_n'') = w_i'',$$

and we hence obtain, in place of the transformation equations:

$$x_i' = f_i(x_1,\dots,x_n,\, a_1,\dots,a_r),$$

the following ones:

$$w_i(y_1',\dots,y_n') = f_i(w_1,\dots,w_n,\, a_1,\dots,a_r),$$

or, by resolution:

$$y_i' = \omega_i\big(f_1(w,a),\,\dots,\, f_n(w,a)\big) = \mathfrak{F}_i(y_1,\dots,y_n,\, a_1,\dots,a_r).$$

It is easy to prove that the equations:

$$y_i' = \mathfrak{F}_i(y_1,\dots,y_n,\, a_1,\dots,a_r) \qquad (i=1\cdots n)$$

with the r essential parameters $a_1,\dots,a_r$ again represent an r-term group. In fact, the known equations:

$$f_i\big(f_1(x,a),\,\dots,\, f_n(x,a),\, b_1,\dots,b_r\big) = f_i(x_1,\dots,x_n,\, c_1,\dots,c_r)$$

are transferred, after the introduction of the new variables, to:

$$f_i\big(f(w,a),\, b\big) = f_i(w_1,\dots,w_n,\, c_1,\dots,c_r),$$

which can also be written:

$$f_i(w_1',\dots,w_n',\, b_1,\dots,b_r) = f_i(w_1,\dots,w_n,\, c_1,\dots,c_r) = w_i'';$$

but from this, it follows by resolution with respect to $y_1'',\dots,y_n''$ that:

$$y_\nu'' = \omega_\nu\big(f_1(w',b),\,\dots,\, f_n(w',b)\big) = \omega_\nu\big(f_1(w,c),\dots,f_n(w,c)\big),$$

or, what is the same:

$$y''_\nu = \mathfrak{F}_\nu(y'_1,\dots,y'_n, b_1,\dots,b_r) = \mathfrak{F}_\nu(y_1,\dots,y_n, c_1,\dots,c_r),$$

that is to say: there exist the equations:

$$\mathfrak{F}_\nu\big(\mathfrak{F}_1(y,a),\dots,\mathfrak{F}_n(y,a),\, b_1,\dots,b_r\big) = \mathfrak{F}_\nu(y_1,\dots,y_n, c_1,\dots,c_r),$$

whence it is indeed proved that the equations $y'_i = \mathfrak{F}_i(y,a)$ represent a group.

Lastly, we can naturally introduce at the same time new parameters and new variables in a given group; it is clear that in this way, we likewise obtain a new group from the original group.

Now we set up the following definition:

Definition. Two r-term groups:

$$x'_i = f_i(x_1,\dots,x_n, a_1,\dots,a_r) \qquad {\scriptstyle (i=1\cdots n)}$$
$$y'_i = \mathfrak{f}_i(y_1,\dots,y_n, b_1,\dots,b_r) \qquad {\scriptstyle (i=1\cdots n)}$$

in the same number of variables are *similar* [ÄHNLICH] to each other as soon as one converts into the other by the introduction of appropriate new variables and of appropriate new parameters.

Obviously, there is an unbounded number of groups which are similar to a given one; but all these unboundedly numerous groups are known simultaneously with the given one. For this reason, as it shall also happen in the sequel, we can consider that two mutually similar groups are not essentially distinct from each other.

Above, we spoke about the introduction of new parameters and of new variables without dealing with the assumptions by which we can ascertain that all group-theoretic properties essential for us are preserved here. We offer a few words concerning this point.

For it to be permitted to introduce, in the group $x'_i = f_i(x_1,\dots,x_n, a_1,\dots,a_r)$, the new parameters $\overline{a}_k = \beta_k(a_1,\dots,a_r)$ in place of the a, the $\overline{a}_k$ must be univalent functions of the a in the complete region (a) defined earlier on, and they must behave regularly everywhere in it; the functional determinant $\sum \pm \partial\beta_1/\partial a_1 \cdots \partial\beta_r/\partial a_r$ should vanish nowhere in the region (a), and lastly, to two distinct systems of values $a_1,\dots,a_r$ of this region, there must always be associated two distinct systems of values $\overline{a}_1,\dots,\overline{a}_r$. In other words: in the region of the $\overline{a}_k$, one must be able to delimit a region $(\overline{a})$ on which the systems of values of the region (a) are represented in a univalent way by the equations $\overline{a}_k = \beta_k(a_1,\dots,a_r)$.

On the other hand, for the introduction of the new variables $y_i = \omega_i(x_1,\dots,x_n)$ to be allowed, the y must be univalent and regular functions of the x for all systems of values $x_1,\dots,x_n$ which come into consideration after establishing the group-theoretic properties of the equations

$x_i' = f_i(x_1, \dots, x_n, a_1, \dots, a_r)$; inside this region, the functional determinant $\sum \pm \partial\omega_1/\partial x_1 \cdots \partial\omega_n/\partial x_n$ should vanish nowhere, and lastly, to two distinct systems of values $x_1, \dots, x_n$ of this region, there must always be associated two distinct systems of values $y_1, \dots, y_n$. The concerned system of values of the x must therefore be represented univalently onto a certain region of systems of values of the y.

If one introduced, in the group $x_i' = f_i(x,a)$, new parameters or new variables without the requirements just explained being satisfied, then it would be conceivable in any case that important properties of the group, for instance the group composition property itself, would be lost; a group with the identity transformation could convert into a group which does not contain the identity transformation, and conversely.

But in certain circumstances, the matter is only to study the family of transformations $x_i' = f_i(x,a)$ in the neighborhood of a single point $a_1, \dots, a_r$ or $x_1, \dots, x_n$. This study will often be facilitated by introducing new variables or new parameters which satisfy the requirements mentioned above in the *neighborhood of the concerned points*.

In such a case, one does not need to deal with the question of whether the concerned requirements are satisfied in the whole extension of the regions (x) and (a).

Now, if we have a family of ∞^r transformations:

$$x_i' = f_i(x_1, \dots, x_n, a_1, \dots, a_r)$$

which forms an r-term group, then there corresponds to this family a family of ∞^r operations by which the points of the space $x_1, \dots, x_n$ are permuted. Evidently, any two of these ∞^r operations, when executed one after the other, always produce an operation which again belongs to the family.

Thus, if we actually call a family of operations of this sort a group operation [OPERATIONSGRUPPE], or shortly, a group, then we can say: *every given r-term transformation group can be interpreted as the analytic representation of a certain group of ∞^r permutations of the points $x_1, \dots, x_n$.*

Conversely, if a group of ∞^r permutations of the points $x_1, \dots, x_n$ is given, and if it is possible to represent these permutations by *analytic* transformation equations, then the corresponding ∞^r transformations naturally form a transformation group.

Now, if one imagines that a determined group-operation is given, and in addition, that an analytic representation of it is given — hence if one has a transformation group —, then this representation has in itself two obvious incidental characters [ZUFÄLLIGKEIT].

The first incidental character is the choice of the parameters $a_1, \dots, a_r$. It stands to reason that this choice has in itself absolutely no influence on the group-operation when we introduce in place of the a the new parameters

$\overline{a}_k = \beta_k(a_1,\dots,a_r)$. Only the analytic expression for the group-operation will be a different one on this occasion; therefore, this expression represents a transformation group as before.

The second incidental character in the analytic representation of our group-operation is the choice of the coordinates in the space $x_1,\dots,x_n$. Every permutation of the points $x_1,\dots,x_n$ is fully independent of the choice of the system of coordinates to which one refers the points $x_1,\dots,x_n$; only the analytic representation of the permutation changes with the concerned system of coordinates. Naturally, the same holds for any group of permutations. From this, it results that by introducing new variables, that is to say, by a change of the system of coordinates, one obtains, from a transformation group, again a transformation group, for the transformation equations that one receives after the introduction of the new variables represent exactly the same group-operation as that of the initial transformation group, so that they form in turn a transformation group.

Thus, the analytic considerations of the previous paragraphs are now explained in a conceptual way. Above all, it is at present clear why two similar transformation groups are to be considered as not essentially distinct from each other; namely, for the reason that they both represent analytically one and the same group-operation.

3.3 Geometric Introduction of Infinitesimal Transformations

Now, for reasons of clarity, we shall present in advance the basic geometric way in which infinitesimal transformations can be introduced, a way which is knowingly passed over in silence in the great treatise [1].

Letting ε denote either an infinitesimal quantity in the sense of Leibniz, or a small quantity subjected to Weierstrass' rigorous epsilon-delta formalism, for fixed $k \in \{1,2,\dots,r\}$, we consider all the points:

$$\begin{aligned} x_i' &= f_i\big(x;\, e_1,\dots,e_k+\varepsilon,\dots,e_n\big) \\ &= x_i + \frac{\partial f_i}{\partial a_k}(x;e)\,\varepsilon + \cdots \qquad (i=1\cdots n) \end{aligned}$$

that are infinitesimally pushed from the starting points $x = f(x;e)$ by adding the tiny increment ε to only the k-th identity parameter e_k. One may reinterpret this common spatial move by introducing the vector field (and a new notation for its coefficients):

$$X_k^e := \sum_{i=1}^{n} \frac{\partial f_i}{\partial a_k}(x;e)\,\frac{\partial}{\partial x_i} := \sum_{i=1}^{n} \xi_{ki}(x)\,\frac{\partial}{\partial x_i},$$

which is either written as a derivation in modern style, or considered as a column vector:

$$^{\tau}\Big(\tfrac{\partial f_1}{\partial a_k},\dots,\tfrac{\partial f_n}{\partial a_k}\Big)\Big|_x = {}^{\tau}\big(\xi_{k1},\dots,\xi_{kn}\big)\Big|_x$$

based at x, where ${}^\tau(\cdot)$ denotes a transposition, yielding column vectors. Then $x' = x + \varepsilon X_k^e + \cdots$, or equivalently:

$$x'_i = x_i + \varepsilon\, \xi_{ki} + \cdots \qquad (i=1\cdots n),$$

where the omitted terms "$+\cdots$" are of course an $O(\varepsilon^2)$, so that from the geometrical viewpoint, x' is infinitesimally pushed along the vector $X_k^e\big|_x$ up to a length ε.

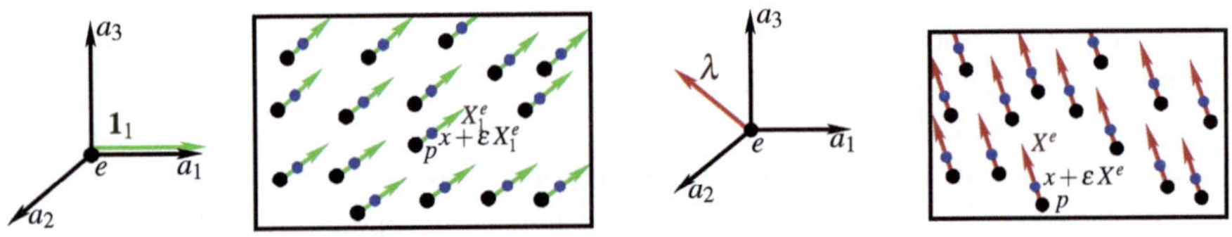

Fig. 3.2 Infinitesimal displacement $x' = x + \varepsilon X^e$ of all points

More generally, still starting from the identity parameter e, when we add to e an arbitrary infinitesimal increment:

$$\big(e_1 + \varepsilon\lambda_1, \ldots, e_k + \varepsilon\lambda_k, \ldots, e_r + \varepsilon\lambda_r\big),$$

where ${}^\tau(\lambda_1, \ldots, \lambda_r)\big|_e$ is a fixed, constant vector based at e in the parameter space, it follows by linearity of the tangential map, or else just by the chain rule in coordinates, that:

$$\begin{aligned} f_i(x;\, e + \varepsilon\lambda) &= x_i + \sum_{k=1}^{n} \varepsilon\,\lambda_k\, \frac{\partial f_i}{\partial a_k}(x;e) + \cdots \\ &= x_i + \varepsilon \sum_{k=1}^{n} \lambda_k\, \xi_{ki}(x) + \cdots, \end{aligned}$$

so that all points $x' = x + \varepsilon X + \cdots$ are infinitesimally and simultaneously pushed along the vector field:

$$X := \lambda_1 X_1^e + \cdots + \lambda_r X_r^e$$

which is the general linear combination of the r previous basic vector fields X_k^e, $k = 1, \ldots, r$.

Occasionally, Lie wrote that such a vector field X *belongs to the group*, $x' = f(x;a)$, to mean that X itself comes with the infinitesimal translation $x' = x + \varepsilon X$ it is supposed to perform (dots should now intuitively be suppressed), and hence accordingly, Lie systematically called such an X an *infinitesimal transformation*, viewing $x' = x + \varepsilon X$ as just a case of $x' = f(x,a)$. Another fundamental and *very* deep reason why Lie said that X belongs to the group $x' = f(x,a)$ is that he showed that local transformation group actions are in one-to-one correspondence with the purely linear vector spaces:

$$\mathrm{Vect}_{\mathbb{K}}(X_1, X_2, \ldots, X_r)$$

of infinitesimal transformations, which in fact also inherit a crucial additional *algebraic* structure directly from the group multiplication law. Without anticipating too much, let us come to the *purely analytic* way how Engel and Lie introduce the infinitesimal transformations.

3.4 Derivation of Fundamental Partial Differential Equations

So we have defined the concept of a purely local Lie transformation group, insisting on the fact that composition and inversion are both represented by some precise local analytic maps defined around the identity. The royal road towards the famous theorems of Lie is to differentiate these finite data, namely to *infinitesimalize*.

We start with the group composition law (1) which we rewrite as follows:

$$x'' = f\big(f(x;a);b\big) = f(x;a\cdot b) =: f(x;c).$$

Here, $c := a\cdot b$ depends on a and b, but instead of a and b, following [1], we want to consider a and c to be the independent parameters, namely we rewrite $b = a^{-1}\cdot c =: b(a,c)$ so that the equations:

$$f_i\big(f(x;a);\, b(a,c)\big) \equiv f_i(x;c) \qquad {\scriptstyle (i=1\cdots n)}$$

hold identically for all x, all a and all c. Next, we differentiate these identities with respect to a_k, defining $f_i' \equiv f_i(x';b)$ and $x_j' \equiv f_j(x;a)$:

$$\frac{\partial f_i'}{\partial x_1'}\frac{\partial x_1'}{\partial a_k}+\cdots+\frac{\partial f_i'}{\partial x_n'}\frac{\partial x_n'}{\partial a_k}+\frac{\partial f_i'}{\partial b_1}\frac{\partial b_1}{\partial a_k}+\cdots+\frac{\partial f_i'}{\partial b_r}\frac{\partial b_r}{\partial a_k}\equiv 0 \qquad {\scriptstyle (i=1\cdots n)}.$$

Here of course again, the argument of f' is $\big(f(x,a);b(a,c)\big)$, the argument of x' is $(x;a)$ and the argument of b is (a,c). Thanks to $x''(x';e)\equiv x'$, the matrix $\frac{\partial f_j'}{\partial x_k'}\big(f(x;e);b(e,e)\big)$ is the identity $I_{n\times n}$. So using Cramer's rule,[2] (Translator's note: — and possibly also shrinking ρ_2 and σ_2 if necessary, and likewise below, without further mention.) for each fixed k, we can solve the preceding n linear equations with respect to the n unknowns $\frac{\partial x_1'}{\partial a_k},\dots,\frac{\partial x_n'}{\partial a_k}$, obtaining expressions of the form:

$$\text{(2)} \qquad \frac{\partial x_\nu'}{\partial a_k}(x;a) = \Xi_{1\nu}(x',b)\,\frac{\partial b_1}{\partial a_k}(a,c)+\cdots+\Xi_{r\nu}(x',b)\,\frac{\partial b_r}{\partial a_k}(a,c)$$

$${\scriptstyle (\nu=1\cdots n;\,k=1\cdots r)},$$

with some analytic functions $\Xi_{j\nu}(x',b)$ that are independent of k.

On the other hand, in order to substitute the $\frac{\partial b_j}{\partial a_k}$, we differentiate with respect to a_k the identically satisfied identities:

$$c_\mu \equiv \mathbf{m}_\mu\big(a, b(a,c)\big) \qquad (\mu = 1 \cdots r).$$

We therefore get:

$$0 \equiv \frac{\partial \mathbf{m}_\mu}{\partial a_k} + \sum_{\pi=1}^{r} \frac{\partial \mathbf{m}_\mu}{\partial b_\pi}\,\frac{\partial b_\pi}{\partial a_k} \qquad (\mu = 1 \cdots r).$$

But since the matrix $\frac{\partial \mathbf{m}_\mu}{\partial b_\pi}$ is the identity $I_{r\times r}$ for $\big(a, b(a,c)\big)\big|_{(a,c)=(e,e)} = (e,e)$, just because $\mathbf{m}(e,b) \equiv b$, Cramer's rule again enables one to solve this system with respect to the r unknowns $\frac{\partial b_\pi}{\partial a_k}$, yielding expressions of the form:

$$\frac{\partial b_\pi}{\partial a_k}(a,c) = \Psi_{k\pi}(a,c),$$

for certain functions $\Psi_{k\pi}$, defined on a possibly smaller parameter product space $\square^r_{\sigma_2} \times \square^r_{\sigma_2}$. Again putting $(a,c) = (e,e)$ in the same system before solving it, we get in fact (notice the minus sign):

$$\left(\frac{\partial b_\pi}{\partial a_k}(e,e)\right)_{1\leqslant k\leqslant r}^{1\leqslant \pi\leqslant r} = -I_{r\times r},$$

so that $\Psi_{k\pi}(e,e) = -\delta_k^\pi$.

We can therefore now insert in (2) the gained value $\Psi_{k\pi}$ of $\frac{\partial b_\pi}{\partial a_k}$, obtaining (and quoting [1], p. 29) the following crucial partial differential equations:

$$\text{(2')} \qquad \frac{\partial x'_\nu}{\partial a_k}(x;a) = \sum_{\pi=1}^{r} \Psi_{k\pi}(a,b)\,\Xi_{\pi\nu}(x',b) \qquad (\nu = 1 \cdots n,\ k = 1 \cdots r)$$

These equations are of utmost importance [ÄUSSERST WICHTIG], *as we will see later.*

Here, we have replaced c by $c = c(a,b) = a \cdot b$, whence $b\big(a, c(a,b)\big) \equiv b$, and we have reconsidered (a,b) as the independent variables.

3.4.1 Restricting Considerations to a Single System of Parameters

Setting $b := e$ in (2') above, the partial derivatives of the group transformation equations $x'_i = x'_i(x;a)$ with respect to the parameters a_k at $(x;a)$:

$$\text{(2'')} \qquad \boxed{\boxed{\frac{\partial x_i'}{\partial a_k}(x;a) = \sum_{j=1}^{r} \psi_{kj}(a)\, \xi_{ji}\big(x'(x;a)\big)}} \qquad {\scriptstyle (i=1\cdots n,\; k=1\cdots r)},$$

appear to be *linear combinations*, with certain coefficients $\psi_{kj}(a) := \Psi_{kj}(a,e)$ *depending only upon* a, of the quantities $\xi_{ji}(x') := \Xi_{ji}(x',b)\big|_{b=e}$. But in fact we already know these quantities.

3.4.2 Comparing Different Frames of Infinitesimal Transformations

Indeed, setting $a = e$ above, thanks to $\psi_{kj}(e) = -\delta_k^j$, we get immediately:

$$\xi_{ki}(x) = \xi_{ki}\big(x'(x;e)\big) = -\frac{\partial x_i'}{\partial a_k}(x;e),$$

whence the $\xi_{ki}(x)$ just coincide with the coefficients of the r infinitesimal transformations already introduced on p. 31 (with an overall opposite sign) and written as derivations:

$$\text{(3)} \qquad \begin{aligned} X_k^e\big|_x &= \frac{\partial f_1}{\partial a_k}(x;e)\,\frac{\partial}{\partial x_1} + \cdots + \frac{\partial f_n}{\partial a_k}(x;e)\,\frac{\partial}{\partial x_n} \\ &=: -\xi_{k1}(x)\,\frac{\partial}{\partial x_1} - \cdots - \xi_{kn}(x)\,\frac{\partial}{\partial x_n} \qquad {\scriptstyle (k=1\cdots r)}. \end{aligned}$$

Now at last, after having reproduced some rather blind computations by which Lie expresses his brilliant synthetic thoughts, the crucial geometric interpretation of the twice-boxed partial differential equations can be unveiled.

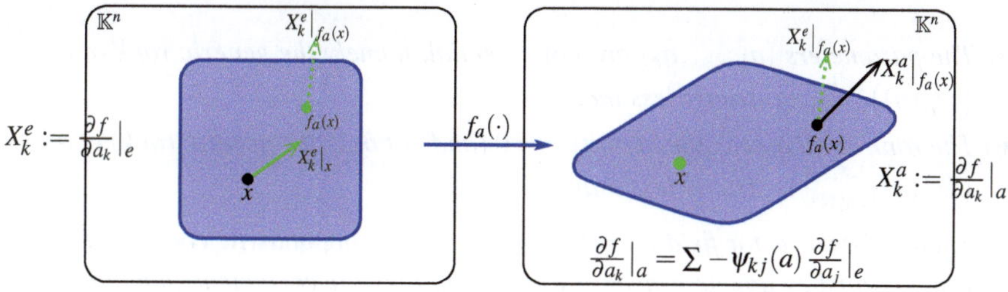

Fig. 3.3 Geometric interpretation of the fundamental differential equations

Instead of differentiating with respect to a_k only at $(x;e)$, we must in principle, for reasons of generality, do the same at any $(x;a)$, which yields the vector fields:

$$X_k^a\big|_{f_a(x)} := \frac{\partial f_1}{\partial a_k}(x;a)\,\frac{\partial}{\partial x_1} + \cdots + \frac{\partial f_n}{\partial a_k}(x;a)\,\frac{\partial}{\partial x_n} \qquad (k=1\cdots r),$$

and then the fundamental partial differential equations say that such new r infinitesimal transformations:

$$\boxed{X_k^a\big|_{f_a(x)} = -\psi_{k1}(a)\,X_1^e\big|_{f_a(x)} - \cdots - \psi_{kr}(a)\,X_r^e\big|_{f_a(x)}}$$

are just linear combinations, *with coefficients depending only upon the group parameters*, of the r infinitesimal transformations $X_1^e,\dots,X_r^e$ computed at the special parameter e, *and considered at the a-pushed points* $f_a(x)$.

Finally, since the matrix $\psi_{kj}(a)$ has an inverse that we will denote, as in [1], by $\alpha_{jk}(a)$ which is analytic in a neighborhood of e, we can also write the fundamental differential equations in the reciprocal form, useful in the sequel:

$$\text{(4)} \qquad \xi_{ji}\big(x'(x;a)\big) = \sum_{k=1}^{r} \alpha_{jk}(a)\,\frac{\partial x_i'}{\partial a_k}(x;a). \qquad (i=1\cdots n;\ j=1\cdots r).$$

3.5 Essentializing the Group Parameters

As we have seen in the previous chapter, the (needed) suppression of illusory parameters in the transformation group equations $x_i' = f_i(x;a) = \sum_{\alpha\in\mathbb{N}^n} \mathcal{U}_\alpha^i(a)\,x^\alpha$ might require us to relocalize considerations to some neighborhood of some generic a^0, *which might possibly not include the identity* e. Fortunately, this does not occur: the group property ensures that the rank of the infinite coefficient map $\mathsf{U}_\infty : a \longmapsto \big(\mathcal{U}_\alpha^i(a)\big)_{\alpha\in\mathbb{N}^n}^{1\leqslant i\leqslant n}$ is *constant* around e.

Proposition 3.1. *For a finite continuous local Lie transformation group* $x_i' = f_i(x;a) = \sum_{\alpha\in\mathbb{N}^n} \mathcal{U}_\alpha^i(a)\,x^\alpha$ *expanded in a power series with respect to* x, *the following four conditions are equivalent:*

(i) *The parameters* $(a_1,\dots,a_r)$ *are* not *essential, namely the generic rank of* $a \longmapsto \big(\mathcal{U}_\alpha^i(a)\big)_{\alpha\in\mathbb{N}^n}^{1\leqslant i\leqslant n}$ *is strictly less than* r.

(ii) *The rank at all* a *near the identity* $e = 0$ *(and not only the generic rank) of* $a \longmapsto \big(\mathcal{U}_\alpha^i(a)\big)_{\alpha\in\mathbb{N}^n}^{1\leqslant i\leqslant n}$ *is strictly less than* r.

(iii) *There exists a vector field* $\mathcal{T} = \sum_{k=1}^r \tau_k(a)\,\frac{\partial}{\partial a_k}$ *having analytic coefficients* $\tau_k(a)$ *which vanishes nowhere near* e *and annihilates all coefficient functions:* $0 \equiv \mathcal{T}\,\mathcal{U}_\alpha^i$ *for all* i *and all* α.

(iv) *There exist constants* $e_1,\dots,e_r \in \mathbb{K}$ *not all zero such that, if*

$$X_k = \frac{\partial f_1}{\partial a_k}(x;e)\,\frac{\partial}{\partial x_1} + \cdots + \frac{\partial f_n}{\partial a_k}(x;e)\,\frac{\partial}{\partial x_n} \qquad (k=1\cdots n)$$

denote the r infinitesimal transformations associated to the group, then:

$$0 \equiv e_1 X_1 + \cdots + e_r X_r.$$

Proof. Suppose that **(iv)** holds, namely via the expressions (3):

$$0 \equiv -e_1\,\xi_{1i}(x) - \cdots - e_r\,\xi_{ri}(x) \qquad (i=1\cdots n).$$

By applying the linear combination $\sum_{j=1}^{r} e_j(\cdot)$ to the partial differential equations (4), the left member thus becomes zero, and we get n equations:

$$0 \equiv \sum_{j=1}^{r}\sum_{k=1}^{r} e_j\,\theta_{jk}(a)\,\frac{\partial x_i'}{\partial a_k}(x;a) \qquad (i=1\cdots n)$$

which just say that the *not identically zero* vector field:

$$\mathscr{T} := \sum_{k=1}^{r}\bigg(\sum_{j=1}^{r} e_j\,\theta_{jk}(a)\bigg)\frac{\partial}{\partial a_k}$$

satisfies $0 \equiv \mathscr{T} f_1 \equiv \cdots \equiv \mathscr{T} f_n$, or equivalently $0 \equiv \mathscr{T}\mathscr{U}_\alpha^i$ for all i and all α. Since the matrix $\theta(a)$ equals $-I_{r\times r}$ at the identity e, the vectors $\mathscr{T}\big|_a$ are *nonzero* for all a in a neighborhood of a. In conclusion, **(iv)** implies **(iii)**, and then straightforwardly **(iii)** $\Rightarrow$ **(ii)** $\Rightarrow$ **(i)**.

It remains to establish the reverse implications: **(i)** $\Rightarrow$ **(ii)** $\Rightarrow$ **(iii)** $\Rightarrow$ **(iv)**. As a key fact, the generic rank in **(i)** happens to be constant.

Lemma 3.1. *There exist two* $\infty\times\infty$ *matrices* $\big(\mathsf{Coeff}_\alpha^\beta(a)\big)_{\alpha\in\mathbb{N}^n}^{\beta\in\mathbb{N}^n}$ *and* $\big(\overline{\mathsf{Coeff}_\alpha^\beta(a)}\big)_{\alpha\in\mathbb{N}^n}^{\beta\in\mathbb{N}^n}$ *of analytic functions of the parameters* $a_1,\dots,a_r$ *which enjoy appropriate Cauchy convergence estimates,*[3] *(Translator's note: — that are automatically satisfied and hence can be passed over in silence —) the second matrix being the inverse of the first, such that:*

$$\mathscr{U}_\alpha^i\big(\mathbf{m}(a,b)\big) \equiv \sum_{\beta\in\mathbb{N}^n} \mathsf{Coeff}_\alpha^\beta(a)\,\mathscr{U}_\beta^i(b) \qquad (i=1\cdots n;\ \alpha\in\mathbb{N}^n),$$

and inversely:

$$\mathscr{U}_\alpha^i(b) \equiv \sum_{\beta\in\mathbb{N}^n} \overline{\mathsf{Coeff}_\alpha^\beta(a)}\,\mathscr{U}_\beta^i\big(\mathbf{m}(a,b)\big) \qquad (i=1\cdots n;\ \alpha\in\mathbb{N}^n).$$

As a consequence, the rank of the infinite coefficient mapping $a \longmapsto \big(\mathscr{U}_\alpha^i(a)\big)_{\alpha\in\mathbb{N}^n}^{1\leqslant i\leqslant n}$ *is constant in a neighborhood of e.*

Proof. By definition, the composition of $x' = \sum_{\alpha\in\mathbb{N}^n}\mathscr{U}_\alpha(a)\,x^\alpha$ and of $x_i'' = \sum_{\beta\in\mathbb{N}^n}\mathscr{U}_\beta^i(b)\,(x')^\beta$ yields $x_i'' = f_i\big(x;\mathbf{m}(a,b)\big)$ and we must therefore compute the composed Taylor series expansion in question, which we place in the right-hand side:

$$\sum_{\alpha\in\mathbb{N}^n} \mathscr{U}_\alpha^i\big(\mathbf{m}(a,b)\big)\,x^\alpha \equiv \sum_{\beta\in\mathbb{N}^n} \mathscr{U}_\beta^i(b)\bigg(\sum_{\alpha\in\mathbb{N}^n} \mathscr{U}_\alpha(a)\,x^\alpha\bigg)^\beta.$$

To this end, we use the general formula for the expansion of a k-th power:

$$\bigg(\sum_{\alpha\in\mathbb{N}^n} \mathsf{c}_\alpha\, x^\alpha\bigg)^k = \sum_{\alpha\in\mathbb{N}^n} x^\alpha\bigg\{\sum_{\alpha_1+\cdots+\alpha_k=\alpha} \mathsf{c}_{\alpha_1}\cdots\mathsf{c}_{\alpha_k}\bigg\}$$

of a *scalar* power series. Thus splitting $x'^\beta = (x'_1)^{\beta_1}\cdots(x'_n)^{\beta_n}$ into scalar powers, we may start to compute the desired expansion:

$$\begin{aligned}
\mathsf{Composition} &= \sum_{(\beta_1,\dots,\beta_n)} \mathscr{U}_\beta^i(b)\bigg(\sum_{\alpha^1\in\mathbb{N}^n} \mathscr{U}_{\alpha^1}^1(a)\,x^{\alpha^1}\bigg)^{\beta_1}\cdots\bigg(\sum_{\alpha^n\in\mathbb{N}^n} \mathscr{U}_{\alpha^n}^n(a)\,x^{\alpha^n}\bigg)^{\beta_n} \\
&= \sum_{(\beta_1,\dots,\beta_n)} \mathscr{U}_\beta^i(b)\bigg[\sum_{\alpha^1\in\mathbb{N}^n} x^{\alpha^1}\bigg\{\sum_{\alpha_1^1+\cdots+\alpha_{\beta_1}^1=\alpha^1} \mathscr{U}_{\alpha_1^1}^1(a)\cdots\mathscr{U}_{\alpha_{\beta_1}^1}^1(a)\bigg\}\bigg]\cdots \\
&\qquad\cdots\bigg[\sum_{\alpha^n\in\mathbb{N}^n} x^{\alpha^n}\bigg\{\sum_{\alpha_1^n+\cdots+\alpha_{\beta_n}^n=\alpha^n} \mathscr{U}_{\alpha_1^n}^n(a)\cdots\mathscr{U}_{\alpha_{\beta_n}^n}^n(a)\bigg\}\bigg].
\end{aligned}$$

To finish off, if we apply the expansion of a product of n power series:

$$\Big(\textstyle\sum_{\alpha^1} \mathsf{c}_{\alpha^1}^1\, x^{\alpha^1}\Big)\cdots\Big(\textstyle\sum_{\alpha^n} \mathsf{c}_{\alpha^n}^n\, x^{\alpha^n}\Big) = \textstyle\sum x^\alpha\Big\{\textstyle\sum_{\alpha^1+\cdots+\alpha^n=\alpha} \mathsf{c}_{\alpha^1}^1\cdots\mathsf{c}_{\alpha^n}^n\Big\},$$

we then obtain straightforwardly:

$$\begin{aligned}
&\sum_{\alpha\in\mathbb{N}^n} \mathscr{U}_\alpha^i\big(\mathbf{m}(a,b)\big)\,x^\alpha = \sum_{\alpha\in\mathbb{N}^n} x^\alpha\bigg\{\sum_{(\beta_1,\dots,\beta_n)} \mathscr{U}_\beta^i(b)\sum_{\alpha^1+\cdots+\alpha^n=\alpha} \\
&\bigg(\sum_{\alpha_1^1+\cdots+\alpha_{\beta_1}^1=\alpha^1} \mathscr{U}_{\alpha_1^1}^1(a)\cdots\mathscr{U}_{\alpha_{\beta_1}^1}^1(a)\bigg)\cdots\bigg(\sum_{\alpha_1^n+\cdots+\alpha_{\beta_n}^n=\alpha^n} \mathscr{U}_{\alpha_1^n}^n(a)\cdots\mathscr{U}_{\alpha_{\beta_n}^n}^n(a)\bigg)\bigg\}.
\end{aligned}$$

By identifying the coefficients of x^α in both sides and by abbreviating just as $\mathsf{Coeff}_\alpha^\beta(a)$ the $\sum_{\alpha^1+\cdots+\alpha^n=\alpha}$ of the products of the n long sums appearing in the second line, we have thus found the first family of equations. The second one is obtained quite similarly simply by expanding the identities:

$$f_i(x;\, b) \equiv f_i\big(f(x;\, \mathbf{i}(a));\, \mathbf{m}(a,b)\big) \qquad {\scriptstyle (i=1\cdots n)}.$$

The two families of equations thus found must clearly be inverses of each other.

To show that the rank at e of $a \mapsto \mathsf{U}_\infty(a)$ is the same as it is at any a near e, we apply $\frac{\partial}{\partial b_k}\big|_{b=e}$ to both the first and the second family of equations, which gives:

$$\sum_{l=1}^{r} \frac{\partial \mathscr{U}_\alpha^i}{\partial a_l}(a)\, \frac{\partial \mathbf{m}_l}{\partial b_k}(a,e) \equiv \sum_{\beta\in\mathbb{N}^n} \mathsf{Coeff}_\alpha^\beta(a)\, \frac{\partial \mathscr{U}_\beta^i}{\partial b_k}(e)$$

$$\frac{\partial \mathscr{U}_\alpha^i}{\partial b_k}(e) \equiv \sum_{\beta\in\mathbb{N}^n} \overline{\mathsf{Coeff}_\alpha^\beta}(a) \bigg(\sum_{l=1}^{r} \frac{\partial \mathscr{U}_\beta^i(a)}{\partial a_l}\, \frac{\partial \mathbf{m}_l}{\partial b_k}(a,e) \bigg)$$

$$(k=1\cdots r;\; i=1\cdots n;\; \alpha\in\mathbb{N}^n).$$

Of course, the appearing $r\times r$ matrix:

$$\mathsf{M}(a) := \left(\frac{\partial \mathbf{m}_l}{\partial b_k}(a,e) \right)_{1\leqslant k\leqslant r}^{1\leqslant l\leqslant r}$$

has nonzero determinant at every a, because left translations $b\mapsto \mathbf{m}(a,b)$ of the group are diffeomorphisms. Thus, if for each fixed i we denote by $\operatorname{Jac}\mathsf{U}_\infty^i(a)$ the $r\times\infty$ Jacobian matrix $\left(\frac{\partial \mathscr{U}_\alpha^i}{\partial a_k}(a)\right)_{1\leqslant k\leqslant r}^{\alpha\in\mathbb{N}^n}$, the previous two families of identities when written in matrix form:

$$\mathsf{M}(a)\operatorname{Jac}\mathsf{U}_\infty^i(a) \equiv \mathsf{Coeff}(a)\operatorname{Jac}\mathsf{U}_\infty^i(a)$$

$$\operatorname{Jac}\mathsf{U}_\infty^i(e) \equiv \overline{\mathsf{Coeff}}(a)\,\mathsf{M}(a)\operatorname{Jac}\mathsf{U}_\infty^i(a)$$

show that the rank of $\operatorname{Jac}_\infty^i(a)$ must be equal to the rank of $\operatorname{Jac}_\infty^i(e)$. This completes the proof of the auxiliary lemma. □

Thus, **(i)** of the proposition implies **(ii)** and moreover, the last part of the theorem on p. 15 yields *annihilating* analytic vector fields $\mathscr{T}_1,\dots,\mathscr{T}_{r-\rho_\infty}$ on the parameter space with $\dim \operatorname{Vect}\big(\mathscr{T}_1\big|_a,\dots,\mathscr{T}_{\rho_\infty}\big|_a\big) = r-\rho_\infty$ constant for all a near e. So **(i)** $\Rightarrow$ **(ii)** $\Rightarrow$ **(iii)**. Finally, assuming **(iii)**, namely:

$$0\equiv \mathscr{T} f_i(x;a) = \sum_{k=1}^{r} \tau_k(a)\, \frac{\partial x_i'}{\partial a_k}(x;a) \qquad (i=1\cdots n)$$

with $\mathscr{T}\big|_e\neq 0$, and replacing $\frac{\partial x_i'}{\partial a_k}$ by its value (2") given by the fundamental differential equations, we get:

$$0\equiv \sum_{j=1}^{r}\sum_{k=1}^{r} \tau_k(a)\,\psi_{kj}(a)\,\xi_{ji}\big(x'(x;a)\big) \qquad (i=1\cdots n).$$

Setting a to be the identity element in these equations and introducing the r constants (recall here that $\psi_{kj}(e) = -\delta_k^j$):

$$c_j := \textstyle\sum_{k=1}^{r} \tau_k(e)\,\psi_{kj}(e) = -\tau_j(e) \qquad (j=1\cdots r)$$

that are not all zero, since $\mathscr{T}\big|_e\neq 0$, we get equations:

$$0\equiv c_1\,\xi_{1i}(x) + \cdots + c_r\,\xi_{ri}(x) \qquad (i=1\cdots n)$$

which, according to (3), are the coordinatewise expression of **(iv)**. This completes the proof of the proposition. □

3.6 The First Fundamental Theorem

Thanks to this proposition and to the corollary on p. 15, *no relocalization is necessary to get rid of superfluous parameters in finite continuous transformation groups* $x' = f(x; a_1, \dots, a_r)$. Thus, without loss of generality parameters can (and surely will) *always* be assumed to be essential. We can now translate Theorem 3 on pp. 33–34 of [1] which summarizes all the preceding considerations, and we add some technical details if this theorem is to be interpreted as a *local* statement (though Lie has something different in mind, cf. the next section). As above, we assume implicitly for simplicity that the group $x' = f(x; a)$ is local and contains the identity transformation $x' = f(x; e) = x$, but the theorem is in fact valid with less restrictive assumptions, *see* Sect. 3.9 and especially Proposition 3.4.

Theorem 3. *If the n equations:*

$$x_i' = f_i(x_1, \dots, x_n, a_1, \dots, a_r) \qquad (i=1\cdots n)$$

represent a finite continuous group, whose parameters $a_1, \dots, a_r$ are all essential, then $x_1', \dots, x_n'$, considered as functions of $a_1, \dots, a_r, x_1, \dots, x_n$ satisfy certain differential equations of the form:

$$(5) \qquad \frac{\partial x_i'}{\partial a_k} = \sum_{j=1}^{r} \psi_{kj}(a_1, \dots, a_r)\, \xi_{ji}(x_1', \dots, x_n') \qquad (i=1\cdots n;\ k=1\cdots r),$$

which can also be written as:

$$(6) \qquad \xi_{ji}(x_1', \dots, x_n') = \sum_{k=1}^{r} \alpha_{jk}(a_1, \dots, a_r) \frac{\partial x_i'}{\partial a_k} \qquad (i=1\cdots n;\ j=1\cdots r).$$

Here, neither the determinant of the $\psi_{kj}(a)$, nor that of the $\alpha_{jk}(a)$ vanishes identically;[4] *(Translator's note: The functions $\xi_{ji}(x)$ are, up to an overall minus sign, just the coefficients of the r infinitesimal transformations obtained by differentiation with respect to the parameters in the identity: $\xi_{ji}(x) = -\frac{\partial f_i}{\partial a_j}(x; e)$, $i = 1, \dots, n$, $j = 1, \dots, n$. Furthermore, with the purely local assumptions we made above, we in fact have $\psi_{kj}(e) = -\delta_k^j$, so the mentioned determinants do not vanish for all a in a neighborhood of e.) in addition, it is impossible to indicate r quantities $e_1, \dots, e_r$ independent of $x_1', \dots, x_n'$ [constants] and not all vanishing such that the n expressions:*

$$e_1\, \xi_{1i}(x') + \cdots + e_r\, \xi_{ri}(x') \qquad (i=1\cdots n)$$

vanish simultaneously.

Furthermore, the latter property is clearly equivalent to the nonexistence of constants $e_1, \dots, e_r$ not all zero such that:

$$0 \equiv e_1 X_1 + \cdots + e_r X_r,$$

that is to say: the r infinitesimal transformations $X_1, \dots, X_r$ are *linearly* independent. This property was shown to derive from essentiality of parameters, and Theorem 8 p. 75 below will establish a satisfactory converse.

Thus in the above formulation, Engel and Lie implicitly introduce the r infinitesimal transformations:

$$X_k := \sum_{i=1}^{n} \xi_{ki}(x_1, \dots, x_n) \frac{\partial}{\partial x_i} \qquad {\scriptstyle (k=1\cdots r)}$$

of an r-term continuous group *not* as partial derivatives at the identity element with respect to the parameters (which would be the intuitively clearest way):

$$X_k^e := \sum_{i=1}^{n} \frac{\partial f_i}{\partial a_k}(x; e) \frac{\partial}{\partial x_i} \equiv -X_k,$$

but rather indirectly as having coefficients $\xi_{ki}(x)$ stemming from the fundamental differential equations (5); in fact, we already saw in (3) that both definitions agree modulo sign. Slightly later in the treatise, the introduction is given explicitly.

Proposition. ([1], p. 67) *Associated to every r-term group $x_i' = f_i(x_1, \dots, x_n, a_1, \dots, a_r)$, there are r infinitesimal transformations:*

$$X_k(f) = \sum_{i=1}^{n} \xi_{ki}(x_1, \dots, x_n) \frac{\partial f}{\partial x_i} \qquad {\scriptstyle (k=1\cdots r)}$$

which stand in such a relationship that equations of the form:

$$\frac{\partial x_i'}{\partial a_k} = \sum_{j=1}^{r} \psi_{kj}(a_1, \dots, a_r)\, \xi_{ji}(x_1', \dots, x_n')$$

hold true, that can be solved with respect to the ξ_{ji}:

$$\xi_{ji}(x_1', \dots, x_n') = \sum_{k=1}^{r} \alpha_{jk}(a_1, \dots, a_r) \frac{\partial x_i'}{\partial a_k}.$$

The gist of Lie's theory is to show that the datum of an r-term continuous (local) transformation group is equivalent the datum of r infinitesimal transformations

$X_1, \dots, X_r$ associated in this way. The next few chapters are devoted to exposing this (well known) one-to-one correspondence between local Lie groups and Lie algebras of local analytic vector fields (Chap. 9), by following, by summarizing and by adapting the original presentation, but without succumbing to the temptation of overformalizing some alternative coordinate-free reasonings with the help of some available contemporary views, because this would certainly impoverish the depth of Lie's original thought.

3.7 Fundamental Differential Equations for the Inverse Transformations

According to fundamental Theorem 3 on p. 40 above, a general r-term continuous transformation group $x_i' = f_i(x; a_1, \dots, a_r)$ satisfies partial differential equations: $\partial f_i / \partial a_k = \sum_{j=1}^r \psi_{kj}(a)\, \xi_{ji}(f_1, \dots, f_n)$ that are used everywhere in basic Lie theory. For the study of the adjoint group in Chap. 16 below, we also need to know how to precisely write the fundamental differential equations that are satisfied by the group of *inverse* transformations:

$$x_i = f_i\big(x'; \mathbf{i}(a)\big) \qquad (i=1 \cdots n),$$

and this is easy. Following a well known path, we begin by differentiating these equations with respect to the parameters a_k:

$$\frac{\partial x_i}{\partial a_k} = \sum_{l=1}^r \frac{\partial f_i}{\partial a_l}\big(x'; \mathbf{i}(a)\big)\, \frac{\partial \mathbf{i}_l}{\partial a_k}(a) \qquad (i=1 \cdots n;\; k=1 \cdots r).$$

Naturally, here we replace the $\partial f_i / \partial a_l$ by their values $\sum_{j=1}^r \psi_{lj}\, \xi_{ji}$ given by the fundamental differential equations of Theorem 3, and we obtain a double sum:

$$\begin{aligned} \frac{\partial x_i}{\partial a_k} &= \sum_{l=1}^r \sum_{j=1}^r \psi_{lj}\big(\mathbf{i}(a)\big)\, \xi_{ji}\big(f(x'; \mathbf{i}(a)\big)\, \frac{\partial \mathbf{i}_l}{\partial a_k}(a) \\ &=: \sum_{j=1}^r \vartheta_{kj}(a)\, \xi_{ji}(x) \qquad (i=1 \cdots n;\; k=1 \cdots r), \end{aligned}$$

which we contract to a single sum by simply introducing the following new $r \times r$ auxiliary matrix of parameter functions:

$$\vartheta_{kj}(a) := \sum_{l=1}^r \psi_{lj}\big(\mathbf{i}(a)\big)\, \frac{\partial \mathbf{i}_l}{\partial a_k}(a) \qquad (k, j=1 \cdots r),$$

whose precise expression in terms of $\mathbf{i}(a)$ will not matter anymore. It now remains to check that this matrix $\big(\vartheta_{kj}(a)\big)_{1 \leqslant k \leqslant r}^{1 \leqslant j \leqslant r}$ is invertible for all a in a neighborhood of the identity element $e = (e_1, \dots, e_r)$. In fact, we claim that:

$$\vartheta_{kj}(e) = \delta_k^j,$$

which will clearly assure the invertibility in question. Firstly, we remember from Theorem 3 on p. 40 that $\psi_{lj}(e) = -\delta_l^j$. Thus secondly, it now only remains to check that $\frac{\partial \mathbf{i}_l}{\partial a_k}(e) = -\delta_k^l$.

To check this, we differentiate with respect to a_k the identities: $e_j \equiv \mathbf{m}_j\big(a, \mathbf{i}(a)\big)$, $j = 1, \dots, r$, which hold by definition, and we get:

$$0 \equiv \frac{\partial \mathbf{m}_j}{\partial a_k}(e,e) + \sum_{l=1}^{r} \frac{\partial \mathbf{m}_j}{\partial b_l}(e,e) \frac{\partial \mathbf{i}_l}{\partial a_k}(e) \qquad (j=1\cdots r).$$

Alternatively, by differentiating the two families of r identities $a_j \equiv \mathbf{m}_j(a,e)$ and $b_j \equiv \mathbf{m}_j(e,b)$ with respect to a_k and with respect to b_l, we immediately get two expressions:

$$\frac{\partial \mathbf{m}_j}{\partial a_k}(e,e) = \delta_k^j \qquad \text{and} \qquad \frac{\partial \mathbf{m}_j}{\partial b_l}(e,e) = \delta_l^j$$

which, when inserted into the above, yield the announced $\frac{\partial \mathbf{i}_l}{\partial a_k}(e) = -\delta_k^l$. Sometimes, we will write $g(x;a)$ instead of $f\big(x; \mathbf{i}(a)\big)$. As a result:

Lemma 3.2. *The finite continuous transformation group $x_i' = f_i(x;a)$ and its inverse transformations $x_i = g_i(x';a) := f_i\big(x'; \mathbf{i}(a)\big)$ both satisfy fundamental partial differential equations of the form:*

$$\begin{cases} \dfrac{\partial x_i'}{\partial a_k}(x;a) = \displaystyle\sum_{j=1}^{r} \psi_{kj}(a)\, \xi_{ji}\big(x'(x;a)\big) & (i=1\cdots n;\ k=1\cdots r), \\[2ex] \dfrac{\partial x_i}{\partial a_k}(x';a) = \displaystyle\sum_{j=1}^{r} \vartheta_{kj}(a)\, \xi_{ji}\big(x(x';a)\big) & (i=1\cdots n;\ k=1\cdots r), \end{cases}$$

where ψ and ϑ are two $r \times r$ matrices of analytic functions with $-\psi_{kj}(e) = \vartheta_{kj}(e) = \delta_k^j$, and where the functions ξ_{ji} appearing in both *systems of equations:*

$$\xi_{ji}(x) := -\frac{\partial f_i}{\partial x_j}(x;e) \qquad (i=1\cdots n;\ j=1\cdots r)$$

are, up to an overall minus sign, just the coefficients of the r infinitesimal transformations

$$X_1^e = \frac{\partial f}{\partial a_1}(x;e), \dots\dots, X_r^e = \frac{\partial f}{\partial a_r}(x;e)$$

obtained by differentiating the finite equations with respect to the parameters at the identity element.

Theorem 4. *If, in the equations:*

$$x_i' = f_i(x_1,\dots,x_n,a_1,\dots,a_r) \qquad (i=1\cdots n)$$

of a group with the r essential parameters $a_1,\dots,a_r$, one considers the x_i as functions of $a_1,\dots,a_r$ and of $x_1',\dots,x_n'$, then there exist differential equations of the form:

$$\frac{\partial x_i}{\partial a_k} = \sum_{j=1}^{r} \vartheta_{kj}(a_1,\dots,a_r)\,\xi_{ji}(x_1,\dots,x_n) \qquad (i=1\cdots n;\; k=1\cdots r).$$

3.8 Transfer of Individual Infinitesimal Transformations by the Group

With $x = g(x';a)$ denoting the inverse of $x' = f(x;a)$, we now differentiate with respect to a_k the identically satisfied equations:

$$x_i' \equiv f_i\big(g(x';a);a\big) \qquad (i=1\cdots n),$$

which just say that an arbitrary transformation of the group followed by its inverse gives the identity transformation, and we immediately get:

$$0 \equiv \sum_{\nu=1}^{n} \frac{\partial f_i}{\partial x_\nu}\frac{\partial g_\nu}{\partial a_k} + \frac{\partial f_i}{\partial a_k} \qquad (i=1\cdots n;\; k=1\cdots r).$$

Thanks to the above two systems of partial differential equations, we may then replace $\partial g_\nu / \partial a_k$ by its value from the second equation of the lemma above, and also $\partial f_i / \partial a_k$ by its value from the first equation in the same lemma:

(7)
$$0 \equiv \sum_{\nu=1}^{n} \Big\{ \sum_{j=1}^{r} \vartheta_{kj}(a)\,\xi_{j\nu}(g) \Big\} \frac{\partial f_i}{\partial x_\nu} + \sum_{j=1}^{r} \psi_{kj}(a)\,\xi_{ji}(f)$$
$$(i=1\cdots n;\; k=1\cdots r).$$

In order to bring these equations to a more symmetric form, following [1] pp. 44–45, we fix k and we multiply, for $i=1$ to n, the i-th equation by $\frac{\partial}{\partial x_i'}$, we apply the summation $\sum_{i=1}^{n}$, we use the fact that, via the diffeomorphism $x \mapsto f_a(x) = x'$, the coordinate vector fields transform as:

$$\frac{\partial}{\partial x_\nu} = \sum_{i=1}^{n} \frac{\partial f_i}{\partial x_\nu}\frac{\partial}{\partial x_i'},$$

which just means in contemporary notation that:

$$(f_a)_*\big(\tfrac{\partial}{\partial x_\nu}\big) = \textstyle\sum_{i=1}^n \frac{\partial f_i}{\partial x_\nu}\frac{\partial}{\partial x_i'} \qquad (\nu = 1 \cdots n),$$

and we obtain, thanks to this observation, completely symmetric equations:

$$0 \equiv \sum_{j=1}^n \vartheta_{kj}(a) \sum_{\nu=1}^n \xi_{j\nu}(x)\frac{\partial}{\partial x_\nu} + \sum_{j=1}^r \psi_{kj}(a) \sum_{\nu=1}^r \xi_{j\nu}(x')\frac{\partial}{\partial x_\nu'} \qquad (k = 1 \cdots r),$$

in which the push-forwards $(f_a)_*(\partial/\partial x_\nu)$ are now implicitly understood. It is easy to see that exactly the same equations, but with the opposite push-forwards $(g_a)_*(\partial/\partial x_\nu')$, can be obtained by subjecting to similar calculations the reverse, identically satisfied equations: $x_i \equiv g_i\big(f(x;a);a\big)$. Consequently, we have obtained two families of equations:

$$(8)\quad \begin{cases} 0 \equiv \displaystyle\sum_{j=1}^n \vartheta_{kj}(a) \sum_{\nu=1}^n \xi_{j\nu}(x)\frac{\partial}{\partial x_\nu}\bigg|_{x\mapsto g_a(x')} + \sum_{j=1}^r \psi_{kj}(a) \sum_{\nu=1}^r \xi_{j\nu}(x')\frac{\partial}{\partial x_\nu'}, \\ 0 \equiv \displaystyle\sum_{\nu=1}^n \vartheta_{kj}(a) \sum_{\nu=1}^n \xi_{j\nu}(x)\frac{\partial}{\partial x_\nu} + \sum_{j=1}^r \psi_{kj}(a) \sum_{\nu=1}^r \xi_{j\nu}(x')\frac{\partial}{\partial x_\nu'}\bigg|_{x'\mapsto f_a(x)} \\ \qquad\qquad (k = 1 \cdots r), \end{cases}$$

in which we represent push-forwards of vector fields by the symbol of variable replacement $x \mapsto g_a(x')$ in the first line, and similarly in the second line, by $x' \mapsto f_a(x)$.

Theorem 5. *If the equations:*

$$x_i' = f_i(x_1,\dots,x_n,a_1,\dots,a_r) \qquad (i = 1 \cdots n)$$

with the r essential parameters $a_1,\dots,a_r$ represent an r-term transformation group, if, moreover, $F(x_1',\dots,x_n')$ denotes an arbitrary function of $x_1',\dots,x_n'$ and lastly, if the $\xi_{j\nu}(x)$, $\psi_{kj}(a)$, $\vartheta_{kj}(a)$ denote the same functions of their arguments as in the two Theorems 3 and 4, then the relations:

$$\sum_{j=1}^r \vartheta_{kj}(a) \sum_{\nu=1}^n \xi_{j\nu}(x)\frac{\partial F}{\partial x_\nu} + \sum_{j=1}^r \psi_{kj}(a) \sum_{\nu=1}^n \xi_{j\nu}(x')\frac{\partial F}{\partial x_\nu'} = 0$$

$$(k = 1 \cdots r)$$

hold true after the substitution $x_1' = f_1(x,a), \dots, x_n' = f_n(x,a)$.

3.8.1 A Synthetic, Geometric Counterpart of the Computations

To formulate an adequate interpretation of the above considerations, we must introduce the two systems of r infinitesimal transformations ($1 \leqslant k \leqslant r$):

$$X_k := \sum_{i=1}^{n} \xi_{ki}(x)\, \frac{\partial}{\partial x_i} \qquad \text{and} \qquad X_k' := \sum_{i=1}^{n} \xi_{ki}(x')\, \frac{\partial}{\partial x_i'},$$

where the second ones are defined to be *exactly the same vector fields* as the first ones, but considered on x'-space. In fact, this target, auxiliary space x' has to be considered to be the *same* space as x-space, because the considered transformation group acts on a single space. So we can also assume that X_k' coincides with the value of X_k at x' and we shall sometimes switch to another notation:

$$\boxed{X_k' \equiv X_k\big|_{x'}}\,.$$

Letting now α and $\widetilde{\vartheta}$ be the inverse matrices of ψ and of ϑ, namely:

$$\sum_{k=1}^{r} \alpha_{lk}(a)\, \psi_{kj}(a) = \delta_l^j, \qquad \sum_{k=1}^{r} \widetilde{\vartheta}_{lk}(a)\, \vartheta_{kj}(a) = \delta_l^j,$$

we can multiply the first (resp. the second) line of (8) by $\alpha_{lk}(a)$ (resp. by $\widetilde{\vartheta}_{lk}(a)$) and then sum over $k = 1, \ldots, r$ in order to obtain the resolved equations:

$$\begin{cases} 0 \equiv \sum_{k=1}^{r} \sum_{j=1}^{r} \alpha_{lk}(a)\, \vartheta_{kj}(a)\, X_j + X_l' & (k=1\cdots r), \\ 0 \equiv X_l + \sum_{k=1}^{r} \sum_{j=1}^{r} \widetilde{\vartheta}_{lk}(a)\, \psi_{kj}(a)\, X_j' & (k=1\cdots r), \end{cases}$$

in which we have suppressed the push-forward symbols. We can readily rewrite such equations in the contracted form:

$$X_k = \sum_{j=1}^{r} \rho_{jk}(a)\, X_j' \qquad \text{and} \qquad X_k' = \sum_{j=1}^{r} \widetilde{\rho}_{jk}(a)\, X_j \qquad (k=1\cdots r),$$

by introducing two appropriate auxiliary $r \times r$ matrices $\rho_{jk}(a) := -\sum_{l=1}^{r} \widetilde{\vartheta}_{kl}(a)\, \psi_{lj}(a)$ and $\widetilde{\rho}_{jk}(a) := -\sum_{l=1}^{r} \alpha_{kl}(a)\, \vartheta_{lj}(a)$ of analytic functions (whose precise expression does not matter here) which depend only upon a and which, naturally, are inverses of each other. A diagram illustrating what we have achieved at this point is welcome and intuitively helpful.

Proposition 3.2. *If, in each one of the r basic infinitesimal transformations of the finite continuous transformation group $x' = f(x; a) = f_a(x)$ having the inverse trans-*

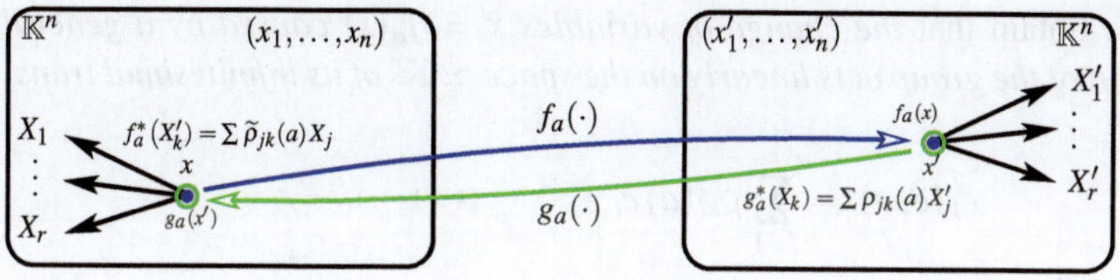

Fig. 3.4 Transfer of infinitesimal transformations by the group

formations $x = g_a(x')$, namely if in the vector fields:

$$X_k = \sum_{i=1}^{n} \xi_{ki}(x) \frac{\partial}{\partial x_i} \qquad {\scriptstyle (k=1\cdots r)}, \qquad \xi_{ki}(x) := -\tfrac{\partial f_i}{\partial a_k}(x; e),$$

one introduces *the new variables $x' = f_a(x)$, that is to say:* replaces *x by $g_a(x')$ and $\frac{\partial}{\partial x_i}$ by $\sum_{\nu=1}^{n} \frac{\partial f_\nu}{\partial x_i}(x; a) \frac{\partial}{\partial x'_\nu}$, then one necessarily obtains a linear combination of the same infinitesimal transformations $X'_l = \sum_{i=1}^{n} \xi_{ki}(x') \frac{\partial}{\partial x'_i}$ at the point x' with coefficients depending only upon the parameters $a_1,\dots,a_r$:*

$$(f_a)_*\big(X_k\big|_x\big) = (g_a)^*\big(X_k\big|_{g_a(x')}\big) = \sum_{l=1}^{r} \rho_{lk}(a_1,\dots,a_r) X_l\big|_{x'} \qquad {\scriptstyle (k=1\cdots r)}.$$

Of course, through the inverse change of variable $x' \mapsto f_a(x)$, the infinitesimal transformations X'_k are subjected to similar linear substitutions:

$$(g_a)_*\big(X'_k\big|_{x'}\big) = (f_a)^*\big(X'_k\big|_{f_a(x)}\big) = \sum_{l=1}^{r} \widetilde{\rho}_{lk}(a) X_l\big|_x \qquad {\scriptstyle (k=1\cdots r)}.$$

3.8.2 Transfer of General Infinitesimal Transformations

Thanks to the linearity of the tangent map, we deduce that the general infinitesimal transformation of our group:

$$X := e_1 X_1 + \cdots + e_r X_r,$$

coordinatized in the basis $\big(X_k\big)_{1 \leqslant k \leqslant r}$ by means of some r arbitrary constants $e_1,\dots,e_r \in \mathbb{K}$, then transforms as:

$$\begin{aligned}(g_a)^*\Big(e_1 X_1 + \cdots + e_r X_r\big|_{g_a(x')}\Big) &= \sum_{k=1}^{r} e_k \sum_{l=1}^{r} \rho_{lk}(a) X_l\big|_{x'} \\ &=: e'_1(e; a) X_1\big|_{x'} + \cdots + e'_r(e; a) X_r\big|_{x'}\end{aligned}$$

and hence we obtain that *the change of variables $x' = f_a(x)$ caused by a general transformation of the group acts linearly on the space $\simeq \mathbb{K}^r$ of its infinitesimal transformations*:

$$e'_k(e;a) := \sum_{l=1}^{r} \rho_{kl}(a)\, e_l \qquad\qquad (k=1\cdots r),$$

by just multiplying the coordinates e_l by the matrix $\rho_{kl}(a)$. Inversely, we have:

$$e_k(e';a) = \sum_{l=1}^{r} \widetilde{\rho}_{kl}(a)\, e'_l \qquad\qquad (k=1\cdots r),$$

where $\widetilde{\rho}(a)$ is the inverse matrix of $\rho(a)$.

Proposition 4. ([1], p. 81) *If the equations $x'_i = f_i(x_1,\dots,x_n,a_1,\dots,a_r)$ represent an r-term group and if this group contains the r independent infinitesimal transformations:*

$$X_k(f) = \sum_{i=1}^{n} \xi_{ki}(x_1,\dots,x_n)\,\frac{\partial f}{\partial x_i} \qquad\qquad (k=1\cdots r),$$

then after the introduction of the new variables $x'_i = f_i(x,a)$, the general infinitesimal transformation:

$$e_1 X_1(f) + \cdots + e_r X_r f$$

keeps its form in so far as, for every system of values $e_1,\dots,e_r$, there is a relation of the form:

$$\sum_{k=1}^{r} e_k X_k(f) = \sum_{k=1}^{r} e'_k X'_k(f),$$

where $e'_1,\dots,e'_r$ are independent linear homogeneous functions of $e_1,\dots,e_r$ with coefficients which are functions of $a_1,\dots,a_r$.

3.8.3 Towards the Adjoint Action

Thanks to the linearity of the tangent map, we deduce that the general transformation of our group:

$$X := e_1 X_1 + \cdots + e_r X_r,$$

coordinatized in the basis $(X_k)_{1 \leqslant k \leqslant r}$ by means of some r arbitrary constants $e_1,\dots,e_r \in \mathbb{K}$, then transforms as:

$$(g_a)^*\Big(e_1X_1+\cdots+e_rX_r\big|_{g_a(x')}\Big)=\sum_{k=1}^r e_k\sum_{l=1}^r \rho_{lk}(a)X_l\big|_{x'}$$
$$=: e_1'(e;a)X_1\big|_{x'}+\cdots+e_r'(e;a)X_r\big|_{x'}$$

and hence we obtain that *the change of variables $x'=f_a(x)$ performed by a general transformation of the group then acts linearly on the space $\simeq \mathbb{K}^r$ of its infinitesimal transformations*:

$$\boxed{e_k'(e;a):=\sum_{l=1}^r \rho_{kl}(a)\cdot e_l} \qquad (k=1\cdots r),$$

by just multiplying the coordinates e_l by the matrix $\rho_{kl}(a)$.

Nowadays, the adjoint action is defined as an action of an abstract Lie group on its Lie algebra. But in Chap. 16 below, Lie defines it in the more general context of a transformation group, as follows. Employing the present terminology, one considers the general infinitesimal transformation $X\big|_{x'}=e_1X_1+\cdots+e_rX_r\big|_{x'}$ of the group as being based at the point x', and one computes the adjoint action $\mathrm{Ad}\,f_a\big(X\big|_{x'}\big)$ of f_a on $X\big|_{x'}$ by differentiating at $t=0$ the composition:[5] (Translator's note: One term groups $(t,x)\longmapsto \exp(tX)(x)$ are introduced in Chap. 4.)

$$f_a\circ\exp(tX)\circ f_a^{-1}$$

which represents the action of the interior automorphism associated to f_a on the one-parameter subgroup $(t,w)\longmapsto \exp(tX)(x)$ generated by X:

$$\begin{aligned}
\mathrm{Ad}\,f_a\big(X\big|_{x'}\big) &:= \frac{\mathrm{d}}{\mathrm{d}t}\Big(f_a\circ\exp(tX)(\cdot)\circ f_a^{-1}(x')\Big)\Big|_{t=0}\\
&=(f_a)_*\frac{\mathrm{d}}{\mathrm{d}t}\Big(\exp(tX)\big(f_a^{-1}(x')\big)\Big)\Big|_{t=0}\\
&=(f_a)_*\big(X\big|_{f_a^{-1}(x')}\big)\\
&=(g_a)^*\big(X\big|_{g_a(x')}\big)\\
&=(g_a)^*\Big(e_1X_1+\cdots+e_rX_r\big|_{g_a(x')}\Big)\\
&=e_1'(e;a)X_1\big|_{x'}+\cdots+e_r'(e;a)X_r\big|_{x'}.
\end{aligned}$$

We thus recover *exactly* the linear action $e_k'=e_k'(e;a_1,\dots,a_r)$ boxed above. A diagram is welcome: from the left, the point x is sent by f_a to $x'=f_a(x)$ — or inversely $x=g_a(x')$ —, the vector $X|_x=\sum e_kX_k\big|_{g_a(x')}$ is sent to $(g_a)^*\big(X\big|_{g_a(x')}\big)$ by means of the differential $(f_a)_*=(g_a)^*$ and Lie's theorem says that this transferred vector is a linear combination of the existing vectors $X_k|_{x'}$ based at x', with coefficients depending on the e_j and on the a_j.

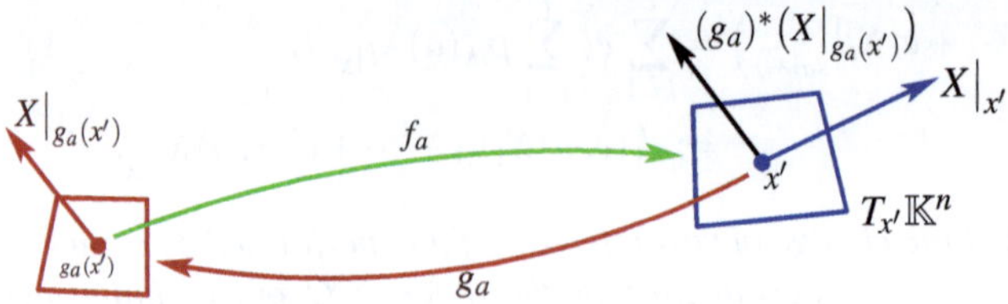

Fig. 3.5 Differentiating the action of an interior automorphism

3.9 Substituting the Axiom of Inverse for a Differential Equations Assumption

Notwithstanding the counterexample discovered by Engel in 1884 (p. 21), in his systematic treatise Lie wanted to avoid employing both the axiom of inverse and the axiom of identity as much as possible. As a result, the first nine fundamental chapters of [1] regularly emphasize what can be derived from only the axiom of composition.

In this section, instead of purely local hypotheses valuable in small polydiscs (Sect. 3.1), we present the semi-global topological hypotheses accurately made by Lie and Engel at the very beginning of their book. After a while, they emphasize that the fundamental differential equations $\frac{\partial x_i'}{\partial a_k} = \sum_{j=1}^r \psi_{kj}(a)\, \xi_{ji}(x')$, which can be deduced from only the composition axiom, should assume the position of being the main continuous group assumption. A central technical theorem states that if some transformation equations $x' = f(x; a)$ with essential parameters satisfy differential equations $\frac{\partial x_i'}{\partial a_k} = \sum_{j=1}^r \psi_{kj}(a)\, \xi_{ji}(x')$, for x and a running in appropriate domains $\mathscr{X} \subset \mathbb{K}^n$ and $\mathscr{A} \subset \mathbb{K}^r$, then every transformation $x' = f(x; a)$ whose parameter a lies in a small neighborhood of some fixed $a^0 \in \mathscr{A}$ can be obtained by firstly performing the initial transformation $\overline{x} = f(x; a^0)$ and then secondly by performing a certain transformation:

$$x_i' = \exp\big(t\lambda_1 X_1 + \cdots + t\lambda_r X_r\big)(\overline{x}_i) \qquad (i=1\cdots n)$$

of the one-term group (Chap. 4 below) generated by some suitable linear combination $\lambda_1 X_1 + \cdots + \lambda_r X_r$ of the n infinitesimal transformations $X_k := \sum_{i=1}^n \xi_{ki}(x) \frac{\partial}{\partial x_i}$. This theorem will be of crucial use when establishing the so-called *Second Fundamental Theorem*: *To any Lie algebra of local analytic vector fields is associated a local Lie transformation group containing the identity element* (Chap. 9 below).

Also postponed to Chap. 9 below, Lie's answer to Engel's counterexample will show that, starting from transformation equations $x' = f(x; a)$ that are only assumed to be closed under composition, one can always capture the identity element and all the inverses of transformations near the identity by appropriately changing coordinates in the parameter space (Theorem 26, p. 177).

3.9.1 Specifying Domains of Existence

Thus, we consider $\mathbb{K}$-analytic transformation equations $x_i' = f_i(x; a)$ defined on a more general domain than a product $\Delta_\rho^n \times \square_\sigma^r$ of two small polydiscs centered at the origin. Here is how Lie and Engel specify their domains of existence on p. 14 of [1], and these domains might be global.

§ **2**. In the transformation equations:

$$x_i' = f_i(x_1, \dots, x_n; a_1, \dots, a_r) \qquad (i=1\cdots n), \tag{1}$$

let all the parameters $a_1, \dots, a_r$ now be essential.

Since the f_i are analytic functions of their arguments, in the domain [GEBIETE] of all systems of values $x_1, \dots, x_n$ and in the domain of all systems of values $a_1, \dots, a_r$, we can choose a region [BEREICH] (x) and, respectively, a region (a) such that the following holds:

Firstly. The $f_i(x, a)$ are single-valued [EINDEUTIG] functions of the $n+r$ variables $x_1, \dots, x_n$, $a_1, \dots, a_r$ in the complete extension [AUSDEHNUNG] of the two regions (x) and (a).

Secondly. The $f_i(x, a)$ behave regularly in the neighborhood of every system of values $x_1^0, \dots, x_n^0$, $a_1^0, \dots, a_r^0$, hence are expandable in ordinary power series with respect to $x_1 - x_1^0$, ..., $x_n - x_n^0$, as soon as $x_1^0, \dots, x_n^0$ lies arbitrarily in the domain (x), and $a_1^0, \dots, a_r^0$ lies arbitrarily in the domain (a).

Thirdly. The functional determinant:

$$\sum \pm \frac{\partial f_1}{\partial x_1} \cdots \frac{\partial f_n}{\partial x_n}$$

vanishes for no combination of systems of values of x_i and of a_k in the two domains (x) and (a), respectively.

Fourthly. If one gives to the parameters a_k in the equations $x_i' = f_i(x, a)$ any of the values a_k^0 in the domain (a), then the equations:

$$x_i' = f_i(x_1, \dots, x_n; a_1^0, \dots, a_r^0) \qquad (i=1\cdots n)$$

always produce two different systems of values $x_1', \dots, x_n'$ for two different systems of values $x_1, \dots, x_n$ of the domain (x).

We assume that the two regions (x) and (a) are chosen in such a way that these four conditions are satisfied. If we give to the variables x_i in the equations $x_i' = f_i(x, a)$ all possible values in (x) and to the parameters a_k all possible values in (a), then in their domain, the x_i' run throughout a certain region, which we can denote symbolically by the equation $x' = f\big((x)(a)\big)$. This new domain has the following properties:

Firstly. If $a_1^0, \dots, a_r^0$ is an arbitrary system of values of (a) and $x_1'^0, \dots, x_n'^0$ an arbitrary system of values of the subregion $x' = f\big((x)(a^0)\big)$, then in the neighborhood of the system of values $x_i'^0$, a_k^0, the $x_1, \dots, x_n$ can be expanded in ordinary power series with respect to $x_1' - x_1'^0, \dots, x_n' - x_n'^0$, $a_1 - a_1^0, \dots, a_n - a_n^0$.

Secondly. If one gives to the a_k fixed values a_k^0 in the domain (a), then in the equations:

$$x_i' = f_i(x_1, \dots, x_n; a_1^0, \dots, a_r^0) \qquad {\scriptstyle (i=1\cdots n)},$$

the quantities $x_1, \dots, x_n$ will be single-valued functions of $x_1', \dots, x_n'$, which behave regularly in the complete extension of the region $x' = f\big((x)(a^0)\big)$.

Referring to the excerpt p. 26, we may rephrase these basic assumptions as follows. The $f_i(x; a)$ are defined for (x, a) belonging to the product:

$$\mathscr{X} \times \mathscr{A} \subset \mathbb{K}^n \times \mathbb{K}^r$$

of two domains $\mathscr{X} \subset \mathbb{K}^n$ and $\mathscr{A} \subset \mathbb{K}^r$. These functions are $\mathbb{K}$-analytic in both variables, hence expandable in Taylor series at every point (x^0, a^0). Furthermore, for every fixed a^0, the map $x \mapsto f(x; a^0)$ is assumed to constitute a $\mathbb{K}$-analytic diffeomorphism of $\mathscr{X} \times \{a^0\}$ onto its image. Of course, the inverse map is also locally expandable in power series, by virtue of the $\mathbb{K}$-analytic inverse function theorem.

To insure that the composition of two transformations exists, one requires that there exist nonempty subdomains:

$$\mathscr{X}^1 \subset \mathscr{X} \qquad \text{and} \qquad \mathscr{A}^1 \subset \mathscr{A},$$

with the property that for every fixed $a^1 \in \mathscr{A}^1$:

$$f\big(\mathscr{X}^1 \times \{a^1\}\big) \subset \mathscr{X},$$

so that for every such an $a^1 \in \mathscr{A}^1$ and for every fixed $b \in \mathscr{A}$, the composed map:

$$x \longmapsto f\big(f(x; a^1); b\big)$$

is well defined for all $x \in \mathscr{X}^1$ and moreover, is a $\mathbb{K}$-analytic diffeomorphism onto its image. In fact, it is even $\mathbb{K}$-analytic with respect to all the variables (x, a^1, b) in $\mathscr{X}^1 \times \mathscr{A}^1 \times \mathscr{A}$. Lie's fundamental and unique *group composition axiom* may then be expressed as follows.

(A1) There exists a $\mathbb{K}^r$-valued $\mathbb{K}$-analytic map $\varphi = \varphi(a, b)$ defined in $\mathscr{A}^1 \times \mathscr{A}^1$ with $\varphi\big(\mathscr{A}^1 \times \mathscr{A}^1\big) \subset \mathscr{A}$ such that:

$$f\big(f(x; a); b\big) \equiv f\big(x; \varphi(a, b)\big) \qquad \text{for all } x \in \mathscr{X}^1,\ a \in \mathscr{A}^1,\ b \in \mathscr{A}^1.$$

Here are two further specific unmentioned assumptions that Lie presupposes, still with the goal of admitting neither the identity element, nor the existence of inverses.

(A2) There is a $\mathbb{K}^r$-valued $\mathbb{K}$-analytic map $(a,c) \longmapsto \mathsf{b} = \mathsf{b}(a,c)$ defined for a running in a certain (nonempty) subdomain $\mathscr{A}^2 \subset \mathscr{A}^1$ and for c running in a certain (nonempty) subdomain $\mathscr{C}^2 \subset \mathscr{A}^1$ with $\mathsf{b}(\mathscr{A}^2 \times \mathscr{C}^2) \subset \mathscr{A}^1$ which solves b in terms of (a,c) in the equations $c_k = \varphi_k(a,b)$, namely which satisfies identically:

$$c \equiv \varphi\big(a,\, \mathsf{b}(a,c)\big) \qquad \text{for all } a \in \mathscr{A}^2,\ c \in \mathscr{C}^2.$$

Conversely, $\varphi(a,b)$ solves c in terms of (a,b) in the equations $b_k = \mathsf{b}_k(a,c)$, namely more precisely: there exists a certain (nonempty) subdomain $\mathscr{A}^3 \subset \mathscr{A}^2$ ($\subset \mathscr{A}^1$) and a certain (nonempty) subdomain $\mathscr{B}^3 \subset \mathscr{A}^1$ with $\varphi\big(\mathscr{A}^3 \times \mathscr{B}^3\big) \subset \mathscr{C}^2$ such that one has identically:

$$b \equiv \mathsf{b}\big(a,\, \varphi(a,b)\big) \qquad \text{for all } a \in \mathscr{A}^3,\ b \in \mathscr{B}^3.$$

Example. In Engel's counterexample of the group $x' = \chi(\lambda)x$ with a Riemann uniformizing map $\omega : \Delta \to \Lambda$ as on p. 21 having inverse $\chi : \Lambda \to \Delta$, these three requirements are satisfied, and in addition, we claim that one may even take $\mathscr{X} = \mathbb{K}$ and $\mathscr{A}^1 = \mathscr{A} = \Lambda$, with no shrinking, for composition happens to hold in fact without restriction in this case. Indeed, starting from the general composition:

$$x'' = \chi(\lambda_2)x' = \chi(\lambda_2)\,\chi(\lambda_1)\,x,$$

that is to say, from $x'' = \chi(\lambda_2)\chi(\lambda_1)x$, in order to represent it in the specific form $x'' = \chi(\lambda_3)x$, it is necessary and sufficient to solve $\chi(\lambda_3) = \chi(\lambda_1)\chi(\lambda_2)$, hence we may take for φ:

$$\lambda_3 = \omega\big(\chi(\lambda_1)\chi(\lambda_2)\big) =: \varphi(\lambda_1, \lambda_2),$$

without shrinking the domains, for the two inequalities $|\chi(\lambda_1)| < 1$ and $|\chi(\lambda_2)| < 1$ readily imply that $|\chi(\lambda_1)\chi(\lambda_2)| < 1$ too so that $\omega\big(\chi(\lambda_1)\chi(\lambda_2)\big)$ is defined. On the other hand, for solving λ_2 in terms of (λ_1, λ_3) in the above equation, we are naturally led to define:

$$\mathsf{b}(\lambda_1, \lambda_3) := \omega\big(\chi(\lambda_3)/\chi(\lambda_1)\big),$$

and then $\mathsf{b} = \mathsf{b}(\lambda_1, \lambda_3)$ is defined under the specific restriction that $|\chi(\lambda_3)| < |\chi(\lambda_1)|$.

However, the axiom **(A2)** still happens to be incomplete for later use, and one should add the following axiom in order to also be able to solve a in $c = \varphi(a,b)$.

(A3) There is a $\mathbb{K}^m$-valued $\mathbb{K}$-analytic map $(b,c) \longmapsto \mathsf{a} = \mathsf{a}(b,c)$ defined in $\mathscr{B}^4 \times \mathscr{C}^4$ with $\mathscr{B}^4 \subset \mathscr{A}^1$ and $\mathscr{C}^4 \subset \mathscr{A}^1$, and with $\mathsf{a}\big(\mathscr{B}^4 \times \mathscr{C}^4\big) \subset \mathscr{A}^1$, such that one has identically:

$$c \equiv \varphi\big(\mathsf{a}(b,c),\, b\big) \qquad \text{for all } b \in \mathscr{B}^4,\ c \in \mathscr{C}^4.$$

Conversely, $\varphi(a,b)$ solves c in the equations $a_k = \mathsf{a}_k(b,c)$, namely more precisely: there exist $\mathscr{B}^5 \subset \mathscr{B}^4$ and $\mathscr{A}^5 \subset \mathscr{A}^1$ with $\varphi\big(\mathscr{A}^5 \times \mathscr{B}^5\big) \subset \mathscr{C}^4$ such that one has identically:

$$a \equiv \mathsf{a}\big(b,\, \varphi(a,b)\big) \qquad \text{for all } a \in \mathscr{A}^5,\ b \in \mathscr{B}^5,$$

and furthermore in addition, with $\mathsf{a}\big(\mathscr{B}^5 \times \mathscr{C}^4\big) \subset \mathscr{A}^2$ and with $\mathsf{b}\big(\mathscr{A}^5 \times \mathscr{C}^4\big) \subset \mathscr{B}^4$ such that one also has identically:

$$b \equiv \mathsf{b}\big(\mathsf{a}(b,c),\, c\big) \qquad \text{for all } b \in \mathscr{B}^5,\ c \in \mathscr{C}^4$$
$$a \equiv \mathsf{a}\big(\mathsf{b}(a,c),\, a\big) \qquad \text{for all } a \in \mathscr{A}^5,\ c \in \mathscr{C}^4.$$

The introduction of the numerous (nonempty) domains $\mathscr{A}^2$, $\mathscr{C}^2$, $\mathscr{A}^3$, $\mathscr{B}^3$, $\mathscr{B}^4$, $\mathscr{C}^4$, $\mathscr{A}^5$, $\mathscr{B}^5$ which appears slightly unnatural and seems to depend upon the order in which the solving maps a and b are considered can be avoided by requiring from the beginning only that there exist two subdomains $\mathscr{A}^3 \subset \mathscr{A}^2 \subset \mathscr{A}^1$ such that one has uniformly:

$$\begin{aligned}
c &\equiv \varphi\big(a,\, \mathsf{b}(a,c)\big) && \text{for all } a \in \mathscr{A}^3,\ c \in \mathscr{A}^3\\
c &\equiv \varphi\big(\mathsf{a}(b,c),\, b\big) && \text{for all } b \in \mathscr{A}^3,\ c \in \mathscr{A}^3\\
b &\equiv \mathsf{b}\big(\mathsf{a}(b,c),\, c\big) && \text{for all } b \in \mathscr{A}^3,\ c \in \mathscr{A}^3\\
b &\equiv \mathsf{b}\big(a,\, \varphi(a,b)\big) && \text{for all } a \in \mathscr{A}^3,\ b \in \mathscr{A}^3\\
a &\equiv \mathsf{a}\big(\mathsf{b}(a,c),\, c\big) && \text{for all } a \in \mathscr{A}^3,\ c \in \mathscr{A}^3\\
a &\equiv \mathsf{a}\big(b,\, \varphi(a,b)\big) && \text{for all } a \in \mathscr{A}^3,\ b \in \mathscr{A}^3.
\end{aligned}$$

We will adopt these axioms in the next subsection. Importantly, we would like to point out that, although $\mathsf{b}(a,c)$ seems to represent the group product $a^{-1} \cdot c = \mathbf{m}\big(\mathbf{i}(a),\, c\big)$, the assumption **(A2)** neither reintroduces inverses, nor the identity element, it just means that one may solve b by means of the implicit function theorem in the parameter composition equations $c_k = \varphi_k(a_1, \dots, a_r, b_1, \dots, b_r)$.

3.9.2 *The Group Composition Axiom and Fundamental Differential Equations*

As mentioned earlier, the fundamental Theorem 3, p. 40, concerning differential equations satisfied by a transformation group was in fact stated and proved in [1] under semi-global assumptions essentially equivalent to the ones we just formulated above with $\mathscr{A}^3 \subset \mathscr{A}^2 \subset \mathscr{A}^1$, so now, we can restitute it.

Proposition 3.4. *Under these assumptions, there is an $r \times r$ matrix of functions $\big(\psi_{kj}(a)\big)_{1 \leqslant k \leqslant r}^{1 \leqslant j \leqslant r}$ which is $\mathbb{K}$-analytic and invertible in $\mathscr{A}^3$, and there are certain functions $\xi_{ji}(x)$, $\mathbb{K}$-analytic in $\mathscr{X}$, such that the following differential equations:*

$$\text{(2)} \qquad \frac{\partial x_i'}{\partial a_k}(x;a) = \sum_{j=1}^{r} \psi_{kj}(a)\,\xi_{ji}(x') \qquad {\scriptstyle (i=1\cdots n;\; k=1\cdots r)}$$

are identically satisfied for all $x \in \mathscr{X}^1$ and all $a \in \mathscr{A}^3$ after replacing x' by $f(x;a)$. Here, the functions $\xi_{ji}(x')$ are defined by choosing arbitrarily some fixed $b^0 \in \mathscr{A}^3$ and by setting:

$$\xi_{ji}(x_1',\dots,x_n') = \left[\frac{\partial x_i'}{\partial b_j}\right]_{b=b^0} = \left[\sum_{k=1}^{r} \frac{\partial x_i'}{\partial a_k}\,\frac{\partial \mathsf{a}_k}{\partial b_j}\right]_{b=b^0},$$

and moreover, in the equations inverse to (2)*:*

$$\xi_{ji}(x') = \sum_{k=1}^{r} \alpha_{jk}(a)\,\frac{\partial x_i'}{\partial a_k}(x;a),$$

the (inverse) coefficients $\alpha_{jk}(a)$ are defined by:

$$\alpha_{jk}(a) = \left[\frac{\partial \mathsf{a}_k}{\partial b_j}\right]_{b=b^0}.$$

Proof. The computations which we have already conducted on p. 33 for a purely local transformation group can be performed in a similar way here. In brief, using $b = \mathsf{b}(a,c)$ from **(A2)**, we consider the identities:

$$f_i\big(f(x;a);\,\mathsf{b}(a,c)\big) \equiv f_i(x;c) \qquad {\scriptstyle (i=1\cdots n)}$$

and we differentiate them with respect to a_k; if we use the shorthand $f_i' \equiv f_i\big(x';\mathsf{b}(a,c)\big)$ and $x_j' \equiv f_j(x;a)$, this gives:

$$\sum_{\nu=1}^{n} \frac{\partial f_i'}{\partial x_\nu'}\,\frac{\partial x_\nu'}{\partial a_k} + \sum_{j=1}^{r} \frac{\partial f_i'}{\partial b_j}\,\frac{\partial \mathsf{b}_j}{\partial a_k} \equiv 0 \qquad {\scriptstyle (i=1\cdots n).}$$

By the diffeomorphism assumptions, the matrix $\big(\frac{\partial f_i'}{\partial x_\nu'}(x';b)\big)_{1\leqslant i\leqslant n}^{1\leqslant \nu\leqslant n}$ has a $\mathbb{K}$-analytic inverse for all $(x',b) \in \mathscr{X} \times \mathscr{A}$, so an application of Cramer's rule yields a resolution of the form:

$$\frac{\partial x_\nu'}{\partial a_k}(x;a) = \Xi_{1\nu}\big(x',\mathsf{b}(a,c)\big)\,\frac{\partial \mathsf{b}_1}{\partial a_k}(a,c) + \cdots + \Xi_{r\nu}\big(x',\mathsf{b}(a,c)\big)\,\frac{\partial \mathsf{b}_r}{\partial a_k}(a,c)$$
$${\scriptstyle (\nu=1\cdots n;\; k=1\cdots r).}$$

Here of course, $a, c \in \mathscr{A}^3$. Then we replace c by $\varphi(a,b)$, with $a, b \in \mathscr{A}^3$, we confer to b any fixed value, say b^0 (recall that the identity e is not available), and we get the desired differential equations with $\xi_{ji}(x') := \Xi_{ji}(x',b^0)$ and with $\psi_{kj}(a) := \frac{\partial \mathsf{b}_j}{\partial a_k}(a,b^0)$. Naturally, the invertibility of the matrix $\psi_{kj}(a)$ comes from **(A2)**.

Next, we multiply each equation just obtained (changing indices):

(2')
$$\frac{\partial x_i'}{\partial a_k}(x;a) \equiv \sum_{\nu=1}^{r} \frac{\partial \mathsf{b}_\nu}{\partial a_k}(a,c)\,\Xi_{\nu i}\big(x', \mathsf{b}(a,c)\big)$$

by $\frac{\partial \mathsf{a}_k}{\partial b_l}(b,c)$, where $l \in \{1,\dots,r\}$ is fixed, we consider (b,c), instead of (a,c), as the $2r$ independent variables while $a = \mathsf{a}(b,c)$, and we sum with respect to k for $k = 1,\dots,r$:

$$\sum_{k=1}^{r} \frac{\partial x_i'}{\partial a_k}\frac{\partial \mathsf{a}_k}{\partial b_l} = \sum_{\nu=1}^{r}\sum_{k=1}^{r} \frac{\partial \mathsf{a}_k}{\partial b_l}\frac{\partial \mathsf{b}_\nu}{\partial a_k}\,\Xi_{\nu i}.$$

Now, the chain rule and the fact that $\partial \mathsf{a}_k/\partial b_l$ is the inverse matrix of $\partial \mathsf{b}_\nu/\partial a_k$ enables us to simplify both sides (interchanging members):

(3)
$$\sum_{k=1}^{r} \frac{\partial x_i'}{\partial a_k}\frac{\partial \mathsf{a}_k}{\partial b_l} = \frac{\partial x_i'}{\partial b_l} = \Xi_{li}.$$

Specializing $b := b^0 \in \mathscr{A}^3$, we get the announced representation:

$$\xi_{ji}(x') = \Xi_{ji}(x', b^0) = \partial x_i'/\partial b_j\big|_{b=b_0},$$

and by identification, we also obtain at the same time the representation $\alpha_{jk}(a) = \partial \mathsf{a}_k/\partial b_j\big|_{b=b_0}$. □

We end up by observing that, similarly as we did on p. 33, one could proceed to some further computations, although it would not really be needed for the proposition. We may indeed differentiate the equations:

$$c_\mu \equiv \varphi_\mu\big(\mathsf{a}(b,c), b\big) \qquad (\mu=1\cdots r)$$

with respect to b_l, and for this, we translate a short passage of [1], p. 20.

Hence one has:

$$\sum_{k=1}^{r} \frac{\partial \varphi_\mu}{\partial a_k}\frac{\partial \mathsf{a}_k}{\partial b_l} + \frac{\partial \varphi_\mu}{\partial b_l} = 0 \qquad (\mu, l=1\cdots r),$$

whence we obtain:

$$\frac{\partial \mathsf{a}_k}{\partial b_l} = -\frac{\sum \pm \frac{\partial \varphi_1}{\partial a_1} \cdots \frac{\partial \varphi_{k-1}}{\partial a_{k-1}}\frac{\partial \varphi_k}{\partial b_l}\frac{\partial \varphi_{k+1}}{\partial a_{k+1}} \cdots \frac{\partial \varphi_r}{\partial a_r}}{\sum \pm \frac{\partial \varphi_1}{\partial a_1} \cdots \frac{\partial \varphi_r}{\partial a_r}} = A_{lk}(a_1,\dots,a_r, b_1,\dots,b_r).$$

When we insert these values into (3), we obtain the equations:

(3')
$$\Xi_{li}(x',b) = \sum_{k=1}^{r} A_{lk}(a,b)\,\frac{\partial x_i'}{\partial a_k} \qquad (i=1\cdots n;\ l=1\cdots r).$$

Of course by identification with (3), we must then have $A_{lk}(a,b) = \partial \mathrm{a}_k / \partial b_l$ here. In conclusion, we have presented the complete thought of Lie and Engel, who did not necessarily consider the axioms of groups to be strictly local.

3.9.3 The Differential Equations Assumption and its Consequences

Now, we would like to emphasize that instead of axioms (**A1-2-3**), in his answer to Engel's counterexample and in several other places as well, Lie says that he wants to set as a fundamental hypothesis the existence of a system of differential equations as the one above, with an invertible matrix $\psi_{kj}(a)$.

§ **17**. ([1], pp. 67–68) For the time being, we want to refrain from assuming that the equations $x_i' = f_i(x_1, \dots, x_n, a_1, \dots, a_r)$ should represent an r-term group. Rather, concerning the equations (1), we want only to assume: *firstly*, that they represent a family of ∞^r different transformations, hence that the r parameters $a_1, \dots, a_r$ are all essential, and *secondly*, that they satisfy differential equations of the specific form (2).

So renaming the domain $\mathscr{A}^3$ considered above simply as $\mathscr{A}^1$, we will fundamentally assume in Sect. 4.6 of Chap. 4 that differential equations of the specific form (2) hold for all $x \in \mathscr{X}^1$ and all $a \in \mathscr{A}^1$, *forgetting completely about composition*, and most importantly, *assuming neither the existence of the identity element, nor the existence of inverse transformations*. At first, as explained by Engel and Lie, one can easily deduce from such new economical assumptions two basic nondegeneracy conditions.

Lemma 3.3. *Consider transformation equations $x_i' = f_i(x; a)$ defined for $x \in \mathscr{X}$ and $a \in \mathscr{A}$ having essential parameters $a_1, \dots, a_r$ which satisfy differential equations of the form:*

$$(2) \qquad \frac{\partial x_i'}{\partial a_k}(x; a) = \sum_{j=1}^{r} \psi_{kj}(a)\, \xi_{ji}(x') \qquad (i=1\cdots n;\ k=1\cdots r),$$

for all $x \in \mathscr{X}^1$ and all $a \in \mathscr{A}^1$. Then the determinant of the $\psi_{kj}(a)$ does not vanish identically and furthermore, the r infinitesimal transformations:

$$X_k' = \sum_{i=1}^{n} \xi_{ki}(x') \frac{\partial}{\partial x_i'} \qquad (k=1\cdots r)$$

are independent of each other.

Proof. If the determinant of the $\psi_{kj}(a)$ vanished identically, there would exist a locally defined (for a running in the locus where $\psi_{kj}(a)$ is of maximal, locally constant rank) nonzero $\mathbb{K}$-analytic vector $\big(\tau_1(a),\dots,\tau_r(a)\big)$ in the kernel of $\psi_{kj}(a)$. Consequently, after multiplying (2) by $\tau_k(a)$, we would derive the equations: $\sum_{k=1}^r \tau_k(a)\,\frac{\partial x_i'}{\partial a_k}(x;a) \equiv 0$ which would then contradict the essentiality of parameters, according to the theorem on p. 15.

As a result, for any a belonging to the open set where $\det \psi_{kj}(a) \neq 0$, we can locally invert the differential equations (2) and write them down in the form:

$$\xi_{ji}(x') = \sum_{k=1}^{r} \alpha_{jk}(a)\,\frac{\partial x_i'}{\partial a_k}(x;a) \qquad (i=1\cdots n;\; j=1\cdots r).$$

If there existed constants $e_1',\dots,e_r'$ not all zero with $e_1'X_1'+\cdots+e_r'X_r'=0$, we would then deduce the relation:

$$0 \equiv \sum_{k=1}^{r}\sum_{j=1}^{r} e_j'\,\alpha_{jk}(a)\,\frac{\partial x_i'}{\partial a_k}(x;a)$$

which would again contradict the essentiality of parameters. □

3.9.4 Towards Theorem 26

At the end of the next chapter in Sect. 4.6, we shall be in a position to pursue the restoration of further refined propositions towards Lie's Theorem 26 (translated in Chap. 9, p. 177) which will fully answer Engel's counterexample. In brief, and by anticipation, this theorem states the following.

Let $x_i' = f_i(x; a_1,\dots,a_r)$ be a family of transformation equations which is only assumed to be closed under composition, a *finite continuous transformation group*, in the sense of Lie. According to Proposition 3.4 above, there exists a system of fundamental differential equations of the form:

$$\frac{\partial x_i'}{\partial a_k} = \sum_{j=1}^{r} \psi_{kj}(a)\cdot \xi_{ji}(x') \qquad (i=1\cdots n; k=1\cdots r)$$

which is identically satisfied by the functions $f_i(x,a)$, where the ψ_{kj} are certain analytic functions of the parameters $(a_1,\dots,a_r)$. If one introduces the r infinitesimal transformations:

$$\sum_{i=1}^{n} \xi_{ki}(x)\,\frac{\partial f}{\partial x_i} =: X_k(f) \qquad (k=1\cdots r),$$

and if one forms the so-called *canonical* finite equations:

$$\begin{aligned} x_i' &= \exp\big(\lambda_1 X_1 + \cdots + \lambda_k X_k\big)(x) \\ &=: g_i(x;\lambda_1,\dots,\lambda_r) \qquad (i=1\cdots n) \end{aligned}$$

of the r-term group which is generated by these r infinitesimal transformations—*see* the next Chap. 4 for $\exp X(x)$—then this group contains the identity element, namely $g(x; 0)$, and its transformations are ordered as inverses by pairs, namely: $g(x; -\lambda) = g(x; \lambda)^{-1}$. Lastly, the Theorem 26 in question states that in these finite equations $x_i' = g_i(x; \lambda)$, it is possible to introduce *new* local parameters $\overline{a}_1, \dots, \overline{a}_r$ in place of $\lambda_1, \dots, \lambda_r$ so that the resulting transformation equations:

$$\begin{aligned} x_i' &= g_i\big(x; \lambda_1(\overline{a}), \dots, \lambda_r(\overline{a})\big) \\ &=: \overline{f}_i(x_1, \dots, x_n, \overline{a}_1, \dots, \overline{a}_r) \qquad (i=1\cdots n) \end{aligned}$$

represent a family of ∞^r transformations which embraces, *possibly after analytic extension*, all the ∞^r initial transformations:

$$x_i' = f_i(x_1, \dots, x_n, a_1, \dots, a_r).$$

In this way, Lie not only answers Engel's counterexample $x' = \chi(\lambda)x$ by saying that one has to restitute the plain transformations $x' = \zeta x$ by appropriately changing coordinates in the parameter space, but also, Lie really establishes the conjecture he suspected (quotation p. 20), modulo the fact that the conjecture was not true without a suitable change of coordinates in the parameter space. To our knowledge, no modern treatise restitutes this theorem of Lie, although a great deal of the first 170 pages of the Theorie der Transformationsgruppen is devoted to economizing the axiom of inverse. We finish this chapter with a quotation ([1], pp. 81–82) motivating the introduction of one-term groups $\exp(tX)(x)$ in the next chapter.

3.9.5 *Metaphysical Links with Substitution Theory*

We conclude this chapter with a brief quotation motivating what will follow.

> The concepts and the propositions of the theory of continuous transformation groups often have their analogues in the *theory of substitutions*,[1] that is to say, in the theory of discontinuous groups. In the course of our studies, we will not emphasize this analogy every time, but we will more often remember it by translating the terminology of the theory of substitutions into the theory of transformation groups, and this shall take place as far as possible.
>
> Here, we want to point out that the *one-term groups* in the theory of transformation groups play the same rôle as the *groups generated by a single substitution* in the theory of substitutions.
>
> In a way, we shall consider the one-term groups, or their infinitesimal transformations, as the elements of the r-term group. In the studies concerning r-term groups, it is, almost in all circumstances, advantageous to direct at first the at-

tention towards the infinitesimal transformations of the concerned group and to choose them as the object of study.

References

1. Engel, F., Lie, S.: Theorie der Transformationsgruppen. Erster Abschnitt. Unter Mitwirkung von Prof. Dr. Friedrich Engel, bearbeitet von Sophus Lie, Verlag und Druck von B.G. Teubner, Leipzig und Berlin, xii+638 pp. (1888). Reprinted by Chelsea Publishing Co., New York, N.Y. (1970)

[1] C. JORDAN, Traité des substitutions, Paris 1870.

Chapter 4
One-Term Groups and Ordinary Differential Equations

Abstract The flow $x' = \exp(tX)(x)$ of a single, arbitrary vector field $X = \sum_{i=1}^{n} \xi_i(x) \frac{\partial}{\partial x_i}$ with analytic coefficients $\xi_i(x)$ always generates a one-term (local) continuous transformation group satisfying:

$$\exp\left(t_1 X\right)\Big(\exp\left(t_2 X\right)(x)\Big) = \exp\left((t_1+t_2)X\right)(x),$$

and:

$$\left[\exp(tX)(\cdot)\right]^{-1} = \exp(-tX)(\cdot).$$

In a neighborhood of any point at which X does not vanish, an appropriate local diffeomorphism $x \mapsto y$ may straighten X to just $\frac{\partial}{\partial y_1}$, hence its flow becomes $y_1' = y_1 + t,\, y_2' = y_2, \dots, y_n' = y_n$.

In fact, in the analytic category (only), computing a general flow $\exp(tX)(x)$ amounts to adding the differentiated terms appearing in the formal expansion of *Lie's exponential series*:

$$\exp(tX)(x_i) = \sum_{k\geqslant 0} \frac{(tX)^k}{k!}(x_i) = x_i + t\,X(x_i) + \cdots + \frac{t^k}{k!} \underbrace{X\big(\cdots\big(X\big(X(x_i)\big)\big)\cdots\big)}_{k \text{ times}} + \cdots,$$

that have been studied extensively by Gröbner in [3].

The famous Lie bracket is introduced by looking at the way a vector field $X = \sum_{i=1}^{n} \xi_i(x) \frac{\partial}{\partial x_i}$ is perturbed, to first order, while introducing the new coordinates $x' = \exp(tY)(x) =: \varphi(x)$ provided by the flow of another vector field Y:

$$\varphi_*(X) = X' + t\left[X', Y'\right] + \cdots,$$

with $X' = \sum_{i=1}^{n} \xi_i(x') \frac{\partial}{\partial x_i'}$ and $Y' = \sum_{i=1}^{n} \eta_i(x') \frac{\partial}{\partial x_i'}$ denoting the two vector fields in the target space x' having the *same* coefficients as X and Y. Here, the analytical expression of the Lie bracket is:

S. Lie, *Theory of Transformation Groups I*,
DOI 10.1007/978-3-662-46211-9_4

$$[X',Y'] = \sum_{i=1}^{n} \left(\sum_{l=1}^{n} \xi_l(x') \frac{\partial \eta_i}{\partial x'_l}(x') - \eta_l(x') \frac{\partial \xi_i}{\partial x'_l}(x') \right) \frac{\partial}{\partial x'_i}.$$

An r-term group $x' = f(x; a)$ satisfying his fundamental differential equations $\frac{\partial x'_i}{\partial a_k} = \sum_{j=1}^{r} \psi_{kj}(a)\, \xi_{ji}(x')$ can, alternatively, be viewed as being generated by its infinitesimal transformations $X_k = \sum_{i=1}^{n} \xi_{ki}(x) \frac{\partial}{\partial x_i}$ in the sense that the totality of the transformations $x' = f(x; a)$ is identical with the totality of all transformations:

$$\begin{aligned} x'_i &= \exp\left(\lambda_1 X_1 + \cdots + \lambda_r X_r\right)(x_i) \\ &= x_i + \sum_{k=1}^{r} \lambda_k\, \xi_{ki}(x) + \sum_{k,j}^{1\ldots r} \frac{\lambda_k \lambda_j}{1 \cdot 2} X_k(\xi_{ji}) + \cdots \qquad (i = 1 \cdots n) \end{aligned}$$

obtained as the time-one map of the one-term group $\exp\left(t \sum \lambda_i X_i\right)(x)$ generated by the general linear combination of the infinitesimal transformations.

A beautiful idea of analyzing the (diagonal) action $x^{(\mu)\prime} = f\left(x^{(\mu)}; a\right)$ induced on r-tuples of points $\left(x^{(1)}, \ldots, x^{(r)}\right)$ in general position enables Lie to show that for every collection of r linearly independent vector fields $X_k = \sum_{i=1}^{n} \xi_{ki}(x) \frac{\partial}{\partial x_i}$, the parameters $\lambda_1, \ldots, \lambda_r$ in the finite transformation equations $x' = \exp\left(\lambda_1 X_1 + \cdots + \lambda_r X_r\right)(x)$ are all essential.

4.1 Mechanical and Mental Images

§ **15**. ([2]) The concept of infinitesimal transformation and likewise that of one-term group gain a certain graphical nature when one makes use of geometric and mechanical images.

In the infinitesimal transformation:

$$x'_i = x_i + \xi_i(x_1, \ldots, x_n)\, \delta t \qquad (i = 1 \cdots n),$$

or in:

$$X(f) = \sum_{i=1}^{n} \xi_i \frac{\partial f}{\partial x_i},$$

we interpret the variables x as Cartesian coordinates of an n-times extended space. Then the transformation can obviously be interpreted in such a way that each point of coordinates $x_1, \ldots, x_n$ is transferred to an infinitesimally neighbouring point of coordinates $x'_1, \ldots, x'_n$. *So the transformation attaches to every point at which not all the* ξ_i *vanish a certain direction of progress* [FORTSCHREITUNGSRICHTUNG], *and along this direction of progress, a cer-*

tain infinitely small line [STRECKE]; the direction of progress is determined by the proportion: $\xi_1(x) : \xi_2(x) : \cdots : \xi_n(x)$; the infinitely small line has the length: $\sqrt{\xi_1^2 + \cdots + \xi_n^2}\,\delta t$. If at a point $x_1, \dots, x_n$ all the ξ are zero, then no direction of progress is attached to the point by the infinitesimal transformation.

If we imagine that the whole space is filled with a compressible fluid, then we can interpret the infinitesimal transformation $X(f)$ simply as an infinitely small movement [BEWEGUNG] of this fluid, and δt as the infinitely small time interval [ZEITABSCHNITT] during which this movement proceeds. Then evidently, the quantities $\xi_1(x) : \cdots : \xi_n(x)$ are the components of the velocity of the fluid particle [FLÜSSIGKEITSTHEILCHENS] which is located precisely at the point $x_1, \dots, x_n$.

The finite transformations of the one-term group $X(f)$ come into being [ENTSTEHEN] by repeating infinitely many times [DURCH UNENDLICHMALIGE WIEDERHOLUNG] the infinitesimal transformation $x_i' = x_i + \xi_i\,\delta t$. In order to arrive at a finite transformation, we must hence imagine that the infinitely small movement represented by the infinitesimal transformation is repeated during infinitely many time intervals δt; in other words, we must follow the movement of the fluid particle during a finite time interval. To this end, it is necessary to integrate the differential equations of this movement, that is to say the simultaneous system:

$$\frac{\mathrm{d}x_1'}{\mathrm{d}\xi_1(x')} = \cdots = \frac{\mathrm{d}x_n'}{\xi_n(x')} = \mathrm{d}t.$$

A fluid particle which, at the time $t = 0$, is located at the point $x_1, \dots, x_n$ will, after a lapse of time t, reach the point $x_1', \dots, x_n'$; so the integration must be executed in such a way that for $t = 0$, one has: $x_i' = x_i$. In reality, we found earlier the general form of a finite transformation of our group exactly in this manner.

The movement of our fluid is one which is a so-called stationary movement, because the velocity components $\mathrm{d}x_i'/\mathrm{d}t$ are free of t. From this, it follows that in one and the same point, the movement is the same at each time, and consequently that the whole process of movement always takes the same course, whatever the time at which one starts to consider it. The proof that the finite transformations produced by the infinitesimal transformation $X(f)$ constitute a group already lies fundamentally in this observation.

If we consider a determined fluid particle, say the one which, at the time $t = 0$ is located at the point $x_1^0, \dots, x_n^0$, then we see that it moves on a curve which passes through the point $x_1^0, \dots, x_n^0$. So the entire space is decomposed only in curves of a constitution such that each particle remains on the curve on which it is located. We want to call these curves the *integral curves* [BAHNCURVEN] *of the infinitesimal transformation* $X(f)$. Obviously, there are ∞^{n-1} such integral curves.

For a given point $x_1^0, \dots, x_n^0$, it is easy to display the equations of the integral curve passing through it. The *integral curve is nothing but the locus of all points in which the point* $x_1^0, \dots, x_n^0$ *is transferred, when the* ∞^1 *transformations:*

$$x_i' = x_i + \frac{t}{1}\,\xi_i(x) + \cdots \qquad (i=1\cdots n)$$

of the one-term group are executed on it. Consequently, the equations:

$$x_i' = x_i^0 + \frac{t}{1}\,\xi_i(x^0) + \frac{t^2}{1\cdot 2}\,\big(X(\xi_i)\big)_{x=x_0} + \cdots \qquad (i=1\cdots n)$$

represent the integral curve in question, when one considers t as an independent variable.

4.2 Straightening of Flows and the Exponential Formula

Let $x_i' = f_i(x; a)$ be a *one-term* local transformation group, with $a \in \mathbb{K}$ $(= \mathbb{R}$ or $\mathbb{C})$ a scalar and with the identity e corresponding to the origin $0 \in \mathbb{K}$ as usual. Its fundamental differential equations (p. 35):

$$\frac{\mathrm{d}x_i'}{\mathrm{d}a} = \psi(a)\,\xi_i(x_1', \dots, x_n') \qquad (i=1\cdots n)$$

then consist of a complete first order PDE system. But by introducing the new parameter:

$$t = t(a) := \int_0^a \psi(a_1)\,\mathrm{d}a_1,$$

we immediately transfer these fundamental differential equations to the *time-independent* system of n ordinary differential equations:

$$\text{(1)} \qquad \frac{\mathrm{d}x_i'}{\mathrm{d}t} = \xi_i(x_1', \dots, x_n') \qquad (i=1\cdots n),$$

the integration of which amounts to computing the so-called *flow* of the vector field $X := \sum_{i=1}^n \xi_i(x)\,\frac{\partial}{\partial x_i}$.

With the same letters f_i, we will write $f_i(x;t)$ instead of $f_i(x; a(t))$. Of course, the (unique) solution of the system (1) with the initial condition $x_i'(x; 0) = x_i$ is nothing but $x_i' = f_i(x;t)$: the flow was in fact known from the beginning. Furthermore, uniqueness of the flow and the fact that the ξ_i are independent of t both imply that the group composition property corresponds just to addition of time parameters ([1, 4]):

$$f_i\big(f(x;t_1);t_2\big) \equiv f_i(x; t_1 + t_2) \qquad (i=1\cdots n).$$

It is classical that one may (locally) straighten X to $\frac{\partial}{\partial y_n}$.

Theorem 4.1. *Every one-term continuous transformation group:*

$$x_i' = f_i(x_1, \dots, x_n; t) \qquad (i=1\cdots n)$$

satisfying differential equations of the form:

$$\frac{\mathrm{d}f_i}{\mathrm{d}t} = \xi_i(f_1,\ldots,f_n) \qquad {\scriptstyle (i=1\cdots n)}$$

is locally equivalent, through a suitable change of variables $y_i = y_i(x)$, *to a group of translations:*

$$y_1' = y_1 + t, \ldots\ldots, \quad y_2' = y_2, \quad y_n' = y_n.$$

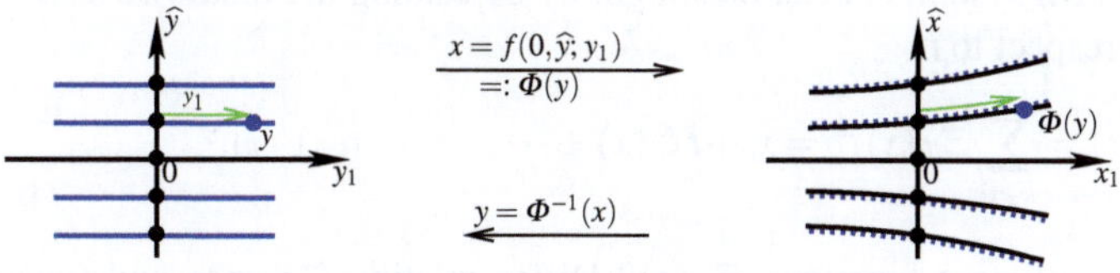

Fig. 4.1 Straightening a flow by means of a diffeomorphism

Proof. We may suppose from the beginning that the coordinates $x_1,\ldots,x_n$ had been chosen so that $\xi_1(0) = 1$ and $\xi_2(0) = \cdots = \xi_n(0) = 0$. In an auxiliary space $y_1,\ldots,y_n$ drawn on the left of the figure, we consider all points $(0,\widehat{y}) := (0,y_2,\ldots,y_n)$ near the origin lying on the coordinate hyperplane which is complementary to the y_1-axis, and we introduce the diffeomorphism:

$$y \longmapsto x = x(y) := f\big(0,\widehat{y}; y_1\big) =: \Phi(y),$$

defined by following the flow up to time y_1 from $\big(0,\widehat{y}\big)$; this is indeed a diffeomorphism fixing the origin thanks to $\frac{\partial \Phi_1}{\partial y_1}(0) = \frac{\partial f_1}{\partial t}(0) = \xi_1(0) = 1$, to $\frac{\partial \Phi_k}{\partial y_1}(0) = \frac{\partial f_k}{\partial t}(0) = \xi_k(0) = 0$ and to $\Phi_k(0,\widehat{y}) \equiv \widehat{y}_k$ for $k = 2,\ldots,n$. Consequently, we get that the (wavy) flow represented on the right figure side has been straightened on the left side to be just a uniform translation directed by the y_1-axis, because by substituting:

$$x' = f\big(0,\widehat{y}'; y_1'\big) = f\big(0,\widehat{y}; y_1 + t\big) = f\big(f\big(0,\widehat{y}; y_1\big); t\big) = f(x;t),$$

we recover the uniquely defined flow $x' = f(x;t)$ when assuming that $\widehat{y}' = \widehat{y}$ and $y_1' = y_1 + t$. □

Theorem 6. *If a one-term group:*

$$x_i' = f_i(x_1,\ldots,x_n,a) \qquad {\scriptstyle (i=1\cdots n)}$$

contains the identity transformation, then its transformations are interchangeable [VERTAUSCHBAR] *with one another and they can be ordered as inverses by pairs. Every group of this sort is equivalent to a group of translations:*

$$y_1' = y_1, \ \ldots, \ y_{n-1}' = y_{n-1}, \ y_n' = y_n + t.$$

4.2.1 The Exponential Analytic Flow Formula

In fact, because we have universally assumed analyticity of all data, the solution $x' = x'(x;t)$ to the PDE system (1) can be sought by expanding the unknown x' in a power series with respect to t:

$$x_i'(x;t) = \sum_{k\geqslant 0} \Xi_{ik}(x)\, t^k = x_i + t\,\xi_i(x) + \cdots \qquad (i=1\cdots n).$$

So, what are the coefficient functions $\Xi_{ik}(x)$? Differentiating (1) once and twice more, we get for instance:

$$\frac{\mathrm{d}^2 x_i'}{\mathrm{d}t^2} = \sum_{k=1}^{n} \frac{\partial \xi_i}{\partial x_k}\frac{\mathrm{d}x_k'}{\mathrm{d}t} = \sum_{k=1}^{n} \frac{\partial \xi_i}{\partial x_k}\,\xi_k = X(\xi_i) \qquad (i=1\cdots n)$$

$$\frac{\mathrm{d}^3 x_i'}{\mathrm{d}t^3} = \sum_{k=1}^{n} \frac{\partial X(\xi_i)}{\partial x_k}\frac{\mathrm{d}x_k'}{\mathrm{d}t} = \sum_{k=1}^{n} \frac{\partial X(\xi_i)}{\partial x_k}\,\xi_k = X\big(X(\xi_i)\big),$$

etc., and hence generally by a straightforward induction:

$$\frac{\mathrm{d}^k x_i'}{\mathrm{d}t^k} = \underbrace{X\big(\cdots\big(X}_{k-1 \text{ times}}(\xi_i)\big)\cdots\big) = \underbrace{X\big(\cdots\big(X\big(X}_{k \text{ times}}(x_i')\big)\big)\cdots\big),$$

for every nonnegative integer k, with the convention $X^0 x_i = x_i$. Setting $t = 0$, we therefore get:

$$k!\ \Xi_{ik}(x) \equiv \underbrace{X\big(\cdots\big(X\big(X}_{k \text{ times}}(x_i)\big)\big)\cdots\big).$$

Thus rather strikingly, computing a flow boils down in the analytic category to summing up differentiated terms.

Proposition 4.1. *The unique solution $x'(x;t)$ to a local analytic system of ordinary differential equations $\frac{\mathrm{d}x_i'}{\mathrm{d}t} = \xi_i(x_1', \ldots, x_n')$ with initial condition $x_i'(x;0) = x_i$ is provided by the power series expansion:*

$$\text{(2)} \quad x_i'(x;t) = x_i + t\,X(x_i) + \cdots + \frac{t^k}{k!}\underbrace{X\big(\cdots\big(X\big(X}_{k \text{ times}}(x_i)\big)\big)\cdots\big) + \cdots \qquad (i=1\cdots n),$$

which can also be written, quite adequately, by means of an exponential denotation:

(2') $$x_i' = \exp(tX)(x_i) = \sum_{k\geqslant 0} \frac{(tX)^k}{k!}(x_i) \qquad {\scriptstyle (i=1\cdots n)}.$$

4.2.2 Action on Functions

Letting $f = f(x_1,\dots,x_n)$ be an arbitrary analytic function, we might compose f with the above flow:

$$f' := f(x_1',\dots,x_n') = f\big(x_1'(x;t),\dots,x_n'(x;t)\big),$$

and we should then expand the result in a power series with respect to t:

$$f' = \big(f'\big)_{t=0} + \tfrac{t}{1!}\big(\tfrac{\mathrm{d}f'}{\mathrm{d}t}\big)_{t=0} + \tfrac{t^2}{2!}\big(\tfrac{\mathrm{d}^2 f'}{\mathrm{d}t^2}\big)_{t=0} + \cdots.$$

Consequently, we need to compute the differential quotients $\frac{\mathrm{d}f'}{\mathrm{d}t}$, $\frac{\mathrm{d}^2 f'}{\mathrm{d}t^2}$, ..., and if we set $\xi_i' := \xi_i(x_1',\dots,x_n')$ and $X' := \sum_{i=1}^n \xi_i' \frac{\partial}{\partial x_i'}$,

$$\frac{\mathrm{d}f'}{\mathrm{d}t} = \sum_{i=1}^n \xi_i' \frac{\partial f'}{\partial x_i'} = X'(f'),$$

$$\frac{\mathrm{d}^2 f'}{\mathrm{d}t^2} = X'\bigg(\sum_{i=1}^n \xi_i' \frac{\partial f'}{\partial x_i'}\bigg) = X'(X'(f')),$$

and so on. After setting $t = 0$, the x_i' become x_i, the f' becomes f, the $X'(f')$ becomes $X(f)$, and so on, whence we obtain the expansion:[1] (Translator's note: Changing t to $-t$ exchanges the rôles of $x' = \exp(tX)(x)$ and of $x = \exp(-tX)(x')$, hence we also have:

(3a) $$f(x_1,\dots,x_n) = f(x_1',\dots,x_n') - \frac{t}{1!}X(f) + \cdots + (-1)^k \frac{t^k}{k!} X(\cdots(X(f))\cdots) + \cdots.$$

(3) $$f(x_1',\dots,x_n') = f(x_1,\dots,x_n) + \frac{t}{1!}X(f) + \cdots + \frac{t^k}{k!} \underbrace{X(\cdots(X(f))\cdots)}_{k \text{ times}} + \cdots.$$

§ **13**. ([2]) Amongst the ∞^1 transformations of the one-term group (2), those whose parameter t has an infinitely small value, say the value δt, play an important role. We now want to consider more precisely these "*infinitely small*" or "*infinitesimal*" transformations of the group.

If we take into account only the first power of δt, whereas we ignore the second and all the higher ones, then we obtain from (2) the desired infinitesimal transformation in the form:

$$x_i' = x_i + \xi_i(x_1,\dots,x_n)\,\delta t \qquad (i=1\cdots n); \tag{4}$$

on the other hand, if we use the equation (3), we get the last n equations condensed in the single one:

$$f' = f + X(f)\,\delta t,$$

or written in greater length:

$$f(x_1',\dots,x_n') = f(x_1,\dots,x_n) + \delta t \sum_{i=1}^{n} \xi_i \frac{\partial f}{\partial x_i}.$$

It is convenient to introduce a specific name for the difference $x_i' - x_i$, that is to say, for the expression $\xi_i\,\delta t$. Occasionally, we want to call $\xi_i \delta t$ the "*increase*" [ZUWACHS], or the "*increment*" [INCREMENT], or also the "*variation*" [VARIATION] of x_i, and write: δx_i. Then we can also represent the infinitesimal transformation in the form:

$$\delta x_1 = \xi_1\,\delta t, \dots, \delta x_n = \xi_n\,\delta t.$$

Correspondingly, we will call the difference $f' - f$, or the expression $X(f)\,\delta t$ the *increase*, or the *variation* of the function $f(x_1,\dots,x_n)$ and we shall write:

$$f' - f = X(f)\,\delta t = \delta f.$$

It stands to reason that the expression, completely alone:

$$X(f) = \sum_{i=1}^{n} \xi_i \frac{\partial f}{\partial x_i}$$

already fully determines the infinitesimal transformation $\delta x_i = \xi_i\,\delta t$, when one understands by $f(x_1,\dots,x_n)$ some undetermined function of its arguments. Indeed, all the n functions $\xi_1,\dots,\xi_n$ are individually given at the same time with $X(f)$.

This is why we shall introduce the expression $X(f) = \frac{\delta f}{\delta t}$ *as being the symbol of the infinitesimal transformation* (4), so we will really speak of the "*infinitesimal transformation* $X(f)$". However, we want to point out that the symbol of the infinitesimal transformation (4) is essentially determined only up to an arbitrary remaining constant factor. In fact, when we multiply the expression $X(f)$ by any finite constant c, then the resulting expression $cX(f)$ is also to be considered as the symbol of the infinitesimal transformation (4). Indeed, according to the concept of an infinitely small quantity, it makes no difference when we substitute in (4) the infinitely small quantity δt by $c\,\delta t$.

The introduction of the symbol $X(f)$ for the infinitesimal transformation (4) presents many advantages. *Firstly*, it is very convenient that the n equations $x_i' = x_i + \xi_i\,\delta t$ of the transformation are replaced by the single expression $X(f)$. *Secondly*, it is convenient that in the symbol $X(f)$, we have to deal with only *one* series of variables, not with the two series: $x_1, \dots, x_n$ and $x_1', \dots, x_n'$. Lastly and *thirdly*, the symbol $X(f)$ establishes the connection between infinitesimal transformations and linear partial differential equations; because in the latter theory, expressions such as $X(f)$ do indeed play an important role. We shall go into more detail later concerning this connection (cf. Chap. 6).

The preceding developments show that a one-term group with the identity transformation always comprises a well determined infinitesimal transformation $x_i' = x_i + \xi_i\,\delta t$, or briefly $X(f)$. But it is also clear that conversely, the one-term group in question is perfectly determined, as soon as one knows its infinitesimal transformation. Indeed, the infinitesimal transformation $x_i' = x_i + \xi_i\,\delta t$ is, so to speak, only another way of writing the simultaneous system (1) from which are derived the equations (2) of the one-term group.

Thus, since every one-term group with the identity transformation is completely determined by its infinitesimal transformation, we shall, for the sake of convenience, introduce the following terminology.

Every transformation of the one-term group:

$$x_i' = x_i + \frac{t}{1}\,\xi_i + \frac{t^2}{1\cdot 2}\,X(\xi_i) + \cdots \qquad (i=1\cdots n)$$

is obtained by repeating infinitely many times the infinitesimal transformation:

$$x_i' = x_i + \xi_i\,\delta t \quad \text{ or } \quad X(f) = \xi_1\,\frac{\partial f}{\partial x_1} + \cdots + \xi_n\,\frac{\partial f}{\partial x_n}.$$

Or yet more briefly:

The one-term group in question is generated by its infinitesimal transformations.

In contrast to the infinitesimal transformation $X(f)$, we call the equations:

$$x_i' = x_i + \frac{t}{1}\,\xi_i + \frac{t^2}{1\cdot 2}\,X(\xi_i) + \cdots$$

the *finite* equations of the one-term group in question.

Now, we may enunciate briefly as follows the connection found earlier between the simultaneous system (1) and the one-term group (2'):

Proposition 1. *Every one-term group which contains the identity transformation is generated by a well determined infinitesimal transformation.*

And conversely:

Proposition 2. *Every infinitesimal transformation generates a completely determined one-term group.*

4.3 Exponential Change of Coordinates and the Lie Bracket

Let $X = \sum_{i=1}^n \xi_i(x) \frac{\partial}{\partial x_i}$ be an infinitesimal transformation which generates the one-term group $x' = \exp(tX)(x)$. What happens if the variables $x_1, \ldots, x_n$ are subjected to an analytic diffeomorphism $y_\nu = \varphi_\nu(x)$ which naturally transfers X to the vector field:

$$Y := \varphi_*(X) = \sum_{\nu=1}^n X(y_\nu) \frac{\partial}{\partial y_\nu} =: \sum_{\nu=1}^n \eta_\nu(y_1, \ldots, y_n) \frac{\partial}{\partial y_\nu}$$

in the new variables $y_1, \ldots, y_n$?

Proposition 4.4. *The new one-term group* $y' = \exp(tY)(y)$ *associated to* $Y = \varphi_*(X)$ *can be recovered from the old one* $x' = \exp(tX)(x)$ *thanks to the formula:*

$$\exp(tY)(y) = \varphi\big(\exp(tX)(x)\big)\big|_{x=\varphi^{-1}(y)}.$$

Proof. Since we work in the analytic category, we are allowed to deal with power series expansions. Through the introduction of the new variables $y_\nu = \varphi_\nu(x)$, an arbitrary function $f(x_1, \ldots, x_n)$ is transformed to the function $F = F(y)$ defined by the identity:

$$f(x) \equiv F\big(\varphi(x)\big).$$

With $x' = \exp(tX)(x)$ by assumption, we may also define $y'_\nu := \varphi_\nu(x')$ so that:

$$F(y')\big|_{y'=\varphi(x')} = f(x').$$

On the other hand, the Jacobian matrix of φ induces a transformation between vector fields; equivalently, this transformation $X \mapsto \varphi_*(X) =: Y$ can be defined by the requirement that for any function f:

$$Y(F)\big|_{y=\varphi(x)} = X(f).$$

By a straightforward induction, it follows for any integer $k \geqslant 1$ that we have:

$$Y\big(Y(F)\big)\big|_{y=\varphi(x)} = X\big(X(f)\big), \ldots\ldots, \; Y^k(F)\big|_{y=\varphi(x)} = X^k(f).$$

Consequently, in the expansion (3) of $f(x') = f\big(\exp(tX)(x)\big)$ with respect to the powers of t, we may perform replacements:

$$F(y')\big|_{y'=\varphi(x')} = f(x') = \sum_{k\geqslant 0} \frac{t^k}{k!} X^k(f) = \sum_{k\geqslant 0} \frac{t^k}{k!} Y^k(F)\big|_{y=\varphi(x)}.$$

Removing the two replacements $|_{y'=\varphi(x')}$ and $|_{y=\varphi(x)}$, we get an identity in terms of the variables y and y':

$$F(y') = \sum_{k\geqslant 0} \frac{t^k}{k!} Y^k(F),$$

which, because the function f — and hence F too — was arbitrary, shows that y' must coincide with $\exp(tY)(y)$, namely:

$$y_i' = \sum_{k\geqslant 0} \frac{t^k}{k!} Y^k(y_i) \qquad (i=1\cdots n).$$

Consequently, after replacing here y' by $\varphi(x')$ and y by $\varphi(x)$, we obtain:

$$\varphi\big(\exp(tX)(x)\big) = \varphi(x') = y' = \sum_{k\geqslant 0} \frac{t^k}{k!} Y^k(y)\big|_{y=\varphi(x)} = \exp(tY)(y)\big|_{y=\varphi(x)},$$

and this is the same relation between flows as the one stated in the proposition, but viewed in x-space. □

Proposition 3. *If, after the introduction of the new independent variables:*

$$y_i = \varphi_i(x_1, \dots, x_n) \qquad (i=1\cdots n),$$

the symbol:

$$X(f) = \sum_{i=1}^{n} \xi_i \frac{\partial f}{\partial x_i}$$

of the infinitesimal transformation:

$$x_i' = x_i + \xi_i\, \delta t \qquad (i=1\cdots n)$$

receives the form:

$$X(f) = \sum_{\nu=1}^{n} X(y_\nu) \frac{\partial f}{\partial y_\nu} = \sum_{\nu=1}^{n} \eta_\nu(y_1, \dots, y_n) = Y(f),$$

then the finite transformations generated by $X(f)$:

$$x_i' = x_i + \frac{t}{1}\xi_i + \frac{t^2}{1\cdot 2} X(\xi_i) + \cdots \qquad (i=1\cdots n)$$

are given, in the new variables, the form:

$$y_i' = y_i + \frac{t}{1}\,\eta_i + \frac{t^2}{1\cdot 2}\,Y(\eta_i) + \cdots \qquad (i=1\cdots n),$$

where the parameter t has the same value in both cases.

4.3.1 Flows as Changes of Coordinates

Suppose we are given an arbitrary infinitesimal transformation:

$$X = \sum_{i=1}^{n} \xi_i(x)\,\frac{\partial}{\partial x_i}.$$

For later use, we want to know how X transforms through the change of coordinates represented by the flow diffeomorphism:

$$x_i' = \exp(tY)(x_i) = x_i + \frac{t}{1}\,\eta_i + \frac{t^2}{1\cdot 2}\,Y(\eta_i) + \cdots \qquad (i=1\cdots n)$$

of *another* infinitesimal transformation:

$$Y = \sum_{i=1}^{n} \eta_i(x_1,\dots,x_n)\,\frac{\partial}{\partial x_i}.$$

At least, we would like to control how X transforms modulo second order terms in t. By definition, X is transformed to the vector field:

$$\varphi_*(X) = \sum_{i=1}^{n} X(x_i')\,\frac{\partial}{\partial x_i'},$$

where the expressions $X(x_i')$ should still be expressed in terms of the target coordinates $x_1',\dots,x_n'$. Replacing x_i' by its above expansion and neglecting the second and higher powers of t, we thus get:

$$X(x_i') = X(x_i) + t\,X(\eta_i) + \cdots \qquad (i=1\cdots n).$$

To express the first term of the right-hand side in terms of the target coordinates, we begin by inverting $x' = \exp(tY)(x)$, getting $x = \exp(-tY')(x')$, with of course $Y' := \sum_{i=1}^{n} \eta_i(x')\,\frac{\partial}{\partial y_i'}$ denoting the same field as Y but written in x'-space, and we compute $X(x_i)$ as follows:

$$\begin{aligned} X(x_i) &= \xi_i(x) = \xi_i\big(\exp(-tY')(x')\big) \\ &= \xi_i\big(x_1' - t\,\eta_1(x'), \dots\dots, x_n' - t\,\eta_n(x')\big) \\ &= \xi_i(x') - t\sum_{l=1}^{n} \frac{\partial \xi_i}{\partial x_l'}(x')\,\eta_l(x') + \cdots \\ &= \xi_i(x') - t\,Y'\big(\xi_i(x')\big) + \cdots \qquad {\scriptstyle (i=1\cdots n)}. \end{aligned}$$

If we now abbreviate $\xi_i' := \xi_i(x')$ and $\eta_i' := \eta_i(x')$, because $\mathrm{O}(t^2)$ is neglected, no computation is needed to express the second term in terms of x':

$$t\,X(\eta_i) = t\,X'(\eta_i') - \cdots \qquad {\scriptstyle (i=1\cdots n)},$$

and consequently by adding we get:

$$X(x_i') = \xi_i' + t\,\big(X'(\eta_i') - Y'(\xi_i')\big) + \cdots \qquad {\scriptstyle (i=1\cdots n)}.$$

As the n coefficients $X'(\eta_i') - Y'(\xi_i')$, either one *recognizes* the coefficients of the *Lie bracket*:

$$\big[X', Y'\big] = \left[\sum_{i=1}^{n} \xi_i(x')\frac{\partial}{\partial x_i'}, \sum_{i=1}^{n} \eta_i(x')\frac{\partial}{\partial x_i'}\right] := \sum_{i=1}^{n}\Big(X'(\eta_i') - Y'(\xi_i')\Big)\frac{\partial}{\partial x_i'},$$

between $X' := \sum_{i=1}^{n} \xi_i(x')\,\frac{\partial}{\partial x_i'}$ and $Y' := \sum_{i=1}^{n} \eta_i(x')\,\frac{\partial}{\partial x_i'}$, or one chooses to *define* once and for all the Lie bracket in such a way. In fact, Engel and Lie mainly introduce brackets in the context of the Clebsch-Frobenius theorem, Chap. 5 below. Now let us state the obtained result in a self-contained manner.

Lemma 4.1. ([2], p. 141) *If, in the infinitesimal transformation* $X = \sum_{i=1}^{n} \xi_i(x)\,\frac{\partial}{\partial x_i}$ *of the space* $x_1, \dots, x_n$*, one introduces as new variables:*

$$x_i' = \exp(tY)(x_i) = x_i + t\,Y(x_i) + \cdots \qquad {\scriptstyle (i=1\cdots n)},$$

those induced by the one-term group generated by another infinitesimal transformation $Y = \sum_{i=1}^{n} \eta_i(x)\,\frac{\partial}{\partial x_i}$*, then setting* $X' := \sum_{i=1}^{n} \xi_i(x')\,\frac{\partial}{\partial x_i'}$ *and* $Y' := \sum_{i=1}^{n} \eta_i(x')\,\frac{\partial}{\partial x_i'}$*, one obtains a transformed vector field:*

$$\varphi_*(X) = X' + t\,\big[X', Y'\big] + \cdots,$$

with a first order perturbation which is the Lie bracket $\big[X', Y'\big]$.

4.4 Essentiality of Multiple Flow Parameters

Now we again consider r arbitrary vector fields with analytic coefficients defined on a certain, unnamed domain of $\mathbb{K}^n$ which contains the origin:

$$X_k = \sum_{i=1}^{n} \xi_{ki}(x_1,\dots,x_n)\,\frac{\partial}{\partial x_i} \qquad {\scriptstyle (k=1\cdots r)}.$$

Although the collection $X_1,\dots,X_k$ does not necessarily stem from an r-term group, each individual X_k nonetheless generates the one-term continuous transformation group $x' = \exp(tX_k)(x)$ with corresponding infinitesimal transformations $x_i' = x_i + \varepsilon\,\xi_{ki}(x)$; this is the reason why we shall henceforth sometimes alternatively refer to such general vector fields X_k as being *infinitesimal transformations*.

Definition 4.1. The r infinitesimal transformations $X_1,\dots,X_r$ will be called *independent* (*of each other*) if they are linearly independent, namely if the n equations:

$$0 \equiv e_1\,\xi_{1i}(x) + \cdots + e_r\,\xi_{ri}(x) \qquad {\scriptstyle (i=1\cdots n)}$$

in which $e_1,\dots,e_r$ are *constants*, imply $e_1 = \cdots = e_r = 0$.

For instance, Theorem 3 on p. 40 states that an r-term continuous local transformation group $x_i' = f_i(x;a_1,\dots,a_r)$ whose parameters a_k are all *essential* always gives rise to the r infinitesimal transformations $X_k := -\frac{\partial f}{\partial a_k}(x;e)$, $k = 1,\dots,r$, which are *independent of each other*.

Introducing r arbitrary auxiliary constants $\lambda_1,\dots,\lambda_r$, one may consider the one-term group generated by the general linear combination:

$$C := \lambda_1 X_1 + \cdots + \lambda_r X_r,$$

of $X_1,\dots,X_r$, namely the flow:

$$\begin{aligned} x_i' = \exp(tC)(x_i) &= x_i + \frac{t}{1}\,C(x_i) + \frac{t^2}{1\cdot 2}\,C\big(C(x_i)\big) + \cdots \\ &= x_i + t\sum_{k=1}^{r} \lambda_k\,\xi_{ki} + t^2 \sum_{k,j}^{1\dots r} \frac{\lambda_k\,\lambda_j}{1\cdot 2}\,X_k(\xi_{ji}) + \cdots \\ &=: h_i(x;t,\lambda_1,\dots,\lambda_r) \qquad {\scriptstyle (i=1\cdots n)}. \end{aligned}$$

If the $X_k \equiv -\frac{\partial f}{\partial a_k}(x;e)$ stem from an r-term continuous group $x_i' = f_i(x;a_1,\dots,a_r)$, a natural question is then to compare the above integrated finite equations $h_i(x;t,\lambda_1,\dots,\lambda_r)$ to the original transformation equations $f_i(x;a_1,\dots,a_r)$. Before studying this question together with Lie and Engel, we focus our attention on a subquestion whose proof shows a beautiful, synthetical, geometrical idea: that of extending the action jointly to finite sets of points.

At first, without assuming that the X_k stem from an r-term continuous group, it is to be asked (subquestion) whether the parameters $\lambda_1,\dots,\lambda_r$ in the above integrated transformation equations $x_i' = h_i(x;\lambda_1,\dots,\lambda_r)$ are all essential. In the formula above, we observe that the $r+1$ parameters only appear in the form $t\,\lambda_1, \dots, t\,\lambda_r$, hence because the λ_k are arbitrary, there is no loss of generality in setting $t = 1$. We shall then simply write $h_i(x;\lambda_1,\dots,\lambda_r)$ instead of $h_i(x;1,\lambda_1,\dots,\lambda_r)$.

Theorem 8. *If the r independent infinitesimal transformations:*

$$X_k(f) = \sum_{i=1}^{n} \xi_{ki}(x_1,\dots,x_n)\,\frac{\partial f}{\partial x_i} \qquad (k=1\cdots r)$$

are independent of each other, and if furthermore $\lambda_1,\dots,\lambda_r$ *are arbitrary parameters, then the totality of all one-term groups* $\lambda_1 X_1(f)+\cdots+\lambda_r X_r(f)$ *forms a family of transformations:*

$$(5) \qquad x_i' = x_i + \sum_{k=1}^{r} \lambda_k\,\xi_{ki} + \sum_{k,\,j}^{1\dots r} \frac{\lambda_k\,\lambda_k}{1\cdot 2}\,X_k(\xi_{ji}) + \cdots \qquad (i=1\cdots n),$$

in which the r parameters $\lambda_1,\dots,\lambda_r$ *are all* essential, *hence a family of* ∞^r *different transformations.*

Proof. Here exceptionally, we observed a harmless technical error in Engel-Lie's proof ([2], pp. 62–65) concerning the link between the generic rank of $X_1\big|_x,\dots,X_r\big|_x$ and a lower bound for the number of essential parameters.[2] (Translator's note: On page 63, it is said that if the number r of the independent infinitesimal transformations X_k is $\leqslant n$, then the $r\times n$ matrix $\big(\xi_{ki}(x)\big)_{1\leqslant k\leqslant r}^{1\leqslant i\leqslant n}$ of their coefficients is of generic rank equal to r, although this claim is contradicted with $n=r=2$ by the two vector fields $x\,\frac{\partial}{\partial x}+y\,\frac{\partial}{\partial y}$ and $xx\,\frac{\partial}{\partial x}+xy\,\frac{\partial}{\partial y}$. Nonetheless, the ideas and the arguments of the written proof (which does not really need such a fact) are perfectly correct.)

However, Lie's main idea is clever and pertinent: it consists in the introduction of exactly r (the number of λ_k's) copies of the same space $x_1,\dots,x_n$ whose coordinates are labelled as $x_1^{(\mu)},\dots,x_n^{(\mu)}$ for $\mu=1,\dots,r$ and to consider the family of transformation equations induced by the *same transformation equations*:

$$x_i^{(\mu)\prime} = \exp(C)\big(x_i^{(\mu)}\big) = h_i\big(x^{(\mu)};\,\lambda_1,\dots,\lambda_r\big) \qquad (i=1\cdots n;\ \mu=1\cdots r)$$

on each copy of space, again with $t=1$. Geometrically, one thus observes how the initial transformation equations $x_i'=h_i(x;\,\lambda_1,\dots,\lambda_r)$ act *simultaneously* on r-tuples of points. Written in greater length, these transformations read:

$$(5') \qquad x_i^{(\mu)\prime} = x_i^{(\mu)} + \sum_{k=1}^{r} \lambda_k\,\xi_{ki}^{(\mu)} + \sum_{k,\,j}^{1\dots r} \frac{\lambda_k\,\lambda_j}{1\cdot 2}\,X_k^{(\mu)}\big(\xi_{ji}^{(\mu)}\big) + \cdots$$

$$(i=1\cdots n;\ \mu=1\cdots r),$$

where we have of course set: $\xi_{ki}^{(\mu)} := \xi_{ki}(x^{(\mu)})$ and $X_k^{(\mu)} := \sum_{i=1}^n \xi_{ki}(x^{(\mu)})\,\frac{\partial}{\partial x_i^{(\mu)}}$. Such an idea also reveals itself to be fruitful in other contexts.

According to the theorem stated on p. 15, in order to check that the parameters $\lambda_1,\dots,\lambda_r$ are essential, one only has to expand x' in a power series with respect to the powers of x at the origin:

$$x'_i = \sum_{\alpha \in \mathbb{N}^n} \mathscr{U}^i_\alpha(\lambda)\, x^\alpha \qquad (i=1\dots n),$$

and to show that the generic rank of the infinite coefficient mapping $\lambda \longmapsto \big(\mathscr{U}^i_\alpha(\lambda)\big)^{1 \leqslant i \leqslant n}_{\alpha \in \mathbb{N}^n}$ is maximal, possibly equal to r. Correspondingly and immediately, we get the expansion of the r-times copied transformation equations:

$$\text{(5'')} \qquad x_i^{(\mu)\prime} = \sum_{\alpha \in \mathbb{N}^n} \mathscr{U}^{i,(\mu)}_\alpha(\lambda)\, (x^{(\mu)})^\alpha \qquad (i=1\dots n;\ \mu=1\dots r),$$

with, for each $\mu = 1, \dots, r$, the *same* coefficient functions:

$$\mathscr{U}^{i,(\mu)}_\alpha(\lambda) \equiv \mathscr{U}^i_\alpha(\lambda) \qquad (i=1\dots n;\ \alpha \in \mathbb{N}^n;\ \mu=1\dots r).$$

So the generic rank of the corresponding infinite coefficient matrix, which is just an r-times copy of the same mapping $\lambda \longmapsto \big(\mathscr{U}^i_\alpha(\lambda)\big)^{1 \leqslant i \leqslant n}_{\alpha \in \mathbb{N}^n}$, is neither increasing nor decreasing.

Thus, the parameters $\lambda_1, \dots, \lambda_r$ for the transformation equations $x' = h(x; \lambda)$ are essential if and only if they are essential for the diagonal transformation equations $x^{(\mu)\prime} = h\big(x^{(\mu)}; \lambda\big)$, $\mu = 1, \dots, r$, induced on the r-fold copy of the space $x_1, \dots, x_n$.

Therefore, we are left with the task of showing that the generic rank of the r-times copy of the infinite coefficient matrix $\lambda \longmapsto \big(\mathscr{U}^{i,(\mu)}_\alpha(\lambda)\big)^{1 \leqslant i \leqslant n,\, 1 \leqslant \mu \leqslant r}_{\alpha \in \mathbb{N}^n}$ is equal to r. We shall in fact establish more, namely that the rank at $\lambda = 0$ of this map already equals r, or equivalently, that the infinite constant matrix:

$$\left(\frac{\partial \mathscr{U}^{i,(\mu)}_\alpha}{\partial \lambda_k}(0)\right)^{1 \leqslant i \leqslant n,\, \alpha \in \mathbb{N}^n,\, 1 \leqslant \mu \leqslant r}_{1 \leqslant k \leqslant r},$$

whose r lines are labelled with respect to partial derivatives, has rank equal to r.

To prepare this infinite matrix, if we differentiate the expansions (5') which identify to (5") with respect to λ_k at $\lambda = 0$, and if we expand the coefficients of our infinitesimal transformations:

$$\xi_{ki}(x^{(\mu)}) = \sum_{\alpha \in \mathbb{N}^n} \xi_{ki\alpha}\, (x^{(\mu)})^\alpha \qquad (i=1\dots n;\ k=1\dots r;\ \mu=1\dots r)$$

with respect to the powers of $x_1, \dots, x_n$, we obtain a more suitable expression of it:

$$\left(\frac{\partial \mathscr{U}^{i,(\mu)}_\alpha}{\partial \lambda_k}(0)\right)^{1 \leqslant i \leqslant n,\, \alpha \in \mathbb{N}^n,\, 1 \leqslant \mu \leqslant r}_{1 \leqslant k \leqslant r} \equiv \Big(\big(\xi_{ki\alpha}\big)^{1 \leqslant i \leqslant n,\, \alpha \in \mathbb{N}^n}_{1 \leqslant k \leqslant r} \cdots \big(\xi_{ki\alpha}\big)^{1 \leqslant i \leqslant n,\, \alpha \in \mathbb{N}^n}_{1 \leqslant k \leqslant r}\Big)$$

$$=: \Big(\mathsf{J}^\infty \Xi(0) \cdots \mathsf{J}^\infty \Xi(0)\Big).$$

As argued up to now, it thus suffices to show that this matrix has rank r. Also, we observe that this matrix identifies with the *infinite jet matrix* $\mathsf{J}^\infty \Xi(0)$ of Taylor coefficients of the r-fold copy of the same $r \times n$ matrix of coefficients of the vector fields X_k:

$$\Xi(x) := \begin{pmatrix} \xi_{11}(x) & \cdots & \xi_{1n}(x) \\ \cdots & \cdots & \cdots \\ \xi_{r1}(x) & \cdots & \xi_{rn}(x) \end{pmatrix}.$$

This justifies the symbol J^∞ introduced above. Now, we can formulate an auxiliary lemma which will enable us to conclude the result.

Lemma 4.2. *Let $n \geqslant 1$, $q \geqslant 1$, $m \geqslant 1$ be integers, let $\mathsf{x} \in \mathbb{K}^n$ and let:*

$$\mathsf{A}(\mathsf{x}) = \begin{pmatrix} a_{11}(\mathsf{x}) & \cdots & a_{1m}(\mathsf{x}) \\ \cdots & \cdots & \cdots \\ a_{q1}(\mathsf{x}) & \cdots & a_{qm}(\mathsf{x}) \end{pmatrix}$$

be an arbitrary $q \times m$ matrix of analytic functions:

$$a_{ij}(\mathsf{x}) = \sum_{\alpha \in \mathbb{N}^n} a_{ij\alpha}\, \mathsf{x}^\alpha \qquad {\scriptstyle (i=1\ldots q;\ j=1\ldots m)}$$

that are all defined in a fixed neighborhood of the origin in $\mathbb{K}^n$, and introduce the $q \times \infty$ constant matrix of Taylor coefficients:

$$\mathsf{J}^\infty \mathsf{A}(0) := \big(a_{ij\alpha}\big)_{1 \leqslant i \leqslant q}^{1 \leqslant j \leqslant m,\, \alpha \in \mathbb{N}^n}$$

whose q lines are labelled by the index i. Then the following inequality between (generic) ranks holds true:

$$\operatorname{rk} \mathsf{J}^\infty \mathsf{A}(0) \geqslant \operatorname{genrk} \mathsf{A}(\mathsf{x}).$$

Proof. Here, our infinite matrix $\mathsf{J}^\infty \mathsf{A}(0)$ will be considered as acting by *left* multiplication on *horizontal vectors* $u = (u_1, \ldots, u_q)$, so that $u\, \mathsf{J}^\infty \mathsf{A}(0)$ is an $\infty \times 1$ matrix, namely an infinite horizontal vector. Similarly, $\mathsf{A}(\mathsf{x})$ will act on horizontal vectors of analytic functions $(u_1(\mathsf{x}), \ldots, u_r(\mathsf{x}))$.

Supposing that $u = (u_1, \ldots, u_q) \in \mathbb{K}^q$ is any nonzero vector in the kernel of $\mathsf{J}^\infty \mathsf{A}(0)$, namely: $0 = u\, \mathsf{J}^\infty \mathsf{A}(0)$, or else in greater length:

$$0 = u_1\, a_{1j\alpha} + \cdots + u_q\, a_{qj\alpha} \qquad {\scriptstyle (j=1\ldots m;\ \alpha \in \mathbb{N}^n)},$$

we then immediately deduce, after multiplying each such equation by x^α and by summing over all $\alpha \in \mathbb{N}^n$:

$$0 \equiv u_1\, a_{1j}(\mathsf{x}) + \cdots + u_q\, a_{qj}(\mathsf{x}) \qquad {\scriptstyle (j=1\ldots m)},$$

so that the same constant vector $u = (u_1, \ldots, u_q)$ also satisfies $0 \equiv u A(\mathsf{x})$. It follows that the dimension of the kernel of $\mathsf{J}^\infty \mathsf{A}(0)$ is smaller than or equal to the dimension of the kernel of $A(\mathsf{x})$ (at a generic x): this is just equivalent to the above inequality between (generic) ranks. □

Now, for each $q = 1, 2, \dots, r$, we want to apply the lemma with the matrix $\mathsf{A}(\mathsf{x})$ being the q-fold copy of matrices $\big(\Xi(x^{(1)}) \cdots \Xi(x^{(q)})\big)$, or equivalently in greater length:

$$\Xi_q(\widetilde{\mathsf{x}}_q) := \begin{pmatrix} \xi_{11}^{(1)} & \cdots & \xi_{1n}^{(1)} & \xi_{11}^{(2)} & \cdots & \xi_{1n}^{(2)} & \cdots\cdots & \xi_{11}^{(q)} & \cdots & \xi_{1n}^{(q)} \\ \cdots & \cdots & \cdots & \cdots & \cdots & \cdots & \cdots\cdots & \cdots & \cdots & \cdots \\ \xi_{r1}^{(1)} & \cdots & \xi_{rn}^{(1)} & \xi_{r1}^{(2)} & \cdots & \xi_{rn}^{(2)} & \cdots\cdots & \xi_{r1}^{(q)} & \cdots & \xi_{rn}^{(q)} \end{pmatrix},$$

where we have used the abbreviation:

$$\widetilde{\mathsf{x}}_q := \big(x^{(1)}, \dots, x^{(q)}\big).$$

Lemma 4.3. *It is a consequence of the fact that $X_1, X_2, \dots, X_r$ are linearly independent of each other that for every $q = 1, 2, \dots, r$, one has:*

$$\operatorname{genrk}\Big(\Xi(x^{(1)}) \quad \Xi(x^{(2)}) \quad \cdots \quad \Xi(x^{(q)})\Big) \geqslant q.$$

Proof. Indeed, for $q = 1$, it is at first clear that $\operatorname{genrk}\big(\Theta(x^{(1)})\big) \geqslant 1$, just because not all the $\xi_{ki}(x)$ vanish identically.

We next establish by induction that, as long as they remain $< r$, generic ranks increase by at least one unit at each step:

$$\operatorname{genrk}\Big(\Xi_{q+1}(\widetilde{\mathsf{x}}_{q+1})\Big) \geqslant 1 + \operatorname{genrk}\Big(\Xi_q(\widetilde{\mathsf{x}}_q)\Big),$$

a fact which will immediately yield the lemma.

Indeed, if on the contrary, the generic ranks stabilized, and were still $< r$, then locally in a neighborhood of a generic, fixed $\widetilde{\mathsf{x}}_{q+1}^0$, both matrices Ξ_{q+1} and Ξ_q would have the same, locally constant rank. Consequently, the solutions $\big(\vartheta_1(\widetilde{\mathsf{x}}_q) \cdots \vartheta_r(\widetilde{\mathsf{x}}_q)\big)$ to the (kernel-like) system of linear equations written in matrix form:

$$0 \equiv \big(\vartheta_1(\widetilde{\mathsf{x}}_q) \cdots \vartheta_r(\widetilde{\mathsf{x}}_q)\big)\, \Xi_q(\widetilde{\mathsf{x}}_q),$$

which are analytic near $\widetilde{\mathsf{x}}_q^0$ thanks to an application of Cramer's rule and thanks to constancy of rank, would automatically also be solutions of the extended system:

$$0 \equiv \big(\vartheta_1(\widetilde{\mathsf{x}}_q) \cdots \vartheta_r(\widetilde{\mathsf{x}}_q)\big)\big(\Xi_q(\widetilde{\mathsf{x}}_q) \;\; \Xi(x^{(q+1)})\big),$$

whence there would exist *nonzero* solutions $(\vartheta_1, \dots, \vartheta_r)$ to the linear dependence equations:

$$0 = \big(\vartheta_1 \cdots \vartheta_r\big)\, \Xi\big(x^{(q+1)}\big)$$

which are *constant* with respect to the variable $x^{(q+1)}$, since they only depend upon $\widetilde{x}_q$. This contradicts the assumption that $X_1^{(q+1)}, \dots, X_r^{(q+1)}$ are independent of each other. □

Lastly, we may form a chain of (in)equalities that are now obvious consequences of the lemma and of the assertion:

$$\text{rank}\Big(\mathsf{J}^\infty \Xi(0) \cdots \mathsf{J}^\infty \Xi(0)\Big) = \text{rank}\, \mathsf{J}^\infty \Xi_r(0) \geqslant \text{genrk}\, \Xi_r(\widetilde{x}_r) = r,$$

and since all ranks are $\leqslant r$, we get the promised rank estimate:

$$r = \text{rank}\Big(\mathsf{J}^\infty \Xi(0) \ \cdots \ \mathsf{J}^\infty \Xi(0)\Big),$$

which finally completes the proof of the theorem. □

In order to help fix in our memory the trick of extending the group action to an r-fold product of the base space, we also translate a summarizing proposition which is formulated on p. 66 of [2].

Proposition 5. *If the r infinitesimal transformations:*

$$X_k(f) = \sum_{i=1}^{n} \xi_{ki}(x_1, \dots, x_n) \frac{\partial f}{\partial x_i} \qquad (k = 1 \dots r)$$

are independent of each other, if furthermore:

$$x_1^{(\mu)}, \dots, x_n^{(\mu)} \qquad (\mu = 1 \dots r)$$

are r different systems of n variables, and if lastly one uses the abbreviation:

$$X_k^{(\mu)}(f) = \sum_{i=1}^{n} \xi_{ki}\big(x_1^{(\mu)}, \dots, x_n^{(\mu)}\big) \frac{\partial f}{\partial x_i^{(\mu)}} \qquad (k, \mu = 1 \dots r),$$

then the r infinitesimal transformations:

$$W_k(f) = \sum_{\mu=1}^{r} X_k^{(\mu)}(f) \qquad (k = 1 \dots r)$$

in the nr variables $x_i^{(\mu)}$ satisfy no *relation of the form:*

$$\sum_{k=1}^{n} \chi_k\big(x_1^{(1)}, \dots, x_n^{(1)}, \dots\dots, x_1^{(r)}, \dots, x_n^{(r)}\big)\, W_k(f) \equiv 0.$$

Also, we remark for later use as in [2], p. 65, that during the proof of Theorem 8 above, it did not really matter that the equations (5) represented the finite equations of a family of one-term groups. In fact, we only considered the terms of first order with respect to $\lambda_1, \dots, \lambda_r$ in the finite equations (5), and the crucial Lemma 4.3 emphasized during the proof was true under the only assumption that the infinitesimal transformations $X_1, \dots, X_r$ were mutually independent. Consequently, Theorem 8 can be somewhat generalized as follows.

Proposition 4. *If a family of transformations contains the r arbitrary parameters $e_1, \dots, e_r$ and if its equations, when they are expanded with respect to powers of $e_1, \dots, e_r$, are of the form:*

$$x_i' = x_i + \sum_{k=1}^{r} e_k \, \xi_{ki}(x_1, \dots, x_n) + \cdots \qquad (i=1 \cdots n),$$

where the neglected terms in $e_1, \dots, e_r$ are of second and of higher order, and lastly, if the functions $\xi_{ki}(x)$ have the property that the r infinitesimal transformations made up with them:

$$X_k(f) = \sum_{i=1}^{n} \xi_{ki}(x_1, \dots, x_n) \frac{\partial f}{\partial x_i} \qquad (k=1 \cdots r)$$

are independent of each other, then those transformation equations represent ∞^r different transformations, or what is the same: the r parameters $e_1, \dots, e_r$ are essential.

4.5 Generation of an r-Term Group by its One-Term Subgroups

After these preparations, we may now come back to our question formulated on p. 74: how to compare the equations $x_i' = f_i(x; a)$ of a given finite continuous transformation group to the equations:

$$\begin{aligned} x_i' &= \exp\big(t \lambda_1 X_1 + \cdots + t \lambda_r X_r\big)(x_i) \\ &=: h_i\big(x; t, \lambda_1, \dots, \lambda_r\big) \qquad (i=1 \dots n) \end{aligned}$$

obtained by integrating the general linear combination of its r infinitesimal transformations $X_k = -\frac{\partial f_i}{\partial a_k}(x; e)$? Sometimes, such equations will be called as in [2] the *canonical finite equations* of the group.

Abbreviating $\lambda_1 X_1 + \cdots + \lambda_r X_r$ as the infinitesimal transformation $C := \sum_{i=1}^{n} \xi_i(x) \frac{\partial}{\partial x_i}$, whose coefficients are given by:

$$\xi_i(x) := \sum_{j=1}^{n} \lambda_j \, \xi_{ji}(x) \qquad (i=1 \dots n),$$

then by definition of the flow $x_i' = \exp(tC)(x)$, the functions h_i satisfy the first order system of ordinary differential equations $\frac{dh_i}{dt} = \xi_i(h_1,\dots,h_n)$, or equivalently:

$$(6) \qquad \frac{\mathrm{d}h_i}{\mathrm{d}t} = \sum_{j=1}^{r} \lambda_j\, \xi_{ji}(h_1,\dots,h_n) \qquad\qquad {\scriptstyle (i=1\dots n)},$$

with of course the initial condition $h(x;0,\lambda) = x$ at $t = 0$. On the other hand, according to Theorem 3 on p. 40, we remember that the f_i satisfy the fundamental differential equations:

$$(7) \qquad \xi_{ji}(f_1,\dots,f_n) = \sum_{k=1}^{r} \alpha_{jk}(a)\, \frac{\partial f_i}{\partial a_k} \qquad\qquad {\scriptstyle (i=1\dots n;\; j=1\dots r)}.$$

Proposition 4.8. *If the parameters $a_1,\dots,a_r$ are the unique solutions $a_k(t,\lambda)$ to the system of first order ordinary differential equations:*

$$\frac{\mathrm{d}a_k}{\mathrm{d}t} = \sum_{j=1}^{r} \lambda_j\, \alpha_{jk}(a) \qquad\qquad {\scriptstyle (k=1\dots r)}$$

with initial condition $a(0,\lambda) = e$ being the identity element, then the following identities hold:

$$f_i\big(x;a(t,\lambda)\big) \equiv \exp\big(t\,\lambda_1 X_1 + \cdots + t\,\lambda_r X_r\big)(x_i) = h_i\big(x;t,\lambda_1,\dots,\lambda_r\big) \qquad {\scriptstyle (i=1\dots n)}$$

and they show how the h_i are recovered from the f_i.

Proof. Indeed, multiplying the equation (7) by λ_j and summing over j for j equals 1 up to r, we get:

$$\sum_{k=1}^{r} \frac{\partial f_i}{\partial a_k} \sum_{j=1}^{r} \lambda_j\, \alpha_{jk}(a) = \sum_{j=1}^{r} \lambda_j\, \xi_{ji}(f_1,\dots,f_n) \qquad\qquad {\scriptstyle (i=1\dots n)}.$$

Thanks to the assumption concerning the a_k, we can replace the second sum of the left-hand side by $\frac{\mathrm{d}a_k}{\mathrm{d}t}$, which yields identities:

$$\sum_{k=1}^{r} \frac{\partial f_i}{\partial a_k} \frac{\mathrm{d}a_k}{\mathrm{d}t} \equiv \sum_{j=1}^{r} \lambda_j\, \xi_{ji}(f_1,\dots,f_n) \qquad\qquad {\scriptstyle (i=1\dots n)}$$

in the left-hand side of which we recognize just a plain derivation with respect to t:

$$\frac{\mathrm{d}f_i}{\mathrm{d}t} = \frac{\mathrm{d}}{\mathrm{d}t}\big[f_i\big(x;a(t,\lambda)\big)\big] \equiv \sum_{j=1}^{r} \lambda_j\, \xi_{ji}(f_1,\dots,f_n) \qquad\qquad {\scriptstyle (i=1\dots n)}.$$

But since $f\big(x;a(0,\lambda)\big) = f(x;e) = x$ has the same initial condition x at $t = 0$ as the solution $h\big(x;t,\lambda\big)$ to (6), the uniqueness of solutions to systems of first order ordinary differential equations immediately gives the asserted coincidence $f(x; a(t, \lambda)) \equiv h(x; t, \lambda)$. □

4.6 Applications to the Economy of Axioms

We now come back to the end of Chap. 3, Sect. 3.9, where the three standard group axioms (composition; identity element; existence of inverses) were superseded by the hypothesis of existence of differential equations. We recall that in the proof of Lemma 3.3, a relocalization was needed to assure that $\det \psi_{kj}(a) \neq 0$. For the sake of clarity and of rigor, we will, in the hypotheses, explicitly mention the subdomain $\mathscr{A}^1 \subset \mathscr{A}$ where the determinant of the $\psi_{kj}(a)$ does not vanish.

The following (apparently technical) theorem which is a mild modification of the Theorem 9 on p. 72 of [2], will be used in an essential way by Lie to derive his famous three fundamental theorems in Chap. 9 below.

Theorem 9. *If, in the transformation equations defined for* $(x, a) \in \mathscr{X} \times \mathscr{A}$:

(1) $$x_i' = f_i(x_1, \dots, x_n; a_1, \dots, a_r) \qquad {\scriptstyle (i=1\cdots n)},$$

the r parameters $a_1, \dots, a_r$ *are all essential and if in addition, certain differential equations of the form:*

(2) $$\frac{\partial x_i'}{\partial a_k} = \sum_{j=1}^{r} \psi_{kj}(a_1, \dots, a_r)\, \xi_{ji}(x_1', \dots, x_n') \qquad {\scriptstyle (i=1\cdots n\,;\; k=1\cdots r)}$$

are identically satisfied by $x_1' = f_1(x; a), \dots, x_n' = f_n(x; a)$*, where the matrix* $\psi_{kj}(a)$ *is holomorphic and invertible in some nonempty subdomain* $\mathscr{A}^1 \subset \mathscr{A}$*, and where the functions* $\xi_{ji}(x')$ *are holomorphic in* $\mathscr{X}$*, then by introducing the r infinitesimal transformations:*

$$X_k := \sum_{i=1}^{n} \xi_{ki}(x) \frac{\partial}{\partial x_i},$$

it holds true that every transformation $x_i' = f_i(x; a)$ *whose parameters* $a_1, \dots, a_r$ *lie in a small neighborhood of some fixed* $a^0 \in \mathscr{A}^1$ *can be obtained by firstly performing the transformation:*

$$\overline{x}_i = f_i(x_1, \dots, x_n; a_1^0, \dots, a_r^0) \qquad {\scriptstyle (i=1\cdots n)},$$

and then secondly, by performing a certain transformation:

$$x_i' = \exp\big(t\lambda_1 X_1 + \cdots + t\lambda_r X_r\big)(\overline{x}_i) \qquad {\scriptstyle (i=1\cdots n)}$$

of the one-term group generated by some suitable linear combination of the X_k, where t and $\lambda_1, \dots, \lambda_r$ are small complex numbers.

This technical statement will be especially useful later to show that whenever r infinitesimal transformations $X_1, \dots, X_r$ form a Lie algebra, the composition of two transformations of the form $x' = \exp\left(t\lambda_1 X_1 + \cdots + t\lambda_r X_r\right)$ is again of the same form, hence the totality of these transformations truly constitutes a group.

Proof. The arguments are essentially the same as those developed at the end of the previous section (p. 81) for a genuinely local continuous transformation group, except that the identity parameter e (which does not necessarily exist here) should be replaced by a^0.

For the sake of completeness, let us give the proof. First, we fix $a^0 \in \mathscr{A}^1$ and we introduce the solutions $a_k = a_k(t, \lambda_1, \dots, \lambda_r)$ of the following system of ordinary differential equations:

$$\frac{\mathrm{d}a_k}{\mathrm{d}t} = \sum_{j=1}^{r} \lambda_j \, \alpha_{jk}(a) \qquad (k=1\cdots r),$$

with initial condition $a_k(0, \lambda_1, \dots, \lambda_r) = a_k^0$, where $\lambda_1, \dots, \lambda_r$ are small complex parameters and where, as before, $\alpha_{jk}(a)$ denotes the inverse matrix of $\psi_{jk}(a)$, which is holomorphic in the whole of $\mathscr{A}_1$.

Secondly, we introduce the local flow:

$$\exp\left(t\lambda_1 X_1 + \cdots + t\lambda_r X_r\right)(\overline{x}) =: h(\overline{x}; t, \lambda)$$

of the general linear combination $\lambda_1 X_1 + \cdots + \lambda_r X_r$ of the r infinitesimal transformations $X_k = \sum_{i=1}^{n} \xi_{ki}(x) \frac{\partial}{\partial x_i}$, where $\overline{x}$ is assumed to run in $\mathscr{A}^1$. Thus by its very definition, this flow integrates the ordinary differential equations:

$$\frac{\mathrm{d}h_i}{\mathrm{d}t} = \sum_{j=1}^{r} \lambda_j \, \xi_{ji}(h_1, \dots, h_n) \qquad (i=1\cdots n)$$

with the initial condition $h(\overline{x}; 0, \lambda) = \overline{x}$.

Thirdly, we first solve the ξ_{ji} in the fundamental differential equations (2) using the inverse matrix α:

$$\xi_{ji}(f_1, \dots, f_n) = \sum_{k=1}^{r} \alpha_{jk}(a) \frac{\partial f_i}{\partial a_k} \qquad (i=1\cdots n;\ j=1\cdots r).$$

Then we multiply by λ_j, we sum and we recognize $\frac{\mathrm{d}a_k}{\mathrm{d}t}$, which we then substitute:

$$\begin{aligned}\sum_{j=1}^{r}\lambda_j\,\xi_{ji}(f_1,\dots,f_n)&=\sum_{k=1}^{r}\frac{\partial f_i}{\partial a_k}\sum_{j=1}^{r}\lambda_j\,\alpha_{jk}(a)\\&=\sum_{k=1}^{r}\frac{\partial f_i}{\partial a_k}\frac{\mathrm{d}a_k}{\mathrm{d}t}\\&=\frac{\mathrm{d}}{\mathrm{d}t}\big[f_i\big(x;\,a(t,\lambda)\big)\big]\qquad\qquad (i=1\cdots n).\end{aligned}$$

So the $f_i\big(x;\,a(t,\lambda)\big)$ satisfy the *same* differential equations as the $h_i\big(\overline{x};\,t,\lambda\big)$, and in addition, if we set $\overline{x}$ equal to $f(x;\,a^0)$, both collections of solutions will have the *same* initial value for $t=0$, namely $f(x;\,a^0)$. In conclusion, by observing that the f_i and the h_i satisfy the same equations, the uniqueness property enjoyed by first order ordinary differential equations yields the identity:

$$\boxed{f\big(x;\,a(t,\lambda)\big)\equiv\exp\big(t_1\lambda_1X_1+\cdots+t\lambda_rX_r\big)\big(f(x;\,a^0)\big)}$$

expressing that every transformation $x'=f(x;\,a)$ for a in a neighborhood of a^0 appears to be the composition of the fixed transformation $\overline{x}=f(x;\,a^0)$ followed by a certain transformation of the one-term group $\exp\big(t\lambda_1X_1+\cdots+t\lambda_rX_r\big)(\overline{x})$. □

§ **18**. We now apply the preceding general developments to the peculiar case where the ∞^r transformations $x_i'=f_i(x_1,\dots,x_n,a_1,\dots,a_r)$ constitute an r-term group.

If the equations (1) represent an r-term group, then according to Theorem 3 p. 40, there are always differential equations of the form (2); so we do not need to specially enunciate this requirement.

Moreover, we observe that all infinitesimal transformations of the form $\sum_{i=1}^{n}\left\{\sum_{j=1}^{r}\lambda_j\xi_{ji}(x)\right\}\frac{\partial f}{\partial x_i}$ can be linearly expressed by means of the following r infinitesimal transformations:

$$X_k(f)=\sum_{i=1}^{n}\xi_{ki}(x_1,\dots,x_n)\frac{\partial f}{\partial x_i}\qquad\qquad (k=1\cdots r),$$

since all the infinitesimal transformations (9) are contained in the expression:

$$\sum_{k=1}^{r}\lambda_k X_k(f).$$

Here, as we have underscored already in the introduction of this chapter, the infinitesimal transformations $X_1(f),\dots,X_r(f)$ are independent of each other.

Consequently, we can state the following theorem concerning arbitrary r-term groups:

Theorem 10. *To every r-term group:*

$$x_i' = f_i(x_1, \dots, x_n, a_1, \dots, a_r) \qquad (i=1\cdots n)$$

are associated r independent infinitesimal transformations:

$$X_k(f) = \sum_{i=1}^{n} \xi_{ki}(x_1, \dots, x_n) \frac{\partial f}{\partial x_i} \qquad (i=1\cdots n),$$

which stand in the following relationship to the finite transformations of the group: if $x_i' = f_i(x_1, \dots, x_n; a_1^0, \dots, a_r^0)$ *is any transformation of the group, then every transformation* $x_i' = f_i(x, a)$ *whose parameter lies in a certain neighborhood of* $a_1^0, \dots, a_r^0$ *can be obtained by firstly executing the transformation* $\overline{x}_i = f_i(x_1, \dots, x_n, a_1^0, \dots, a_r^0)$ *and secondly a certain transformation* $x_i' = \omega_i(\overline{x}_1, \dots, \overline{x}_n)$ *of a one-term group, the infinitesimal transformation of which has the form* $\lambda_1 X_1(f) + \cdots + \lambda_r X_r(f)$*, where* $\lambda_1, \dots, \lambda_r$ *denote certain suitably chosen constants.*

If we not only know that the equations $x_i' = f_i(x, a)$ represent an r-term group, but also that this group contains the identity transformation, and lastly also that the parameters a_k^0 represent a system of values of the identity transformation in the domain $((a_k))$, then we can still say more. Indeed, if we in particular choose for the transformation $\overline{x}_i = f_i(x, a^0)$ the identity transformation, we then realize immediately that the transformations of our group are nothing but the transformations of those one-term groups that are generated by the infinitesimal transformations:

$$\lambda_1 X_1(f) + \cdots + \lambda_r X_r(f).$$

Thus, if we use the abbreviation:

$$\sum_{k=1}^{r} \lambda_k X_k(f) = C_k(f),$$

then the equations:

$$x_i' = x_i + \frac{t}{1} C(x_i) + \frac{t^2}{1 \cdot 2} C\big(C(x_i)\big) + \cdots \qquad (i=1\cdots n)$$

represent the ∞^r transformations of the group. The fact that the $r+1$ parameters: $\lambda_1, \dots, \lambda_r, t$ appear is just fictitious here, for they are indeed only found in the r combinations $\lambda_1 t, \dots, \lambda_r t$. We can therefore quietly set t equal to 1. In addition, if we remember the representation of a one-term group by a single equation given in eq. (3) on p. 67, then we realize that the equations of our r-term group may be condensed into the single equation:

$$f(x_1', \dots, x_n') = f(x_1, \dots, x_n) + C(f) + \frac{1}{1 \cdot 2} C\big(C(f)\big) + \cdots.$$

That it is possible to order the transformations of the group as inverses in pairs thus hardly requires a mention.

We can briefly state the above result on r-term groups with identity transformation as follows.

Theorem 11. *If an r-term group:*

$$x_i' = f_i(x_1,\dots,x_n,a_1,\dots,a_r) \qquad (i=1\cdots n)$$

contains the identity transformation, then its ∞^r transformations can be organized in ∞^{r-1} families of ∞^1 transformations in such a way that each family amongst these ∞^{r-1} families consists of all the transformations of a certain one-term group with the identity transformation. In order to find these one-term groups, one forms the known equations:

$$\frac{\partial x_i'}{\partial a_k} = \sum_{j=1}^{r} \psi_{kj}(a_1,\dots,a_r)\,\xi_{ji}(x_1',\dots,x_n') \qquad (i=1\cdots n\,;\,k=1\cdots r),$$

which are identically satisfied after substituting $x_i' = f_i(x,a)$. One further sets:

$$\sum_{i=1}^{n} \xi_{ki}(x_1,\dots,x_n)\,\frac{\partial f}{\partial x_i} = X_k(f) \qquad (k=1\cdots r),$$

hence the expression:

$$\sum_{k=1}^{r} \lambda_k X_k(f)$$

where the r arbitrary parameters $\lambda_1,\dots,\lambda_r$ represent the infinitesimal transformations of these ∞^{r-1} one-term groups, and their finite equations have the form:

$$x_i' = x_i + \sum_{k=1}^{r} \lambda_k\,\xi_{ki}(x) + \sum_{k,j}^{1\dots r} \frac{\lambda_k\lambda_j}{1\cdot 2} X_k(\xi_{ji}) + \cdots \qquad (i=1\cdots n).$$

The totality of all these finite transformations is identical with the totality of all transformations of the group $x_i' = f_i(x,a)$. Furthermore, the transformations of this group can be ordered as inverses in pairs.

§ **19**. In general, if an r-term group contains all transformations of some one-term group and if, in the sense discussed earlier, this one-term group is generated by the infinitesimal transformation $X(f)$, then we say that the r-term group *contains the infinitesimal transformation* $X(f)$. Now, we have just seen that every r-term group with the identity transformation can be brought to the form:

$$f(x'_1,\dots,x'_n) = f(x_1,\dots,x_n) + \sum_{k=1}^{r} \lambda_k X_k(f)$$
$$+\frac{1}{1\cdot 2}\sum_{k,j}^{1\cdots r} \lambda_k \lambda_j X_k\big(X_j(f)\big) + \cdots,$$

where $\lambda_1,\dots,\lambda_r$ denote arbitrary constants, while $X_1(f),\dots,X_r(f)$ stand for mutually independent infinitesimal transformations. So we can say that *every such r-term group contains r independent infinitesimal transformations.*

We are very close to showing that an r-term group cannot contain more than r independent infinitesimal transformations.

In order to clarify this point, we want to consider directly the question of when the infinitesimal transformation:

$$Y(f) = \sum_{i=1}^{n} \eta_i(x_1,\dots,x_n)\frac{\partial f}{\partial x_i}$$

is contained in the r-term group with the r independent infinitesimal transformations $X_1(f),\dots,X_r(f)$.

If $Y(f)$ belongs to the r-term group in question, then the same is also true of the transformations:

$$x'_i = x_i + \frac{\tau}{1}\eta_i(x) + \frac{\tau^2}{1\cdot 2}Y(\eta_i) + \cdots \qquad (i=1\cdots n)$$

of the one-term group generated by $Y(f)$. Hence if we first execute an arbitrary transformation of this one-term group and then an arbitrary transformation:

$$x''_i = x'_i + \sum_{k=1}^{r} \lambda_k \xi'_{ki} + \sum_{k,j}^{1\cdots r} \frac{\lambda_k \lambda_j}{1\cdot 2} X'_j(\xi'_{ki}) + \cdots$$

of the r-term group, we must then obtain a transformation which also belongs to the r-term group. By a calculation, we find that this new transformation has the form:

$$x''_i = x_i + \tau\,\eta_i(x) + \sum_{k=1}^{r} \lambda_k \xi_{ki}(x) + \cdots \qquad (i=1\cdots n),$$

where all the omitted terms are of second and of higher order with respect to $\lambda_1,\dots,\lambda_r,\tau$. For arbitrary $\lambda_1,\dots,\lambda_r,\tau$, this transformation must belong to the r-term group. Now, if the $r+1$ infinitesimal transformations $X_1(f),\dots,X_r(f),Y(f)$ were independent of each other, then according to Proposition 4, p. 80, the last written equations would represent ∞^{r+1} transformations; but this is impossible, for the r-term group contains in general only ∞^r transformations. Consequently, $X_1(f),\dots,X_r(f),Y(f)$ are not independent of each other, but since $X_1(f),\dots,X_r(f)$ are so, then $Y(f)$ must be a linear combination of $X_1(f),\dots,X_r(f)$, hence have the form:

$$Y(f) = \sum_{k=1}^{r} l_k X_k(f),$$

where $l_1,\dots,l_r$ denote appropriate constants. That is to say, the following holds:

Proposition 2. *If an r-term group contains the identity transformation, then it contains r independent infinitesimal transformations $X_1(f),\dots,X_r(f)$ and every infinitesimal transformation contained in it is of the form $\lambda_1 X_1(f) + \cdots + \lambda_r X_r(f)$, where $\lambda_1,\dots,\lambda_r$ denote constants.*

We even saw above that *every* infinitesimal transformation of the form $\lambda_1 X_1(f) + \cdots + \lambda_r X_r(f)$ belongs to the group; hence in future, we shall call the expression $\lambda_1 X_1(f) + \cdots + \lambda_r X_r(f)$ with the r arbitrary constants $\lambda_1,\dots,\lambda_r$ *the general infinitesimal transformation* of the r-term group in question.

From the preceding considerations we also obtain the following, certainly special, but nevertheless important:

Proposition 2. *If an r-term group contains the $m \leqslant r$ mutually independent infinitesimal transformations $X_1(f),\dots,X_r(f)$, then it also contains every infinitesimal transformation of the following form: $\lambda_1 X_1(f) + \cdots + \lambda_m X_m(f)$, where $\lambda_1,\dots,\lambda_m$ denote completely arbitrary constants.*

Of course, the work done so far gives us the means to determine the infinitesimal transformations of an r-term group $x_i' = f_i(x,a)$ with the identity transformation. But it is possible to reach the objective more rapidly.

Let the identity transformation of our group go with the parameters: $a_1^0,\dots,a_r^0$, and let $a_1^0,\dots,a_r^0$ lie in the domain $((a))$, so that the determinant $\sum \pm \psi_{11}(a^0)\cdots\psi_{rr}(a^0)$ is hence certainly different from zero. Now, we have:

$$\sum_{j=1}^{r} \psi_{kj}(a_1,\dots,a_r)\,\xi_{ji}(f_1(x,a),\dots,f_n(x,a)) \equiv \frac{\partial}{\partial a_k} f_i(x,a),$$

hence when one sets $a_k = a_k^0$:

$$\sum_{j=1}^{r} \psi_{kj}(a^0)\,\xi_{ji}(x_1,\dots,x_n) \equiv \left[\frac{\partial}{\partial a^k} f_i(x,a)\right]_{a=a^0}.$$

We multiply this equation by $\frac{\partial f}{\partial x_i}$ and we sum for i from 1 to n, which then gives:

$$\sum_{j=1}^{r} \psi_{kj}(a^0) X_j(f) = \sum_{i=1}^{n} \left[\frac{\partial}{\partial a_k} f_i(x,a)\right]_{a=a^0} \frac{\partial f}{\partial x_i} \qquad (k=1\cdots r)$$

Now, since $X_1(f),\dots,X_r(f)$ are independent infinitesimal transformations and since in addition the determinant of the $\psi_{kj}(a^0)$ is different from zero, the right-hand sides of the latter equations represent r independent infinitesimal transformations of our group.

The following method is even somewhat simpler.

One sets $a_k = a_k^0 + \delta t_k$, where it is understood that $\delta t_1,\dots,\delta t_r$ are infinitely small quantities. Then:

$$x_i' = f_i(x_1,\dots,x_n,\, a_1^0+\delta t_1,\dots,a_r^0+\delta t_r)$$
$$= x_i + \sum_{k=1}^{r} \left[\frac{\partial}{\partial a_k} f_i(x,a)\right]_{a=a^0} \delta t_k + \cdots,$$

where the omitted terms are of second and of higher order with respect to the δt_k. Here, it is now immediately apparent that our group contains the r infinitesimal transformations:

$$x_i' = x_i + \left[\frac{\partial}{\partial a_k} f_i(x,a)\right]_{a=a_0} \delta t_k \qquad (i=1\cdots n)$$
$$(k=1\cdots r)$$

However, the question of whether these r infinitesimal transformations are independent of each other requires in each individual case a specific examination, if one does not know from the beginning that the determinant of the ψ does not vanish for $a_k = a_k^0$.

Example. We consider the general projective group:

$$x' = \frac{x+a_1}{a_2x+a_3}$$

of the once-extended manifold.

The infinitesimal transformations of this group are obtained very easily by means of the following method. Indeed, one has $a_1^0=0$, $a_2^0=0$, $a_3^0=1$, hence we have:

$$x' = \frac{x+\delta t_1}{x\,\delta t_2 + 1 + \delta t_3} = x + \delta t_1 - x\,\delta t_3 - x^2\,\delta t_2 + \cdots,$$

that is to say, our group contains the three mutually independent infinitesimal transformations:

$$X_1(f) = \frac{\mathrm{d}f}{\mathrm{d}x}, \qquad X_2(f) = x\frac{\mathrm{d}f}{\mathrm{d}x}, \qquad X_3(f) = x^2\frac{\mathrm{d}f}{\mathrm{d}x}.$$

The general infinitesimal transformation of our group has the form:

$$(\lambda_1+\lambda_2x+\lambda_3x^2)\,\frac{\mathrm{d}f}{\mathrm{d}x},$$

hence we obtain its finite transformations by integrating the ordinary differential equation:

$$\frac{\mathrm{d}x'}{\lambda_1+\lambda_2x'+\lambda_3x'^2}=\mathrm{d}t,$$

adding the initial condition: $x'=x$ for $t=0$.

In order to carry out this integration, we bring the differential equation to the form:

$$\frac{\mathrm{d}x'}{x'-\alpha}-\frac{\mathrm{d}x'}{x'-\beta}=\gamma\,\mathrm{d}t,$$

by setting:

$$\lambda_1=\frac{\alpha\beta\gamma}{\alpha-\beta},\qquad \lambda_2=-\frac{\alpha+\beta}{\alpha-\beta}\,\gamma,\qquad \lambda_3=\frac{\gamma}{\alpha-\beta},$$

whence:

$$2\alpha=-\frac{\lambda_2}{\lambda_3}+\frac{\sqrt{\lambda_2^2-4\lambda_1\lambda_3}}{\lambda_3},\qquad 2\beta=-\frac{\lambda_2}{\lambda_3}-\frac{\sqrt{\lambda_2^2-4\lambda_1\lambda_3}}{\lambda_3},$$

$$\gamma=\sqrt{\lambda_2^2-4\lambda_1\lambda_3}.$$

By integration, we find:

$$l(x'-\alpha)-l(x'-\beta)=\gamma t+l(x-\alpha)-l(x-\beta),$$

or:

$$\frac{x'-\alpha}{x'-\beta}=e^{\gamma t}\,\frac{x-\alpha}{x-\beta},$$

and now there is absolutely no difficulty in expressing α,β,γ in terms of $\lambda_1,\lambda_2,\lambda_3$, in order to obtain the ∞^3 transformations of our three-term group arranged in ∞^2 one-term groups, exactly as enunciated in Theorem 11.

Furthermore, a simple known form of our group is obtained if one keeps the two parameters α and β, while one introduces the new parameter $\overline{\gamma}$ instead of $e^{\gamma t}$; then our group appears in the form:

$$\frac{x'-\alpha}{x'-\beta}=\overline{\gamma}\,\frac{x-\alpha}{x-\beta}.$$

References

1. Arnol'd, V.I.: Ordinary differential equations. Translated from the Russian and edited by R.A. Silverman, MIT Press, Cambridge, Mass.-London (1978)
2. Engel, F., Lie, S.: Theorie der Transformationsgruppen. Erster Abschnitt. Unter Mitwirkung von Prof. Dr. Friedrich Engel, bearbeitet von Sophus Lie, Verlag und Druck von B.G. Teubner, Leipzig und Berlin, xii+638 pp. (1888). Reprinted by Chelsea Publishing Co., New York, N.Y. (1970)
3. Gröbner, W.: Die Lie-Reihen und ihre Anwendungen. Math. Monog. Veb Deutschen Verlag der Wissenschaften (1960)
4. Rao, M.R.M.: Ordinary differential equations, theory and applications. Edward Arnold, London (1981)

Part II
English Translation

Chapter 5
Complete Systems of Partial Differential Equations

Abstract Any infinitesimal transformation $X = \sum_{i=1}^{n} \xi_i(x) \frac{\partial}{\partial x_i}$ can be considered as the first order analytic partial differential equation $X\omega = 0$ with the unknown ω. After a relocalization, a renumbering and a rescaling, one may suppose $\xi_n(x) \equiv 1$. Then the general solution ω happens to be any (local, analytic) function $\Omega\big(\omega_1, \dots, \omega_{n-1}\big)$ of the $(n-1)$ functionally independent solutions defined by the formula:

$$\omega_k(x) := \exp\big(-x_n X\big)(x_k) \qquad\qquad {\scriptstyle (k=1\cdots n-1)}.$$

What about first order *systems* $X_1\omega = \cdots = X_q\omega = 0$ of such differential equations? Any solution ω also trivially satisfies $X_i\big(X_k(\omega)\big) - X_k\big(X_i(\omega)\big) = 0$. But it appears that the subtraction in the Jacobi commutator $X_i\big(X_k(\cdot)\big) - X_k\big(X_i(\cdot)\big)$ kills all the second-order differentiation terms, so that one may freely add such supplementary first-order differential equations to the original system, continuing again and again, until the system, still denoted by $X_1\omega = \cdots = X_q\omega = 0$, becomes *complete* in the sense of Clebsch, namely satisfies, locally in a neighborhood of a generic point x^0:

(i) for all indices $i, k = 1, \dots, q$, there are appropriate functions $\chi_{ik\mu}(x)$ such that $X_i\big(X_k(f)\big) - X_k\big(X_i(f)\big) = \chi_{ik1}(x)X_1(f) + \cdots + \chi_{ikq}(x)X_q(f)$;

(ii) the rank of the vector space generated by the q vectors $X_1\big|_x, \dots, X_q\big|_x$ is constant equal to q for all x near the central point x^0.

Under these assumptions, it is shown in this chapter that there are $n-q$ functionally independent solutions $x_1^{(q)}, \dots, x_{n-q}^{(q)}$ of the system that are analytic near x_0 such that any other solution is a suitable function of these $n-q$ fundamental solutions.

S. Lie, *Theory of Transformation Groups I*,
DOI 10.1007/978-3-662-46211-9_5

A First Order Scalar Partial Differential Equation

As a prologue, we ask what are the general solutions ω of a first order partial differential equation $X\omega = 0$ naturally associated to a local analytic vector field $X = \sum_{i=1}^{n} \xi_i(x) \frac{\partial}{\partial x_i}$. Free relocalization always being allowed in Lie's theory, we may assume, after possibly renumbering the variables, that ξ_n does not vanish in a small neighborhood of some point at which we center the origin of the coordinates. Dividing then by $\xi_n(x)$, it is equivalent to seek functions ω that are annihilated by the differential operator:

$$X = \sum_{i=1}^{n-1} \frac{\xi_i(x)}{\xi_n(x)} \frac{\partial}{\partial x_i} + \frac{\partial}{\partial x_n},$$

still denoted by X and which now satisfies $X(x_n) \equiv 1$. We recall that the corresponding system of ordinary differential equations which defines curves that are everywhere tangent to X, namely the system:

$$\frac{d\mathrm{x}_1}{dt} = \frac{\xi_1(\mathrm{x}(t))}{\xi_n(\mathrm{x}(t))}, \quad \dots\dots, \quad \frac{d\mathrm{x}_{n-1}}{dt} = \frac{\xi_{n-1}(\mathrm{x}(t))}{\xi_n(\mathrm{x}(t))}, \quad \frac{d\mathrm{x}_n(t)}{dt} = 1,$$

with initial condition for $t = 0$ being an arbitrary point of the hyperplane $\{x_n = 0\}$:

$$\mathrm{x}_1(0) = x_1, \dots\dots, \mathrm{x}_{n-1}(0) = x_{n-1}, \quad \mathrm{x}_n(0) = 0$$

is *solvable* and has a unique vectorial solution $(\mathrm{x}_1, \dots, \mathrm{x}_{n-1}, \mathrm{x}_n)$ which is analytic in a neighborhood of the origin. In fact, $\mathrm{x}_n(t) = t$ by an obvious integration, and the $(n-1)$ other $\mathrm{x}_k(t)$ are given by the marvelous exponential formula already shown on p. 67:

$$\mathrm{x}_k(t) = \exp(tX)(x_k) = \sum_{l \geqslant 0} \frac{t^l}{l!} X^l(x_k) \qquad (k = 1 \cdots n-1).$$

We then set $t := -x_n$ in this formula (the minus sign will be crucial) and we define the $(n-1)$ functions that are relevant to us:

$$\begin{aligned} \omega_k(x_1, \dots, x_n) &:= \mathrm{x}_k(-x_n) = \exp\big(-x_n X\big)(x_k) \\ &= \sum_{l \geqslant 0} (-1)^l \frac{(x_n)^l}{l!} X^l(x_k). \end{aligned}$$

Proposition 5.1. *The $(n-1)$ so defined functions $\omega_1, \dots, \omega_{n-1}$ are functionally independent solutions of the partial differential equation $X\omega = 0$ with the rank of their Jacobian matrix $\big(\frac{\partial \omega_k}{\partial x_i}\big)_{1 \leqslant i \leqslant n}^{1 \leqslant k \leqslant n-1}$ being equal to $n-1$ at the origin. Furthermore, for every other solution ω of $X\omega = 0$, there exists a local analytic function $\Omega = \Omega\big(\omega_1, \dots, \omega_{n-1}\big)$ defined in a neighborhood of the origin in $\mathbb{K}^{n-1}$*

such that:

$$\omega(x) \equiv \Omega\big(\omega_1(x), \dots, \omega_{n-1}(x)\big).$$

Proof. Indeed, when applying X to the above series defining the ω_k, we see that all the terms cancel out, thanks to an application of Leibniz' formula in the form:

$$X\big[(x_n)^l X^l(x_k)\big] = l\,(x_n)^{l-1} X^l(x_k) + (x_n)^l X^{l+1}(x_k).$$

Next, the assertion that the map $x \longmapsto \big(\omega_1(x), \dots, \omega_{n-1}(x)\big)$ has rank $n-1$ is clear, for $\omega_k(x_1, \dots, x_{n-1}, 0) \equiv x_k$ by construction. Finally, after straightening X to $X' := \frac{\partial}{\partial x_n'}$ in some new coordinates $(x_1', \dots, x_n')$ thanks to the theorem on p. 64, the general solution $\omega'(x')$ to $X'\omega' = 0$ is trivially any function $\Omega'(x_1', \dots, x_{n-1}')$ of $x_1' \equiv \omega_1', \dots, x_{n-1}' \equiv \omega_{n-1}'$. □

Chapter 5

The Complete Systems

We assume that the theory of the integration of an individual first order linear partial differential equation:

$$X(f) = \sum_{i=1}^{n} \xi_i(x_1, \dots, x_n)\,\frac{\partial f}{\partial x_i} = 0,$$

or of the equivalent simultaneous system of ordinary differential equations:

$$\frac{\mathrm{d}x_1}{\xi_1} = \cdots = \frac{\mathrm{d}x_n}{\xi_n},$$

is known; nonetheless, as an introduction, we compile without demonstration a few related propositions. Based on these propositions, we shall very briefly derive the theory of the integration of simultaneous first order linear partial differential equations. In the next chapter, we shall place in a new light [IN EIN NEUES LICHT SETZEN] this theory due in the main wholly to JACOBI and to CLEBSCH, by explaining more closely the connection between the concepts [BEGRIFFEN] of "linear partial differential equation" and of "infinitesimal transformation", a connection that we have already mentioned earlier (Chap. 4, p. 69).

§ 21. One can suppose that $\xi_1, \dots, \xi_n$ behave regularly in the neighborhood of a determinate system of values $x_1^0, \dots, x_n^0$, and also that $\xi_n(x_1^0, \dots, x_n^0)$ is different from zero. Under these assumptions, one can determine $x_1, \dots, x_{n-1}$ as analytic functions of x_n in such a way that by substitution of these functions, the simultaneous system:

$$\frac{dx_1}{dx_n} = \frac{\xi_1}{\xi_n}, \ldots, \frac{dx_{n-1}}{dx_n} = \frac{\xi_{n-1}}{\xi_n}$$

is identically satisfied, and that in addition $x_1, \ldots, x_{n-1}$ for $x_n = x_n^0$ take certain prescribed *initial values* [ANFANGSWERTHE] $x_1', \ldots, x_{n-1}'$. These initial values have to be interpreted as the integration constants.

The equations which, in the concerned way, represent $x_1, \ldots, x_{n-1}$ as functions of x_n are called the *complete integral equations of the simultaneous system*; they can receive the form:

$$x_k = x_k' + (x_n^0 - x_n)\,\mathfrak{P}_k(x_1' - x_1^0, \ldots, x_{n-1}' - x_{n-1}^0, x_n^0 - x_n)$$
$$(k=1 \cdots n-1),$$

where the $\mathfrak{P}_k$ denote ordinary power series in their arguments. By inverting the relation between $x_1, \ldots, x_{n-1}, x_n$ and $x_1', \ldots, x_{n-1}', x_n^0$, one again obtains the integral equations, resolved with respect to only the initial values $x_1', \ldots, x_{n-1}'$:

$$x_k' = x_k + (x_n - x_n^0)\,\mathfrak{P}_k(x_1 - x_1^0, \ldots, x_n - x_n^0) = \omega_k(x_1, \ldots, x_n)$$
$$(k=1 \cdots n-1).$$

Here, the functions ω_k are the so-called *integral functions* of the simultaneous system, since the differentials of these functions:

$$d\,\omega_k = \sum_{i=1}^{n} \frac{\partial \omega_k}{\partial x_i}\, dx_i \qquad (k=1 \cdots n-1)$$

all vanish identically by virtue of the simultaneous system, and every function of this sort is called an integral function of the simultaneous system. But every such integral function is at the same time a solution of the linear partial differential equation $X(f) = 0$, whence $\omega_1, \ldots, \omega_{n-1}$ are solutions of $X(f) = 0$, and in fact, they are obviously independent. In a certain neighborhood of $x_1^0, \ldots, x_n^0$ these solutions behave regularly; in addition, they reduce for $x_n = x_n^0$ to $x_1, \ldots, x_{n-1}$ respectively; that is why they are called the *general solutions of the equation* $X(f) = 0$ *relative to* $x_n = x_n^0$.

If one knows altogether $n-1$ independent solutions:

$$\psi_1(x_1, \ldots, x_n), \ldots\ldots, \psi_{n-1}(x_1, \ldots, x_n)$$

of the equation $X(f) = 0$, then the most general solution of it has the form $\Omega(\psi_1, \ldots, \psi_{n-1})$, where Ω denotes an arbitrary analytic function of its arguments.

§ 22. If a function $\psi(x_1, \ldots, x_n)$ satisfies the two equations:

$$X_1(f) = 0, \qquad X_2(f) = 0,$$

then it naturally also satisfies the two differential equations of second order:

$$X_1\big(X_2(f)\big) = 0, \qquad X_2\big(X_1(f)\big) = 0,$$

and in consequence of that, also the equation:

$$X_1\big(X_2(f)\big) - X_2\big(X_1(f)\big) = 0,$$

which is obtained by subtraction from the last two written ones.

If now one lets:

$$X_k(f) = \sum_{i=1}^{n} \xi_{ki}(x_1,\dots,x_n)\,\frac{\partial f}{\partial x_i} \qquad (k=1,2),$$

then:

$$X_1\big(X_2(f)\big) - X_2\big(X_1(f)\big) = \sum_{i=1}^{n} \{X_1(\xi_{2i}) - X_2(\xi_{1i})\}\,\frac{\partial f}{\partial x_i},$$

because all terms which contain second order differential quotients are canceled. Thus, the following holds:

Proposition 1. *If a function $\psi(x_1,\dots,x_n)$ satisfies the two differential equations of first order:*

$$X_k(f) = \sum_{i=1}^{n} \xi_{ki}(x_1,\dots,x_n)\,\frac{\partial f}{\partial x_i} = 0 \qquad (k=1,2),$$

then it also satisfies the equation:

$$X_1\big(X_2(f)\big) - X_2\big(X_1(f)\big) = \sum_{i=1}^{n} \{X_1(\xi_{2i}) - X_2(\xi_{1i})\}\,\frac{\partial f}{\partial x_i} = 0,$$

which, likewise, is of first order.

It is of great importance to know how the expression $X_1\big(X_2(f)\big) - X_2\big(X_1(f)\big)$ behaves when, in place of $x_1,\dots,x_n$, new independent variables $y_1,\dots,y_n$ are introduced.

We agree that by introduction of the y, we obtain:

$$X_k(f) = \sum_{i=1}^{n} X_k(y_i)\,\frac{\partial f}{\partial y_i} = \sum_{i=1}^{n} \eta_{ki}(y_1,\dots,y_n)\,\frac{\partial f}{\partial y_i} = Y_k(f) \qquad (k=1,2).$$

Since f denotes here a completely arbitrary function of $x_1,\dots,x_n$, we can substitute $X_1(f)$ or $X_2(f)$ in place of f, so we have:

$$X_1\big(X_2(f)\big) = Y_1\big(X_2(f)\big) = Y_1\big(Y_2(f)\big)$$
$$X_2\big(X_1(f)\big) = Y_2\big(X_1(f)\big) = Y_2\big(Y_1(f)\big),$$

and consequently:

$$X_1\big(X_2(f)\big)-X_2\big(X_1(f)\big)=Y_1\big(Y_2(f)\big)-Y_2\big(Y_1(f)\big).$$

Thus we have the

Proposition 2. *If, by the introduction of a new independent variable, the expressions $X_1(f)$ and $X_2(f)$ are transferred to $Y_1(f)$ and respectively to $Y_2(f)$, then the expression $X_1\big(X_2(f)\big)-X_2\big(X_1(f)\big)$ is transferred to $Y_1\big(Y_2(f)\big)-Y_2\big(Y_1(f)\big)$.*

This property of the expression $X_1\big(X_2(f)\big)-X_2\big(X_1(f)\big)$ will frequently be used in the course of our study. The same proposition can be stated more briefly: *the expression $X_1\big(X_2(f)\big)-X_2\big(X_1(f)\big)$ behaves invariantly through the introduction of a new variable.*

We now consider the q equations:

$$(1) \qquad X_1(f)=0,\dots\dots,X_q(f)=0,$$

and we ask about its possible joint solutions.

It is conceivable that between the expressions $X_k(f)$, there are relations of the form:

$$(2) \qquad \sum_{k=1}^{q}\chi_k(x_1,\dots,x_n)\,X_k(f)\equiv 0.$$

If this were the case, then certain amongst our equations would be a consequence of the remaining ones, and they could easily be left out while taking for granted the solution of the stated problem. Therefore it is completely legitimate to make the assumption that there are no relations of the form (2), hence that the equations (1) are solvable with respect to q of the differential quotients $\frac{\partial f}{\partial x_i}$. When we refer to the *equations* (1) as *independent* of each other, it is to be understood in this sense.[1] (Translator's note: This is a typical place where a relocalization is in general required in order to insure that the vectors $X_1\big|_x,\dots,X_q\big|_x$ are locally linearly independent, and the authors, as usual, understand it mentally.)

According to what has been said above about the two equations $X_1(f)=0$ and $X_2(f)=0$, it is clear that the possible joint solutions of our q equations also satisfy all equations of the form:

$$X_i\big(X_k(f)\big)-X_k\big(X_i(f)\big)=0.$$

Two cases can occur.

Firstly, the equations obtained this way can be a consequence of the former, when for every i and $k\leqslant q$, a relation of the following form:

$$X_i\big(X_k(f)\big)-X_k\big(X_i(f)\big)=$$
$$=\chi_{ik1}(x_1,\dots,x_n)\,X_1(f)+\cdots+\chi_{ikq}(x_1,\dots,x_n)\,X_q(f)$$

holds. With CLEBSCH, we then say that *the q independent equations $X_1(f)=0, \dots, X_q(f)=0$ form a q-term complete system* [q-GLIEDRIG VOLLSTÄNDIG SYSTEM].

However, in general the second possible case will occur; amongst the new formed equations:

$$X_i\big(X_k(f)\big)-X_k\big(X_i(f)\big)=0,$$

will be found a certain number which are independent of each other, and from the q presented ones. We add them, say:

$$X_{q+1}(f)=0,\dots\dots,X_{q+s}(f)=0$$

to the q initial equations and we now treat the obtained $q+s$ equations exactly as the original q equations. So we continue; but since we cannot come to more than n equations $X_i(f)=0$ that are independent of each other, we must finally reach a complete system which consists of n or fewer independent equations. Therefore we have the proposition:

Proposition 3. *The determination of the joint solutions of q given linear partial differential equations of first order $X_1(f)=0,\dots,X_q(f)=0$ can always be reduced, by differentiation and elimination, to the integration of a complete system.*

We now assume that $X_1(f)=0, \dots, X_q(f)=0$ form a q-term complete system. Obviously, these equations may be replaced by q other equations:

$$Y_k(f)=\sum_{j=1}^{q}\psi_{kj}(x_1,\dots,x_n)\,X_j(f)=0 \qquad (k=1\cdots q).$$

In the process, it is only required that the determinant of the ψ_{kj} does not vanish identically, whence the $X_j(f)$ can also be expressed linearly in terms of the $Y_k(f)$. Visibly, relations of the form:

$$Y_i\big(Y_k(f)\big)-Y_k\big(Y_i(f)\big)=\sum_{j=1}^{q}\omega_{ikj}(x_1,\dots,x_n)\,Y_j(f) \tag{3}$$

then hold true; so the equations $Y_k(f)=0$ also form a q-term complete system and hence are totally equivalent to the equations $X_k(f)=0$.

As has been pointed out for the first time by CLEBSCH, some ψ_{kj} are always available for which all the ω_{ikj} vanish. In order to show this in the simplest way, we select with A. MAYER the ψ_{kj} so that the $Y_k(f)=0$ appear to be resolved with respect to q of the differential quotients, for instance with respect to $\frac{\partial f}{\partial x_n}, \dots, \frac{\partial f}{\partial x_{n-q+1}}$:

$$Y_k(f)=\frac{\partial f}{\partial x_{n-q+k}}+\sum_{i=1}^{n-q}\eta_{ki}\,\frac{\partial f}{\partial x_i} \qquad (k=1\cdots q). \tag{4}$$

Then the expressions $Y_i\big(Y_k(f)\big)-Y_k\big(Y_i(f)\big)$ will all be free of $\frac{\partial f}{\partial x_{n-q+1}}, \dots, \frac{\partial f}{\partial x_n}$, and consequently they can have the form $\sum_j \omega_{ikj}\,Y_j(f)$ only if all the ω_{ikj} vanish. Thus:

Proposition 4. *If one solves a q-term complete system:*

$$X_1(f) = 0, \ldots\ldots, X_q(f) = 0$$

with respect to q of the differential quotients, then the resulting equations:

$$(4) \qquad Y_k(f) = \frac{\partial f}{\partial x_{n-q+k}} + \sum_{i=1}^{n-q} \eta_{ki} \frac{\partial f}{\partial x_i} = 0 \qquad (k=1\cdots q)$$

stand pairwise in the relationships:

$$(5) \qquad Y_i\big(Y_k(f)\big) - Y_k\big(Y_i(f)\big) = 0 \qquad (i,k=1\cdots q).$$

§ 23. We now imagine that a given q-term complete system is brought to the above-mentioned form:

$$(4) \qquad Y_k(f) = \frac{\partial f}{\partial x_{n-q+k}} + \sum_{i=1}^{n-q} \eta_{ki} \frac{\partial f}{\partial x_i} = 0 \qquad (k=1\cdots q).$$

It will be shown that this system possesses $n-q$ independent solutions, for the determination of which it suffices to integrate q individual linear partial differential equations one after the other.

At first, we integrate the differential equation:

$$Y_q(f) = \frac{\partial f}{\partial x_n} + \sum_{i=1}^{n-q} \eta_{qi} \frac{\partial f}{\partial x_i} = 0;$$

amongst its $n-1$ independent solutions $x_1', \ldots, x_n'$, the following $q-1$, namely:

$$x_{n-q+1}' = x_{n-q+1}, \ldots\ldots, x_{n-1}' = x_{n-1}$$

are known at once. If the $n-1$ expressions $x_1', \ldots, x_{n-1}'$ together with the quantity x_n which is independent of them, are introduced as new variables, then our complete system receives the form:

$$Y_q(f) = \frac{\partial f}{\partial x_n} = 0, \qquad Y_k(f) = \frac{\partial f}{\partial x_{n-q+1}'} + \sum_{i=1}^{n-q} \eta_{ki}' \frac{\partial f}{\partial x_i'} = 0$$

$$(k=1\cdots q-1).$$

Now, since the expressions $Y_i\big(Y_k(f)\big) - Y_k\big(Y_i(f)\big)$ behave invariantly through the introduction of new variables (cf. Proposition 2 of this chapter), they must now again also vanish, from which it follows that all η_{ki}' must be functions of only $x_1', \ldots, x_{n-1}'$, and be free of x_n. The initial problem of integration is therefore reduced to finding the joint solutions of the $q-1$ equations:

$$Y_k'(f) = \frac{\partial f}{\partial x'_{n-q+k}} + \sum_{i=1}^{n-q} \eta'_{ki}(x'_1, \dots, x'_{n-1}) \frac{\partial f}{\partial x'_i} = 0$$
$$(k = 1 \cdots q-1),$$

and in fact these equations, which depend upon only $n-1$ independent variables, namely $x'_1, \dots, x'_{n-1}$, again stand pairwise in the relationships: $Y_i'(Y_k'(f)) - Y_k'(Y_i'(f)) = 0$.

We formulate this result in the following way.

Proposition 5. *The joint solutions of the equations of a q-term complete system in n variables can also be defined as the joint solutions of the equations of a $(q-1)$-term complete system in $n-1$ variables. In order to be able to set up this new complete system, one only has to integrate a single linear partial differential equation of first order in $(n-q+1)$ variables.*

The new complete system again appears in resolved form; we can hence at once continue our above-mentioned process and by applying it $(q-1)$ times, we obtain the following:

Proposition 6. *The joint solutions of the equations of a q-term complete system in n variables can also be defined as the solutions of a single linear partial differential equation of first order in $n-q+1$ variables. In order to be able to set up this equation, it suffices to integrate $q-1$ individual equations of this sort one after the other.*

From this it follows easily:

Proposition 7. *A q-term complete system in n independent variables always possesses $n-q$ independent solutions.*

But conversely, the following also holds:

Proposition 8. *If q independent linear partial differential equations of first order in n independent variables:*

$$X_1(f) = 0, \dots\dots, X_q(f) = 0$$

have exactly $n-q$ independent solutions in common, then they form a q-term complete system.

For the proof, we remark that according to the preceding, the equations:

$$X_1(f) = 0, \dots\dots, X_q(f) = 0$$

determine a complete system with q or more terms; now, if this complete system contained more than q independent equations, then it would possess not $n-q$, but only a smaller number of independent solutions; under the assumptions of the proposition, it is therefore q-term, that is to say, it is constituted by the equations $X_1(f) = 0, \dots, X_q(f) = 0$ themselves.

If $n-q$ independent functions $\omega_1,\dots,\omega_{n-q}$ of $x_1,\dots,x_n$ are presented, then one can always set up q independent linear partial differential equations which are identically satisfied by all ω. Indeed, if we take for granted that the determinant:

$$\sum \pm \frac{\partial \omega_1}{\partial x_1} \cdots \frac{\partial \omega_{n-q}}{\partial x_{n-q}}$$

does not vanish identically, which we can achieve without restriction, then the q equations:

$$\begin{vmatrix} \frac{\partial f}{\partial x_1} & \cdots & \frac{\partial f}{\partial x_{n-q}} & \frac{\partial f}{\partial x_{n-q+k}} \\ \frac{\partial \omega_1}{\partial x_1} & \cdots & \frac{\partial \omega_1}{\partial x_{n-q}} & \frac{\partial \omega_1}{\partial x_{n-q+k}} \\ \cdot & \cdots & \cdot & \cdot \\ \frac{\partial \omega_{n-q}}{\partial x_1} & \cdots & \frac{\partial \omega_{n-q}}{\partial x_{n-q}} & \frac{\partial \omega_{n-q}}{\partial x_{n-q+k}} \end{vmatrix} = 0 \qquad (k=1\cdots q)$$

are differential equations of that kind; because after the substitution $f = \omega_j$, they go to an identity, and because they are independent of each other, as their form shows. According to the last proposition, these equations form a q-term complete system. Thus:

Proposition 9. *If $n-q$ independent functions of n variables $x_1,\dots,x_n$ are presented, then there always exists a determinate q-term complete system in $x_1,\dots,x_n$ of which these functions constitute a system of solutions.*

§ 24. However, for our goal, it does not suffice to have proved the existence of the solutions to a complete system, it is on the contrary rather necessary to deal more closely with the analytic properties of these solutions.

For this purpose, we shall assume that in the presented q-term complete system:

$$Y_k(f) = \frac{\partial f}{\partial x_{n-q+k}} + \sum_{i=1}^{n-q} \eta_{ki} \frac{\partial f}{\partial x_i} = 0 \qquad (k=1\cdots q),$$

the analytic functions η_{ki} behave regularly in the neighborhood of:

$$x_1 = \cdots = x_n = 0.$$

As quantities $x'_1,\dots,x'_{n-1}$, we now choose amongst the solutions of the equation:

$$\frac{\partial f}{\partial x_n} + \sum_{i=1}^{n-q} \eta_{qi} \frac{\partial f}{\partial x_i} = 0$$

the formerly defined general solutions of this equation, relative to $x_n = 0$.

We know that these general solutions $x'_1,\dots,x'_{n-1}$ are ordinary power series with respect to $x_1,\dots,x_n$ in a certain neighborhood of $x_1 = \cdots = x_n = 0$, and that for

$x_n = 0$, they reduce to $x_1,\ldots,x_{n-1}$ respectively. The solutions already mentioned above:

$$x'_{n-q+1} = x_{n-q+1},\ldots\ldots,x'_{n-1} = x_{n-1}$$

are therefore general solutions.

Conversely, according to the observations in § 21, $x_1,\ldots,x_{n-1}$ are also analytic functions of $x'_1,\ldots,x'_{n-1},x_n$ and they behave regularly in the neighborhood of the system of values $x'_1 = \cdots = x'_{n-1} = x_n = 0$. Now, since in the new complete system $Y'_k(f) = 0$, the coefficients η'_{ki} behave regularly as functions of $x_1,\ldots,x_n$, they will also be ordinary power series with respect to $x'_1,\ldots,x'_{n-1},x_n$ in a certain neighborhood of $x'_1 = 0,\ldots,x'_{n-1} = 0$, $x_n = 0$, and in fact they will be, as we know, free of x_n.

Next, we determine the general solutions of the equation:

$$\frac{\partial f}{\partial x'_{n-1}} + \sum_{i=1}^{n-q} \eta'_{q-1,i} \frac{\partial f}{\partial x'_i} = 0$$

relative to $x'_{n-1} = 0$. These solutions, which we may call $x''_1,\ldots,x''_{n-2}$, behave regularly as functions of $x'_1,\ldots,x'_{n-1}$ in the neighborhood of $x'_1 = \cdots = x'_{n-1} = 0$ and are hence also regular in the neighborhood of $x_1 = \cdots = x_n = 0$, as functions of $x_1,\ldots,x_n$. After the substitution $x'_{n-1} = 0$, the functions $x''_1,\ldots,x''_{n-2}$ reduce to $x'_1,\ldots,x'_{n-2}$, whence they reduce to $x_1,\ldots,x_{n-2}$ after the substitution $x_{n-1} = x_n = 0$. The coefficients of the next $(q-2)$-term complete system are naturally ordinary power series in $x''_1,\ldots,x''_{n-2}$.

After iterating these considerations q times, we obtain at the end $n-q$ independent solutions: $x_1^{(q)},\ldots,x_{n-q}^{(q)}$ of our complete system. These are ordinary power series with respect to the $x_i^{(q-1)}$ in a certain neighborhood of $x_1^{(q-1)} = \cdots = x_{n-q+1}^{(q)} = 0$, and just in the same way, also series with respect to the x_i in a certain neighborhood of $x_1 = \cdots = x_n = 0$. For $x_{n-q+1}^{(q-1)} = 0$, the $x_1^{(q)},\ldots,x_{n-q}^{(q)}$ reduce respectively to $x_1^{(q-1)},\ldots,x_{n-q}^{(q-1)}$ and hence to $x_1,\ldots,x_{n-q}$ for $x_{n-q+1} = \cdots = x_n = 0$. Therefore one has:

$$x_i^{(q)} = x_i + \mathfrak{P}_i(x_1,\ldots,x_n) \qquad (i=1\cdots n-q),$$

where the $\mathfrak{P}_i$ all vanish for $x_{n-q+1} = \cdots = x_n = 0$.

We call the solutions $x_1^{(q)},\ldots,x_{n-q}^{(q)}$ of our complete system its *general solutions relative to* $x_{n-q+1} = 0,\ldots,x_n = 0$.

We may state the gained result in a somewhat more general form by introducing a general system of values: $x_1^0,\ldots,x_n^0$ in place of the special one:

$$x_1 = x_2 = \cdots = x_n = 0.$$

Then we can say:

Theorem 12. *Every q-term complete system:*

$$\frac{\partial f}{\partial x_{n-q+k}} + \sum_{i=1}^{n-q} \eta_{ki}(x_1, \ldots, x_n) \frac{\partial f}{\partial x_i} = 0 \qquad (k=1\cdots q)$$

whose coefficients η_{ki} behave regularly in the neighborhood of $x_1 = x_1^0$, ..., $x_n = x_n^0$ possesses $n-q$ independent solutions $x_1^{(q)}, \ldots, x_{n-q}^{(q)}$ which behave regularly in a certain neighborhood of $x_1 = x_1^0$, ..., $x_n = x_n^0$ and which in addition reduce to $x_1, \ldots, x_{n-q}$ respectively after the substitution $x_{n-q+1} = x_{n-q+1}^0$, ..., $x_n = x_n^0$.

The main theorem of the theory of complete systems has not been stated in this precise version, either by JACOBI, or by CLEBSCH. Nevertheless, this theorem is implicitly contained in the work of CAUCHY, WEIERSTRASS, BRIOT and BOUQUET, KOWALEVSKY and DARBOUX, which treat the question of existence of solutions to given differential equations.

§ 25. The theory of a single linear partial differential equation:

$$X(f) = \sum_{i=1}^{n} \xi_i(x_1, \ldots, x_n) \frac{\partial f}{\partial x_i} = 0$$

stands, as already said, in the closest connection with the theory of the simultaneous system:

$$\frac{\mathrm{d}x_1}{\xi_1} = \cdots = \frac{\mathrm{d}x_n}{\xi_n}.$$

Moreover, something completely analogous also takes place for systems of linear partial differential equations.[1]

Consider q independent linear partial differential equations of first order which, however, need not constitute a complete system. For the sake of simplicity, we imagine that the equations are solved with respect to q of the differential quotients:

$$\text{(4')} \qquad Y_k(f) = \frac{\partial f}{\partial x_{n-q+k}} + \sum_{i=1}^{n-q} \eta_{ki}(x_1, \ldots, x_n) \frac{\partial f}{\partial x_i} = 0 \qquad (k=1\cdots q).$$

If $\omega(x_1, \ldots, x_n)$ is a general solution of these equations, then:

$$\frac{\partial \omega}{\partial x_{n-q+k}} \equiv -\sum_{i=1}^{n-q} \eta_{ki} \frac{\partial \omega}{\partial x_i} \qquad (k=1\cdots q),$$

whence:

$$\mathrm{d}\omega = \sum_{i=1}^{n-q} \frac{\partial \omega}{\partial x_i} \left\{ dx_i - \sum_{k=1}^{q} \eta_{ki}\, dx_{n-q+k} \right\}.$$

[1] This connection has been explained in detail for the first time by BOOLE. Cf. also A. MAYER concerning unrestricted integrable differential equations, Math. Ann. Vol. V.

Consequently, the differential $d\omega$ vanishes identically by virtue of the $n-q$ total differential equations:

$$\text{(6)} \qquad dx_i - \sum_{k=1}^{q} \eta_{ki}\, dx_{n-q+k} = 0 \qquad (i=1\cdots n-q).$$

But one calls every function of this nature an *integral function* of this system of total differential equations. Thus we can say:

Every joint solution of the q linear partial differential equations (4') *is an integral function of the system of n − q total differential equations* (6).

But conversely also, every integral function of the system (6) *is a joint solution of the equations* (4').

Indeed, if $w(x_1,\ldots,x_n)$ is an integral function of (6), then the expression:

$$dw = \sum_{i=1}^{n} \frac{\partial w}{\partial x_i}\, dx_i$$

vanishes identically by virtue of (6), that it is to say, one has:

$$\sum_{k=1}^{q} \left\{ \frac{\partial w}{\partial x_{n-q+k}} + \sum_{i=1}^{n-q} \eta_{ki} \frac{\partial w}{\partial x_i} \right\} dx_{n-q+k} \equiv 0,$$

and from this it is evident that w satisfies the equations (4) identically.

But now, the integration of the system (6) is synonymous to the finding [AUFFINDUNG] of all its integral functions; for, what does it mean to integrate the system (6)? Nothing else, as is known, but to determine all possible functions $\rho_1,\ldots,\rho_{n-q}$ of $x_1,\ldots,x_n$ which make the expression:

$$\sum_{i=1}^{n-q} \rho_i(x_1,\ldots,x_n) \left\{ dx_i - \sum_{k=1}^{q} \eta_{ki}\, dx_{n-q+k} \right\}$$

become a complete differential, hence to be the differential of an integral function. In other words:

The integration of the system (4') *of q linear partial differential equations is accomplished if one integrates the system of the n − q total differential equations* (6), *and conversely.*

This connection between the two systems (4') and (6) naturally presupposes that they (respectively) possess integral functions; however, there still exists a certain connection also when there are no joint solutions to (4'), and hence too, no integral functions for (6). This will be considered on another occasion (Chap. 6, p. 120).

Since the equations (4') have at most $n-q$ independent solutions in common, the system (6) has at most $n-q$ integral functions. If it possess precisely $n-q$ such independent solutions, the system (6) is called *unrestrictedly inte-*

grable [UNBESCHRÄNKT INTEGRABEL]; so this case occurs only when the q equations (4') form a q-term complete system.

We suppose that the system (6) is unrestrictedly integrable, or what is the same, that all the expressions $Y_k\big(Y_j(f)\big) - Y_j\big(Y_k(f)\big)$ vanish identically. Furthermore, we imagine that the $n-q$ general solutions:

$$\omega_1(x_1,\dots,x_n),\ \dots,\ \omega_{n-q}(x_1,\dots,x_n)$$

of the complete system (4') relative to:

$$x_{n-q+1} = x_{n-q+1}^0, \dots, x_n = x_n^0$$

are determined. Then the equations:

$$\omega_1(x_1,\dots,x_n) = a_1,\ \dots,\ \omega_{n-q}(x_1,\dots,x_n) = a_{n-q}$$

with the $n-q$ arbitrary constants $a_1,\dots,a_{n-q}$ are called the *complete integral equations* of the system (6). These integral equations are obviously solvable with respect to $x_1,\dots,x_{n-q}$, for $\omega_1,\dots,\omega_{n-q}$ reduce to $x_1,\dots,x_{n-q}$ (respectively) for:

$$x_{n-q+1} = x_{n-q+1}^0,\ \dots,\ x_n = x_n^0.$$

Hence we obtain:

$$x_i = \psi_i\big(x_{n-q+1},\dots,x_n,\,a_1,\dots,a_{n-q}\big) \qquad (i=1\cdots n-q).$$

One can easily see that the equations (6) become identities after the substitution $x_1 = \psi_1,\dots,x_{n-q} = \psi_{n-q}$. Indeed, we at first have:

$$x_i - \psi_i\big(x_{n-q+1},\dots,x_n,\,\omega_1,\dots,\omega_{n-q}\big) \equiv 0 \qquad (i=1\cdots n-q);$$

so if we introduce $x_i - \psi_i$ in place of f in $Y_k(f)$, we naturally again obtain an identically vanishing expression, and since all $Y_k(\omega_1),\dots,Y_k(\omega_{n-q})$ are identically zero, we get:

$$\eta_{ki}(x_1,\dots,x_n) - \frac{\partial}{\partial x_{n-q+k}}\,\psi_i\big(x_{n-q+1},\dots,x_n,\,\omega_1,\dots,\omega_{n-q}\big) \equiv 0$$
$$(k=1\cdots q;\, i=1\cdots n-q).$$

If we make the substitution $x_1 = \psi_1,\dots,\, x_{n-q} = \psi_{n-q}$ here, we obtain:

$$\eta_{ki}\big(\psi_1,\dots,\psi_{n-q},\,x_{n-q+1},\dots,x_n\big) - \frac{\partial}{\partial x_{n-q+k}}\,\psi_i\big(x_{n-q+1},\dots,x_n,\, a_1,\dots,a_{n-q}\big) \equiv 0.$$

We multiply this by dx_{n-q+k}, we sum with respect to k from 1 to q and we then realize that the expression:

$$\mathrm{d}x_i - \sum_{k=1}^{q} \eta_{ki}(x_1,\dots,x_n)\,\mathrm{d}x_{n-q+k}$$

effectively vanishes identically after the substitution $x_1 = \psi_1, \dots, x_{n-q} = \psi_{n-q}$. If we add that, from the equations $\omega_i = a_i$, we can always determine the a_i so that the variables $x_1,\dots,x_{n-q}$ take prescribed initial values $\overline{x}_1,\dots,\overline{x}_{n-q}$ for $x_{n-q+1} = x^0_{n-q+1}, \dots, x_n = x^0_n$, then we can say:

If the system of total differential equations (6) *is unrestrictedly integrable, then it is always possible to determine analytic functions* $x_1,\dots,x_{n-q}$ *of* $x_{n-q+1},\dots,x_n$ *in such a way that the system* (6) *is identically satisfied and that* $x_1,\dots,x_{n-q}$ *take prescribed initial values for* $x_{n-q+1} = x^0_{n-q+1}, \dots, x_n = x^0_n$.

§ 26. For reasons of convenience, we introduce a few abbreviations which will be useful in the future.

Firstly, the parentheses around the f in $X(f)$ shall from now on be frequently omitted.

Further, since expressions of the form $X\big(Y(f)\big) - Y\big(X(f)\big)$ will always occur more frequently [IMMER HAÜFIGER], we want to write:

$$X\big(Y(f)\big) - Y\big(X(f)\big) = XYf - YXf = \big[X,Y\big];$$

also, we shall often employ the following language [REDEWEISE BEDIENEN]: the expression, or the infinitesimal transformation $\big[X,Y\big]$ arises as the "*composition*" [ZUSAMMENSETZUNG], or the "*combination*" [COMBINATION], of Xf and Yf.

The following observation is also pertinent here:

Between any three expressions Xf, Yf, Zf there always exists the following identity:

$$\big[\big[X,Y\big],Z\big] + \big[\big[Y,Z\big],X\big] + \big[\big[Z,X\big],Y\big] \equiv 0. \tag{7}$$

This identity is a special case of the so-called *Jacobi identity*, which we shall get to know later. Here, we want to content ourselves with verifying the correctness of the special identity (7); later at the concerned place (cf. Volume 2), we shall enter into the sense of the Jacobi identity.

One has obviously:

$$\big[\big[X,Y\big],Z\big] = XYZf - YXZf - ZXYf + ZYXf;$$

if one permutes Xf, Yf, Zf circulary and then sums the three obtained relations, one gets rid of all the terms on the right-hand side, and one obtains the identity pointed out above.

This special Jacobi identity turns out to be extremely important in all research on transformation groups.

The above simple verification of the *special* Jacobi identity was given for the first time by ENGEL.

Chapter 6
New Interpretation of the Solutions of a Complete System

If, in the developments of the preceding chapter, we interpret the expression $X(f)$ as the symbol of an infinitesimal transformation, or what is the same, as the symbol of a one-term group, then all that has been said receives a new sense. If on the other hand, we interpret the variables $x_1, \dots, x_n$ as point coordinates [PUNKTCOORDINATEN] in a space of n dimensions, the results obtained also receive a certain graphical nature [ANSCHAULICHKEIT].

The goal of the present chapter is to present these two aspects in detail and then to put them in association; but for that, the introduction of various new concepts turns out to be necessary.[1]

§ 27. Let $x_i' = f_i(x_1, \dots, x_n)$ be a transformation in the variables $x_1, \dots, x_n$ and let $\Phi(x_1, \dots, x_n)$ be an arbitrary function; now, if by chance this function is constituted in such a way that the relation:

$$\Phi\big(f_1(x), \dots, f_n(x)\big) = \Phi(x_1, \dots, x_n)$$

holds identically, then we say: *the function* $\Phi(x_1, \dots, x_n)$ *admits* [GESTATTET] *the transformation* $x_i' = f_i(x_1, \dots, x_n)$, or: it *allows* [ZULÄSST] this transformation; we also express ourselves as follows: *the function* $\Phi(x_1, \dots, x_n)$ *remains invariant through the mentioned transformation, it behaves as an invariant with respect to this transformation.*

If a function $\Phi(x_1, \dots, x_n)$ admits all the ∞^r transformations $x_i' = f_i(x_1, \dots, x_n, a_1, \dots, a_r)$ of an r-term group, we say that it *remains invariant by this group*, and that *it admits this group*; at the same time, we call Φ an *absolute invariant*, or briefly an *invariant* of the group.

Here, we restrict ourselves to one-term groups. So consider a one-term group:

[1] The formations of concepts [BEGRIFFSBILDUNGEN] presented in this chapter have been developed by Lie in the Memoirs of the Scientific Society of Christiania 1872, 1873, 1874 and 19 February 1875. Cf. also Math. Ann. Vols. VIII, IX and XI.

S. Lie, *Theory of Transformation Groups I*,
DOI 10.1007/978-3-662-46211-9_6

$$X(f) = \sum_{i=1}^{n} \xi_i(x_1,\dots,x_n)\,\frac{\partial f}{\partial x_i},$$

the finite transformations of which have the form:

$$x_i' = x_i + \frac{t}{1}\,\xi_i + \frac{t^2}{1\cdot 2}\,X(\xi_i) + \cdots \qquad (i=1\cdots n).$$

We ask for all invariants of this group.

If $\Phi(x_1,\dots,x_n)$ is such an invariant, then for every t, the equation:

$$\Phi\big(x_1 + t\xi_1 + \cdots,\ \dots\dots,\ x_n + t\xi_n + \cdots\big) = \Phi(x_1,\dots,x_n)$$

must hold identically. If we expand here the left-hand side in a series of powers of t according to the general formula (7) in Chap. 4, p. 67, then we obtain the condition:

$$\Phi(x_1,\dots,x_n) + \frac{t}{1}\,X(\Phi) + \frac{t^2}{1\cdot 2}\,X\big(X(\Phi)\big) + \cdots \equiv \Phi(x_1,\dots,x_n)$$

for every t. From this, it immediately follows that the expression $X(\Phi)$ must vanish identically, if Φ admits our one-term group; as a result, we thus have a *necessary* criterion for the invariance of the function Φ by the one-term group $X(f)$.

As we saw earlier (cf. p. 68), the expression $X(\Phi)$ determines the increase [ZUWACHS] that the function Φ undergoes by the infinitesimal transformation $X(f)$. Now, since this increase $\delta\Phi = X(\Phi)\delta t$ vanishes together with $X(\Phi)$, it is natural to introduce the following language: *when the expression $X(\Phi)$ vanishes identically, we say that the function Φ admits the infinitesimal transformation $X(f)$.*

Thus, our result above can also be enunciated as follows:

For a function Φ of $x_1,\dots,x_n$ to admit all transformations of the one-term group $X(f)$, it is a necessary condition that it admits the infinitesimal transformation $X(f)$ of the concerned group.

However, it is easy to see that this necessary condition is at the same time sufficient. Indeed, together with $X(\Phi)$, all the expressions $X\big(X(\Phi)\big), X\big(X\big(X(\Phi)\big)\big)$, etc. also vanish identically, and consequently, the equation:

$$\Phi(x_1',\dots,x_n') = \Phi(x_1,\dots,x_n) + \frac{t}{1}\,X(\Phi) + \cdots$$

reduces to $\Phi(x') = \Phi(x)$ for every value of t, hence with that, it is proved that the function $\Phi(x)$ admits all transformations of the one-term group $X(f)$. Now, since the functions $\Phi(x_1,\dots,x_n)$ for which the expression $X(\Phi)$ vanishes identically are nothing but the solutions of the linear partial differential equation $X(f) = 0$, we can now state the following theorem.

Theorem 13. *The solutions of the linear partial differential equation:*

$$X(f) = \sum_{i=1}^{n} \xi_i(x_1,\ldots,x_n)\frac{\partial f}{\partial x_i} = 0$$

are invariant by the one-term group $X(f)$, and in fact, they are the only invariants by $X(f)$.

Of course, it should not be forgotten that the invariants of the one-term group $X(f)$ are also at the same time the invariants of any one-term group of the form:

$$\rho(x_1,\ldots,x_n)\sum_{i=1}^{n} \xi_i(x_1,\ldots,x_n)\frac{\partial f}{\partial x_i} = \rho X(f),$$

whatever ρ is as a function of its arguments.

This simply follows from the fact that one can multiply the equation $X(f) = 0$ which defines the invariants in question by any arbitrary function ρ of $x_1,\ldots,x_n$. We can also express this fact as follows: if a function of $x_1,\ldots,x_n$ admits the infinitesimal transformation $X(f)$, then at the same time, it admits every infinitesimal transformation $\rho(x_1,\ldots,x_n)X(f)$.

One sees that the two concepts of one-term group $X(f)$ and of infinitesimal transformation $X(f)$ are more special [SPECIELLER] than the concept of linear partial differential equation $X(f) = 0$.

On the basis of the above developments, we can now express in the following way our former observation that the joint solutions of two equations $X_i(f) = 0$ and $X_k(f) = 0$ simultaneously satisfy the third equation $X_i(X_k(f)) - X_k(X_i(f)) = 0$.

Proposition 1. *If a function of $x_1,\ldots,x_n$ admits the two infinitesimal transformations $X_i(f)$ and $X_k(f)$ in these variables, then it also admits the infinitesimal transformation $X_i(X_k(f)) - X_k(X_i(f))$.*

Expressed differently:

Proposition 2. *If a function of $x_1,\ldots,x_n$ admits the two one-term groups $X_i(f)$ and $X_k(f)$, then it also admits the one-term group $X_i(X_k(f)) - X_k(X_i(f))$.*

If $\psi_1,\ldots,\psi_{n-q}$ is a system of independent solutions of the q-term complete system $X_1(f) = 0,\ldots,X_q(f) = 0$, then $\Omega(\psi_1,\ldots,\psi_{n-q})$ is the general form of a solution of this complete system, whence $\Omega(\psi_1,\ldots,\psi_{n-q})$ generally admits every infinitesimal transformation of the form:

$$\chi_1(x_1,\ldots,x_n)X_1(f) + \cdots + \chi_q(x_1,\ldots,x_n)X_q(f), \tag{1}$$

whatever functions $\chi_1,\ldots,\chi_q$ one can choose. It is even clear that aside from the ones written just now, there are no infinitesimal transformations by which all functions of the form $\Omega(\psi_1,\ldots,\psi_{n-q})$ remain invariant; for we know that the q-term complete system $X_1(f) = 0,\ldots,X_q(f) = 0$ is characterized by its solutions $\psi_1,\ldots,\psi_{n-q}$.

Naturally, the functions Ω also admit all finite transformations of the one-term group (1). Furthermore, one can indicate all finite transformations through which all functions $\Omega(\psi_1,\dots,\psi_{n-q})$ remain simultaneously invariant. The form of these transformations is obviously:

$$\psi_k(x_1',\dots,x_n') = \psi_k(x_1,\dots,x_n) \qquad {\scriptstyle (k=1\cdots n-q)}$$
$$\mathfrak{U}_j(x_1',\dots,x_n',x_1,\dots,x_n) = 0 \qquad {\scriptstyle (j=1\cdots q)},$$

where the $\mathfrak{U}_j$ are subjected to the condition that a transformation really arises. Moreover, the x_k' and the x_k must also stay inside a certain region.

We want to say that *the system of equations:*

$$\pi_1(x_1,\dots,x_n) = 0,\ \dots,\ \pi_m(x_1,\dots,x_n) = 0$$

admits the transformation $x_i' = f_i(x_1,\dots,x_n)$ *when the system of equations:*

$$\pi_1(x_1',\dots,x_n') = 0,\ \dots,\ \pi_m(x_1',\dots,x_n') = 0$$

is equivalent to $\pi_1(x) = 0,\ \dots,\ \pi_m(x) = 0$ *after the substitution* $x_i' = f_i(x_1,\dots,x_n)$, *hence when every system of values* $x_1,\dots,x_n$ *which satisfies the* m *equations* $\pi_\mu(x) = 0$ *also satisfies the* m *equations:*

$$\pi_1\big(f_1(x),\dots,f_n(x)\big) = 0,\ \dots,\ \pi_m\big(f_1(x),\dots,f_m(x)\big) = 0.$$

With the introduction of this definition, it is not even necessary to assume that the m equations $\pi_1 = 0,\dots,\pi_m = 0$ are independent of each other, though this assumption will always be made in the sequel, unless the contrary is expressly stated.

From the preceding, we immediately get the following

Proposition 3. *If* $W_1, W_2,\dots,W_m$ $(m \leqslant n-q)$ *are arbitrary solutions of the q-term complete system* $X_1(f) = 0,\dots,X_q(f) = 0$ *in the n independent variables* $x_1,\dots,x_n$ *and if furthermore* $a_1,\dots,a_m$ *are arbitrarily chosen constants, then the system of equation:*

$$W_1 = a_1,\ \dots,\ W_m = a_m$$

admits any one-term group of the form:

$$\sum_{k=1}^{q} \chi_k(x_1,\dots,x_n)\, X_k(f),$$

where it is understood that $\chi_1,\dots,\chi_q$ *are arbitrary functions of their arguments.*

§ 28. In the previous section, we have shown in a new light the theory of the integration of linear partial differential equations, in such a way that we brought to connection the infinitesimal transformations and the one-term groups. At present, we want to take another route, we want to attempt to make accessible the clear, illustrated conception of this theory of integration (and of what is linked with it), by means of manifold considerations [MANNIGFALTIGKEITSBETRACHTUNGEN].

If we interpret $x_1,\ldots,x_n$ as coordinates in an n-times extended space R_n, then the simultaneous system:

$$\frac{\mathrm{d}x_1}{\xi_1} = \cdots = \frac{\mathrm{d}x_n}{\xi_n}$$

receives a certain illustrative sense; namely it attaches to each point $x_1,\ldots,x_n$ of the R_n a certain direction [RICHTUNG].

The integral equations of the simultaneous system determine $n-1$ of the variables $x_1,\ldots,x_n$, hence for instance $x_1,\ldots,x_{n-1}$, as functions of the n-th: x_n and of the initial values $\overline{x}_1,\ldots,\overline{x}_n$; consequently, after a definite choice of initial values, these integral equations represent a determinate once-extended manifold which we call an *integral curve* of the simultaneous system. *Every such integral curve comes into contact* [BERÜHRT], *in each one of its points, with the direction attached to the point.*

There is in total ∞^{n-1} different integral curves of the simultaneous system, and in fact, through every point of the R_n, there passes in general one integral curve.

If $\psi_1,\ldots,\psi_{n-1}$ are independent integral functions of the simultaneous system, then all the integral curves are also represented by means of the $n-1$ equations:

$$\psi_k(x_1,\ldots,x_n) = C_k \qquad (k=1\cdots n-1),$$

with the $n-1$ arbitrary constants $C_1,\ldots,C_{n-1}$. If one sets an arbitrary integral function $\Omega(\psi_1,\ldots,\psi_{n-1})$ to be equal to an arbitrary constant:

$$\Omega(\psi_1,\ldots,\psi_{n-1}) = A,$$

then one obtains the equation of ∞^1 $(n-1)$-times extended manifolds, which are entirely constituted of integral curves and in fact, every such manifold is constituted of ∞^{n-2} different integral manifolds. Lastly, if one sets in general $m \leqslant n-1$ independent integral functions $\Omega_1,\ldots,\Omega_m$ to be equal to arbitrary constants:

$$\Omega_\mu(\psi_1,\ldots,\psi_{n-1}) = A_\mu \qquad (\mu=1\cdots m),$$

then one obtains the analytic expression of a family of ∞^m $(n-m)$-times extended manifolds, each one of which consists of ∞^{n-m-1} integral curves.

The integral functions of our simultaneous system are at the same time the solutions of the linear partial differential equations:

$$\sum_{i=1}^{n} \xi_i(x_1,\ldots,x_n)\,\frac{\partial f}{\partial x_i} = X(f) = 0.$$

Occasionally, we also call the integral curves of the simultaneous system the *characteristics of the linear partial differential equation* $X(f)=0$, if we again take up the terminology introduced by MONGE for $n=3$. Using this terminology, we can also say: every solution of the linear partial differential equation $X(f)=0$ represents, when it is set equal to a constant, a family of ∞^1 manifolds which consists of ∞^{n-2} characteristics of $X(f)=0$.

Now, we imagine that we are given two linear partial differential equations, say $X_1(f) = 0$ and $X_2(f) = 0$.

It is possible that the two equations have characteristics in common. This case happens when there is an identity of the form:

$$\chi_1(x_1, \dots, x_n) X_1(f) + \chi_2(x_1, \dots, x_n) X_2(f) \equiv 0,$$

without χ_1 and χ_2 both vanishing. Then obviously every solution of $X_1(f) = 0$ also satisfies $X_2(f) = 0$ and conversely.

If the two equations $X_1(f) = 0$ and $X_2(f) = 0$ have distinct characteristics, then they do not have all their solutions in common; then the question is whether they in general possess solutions in common, or, as we can now express: whether the characteristics of $X_1(f) = 0$ can be gathered as manifolds which consist of characteristics of $X_2(f) = 0$.

This question can be answered directly when one knows the characteristics of the two equations $X_1(f) = 0$ and $X_2(f) = 0$; however, we do not want to halt here. In the sequel, we shall restrict ourselves to expressing in the language of the theory of manifolds [MANNIGFALTIGKEITSLEHRE] the former results which have been deduced by means of the analytic method [DURCH ANALYTISCHE METHODEN].

Let the q mutually independent equations:

$$X_1(f) = 0, \dots, X_q(f) = 0$$

form a q-term complete system and let $\psi_1, \dots, \psi_{n-q}$ be independent solution of it. Then the equations:

$$\psi_1(x_1, \dots, x_n) = C_1, \dots, \psi_{n-q}(x_1, \dots, x_n) = C_{n-q}$$

with the $n - q$ arbitrary constants C_k represent a family of ∞^{n-q} q-times extended manifolds, each one of which consists of ∞^{q-1} characteristics of each individual equation amongst the q equations $X_1(f) = 0, \dots, X_q(f) = 0$. We call these ∞^{n-q} manifolds the *characteristic manifolds of the complete system.*

If one sets any $n - q - m$ independent functions of $\psi_1, \dots, \psi_{n-q}$ equal to arbitrary constants:

$$\Omega_1(\psi_1, \dots, \psi_{n-q}) = A_1, \dots, \Omega_{n-q-m}(\psi_1, \dots, \psi_{n-q}) = A_{n-q-m},$$

then one gets the analytic expression of a family of ∞^{n-q-m} $(q+m)$-times extended manifolds, amongst which each individual one consists of ∞^m characteristic manifolds.

The equations of the ∞^{n-q} characteristic manifolds show that every point of R_n belongs to one and to only one characteristic manifold. Consequently, we can say that *the whole R_n is decomposed* [ZERLEGT] *into ∞^{n-q} q-times extended manifolds, hence that our complete system defines a decomposition* [ZERLEGUNG] *of the space.*

Conversely, every decomposition of R_n into ∞^{n-q} q-times extended manifolds:

$$\varphi_1(x_1,\ldots,x_n) = A_1,\ \ldots,\ \varphi_{n-q}(x_1,\ldots,x_n) = A_{n-q}$$

can be defined by means of a q-term complete system; for $\varphi_1,\ldots,\varphi_{n-q}$ are necessarily independent functions, whence according to Proposition 9, p. 104, there is a q-term complete system the most general solution of which is an arbitrary function of $\varphi_1,\ldots,\varphi_{n-q}$; this complete system then defines the decomposition in question.

An individual linear partial differential equation $X(f) = 0$ attaches to every point of R_n a certain direction. If one has several such equations, for instance the following ones, which can be chosen in a completely arbitrary way:

$$X_k(f) = \sum_{i=1}^{n} \xi_{ki}(x_1,\ldots,x_n)\,\frac{\partial f}{\partial x_i} = 0 \qquad {\scriptstyle (k=1\cdots q)},$$

then each one of these equations associates to every point of the space a direction of progress [FORTSCHREITUNGSRICHTUNG]. For instance, the q directions associated to the point $x_1^0,\ldots,x_n^0$ are determined by:

$$\delta x_1^0 : \delta x_2^0 : \cdots : \delta x_n^0 = \xi_{k1}(x^0) : \xi_{k2}(x^0) : \cdots : \xi_{kn}(x^0)$$
$$ {\scriptstyle (k=1\cdots q)}.$$

We call these q directions in the chosen point *independent of each other* if none of them can be linearly deduced from the remaining ones, that is to say: if it is not possible to indicate q numbers $\lambda_1,\ldots,\lambda_q$ which do not all vanish although the n equations:

$$\lambda_1\xi_{1i}^0 + \cdots + \lambda_q\xi_{qi}^0 = 0 \qquad {\scriptstyle (i=1\cdots n)}$$

are satisfied.

From this, it follows that the q equations $X_1(f) = 0,\ldots,X_q(f) = 0$ associate to *every point* in general position q independent directions when they are themselves mutually independent, thus when the equation:

$$\sum_{k=1}^{q} \chi_k(x_1,\ldots,x_n)\,X_k(f) = 0$$

can be identically satisfied only for $\chi_1 = 0,\ \ldots,\ \chi_q = 0$.

If one wants to visualize geometrically what is understood by "independent directions", one should best start in ordinary, thrice-extended space R_3; for it is then obvious. At a point of R_3 one calls two directions independent of each other when they are generally distinct; three directions are independent when they do not fall in the same plane passing through the point; there are in general no more than three mutually independent directions at a point of R_3.

Accordingly, q directions at a point of R_n are independent of each other if and only if, when collected together, they are not contained in any smooth manifold through this point which has fewer than q dimensions.

Every possible common solution of the q equations:

$$X_1(f) = 0, \ldots, X_q(f) = 0$$

also satisfies all equations of the form:

$$\sum_{k=1}^{q} \chi_k(x_1, \ldots, x_n)\, X_k(f) = 0. \tag{2}$$

The totality of all these equations associates a whole family of directions to any point of R_n. If we assume from the beginning that the equations $X_1(f) = 0, \ldots, X_q(f) = 0$ are mutually independent, we do not restrict the generality of the investigation; thus through the equations (2), we are given a family of ∞^{q-1} different directions which are attached to each point $x_1, \ldots, x_n$. One easily sees that these ∞^{q-1} directions at every point form a smooth bundle, and hence determine a smooth q-times extended manifold passing through this point, namely the smallest smooth manifold through the point which contains the q independent directions of the q equations $X_1(f) = 0, \ldots, X_q(f) = 0$.

Every possible joint solution of the equations:

$$X_1(f) = 0, \ldots, X_q(f) = 0$$

also satisfies all equations of the form:

$$X_k\big(X_j(f)\big) - X_j\big(X_k(f)\big) = 0.$$

These equations also attach to any point $x_1, \ldots, x_n$ certain directions, but in general, the directions in question will only exceptionally belong to the above-mentioned bundle of ∞^{q-1} directions at the point $x_1, \ldots, x_n$. It is only in one case that at each point of the space, the directions attached to all equations:

$$X_k\big(X_j(f)\big) - X_j\big(X_k(f)\big) = 0$$

belong to the bundle in question, namely only if for every k and j there exists a relation of the form:

$$X_k\big(X_j(f)\big) - X_j\big(X_k(f)\big) = \sum_{s=1}^{q} \omega_{kjs}(x_1, \ldots, x_n)\, X_s(f),$$

that is to say, when the equations $X_1(f) = 0, \ldots, X_q(f) = 0$ coincidentally form a q-term complete system. —

We can also define by the equations:

$$\delta x_1 : \cdots : \delta x_n = \sum_{k=1}^{q} \chi_k(x)\, \xi_{k1}(x) : \cdots : \sum_{k=1}^{q} \chi_k(x)\, \xi_{kn}(x)$$

$$(k = 1 \cdots q)$$

the bundle of directions which is determined at every point $x_1,\dots,x_n$ by the equations $X_1 f = 0, \dots, X_q f = 0$. By eliminating from this the arbitrary functions $\chi_1,\dots,\chi_q$, or, what is the same, by setting equal to zero all the determinants in $q+1$ columns of the matrix:

$$\begin{vmatrix} dx_1 & dx_2 & \cdots & dx_n \\ \xi_{11} & \xi_{12} & \cdots & \xi_{1n} \\ \cdot & \cdot & & \cdot \\ \xi_{q1} & \xi_{q2} & \cdots & \xi_{qn} \end{vmatrix}$$

we obtain a system of $n-q$ independent total differential equations. This system attaches to every point $x_1,\dots,x_n$ exactly the same smooth bundle of ∞^{q-1} directions as did the equations (2). It follows conversely that the totality of the linear partial differential equations (2) is also completely determined by the system of total differential equations just introduced.

The formulas become particularly convenient when one replaces the q equations $X_1(f) = 0, \dots, X_q(f) = 0$ by q other equations which are resolved with respect to q of the differential quotients $\frac{\partial f}{\partial x_i}$, for instance by the following q equations:

$$Y_k(f) = \frac{\partial f}{\partial x_{n-q+k}} + \sum_{i=1}^{n-q} \eta_{ki}(x_1,\dots,x_n)\,\frac{\partial f}{\partial x_i} = 0 \qquad (k=1\cdots q).$$

The totality of all the equations (2) is equivalent to the totality of all the equations:

$$\sum_{k=1}^{q} \chi_k(x_1,\dots,x_n)\,Y_k(f) = 0;$$

hence the directions which are attached to the point $x_1,\dots,x_n$ are also represented by the equations:

$$\begin{aligned} dx_1 : \cdots : dx_{n-q} : dx_{n-q+1} : \cdots : dx_n = \\ = \sum_{k=1}^{q} \chi_k\,\eta_{k1} : \cdots : \sum_{k=1}^{q} \chi_k\,\eta_{k,n-q} : \chi_1 : \cdots : \chi_q. \end{aligned}$$

If we therefore eliminate $\chi_1,\dots,\chi_q$, we obtain the following system of total differential equations:

$$\text{(3)} \qquad dx_i - \sum_{k=1}^{q} \eta_{ki}(x)\,dx_{n-q+k} = 0 \qquad (i=1\cdots n-q).$$

We have already seen in the preceding chapter, from p. 106 up to p. 107, that there is a connection between the system of linear partial differential equations $Y_1(f) = 0, \dots, Y_q(f) = 0$ and the above system of total differential equations. But at that time, we limited ourselves to the special case where the q equations:

$$Y_1(f) = 0, \;\dots,\; Y_q(f) = 0$$

have solutions in common, and we showed that the determination of these joint solutions amounts to the integration of the above total differential equations.

However, in the developments carried out just now, the integrability of the concerned system of differential equations is out of the question. With that, the connection between the system of linear partial differential equations $Y_1(f) = 0, \ldots, Y_q(f) = 0$ and the system of total differential equations (3) is completely independent of the integrability of these two systems; this connection is just based on the fact that the two systems attach a single and the same smooth bundle of ∞^{q-1} directions.

To conclude, we make the following remark which seems most certainly obvious, but nevertheless has to be done: if the q equations $X_1 f = 0, \ldots, X_q f = 0$ constitute a q-term complete system, then at each point $x_1, \ldots, x_n$, the characteristic manifold of the complete system comes into contact with the ∞^{q-1} directions that all equations of the form (2) attach to this point.

§ **29.** Lastly, it is advisable to combine the manifold-type considerations [MANNIGFALTIGKEITSBETRACHTUNGEN] of the previous section with the developments of § 27. Before we do that, we continue to pursue some further manifold considerations.

Every transformation $x_i' = f_i(x_1, \ldots, x_n)$ can be interpreted as an operation which exchanges [VERTAUSCHT] the points of R_n, as it transfers each point $x_1, \ldots, x_n$ to the new position $x_1' = f_1(x), \ldots, x_n' = f_n(x)$ (Chap. 1, p. 3).

A system of m independent equations:

$$\Omega_1(x_1, \ldots, x_n) = 0, \ldots, \Omega_m(x_1, \ldots, x_n) = 0$$

represents an $(n-m)$-times extended manifold of R_n. We say that this manifold *admits the transformation* $x_i' = f_i(x_1, \ldots, x_n)$ if the system of equations $\Omega_1 = 0, \ldots, \Omega_m = 0$ admits this transformation. According to § 27 this is the case when every system of values $x_1, \ldots, x_n$ which satisfies the equations $\Omega_1(x) = 0, \ldots, \Omega_m(x) = 0$ satisfies at the same time the equations:

$$\Omega_1\big(f_1(x), \ldots, f_n(x)\big) = 0, \ldots, \Omega_m\big(f_1(x), \ldots, f_n(x)\big) = 0.$$

Hence we can also express ourselves as follows:

The manifold:

$$\Omega_1(x_1, \ldots, x_n) = 0, \ldots, \Omega_m(x_1, \ldots, x_n) = 0$$

admits the transformation:

$$x_1' = f_1(x_1, \ldots, x_n), \ldots, x_n' = f_n(x_1, \ldots, x_n)$$

if every point $x_1, \ldots, x_n$ of the manifold is transferred by this transformation to a point $x_1', \ldots, x_n'$ which likewise belongs to the manifold.

In Chapter 3, p. 62 sq., we saw that through every point $x_1,\dots,x_n$ of R_n in general position there passes an integral curve [BAHNCURVE] of the infinitesimal transformation:

$$X(f) = \sum_{i=1}^{n} \xi_i(x_1,\dots,x_n)\,\frac{\partial f}{\partial x_i}.$$

There, we defined this integral curve as the totality of all the positions that the point $x_1,\dots,x_n$ can take by means of all the ∞^1 transformations of the one-term group $X(f)$; furthermore, the point $x_1,\dots,x_n$ on the integral curve in question could be chosen completely arbitrarily. From this, it results that, through every transformation of the one-term group $X(f)$, every point $x_1,\dots,x_n$ stays on the integral curve passing through it, whence *every integral curve of the infinitesimal transformation $X(f)$ remains invariant by the ∞^1 transformations of the one-term group $X(f)$. The same naturally holds true for every manifold which consists of integral curves.*

But the integral curves of the infinitesimal transformation $X(f)$ are nothing else than the integral curves of the simultaneous system:

$$\frac{\mathrm{d}x_1}{\xi_1} = \cdots = \frac{\mathrm{d}x_n}{\xi_n}$$

hence from this it again follows that the above-mentioned characteristics of the linear partial differential equation $X(f) = 0$ coincide with the integral curves of the infinitesimal transformation $X(f)$.

Earlier on (Chap. 4, p. 62), we emphasized that an infinitesimal transformation Xf attaches to every point $x_1,\dots,x_n$ in general position a certain direction of progress, namely the one which comes into contact with the integral curve passing through. Obviously this direction of progress coincides with the direction that the equation Xf associates to the point in question.

Several, say q, infinitesimal transformations:

$$X_k f = \sum_{i=1}^{n} \xi_{ki}(x_1,\dots,x_n)\,\frac{\partial f}{\partial x_i} \qquad (k=1\cdots q)$$

determine q different directions of progress at each point $x_1,\dots,x_n$ in general position; in accordance with the preceding, we call these directions of progress *independent of each other* if the equations $X_1 f = 0, \dots, X_q f = 0$ are mutually independent.

From this, it follows immediately that the q infinitesimal transformations $X_1 f,\dots,X_q f$ attach precisely $h \leqslant q$ independent directions of progress when all the $(h+1)\times(h+1)$ determinants of the matrix:

$$\begin{vmatrix} \xi_{11} & \cdot\,\cdot & \xi_{1n} \\ \cdot & \cdot\,\cdot & \cdot \\ \xi_{q1} & \cdot\,\cdot & \xi_{qn} \end{vmatrix}$$

vanish identically, without all its $h \times h$ determinants doing so. —

What we have to say in addition here can be summarized as a statement.

Proposition 4. *If q infinitesimal transformations $X_1 f, \dots, X_q f$ of the n-times extended space $x_1, \dots, x_n$ are constituted in such a way that the q equations $X_1 f = 0, \dots, X_q f = 0$ are independent of each other, then $X_1 f, \dots, X_q f$ attach to every point $x_1, \dots, x_n$ in general position q independent directions of progress; moreover, if the equations $X_1 f = 0, \dots, X_q f = 0$ form a q-term complete system, then $X_1 f, \dots, X_q f$ determine a decomposition of the space into ∞^{n-q} q-times extended manifolds, the characteristic manifolds of the complete system. Each one of these manifolds comes into contact in each of its points with the directions that $X_1 f, \dots, X_q f$ associate to the point; each such manifold can be engendered by the ∞^{q-1} integral curves of an arbitrary infinitesimal transformation of the form:*

$$\chi_1(x_1, \dots, x_n) X_1 f + \cdots + \chi_q(x_1, \dots, x_n) X_q f,$$

where it is understood that $\chi_1, \dots, \chi_q$ are arbitrary functions of their arguments; lastly, each one of the discussed manifolds admits all the transformations of any one-term group:

$$\chi_1 X_1 f + \cdots + \chi_q X_q f.$$

Chapter 7
Determination of All Systems of Equations Which Admit Given Infinitesimal Transformations

First, we shall define what we mean by the phrase that the system of equations:

$$\Omega_1(x_1,\dots,x_n)=0,\;\dots,\;\Omega_{n-m}(x_1,\dots,x_n)=0$$

admits the infinitesimal transformation $X(f)$. Then, we shall settle the extremely important problem of determining all systems of equations which admit given infinitesimal transformations.[1]

But beforehand, we observe the following:

We naturally consider only such system of equations:

$$\Omega_1=0,\;\dots,\;\Omega_{n-m}=0$$

that are satisfied by certain systems of values $x_1,\dots,x_n$; at the same time, we *always* restrict ourselves to systems of values $x_1,\dots,x_n$ in the neighborhood of which the functions $\Omega_1,\dots,\Omega_{n-m}$ behave regularly. In addition, we want *once and for all* to agree on the following: unless the contrary is expressly allowed, every system of equations $\Omega_1=0,\dots,\Omega_{n-m}=0$ which we consider should be constituted in such a way that not all $(n-m)\times(n-m)$ determinants of the matrix:

$$\begin{vmatrix} \frac{\partial\Omega_1}{\partial x_1} & \cdot\cdot & \frac{\partial\Omega_1}{\partial x_n} \\ \cdot & \cdot\cdot & \cdot \\ \frac{\partial\Omega_{n-m}}{\partial x_1} & \cdot\cdot & \frac{\partial\Omega_{n-m}}{\partial x_n} \end{vmatrix} \tag{1}$$

vanish by means of $\Omega_1=0,\dots,\Omega_{n-m}=0$. It is permitted to make this assumption, since a system of equations which does not have the demanded property can always be brought to a form which does satisfy the stated requirement.

§ **30.** In the variables $x_1,\dots,x_n$, let an infinitesimal transformation:

[1] Cf. LIE, Scientific Society of Christiania 1872–74, and also Math. Ann. Vol. XI, Vol. XXIV, pp. 542–544.

S. Lie, *Theory of Transformation Groups I*,
DOI 10.1007/978-3-662-46211-9_7

$$Xf = \sum_{i=1}^{n} \xi_i(x_1,\dots,x_n)\,\frac{\partial f}{\partial x_i}$$

be given. In our studies which relate to such an infinitesimal transformation, *we shall always restrict ourselves to systems of values for which the ξ_i behave regularly.*

If a system of equations $\Omega_1 = 0, \dots, \Omega_{n-m} = 0$ admits all finite transformations:

$$x_i' = x_i + eXx_i + \cdots \qquad (i=1\cdots n)$$

of the one-term group Xf, then the system of equations:

$$\Omega_k\big(x_1 + eXx_1 + \cdots,\ \dots,\ x_n + eXx_n + \cdots\big) = 0$$
$$(k=1\cdots n-m),$$

or, what is the same, the system:

$$\Omega_k + eX\Omega_k + \cdots = 0 \qquad (k=1\cdots n-m)$$

must be equivalent to the system of equations:

$$\Omega_1 = 0,\ \dots,\ \Omega_{n-m} = 0$$

for all values of e. To this end it is obviously *necessary* that all $X\Omega_k$ vanish for the system of values of $\Omega_1 = 0, \dots, \Omega_{n-m} = 0$, hence that the increment $X\Omega_k\,\delta t$ that Ω_k undergoes by the infinitesimal transformation $x_i' = x_i + \xi_i\,\delta t$ vanishes by means of $\Omega_1 = 0, \dots, \Omega_{n-m} = 0$.

These considerations lead us to set up the following definition:

A system of equations:

$$\Omega_1(x_1,\dots,x_n) = 0,\ \dots,\ \Omega_{n-m}(x_1,\dots,x_n) = 0$$

admits the infinitesimal transformation Xf as soon as all the $n-m$ expressions $X\Omega_k$ vanish by means[1] *(Translator's note: An example appearing on p. 128 below illustrates this condition.) of the system of equations.*

Then on the basis of this definition, the following obviously holds:

Proposition 1. *If a system of equations admits all transformations of the one-term group Xf, then in any case, it must admit the infinitesimal transformation Xf.*

In addition, we immediately realize that *a system of equations which admits the infinitesimal transformation Xf allows at the same time every infinitesimal transformation of the form $\chi(x_1,\dots,x_n)Xf$, provided of course that the function χ behaves regularly for the system of values which comes into consideration in the concerned system of equations.*

Without difficulty, one can see that the above definition is independent of the choice of coordinates, so that every system of equations $\Omega_1 = 0, \dots, \Omega_{n-m} = 0$

which admits the infinitesimal transformation Xf in the sense indicated above, must also admit it, when new independent variables $y_1,\dots,y_n$ are introduced in place of the x. The fact that this really holds true follows immediately from the behavior of the symbol Xf after the introduction of new variables. Here as always, it is assumed that the y are ordinary power series in the x, and that the x are ordinary power series in the y, for all the systems of values $x_1,\dots,x_n$ and $y_1,\dots,y_n$ coming into consideration.

It now remains to show that the definition set up above is also independent of the form of the system of equations $\Omega_1 = 0, \dots, \Omega_{n-m} = 0$. Only when we will have proved this fact shall the legitimacy of the definition really be established.

Now, in order to be able to carry out this proof, we first provide a few general results which are already important and which will later find several applications.

Suppose that the system of equations $\Omega_1 = 0, \dots, \Omega_{n-m} = 0$ admits the infinitesimal transformation Xf. Since not all $m \times m$ determinants of the matrix (1) vanish by means of $\Omega_1 = 0, \dots, \Omega_{n-m} = 0$, we can assume that the determinant:

$$\sum \pm \frac{\partial \Omega_1}{\partial x_1} \cdots \frac{\partial \Omega_{n-m}}{\partial x_{n-m}}$$

belongs to the nonvanishing ones. Then it is possible to resolve the equations $\Omega_k = 0$ with respect to $x_1,\dots,x_m$, and this naturally delivers a system of equations:

$$x_1 = \varphi_1(x_{n-m+1},\dots,x_n), \dots, x_{n-m} = \varphi_{n-m}(x_{n-m+1},\dots,x_n)$$

which is analytically equivalent to the system $\Omega_1 = 0, \dots, \Omega_{n-m} = 0$. Therefore, if by the symbol $[\,]$ we denote the substitution $x_1 = \varphi_1, \dots, x_{n-m} = \varphi_{n-m}$, we have:

$$[\Omega_1] \equiv 0, \dots, [\Omega_{n-m}] \equiv 0;$$

and moreover, the fact that the system of equations $\Omega_1 = 0, \dots, \Omega_{n-m} = 0$ admits the infinitesimal transformation Xf is expressed by the identities:

$$[X\Omega_1] \equiv 0, \dots, [X\Omega_{n-m}] \equiv 0.$$

Now, let $\Phi(x_1,\dots,x_n)$ be an arbitrary function which behaves regularly for the system of values $x_1,\dots,x_n$ coming into consideration. Then one has:

$$[X\Phi] = \sum_{i=1}^{n} [Xx_i] \left[\frac{\partial \Phi}{\partial x_i}\right],$$

and on the other hand:

$$[X[\Phi]] = \sum_{k=1}^{n-m} [X\varphi_k] \left[\frac{\partial \Phi}{\partial x_k}\right] + \sum_{\mu=1}^{m} [Xx_{n-m+\mu}] \left[\frac{\partial \Phi}{\partial x_{n-m+\mu}}\right],$$

whence:

(2) $$\left[X\Phi\right] = \left[X[\Phi]\right] + \sum_{k=1}^{n-m} \left[X(x_k - \varphi_k)\right] \left[\frac{\partial \Phi}{\partial x_k}\right].$$

If in place of Φ we insert one after the other the functions $\Omega_1, \dots, \Omega_{n-m}$, and if we take into account that $\left[\Omega_j\right]$ and also $X\left[\Omega_j\right]$ plus $\left[X[\Omega_j]\right]$ vanish identically, then we find:

$$\left[X\Omega_j\right] = \sum_{k=1}^{n-m} \left[X(x_k - \varphi_k)\right] \left[\frac{\partial \Omega_j}{\partial x_k}\right]$$
$$(j = 1 \cdots n-m).$$

Now, because $\left[X\Omega_j\right]$ vanishes identically in any case, while the determinant:

$$\sum \pm \left[\frac{\partial \Omega_1}{\partial x_1}\right] \cdots \left[\frac{\partial \Omega_{n-m}}{\partial x_{n-m}}\right]$$

does not vanish identically, it follows that:

$$\left[X(x_k - \varphi_k)\right] \equiv 0 \qquad (k = 1 \cdots n-m),$$

whence the equation (2) takes the form:

(3) $$\left[X\Phi\right] \equiv \left[X[\Phi]\right].$$

This formula, which is valid for any function $\Phi(x_1, \dots, x_n)$, will be very useful later. Here, we need it only in the special case where Φ vanishes by means of $\Omega_1 = 0, \dots, \Omega_{n-m} = 0$; then $\left[\Phi\right]$ is identically zero and likewise $\left[X[\Phi]\right]$; our formula hence shows that $\left[X\Phi\right]$ also vanishes identically. We can express this result as follows:

Proposition 2. *If a system of equations:*

$$\Omega_1(x_1, \dots, x_n) = 0, \dots, \Omega_{n-m}(x_1, \dots, x_n) = 0$$

admits the infinitesimal transformation:

$$Xf = \sum_{i=1}^{n} \xi_i(x_1, \dots, x_n) \frac{\partial f}{\partial x_i}$$

and if $V(x_1, \dots, x_n)$ is a function which vanishes by means of this system of equations, then the function XV also vanishes by means of $\Omega_1 = 0, \dots, \Omega_{n-m} = 0$.

Now, if $V_1 = 0, \dots, V_{n-m} = 0$ is an arbitrary system analytically equivalent to:

$$\Omega_1 = 0, \dots, \Omega_{n-m} = 0,$$

then according to the proposition just stated, all the $n-m$ expressions XV_k vanish by means of $\Omega_1 = 0, \dots, \Omega_{n-m} = 0$ and hence they also vanish by means of $V_1 =$

$0, \dots, V_{n-m} = 0$; in other words: the system of equations $V_1 = 0, \dots, V_{n-m} = 0$ also admits the infinitesimal transformation Xf.

Finally, as a result, it is established that our above definition for the invariance of a system of equations by an infinitesimal transformation is also independent of the form of this system of equations. Therefore, the introduction of this definition is completely natural [NATURGEMÄSS].

We know that a system of equations can admit all transformations of the one-term group Xf only when it admits the infinitesimal transformation Xf. But this necessary condition is at the same time sufficient; indeed, it can be established that every system of equations which admits the infinitesimal transformation Xf generally allows all the transformations of the one-term group Xf.

In fact, let the system of equation $\Omega_1 = 0, \dots, \Omega_{n-m} = 0$ admit the infinitesimal transformation Xf; moreover, let $x_1 = \varphi_1, \dots, x_{n-m} = \varphi_{n-m}$ be a resolved form for the system of equations $\Omega_1 = 0, \dots, \Omega_{n-m} = 0$; lastly, let the substitution $x_\mu = \varphi_\mu$ again be denoted by the symbol $[\,]$.

Under these assumptions, we have at first $[\Omega_k] \equiv 0$, then $[X\Omega_k] \equiv 0$ and from Proposition 2 just stated we obtain furthermore:

$$[XX\Omega_k] \equiv 0, \qquad [XXX\Omega_k] \equiv 0, \cdots.$$

Consequently, the infinite series:

$$\Omega_k + \frac{e}{1} X\Omega_k + \frac{e^2}{1 \cdot 2} XX\Omega_k + \cdots$$

vanishes identically after the substitution $x_\mu = \varphi_\mu$, whichever value the parameter e can have. So for any e, the system of equations:

$$\Omega_k + eX\Omega_k + \cdots = 0 \qquad {\scriptstyle (k=1\cdots n-m)}$$

will be satisfied by the systems of values of the system of equations $\Omega_1 = 0, \dots, \Omega_{n-m} = 0$, and according to what has been said earlier, this simply means that the system of equations:

$$\Omega_1 = 0, \dots, \Omega_{n-m} = 0$$

admits all transformations:

$$x_i' = x_i + eXx_i + \cdots \qquad {\scriptstyle (i=1\cdots n)}$$

of the one-term group Xf. With that, the assertion made above is proved; as a result, we have the

Theorem 14. *The system of equations:*

$$\Omega_1(x_1, \dots, x_n) = 0, \dots, \Omega_{n-m}(x_1, \dots, x_n) = 0$$

admits all transformations of the one-term group Xf *if and only if it admits the infinitesimal transformation* Xf*, that is to say, when all the* $n-m$ *expressions* $X\Omega_k$ *vanish by means of* $\Omega_1=0,\dots,\Omega_{n-m}=0$.

This theorem is proved under the assumption, which we always make unless otherwise stated, that not all $(n-m)\times(n-m)$ determinants of the matrix (1) vanish by means of $\Omega_1=0,\dots,\Omega_{n-m}=0$. In addition, as already said, both the Ω_k and the ξ_i must behave regularly for the systems of values $x_1,\dots,x_n$ coming into consideration.

It can be seen that Theorem 14 no longer holds true when this assumption on the determinant of the matrix (1) is not fulfilled. Indeed, we consider for instance the system of equations:

$$\Omega_1=x_1^2=0,\ \dots,\ \Omega_{n-m}=x_{n-m}^2=0,$$

by means of which all the $(n-m)\times(n-m)$ determinants of the matrix (1) vanish. We find for these equations: $X\Omega_k=2x_kXx_k$, hence all the $X\Omega_k$ vanish by means of $\Omega_1=0,\dots,\Omega_{n-m}=0$, whatever form Xf has. Consequently, if Theorem 14 were also true here, then the system of equations:

$$x_1^2=0,\ \dots,\ x_{n-m}^2=0$$

would admit any arbitrary one-term group Xf, which obviously is not the case.

From this we conclude the following: when the system of equations $\Omega_1=0,\dots,\Omega_{n-m}=0$ brings to zero all the $(n-m)\times(n-m)$ determinants of the matrix (1), the vanishing of all the $X\Omega_k$ by means of $\Omega_1=0,\dots,\Omega_{n-m}=0$ is of course necessary in order that this system of equations admits the one-term group Xf, but however, it is not sufficient.

Nevertheless, in the course of research we are often led to systems of equations for which one has no means to decide whether the repeatedly mentioned requirement is met. Then how should one recognize that the system of equations in question admits, or does not admit, a given one-term group?

In such circumstances, there is a criterion which is frequently of great help and which we now want to develop.

Let:

$$\Delta_1(x_1,\dots,x_n)=0,\ \dots,\ \Delta_s(x_1,\dots,x_n)=0$$

be a system of equations. We assume that the functions $\Delta_1,\dots,\Delta_s$ behave regularly inside a certain region B, in the neighborhood of those systems of values $x_1,\dots,x_n$ which satisfy the system of equations. However, we assume nothing about the behavior of the functional determinants of the Δ's; we no longer demand that our s equations are independent of each other, so the number s can even be larger than n in certain circumstances.

Moreover, let:

$$Xf=\sum_{i=1}^{n}\xi_i(x_1,\dots,x_n)\,\frac{\partial f}{\partial x_i}$$

be an infinitesimal transformation and suppose that amongst the system of values $x_1, \dots, x_n$ for which $\xi_1, \dots, \xi_n$ behave regularly, there exist some which satisfy the equations $\Delta_1 = 0, \dots, \Delta_s = 0$ and which in addition belong to the domain B.

Now, if under these assumptions the s expressions $X\Delta_\sigma$ can be represented as:

$$X\Delta_\sigma \equiv \sum_{\tau=1}^{s} \rho_{\sigma\tau}(x_1, \dots, x_n)\, \Delta_\tau(x_1, \dots, x_n) \qquad (\sigma = 1 \cdots s),$$

and if at the same time all the $\rho_{\sigma\tau}$ behave regularly for the concerned system of values of:

$$\Delta_1 = 0, \dots, \Delta_s = 0,$$

then our system of equations admits every transformation:

$$x_i' = x_i + \frac{e}{1} X x_i + \cdots \qquad (i = 1 \cdots n)$$

of the one-term group Xf.

The proof of that is very simple. We have:

$$\Delta_\sigma(x_1', \dots, x_n') = \Delta_\sigma(x_1, \dots, x_n) + \frac{e}{1} X\Delta_\sigma + \cdots;$$

but we have:

$$XX\Delta_\sigma \equiv \sum_{\tau=1}^{s} \left\{ X\rho_{\sigma\tau} + \sum_{\pi=1}^{s} \rho_{\sigma\pi}\, \rho_{\pi\tau} \right\} \Delta_\tau,$$

where in the right-hand side the coefficients of the Δ again behave regularly for the system of values of $\Delta_1 = 0, \dots, \Delta_s = 0$. In the same way, the $XXX\Delta_\sigma$ express linearly in terms of $\Delta_1, \dots, \Delta_s$, and so on. In brief, we find:

$$\Delta_\sigma(x_1', \dots, x_n') = \sum_{\tau=1}^{s} \psi_{\sigma\tau}(x_1, \dots, x_n, e)\, \Delta_\tau(x) \qquad (\sigma = 1 \cdots s),$$

where the $\psi_{\sigma\tau}$ are ordinary power series in e and behave regularly for the system of values of $\Delta_1 = 0, \dots, \Delta_s = 0$. From this, it results that every system of values $x_1, \dots, x_n$ which satisfies the equations $\Delta_1(x) = 0, \dots, \Delta_s(x) = 0$ also satisfies the equations:

$$\Delta_\sigma\Big(x_1 + \frac{e}{1} X x_1 + \cdots, \; \dots, \; x_n + \frac{e}{1} X x_n + \cdots \Big) \qquad (\sigma = 1 \cdots s)$$

so that the system of equations $\Delta_1 = 0, \dots, \Delta_s = 0$ really admits the one-term group Xf.

As a result, we have the

Proposition 3. *If, in the variables $x_1, \dots, x_n$, a system of equations:*

$$\Delta_1(x_1, \dots, x_n) = 0, \dots, \Delta_s(x_1, \dots, x_n) = 0$$

is given, for which it is not assumed that its equations are mutually independent, and even less that the $s \times s$ determinants of the matrix:

$$\begin{vmatrix} \frac{\partial \Delta_1}{\partial x_1} & \cdot\cdot & \frac{\partial \Delta_1}{\partial xi_1} \\ \cdot & \cdot\cdot & \cdot \\ \frac{\partial \Delta_s}{\partial x_1} & \cdot\cdot & \frac{\partial \Delta_s}{\partial x_n} \end{vmatrix}$$

vanish or do not vanish by means of $\Delta_1 = 0, \dots, \Delta_s = 0$, then this system of equations surely admits all the transformations of the one-term group Xf when the s expressions $X\Delta_\sigma$ can be represented in the form:

$$X\Delta_\sigma \equiv \sum_{\tau=1}^{s} \rho_{\sigma\tau}(x_1,\dots,x_n)\,\Delta_\tau \qquad (\sigma=1\cdots s),$$

and when at the same time the $\rho_{\sigma\tau}$ behave regularly for those systems of values $x_1,\dots,x_n$ which satisfy the system of equations $\Delta_1 = 0, \dots, \Delta_s = 0$.

§ 31. In the preceding section, we have shown that the determination of all systems of equations which admit the one-term group Xf amounts to determining all systems of equations which admit the infinitesimal transformation Xf. Hence the question arises to ask for [Es ENTSTEHT DAHER DIE FRAGE NACH] all systems of equations $\Omega_1 = 0, \dots, \Omega_{n-m} = 0$ $(m \leqslant n)$ which admit the infinitesimal transformation:[2] (Translator's note: (in the sense of the definition on p. 124))

$$Xf = \sum_{i=1}^{n} \xi_i(x_1,\dots,x_n)\,\frac{\partial f}{\partial x_i}.$$

This question will be answered in the present section.

Two cases must be distinguished, namely either not all functions $\xi_1,\dots,\xi_n$ vanish by means of:

$$\Omega_1 = 0, \dots, \Omega_{n-m} = 0$$

or the equations $\xi_1 = 0, \dots, \xi_n = 0$ are a consequence of:

$$\Omega_1 = 0, \dots, \Omega_{n-m} = 0.$$

To begin with, we treat the first case.

Suppose, to fix ideas, that ξ_n does not vanish by means of $\Omega_1 = 0, \dots, \Omega_{n-m} = 0$. Then the concerned system of equations also admits the infinitesimal transformation:

$$Yf = \frac{1}{\xi_n}Xf = \frac{\xi_1}{\xi_n}\frac{\partial f}{\partial x_1} + \cdots + \frac{\xi_{n-1}}{\xi_n}\frac{\partial f}{\partial x_{n-1}} + \frac{\partial f}{\partial x_n}.$$

If now $x_1^0,\dots,x_n^0$ is a system of values which satisfies the equations $\Omega_k = 0$ and for which ξ_n does not vanish, then we may think that the general solutions of $Xf = 0$ relative to $x_n = x_n^0$, or, what is the same, of Yf, are determined; these general solutions, that we may call $y_1, \dots, y_{n-1}$, behave regularly in the neighborhood of $x_1^0, \dots, x_n^0$ and are independent of x_n. Hence if we introduce the new independent variables $y_1, \dots, y_{n-1}, y_n = x_n$ in place of the x, this will be an allowed transformation. Doing so, Yf receives the form $\frac{\partial f}{\partial y_n}$, and the system of equations $\Omega_1 = 0, \dots, \Omega_{n-m} = 0$ is transferred to a new one:

$$\overline{\Omega}_1(y_1, \dots, y_n) = 0, \dots, \overline{\Omega}_{n-m}(y_1, \dots, y_n) = 0$$

which admits the infinitesimal transformation $\frac{\partial f}{\partial y_n}$.

From this, it follows that the system of equations $\overline{\Omega}_k = 0$ is not solvable with respect to y_n. Indeed, if it were solvable with respect to y_n, and hence yielded $y_n - \varphi(y_1, \dots, y_{n-1}) = 0$, then the expression:

$$Y(y_n - \psi) = \frac{\partial}{\partial y_n}(y_n - \psi) = 1$$

would vanish by means of $\overline{\Omega}_1 = 0, \dots, \overline{\Omega}_{n-m} = 0$, which is nonsensical. Consequently, y_n can at most appear purely formally in the equations $\overline{\Omega}_k = 0$, that is to say, these equations can in all circumstances be brought to a form such as they represent relations between $y_1, \dots, y_{n-1}$ alone.[3] (Translator's note: For instance, the system of two equations $y_1 y_n = 0$ and $y_1 = 0$ invariant by $\frac{\partial}{\partial y_n}$ amounts to just $y_1 = 0$.) Here, the form of these relations is subjected to no further restriction.

If we now return to the initial variables, we immediately realize that the system of equations $\Omega_k = 0$ can be expressed by means of relations between the $n-1$ independent solutions $y_1, \dots, y_{n-1}$ of the equation $Xf = 0$. This outcome is obviously independent of the assumption that ξ_n itself should not vanish by means of $\Omega_k = 0$; we therefore see that every system of equations which admits the infinitesimal transformation Xf and which does not annihilate $\xi_1, \dots, \xi_n$ is represented by relations between the solutions of $Xf = 0$. On the other hand, we know that completely arbitrary relations between the solutions of $Xf = 0$ represent a system of equations which admits not only the infinitesimal transformation, but also all transformations of the one-term group Xf (Chap. 6, Proposition 3, p. 113). Consequently, this confirms the previously established result that our system of equations $\Omega_k = 0$ admits the one-term group Xf.

We now come to the second of the above two distinguished cases; naturally, this case can occur only when there are in general systems of values $x_1, \dots, x_n$ for which all the n functions ξ_i vanish.

If $x_1^0, \dots, x_n^0$ is an arbitrary system of values for which all ξ_i vanish, then the transformation of our one-term group:

$$x_i' = x_i + \frac{e}{1}\xi_i + \frac{e^2}{1 \cdot 2} X\xi_i + \cdots \qquad (i=1\cdots n)$$

reduces after the substitution $x_i = x_i^0$ to:

$$x_1' = x_1^0, \dots, x_n = x_n^0,$$

and we express this as: the system of values $x_1^0, \dots, x_n^0$ remains invariant by all transformations of the one-term group Xf. For this reason, it follows that every system of equations of the form:

$$\xi_1 = 0, \dots, \xi_n = 0, \qquad \psi_1(x_1, \dots, x_n) = 0, \quad \psi_2(x_1, \dots, x_n) = 0, \dots,$$

admits the one-term group Xf, whatever systems of values $x_1, \dots, x_n$ are involved in it. But such a form embraces[4] (Translator's note: In Lie's thought, a first system of equations *embraces* (verb: UMFASSEN) a second system of equations when the first zero-set is larger than the second one, so that the first system *implies* the second one, at least locally and generically, and perhaps after some allowed algebraic manipulations. Nothing more precise concerning this notion will appear later, and certainly nothing approaching either the Nullstellensatz or some of the concepts of the so-called theory of complex spaces.) [UMFASST] all systems of equations which bring $\xi_1, \dots, \xi_n$ to zero; so as a result, the second of the two previously distinguished cases is settled.

We summarize the obtained result in the following

Theorem 15. *There are two sorts of systems of equations which admit the infinitesimal transformation:*

$$Xf = \sum_{i=1}^{n} \xi_i(x_1, \dots, x_n) \frac{\partial f}{\partial x_i},$$

and hence in general admit all transformations of the one-term group Xf. The systems of equations of the first sort are represented by completely arbitrary relations between the solutions of the linear partial differential equation $Xf = 0$. The systems of equations of the second sort have the form:

$$\xi_1 = 0, \dots, \xi_n = 0, \qquad \psi_1(x_1, \dots, x_n) = 0, \quad \psi_2(x_1, \dots, x_n) = 0, \dots,$$

in which the ψ are absolutely arbitrary, provided of course that there are systems of values $x_1, \dots, x_n$ which satisfy the equations in question.

We now make a brief remark on this.

Let $C_1, \dots, C_{n-m}$ be arbitrary constants, and let $\Omega_1, \dots, \Omega_{n-m}$ be functions of x, which however are free of the C; lastly, suppose that each system of equations of the form:

$$\Omega_1(x_1, \dots, x_n) = C_1 \dots, \Omega_{n-m}(x_1, \dots, x_n) = C_{n-m}$$

admits the infinitesimal transformation Xf, which is free of C. Under these assumptions, the $n-m$ expression $X\Omega_k$ must vanish by means of:

$$\Omega_1 = C_1, \dots, \Omega_{n-m} = C_{n-m},$$

and certainly, for all values of the C. But since the $X\Omega_k$ are all free of the C, this is only possible when the $X\Omega_k$ vanish identically, that is to say, when the Ω_k are solutions of the equation $Xf = 0$. Thus, the following holds.

Proposition 4. *If the equations:*

$$\Omega_1(x_1,\dots,x_n) = C_1, \dots, \Omega_{n-m}(x_1,\dots,x_n) = C_{n-m}$$

with the arbitrary constants $C_1,\dots,C_{n-m}$ represent a system of equations which admits the infinitesimal transformation Xf, then $\Omega_1,\dots,\Omega_{n-m}$ are solutions of the differential equation Xf, which is free of the C.

§ 32. *At present, we consider q arbitrary infinitesimal transformations:*

$$X_k f = \sum_{i=1}^{n} \xi_{ki}(x_1,\dots,x_n)\frac{\partial f}{\partial x_i} \qquad (k=1\cdots q)$$

and ask for the systems of equations $\Omega_1 = 0, \Omega_2 = 0, \dots$ which admit all these infinitesimal transformations.

As always, we restrict ourselves here to systems of values $x_1,\dots,x_n$ for which all the ξ_{ki} behave regularly.

At first, it is clear that every sought system of equations also admits all infinitesimal transformations of the form:

$$\sum_{k=1}^{q} \chi_k(x_1,\dots,x_n)X_k f,$$

provided that $\chi_1,\dots,\chi_q$ behave regularly for the systems of values $x_1,\dots,x_n$ which satisfy the system of equations. From the previous section, we moreover see that each such system of equations also admits all finite transformations of the one-term groups that arise from the discussed infinitesimal transformations.[5] (Translator's note: In fact, Proposition 3, p. 129, gives a more precise statement.)

Furthermore, we remember Chap. 6, Proposition 1, p. 113. At that time, we saw that every function of $x_1,\dots,x_n$ which admits the two infinitesimal transformations $X_1 f$ and $X_2 f$ also admits the transformation $X_1X_2f - X_2X_1f$. Exactly the same property also holds true for every system of equations which admits the two infinitesimal transformations $X_1 f$ and $X_2 f$.

In fact, assume that the system of equations:

$$\Omega_k(x_1,\dots,x_n) = 0 \qquad (k=1\cdots n-m)$$

admits the two infinitesimal transformations $X_1 f$ and $X_2 f$, so that all the expressions $X_1\Omega_j$ and $X_2\Omega_j$ vanish by means of the system $\Omega_1 = 0, \dots, \Omega_{n-m} = 0$. Then according to Proposition 2, p. 126, the same holds true for all expressions $X_1X_2\Omega_j$ and $X_2X_1\Omega_j$, so that each $X_1X_2\Omega_j - X_2X_1\Omega_j$ vanishes by means of the system of equations $\Omega_1 = 0, \dots, \Omega_{n-m} = 0$. As a result, we have the

Proposition 5. *If a system of equations:*

$$\Omega_1(x_1,\dots,x_n) = 0, \dots, \Omega_{n-m}(x_1,\dots,x_n) = 0$$

admits the two infinitesimal transformations $X_1 f$ and $X_2 f$, then it also admits the infinitesimal transformation $X_1X_2f - X_2X_1f$.

We now apply this proposition similarly as we did earlier in Proposition 1 on p. 99, where the question was to determine the joint solutions of q given equations $X_1 f = 0, \dots, X_q f = 0$. At that time, we reduced the stated problem to the determination of the solutions of a complete system. Now, we proceed as follows.

We form all infinitesimal transformations:

$$X_kX_jf - X_jX_kf = [X_k, X_j]$$

and we ask whether the linear partial differential equations $[X_k, X_j] = 0$ are a consequence of $X_1 f = 0, \dots, X_q f = 0$. If this is not the case, then we add all transformations $[X_k, X_j]$ to the infinitesimal transformations $X_1 f, \dots, X_q f$,[2] which is permitted, since every system of equations which admits $X_1 f, \dots, X_q f$ also admits $[X_k, X_j]$. At present, we treat the infinitesimal transformations taken together:

$$X_1 f, \dots, X_q f, [X_k, X_j] \qquad {\scriptstyle (k,\, j = 1 \cdots q)}$$

exactly as we did at first with $X_1 f, \dots, X_q f$, that is to say, we form all infinitesimal transformations:

$$[[X_k, X_j], X_l], \qquad [[X_k, X_j], [X_h, X_l]],$$

and we ask whether the equations obtained by setting these expressions equal to zero are a consequence of $X_k f = 0$, $[X_k, X_j] = 0$ $(k, j = 1, \dots, q)$. If this is not the case, we add all the found infinitesimal transformations to $X_k f$, $[X_k, X_j]$ $(k, j = 1, \dots, q)$.

We continue in this way, and so at the end we must obtain a series of infinitesimal transformations:

$$X_1 f, \dots, X_q f, X_{q+1} f, \dots, X_{q'} f \qquad {\scriptstyle (q' \geqslant q)}$$

which is constituted in such a way that every equation:

$$[X_k, X_j] = 0 \qquad {\scriptstyle (k,\, j = 1 \cdots q')}$$

is a consequence of $X_1 f = 0, \dots, X_{q'} f = 0$. The equations:

$$X_1 f = 0, \dots, X_{q'} f = 0$$

[2] In practice, one will in general not add all infinitesimal transformations $[X_k, X_j]$, but only a certain number amongst them.

then define a complete system with q' or fewer terms. Hence we have the

Theorem 16. *The problem of determining all systems of equations which admit q given infinitesimal transformations $X_1 f, \dots, X_q f$ can always be led back to the determination of all systems of equations which, aside from $X_1 f, \dots, X_q f$, admit certain further infinitesimal transformations:*

$$X_{q+1} f, \dots, X_{q'} f \qquad (q' \geqslant q),$$

where now the equations:

$$X_1 f = 0, \dots, X_q f = 0, \;\; X_{q+1} f = 0, \dots, X_{q'} f = 0$$

define a complete system which has as many terms as there are independent equations in it.

Thus we can from now on limit ourselves to the following more special problem: *Consider q infinitesimal transformations in the variables $x_1, \dots, x_n$:*

$$X_k f = \sum_{i=1}^{n} \xi_{ki}(x_1, \dots, x_n) \frac{\partial f}{\partial x_i} \qquad (k = 1 \cdots q)$$

with the property that amongst the q equations:

$$X_1 f = 0, \dots, X_q f = 0,$$

there are exactly $p \leqslant q$ equations which are mutually independent, and with the property that any p independent ones amongst these q equations form a p-term complete system which belongs to $X_1 f = 0, \dots, X_q f = 0$. We seek all systems of equations in $x_1, \dots, x_n$ which admit the infinitesimal transformations $X_1 f, \dots, X_q f$.

It is clear that we can suppose without loss of generality that $X_1 f, \dots, X_q f$ are independent of each other as *infinitesimal transformations*. Furthermore, we notice that under the assumptions of the problem, the $(p+1) \times (p+1)$ determinants of the matrix:

$$\begin{vmatrix} \xi_{11} & \cdot\cdot & \xi_{1n} \\ \cdot & \cdot\cdot & \cdot \\ \xi_{q1} & \cdot\cdot & \xi_{qn} \end{vmatrix} \tag{4}$$

all vanish identically, whereas not all $p \times p$ determinants do.

The first step towards the solution of our problem is to distribute the systems of equations which admit the q infinitesimal transformations $X_1 f, \dots, X_q f$ in two separate classes [IN ZWEI GETRENNTE CLASSEN]; here, as a principle of classification, we take the behavior of the $p \times p$ determinants of (4).

In the first class, we include all systems of equations by means of which not all $p \times p$ determinants of the matrix (4) *vanish.*

In the second class, we include all systems of equations by means of which the $p \times p$ determinants in question all vanish.

We now examine the two classes one after the other.

§ 33. Amongst the q equations $X_1 f = 0, \dots, X_q f = 0$, we choose any p equations which are independent of each other; to fix ideas, let $X_1 f = 0, \dots, X_p f = 0$ be such equations, so that not all $p \times p$ determinants of the matrix:

$$\begin{vmatrix} \xi_{11} & \cdot\cdot & \xi_{1n} \\ \cdot & \cdot\,\cdot & \cdot \\ \xi_{p1} & \cdot\cdot & \xi_{pn} \end{vmatrix} \tag{5}$$

vanish identically. We now seek, amongst the systems of equations $\Omega_1 = 0, \dots, \Omega_{n-m} = 0$ of the first species, those which do not cancel the independence of the equations $X_1 f = 0, \dots, X_p f = 0$, hence those which do not annihilate all $p \times p$ determinants of the matrix (5). Here, we want at first to make the special assumption that a definite $p \times p$ determinant of the matrix (5), say the following one:

$$D = \begin{vmatrix} \xi_{1,n-p+1} & \cdot\cdot & \xi_{1n} \\ \cdot & \cdot\,\cdot & \cdot \\ \xi_{p,n-p+1} & \cdot\cdot & \xi_{pn} \end{vmatrix}$$

neither vanishes identically, nor vanishes by means of $\Omega_1 = 0, \dots, \Omega_{n-m} = 0$.

Under the assumptions made, there are identities of the form:

$$X_{p+j} f \equiv \sum_{\pi=1}^{p} \chi_{j\pi}(x_1, \dots, x_n) X_\pi f \qquad (j=1 \cdots q-p).$$

For the determination of the functions $\chi_{j\pi}$, we have here the equations:

$$\sum_{\pi=1}^{p} \xi_{\pi\nu}\, \chi_{j\pi} = \xi_{p+j,\nu} \qquad (\nu=1 \cdots n;\, j=1 \cdots q-p);$$

now since the determinant D does not vanish by means of $\Omega_1 = 0, \dots, \Omega_{n-m} = 0$, we realize that the $\chi_{j\pi}$ behave regularly for the system of values of $\Omega_1 = 0, \dots, \Omega_{n-m} = 0$, so that we can, under the assumptions made, leave out the infinitesimal transformations $X_{p+1} f, \dots, X_q f$; because, if the system of equations $\Omega_1 = 0, \dots, \Omega_{n-m} = 0$ admits the transformations $X_1 f, \dots, X_p f$, then it also automatically admits $X_{p+1} f, \dots, X_q f$.

We replace the infinitesimal transformations $X_1 f, \dots, X_p f$ by p other infinitesimal transformations of the specific form:

$$Y_\pi f = \frac{\partial f}{\partial x_{n-p+\pi}} + \sum_{i=1}^{n-p} \eta_{\pi i}(x_1, \dots, x_n) \frac{\partial f}{\partial x_i} \qquad (\pi=1 \cdots p).$$

We are allowed to do this, because for the determination of $Y_1 f, \dots, Y_p f$ we obtain the equations:

$$\sum_{\pi=1}^{p} \xi_{j,n-p+\pi} Y_\pi f = X_j f \qquad (j=1\cdots p),$$

which are solvable, and which provide for the Y_π expressions of the form:

$$Y_\pi f = \sum_{j=1}^{p} \rho_{\pi j}(x_1, \dots, x_n) X_j f \qquad (\pi=1\cdots p),$$

where the coefficients $\rho_{\pi j}$ behave regularly for the systems of values of:

$$\Omega_1 = 0, \dots, \ \Omega_{n-m} = 0.$$

Hence if the system of equations $\Omega_k = 0$ admits the infinitesimal transformations $X_1 f, \dots, X_p f$, it also admits $Y_1 f, \dots, Y_p f$, and conversely.

Let $x_1^0, \dots, x_n^0$ be any system of values which satisfies the equations $\Omega_1 = 0, \dots, \Omega_{n-m} = 0$ and for which the determinant D is nonzero. Then the $\eta_{\pi i}$ behave regularly in the neighborhood of $x_1^0, \dots, x_n^0$. Now, the equations $Y_1 f = 0, \dots, Y_p f = 0$ constitute a p-term complete system just as the equations $X_1 f = 0, \dots, X_p f = 0$, hence according to Theorem 12, p. 105, they have $n-p$ general solutions $y_1, \dots, y_{n-p}$ which behave regularly in the neighborhood of $x_1^0, \dots, x_n^0$ and which reduce to $x_1, \dots, x_{n-p}$ (respectively) for $x_{n-p+1} = x_{n-p+1}^0, \dots, x_n = x_n^0$.

Thus, if we set $y_{n-p+1} = x_{n-p+1}, \dots, y_n = x_n$, we can introduce $y_1, \dots, y_n$ as new variables in place of the x. At the same time, the infinitesimal transformations $Y_\pi f$ receive the form:

$$Y_1 f = \frac{\partial f}{\partial y_{n-p+1}}, \dots, \ Y_p f = \frac{\partial f}{\partial y_n};$$

but the system of equations $\Omega_k = 0$ is transferred to:

$$\overline{\Omega}_1(y_1, \dots, y_n) = 0, \dots, \ \overline{\Omega}_{n-m}(y_1, \dots, y_n) = 0,$$

and now these new equations must admit the infinitesimal transformations $\frac{\partial f}{\partial y_{n-p+1}}, \dots, \frac{\partial f}{\partial y_n}$. From this, it follows that the equations $\overline{\Omega}_k = 0$ are not solvable with respect to any of the variables $y_{n-p+1}, \dots, y_n$, that they contain these variables at most fictitiously and that they can be reshaped so as to represent only relations between $y_1, \dots, y_{n-p}$ alone.

If we now return to the original variables $x_1, \dots, x_n$, we then see that the equations $\Omega_1 = 0, \dots, \Omega_{n-m} = 0$ are nothing but relations between the solutions of the complete system $Y_1 f = 0, \dots, Y_p f = 0$, or, what is the same, of the complete system $X_1 f = 0, \dots, X_p f = 0$. Moreover, this result is independent of the assumption that the determinant D should not vanish by means of $\Omega_1 = 0, \dots, \Omega_{n-m} = 0$; thus it always holds true when not all $p \times p$ determinants of the matrix (5) vanish by means of $\Omega_1 = 0, \dots, \Omega_{n-m} = 0$.

Consequently, we can state the following theorem:

Theorem 17. *If q infinitesimal transformations $X_1f,\dots,X_qf$ in the variables $x_1,\dots,x_n$ provide exactly $p \leqslant q$ independent equations when equated to zero, say $X_1f = 0,\dots,X_pf = 0$, and if the latter equations form a p-term complete system, then every system of equations:*

$$\Omega_1(x_1,\dots,x_n) = 0,\dots,\ \Omega_{n-m}(x_1,\dots,x_n) = 0$$

which admits the q infinitesimal transformations $X_1f,\dots,X_qf$ without cancelling the independence of the equations:

$$X_1f = 0,\dots,\ X_pf = 0,$$

is represented by relations between the solutions of the p-term complete system $X_1f = 0,\dots,X_pf = 0$.

Thanks to this theorem, the determination of all systems of equations which belong to our first class is accomplished. In place of $X_1f = 0,\dots,X_pf = 0$, we only have to insert in the theorem one after the other all the systems of p independent equations amongst the q equations $X_1f = 0,\dots,X_qf = 0$.

§ 34. We now come to the second class of systems of equations which admit the infinitesimal transformations $X_1f,\dots,X_qf$, namely to the systems of equations which bring to zero all $p\times p$ determinants of the matrix (4).

Here, a series of subcases must immediately be distinguished. Namely it is possible that aside from the $p\times p$ determinants of the matrix (4)*, the system of equations:*

$$\Omega_1 = 0,\dots,\ \Omega_{n-m} = 0$$

also brings to zero all $(p-1)\times(p-1)$ determinants, all $(p-2)\times(p-2)$ determinants, and so on.

Hence we see that to every sought system of equations is associated a determinate number $h < p$ with the property that the concerned system of equations brings to zero all $p\times p$, all $(p-1)\times(p-1)$, ..., all $(h+1)\dots(h+1)$ determinants of the matrix (4), but not all $h\times h$ determinants. Thus, we must go in detail through all the various possible values $1,2,\dots,p-1$ of h and for each one of these values, we must set up the corresponding systems of equations that are admitted by $X_1f,\dots,X_qf$.

Let h be any of the numbers $1,2,\dots,p-1$. A system of equations which brings to zero all $(h+1)\times(h+1)$ determinants of the matrix (4), but not all the $h\times h$ determinants, contains in any case all equations which are obtained by equating to zero all $(h+1)\times(h+1)$ determinants $\Delta_1,\dots,\Delta_s$ of (4). Now, if the equations $\Delta_1 = 0,\dots,\Delta_s = 0$ were not satisfied by the systems of values $x_1,\dots,x_n$ for which the $\xi_{ki}(x)$ behave regularly, or else, if the equations $\Delta_1 = 0,\dots,\Delta_s = 0$ brought to zero all $h\times h$ determinants of (4), then this would be a sign that there is absolutely no system of the demanded nature to which is associated the chosen number h. So we assume that neither of these two cases occurs.

At first, it is to be examined whether the system of equations $\Delta_1 = 0, \dots, \Delta_s = 0$ is reducible. When this is the case, one should consider every irreducible[6] (Translator's note: This means smooth after stratification and not decomposable further in an invariant way.) [IRREDUCIBEL] system of equations which comes from $\Delta_1 = 0, \dots, \Delta_s = 0$, and which does not bring to zero all $h \times h$ determinants of (4).

Let $W_1 = 0, \dots, W_l = 0$ be one of the found irreducible systems of equations, and suppose that it has already been brought to a form such that not all $l \times l$ determinants of the matrix:

$$\begin{vmatrix} \frac{\partial W_1}{\partial x_1} & \cdot\cdot & \frac{\partial W_1}{\partial x_n} \\ \cdot & \cdot\cdot & \cdot \\ \frac{\partial W_l}{\partial x_1} & \cdot\cdot & \frac{\partial W_l}{\partial x_n} \end{vmatrix}$$

vanish by means of $W_1 = 0, \dots, W_l = 0$. Thus we must determine all systems of equations which are admitted by $X_1 f, \dots, X_q f$, which contain the equations $W_1 = 0, \dots, W_l = 0$ and which at the same time do not bring to zero all $h \times h$ determinants of (4). When we execute this for each individual irreducible system obtained from $\Delta_1 = 0, \dots, \Delta_s = 0$, we find all systems of equations which admit $X_1 f, \dots, X_q f$ and to which is associated the number h.

In general, the system of equations $W_1 = 0, \dots, W_l = 0$ will in fact not admit the infinitesimal transformations $X_1 f, \dots, X_q f$. To a system of equations which contains $W_1 = 0, \dots, W_l = 0$ and which in addition admits $X_1 f, \dots, X_q f$ there must belong in any case the equations:

$$X_k W_\lambda = 0, \qquad X_j X_k W_\lambda = 0 \qquad {\scriptstyle (k,\, j = 1 \cdots q;\ \lambda = 1 \cdots l)},$$

and so on. We form these equations and we examine whether they are contradictory with each other, or with $W_1 = 0, \dots, W_l = 0$, and whether they possibly bring to zero all $h \times h$ determinants of (4). If one of these two cases occurs, then there is no system of equations of the demanded constitution; if none occurs, then the independent equations amongst the equations:

$$W_\lambda = 0, \quad X_k W_\lambda = 0, \quad X_j X_k W_\lambda = 0, \dots \quad {\scriptstyle (\lambda = 1 \cdots l;\ j,k = 1 \cdots q)}$$

represent a system of equations which is admitted by $X_1 f, \dots, X_q f$. Into this system of equations, which can naturally be reducible, we put all systems of equations which admit $X_1 f, \dots, X_q f$, which contain the equations $W_1 = 0, \dots, W_l = 0$, but which embrace no smaller system of equations of the same nature.

The corresponding systems of equations are obviously the *smallest* systems of equations which are admitted by $X_1 f, \dots, X_q f$ and which bring to zero all $(h+1) \times (h+1)$ determinants of (4), though not all the $h \times h$ determinants.

Let now:

$$\text{(6)} \qquad W_1 = 0, \dots, W_{n-m} = 0 \qquad {\scriptstyle (n-m \geqslant l)}$$

be one of the found irreducible systems of equations; then the question is to add, in the most general way, new equations to this system so that one obtains a system of equations which is admitted by $X_1 f, \dots, X_q f$ but which does not bring to zero all $h \times h$ determinants of (4).

Since not all $h \times h$ determinants of (4) vanish by means of:

$$W_1 = 0, \dots, W_{n-m} = 0, \tag{6}$$

we can assume that for instance the determinant:

$$\Delta = \begin{vmatrix} \xi_{1,n-h+1} & \cdot\cdot & \xi_{1n} \\ \cdot & \cdot\cdot & \cdot \\ \xi_{h,n-h+1} & \cdot\cdot & \xi_{hn} \end{vmatrix}$$

belongs to the nonvanishing ones and we can set ourselves the problem of determining all systems of equations which are admitted by $X_1 f, \dots, X_q f$ and which at the same time do not make Δ equal to zero. When we carry out this task for every individual $h \times h$ determinant of (4) which does not vanish already by means of (6), we then evidently obtain all the sought systems of equations which embrace $W_1 = 0, \dots, W_{n-m} = 0$.

The independence of the equations $X_1 f = 0, \dots, X_h f = 0$ is, under the assumptions made, not canceled by $W_1 = 0, \dots, W_{n-m} = 0$, and in fact, for the systems of values of (6), there are certain relations of the form:

$$X_{h+j} f = \sum_{k=1}^{h} \psi_{jk}(x_1, \dots, x_n) X_k f \qquad (j=1 \cdots q-h)$$

where the ψ_{jk} are to be determined by the equations:

$$\xi_{h+j,\nu} = \sum_{k=1}^{h} \psi_{jk} \, \xi_{k\nu} \qquad (j=1 \cdots q-h;\ \nu=1 \cdots n).$$

Since all $(h+1) \times (h+1)$ determinants of (4) vanish by means of (6) while Δ does not, the functions ψ_{jk} are perfectly determined and they behave regularly for the systems of values of (6). Without loss of generality, we are hence allowed to leave out the infinitesimal transformations $X_{h+1} f, \dots, X_q f$; because every system of equations which contains the equations (6), which at the same time does not bring to zero the determinant Δ, and lastly, which is admitted by $X_1 f, \dots, X_h f$, is also automatically admitted by $X_{h+1} f, \dots, X_q f$.

One can easily see that no relation between $x_{n-h+1}, \dots, x_n$ alone can be derived from the equations (6). Indeed, if one obtained such a relation, say:

$$x_n - \omega(x_{n-h+1}, \dots, x_{n-1}) = 0,$$

then the h expressions:

$$X_k(x_n - \omega) = \xi_{kn} - \sum_{j=1}^{h-1} \frac{\partial \omega}{\partial x_{n-h+j}} \, \xi_{k,n-h+j} \qquad (k=1 \cdots h)$$

would vanish by means of (6). From this, we draw the conclusion that the number $n-m$ is in any case not larger than $n-h$ and that the equations (6) can be resolved with respect to $n-m$ of the variables $x_1,\dots,x_{n-h}$, say with respect to $x_1,\dots,x_{n-m}$:

$$x_k = \varphi_k(x_{n-m+1},\dots,x_n) \qquad (k=1\cdots n-m) \qquad (n-m \leqslant n-h). \tag{7}$$

We can therefore replace the system (6) by these equations.

Every system of equations which contains the equations (6) or equivalently the equations (7) can obviously be brought to a form such that, aside from the equations (7), it contains a certain number of relations:

$$V_j(x_{n-m+1},\dots,x_n) = 0 \qquad (j=1,2,\cdots)$$

between $x_{n-m+1},\dots,x_n$ alone. Now, if the system of equations in question is supposed to admit the infinitesimal transformations $X_1f,\dots,X_hf$, then all the expressions X_kV_j must vanish by means of (7) and of $V_1=0,\dots$; or, if we again denote by the symbol $[\,]$ the substitution $x_1=\varphi_1,\dots,x_{n-m}=\varphi_{n-m}$: the expressions:

$$\big[X_kV_j\big] = \sum_{\mu=1}^{m} \big[\xi_{k,n-m+\mu}\big] \frac{\partial V_j}{\partial x_{n-m+\mu}} \qquad (k=1\cdots h;\, j=1,2,\cdots)$$

vanish by means of $V_1=0,\dots$. But we can also express this as follows: the system of equations $V_1=0,\dots$ in the variables $x_{n-m+1},\dots,x_n$ must admit the h reduced infinitesimal transformations:

$$\overline{X}_kf = \sum_{\mu=1}^{m} \big[\xi_{k,n-m+\mu}\big] \frac{\partial f}{\partial x_{n-m+\mu}} \qquad (k=1\cdots h)$$

in these variables. In addition, the system of equations $V_1=0,\dots$ should naturally not bring to zero the determinant $\big[\Delta\big]$.

Conversely, every system of equations $V_1=0,\dots$ which possesses the property just discussed provides, together with (7), a system of equations which does not make Δ zero and which in addition admits $X_1f,\dots,X_hf$ and also $X_{h+1}f,\dots,X_qf$ as well.

As a result, our initial problem is reduced to the simpler problem of determining all systems of equations in $m<n$ variables which admit the $h\leqslant m$ infinitesimal transformations $\overline{X}_1f,\dots,\overline{X}_hf$ and for which the determinant:

$$\sum \pm \big[\xi_{1,n-h+1}\big] \cdots \big[\xi_{h,n-h+h}\big]$$

is not made zero.

We summarize what has been done up to now in the following

Theorem 18. *If q infinitesimal transformations:*

$$X_kf = \sum_{i=1}^{n} \xi_{ki}(x_1,\dots,x_n) \frac{\partial f}{\partial x_i} \qquad (k=1\cdots q)$$

are constituted in such a way that all $(p+1)\times(p+1)$ *determinants, but not all* $p\times p$ *determinants of the matrix:*

$$\begin{vmatrix} \xi_{11} & \cdot\cdot & \xi_{1n} \\ \cdot & \cdot\cdot & \cdot \\ \xi_{q1} & \cdot\cdot & \xi_{qn} \end{vmatrix}$$

vanish identically, and that any p independent equations amongst the equations $X_1 f = 0, \dots, X_q f = 0$ *form a p-term complete system, then one finds in the following manner all systems of equations in* $x_1, \dots, x_n$ *which admit* $X_1 f, \dots, X_q f$ *and which at the same time bring to zero all* $(h+1)\times(h+1)$ *determinants of the above matrix, but not all* $h\times h$ *determinants: by forming determinants* [DURCH DETERMINANTENBILDUNG], *one seeks at first the smallest system of equations for which all* $(h+1)\times(h+1)$ *determinants of the matrix vanish, but not all* $h\times h$ *determinants. If there exists a system of this sort, and if* $W_1 = 0, \dots, W_l = 0$ *is one such, then one forms the equations* $X_k W_i = 0$, $X_j X_k W_i = 0, \dots,$ *and one determines in this way the possibly existing smallest system of equations which embraces* $W_1 = 0, \dots, W_l = 0$, *which is admitted by* $X_1 f, \dots, X_q f$ *and which does not make equal to zero all* $h\times h$ *determinants of the matrix; if* $W_1 = 0, \dots, W_{n-m} = 0$ $(n-m \geqslant l)$ *is such a system of equations, which does not bring to zero for instance the determinant:*

$$\Delta = \sum \pm \xi_{1,n-h+1} \cdots \xi_{h,n-h+h},$$

then $h \leqslant m$ *and the equations* $W_1 = 0, \dots, W_{n-m} = 0$ *can be resolved with respect to* $n-m$ *of the variables* $x_1, \dots, x_{n-h}$, *say as follows:*

$$x_k = \varphi_k(x_{n-m+1}, \dots, x_n) \qquad (k=1\cdots n-m).$$

Lastly, one determines all systems of equations in the m variables $x_{n-m+1}, \dots, x_n$ *which admit the h reduced infinitesimal transformations:*

$$\begin{aligned} \overline{X}_k f &= \sum_{\mu=1}^{m} \xi_{k,n-m+\mu}(\varphi_1, \dots, \varphi_{n-m}, x_{n-m+1}, \dots, x_n) \frac{\partial f}{\partial x_{n-m+\mu}} \\ &= \sum_{\mu=1}^{m} \left[\xi_{k,n-m+\mu}\right] \frac{\partial f}{\partial x_{n-m+\mu}} \qquad (k=1\cdots h) \end{aligned}$$

and which do not bring to zero the determinant:

$$[\Delta] = \sum \pm \left[\xi_{1,n-h+1}\right] \cdots \left[\xi_{h,n_h+h}\right].$$

Each one of these systems of equations represents, after adding the equations $x_1 = \varphi_1, \dots, x_{n-m} = \varphi_{n-m}$, *a system of equations of the demanded constitution. By carrying out the indicated developments in all possible cases, one obtains all systems of equations of the demanded constitution.*

It frequently happens that one already knows a system of equations $U_1 = 0$, $U_2 = 0, \dots, U_{n-s} = 0$ which admits the infinitesimal transformations $X_1 f, \dots, X_q f$ and which brings to zero all $(h+1)\times(h+1)$ determinants of the matrix (4), and which nevertheless does not bring to zero the $h\times h$ determinant Δ. Then one can ask for all systems of equations which comprise the equations $U_1=0,\dots,U_{n-s}=0$ and which likewise do not bring Δ to zero.

The determination of all these systems of equations can be carried out exactly as in the special case above, where the smallest system of equations $W_1=0,\dots,W_{n-m}=0$ of the relevant nature was known, and where one was looking for all systems of equations which were admitted by $X_1f,\dots,X_qf$, which comprised the system $W_1=0,\dots,W_{n-m}=0$ and which did not bring Δ to zero.

Indeed, in exactly the same way as above, one first shows that $h\leqslant s$ and that the equations $U_1=0,\dots,U_{n-s}=0$ can be resolved with respect to $n-s$ of the variables $x_1,\dots,x_{n-h}$, say as follows:

$$x_k=\psi_k(x_{n-s+1},\dots,x_n)\qquad (k=1\cdots n-s).$$

Now, in order to find the sought systems of equations, one forms the h reduced infinitesimal transformations:

$$\begin{aligned}\widehat{X}_kf&=\sum_{\sigma=1}^{s}\xi_{k,n-s+\sigma}(\psi_1,\dots,\psi_{n-s},x_{n-s+1},\dots,x_n)\frac{\partial f}{\partial x_{n-s+\sigma}}\\&=\sum_{\sigma=1}^{s}\widehat{\xi}_{k,n-s+\sigma}\frac{\partial f}{\partial x_{n-s+\sigma}}\qquad (k=1\cdots h)\end{aligned}$$

and next, one determines all systems of equations in the s variables $x_{n-s+1},\dots,x_n$ which admit $\widehat{X}_1f,\dots,\widehat{X}_hf$ and which do not bring to zero the determinant:

$$\widehat{\Delta}=\sum\pm\widehat{\xi}_{1,n-h+1}\cdots\widehat{\xi}_{h,n-h+h}.$$

By adding these equations one after the other to the equations $x_1-\psi_1=0,\dots,x_{n-s}-\psi_{n-s}=0$, one obtains all systems of equations of the demanded constitution.

It is not necessary to explain more precisely what has been just said.

§ 35. The problem which was set up at the beginning of § 32 is now basically settled. Indeed, thanks to the latter theorem, this problem is reduced to the determination of all systems of equations in the m variables $x_{n-m+1},\dots,x_n$ which admit the h infinitesimal transformations $\overline{X}_1f,\dots,\overline{X}_hf$. But this is a problem of the same type as the original one, which is simplified only inasmuch as the number m of the variables is smaller than n.

Now, concerning the reduced problem, the same considerations as those regarding the initial problem can be made use of. That is to say: if the equations $\overline{X}_1f=0,\dots,\overline{X}_hf=0$ do not actually form an h-term complete system, then one has to set up for $k,j=1,\dots,h$ the infinitesimal transformations $[\overline{X}_k,\overline{X}_j]$ and to ask whether the independent equations amongst the equations $\overline{X}_kf=0$, $[\overline{X}_k,\overline{X}_j]=0$ for a complete system, and so on, in brief, one proceeds just as for the original problem. The only difference in comparison to the former situation is that from the beginning one looks for systems of equations which do not cancel the independence of the equations:

$$\overline{X}_1 f = 0, \ldots, \overline{X}_h f = 0,$$

and especially, which do not bring the determinant $\left[\Delta\right]$ to zero.

Exactly as above for the original problem, the reduced problem can be dealt with in parts, and be partially reduced to a problem in fewer variables, and so forth. Briefly, one sees that the complete resolution of the original problem can be attained after a finite number of steps.

We therefore need not address this issue further, but we want to consider in more detail a particularly important special case.

Let as above the system of equations:

$$\text{(7')} \qquad x_1 - \varphi_1(x_{n-m+1}, \ldots, x_n) = 0, \ldots, x_{n-m} - \varphi_{n-m}(x_1, \ldots, x_n) = 0$$

be constituted in such a way that it admits the infinitesimal transformations $X_1 f, \ldots, X_q f$ and assume that it brings to zero all $(h+1)\times(h+1)$ determinants of the matrix (4), while it does not bring to zero the determinant Δ. In comparison, by the system of equations (7'), we want to understand *not just the smallest one*, but a *completely arbitrary one* of the demanded constitution.

Amongst the equations $X_1 f = 0, \ldots, X_q f = 0$, assume that any p independent equations amongst them, say for instance $X_1 f = 0, \ldots, X_p f = 0$, form a p-complete system; thus in any case, there are relations of the form:

$$[X_k, X_j] = \sum_{\sigma=1}^{q} \omega_{kj\sigma}(x_1, \ldots, x_n)\, X_\sigma f \qquad {\scriptstyle (k,\, j=1\cdots q)}.$$

For the systems of values of (7') there remain only h equations:

$$X_1 f = 0, \ldots, X_h f = 0$$

that are independent of each other, whereas $X_{h+1} f, \ldots, X_q f$ can be represented in the form:

$$X_{h+j} f = \sum_{\tau=1}^{h} \psi_{j\tau}(x_1, \ldots, x_n)\, X_\tau f \qquad {\scriptstyle (j=1\cdots q-h)},$$

where the ψ_{jk} behave regularly for the concerned system of values.

Now, we want to make the specific assumption that all the coefficients ω_{kjs} also behave regularly for the system of values of (7); then evidently for the concerned system of values, equations of the form:

$$[X_k, X_j] = \sum_{\sigma=1}^{h} \left\{ \omega_{kj\sigma} + \sum_{\tau=1}^{q-h} \omega_{kj\,h+\tau}\, \psi_{\tau\sigma} \right\} X_\sigma f$$

hold true. In this special case one can indicate in one stroke all systems of equations which embrace the equations (7'), which admit $X_1 f, \dots, X_q f$ and which do not bring Δ to zero. This should now be shown.

For all systems of values $x_1, \dots, x_n$ which satisfy the equations (7'), there are relations of the form:

$$[X_k, X_j] = \sum_{\sigma=1}^{h} w_{kj\sigma}(x_1, \dots, x_n) X_\sigma f \qquad (k, j = 1 \cdots h).$$

Here, the $w_{kj\sigma}$ behave regularly and likewise the functions $[w_{kjs}]$, if by the symbol $[\,]$ we denote as before the substitution:

$$x_1 = \varphi_1, \dots, x_{n-m} = \varphi_{n-m}.$$

We decompose the above relations into the following:

$$X_k \xi_{jv} - X_j \xi_{kv} = \sum_{s=1}^{h} w_{kjs}(x_1, \dots, x_n)\, \xi_{sv}$$
$$(k, j = 1 \cdots h;\ v = 1 \cdots n),$$

and they naturally hold identically after the substitution $[\,]$, so that we have:

$$\text{(8)} \qquad \big[X_k \xi_{jv}\big] - \big[X_j \xi_{kv}\big] \equiv \sum_{s=1}^{h} [w_{kjs}]\, [\xi_{sv}].$$

But now the system of equations $x_k - \varphi_k = 0$ admits the infinitesimal transformations $X_1 f, \dots, X_h f$, so the relation (3) derived on p. 126:

$$\big[X_k \Phi\big] \equiv \big[X_k [\Phi]\big]$$

holds true, in which $\Phi(x_1, \dots, x_n)$ is a completely arbitrary function of its arguments. We can write this relation somewhat differently, if we remember the infinitesimal transformations:

$$\overline{X}_k f = \sum_{\mu=1}^{m} [\xi_{k,n-m+\mu}] \frac{\partial f}{\partial x_{n-m+\mu}} \qquad (k = 1 \cdots h);$$

indeed, one evidently has:

$$\big[X_k [\Phi]\big] \equiv \overline{X}_k [\Phi],$$

whence:

$$\big[X_k \Phi\big] \equiv \overline{X}_k [\Phi].$$

From this, it follows that the identities (8) can be replaced by the following ones:

$$\overline{X}_k [\xi_{jv}] - \overline{X}_j [\xi_{kv}] \equiv \sum_{s=1}^{h} [w_{kjs}]\, [\xi_{sv}].$$

In other words, there are identities of the form:

$$[\overline{X}_k, \overline{X}_j] = \overline{X}_k\overline{X}_j f - \overline{X}_j\overline{X}_k f \equiv \sum_{s=1}^{h} [w_{kjs}]\overline{X}_s f,$$

that is to say, the equations $\overline{X}_1 f = 0, \ldots, \overline{X}_h f = 0$ form an h-term complete system in the m independent variables $x_{n-m+1}, \ldots, x_n$.

Now, we recall the observations that we have linked up with Theorem 18. They showed that our problem stated above comes down to determining all systems of equations in $x_{n-m+1}, \ldots, x_n$ which admit $\overline{X}_1 f, \ldots, \overline{X}_h f$ and do not cancel the independence of the equations $\overline{X}_1 f = 0, \ldots, \overline{X}_h f = 0$. But since in our case the equations $\overline{X}_1 f = 0, \ldots, \overline{X}_h f = 0$ form an h-term complete system, we can at once apply Theorem 17. Thanks to it, we see that the sought systems of equations in $x_{n-m+1}, \ldots, x_n$ are represented by relations between the solutions of the complete system $\overline{X}_1 f = 0, \ldots, \overline{X}_h f = 0$.

Consequently, if we add arbitrary relations between the solutions of this complete system to the equations $x_k = \varphi_k$, we obtain the general form of a system of equations which admits the infinitesimal transformations $X_1 f, \ldots, X_q f$, which comprises the equations $x_k = \varphi_k$ and which does not bring Δ to zero.

It appears superfluous to formulate as a proposition the result obtained here in its full generality. By contrast, it is useful for the sequel to expressly state the following theorem, which corresponds to the special case $q = p = h$.

Theorem 19. *If a system of $n-m$ independent equations in the variables $x_1, \ldots, x_n$ admits the h infinitesimal transformations:*

$$X_k f = \sum_{i=1}^{n} \xi_{ki}(x_1, \ldots, x_n) \frac{\partial f}{\partial x_i} \qquad (k = 1 \cdots h)$$

and if at the same time the determinant:

$$\Delta = \sum \pm \xi_{1,n-h+1} \cdots \xi_{h,n-h+h}$$

neither vanishes identically, nor vanishes by means of the system of equations, then $h \leqslant m$ and the system of equations can be resolved with respect to $n-m$ of the variables $x_1, \ldots, x_{n-h}$, say as follows:

$$x_1 = \varphi_1(x_{n-m+1}, \ldots, x_n), \; \ldots, \; x_{n-m} = \varphi_{n-m}(x_{n-m+1}, \ldots, x_n).$$

Now if, for the systems of values $x_1, \ldots, x_n$ of these equations, all the expressions $[X_k, X_j]$ can be represented in the form:

$$[X_k, X_j] = \sum_{s=1}^{h} w_{kjs}(x_1, \ldots, x_n) X_s f \qquad (k, j = 1 \cdots h),$$

where the w_{kjs} behave regularly for the concerned systems of values, then one finds as follows all systems of equations which comprise the equations:

$$x_1 - \varphi_1 = 0, \dots, x_{n-m} - \varphi_{n-m} = 0,$$

which admit the infinitesimal transformations $X_1 f, \dots, X_h f$ *and which do not bring to zero the determinant* Δ*: one sets up the* h *reduced infinitesimal transformations:*

$$\overline{X}_k f = \sum_{\mu=1}^{m} \xi_{k,n-m+\mu}\big(\varphi_1, \dots, \varphi_{n-m}, x_{n-m+1}, \dots, x_n\big)\, \frac{\partial f}{\partial x_{n-m+\mu}}$$
$$(k=1\cdots h);$$

then the h *mutually independent equations* $\overline{X}_1 f = 0, \dots, \overline{X}_h f = 0$ *form an* h*-term complete system in the independent variables* $x_{n-m+1}, \dots, x_n$*; if:*

$$u_1(x_{n-m+1}, \dots, x_n), \dots, u_{m-h}(x_{n-m+1}, \dots, x_n)$$

are independent solutions of this complete system, then:

$$x_1 - \varphi_1 = 0, \dots, x_{n-m} - \varphi_{n-m} = 0, \quad \Phi_i(u_1, \dots, u_{m-h}) = 0$$
$$(i=1,2,\cdots)$$

is the general form of the sought system of equations; here the Φ_i *are understood to be arbitrary functions of their arguments.*

§ 36. The analytical developments of the present chapter receive a certain clarity and especially a better transparency when one applies the ideas and concepts from the theory of manifolds [DIE VORSTELLUNGEN UND BEGRIFFSBILDUNGEN DER MANNIGFALTIGKEITSLEHRE]. We now want to do that. The sequel stands in comparison to §§ 30 to 35 in exactly the same relationship as Chap. 6 stands in comparison to Chap. 5.

Every system of equations in the variables $x_1, \dots, x_n$ represents a manifold of the n-times extended manifold R_n. If the system of equations admits the one-term group Xf, then according to the previously introduced terminology, the corresponding manifold also admits Xf; so if a point $x_1, \dots, x_n$ belongs to the manifold, then all points into which $x_1, \dots, x_n$ is transferred by all the transformations of the one-term group Xf also lie in the same manifold.

If now $x_1^0, \dots, x_n^0$ is any point of the space, two cases can occur: either $\xi_1, \dots, \xi_n$ do not all vanish for $x_i = x_i^0$, or the quantities $\xi_1(x^0), \dots, \xi_n(x^0)$ are all equal to zero. In the first case, the point $x_1^0, \dots, x_n^0$ takes infinitely many positions via the ∞^1 transformations of the one-term group Xf, and as we know, the totality of these positions is invariant by the one-term group Xf and it constitutes an integral curve of the infinitesimal transformation Xf. In the second case, $x_1^0, \dots, x_n^0$ keeps its position through all transformations of the one-term group Xf; the integral curve passing through $x_1^0, \dots, x_n^0$ shrinks to the point itself.

So if a manifold admits the one-term group Xf and if it consists in general of points for which $\xi_1, \dots, \xi_n$ do not all vanish, then it is constituted of integral curves of the infinitesimal transformation Xf. By contrast, if the manifold in question consists only of points for which $\xi_1, \dots, \xi_n$ all vanish, then each point of the manifold keeps its position through all transformations of the one-term group Xf. Evidently, every smaller manifold contained in this manifold then also admits the one-term group Xf.

With these words, the conceptual content [DER BEGRIFFLICHE INHALT] of Theorem 15 is clearly stated, and in fact fundamentally, the preceding considerations can be virtually regarded as providing a new demonstration of Theorem 15.

We still have to explain what it means conceptually for a system of equations $\Omega_1 = 0, \dots, \Omega_{n-m} = 0$ to admit the infinitesimal transformation Xf.

To this end, we remember that the infinitesimal transformation Xf attaches to each point $x_1, \dots, x_n$ at which not all ξ vanish the completely determined direction of progress:

$$\delta x_1 : \cdots : \delta x_n = \xi_1 : \cdots : \xi_n,$$

while it attaches no direction of progress to a point for which $\xi_1 = \cdots = \xi_n = 0$. We can therefore say:

The system of equations $\Omega_1 = 0, \dots, \Omega_{n-m} = 0$ admits the infinitesimal transformation Xf when the latter attaches to each point of the manifold $\Omega_1 = 0, \dots, \Omega_{n-m} = 0$ either absolutely no direction of progress, or a direction of progress which satisfies the $n-m$ equations:

$$\frac{\partial \Omega_k}{\partial x_1}\,\delta x_1 + \cdots + \frac{\partial \Omega_k}{\partial x_n}\,\delta x_n \qquad (k=1 \cdots n-m),$$

hence which comes into contact with the manifold.

This definition is visibly independent both of the choice of the variables and of the form of the system of equations:

$$\Omega_1 = 0, \dots, \Omega_{n-m} = 0.$$

Theorem 14 can now be given the following visual interpretation:

If the infinitesimal transformation Xf attaches to each point of a manifold either absolutely no direction of progress, or a direction of progress which comes into contact with the manifold, then the manifold admits all transformations of the one-term group Xf.

In conclusion, if we introduce the terminology: "the manifold $\Omega_1 = 0, \dots, \Omega_{n-m} = 0$ admits the infinitesimal transformation Xf", we can express Theorem 14 as follows.

A manifold admits all transformations of the one-term group Xf if and only if it admits the infinitesimal transformation Xf.

In §§ 32 and 34, we gave a classification of all systems of equations which admit the q infinitesimal transformations $X_1 f, \dots, X_q f$. There, we took as a starting point the behavior of the determinants of the matrix (4):

$$\begin{vmatrix} \xi_{11} & \cdot\cdot & \xi_{1n} \\ \cdot & \cdot\cdot & \cdot \\ \xi_{q1} & \cdot\cdot & \xi_{qn} \end{vmatrix}.$$

We assumed that all $(p+1)\times(p+1)$ determinants of this matrix vanished identically, whereas the $p\times p$ determinants did not. We then included in the same class all systems of equations by means of which all $(h+1)\times(h+1)$ determinants of the matrix vanish, but not all $h\times h$ determinants; in the process, the integer h could have the $p+1$ different values: $p, p-1, \dots, 2, 1, 0$.

Already in Chap. 6 we observed that, under the assumptions just made, the q infinitesimal transformations:

$$X_1 f, \dots, X_q f$$

attach exactly p independent directions of progress to any point $x_1, \dots, x_n$ in general position. It stands to reason in the corresponding way that $X_1 f, \dots, X_q f$ attach exactly h independent directions of progress to a point $x_1, \dots, x_n$, when the $h\times h$ determinants of the above matrix do not all vanish at the point in question, while by contrast all $(h+1)\times(h+1)$ determinants are zero. Now, since every system of equations which admits the infinitesimal transformations $X_1 f, \dots, X_q f$ represents a manifold having the same property, our previous classification of the systems of equations immediately provides a classification of the manifolds. Indeed, amongst the manifolds which admit $X_1 f, \dots, X_q f$ we include in the same class those systems to the points of which are associated the same number $h \leqslant p$ of independent directions of progress by the infinitesimal transformations $X_1 f, \dots, X_q f$.

If a manifold admits the infinitesimal transformations $X_1 f, \dots, X_q f$, then at each one of its points, it comes into contact with the directions of progress that are attached to the point by $X_1 f, \dots, X_q f$. Now, if $X_1 f, \dots, X_q f$ determine precisely h independent directions at each point of the manifold, then the manifold must obviously have at least h dimensions. Hence we can state the proposition:

Proposition 6. *If q infinitesimal transformations $X_1 f, \dots, X_q f$ attach precisely h independent directions of progress to a special point $x_1^0, \dots, x_n^0$, then there is in any case no manifold of smaller dimension than h which contains the point $x_1^0, \dots, x_n^0$ and which admits the q infinitesimal transformations $X_1 f, \dots, X_q f$.*

This proposition is essentially just another formulation of a former result. Indeed, in § 34 we considered the systems of equations which admit $X_1 f, \dots, X_q f$ and which at the same time leave only h mutually independent equations amongst the q equations $X_1 f = 0, \dots, X_q f = 0$. On that occasion, we saw that such a system of equations consists of at most $n-h$ independent equations, so that it represents a manifold having at least h dimensions.

Chapter 8
Complete Systems Which Admit All Transformations of a One-term Group

If, in a q-term complete system:

$$X_k f = \sum_{i=1}^{r} \xi_{ki}(x_1,\dots,x_n)\,\frac{\partial f}{\partial x_i} \qquad {\scriptstyle (k=1\cdots q)}$$

one introduces new independent variables $x_1' = F_1(x_1,\dots,x_n),\dots,x_n' = F_n(x_1,\dots,x_n)$, then as was already observed earlier (cf. Chap. 5, p. 102), one again obtains a q-term complete system in $x_1',\dots,x_n'$. Naturally, this new complete system has in general a form different from the initial one; nonetheless, it can also happen that the two complete systems do not essentially differ in their form, when relationships of the form:

$$X_k f = \sum_{j=1}^{q} \psi_{kj}(x_1',\dots,x_n') \sum_{i=1}^{n} \xi_{ji}(x_1',\dots,x_n')\,\frac{\partial f}{\partial x_i'} \qquad {\scriptstyle (k=1\cdots q)}$$

hold, where of course the determinant of the ψ_{kj} does not vanish identically. In this case, we say: *the complete system:*

$$X_1 f = 0,\ \dots,\ X_q f = 0$$

admits the transformation $x_i' = F_i(x_1,\dots,x_n)$, or: *it remains invariant through this transformation.*

By making use of the abbreviated notations:

$$\xi_{ki}(x_1',\dots,x_n') = \xi_{ki}',$$
$$\sum_{i=1}^{n} \xi_{ki}'\,\frac{\partial f}{\partial x_i'} = X_k' f,$$

we can set up the following definition:

The q-term complete system:

S. Lie, *Theory of Transformation Groups I*,
DOI 10.1007/978-3-662-46211-9_8

$$X_k f = \sum_{i=1}^{n} \xi_{ki}(x_1,\dots,x_n)\frac{\partial f}{\partial x_i} = 0 \qquad (k=1\cdots q)$$

admits the infinitesimal transformation $x_i' = F_i(x_1,\dots,x_n)$ *if and only if, for every k, there is a relation of the form:*[1]

(1) $$X_k f = \sum_{j=1}^{q} \psi_{kj}(x_1',\dots,x_n')X_j' f.$$

The sense of this important definition is best elucidated [VERDEUTLICHEN] by means of a simple example.

The two equations:

$$X_1 f = \frac{\partial f}{\partial x_1} = 0, \qquad X_2 f = \frac{\partial f}{\partial x_2} = 0$$

in the three variables x_1, x_2, x_3 form a two-term complete system. By introducing the new variables:

$$x_1' = x_1 + x_2, \qquad x_2' = x_1 - x_2, \qquad x_3' = x_3$$

in place of the x, we obtain the new complete system:

$$\frac{\partial f}{\partial x_1} = \frac{\partial f}{\partial x_1'} + \frac{\partial f}{\partial x_2'} = 0, \qquad \frac{\partial f}{\partial x_2} = \frac{\partial f}{\partial x_1'} - \frac{\partial f}{\partial x_2'} = 0$$

which is equivalent to the system $\frac{\partial f}{\partial x_1'} = 0$, $\frac{\partial f}{\partial x_2'} = 0$. So there are relations of the form:

$$X_1 f = X_1' f + X_2' f, \qquad X_2 f = X_1' f - X_2' f,$$

whence the complete system $\frac{\partial f}{\partial x_1} = 0$, $\frac{\partial f}{\partial x_2} = 0$ admits the transformation: $x_1' = x_1 + x_2$, $x_2' = x_1 - x_2$.

§ 37. Let the q-term complete system $X_1 f = 0, \dots, X_q f = 0$ admit the transformation $x_i' = F_i(x_1,\dots,x_n)$, so that for every f there are relations of the form (1). Now, if $\varphi(x_1,\dots,x_n)$ is a solution of the complete system, the right-hand side of (1) vanishes identically after the substitution $f = \varphi(x_1',\dots,x_n')$, whence the left-hand side also vanishes identically after the substitution $f = \varphi\big(F_1(x),\dots,F_n(x)\big)$; consequently, like $\varphi(x)$ itself, $\varphi\big(F_1(x),\dots,F_n(x)\big)$ also constitutes a solution, or, in other terms: the transformation $x_i' = F_i(x)$ transfers every solution of the complete system $X_1 f = 0, \dots, X_q f = 0$ to a solution of the same complete system.

However, conversely the following also holds true: if every solution of the q-term complete system $X_1 f = 0, \dots, X_q f = 0$ is transferred, by means of a transformation $x_i' = F_i(x_1,\dots,x_n)$, to a solution, then the complete system admits the transformation

[1] Lie, Scientific Society of Christiania, February 1875.

in question. Indeed, by the introduction of the variables x_i in place of the variables x_i', the equations $X_1'f = 0, \dots, X_q'f = 0$ convert into a q-term complete system which has all its solutions in common with the q-term system $X_1 f = 0, \dots, X_q f = 0$; but from this it follows that relations of the form (1) hold, hence that the complete system $X_1 f = 0, \dots, X_q f = 0$ really admits the transformation $x_i' = F_i(x)$.

From this, it follows that *a q-term complete system admits the transformation $x_i' = F_i(x_1, \dots, x_n)$ if and only if this transformation transfers every solution of the complete system into a solution. Naturally, for this to hold, it is only necessary that the transformation transfers to solutions any $n-q$ independent solutions of the system.*

Next, the present line of thought will completely correspond to the one followed in Chaps. 6 and 7 if we now ask the question: how can one show that a q-term complete system $X_1 f = 0, \dots, X_q f = 0$ admits all transformations of the one-term group Yf? Indeed, this question actually belongs to general research on differential equations which admit one-term groups, and that is why this question will also be taken up again in a subsequent chapter of this part, in the chapter on differential invariants, and on the basis of the general theory developed there, it will be settled. But first, we need criteria by means of which we can recognize whether or not a given complete system admits all transformations of a given one-term group. This is why we now want to derive such criteria, with somewhat simpler expedients.

Let us denote any $n-q$ independent solutions of the q-term complete system $X_1 f = 0, \dots, X_q f = 0$ by $\varphi_1, \dots, \varphi_{n-q}$. If now the complete system admits all transformations:

$$x_i' = x_i + \frac{t}{1} Y x_i + \frac{t^2}{1 \cdot 2} Y Y x_i + \cdots \qquad (i=1\cdots n)$$

of the one-term group Yf, then the $n-q$ independent functions:

$$\varphi_k\big(x + tYx + \cdots\big) = \varphi_k(x) + \frac{t}{1} Y\varphi_k + \frac{t^2}{1 \cdot 2} YY\varphi_k + \cdots$$
$$(k=1\cdots n-q)$$

must also be solutions of the system, and for every value of t. From this, we deduce that the $n-q$ expressions $Y\varphi_k$ are in any case solutions of the system, hence that relations of the form:

$$(2) \qquad Y\varphi_k = \omega_k(\varphi_1, \dots, \varphi_{n-q}) \qquad (k=1\cdots n-q)$$

must hold. This condition is necessary; but at the same time it is also sufficient, since if it is satisfied, then all $YY\varphi_k$, $YYY\varphi_k$, ..., will be functions of $\varphi_1, \dots, \varphi_{n-q}$ only, whence the expressions $\varphi_k(x + tYx + \cdots)$ will be solutions of the complete system, and from this it follows that this system effectively admits all transformations of the one-term group Yf.

Consequently, the following holds.

Proposition 1. *A q-term complete system $X_1 f = 0, \dots, X_q f = 0$ with the $n-q$ independent solutions $\varphi_1, \dots, \varphi_{n-q}$ admits all the transformations of the one-term group:*

$$Yf = \sum_{i=1}^{n} \eta_i(x_1, \dots, x_n) \frac{\partial f}{\partial x_i},$$

if and only if $n-q$ relations of the form:

$$Y\varphi_k = \omega_k(\varphi_1, \dots, \varphi_{n-q}) \qquad (k=1\cdots n-q)$$

hold.

This said, the criterion obtained is practically applicable only when the complete system is already integrated. However, it is very easy to deduce from this criterion another one which does not presuppose that one knows the solutions of the complete system.

If in the identity:

$$X_k(Y(f)) - Y(X_k(f)) \equiv \sum_{i=1}^{n} (X_k\eta_i - Y\xi_{ki}) \frac{\partial f}{\partial x_i}$$

one sets in place of f any solution φ of the complete system, then one obtains:

$$X_k(Y(\varphi)) \equiv \sum_{i=1}^{n} (X_k\eta_i - Y\xi_{ki}) \frac{\partial \varphi}{\partial x_i}.$$

Now, if the complete system allows all transformations of the one-term group Yf, then by the above, $Y(\varphi)$ is also a solution of the system, hence the left-hand side of the last equation vanishes identically; naturally, the right-hand side does the same, so every solution of the complete system also satisfies the q equations:

$$\sum_{i=1}^{n} (X_k\eta_i - Y\xi_{ki}) \frac{\partial f}{\partial x_i} = 0;$$

but from this it follows that q identities of the form:

$$\sum_{i=1}^{n} (X_k\eta_i - Y\xi_{ki}) \frac{\partial f}{\partial x_i} \equiv \sum_{j=1}^{q} \chi_{kj}(x_1, \dots, x_n) X_j f \qquad (k=1\cdots q)$$

hold.

On the other hand, we assume that identities of this form exist and we again understand by φ any solution of the complete system. Then it follows immediately that the q expressions $X_k(Y(f)) - Y(X_k(f))$ vanish identically after the substitution $f = \varphi$; but from that, we get: $X_k(Y(\varphi)) \equiv 0$, that is to say $Y\varphi$ is a solution of the system, so there exist $n-q$ relations of the form:

$$Y\varphi_k = \omega_k(\varphi_1, \dots, \varphi_{n-q}) \qquad (k=1\cdots n-q), \tag{2}$$

and they show that the complete system admits all transformations of the one-term group Yf.

As a result, we have the

Theorem 20. [2] *A q-term complete system:*

$$X_k f = \sum_{i=1}^{n} \xi_{ki}(x_1,\dots,x_n)\frac{\partial f}{\partial x_i} = 0 \qquad (k=1\cdots q)$$

in the variables $x_1,\dots,x_n$ admits all transformations of the one-term group:

$$Yf = \sum_{i=1}^{n} \eta_i(x_1,\dots,x_n)\frac{\partial f}{\partial x_i}$$

if and only if all $X_k\big(Y(f)\big) - Y\big(X_k(f)\big)$ can be expressed linearly in terms of $X_1 f,\dots,X_q f$:

$$[X_k, Y] = X_k\big(Y(f)\big) - Y\big(X_k(f)\big) = \sum_{j=1}^{q} \chi_{kj}(x_1,\dots,x_n)X_j f \tag{3}$$

$$(k=1\cdots q).$$

The existence of relations of the form (3) is therefore necessary and sufficient in order that the system $X_1 f = 0, \dots, X_q f = 0$ admits all transformations of the one-term group Yf. For various fundamental reasons, it is also desirable to prove the *necessity* of the relation (3) by means of a direct method.

The transformations of the one-term group Yf have the form:

$$x_i' = x_i + \frac{t}{1}\eta_i + \frac{t^2}{1\cdot 2}Y\eta_i + \cdots \qquad (i=1\cdots n).$$

If we now introduce the new variables $x_1',\dots,x_n'$ in the expressions $X_k f$ by means of this formula in place of $x_1,\dots,x_n$, we obtain:

$$X_k f = \sum_{i=1}^{n} X_k x_i' \frac{\partial f}{\partial x_i'}.$$

Here, the $X_k x_i'$ still have to be expressed in terms of $x_1',\dots,x_n'$. By leaving out the second and the higher powers of t, we immediately get:

$$X_k x_i' = X_k x_i + t X_k \eta_i + \cdots. \tag{4}$$

Furthermore, one has (cf. Chap. 4, Eq. (3a), p. 67):

$$X_k x_i = \xi_{ki}(x) = \xi_{ki}' - \frac{t}{1}Y'\xi_{ki}' + \cdots$$
$$t X_k \eta_i = t X_k' \eta_i' - \cdots.$$

Hence we have:

[2] Lie, Gesellschaft d. W. zu Christiania, 1874.

$$X_k x_i' = \xi_{ki}' + t\big(X_k'\eta_i' - Y'\xi_{ki}'\big) + \cdots,$$

and lastly:

$$X_k f = X_k' f + \frac{t}{1}\,\big(X_k'Y'f - Y'X_k'f\big) + \cdots, \tag{5}$$

where the omitted terms are of second or of higher order in t.

Now, if the complete system $X_1 f = 0, \dots, X_q f = 0$ admits all transformations of the one-term group Yf, then the system of q equations:

$$X_k f + \frac{t}{1}\,[X_k, Y] + \cdots = 0 \qquad\qquad (k=1\cdots q)$$

must be equivalent to $X_1 f = 0, \dots, X_q f = 0$ and this, for all values of t. Consequently, the coefficients of t must in any case be expressible by means of $X_1 f, \dots, X_q f$, that is to say, relations of the form (3):

$$X_k\big(Y(f)\big) - Y\big(X_k(f)\big) = [X_k, Y] = \sum_{j=1}^{q} \chi_{kj}(x_1,\dots,x_n)\,X_j f$$

$$(k=1\cdots q)$$

must exist.

As a result, we have shown directly that the existence of these relations is necessary; but, being content with what has already been said, we do not want to prove its sufficiency.

We want to speak of complete systems which admit an infinitesimal transformation Yf in the same way as we spoke of the systems of equations:

$$\Omega_1(x_1,\dots,x_n) = 0, \ \dots, \ \Omega_{n-m}(x_1,\dots,x_n) = 0$$

which do the same. We shall say: *the q-term complete system $X_1 f = 0, \dots, X_q f = 0$ admits the infinitesimal transformation Yf when relations of the form* (3) *hold.*

With the use of this language, we can also express Theorem 20 as follows:

The q-term complete system $X_1 f = 0, \dots, X_q f = 0$ admits all transformations of the one-term group Yf if and only if it admits the infinitesimal transformation Yf.

When a complete system admits all transformations of a one-term group, we say briefly *that it admits this one-term group*.

The conditions of Theorem 20 are in particular satisfied when Yf has the form:

$$Yf = \sum_{j=1}^{q} \rho_j(x_1,\dots,x_n)\,X_j f,$$

where it is understood that the ρ_j are arbitrary functions of the x. Indeed, there always exist relations of the form (3) and in addition, it holds that $Y\varphi \equiv 0$. This is in accordance with the developments of Chap. 6, p. 113, which showed that every solution $\Omega(\varphi_1,\dots,\varphi_{n-q})$ of the complete system $X_1f = 0,\dots,X_qf = 0$ remains invariant by the transformations of all one-term groups of the form $\sum \rho_j X_j f$.

On the other hand, if an infinitesimal transformation Yf which does not have the form $\sum \rho_j(x) X_j f$ satisfies the conditions of Theorem 20, then the finite transformations of the one-term group Yf do not leave invariant each individual solution of the complete system, but instead, they leave invariant the totality of all these solutions.

Amongst the infinitesimal transformations that are admitted by a given complete system $X_1f = 0$, ..., $X_qf = 0$, there are those of the form $\sum \rho_j(x) X_j f$ which are given simultaneously with the complete system, and which, for this reason, have to be considered as trivial. By contrast, one cannot in general indicate the remaining infinitesimal transformations that the system admits before one has integrated the system.

As we have seen above, when the complete system $X_1f = 0$, ..., $X_qf = 0$ admits the infinitesimal transformation Yf, every solution $\Omega(\varphi_1,\dots,\varphi_{n-q})$ of the complete system is transferred to a solution by every transformation:

$$x_i' = x_i + \frac{t}{1}\, Y x_i + \cdots \qquad (i=1\cdots n)$$

of the one-term group Yf. Hence if we interpret the x and the x' as coordinates for the points of an n-times extended space and the transformation just written as an operation by which the point $x_1,\dots,x_n$ takes the new position $x_1',\dots,x_n'$, and if we recall in addition that, with the constants $a_1,\dots,a_{n-q}$, the equations $\varphi_1 = a_1$, ..., $\varphi_{n-q} = a_{n-q}$ represent a characteristic manifold of the complete system $X_1f = 0,\dots,$ $X_qf = 0$ (cf. Chap. 6, p. 116), then we realize immediately that the transformations of the one-term group Yf send every characteristic manifold of the complete system to a characteristic manifold. *The characteristic manifolds of our complete system are therefore permuted with each other by the transformations of the one-term group Yf, hence they form, as we shall express it, a family* [SCHAAR] *which is invariant by the one-term group Yf.* Since the mentioned characteristic manifolds determine, according to Chap. 6, p. 116, a decomposition of the space, we can also say that *this decomposition remains invariant by the one-term group Yf.*

§ 38. We present a few simple statements concerning complete systems which admit one-term groups:

Proposition 2. *If a q-term complete system admits every transformation of the two one-term groups Yf and Zf, then it also admits every transformation of the one-term group $Y(Z(f)) - Z(Y(f)) = [Y,Z]$.*

Let $X_kf = 0$ be the equations of the complete system. One then forms the Jacobi identity:

$$\big[[Y,Z],X_k\big] + \big[[Z,X_k],Y\big] + \big[[X_k,Y],Z\big] = 0,$$

and one takes into consideration that $[Y, X_k]$ and $[Z, X_k]$ can, according to the assumption, be expressed linearly in terms of $X_1 f, \dots, X_q f$, hence one realizes that the same property is enjoyed by $\big[[Y, Z], X_k\big]$. As a result, our proposition is proved.

Proposition 3. *If the equations $A_1 f = 0, \dots, A_q f = 0$ and likewise the equations $B_1 f = 0, \dots, B_s f = 0$ form a complete system, then a complete system is also formed by the totality of all possible equations $Cf = 0$ which the two complete systems share, in the sense that relations of the form:*

$$Cf = \sum_{j=1}^{q} \alpha_j(x_1, \dots, x_n) A_j f = \sum_{k=1}^{s} \beta_k(x_1, \dots, x_n) B_k f$$

hold. If the two complete systems $A_k f = 0$ and $B_k f = 0$ admit a certain one-term group Xf, then the complete system of equations $Cf = 0$ also admits this group.

Proof. Amongst the equations $Cf = 0$, one can select exactly, say m but not more, equations:

$$C_1 f = 0, \dots, C_m f = 0$$

which are independent of each other. Next, there exist for every $\mu = 1, \dots, m$ relations of the form:

$$C_\mu f = \sum_{j=1}^{q} \alpha_{\mu j}(x) A_j f = \sum_{k=1}^{s} \beta_{\mu k}(x) B_k f;$$

consequently, every $C_\mu\big(C_\nu(f)\big) - C_\nu\big(C_\mu(f)\big) = [C_\mu, C_\nu]$ can be expressed both in terms of the Af and in terms of the Bf, that is to say, every $[C_\mu, C_\nu]$ can be expressed linearly in terms of $C_1 f, \dots, C_m f$. As a result, the first part of our proposition is proved. Further, every $[X, C_\mu]$ can be expressed both in terms of the Af and in terms of the Bf, hence there exist relations of the form:

$$[X, C_\mu] = \sum_{\nu=1}^{m} \gamma_{\mu\nu}(x_1, \dots, x_n) C_\nu f \qquad (\mu = 1 \cdots m).$$

This is the second part of the proposition.

If two complete systems $A_1 f = 0, \dots, A_q f = 0$ and $B_1 f = 0, \dots, B_s f = 0$ are given, then all possible solutions that are common to the two systems can be defined by means of a complete system which, under the guidance of Chap. 5, p. 101, one can derive from the equations:

$$A_1 f = 0, \dots, A_q f = 0, \quad B_1 f = 0, \dots, B_s f = 0.$$

What is more, the following holds.

Proposition 3. *If the two complete systems $A_1 f = 0, \dots, A_q f = 0$ and $B_1 f = 0, \dots, B_s f = 0$ admit the one-term group Xf, then the complete system which defines the common solution of all the equations $A_k f = 0$ and $B_k f = 0$ also admits the one-term group Xf.*

Proof. The identity:

$$\big[[A_j, B_k], X\big] + \big[[B_k, X], A_j\big] + \big[[X, A_j], B_k\big] = 0$$

shows that all $\big[[A_j, B_k], X\big]$ can be expressed linearly in terms of the Af, of the Bf and of the $[A, B]$. Hence, when the equations $[A_j, B_k] = 0$, together with the $Af = 0$ and the $Bf = 0$, already produce a complete system, the assertion of our proposition is proved. Otherwise, one treats the system of the equations $Af = 0$, $Bf = 0$, $[A, B] = 0$ exactly in the same way as the system $Af = 0$, $Bf = 0$ just considered, that is to say: one forms the Jacobi identity with Xf for any two expressions amongst the Af, Bf, $[A, B]$, and so on.

Propositions 3 and 4 can be given a simple conceptual sense when $x_1, \dots, x_n$ are interpreted as coordinates for the points of a space R_n.

First, we recall that the ∞^{n-q} q-times extended characteristic manifolds M_q of the complete system $A_1 f = 0, \dots, A_q f = 0$ form a family that remains invariant by the one-term group Xf. Next, we remark that the ∞^{n-s} s-times extended characteristic manifolds M_s of the complete system $B_1 f = 0, \dots, B_s f = 0$ also form such an *invariant family*.

Let us agree that the number s is at least equal to q. Then every M_s in general position will decompose into a family of ∞^{s-q+h} $(q-h)$-extended manifolds, those in which it is cut by the M_q, where one understands that h is a determined number amongst $0, 1, \dots, q$. Therefore in this way, the whole R_n is decomposed into a family of ∞^{n-q+h} $(q-h)$-extended manifolds. Naturally, the totality of these manifolds remain invariant by the one-term group Xf, for it is the cutting of the totality of all M_q with the totality of all M_s, and these two totalities are, as already said, invariant by the group Xf.

The considered $(q-h)$-times extended manifolds are nothing but the characteristic manifolds of the complete system $Cf = 0$ which appears in Proposition 3, and under the assumptions made, this complete system is $(q-h)$-term.

On the other hand, one can ask for the smallest manifolds which consist both of M_q and of M_s. When there are manifolds of this kind, the totality of them naturally remains invariant by the one-term group Xf; they are the characteristic manifolds of the complete system which is defined in Proposition 4.

Chapter 9
Characteristic Relationships Between the Infinitesimal Transformations of a Group

In Chap. 3 and in Chap. 4 (Proposition 1, p. 41) it has been shown that to every r-term group, there belong r independent infinitesimal transformations which stand in a characteristic relationship to the group in question. Now, we want at first to derive certain important relations which exist between these infinitesimal transformations. Afterwards, we shall prove the equally important proposition that r independent infinitesimal transformations which satisfy the concerned relations always determine an r-term group with the identity transformation.

§ 39. Instead of considering an r-term group, we presently want to take the somewhat more general point of view of considering a family of ∞^r different transformations:

$$x_i' = f_i(x_1, \dots, x_n, a_1, \dots, a_r) \qquad (i=1\cdots n)$$

which satisfy differential equations of the form:

$$\frac{\partial x_i'}{\partial a_k} = \sum_{j=1}^{r} \psi_{kj}(a_1, \dots, a_r)\, \xi_{ji}(x_1', \dots, x_n')$$

$$(i=1\cdots n;\; k=1\cdots r).$$

We then know (cf. Chap. 3, p. 57) that the r infinitesimal transformations:

$$X_k'(f) = \sum_{i=1}^{n} \xi_{ki}(x')\, \frac{\partial f}{\partial x_i'} \qquad (k=1\cdots r)$$

are independent of each other, and that the determinant of the $\psi_{kj}(a)$ do not vanish identically; consequently, as we did earlier, we can also write down the above differential equations as:

S. Lie, *Theory of Transformation Groups I*,
DOI 10.1007/978-3-662-46211-9_9

$$\xi_{ji}(x'_1,\dots,x'_n) = \sum_{k=1}^{r} \alpha_{jk}(a_1,\dots,a_r)\,\frac{\partial x'_i}{\partial a_k} \tag{1}$$

$$(i=1\cdots n;\ j=1\cdots r).$$

Here naturally, the determinant of the $\alpha_{jk}(a)$ does not vanish identically.

If on the other hand, we imagine that the equations:

$$x'_i = f_i(x_1,\dots,x_n,\,a_1,\dots,a_r) \qquad (i=1\cdots n)$$

are solved with respect to $x_1,\dots,x_n$:

$$x_i = F_i(x'_1,\dots,x'_n,\,a_1,\dots,a_r) \qquad (i=1\cdots n),$$

then we can very easily derive certain differential equations that are satisfied by $F_1,\dots,F_n$. We simply differentiate the identity:

$$F_i\big(f_1(x,a),\dots,f_n(x,a),\,a_1,\dots,a_r\big) \equiv x_i$$

with respect to a_k; then we have:

$$\sum_{\nu=1}^{n} \frac{\partial F_i(x',a)}{\partial x'_\nu}\,\frac{\partial f_\nu(x,a)}{\partial a_k} + \frac{\partial F_i(x',a)}{\partial a_k} \equiv 0,$$

provided that one sets $x'_\nu = f_\nu(x,a)$ everywhere. We multiply this identity by $\alpha_{jk}(a)$ and we sum it for k from 1 to r; then on account of:

$$\sum_{k=1}^{r} \alpha_{jk}(a)\,\frac{\partial f_\nu(x,a)}{\partial a_k} \equiv \xi_{j\nu}(f_1,\dots,f_n),$$

we obtain the following equations:

$$\sum_{\nu=1}^{n} \xi_{j\nu}(x'_1,\dots,x'_n)\,\frac{\partial F_i}{\partial x'_\nu} + \sum_{k=1}^{r} \alpha_{jk}(a_1,\dots,a_r)\,\frac{\partial F_i}{\partial a_k} = 0$$

$$(i=1\cdots n;\ j=1\cdots r).$$

According to their derivation, these equations hold identically when one makes the substitution $x'_\nu = f_\nu(x,a)$ in them; but since they do not contain $x_1,\dots,x_n$, they must actually hold identically, that is to say: $F_1,\dots,F_n$ are all solutions of the following linear partial differential equations:

$$\Omega_j(F) = \sum_{\nu=1}^{n} \xi_{j\nu}(x')\,\frac{\partial F}{\partial x'_\nu} + \sum_{\mu=1}^{r} \alpha_{j\mu}(a)\,\frac{\partial F}{\partial a_\mu} = 0 \tag{2}$$

$$(j=1\cdots r).$$

These r equations contain $n+r$ variables, namely $x'_1,\dots,x'_n$ and $a_1,\dots,a_r$; in addition, they are independent of each other, for the determinant of the $\alpha_{j\mu}(a)$ does not vanish identically, and hence a resolution with respect to the r differential quotients $\frac{\partial F}{\partial a_1}, \dots, \frac{\partial F}{\partial a_r}$ is possible. But on the other hand, the equations (2) have n independent solutions in common, namely just the functions $F_1(x', a), \dots, F_n(x', a)$ whose functional determinant with respect to the x':

$$\sum \pm \frac{\partial F_1}{\partial x'_1} \cdots \frac{\partial F_n}{\partial x'_n} = \frac{1}{\sum \pm \frac{\partial f_1}{\partial x_1} \cdots \frac{\partial f_n}{\partial x_n}}$$

does not vanish identically, because the equations $x'_i = f_i(x, a)$ represent transformations by assumption. Therefore, the hypotheses of Proposition 8 in Chap. 5, p. 103, are met by the equations (2), that is to say, these equations constitute an r-term complete system.

If we set:

$$\sum_{k=1}^{r} \alpha_{jk}(a) \frac{\partial F}{\partial a_k} = A_j(F)$$

and furthermore:

$$\sum_{\nu=1}^{n} \xi_{j\nu}(x') \frac{\partial F}{\partial x'_\nu} = X'_j(F),$$

in accordance with a designation employed earlier, then the equations (2) receive the form:

$$\Omega_j(F) = X'_j(F) + A_j(F) = 0 \qquad (j=1\cdots r).$$

As we know, the fact that they constitute a complete system is found in their expressions: certain equations of the form:

$$\Omega_k\big(\Omega_j(F)\big) - \Omega_j\big(\Omega_k(F)\big) = \sum_{s=1}^{r} \vartheta_{kjs}(x'_1, \dots, x'_n, a_1, \dots, a_r)\, \Omega_s(F)$$
$$(k, j=1\cdots r)$$

must hold identically,[1] (Translator's note: These are identities between vector fields.) whatever F is as a function of $x'_1, \dots, x'_n, a_1, \dots, a_r$. But since these identities can also be written as:

$$X'_k\big(X'_j(F)\big) - X'_j\big(X'_k(F)\big) + A_k\big(A_j(F)\big) - A_j\big(A_k(F)\big) =$$
$$= \sum_{s=1}^{r} \vartheta_{kjs} X'_s(F) + \sum_{s=1}^{r} \vartheta_{kjs} A_s(F),$$

we can immediately split them into two identities:

$$\text{(3)} \qquad \begin{cases} X'_k\big(X'_j(F)\big) - X'_j\big(X'_k(F)\big) = \sum_{s=1}^{r} \vartheta_{kjs} X'_s(F) \\ A_k\big(A_j(F)\big) - A_j\big(A_k(F)\big) = \sum_{s=1}^{r} \vartheta_{kjs} A_s(F), \end{cases}$$

and here, the second series of equations can yet again be decomposed as:

$$A_k(\alpha_{j\mu}) - A_j(\alpha_{k\mu}) = \sum_{s=1}^{r} \vartheta_{kjs}\, \alpha_{s\mu} \qquad (k, j, \mu = 1 \cdots r).$$

Now, because the determinant of the $\alpha_{s\mu}$ does not vanish identically, then the ϑ_{kjs} in these last conditions are completely determined, and it follows that the ϑ_{kjs} can only depend upon $a_1, \dots, a_r$, whereas they are in any case free of $x'_1, \dots, x'_n$. However, it can be established that the ϑ_{kjs} are also free of $a_1, \dots, a_r$. For, if in the first series of the identities (3), we consider F as an arbitrary function of only $x'_1, \dots, x'_n$, then we obtain by differentiating with respect to a_μ the following identically satisfied equations:

$$0 \equiv \sum_{s=1}^{r} \frac{\partial \vartheta_{kjs}}{\partial a_\mu} X'_s(F) \qquad (k, j, \mu = 1 \cdots r).$$

But since $X'_1(F), \dots, X'_r(F)$ are independent infinitesimal transformations, and since in addition the $\frac{\partial \vartheta_{kjs}}{\partial a_\mu}$ do not depend upon $x'_1, \dots, x'_n$, then all the $\frac{\partial \vartheta_{kjs}}{\partial a_\mu}$ vanish identically; that is to say, the ϑ_{kjs} are also free of $a_1, \dots, a_r$, they are numerical constants.

Thus, we have the

Theorem 21. *If a family of ∞^r transformations:*

$$x'_i = f_i(x_1, \dots, x_n, a_1, \dots, a_r) \qquad (i = 1 \cdots n)$$

satisfies certain differential equations of the specific form:

$$\frac{\partial x'_i}{\partial a_k} = \sum_{j=1}^{r} \psi_{kj}(a_1, \dots, a_r)\, \xi_{ji}(x'_1, \dots, x'_n) \qquad (i = 1 \cdots n;\ k = 1 \cdots r)$$

and if one writes these differential equations in the form:

$$\xi_{ji}(x'_1, \dots, x'_n) = \sum_{k=1}^{r} \alpha_{jk}(a_1, \dots, a_r) \frac{\partial x'_i}{\partial a_k} \qquad (i = 1 \cdots n;\ j = 1 \cdots r),$$

which is always possible, then there exist between the $2r$ independent infinitesimal transformations:

$$X'_j(F) = \sum_{i=1}^{n} \xi_{ji}(x'_1, \dots, x'_n) \frac{\partial F}{\partial x_i} \qquad (j = 1 \cdots r)$$

$$A_j(F) = \sum_{\mu=1}^{r} \alpha_{j\mu}(a_1, \dots, a_r) \frac{\partial F}{\partial a_\mu} \qquad (j = 1 \cdots r)$$

relationships of the form:

$$
(4)\quad \begin{cases} X'_k(X'_j(F)) - X'_j(X'_k(F)) = \sum_{s=1}^{r} c_{kjs} X'_s(F) & (k, j = 1 \cdots r), \\ A_k(A_j(F)) - A_j(A_k(F)) = \sum_{s=1}^{r} c_{kjs} A_s(F) & (k, j = 1 \cdots r), \end{cases}
$$

where the c_{kjs} denote numerical constants. In consequence of that, the r equations:

$$X'_j(F) + A_j(F) = 0 \qquad (k = 1 \cdots r),$$

which are solvable with respect to $\frac{\partial F}{\partial a_1}, \dots, \frac{\partial F}{\partial a_r}$, constitute an r-term complete system in the $n+r$ variables $x'_1, \dots, x'_n, a_1, \dots, a_r$; if one solves the n equations $x'_i = f_i(x, a)$ with respect to $x_1, \dots, x_n$:

$$x_i = F_i(x'_1, \dots, x'_n, a_1, \dots, a_r) \qquad (i = 1 \cdots n),$$

then $F_1(x', a), \dots, F_n(x', a)$ are independent solutions of this complete system.

This theorem can now be immediately applied to all r-term groups, whether or not they contain the identity transformation.

When applied to the case of an r-term group with the identity transformation, the theorem gives us certain relationships which exist between the infinitesimal transformations of this group. We therefore obtain the important

Theorem 22. *If an r-term group in the variables $x_1, \dots, x_n$ contains the r independent infinitesimal transformations:*

$$X_k(f) = \sum_{i=1}^{n} \xi_{ki}(x_1, \dots, x_n) \frac{\partial f}{\partial x_i} \qquad (k = 1 \cdots r),$$

then there exist between these infinitesimal transformations pairwise relationships of the form:

$$X_k(X_j(f)) - X_j(X_k(f)) = \sum_{s=1}^{r} c_{kjs} X_s(f),$$

where the c_{kjs} denote numerical constants.[1]

From this follows in particular the important

Proposition 1. *If a finite continuous group contains the two infinitesimal transformations:*

$$X(f) = \sum_{i=1}^{n} \xi_i(x_1, \dots, x_n) \frac{\partial f}{\partial x_i}, \qquad Y(f) = \sum_{i=1}^{n} \eta_i(x_1, \dots, x_n) \frac{\partial f}{\partial x_i},$$

then it also contains the infinitesimal transformation:

---

[1] Lie, Math. Ann. Vol. 8, p. 303; Göttinger Nachr. 1874.

$$X(Y(f)) - Y(X(f)).$$

§ 40. Conversely, we imagine that r independent infinitesimal transformations in $x'_1, \dots, x'_n$:

$$X'_j(F) = \sum_{i=1}^{n} \xi_{ji}(x'_1, \dots, x'_n) \frac{\partial F}{\partial x'_i} \qquad (j = 1 \cdots r)$$

are presented, which stand pairwise in relationships of the form:

$$X'_k(X'_j(F)) - X'_j(X'_k(F)) = \sum_{s=1}^{r} c_{kjs} X'_s(F),$$

where the c_{kjs} are numerical constants. In addition, we imagine that r infinitesimal transformations in $a_1, \dots, a_r$:

$$A_j(F) = \sum_{\mu=1}^{r} \alpha_{j\mu}(a_1, \dots, a_r) \frac{\partial F}{\partial a_\mu} \qquad (j = 1 \cdots r)$$

are given, which satisfy analogous relations in pairs of the form:

$$A_k(A_j(F)) - A_j(A_k(F)) = \sum_{s=1}^{r} c_{kjs} A_s(F),$$

with the same c_{kjs} and whose determinant $\sum \pm \alpha_{11}(a) \cdots \alpha_{rr}(a)$ does not vanish identically. We will show that under these assumptions, the infinitesimal transformations $X'_1(F), \dots, X'_r(F)$ generate a completely determined r-term group with the identity transformation.

To this end, we form the equations:

$$\Omega_j(F) = X'_j(F) + A_j(F) = 0 \qquad (j = 1 \cdots r),$$

which, according to the assumptions made, constitute an r-term complete system; indeed, there exist relations of the form:

$$\Omega_k(\Omega_j(F)) - \Omega_j(\Omega_k(F)) = \sum_{s=1}^{r} c_{kjs} \Omega_s(F)$$

and in addition, the equations $\Omega_1(F) = 0, \dots, \Omega_r(F) = 0$ are solvable with respect to $\frac{\partial F}{\partial a_1}, \dots, \frac{\partial F}{\partial a_r}$.

Now, let $a_1^0, \dots, a_r^0$ be a system of values of the a, in a neighborhood of which the $\alpha_{jk}(a)$ behave regularly and for which the determinant $\sum \pm \alpha_{11}(a^0) \cdots \alpha_{rr}(a^0)$ is different from zero. Then according to Theorem 12, Chap. 5, p. 105, the complete system $\Omega_j(F) = 0$ possesses n solutions $F_1(x', a), \dots, F_n(x', a)$ which reduce to $x'_1, \dots, x'_n$ respectively for $a_k = a_k^0$; they are the so-called general solutions of the complete system relative to $a_k = a_k^0$. We imagine that these general solutions are given, we form the n equations:

$$x_i = F_i(x_1',\dots,x_n',a_1,\dots,a_r) \qquad {\scriptstyle (i=1\cdots n)},$$

and we resolve them with respect to $x_1',\dots,x_n'$, which is always possible, for $F_1,\dots,F_n$ are obviously independent of each other, as far as $x_1',\dots,x_n'$ are concerned. The equations obtained in this way:

$$x_i' = f_i(x_1,\dots,x_n,a_1,\dots,a_r) \qquad {\scriptstyle (i=1\cdots n)}$$

represent, as we shall now show, an r-term group and in fact naturally, a group with the identity transformation; because for $a_k = a_k^0$, one gets $x_i' = x_i$.

At first, we have identically:

$$\sum_{\nu=1}^{n} \xi_{j\nu}(x')\,\frac{\partial F_i}{\partial x_\nu'} + \sum_{\mu=1}^{r} \alpha_{j\mu}(a)\,\frac{\partial F_i}{\partial a_\mu} = 0 \tag{5}$$

$$(i=1\cdots n;\ j=1\cdots r).$$

On the other hand, by differentiating $x_i = F_i(x',a)$ with respect to a_μ, one obtains the equation:

$$0 = \sum_{\nu=1}^{n} \frac{\partial F_i}{\partial x_\nu'}\,\frac{\partial x_\nu'}{\partial a_\mu} + \frac{\partial F_i}{\partial a_\mu} \qquad {\scriptstyle (i=1\cdots n;\ \mu=1\cdots r)},$$

which holds identically after the substitution $x_\nu' = f_\nu(x,a)$. We multiply this equation by $\alpha_{j\mu}(a)$ and we sum for μ from 1 to r, hence we obtain an equation which, after using (5), becomes:

$$\sum_{\nu=1}^{n} \frac{\partial F_i}{\partial x_\nu'}\left(\sum_{\mu=1}^{r} \alpha_{j\mu}(a)\,\frac{\partial x_\nu'}{\partial a_\mu} - \xi_{j\nu}(x')\right) = 0$$

$$(i=1\cdots n;\ \mu=1\cdots r).$$

But since the determinant $\sum \pm \frac{\partial F_1}{\partial x_1'} \cdots \frac{\partial F_n}{\partial x_n'}$ does not vanish identically, we therefore obtain:

$$\sum_{\mu=1}^{r} \alpha_{j\mu}(a)\,\frac{\partial x_\nu'}{\partial a_\mu} = \xi_{j\nu}(x'),$$

a system that we can again resolve with respect to the $\frac{\partial x_\nu'}{\partial a_\mu}$, for the determinant of the $\alpha_{j\mu}(a)$ does not vanish. Thus, we obtain finally that equations of the form:

$$\frac{\partial x_\nu'}{\partial a_\mu} = \sum_{j=1}^{r} \psi_{\mu j}(a_1,\dots,a_r)\,\xi_{j\nu}(x_1',\dots,x_n') \tag{6}$$

$$(\nu=1\cdots n;\ \mu=1\cdots r)$$

hold true, which naturally reduce to identities after the substitution $x_\nu' = f_\nu(x,a)$.

At this point, the demonstration that the equations $x_i' = f_i(x, a)$ represent an r-term group is not at all difficult.

Indeed, it is at first easy to see that the equations $x_i' = f_i(x, a)$ represent ∞^r distinct transformations, hence that the parameters $a_1, \dots, a_r$ are all essential. Indeed, otherwise all functions $f_1(x, a)$, $\dots$, $f_n(x, a)$ would satisfy a linear partial differential equation of the form (cf. p. 15):

$$\sum_{k=1}^{r} \chi_k(a_1, \dots, a_r) \frac{\partial f}{\partial a_k} = 0,$$

where the χ_k are free of $x_1, \dots, x_n$. On account of (6), we would then have:

$$\sum_{k,\,j}^{1\dots r} \chi_k(a)\, \psi_{kj}(a)\, \xi_{j\nu}(f_1, \dots, f_n) \equiv 0 \qquad (\nu = 1 \cdots n),$$

whence, since $X_1'(F), \dots, X_r'(F)$ are independent infinitesimal transformations:

$$\sum_{k=1}^{r} \chi_k(a)\, \psi_{kj}(a) = 0 \qquad (j = 1 \cdots r);$$

but from this, it follows immediately that: $\chi_1(a) = 0$, $\dots$, $\chi_r(a) = 0$, because the determinant of the $\psi_{kj}(a)$ does not vanish identically.

Thus, the equations $x_i' = f_i(x, a)$ effectively represent a family of ∞^r different transformations. But now this family satisfies certain differential equations of the specific form (6); hence we can immediately apply Theorem 9 of Chap. 4 on p. 82. According to it, the following holds true: if $\overline{a}_1, \dots, \overline{a}_r$ is a system of values of the a for which the determinant $\sum \pm \psi_{11}(\overline{a}) \cdots \psi_{rr}(\overline{a})$ does not vanish, and the $\psi_{kj}(a)$ behave regularly, then every transformation $x_i' = f_i(x, a)$ whose parameters $a_1, \dots, a_r$ lie in a certain neighborhood of $\overline{a}_1, \dots, \overline{a}_r$ can be obtained by firstly executing the transformation:

$$\overline{x}_i = f_i(x_1, \dots, x_n, \overline{a}_1, \dots, \overline{a}_r) \qquad (i = 1 \cdots n)$$

and then a transformation:

$$x_i' = \overline{x}_i + \sum_{k=1}^{r} \lambda_k\, \xi_{ki}(\overline{x}) + \cdots \qquad (i = 1 \cdots n)$$

of a one-term group $\lambda_1 X_1(f) + \cdots + \lambda_r X_r(f)$, where it is understood that $\lambda_1, \dots, \lambda_r$ are appropriate constants. If we set in particular $\overline{a}_k = a_k^0$, we then get $\overline{x}_i = x_i$, hence we see that the family of the ∞^r transformations $x_i' = f_i(x, a)$ coincides, in a certain neighborhood of $a_1^0, \dots, a_r^0$, with the family of the transformations:

$$x_i' = x_i + \sum_{k=1}^{r} \lambda_k\, \xi_{ki}(x) + \cdots \tag{7}$$

$$(i = 1 \cdots n).$$

If, on the other hand, we choose arbitrary $\overline{a}_1,\dots,\overline{a}_r$ in a certain neighborhood of $a_1^0,\dots,a_r^0$, then the transformation $\overline{x}_i = f_i(x,\overline{a})$ always belongs to the family (7). But if we first execute the transformation $\overline{x}_i = f_i(x,\overline{a})$ and then an appropriate transformation:

$$x_i' = \overline{x}_i + \sum_{k=1}^{r} \lambda_k\, \xi_{ki}(\overline{x}) + \cdots$$

of the family (7), then by what has been said above, we obtain a transformation $x_i' = f_i(x,a)$, where $a_1,\dots,a_r$ can take all values in a certain neighborhood of $\overline{a}_1,\dots,\overline{a}_r$. In particular, if we choose $a_1,\dots,a_r$ in the neighborhood of $a_1^0,\dots,a_r^0$ mentioned earlier on, which is always possible, then again the transformation $x_i' = f_i(x,a)$ also belongs to the family (7); Consequently,[2] (Translator's note: The present goal is mainly to establish that if r infinitesimal transformations $X_1,\dots,X_r$ stand pairwise in the relationships $\big[X_k, X_j\big] = \sum_{j=1}^{r} c_{kjs} X_s$, where the c_{kjs} are constants, then the totality of transformations $x' = \exp\big(\lambda_1 X_1 + \cdots + \lambda_r X_r\big)(x)$ constitutes an r-term continuous local Lie group; Theorem 24 below will conclude such a fundamental statement. In particular, the exponential family $x' = \exp\big(\lambda_1 X_1 + \cdots + \lambda_r X_r\big)(x)$ will be shown (after appropriate shrinkings) to be closed under composition, a property that we may abbreviate informally by $\exp \circ \exp \equiv \exp$.

However, Theorem 9 on p. 82 only says that $f \circ \exp \equiv f$, or in greater detail:

$$(*) \qquad \begin{pmatrix} \overline{x} = f(x;\overline{a}) \\ \overline{a} \text{ near } a^0 \end{pmatrix} \circ \begin{pmatrix} x' = \exp(\lambda X)(\overline{x}) \\ \lambda \text{ near } 0 \end{pmatrix} \equiv \begin{pmatrix} x' = f(x;a) \\ a \text{ near } \overline{a} \end{pmatrix}.$$

But if we now apply this statement where $\overline{a} = a^0$ is the system of values introduced while solving the complete system $\Omega_1(F) = \cdots = \Omega_r(f) = 0$, then by construction, a^0 yields the identity transformation $\overline{x} = f(x;a^0) = x$ and we hence get in particular:

$$\begin{pmatrix} x' = \exp(\lambda X)(\overline{x}) \\ \lambda \text{ near } 0 \end{pmatrix} \equiv \begin{pmatrix} x' = f(x;a) \\ a \text{ near } a^0 \end{pmatrix}.$$

We can therefore replace the two occurrences of the family $f(x;a)$ in $(*)$ by exponentials to see that the exponential family is indeed closed under composition, as was claimed in the text.) we see that two transformations of the family (7), when executed one after the other, once again yield a transformation of this family. As a result, this family, and naturally also the family $x_i' = f_i(x,a)$ which identifies with it, forms an r-term group, a group which contains the identity transformation and whose transformations can be ordered as inverses in pairs. We can state the obtained result as follows:

Theorem 23. *If r independent infinitesimal transformations:*

$$X_k'(f) = \sum_{i=1}^{n} \xi_{ki}(x_1',\dots,x_n')\, \frac{\partial f}{\partial x_i'} \qquad {\scriptstyle (k=1\cdots r)}$$

in the variables $x_1',\dots,x_n'$ satisfy conditions in pairs of the form:

$$X_k'(X_j'(f)) - X_j'(X_k'(f)) = \sum_{s=1}^{r} c_{kjs} X_s'(f),$$

if furthermore r independent infinitesimal transformations:

$$A_k(f) = \sum_{\mu=1}^{r} \alpha_{k\mu}(a_1, \dots, a_r) \frac{\partial f}{\partial a_\mu} \qquad (k=1\cdots r)$$

in the variables $a_1, \dots, a_r$ satisfy the analogous conditions:

$$A_k(A_j(f)) - A_j(A_k(f)) = \sum_{s=1}^{r} c_{kjs} A_s(f)$$

with the same c_{kjs}, and if, in addition, the determinant $\sum \pm \alpha_{11}(a) \cdots \alpha_{rr}(a)$ does not vanish identically, then one obtains in the following way the equations of an r-term group: one forms the r-term complete system:

$$X_k'(f) + A_k(f) = 0 \qquad (k=1\cdots r)$$

and one determines its general solutions relative to a suitable system of values $a_k = a_k^0$. If $x_i = F_i(x_1', \dots, x_n', a_1, \dots, a_r)$ are these general solutions, then the equations $x_i' = f_i(x_1, \dots, x_n, a_1, \dots, a_r)$ which arise from resolution represent an r-term continuous transformation group. This group contains the identity transformation and for each one of its transformations, also the inverse transformation; it is generated by the ∞^{r-1} infinitesimal transformations:

$$\lambda_1 X_1'(f) + \cdots + \lambda_r X_r'(f),$$

where $\lambda_1, \dots, \lambda_r$ denote arbitrary constants. By introducing new parameters in place of the a_k, the equations of the group can therefore be brought to the form:

$$x_i' = x_i + \sum_{k=1}^{r} \lambda_k \, \xi_{ki}(x) + \sum_{k,\,j}^{1\dots r} \frac{\lambda_k \lambda_j}{12} X_j(\xi_{ki}) + \cdots \qquad (i=1\cdots n).$$

Obviously, the equations $x_i = F_i(x', a)$ appearing in this theorem represent a group, and in fact, just the group $x_i' = f_i(x, a)$.

§ **41.** The hypotheses which are made in the important Theorem 23 can be simplified in an essential way.

The theorem expresses that the $2r$ infinitesimal transformations $X_k(f)$ and $A_k(f)$ determine a certain r-term group in x-space; but at the same time, there is a representation of this group which is absolutely independent of the $A_k(f)$; indeed, according to the cited theorem, the group in question identifies with the family of ∞^{r-1} one-term groups $\lambda_1 X_1(f) + \cdots + \lambda_r X_r(f)$, and this family is already completely determined by the $X_k(f)$ alone. This circumstance brings us to the conjecture that the family of ∞^{r-1} one-term groups $\lambda_1 X_1(f) + \cdots + \lambda_r X_r(f)$ always forms an r-term group if and only if the independent infinitesimal transformations $X_1(f), \dots, X_r(f)$

stand pairwise in relations of the form:

$$X_k\big(X_j(f)\big) - X_j\big(X_k(f)\big) = [X_k, X_j] = \sum_{s=1}^{r} c_{kjs} X_s(f). \tag{8}$$

According to Theorem 22, this condition is necessary for the ∞^{r-1} one-term groups $\sum \lambda_k X_k(f)$ to form an r-term group. Thus our conjecture amounts to the fact that this necessary condition is also sufficient.

This presumption would be brought to certainty if we could produce, for every system of the discussed nature, r independent infinitesimal transformations in $(a_1, \dots, a_r)$:

$$A_k(f) = \sum_{\mu=1}^{r} \alpha_{k\mu}(a_1, \dots, a_r) \frac{\partial f}{\partial a_\mu} \qquad {\scriptstyle (k=1\cdots r)}$$

which satisfy the corresponding relations:

$$A_k\big(A_j(f)\big) - A_j\big(A_k(f)\big) = \sum_{s=1}^{r} c_{kjs} A_s(f),$$

while, however, the determinant $\sum \pm \alpha_{11} \cdots \alpha_{rr}$ does not vanish identically, or expressed differently, while no relation of the form:

$$\sum_{k=1}^{r} \chi_k(a_1, \dots, a_r) A_k(f) = 0$$

holds identically.

With the help of Proposition 5 in Chap. 4, p. 79, we can now in fact always manage to produce such a system of infinitesimal transformations $A_k(f)$. In a similar way, we set:

$$X_k^{(\mu)}(f) = \sum_{i=1}^{n} \xi_{ki}\big(x_1^{(\mu)}, \dots, x_n^{(\mu)}\big) \frac{\partial f}{\partial x_i^{(\mu)}},$$

and we make up the r infinitesimal transformations:

$$W_k(f) = \sum_{\mu=1}^{r} X_k^{(\mu)}(f).$$

According to the stated proposition, these infinitesimal transformations have the property that no relation of the form:

$$\sum_{k=1}^{r} \psi_k\big(x_1', \dots, x_n', x_1'', \dots, x_n'', \cdots\cdots, x_1^{(r)}, \dots, x_n^{(r)}\big)\, W_k(f) = 0$$

holds. Now, since in addition we have:

$$W_k\big(W_j(f)\big) - W_j\big(W_k(f)\big) = \sum_{s=1}^{r} c_{kjs} W_s(f),$$

the r equations, independent of each other:

$$W_1(f)=0,\cdots\cdots,W_r(f)=0$$

form an r-term complete system in the rn variables $x_1',\dots,x_n',\cdots,x_1^{(r)},\dots,x_n^{(r)}$. This complete system possesses $r(n-1)$ independent solutions, which can be called $u_1,u_2,\dots,u_{rn-r}$. Hence, if we select r functions $y_1,\dots,y_r$ of the rn quantities $x_i^{(\mu)}$ that are independent of each other and independent of $u_1,\dots,u_{rn-r}$, we can introduce the y and the u as new independent variables in place of the $x_i^{(\mu)}$. By this, we obtain:

$$W_k(f)=\sum_{\pi=1}^{r}W_k(y_\pi)\frac{\partial f}{\partial y_\pi}+\sum_{\tau=1}^{rn-r}W_k(u_\tau)\frac{\partial f}{\partial u_\tau},$$

or, since all $W_k(u_\tau)$ vanish identically:

$$W_k(f)=\sum_{\pi=1}^{r}\omega_{k\pi}(y_1,\dots,y_r,u_1,\dots,u_{rn-r})\frac{\partial f}{\partial y_\pi},$$

where $W_1(f),\dots,W_r(f)$ are linked by no relation of the form:

$$\sum_{k=1}^{r}\varphi_k(y_1,\dots,y_r,u_1,\dots,u_{rn-r})W_k(f)=0.$$

This property of the $W_k(f)$ naturally also remains true when we confer to the u_τ appropriate fixed values u_τ^0. If we then set $\omega_{k\pi}(y,u^0)=\omega_{k\pi}^0(y)$, the r independent infinitesimal transformations in the independent variables $y_1,\dots,y_r$:

$$V_k(f)=\sum_{\pi=1}^{r}\omega_{k\pi}^0(y_1,\dots,y_r)\frac{\partial f}{\partial y_\pi}$$

stand pairwise in the relationships:

$$V_k(V_j(f))-V_j(V_k(f))=\sum_{s=1}^{r}c_{kjs}V_s(f)$$

and in addition, are linked by no relation of the form:

$$\sum_{k=1}^{r}\varphi_k(y_1,\dots,y_r)V_k(f)=0.$$

Consequently, the $V_k(f)$ are infinitesimal transformations of the required constitution. So we can immediately apply Theorem 23, p. 169, to the $2r$ infinitesimal transformations $X_1(f),\dots,X_r(f)$, $V_1(f),\dots,V_r(f)$ and as a result, we have proved that the ∞^{r-1} one-term groups $\sum\lambda_kX_k(f)$ constitute an r-term group. Therefore the following holds true:

Theorem 24. *If r independent infinitesimal transformations:*

$$X_k(f) = \sum_{i=1}^{n} \xi_{ki}(x_1,\dots,x_n)\frac{\partial f}{\partial x_i} \qquad (k=1\cdots r)$$

stand pairwise in the relationships:

$$(8) \qquad X_k(X_j(f)) - X_j(X_k(f)) = [X_k, X_j] = \sum_{s=1}^{r} c_{kjs} X_s(f),$$

where the c_{kjs} are constants, then the totality of the ∞^{r-1} one-term groups:

$$\lambda_1 X_1(f) + \cdots + \lambda_r X_r(f)$$

forms an r-term continuous group, which contains the identity transformation and whose transformations are ordered as inverses in pairs.[2]

When we have r independent infinitesimal transformations:

$$X_1(f),\dots,X_r(f)$$

which match the hypotheses of the above theorem, we shall henceforth say that $X_1(f),\dots,X_r(f)$ generate an r-term group, and *we shall also virtually speak of the r-term group* $X_1(f),\dots,X_r(f)$.

§ 42. Let $X_1(f),\dots,X_r(f)$ be an r-term group in the variables $x_1,\dots,x_n$ and $Y_1,\dots Y_m(f)$ be an m-term group in the same variables. The relations between the $X_k(f)$ and the $Y_\mu(f)$ respectively may have the form:

$$X_k(X_j(f)) - X_j(X_k(f)) = \sum_{s=1}^{r} c_{kjs} X_s(f) = [X_k, X_j]$$

$$Y_\mu(Y_\nu(f)) - Y_\nu(Y_\mu(f)) = \sum_{s=1}^{m} c'_{\mu\nu s} Y_s(f) = [Y_\mu, Y_\nu]$$

$$(k,j=1\cdots r;\ \mu,\nu=1\cdots m).$$

Now, it can happen that these two groups have certain infinitesimal transformations in common. We suppose that they have exactly l independent such transformations in common, say:

$$Z_\lambda(f) = \sum_{k=1}^{r} g_{\lambda k} X_k(f) = \sum_{\mu=1}^{m} h_{\lambda\mu} Y_\mu(f) \qquad (\lambda=1\cdots l),$$

where the $g_{\lambda k}$ and the $h_{\lambda\mu}$ denote constants. Then every other infinitesimal transformation contained in the two groups can be linearly deduced from $Z_1(f),\dots,Z_l(f)$.

---

[2] Lie, Math. Annalen Vol. 8, p. 303, 1874; Göttinger Nachrichten, 1874, p. 533 and 540; Archiv for Math. og Naturv. Christiania 1878.

However, if we form the expressions:

$$Z_\lambda\big(Z_\nu(f)\big) - Z_\nu\big(Z_\lambda(f)\big) = [Z_\lambda, Z_\nu],$$

we realize that they can be deduced linearly from $X_1(f), \dots, X_r(f)$ and also from $Y_1(f), \dots, Y_m(f)$, hence they are common to the two groups. Consequently, relations of the form:

$$Z_\lambda\big(Z_\nu(f)\big) - Z_\nu\big(Z_\lambda(f)\big) = [Z_\lambda, Z_\nu] = \sum_{s=1}^{l} d_{\lambda\nu s}\, Z_s(f)$$

hold true, that is to say $Z_1(f), \dots, Z_l(f)$ generate an l-term group.

As a result, we have the

Proposition 2. *If the two continuous groups: $X_1(f), \dots, X_r(f)$ and $Y_1(f), \dots, Y_m(f)$ in the same variables have exactly l and not more independent infinitesimal transformations in common, then these transformations generate, as far as they are concerned, an l-term continuous group.*

§ 43. If the equations $x_i' = f_i(x_1, \dots, x_n, a_1, \dots, a_r)$ represent a family of ∞^r transformations and if in addition they satisfy differential equations of the specific form:

$$\frac{\partial x_i'}{\partial a_k} = \sum_{j=1}^{r} \psi_{kj}(a_1, \dots, a_r)\, \xi_{ji}(x_1', \dots, x_n') \qquad (i=1\cdots n;\ k=1\cdots r),$$

then as we know, the r infinitesimal transformations:

$$X_k(f) = \sum_{i=1}^{n} \xi_{ki}(x_1, \dots, x_n)\, \frac{\partial f}{\partial x_i} \qquad (k=1\cdots r)$$

are independent of each other, and in addition, according to Theorem 21 on p. 164, they are linked together through relations of the form:

$$X_k\big(X_j(f)\big) - X_j\big(X_k(f)\big) = [X_k, X_j] = \sum_{s=1}^{r} c_{kjs}\, X_s(f).$$

Hence the family of the ∞^{r-1} one-term groups:

$$\lambda_1 X_1(f) + \cdots + \lambda_r X_r(f)$$

forms an r-term group with the identity transformation. Consequently, we can state Theorem 9 on p. 82 as follows:

Theorem 25. *If a family of ∞^r transformations:*

$$x_i' = f_i(x_1, \dots, x_n, a_1, \dots, a_r) \qquad (i=1\cdots n)$$

satisfies certain differential equations of the form:

$$\frac{\partial x_i'}{\partial a_k} = \sum_{j=1}^{r} \psi_{kj}(a_1,\dots,a_r)\,\xi_{ji}(x_1',\dots,x_n') \qquad (i=1\cdots n;\ k=1\cdots r),$$

and if $a_1^0,\dots,a_r^0$ is a system of values of the a for which the $\psi_{kj}(a)$ behave regularly and for which in addition, the determinant:

$$\sum \pm \psi_{11}(a)\cdots\psi_{rr}(a)$$

is different from zero, then every transformation $x_i' = f_i(x,a)$ whose parameters $a_1,\dots,a_r$ lie in a certain neighborhood of $a_1^0,\dots,a_r^0$ can be thought to be produced by performing firstly the transformation $\overline{x}_i = f_i(x,a^0)$ and secondly, a completely determined transformation:

$$x_i' = \overline{x}_i + \sum_{k=1}^{r} \lambda_k\,\xi_{ki}(\overline{x}) + \cdots \qquad (i=1\cdots n)$$

of the r-term group which, under the assumptions made, is generated by the r independent infinitesimal transformations:

$$X_k(f) = \sum_{i=1}^{n} \xi_{ki}(x_1,\dots,x_n)\,\frac{\partial f}{\partial x_i} \qquad (k=1\cdots r).$$

The above theorem is then of special interest when the equations $x_i' = f_i(x,a)$ represent an r-term group which does not contain the identity transformation, at least in the domain $((a))$. In this case, we will derive a few important conclusions.

Thus, let $x_i' = f_i(x_1,\dots,x_n,\,a_1,\dots,a_r)$ be an r-term group without identity transformation, and assume that the two transformations:

$$x_i' = f_i(x_1,\dots,x_n,\,a_1,\dots,a_r)$$
$$x_i'' = f_i(x_1',\dots,x_n',\,b_1,\dots,b_r)$$

executed one after the other produce the transformation:

$$x_i'' = f_i(x_1,\dots,x_n,\,c_1,\dots,c_r) = f_i\big(x_1,\dots,x_n,\,\varphi_1(a,b),\dots,\varphi_r(a,b)\big).$$

Here, if we employ the previous notation, $x_1,\dots,x_n$ lie arbitrarily in the domain $((x))$, $a_1,\dots,a_r$ and $b_1,\dots,b_r$ in the domain $((a))$, while the positions of the x_i', x_i'' and of the c_k are determined by the indicated equations. In addition, there are still differential equations of the specific form:

$$\frac{\partial x_i'}{\partial a_k} = \sum_{j=1}^{r} \psi_{kj}(a_1,\dots,a_r)\,\xi_{ji}(x_1',\dots,x_n')$$

$$(i=1\cdots n;k=1\cdots r).$$

In what follows, $a_1^0, \dots, a_r^0$ and likewise $b_1^0, \dots, b_r^0$ should now denote a determined point [STELLE] of the domain $((a))$ and $\varphi_k(a^0, b^0)$ should be equal to c_k^0. By contrast, by $\overline{a}_1, \dots, \overline{a}_r$ we want to understand an arbitrary point in the domain (a), so that the equations:

$$\overline{x}_i = f_i(x_1, \dots, x_n, \overline{a}_1, \dots, \overline{a}_r)$$

should represent any transformation of the given group.

Every transformation of the form $\overline{x}_i = f_i(x, \overline{a})$ can be obtained by performing at first the transformation $x_i' = f_i(x, a^0)$ and afterwards a certain second transformation. In order to find the latter transformation, we solve the equations $x_i' = f_i(x, a^0)$ with respect to $x_1, \dots, x_n$:

$$x_i = F_i(x_1', \dots, x_n', a_1^0, \dots, a_r^0)$$

and we introduce these values of the x_i in $\overline{x}_i = f_i(x, \overline{a})$. In this way, we obtain for the sought transformation an expression of the form:

$$\overline{x}_i = \Phi_i\big(x_1', \dots, x_n', \overline{a}_1, \dots, \overline{a}_r\big) \qquad (i=1\cdots n); \tag{9}$$

we here do not write down the a_k^0, because we want to consider them as numerical constants.

The transformation (9) is well defined for all systems of values $\overline{a}_k$ in the domain (a) and its expression can be analytically continued to the whole domain of such systems of values; this follows from the hypotheses that we have made previously about the nature of the functions f_i and F_i.

We now claim that for certain values of the parameters $\overline{a}_k$, the transformations of the family $\overline{x}_i = \Phi_i(x', \overline{a})$ belong to the initially given group $x_i' = f_i(x, a)$, whereas by contrast, for certain other values of the $\overline{a}_k$, they belong to the group $X_1 f, \dots, X_r f$ with identity transformation.

We establish as follows the first part of the claim just stated. We know that the two transformations:

$$x_i' = f_i(x_1, \dots, x_n, a_1^0, \dots, a_r^0), \qquad \overline{x}_i = f_i(x_1', \dots, x_n', b_1, \dots, b_r)$$

executed one after the other produce the transformation $\overline{x}_i = f_i(x, c)$, where $c_k = \varphi_k(a^0, b)$; here, we may set for $b_1, \dots, b_r$ any system of values of the domain $((a))$, while the system of values $c_1, \dots, c_r$ then lies in the domain (a), in a certain neighborhood of $c_1^0, \dots, c_r^0$. But according to what has been said earlier, the transformation $\overline{x}_i = f_i(x, c)$ is also obtained when the two transformations:

$$x_i' = f_i(x_1, \dots, x_n, a_1^0, \dots, a_r^0), \qquad \overline{x}_i = \Phi_i(x_1', \dots, x_n', \overline{a}_1, \dots, \overline{a}_r)$$

are executed one after the other and when one chooses $\overline{a}_k = c_k$. Consequently, the transformation $\overline{x}_i = \Phi_i(x', \overline{a})$ is identical, after the substitution $\overline{a}_k = \varphi_k(a^0, b)$, to the transformation $\overline{x}_i = f_i(x', b)$, that is to say: *all transformations $\overline{x}_i = \Phi_i(x', \overline{a})$ whose parameters $\overline{a}_k$ lie in a certain neighborhood of $c_1^0, \dots, c_r^0$, defined through the equation $\overline{a}_k = \varphi_k(a^0, b)$, belong to the presented group $x_i' = f_i(x, a)$.*

In order to establish the second part of our claim stated above, we recall Theorem 25. If $\overline{a}_1,\dots,\overline{a}_r$ lie in a certain neighborhood of $a_1^0,\dots,a_r^0$, then according to this theorem, the transformation $\overline{x}_i = f_i(x,\overline{a})$ can be obtained by first performing the transformation:

$$x_i' = f_i(x_1,\dots,x_n,\, a_1^0,\dots,a_r^0)$$

and then a completely determined transformation:

$$\overline{x}_i = x_i' + \sum_{k=1}^{r} \lambda_k\, \xi_{ki}(x') + \cdots \tag{10}$$

of the r-term group that is generated by the r independent infinitesimal transformations:

$$X_k(f) = \sum_{i=1}^{n} \xi_{ki}(x_1,\dots,x_n)\,\frac{\partial f}{\partial x_i} \qquad (k=1\cdots r).$$

From earlier considerations (cf. Chap. 4, proof of Theorem 9, p. 82), we know in addition that one finds the transformation (10) in question when one chooses in an appropriate way $\overline{a}_1,\dots,\overline{a}_r$ as independent functions of $\lambda_1,\dots,\lambda_r$ and when one then determines by resolution $\lambda_1,\dots,\lambda_r$ as functions of $\overline{a}_1,\dots,\overline{a}_r$. But on the other hand, we also obtain the transformation $\overline{x}_i = f_i(x,\overline{a})$ when we at first execute the transformation $x_i' = f_i(x,a^0)$ and afterwards the transformation $\overline{x}_i = \Phi_i(x',\overline{a})$. Consequently, the transformation $\overline{x}_i = \Phi_i(x',\overline{a})$ belongs to the group generated by $X_1f,\dots,X_rf$ as soon as the system of values $\overline{a}_1,\dots,\overline{a}_r$ lies in a certain neighborhood of $a_1^0,\dots,a_r^0$. Expressed differently: the equations $\overline{x}_i = \Phi_i(x',\overline{a})$ are transferred to the equations (10) when $\overline{a}_1,\dots,\overline{a}_r$ are replaced by the functions of $\lambda_1,\dots,\lambda_r$ discussed above.

With these words, our claim stated above is completely proved.

Thus the transformation equations $\overline{x}_i = \Phi_i(x',\overline{a})$ possess the following important property: if in place of the $\overline{a}_k$, the new parameters $b_1,\dots,b_r$ are introduced by means of the equations $\overline{a}_k = \varphi_k(a^0,b)$, then for a certain domain of the variables, the equations $\overline{x}_i = \Phi_i(x',\overline{a})$ take the form $\overline{x}_i = f_i(x',b)$; on the other hand, if in place of the $\overline{a}_k$, the new parameters $\lambda_1,\dots,\lambda_r$ are introduced, then for a certain domain, the equations $\overline{x}_i = \Phi_i(x',\overline{a})$ convert into:

$$x_i = x_i' + \sum_{k=1}^{r} \lambda_k\, \xi_{ki}(x') + \cdots \qquad (i=1\cdots n).$$

Here lies an important feature of the initially given group $x_i' = f_i(x,a)$. Namely, when we introduce in the equations $x_i' = f_i(x,a)$ the new parameters $\overline{a}_1,\dots,\overline{a}_r$ in place of the a_k by means of $\overline{a}_k = \varphi_k(a^0,a)$, then we obtain a system of transformation equations $x_i' = \Phi_i(x,\overline{a})$ which, by performing in addition its analytic continuation, represents a family of transformations to which belong all transformations of some r-term group with identity transformation.

We can also express this as follows.

Theorem 26. *Every r-term group $x_i' = f_i(x_1,\dots,x_n,a_1,\dots,a_r)$ which is not generated by r independent infinitesimal transformations can be derived from an r-term group with r independent infinitesimal transformations in the following way: one first sets up the differential equations:*

$$\frac{\partial x_i'}{\partial a_k} = \sum_{j=1}^{r} \psi_{kj}(a)\,\xi_{ji}(x') \qquad (i=1\cdots n\,;\,k=1\cdots r),$$

which are satisfied by the equations $x_i' = f_i(x,a)$, then one sets:

$$\sum_{i=1}^{n} \xi_{ki}(x)\,\frac{\partial f}{\partial x_i} = X_k(f) \qquad (k=1\cdots r)$$

and one forms the finite equations:

$$x_i' = x_i + \sum_{k=1}^{r} \lambda_k\,\xi_{ki}(x) + \cdots \qquad (i=1\cdots n)$$

of the r-term group with identity transformation which is generated by the r independent infinitesimal transformations $X_1f,\dots,X_rf$. Then it is possible, in these finite equations, to introduce new parameters $\overline{a}_1,\dots,\overline{a}_r$ in place of $\lambda_1,\dots,\lambda_r$ in such a way that the resulting transformation equations:

$$x_i' = \Phi_i(x_1,\dots,x_n,\overline{a}_1,\dots,\overline{a}_r) \qquad (i=1\cdots n)$$

represent a family of ∞^r transformations which embrace, after analytic continuation, all the ∞^r transformations:

$$x_i' = f_i(x_1,\dots,x_n,a_1,\dots,a_r)$$

of the group.

§ 44. Without difficulty, it can be shown that there exist groups which do not contain the identity transformation and also whose transformations are not ordered as inverses by pairs.

The equation:

$$x' = ax$$

with the arbitrary parameter a represents a one-term group. If one executes two transformations:

$$x' = ax, \qquad x'' = bx'$$

of this group one after the other, then one gets the transformation:

$$x'' = abx,$$

which belongs to the group as well. From this, it results that the family of all transformations of the form $x' = ax$, in which the absolute value of a is smaller than 1, constitutes a group too.[3] (Translator's note: In the sought economy of axioms, what matters is only closure under composition.) Obviously, this family does not contain the identity transformation, nor can its transformations be ordered as inverses by pairs.

Hence, if there existed an analytic expression which represented only those transformations of the form $x' = ax$ in which the absolute value of a is smaller than 1, then with it, we would have a finite continuous group without the identity transformation and without inverse transformations.

Now, an analytic expression of the required constitution can effectively be indicated.

It is known that the function:[4] (Translator's note: A presentation of this passage has been anticipated in Sect. 2.3.)

$$\sum_{\nu=1}^{\infty} \frac{a^{\nu}}{1-a^{2\nu}}$$

can be expanded, in the neighborhood of $a = 0$, as an ordinary power series with respect to a which converges as long as the absolute value $|a|$ of a is smaller than 1. The power series in question has the form:

$$\sum_{\mu=1}^{\infty} k_{\mu}\, a^{\mu} = \omega(a),$$

where the k_{μ} denote whole numbers depending on the index μ. Hence if we interpret the complex values of a as points in a plane, then $\omega(a)$, as an analytic function of a, is defined in the interior of the circle of radius 1 which can be described [BESCHREIBEN] around the point $a = 0$.

Furthermore, it is known that the function $\omega(a)$ is not defined for a whose absolute value equals 1, so that the circle in question around the point $a = 0$ constitutes the natural boundary for $\omega(a)$, across which this function cannot be analytically continued.

We now set $\omega(a) = \lambda$, and moreover, let $|a^0| < 1$ and $\omega(a^0) = \lambda^0$; then we can solve the equation $\omega(a) = \lambda$ with respect to a, that is to say, we can represent a as an ordinary power series in $\lambda - \lambda^0$ in such a way that it gives: $a = a^0$ for $\lambda = \lambda^0$ and that the equation $\omega(a) = \lambda$ is identically satisfied after substitution of this expression for a.

Let $a = \chi(\lambda)$; then $\chi(\lambda)$ is an analytic function which takes only values whose absolute value are smaller than 1; this holds true not only for the found function element which is represented in the neighborhood of $\lambda = \lambda^0$ by an ordinary power series in $\lambda - \lambda^0$, but also for every analytic continuation of this function element.

Hence if we set:

$$x' = \chi(\lambda)\, x,$$

we obtain the desired analytic expression for all transformations $x' = ax$ in which $|a|$ is smaller than 1. If now we have:

$$x' = \chi(\lambda_1)x, \qquad x'' = \chi(\lambda_2)x',$$

then:

$$x'' = \chi(\lambda_1)\chi(\lambda_2)x;$$

but this equation can always be brought to the form:

$$x'' = \chi(\lambda_3)x;$$

indeed, we have $|\chi(\lambda_1)\chi(\lambda_2)| < 1$; so if we set $\chi(\lambda_1)\chi(\lambda_2) = \alpha$, we simply obtain: $\lambda_3 = \omega(\alpha)$.

With this, it is shown that the equation $x' = \chi(\lambda)x$ with the arbitrary parameter λ represents a group. This group is continuous and finite, but lastly, it does not contain the identity transformation, nor can its transformations be ordered as inverses by pairs. Our purpose: the proof that there are groups of this kind, is attained with it. Furthermore, it is easy to see that in a similar manner, one can form arbitrarily many groups having this constitution.

Remarks. In his early work on finite continuous transformation groups, LIE attempted to show that *every* r-term group contains the identity transformation plus infinitesimal transformations, and is generated by the latter (cf. notably the two articles in Archiv for Math. og Naturvid., Bd. 1, Christiania 1876). However, he soon realized that in his proof he had made certain implicit assumptions about the constitution of the occurring functions; as a consequence, he restricted himself expressly to groups whose transformations can be ordered as inverses by pairs and he showed that in any case, the mentioned statement was correct for such groups (Math. Ann. Vol. 16, p. 441 sq.).

Later, in the year 1884, ENGEL succeeded in constructing a finite continuous group which does not contain the identity transformation and whose transformations do not order as inverses by pairs; this is the group described in the preceding section.

Finally, LIE found that the equations of an *arbitrary* finite continuous group with r parameters can in any case be derived, after the introduction of new parameters and analytic continuation, from the equations of an r-term group which contains the identity transformation and r independent infinitesimal transformations, while its finite transformations can be ordered as inverses by pairs (Theorem 26).

§ 45. In Chap. 4, p. 86, we found that every r-term group which contains r independent infinitesimal transformations has the property that its finite transformations can be ordered as inverses by pairs. On the other hand, it was mentioned at the end of the previous section that this statement can be reversed, hence that every r-term group whose transformations order as inverses by pairs, contains the identity transformation and is produced by r infinitesimal transformations. We will indicate how the correctness of this assertion can be seen.

Let the equations:

$$x'_i = f_i(x_1,\dots,x_n,\, a_1,\dots,a_n) \qquad {\scriptstyle (i=1\cdots n)}$$

with r essential parameters $a_1,\dots,a_r$ represent an r-term group with inverse transformations by pairs. By resolution with respect to $x_1,\dots,x_n$, one can obtain:

$$x_i = F_i(x'_1,\dots,x'_n,\, a_1,\dots,a_r) \qquad {\scriptstyle (i=1\cdots n)}.$$

When the system of values $\varepsilon_1,\dots,\varepsilon_r$ lies in a certain neighborhood of $\varepsilon_1 = 0,\dots,$ $\varepsilon_r = 0$, we can execute the two transformations:

$$\begin{aligned} x_i &= F_i(x'_1,\dots,x'_n,\, a_1,\dots,a_r) \\ x''_i &= f_i(x_1,\dots,x_n,\, a_1+\varepsilon_1,\, \dots,\, a_r+\varepsilon_r) \end{aligned}$$

one after the other and we thus obtain a transformation:

$$x''_i = f_i\big(F_1(x',a),\dots,F_n(x',a),\, a_1+\varepsilon_1,\, \dots,\, a_r+\varepsilon_r\big)$$

which likewise belongs to our group and which can be expanded in a power series with respect to $\varepsilon_1,\dots,\varepsilon_r$:

$$\text{(11)} \qquad x''_i = x'_i + \sum_{k=1}^{r} \varepsilon_k \left[\frac{\partial f_i(x,a)}{\partial a_k}\right]_{x=F(x',a)} + \cdots.$$

If we set all ε_k equal to zero here, then we get the identity transformation, which hence appears in our group. If on the other hand we choose all the ε_k infinitely small, we obtain transformations of our group which differ infinitely little from the identity transformation.

For brevity, we set:

$$\text{(12)} \qquad \left[\frac{\partial f_i(x,a)}{\partial a_k}\right]_{x=F(x',a)} = \eta_{ki}(x',a),$$

so that the transformation (11) is of the form:

$$x''_i = x'_i + \sum_{k=1}^{r} \varepsilon_k\, \eta_{ki}(x',a) + \cdots \qquad {\scriptstyle (i=1\cdots n)}.$$

Then, in the variables $x'_1,\dots,x'_n$, we form the r infinitesimal transformations:

$$Y'_k f = \sum_{i=1}^{n} \eta_{ki}(x',a)\, \frac{\partial f}{\partial x'_k} \qquad {\scriptstyle (k=1\cdots r)},$$

which are certainly independent from each other for undetermined values of the a_k. Indeed, in the contrary case, there would be r quantities $\rho_1,\dots,\rho_r$, not all vanishing, which would not depend upon $x'_1,\dots,x'_n$ such that the equation:

$$\sum_{k=1}^{r} \rho_k \, Y_k' f = 0$$

would be identically satisfied; then the n relations:

$$\sum_{k=1}^{r} \rho_k \, \eta_{ki}(x', a) = 0 \qquad (i=1\cdots n)$$

would follow and in turn, these relations would, after the substitution $x_i' = f_i(x, a)$, be transferred to:

$$\sum_{k=1}^{r} \rho_k \, \frac{\partial f_i(x,a)}{\partial a_k} = 0 \qquad (i=1\cdots n);$$

but such relations cannot exist, since according to the assumption, the parameters are essential in the equations $x_i' = f_i(x, a)$ (cf. Chap. 2, p. 16).

From this, we then conclude that $Y_1' f, \dots, Y_r' f$ also remain independent from each other, when one inserts for $a_1, \dots, a_r$ some determined system of values in general position. If $\overline{a}_1, \dots, \overline{a}_r$ is such a system of values, we want to write:

$$\eta_{ki}(x', \overline{a}) = \xi_{ki}(x');$$

then the r infinitesimal transformations:

$$X_k' f = \sum_{i=1}^{n} \xi_{ki}(x') \, \frac{\partial f}{\partial x_i'} \qquad (k=1\cdots r)$$

are also independent of each other. It remains to show that our group is generated by the r infinitesimal transformations $X_k' f$.

We execute two transformations of our group one after the other, namely firstly the transformation:

$$x_i'' = x_i' + \sum_{k=1}^{r} \varepsilon_k \, \eta_{ki}(x', a) + \cdots,$$

and secondly the transformation:

$$\begin{aligned} x_i''' &= x_i'' + \sum_{k=1}^{r} \vartheta_k \, \eta_{ki}(x'', \overline{a}) + \cdots \\ &= x_i'' + \sum_{k=1}^{r} \vartheta_k \, \xi_{ki}(x'') + \cdots. \end{aligned}$$

If, as up to now, we only take into consideration the first-order terms, we then obtain in the indicated way the transformation:

$$x_i''' = x_i' + \sum_{k=1}^{r} \varepsilon_k \, \eta_{ki}(x', a) + \sum_{k=1}^{r} \vartheta_k \, \xi_{ki}(x') + \cdots,$$

which belongs naturally to our group, and this, inside a certain region for all values of the parameters a, ε, ϑ.

If there were, amongst the infinitesimal transformations $Y_1'f, \dots, Y_r'f$, also a certain number independent of $X_1'f, \dots, X_r'f$, then the last written equations would, according to Chap. 4, Proposition 4, p. 80, represent at least ∞^{r+1} different transformations, whereas our group nevertheless contains only ∞^r different transformations. Consequently, each transformation $Y_k'f$ must be linearly expressible in terms of $X_1'f, \dots, X_r'f$, whatever values the a assume. By considerations similar to those of Chap. 2, p. 39, one now realizes that r identities of the form:

$$Y_k'f \equiv \sum_{j=1}^{r} \psi_{kj}(a_1, \dots, a_r)\, X_j'f \qquad (k=1\cdots r)$$

hold true, where the ψ_{kj} behave regularly in a certain neighborhood of $a_k = \overline{a}_k$; in addition, the determinant of the ψ_{kj} does not vanish identically, since otherwise $Y_1'f, \dots, Y_r'f$ would no longer be independent infinitesimal transformations.

At present, it is clear that the $\eta_{ki}(x', a)$ can be expressed as follows in terms of the $\xi_{ji}(x')$:

$$\eta_{ki}(x', a) \equiv \sum_{j=1}^{r} \psi_{kj}(a)\, \xi_{ji}(x').$$

Finally, if we recall the equations (12) which define the functions $\eta_{ki}(x', a)$, we realize that the differential equations:

$$\frac{\partial x_i'}{\partial a_k} = \sum_{j=1}^{r} \psi_{kj}(a)\, \xi_{ji}(x') \qquad (13)$$

$$(i=1\cdots n;\; k=1\cdots r)$$

are identically satisfied after the substitution $x_i' = f_i(x, a)$.

As a result, it has been directly shown that every group with inverse transformations by pairs satisfies certain differential equations of the characteristic form (13); thus, we have reached the starting point for the developments of Chap. 3, Sect. 3.9.

Now, if it were possible to prove that, for the parameter values $a_1^0, \dots, a_r^0$ of the identity transformation, the determinant of the $\psi_{kj}(a)$ has a value distinct from zero, then it would follow from the stated developments that the group $x_i' = f_i(x, a)$ is generated by the r infinitesimal transformations $X_1f, \dots, X_rf$. But now, it is not in the nature of things that one can prove that the determinant $\sum \pm \psi_{11}(a^0) \cdots \psi_{rr}(a^0)$ is distinct from zero. One can avoid this difficulty as follows.[5] (Translator's note: For (local) continuous finite transformation groups containing the identity transformation, this property has already been seen in Sect. 3.4.)

One knows that the equations:

$$x_i' = x_i + \sum_{k=1}^{r} \varepsilon_k \, \xi_{ki}(x) + \cdots \tag{14}$$
$$(i=1 \cdots n)$$

represent transformations of our group $x_i' = f_i(x,a)$ as soon as the ε_k lie in a certain neighborhood of $\varepsilon_1 = 0, \ldots, \varepsilon_r = 0$; furthermore, one can show that one obtains in this way *all* transformations $x_i' = f_i(x,a)$, the parameters of which lie in a certain neighborhood of $a_1^0, \ldots, a_r^0$. Now, if one puts the equations (14) at the foundation, one easily realizes that the x', interpreted as functions of the ε and of the x, satisfy differential equations of the form:

$$\frac{\partial x_i'}{\partial \varepsilon_k} = \sum_{j=1}^{r} \chi_{kj}(\varepsilon_1, \ldots, \varepsilon_r) \, \xi_{ji}(x_1', \ldots, x_n')$$
$$(i=1 \cdots n \,;\, k=1 \cdots r),$$

where now the determinant of the $\chi_{kj}(\varepsilon)$ for $\varepsilon_1 = 0, \ldots, \varepsilon_r = 0$ does not vanish. In this way, one finally comes to the following result:

Every r-term group with transformations inverse by pairs contains the identity transformation, and in addition r independent infinitesimal transformations by which it is generated.

§ **46.** Consider r independent infinitesimal transformations:

$$X_k f = \sum_{\nu=1}^{n} \xi_{k\nu}(x_1, \ldots, x_n) \frac{\partial f}{\partial x_\nu} \qquad (k=1 \cdots r)$$

which satisfy relations in pairs of the form:

$$X_i\big(X_k(f)\big) - X_k\big(X_i(f)\big) = \big[X_i, X_k\big] = \sum_{s=1}^{r} c_{iks} X_s(f) \tag{8}$$
$$(i,k=1 \cdots r),$$

with certain constants c_{iks}, so that according to Theorem 24, p. 172, the totality of all one-term groups of the form:

$$\lambda_1 X_1(f) + \cdots + \lambda_r X_r(f)$$

constitutes an r-term group. We will show that, as far as they are concerned, the constants c_{iks} in the above relations are then tied up together with certain equations.

To begin with, we have $\big[X_i, X_k\big] = -\big[X_k, X_i\big]$, from which it follows immediately that: $c_{iks} = -c_{kis}$. When the number r is greater than 2, we find still other relations. Indeed, in this case, there is the Jacobi identity (Chap. 5, § 26, p. 7) which holds between any three $X_i f$, $X_k f$, $X_j f$ amongst the r infinitesimal transformations

$X_1 f, \dots, X_r f$:

$$[[X_i, X_k], X_j] + [[X_k, X_j], X_i] + [[X_j, X_i], X_k] = 0$$
$$(i, k, j = 1 \cdots n).$$

By making use of the above relation (8), we obtain:

$$\sum_{s=1}^{r} \left\{ c_{iks} [X_s, X_j] + c_{kjs} [X_s, X_i] + c_{jis} [X_s, X_k] \right\} = 0,$$

and then by repeated applications of this relation:

$$\sum_{s,\tau}^{1\cdots r} \left\{ c_{iks} c_{sj\tau} + c_{kjs} c_{si\tau} + c_{jis} c_{sk\tau} \right\} X_\tau f = 0.$$

But since the infinitesimal transformations $X_\tau f$ are independent of each other, this equation decomposes into the following r equations:

$$\sum_{s=1}^{r} \left\{ c_{iks} c_{sj\tau} + c_{kjs} c_{si\tau} + c_{jis} c_{sk\tau} \right\} = 0 \tag{15}$$
$$(\tau = 1 \cdots r).$$

Thus the following holds.

Theorem 27. [3] *If r independent infinitesimal transformations $X_1 f, \dots, X_r f$ are constituted in such a way that they satisfy relations in pairs of the form:*

$$X_i(X_k(f)) - X_k(X_i(f)) = [X_i, X_k] = \sum_{s=1}^{r} c_{iks} X_s(f) \tag{8}$$
$$(i, k = 1 \cdots r),$$

with certain constants c_{iks}, so that the totality of all ∞^{r-1} one-term groups of the form:

$$\lambda_1 X_1(f) + \cdots + \lambda_r X_r(f)$$

forms an r-term group, then between the constants c_{iks}, there exist the following relations:

$$\begin{cases} c_{ik\tau} + c_{ki\tau} = 0 \\ \sum_{s=1}^{r} \left\{ c_{iks} c_{sj\tau} + c_{kjs} c_{si\tau} + c_{jis} c_{sk\tau} \right\} = 0 \\ (i, k, j, \tau = 1 \cdots r). \end{cases} \tag{16}$$

[3] Lie, Archiv for Math. og Naturv. Bd. 1, p. 192, Christiania 1876.

The equations (16) are completely independent of the number of the variables $x_1,\dots,x_n$ in the infinitesimal transformations $X_1(f),\dots,X_r(f)$. Hence from the preceding statement, we can still conclude what follows:

Even if the number n of the independent variables $x_1,\dots,x_n$ can be chosen in a completely arbitrary way, one nevertheless cannot associate to every system of constants c_{iks} $(i,k,s=1\cdots r)$ r independent infinitesimal transformations:

$$X_k(f) = \sum_{\nu=1}^{n} \xi_{k\nu}(x_1,\dots,x_n)\,\frac{\partial f}{\partial x_\nu} \qquad (k=1\cdots r)$$

which pairwise satisfy the relations:

$$\big[X_i, X_k\big] = \sum_{s=1}^{r} c_{iks}\, X_s(f) \qquad (i,k=1\cdots r).$$

Rather, for the existence of such infinitesimal transformations, the existence of the equations (16) is necessary; but it is also sufficient, as we will see later.

Unless the contrary is specially notified, in all the subsequent studies, we shall restrict ourselves to the r-term groups which contain r independent infinitesimal transformations:

$$X_k f = \sum_{i=1}^{n} \xi_{ki}(x_1,\dots,x_n)\,\frac{\partial f}{\partial x_i},$$

and so, which are generated by the same transformations. Here, we shall always consider only systems of values $x_1,\dots,x_n$ for which all the ξ_{ki} behave regularly.

Further, we stress once more that in future, we shall often call an r-term group with the independent infinitesimal transformations $X_1 f,\dots,X_r f$ "the group $X_1 f,\dots,X_r f$"*. Amongst the various forms the finite equations of the r-term group $X_1 f,\dots,X_r f$ can be given, we shall call:*

$$x_i' = x_i + \sum_{k=1}^{r} e_k\, \xi_{ki}(x) + \cdots \qquad (i=1\cdots n)$$

"a canonical form" *of the group.*

Chapter 10
Systems of Partial Differential Equations the General Solution of Which Depends Only Upon a Finite Number of Arbitrary Constants

In the variables $x_1, \dots, x_n$, $z_1, \dots, z_m$ let us imagine that a system of partial differential equations of arbitrary order is given. Aside from $x_1, \dots, x_n$, $z_1, \dots, z_m$, such a system can only contain differential quotients of $z_1, \dots, z_m$ with respect to $x_1, \dots, x_n$, so that we have hence to consider $x_1, \dots, x_n$ as independent of each other, while $z_1, \dots, z_m$ are to be determined as functions of $x_1, \dots, x_n$ in such a way that the system be identically satisfied.

Our system of differential equations shall not be completely arbitrary, but it will possess certain special properties. We want to assume that, in the form in which it is presented, it satisfies the following conditions.

Firstly. If s is the order of the highest differential quotient occurring in the system, then by *resolution* of the equations of the system, all s-th order differential quotients of $z_1, \dots, z_m$ with respect to $x_1, \dots, x_n$ are supposed to be expressible in terms of the differential quotients of the first order up to the $(s-1)$-th, and in terms of $z_1, \dots, z_m$, $x_1, \dots, x_n$. By contrast, it shall not be possible to express in this way all differential quotients of order $(s-1)$ in terms of those of lower order and in terms of $z_1, \dots, z_m$, $x_1, \dots, x_n$.

Secondly. By differentiating the given system once with respect to the individual variables $x_1, \dots, x_n$ and by combining the obtained equations, we obtain only relations between $x_1, \dots, x_n$, $z_1, \dots, z_m$ and the differential quotients of order $(s-1)$ which already follow from the given system.

We make these special assumptions about the form of the given system for reasons of convenience. Naturally, all subsequent considerations can be applied on the whole to every system of partial differential equations which can, through differentiations and elimination, be given the form just described.

According to the known theory of differential equations, it follows without difficulty that every system of differential equations which possesses the properties just discussed is integrable and that the most general functions $z_1, \dots, z_m$ of $x_1, \dots, x_n$ which satisfy the system depend only on a finite number of constants.

We now want to give somewhat different reasons to this proposition, by reducing the discussed problem of integration to the problem of finding systems of equations

S. Lie, *Theory of Transformation Groups I*,
DOI 10.1007/978-3-662-46211-9_10

which admit a certain number of given infinitesimal transformations; on the basis of the developments of Chap. 7, we can indeed solve the latter problem straightaway.

On the other hand, we will also show that the mentioned proposition can be inverted: we will show that every system of equations:

$$z_\mu = Z_\mu(x_1, \dots, x_n, a_1, \dots, a_r) \qquad (\mu = 1 \cdots m)$$

which contains a finite number r of arbitrary parameters a_k represents the most general system of solutions for a certain system of partial differential equations.

For us, this second proposition is the more important; we shall use it again in the next chapter.

Therefore, it appears to be desirable to derive the two propositions in an independent way, without marching into the theory of total differential equations.

§ 47. Let the number of differential quotients of order k of $z_1, \dots, z_m$ with respect to $x_1, \dots, x_n$ be denoted by ε_k. For the differential quotients of order k themselves, we introduce the notation $p_i^{(k)}$, where i has to run through the values $1, 2, \dots, \varepsilon_k$; but we reserve the right to simply write $z_1, \dots, z_m$ instead of $p_1^{(0)}, \dots, p_{\varepsilon_0}^{(0)}$. Lastly, we also set:

$$\frac{\partial p_i^{(k)}}{\partial x_j} = p_{ij}^{(k)},$$

so that $p_{ij}^{(k)}$ denotes one of the ε_{k+1} differential quotients of order $(k+1)$.

According to these definitions, we can write as follows:

$$(1) \qquad \begin{cases} W_1(x, z, p^{(1)}, \dots, p^{(s-1)}) = 0, \quad \dots, \quad W_q(x, z, p^{(1)}, \dots, p^{(s-1)}) = 0 \\ p_{ij}^{(s-1)} = P_{ij}(x, z, p^{(1)}, \dots, p^{(s-1)}) \qquad (i = 1 \cdots \varepsilon_{s-1};\ j = 1 \cdots n) \end{cases}$$

the system of differential equations to be studied.

We assume here that the equations $W_1 = 0, \dots, W_q = 0$ are independent of each other. The $n\varepsilon_{s-1}$ equations $p_{ij}^{(s-1)} = P_{ij}$ are independent of the $W = 0$, but they are not independent of each other, because indeed the $n\varepsilon_{s-1}$ expressions $p_{ij}^{(s-1)}$ do not represent only distinct differential quotients of order s. However, for what follows the above notation is more convenient than if we had written the system of equations (1) in the form:

$$W_1 = 0, \ \dots, \ W_q = 0, \quad p_i^{(s)} = P_i(x, z, p^{(1)}, \dots, p^{(s-1)}) \qquad (i = 1 \cdots \varepsilon_s).$$

Now, by hypothesis, our system of equations (1) has the property that by differentiating it once with respect to the x, no new relation between the x, z, $p^{(1)}, \dots, p^{(s-1)}$ is produced. All relations between the x, z, $p^{(1)}, \dots, p^{(s-1)}$ which come out by differentiating (1) once must therefore be a consequence of $W_1 = 0, \dots, W_q = 0$.

Obviously, we find the relations in question by differentiating (1) with respect to the n variables $x_1,\dots,x_n$ and afterwards, by taking away all differential quotients of order $(s+1)$, and by substituting all differential quotients of order s by means of (1).

By differentiation of $W_k = 0$ and then by elimination of the differential quotients of order s, we obtain the equations:

$$\frac{\partial W_k}{\partial x_\nu} + \sum_{i=1}^{m} p_{i\nu}^{(0)} \frac{\partial W_k}{\partial z_i} + \sum_{i=1}^{\varepsilon_1} p_{i\nu}^{(1)} \frac{\partial W_k}{\partial p_i^{(1)}} + \cdots +$$
$$+ \sum_{i=1}^{\varepsilon_{s-2}} p_{i\nu}^{(s-2)} \frac{\partial W_k}{\partial p_i^{(s-2)}} + \sum_{i=1}^{\varepsilon_{s-1}} P_{i\nu} \frac{\partial W_k}{\partial p_i^{(s-1)}} = 0$$
$$(k = 1 \cdots q;\ \nu = 1 \cdots n).$$

By the above, these equations are a consequence of $W_1 = 0, \dots, W_q = 0$. In other words: *the system of equations $W_1 = 0, \dots, W_q = 0$ in the variables x, z, $p^{(1)}, \dots, p^{(s-1)}$ admits the n infinitesimal transformations:*

$$\text{(2)} \quad \begin{cases} \Omega_\nu f = \dfrac{\partial f}{\partial x_\nu} + \displaystyle\sum_{i=1}^{m} p_{i\nu}^{(0)} \frac{\partial f}{\partial z_i} + \cdots + \\ \qquad + \displaystyle\sum_{i=1}^{\varepsilon_{s-2}} p_{i\nu}^{(s-2)} \frac{\partial f}{\partial p_i^{(s-2)}} + \sum_{i=1}^{\varepsilon_{s-1}} P_{i\nu} \frac{\partial f}{\partial p_i^{(s-1)}} \\ (\nu = 1 \cdots n) \end{cases}$$

in these variables (cf. Chap. 7, p. 124).

If on the other hand, one differentiates the equations $p_{ij}^{(s-1)} = P_{ij}$ with respect to x_ν and then eliminates all differential quotients of order s, then one obtains:

$$\frac{\partial}{\partial x_\nu} p_{ij}^{(s-1)} = \frac{\partial^2}{\partial x_\nu \partial x_j} p_i^{(s-1)} = \Omega_\nu(P_{ij}).$$

One still has to take away all the differential quotients of order $(s+1)$ from these equations. One easily realizes that only the following equations come out:

$$\text{(3)} \quad \Omega_\nu(P_{ij}) - \Omega_j(P_{i\nu}) = 0 \qquad (\nu, j = 1 \cdots n;\ i = 1 \cdots \varepsilon_{s-1}),$$

which likewise must therefore be a consequence of $W_1 = 0, \dots, W_q = 0$.

With this, the properties of the system (1) demanded in the introduction are formulated analytically.

Now, we imagine that an arbitrary system of solutions:

$$z_1 = \varphi_1(x_1,\dots,x_n),\ \dots,\ z_m = \varphi_m(x_1,\dots,x_n)$$

of the differential equations (1) is presented. By differentiating this system we obtain that the $p^{(1)}, \dots, p^{(s-1)}, p^{(s)}$ are represented as functions of $x_1, \dots, x_n$:

$$p_{i_1}^{(1)} = \varphi_{i_1}^{(1)}(x_1, \dots, x_n), \; \dots, \; p_{i_{s-1}}^{(s-1)} = \varphi_{i_{s-1}}^{(s-1)}(x_1, \dots, x_n), \; p_{i_{s-1},\nu}^{(s-1)} = \frac{\partial \varphi_{i_{s-1}}^{(s-1)}}{\partial x_\nu}$$

$$(i_k = 1, 2 \cdots \varepsilon_k; \; \nu = 1 \cdots n),$$

and when we insert these expressions, and the expressions for $z_1, \dots, z_m$, into the equations (1), we naturally receive nothing but identities. From this, it follows that the equations $W_1 = 0, \dots, W_q = 0$ in their turn convert into identities after the substitution:

$$\text{(4)} \quad z_\mu = \varphi_\mu(x_1, \dots, x_n), \; p_{i_1}^{(1)} = \varphi_{i_1}^{(1)}(x_1, \dots, x_n), \; \dots, \; p_{i_{s-1}}^{(s-1)} = \varphi_{i_{s-1}}^{(s-1)}(x_1, \dots, x_n)$$

$$(\mu = 1 \cdots m; \; i_k = 1 \cdots \varepsilon_k);$$

clearly, we can also express this as follows: the system of equations (4) embraces [UMFASST] the equations $W_1 = 0, \dots, W_q = 0$.

Furthermore, we claim that the system of equations (4) admits the infinitesimal transformations $\Omega_1 f, \dots, \Omega_n f$ discussed above.

Indeed, at first, all the expressions:

$$\Omega_\nu(z_\mu - \varphi_\mu) = p_{\mu\nu}^{(0)} - \frac{\partial \varphi_\mu}{\partial x_\nu},$$

$$\Omega_\nu(p_{i_1}^{(1)} - \varphi_{i_1}^{(1)}) = p_{i,\nu}^{(1)} - \frac{\partial \varphi_{i_1}^{(1)}}{\partial x_\nu},$$

$$\cdots\cdots\cdots\cdots\cdots\cdots\cdots\cdots$$

$$\Omega_\nu(p_{i_{s-2}}^{(s-2)} - \varphi_{i_{s-2}}^{(s-2)}) = p_{i_{s-2}\nu}^{(s-2)} - \frac{\partial \varphi_{i_{s-2}}^{(s-2)}}{\partial x_\nu}$$

vanish by means of (4), since the equations (4) come from $z_\mu - \varphi_\mu = 0$ by differentiation with respect to $x_1, \dots, x_n$. But the expressions:

$$\Omega_\nu(p_{i_{s-1}}^{(s-1)} - \varphi_{i_{s-1}}^{(s-1)}) = P_{i_{s-1}\nu} - \frac{\partial \varphi_{i_{s-1}}^{(s-1)}}{\partial x_\nu}$$

also vanish by means of (4), since the equations:

$$p_{i_{s-1}\nu}^{(s-1)} = P_{i_{s-1}\nu}$$

are, as already said above, identically satisfied after the substitution:

$$z_\mu = \varphi_\mu, \; p_{i_1}^{(1)} = \varphi_{i_1}^{(1)}, \; \dots, \; p_{i_{s-1}}^{(s-1)} = \varphi_{i_{s-1}}^{(s-1)}, \; p_{i_{s-1}\nu}^{(s-1)} = \frac{\partial \varphi_{i_{s-1}}^{(s-1)}}{\partial x_\nu}.$$

As a result, our claim stated above is proved.

Conversely, imagine now that we are given a system of equations of the form (4) about which moreover we know neither whether it admits the infinitesimal transformations $\Omega_1 f, \dots, \Omega_n f$, nor whether it embraces the equations $W_1 = 0, \dots, W_q = 0$.

Clearly, the equations (4) in question are all created by differentiating the equations $z_\mu = \varphi_\mu$ with respect to $x_1, \dots, x_n$. In addition, since under the given assumptions, the equations:

$$\Omega_\nu\big(p_{i_{s-1}}^{(s-1)} - \varphi_{i_{s-1}}^{(s-1)}\big) = P_{i_{s-1}\nu} - \frac{\partial \varphi_{i_{s-1}}^{(s-1)}}{\partial x_\nu} = 0$$

are a consequence of (4) and likewise, of the equations $W_1 = 0, \dots, W_q = 0$, then the system of equations (1) will be identically satisfied when one executes the substitution (4) in it, and then sets $p_{i\nu}^{(s-1)} = \frac{\partial \varphi_i^{(s-1)}}{\partial x_\nu}$. Consequently, the equations $z_\mu = \varphi_\mu$ represent a system of solutions for the differential equations (1).

From this, we see: every solution $z_\mu = \varphi_\mu$ of the differential equations (1) provides a completely determined system of equations of the form (4) which admits the infinitesimal transformations $\Omega_1 f, \dots, \Omega_n f$ and which embraces the equations $W_1 = 0, \dots, W_q = 0$; conversely, every system of equations of the form (4) which possesses the properties just indicated provides a completely determined system of solutions for the differential equations (1). *Consequently, the problem of determining all systems of solutions of the differential equations* (1) *is equivalent to the problem of determining all systems of equations of the form* (4) *which admit the infinitesimal transformations* $\Omega_1 f, \dots, \Omega_n f$ *and which in addition embrace the equations* $W_1 = 0, \dots, W_q = 0$. *If one knows the most general solution for one of these two problems, then at the same time, the most general solution of the other system is given.*

But we can solve this new problem on the basis of the developments of Chap. 7, p. 133 sq.

According to Chap. 7, Proposition 5, p. 134, every sought system of equations also admits, simultaneously with $\Omega_1 f, \dots, \Omega_n f$, all infinitesimal transformations of the form $\Omega_\nu\big(\Omega_j(f)\big) - \Omega_j\big(\Omega_\nu(f)\big)$. By computation, one verifies that:

$$\Omega_\nu\big(\Omega_j(f)\big) - \Omega_j\big(\Omega_\nu(f)\big) = \sum_{i=1}^{\varepsilon_s - 1} \big\{\Omega_\nu(P_{ij}) - \Omega_j(P_{i\nu})\big\} \frac{\partial f}{\partial p_i^{(s-1)}} \qquad (\nu, j = 1 \cdots n),$$

since the expressions:

$$\Omega_\nu(p_{ij}^{(k)}) - \Omega_j(p_{i\nu}^{(k)})$$

all vanish identically, as long as k is smaller than $s-1$. But now, we have seen above that the equations (3):

$$\Omega_\nu(P_{ij}) - \Omega_j(P_{i\nu}) = 0$$

are a consequence of $W_1 = 0, \dots, W_q = 0$. Consequently, for the systems of values $x, z, p^{(1)}, \dots, p^{(s-1)}$ of the system of equations $W_1 = 0, \dots, W_q = 0$, there exist relations of the form:

$$\Omega_\nu\big(\Omega_j(f)\big)-\Omega_j\big(\Omega_\nu(f)\big)=\sum_{\tau=1}^{n}\omega_{\nu j\tau}\big(x,z,p^{(1)},\dots,p^{(s-1)}\big)\cdot\Omega_\tau f \qquad (\nu,j=1\cdots n),$$

where the functions $\omega_{\nu j\tau}$ are all equal to zero and behave regularly for the systems of values of $W_1=0$, ..., $W_q=0$.

In addition, it is still to be underlined that amongst the $n\times n$ determinants of the matrix:

$$\begin{vmatrix} 1\;0\cdots 0 & p_{11}^{(0)} & \cdots & p_{\varepsilon_{s-2},1}^{(s-2)} & P_{11} & \cdots & P_{\varepsilon_s-1,1} \\ \cdot\;\cdot\cdots\cdot & \cdots & \cdot & \cdot & \cdot & \cdots & \cdot \\ 0\;0\cdots 1 & p_{1n}^{(0)} & \cdots & p_{\varepsilon_{s-2},n}^{(s-2)} & P_{1n} & \cdots & P_{\varepsilon_{s-1},n} \end{vmatrix},$$

one has the value 1, and thus it cannot be brought to zero for any of the sought systems of equations.

From this, we see that we have in front of us the special case that is handled by Theorem 19 in Chap. 7, p. 146. With the help of this theorem, we can actually set up all systems of equations which admit $\Omega_1 f,\dots,\Omega_n f$ and which embrace the equations $W_1=0$, ..., $W_q=0$. There is no difficulty in identifying such systems of equations which can be given the form (4).

From one system of equations which admits the infinitesimal transformations $\Omega_1 f,\dots,\Omega_n f$, one can never derive a relation between the variables $x_1,\dots,x_n$. This follows from the mentioned theorem and can also easily be seen directly. As a result, it is possible to solve the equations $W_1=0$, ..., $W_q=0$ with respect to q, amongst $\varepsilon_0+\varepsilon_1+\cdots+\varepsilon_{s-1}$, of the quantities z, $p^{(1)}$, ..., $p^{(s-1)}$. When we do that, we receive q of the $\sum\varepsilon_k$ variables z, $p^{(1)}$, ..., $p^{(s-1)}$ expressed by means of the $\sum\varepsilon_k-q$ remaining variables and by means of $x_1,\dots,x_n$.

Thanks to these preparations, we can form the reduced infinitesimal transformations that are considered in the mentioned theorem. We obtain them when we leave out all differential quotients of f with respect to the q considered variables amongst z, $p^{(1)}$, ..., $p^{(s-1)}$ and afterwards, by substituting, in the remaining terms, each one of the q variables with its expression by means of the $\sum\varepsilon_k-q$ remaining variables and the x.

The n reduced infinitesimal transformations so obtained, which we can denote by $\overline{\Omega}_1 f,\dots,\overline{\Omega}_n f$, contain $n-q+\sum\varepsilon_k$ independent variables and in addition stand pairwise in the relationships:

$$\overline{\Omega}_\nu\big(\overline{\Omega}_j(f)\big)-\overline{\Omega}_j\big(\overline{\Omega}_\nu(f)\big)\equiv 0 \qquad (\nu,j=1\cdots n),$$

according to the mentioned theorem.

Consequently, the n mutually independent equations $\overline{\Omega}_1 f=0$, ..., $\overline{\Omega}_n f=0$ form an n-term complete system with the $\sum\varepsilon_k-q$ independent solutions: u_1, u_2, ..., $u_{\sum\varepsilon_k-q}$. If these solutions are determined, then one can indicate all systems of equations which admit the infinitesimal transformations $\Omega_1 f,\dots,\Omega_n f$ and which embrace the equations $W_1=0,\dots,W_q=0$. The general form of a system of this kind is: $W_1 = 0, \ldots, W_q = 0$ together with arbitrary relations between the $\sum \varepsilon_k - q$ solutions u.

However now, the matter is not about all systems of equations of this kind, but only about those which can be brought to the form (4). Every system of equations of this constitution contains exactly $\sum \varepsilon_k$ equations, hence can be given the form:

$$\text{(5)} \qquad W_1 = 0, \; \ldots, \; W_q = 0, \quad u_1 = a_1, \; \ldots, \; u_{\sum \varepsilon_k - q} = a_{\sum \varepsilon_k - q},$$

where the a denotes constants. But every system of the form just indicated can be solved with respect to the $\sum \varepsilon_k$ variables $z, p^{(1)}, \ldots, p^{(s-1)}$, since it admits the infinitesimal transformations $\Omega_1 f, \ldots, \Omega_n f$. If we perform the resolution in question, we then obtain a system of equations:

$$\text{(5')} \qquad \begin{gathered} z_\mu = Z_\mu(x_1, \ldots, x_n, a_1, a_2, \ldots), \quad p_{i_1}^{(1)} = \Pi_{i_1}^{(1)}(x_1, \ldots, x_n, a_1, a_2, \ldots), \\ \cdots \; p_{i_{s-1}}^{(s-1)} = \Pi_{i_{s-1}}(x_1, \ldots, x_n, a_1, a_2, \ldots) \\ (\mu = 1 \cdots m; \; i_k = 1 \cdots \varepsilon_k) \end{gathered}$$

which satisfies all the stated requirements: it admits the infinitesimal transformations $\Omega_1 f, \ldots, \Omega_n f$, it embraces the equations $W_1 = 0, \ldots, W_q = 0$, and it is of the form (4). At present, when we consider the $\sum \varepsilon_k - q$ constants a as arbitrary constants, we therefore obviously have the most general system of equations of the demanded constitution.

From this, it follows that according to what has been said, the equations:

$$\text{(6)} \qquad z_\mu = Z_\mu(x_1, \ldots, x_n, a_1, a_2, \ldots) \qquad (\mu = 1 \cdots m)$$

represent a system of solutions for the differential equations (1) and in fact, the most general system of solutions. We now claim that in this system of solutions, the $\sum \varepsilon_k - q$ arbitrary constants a are all essential.

To prove our claim, we recall that the equations (5') can be obtained from the equations $z_\mu = Z_\mu$ by differentiation with respect to $x_1, \ldots, x_n$, provided that the $p^{(1)}, p^{(2)}, \ldots$ are again interpreted as differential quotients of the z with respect to the x. Now, if the $\sum \varepsilon_k - q$ parameters a in the equations (6) were not essential, then the number of parameters could be lowered by introducing appropriate functions of them. But with this, according to the preceding, the number of parameters in the equations (5') would at the same time be lowered, and this is impossible, since the equations (5') can be brought to the form (5), from which it follows immediately that the parameters a in (5') are all essential. This is a contradiction, so the assumption made a short while ago is false and the parameters a in the equations (6) are all essential.

We can hence state the following proposition.

Proposition 1. *If a system of partial differential equations of order s of the form:*

$$F_\sigma\left(x_1, \ldots, x_n, \, z_1, \ldots, z_m, \, \frac{\partial z_1}{\partial x_1}, \ldots, \frac{\partial^2 z_1}{\partial x_1^2}, \ldots, \frac{\partial^s z_m}{\partial x_n^s}\right) = 0 \qquad (\sigma = 1, 2, \cdots)$$

possesses the property that all differential quotients of order s of the z with respect to the x can be expressed by means of the differential quotients of lower order, by means of $z_1, \ldots, z_m$ *and by means of* $x_1, \ldots, x_n$*, while the corresponding property does not hold in any case for the differential quotients of order* $(s-1)$*, and if in addition, by differentiating it once with respect to the x, the system produces only relations between* $x_1, \ldots, x_n$*,* $z_1, \ldots, z_n$ *and the differential quotients of the first order up to the* $(s-1)$*-th which follow from the system itself, then the most general system of solutions:*

$$z_\mu = \varphi_\mu(x_1, \ldots, x_n) \qquad {\scriptstyle (\mu = 1 \cdots m)}$$

of the concerned system of differential equations contains only a finite number of arbitrary constants. The number of these arbitrary constants is equal to $\varepsilon_0 + \varepsilon_1 + \cdots + \varepsilon_{s-1} - q$*, where* ε_k *denotes the number of differential quotients of order k of* $z_1, \ldots, z_m$ *with respect to* $x_1, \ldots, x_n$ *and where q is the number of independent relations which the system in question yields between* $x_1, \ldots, x_n$*,* $z_1, \ldots, z_n$ *and the differential quotients up to order* $(s-1)$*. One finds the most general system of solutions* $z_\mu = \varphi_\mu$ *itself by integrating an n-term complete system in* $n - q + \varepsilon_0 + \cdots + \varepsilon_{s-1}$ *independent variables.*

For the proof of the above proposition, we imagined the equations $W_1 = 0$, ..., $W_q = 0$ solved with respect to q of the quantities z, $p^{(1)}$, ..., $p^{(s-1)}$. In principle, it is completely indifferent with respect to which amongst these quantities the equations are solved; but now, since the equations $W_1 = 0$, ..., $W_q = 0$ are differential equations and since the $p^{(1)}, p^{(2)}, \ldots$ denote differential quotients, it is advisable to undertake the concerned resolution in a specific way, which we now explain.

At first, we eliminate all differential quotients $p^{(1)}$, $p^{(2)}$, ..., $p^{(s-1)}$ from the equations $W_1 = 0$, ..., $W_q = 0$; then we obtain, say, ν_0 independent equations between the x and z alone, and so we can represent ν_0 of the z as functions of the $\varepsilon_0 - \nu_0 = m - \nu_0$ remaining ones, and of the x.

We insert the expressions for these ν_0 quantities z in the equations $W_1 = 0$, ..., $W_q = 0$, which now reduce to $q - \nu_0$ mutually independent equations. Afterwards, from these $q - \nu_0$ equations, we remove all the differential quotients $p^{(2)}, \ldots, p^{(s-1)}$ and we obtain, say ν_1 mutually independent equations by means of which we can express ν_1 of the quantities $p^{(1)}$ in terms of the $\varepsilon_1 - \nu_1$ remaining ones, in terms of the $\varepsilon_0 - \nu_0$ of the z and in terms of $x_1, \ldots, x_n$.

If we continue in the described way, then at the end, the system of equations $W_1 = 0$, ..., $W_q = 0$ will be resolved, and to be precise, it will be resolved with respect to ν_k amongst the ε_k differential quotients $p^{(k)}$, where it is understood that k is any of the numbers 0, 1, 2, ..., $s-1$. Here, the concerned ν_k amongst the $p^{(k)}$ are each time expressed in terms of the $\varepsilon_k - \nu_k$ remaining ones, in terms of certain of the differential quotients: $p^{(k-1)}, \ldots, p^{(1)}, p^{(0)}$, and in terms of the x. Naturally, the sum $\nu_0 + \nu_1 + \cdots + \nu_{s-1}$ has the value q. Lastly, it can be shown that ν_k is always smaller than ε_k. Indeed at first, ν_{s-1} is certainly smaller than ε_{s-1}, because we have assumed that in our system of differential equations, not all differential quotients of order $(s-1)$ can be expressed in terms of those of lower order, and in terms of the x. But if any other numbers ν_k were equal to ε_k, we would have:

$$p_i^{(k)} = F_i(x, z, p^{(1)}, \dots, p^{(k-1)}) \qquad (i=1,2\cdots\varepsilon_k);$$

now, since the system of equations $W_1 = 0, \dots, W_q = 0$ admits the infinitesimal transformations $\Omega_1 f, \dots, \Omega_n f$, then all systems of equations:

$$p_{iv}^{(k)} = \Omega_v(F_i) \qquad (i=1\cdots\varepsilon_k;\ v=1\cdots n)$$

would be consequences of $W_1 = 0, \dots, W_q = 0$, so one would also have $v_{k+1} = \varepsilon_{k+1}$ and in the same way $v_{k+2} = \varepsilon_{k+2}, \dots, v_{s-1} = \varepsilon_{s-1}$, and this is impossible according to what has been said.

Once we have resolved the equations $W_1 = 0, \dots, W_q = 0$, then with this at the same time, the system of differential equations (1) is resolved in a completely determined way, namely with respect to v_k of the ε_k differential quotients of order k, where it is understood that k is any of the numbers 0, 1, 2, ..., s. At each time, the v_k differential quotients of order k are expressed in terms of the $\varepsilon_k - v_k$ remaining ones of order k, in terms of those of lower order, and in terms of the x.

If one knows all the numbers ε_k and v_k, then one can immediately indicate the number of arbitrary constants in the most general system of solutions of the differential equations (1). Indeed, according to the above proposition, this number is equal to:

$$\varepsilon_0 + \varepsilon_1 + \cdots + \varepsilon_{s-1} - q = (\varepsilon_0 - v_0) + (\varepsilon_1 - v_1) + \cdots + (\varepsilon_{s-1} - v_{s-1}).$$

§ **48.** Conversely, consider a system of equations of the form:

$$z_\mu = Z_\mu(x_1, \dots, x_n, a_1, \dots, a_r) \qquad (\mu=1\cdots m), \tag{7}$$

in which $x_1, \dots, x_n$ are interpreted as independent variables, and $a_1, \dots, a_r$ as arbitrary parameters. The r parameters $a_1, \dots, a_r$ whose number is finite can all be assumed to be essential.

We will prove that there is a system of partial differential equations which is free of $a_1, \dots, a_r$ whose most general solution is represented by the equations (7).

By differentiating the equations (7) with respect to $x_1, \dots, x_n$, we obtain the equations of the form:

$$p_i^{(1)} = Z_i^{(1)}(x_1, \dots, x_n, a_1, \dots, a_r) \qquad (i=1\cdots\varepsilon_1) \tag{7_1}$$

$$p_i^{(2)} = Z_i^{(2)}(x_1, \dots, x_n, a_1, \dots, a_r) \qquad (i=1\cdots\varepsilon_2), \tag{7_2}$$

and so on.

Now, by means of the equations (7), a certain number, say v_0, of the a can be expressed in terms of the $r - \mu_0$ remaining ones, and in terms of the x and the z.

By taking the equations (7) and (7_1) together, we can express $\mu_0+\mu_1$ of the a in terms of the $r-\mu_0-\mu_1$ remaining ones, in terms of the $p^{(1)}$, the z and the x. In general, if we take the equations (7), (7_1), ..., (7_k) together, we can then express $\mu_0+\mu_1+\cdots+\mu_k$ of the quantities a in terms of the $r-\mu_0-\cdots-\mu_k$ remaining ones and in terms of the $p^{(k)}$, $p^{(k-1)}$, ..., $p^{(1)}$, z, x. Here, μ_0, μ_1, ... are entirely determined positive whole numbers.

Naturally, the sum $\mu_0+\mu_1+\cdots+\mu_k$ is at most equal to r; furthermore, the number μ_0 is in any case different from zero, because we can assume that r is bigger than zero. Consequently, there must exist a whole number $s \geqslant 1$ having the property that μ_0, μ_1, ..., μ_{s-1} are all different from zero, but the number μ_s vanishes. Then if the said quantities amongst $a_1,\dots,a_r$ are determined from the equations (7), (7_1), ..., (7_{s-1}) and if their values are inserted into the equations (7_s), then one obtains only equations which are free of $a_1,\dots,a_r$; because in the contrary case, one could determine more than $\mu_0+\cdots+\mu_{s-1}$ of the quantities a from the equations (7), (7_1), ..., (7_s), so one would have $\mu_s>0$, which is not the case according to the above.

As a result, by eliminating $a_1,\dots,a_r$ from the equations (7), (7_1), ..., (7_s), we obtain the following equations:

Firstly, ε_s equations which express all the ε_s differential quotients $p^{(s)}$ in terms of the $p^{(s-1)}$, ..., $p^{(1)}$, z, x, and:

Secondly: $\varepsilon_0-\mu_0+\varepsilon_1-\mu_1+\cdots+\varepsilon_{s-1}-\mu_{s-1}$ mutually independent equations between the x, z, $p^{(1)}$, ..., $p^{(s-1)}$.

It is easy to see that the so obtained system of differential equations of order s possesses all properties which were ascribed in Proposition 1, p. 193, to the differential equations $F_\sigma\big(x, z, \frac{\partial z}{\partial x}, \dots\big)$.

Indeed, all ε_s differential quotients of order s are functions of the differential quotients of lower order, and of the x; however, the corresponding property does not hold true for the differential quotients of order $(s-1)$, because the number μ_{s-1} discussed above is indeed different from zero. Lastly, by differentiation with respect to the x and by combination of the obtained equations, we obtain only relations between the x, z, $p^{(1)}$, ..., $p^{(s-1)}$ which follow from the above-mentioned relations. Indeed, the equations (7), (7_1), ..., (7_{s-1}) and the ones which follow from them are the only finite relations through which the quantities $a_1,\dots,a_r$, x, z, $p^{(1)}$, ..., $p^{(s-1)}$ are linked; hence when we eliminate the a, we obtain the only finite relations which exist between the x, z, $p^{(1)}$, ..., $p^{(s-1)}$, namely the $\varepsilon_0-\mu_0+\cdots+\varepsilon_{s-1}-\mu_{s-1}$ relations mentioned above.

Proposition 1, p. 193, can therefore easily be applied to our system of differential equations of order s. The numbers ν_k defined at that time are equal to $\varepsilon_k-\mu_k$, and hence in the present case the number q has the value:

$$\varepsilon_0-\mu_0+\cdots+\varepsilon_{s-1}-\mu_{s-1},$$

and therefore the most general system of solutions of our differential equations contains precisely $\mu_0+\mu_1+\cdots+\mu_{s-1}$ arbitrary constants. Now, since on the other hand the equations (7) also represent a system of solutions for our differential equations, and a system with the r essential parameters $a_1,\dots,a_r$ as arbitrary constants, it follows that the number $\mu_0 + \cdots + \mu_{s-1}$, which is at most equal to r, must be precisely equal to r. In other words: the equations (7) represent the most general system of solutions of our differential equations.

As a result, we have the

Proposition 2. *If $z_1, \dots, z_m$ are given functions of the variables $x_1, \dots, x_n$ and of a finite number of parameters $a_1, \dots, a_r$:*

$$z_\mu = Z_\mu(x_1, \dots, x_n, a_1, \dots, a_r) \qquad (\mu = 1 \cdots m),$$

then there always exists an integrable system of partial differential equations which determines the z as functions of the x, and whose most general solutions are represented by the equations $z_\mu = Z_\mu(x, a)$.

If we compare this proposition with Proposition 1, p. 193, we immediately obtain the

Proposition 3. *If an integrable system of partial differential equations:*

$$F_\sigma\left(x_1, \dots, x_n, z_1, \dots, z_m, \frac{\partial z_1}{\partial x_1}, \dots, \frac{\partial^2 z_1}{\partial x_1^2}, \dots\right) = 0 \qquad (\sigma = 1, 2 \cdots)$$

is constituted in such a way that its most general solutions depend only on a finite number of arbitrary constants, then by means of differentiation and of elimination, it can always be brought to a form which possesses the following two properties: firstly, all differential quotients of a certain order, say s, can be expressed in terms of those of lower order, and in terms of $z_1, \dots, z_m$, $x_1, \dots, x_n$, whereas the corresponding property does not, in any case, hold true for all differential quotients of order $(s-1)$. Secondly, by differentiating once with respect to the x, we only obtain relations between the x, z and the differential quotients of orders 1 up to $(s-1)$ which follow from the already extant equations.

Furthermore, the developments of the present section provide a simple method for answering the question of how many parameters $a_1, \dots, a_r$ are essential amongst the ones of a given system of equations:

$$z_\mu = Z_\mu(x_1, \dots, x_n, a_1, \dots, a_r) \qquad (\mu = 1 \cdots m).$$

Indeed, in order to be able to answer this question, we only need to compute the whole numbers μ_0, μ_1, ..., μ_{s-1} defined above; the sum $\mu_0 + \mu_1 + \cdots + \mu_{s-1}$ then identifies the number of essential parameters amongst $a_1, \dots, a_r$, because the equations $z_\mu = Z_\mu(x, a)$ represent the most general solutions of a system of differential equations the most general solutions of which, according to the above developments, contain precisely $\mu_0 + \cdots + \mu_{s-1}$ essential parameters.

Chapter 11
The Defining Equations for the Infinitesimal Transformations of a Group

In Chap. 9, we have reduced the finding of all r-term groups to the determination of all systems of r independent infinitesimal transformations $X_1 f, \dots, X_r f$ which satisfy relations of the form:

$$X_i X_k f - X_k X_i f = [X_i, X_k] = \sum_{s=1}^{r} c_{iks} X_s f,$$

with certain constants c_{iks}. Only later will we find means of treating this reduced problem; temporarily, we must restrict ourselves to admitting systems $X_1 f, \dots, X_r f$ of the concerned nature and to studying their properties.

In the present chapter we begin with an application of the developments of the preceding chapter; from this chapter, we conclude that the general infinitesimal transformation:

$$e_1 X_1 f + \cdots + e_r X_r f$$

of a given r-term group $X_1 f, \dots, X_r f$ can be defined by means of certain linear partial differential equations, which we call the *defining equations* of the group. From that, further conclusions will then be drawn.[1]

§ 49. Consider r infinitesimal transformations:

$$X_k f = \sum_{i=1}^{n} \xi_{ki}(x_1, \dots, x_n) \frac{\partial f}{\partial x_i} \qquad (k=1 \cdots r)$$

which generate an r-term group. Then the general infinitesimal transformation of this group has the form:

---

[1] LIE, Archiv for Mathematik og Naturvidenskab Vol. 3, Christiania 1878 and Vol. 8, 1883; Gesellschaft der Wissenschaften zu Christiania 1883, No. 12.

S. Lie, *Theory of Transformation Groups I*,
DOI 10.1007/978-3-662-46211-9_11

$$\sum_{i=1}^{n} \xi_i \frac{\partial f}{\partial x_i} = \sum_{i=1}^{n} \sum_{k=1}^{r} e_k \, \xi_{ki} \frac{\partial f}{\partial x_i},$$

where the e_k denote arbitrary constants. For the ξ_i, we obtain from this the expressions:

(1) $$\xi_i = \sum_{k=1}^{r} e_k \, \xi_{ki}(x) \qquad (i=1\cdots n),$$

in which the $\xi_{ki}(x)$ are given functions of $x_1, \dots, x_n$.

Now, according to Proposition 2 of the preceding chapter, the expressions $\xi_1, \dots, \xi_n$ just written are the most general system of solutions of a certain system of partial differential equations which is free of the arbitrary constants $e_1, \dots, e_r$. In order to set up this system, we proceed as in the introduction to § 48; we differentiate the equations (1) with respect to every variable $x_1, \dots, x_n$, next we differentiate in the same way the obtained equations with respect to $x_1, \dots, x_n$, and so on. Then when we have computed all differential quotients of the ξ up to a certain order (considered in more detail in § 48), we take away from the found equations the arbitrary constants $e_1, \dots, e_r$ and we obtain in this way the desired system of differential equations, the most general solutions of which are just the expressions $\xi_1, \dots, \xi_n$.

We can always arrange that all equations of the discussed system in the ξ_i and in their differential quotients are linear homogeneous; indeed, from any two systems of solutions:

$$\xi_{k\nu}, \quad \xi_{j\nu} \qquad (\nu=1\cdots n),$$

one can always derive another system of solutions $e_k \xi_{k\nu} + e_j \xi_{j\nu}$ of the concerned differential equations with two arbitrary constants e_k and e_j.

The system of differential equations which define $\xi_1, \dots, \xi_n$ therefore has the form:

$$\sum_{\nu=1}^{n} A_{\mu\nu}(x_1, \dots, x_n) \, \xi_\nu + \sum_{\nu,\,\pi}^{1\cdots n} B_{\mu\nu\pi}(x_1, \dots, x_n) \frac{\partial \xi_\nu}{\partial x_\pi} + \cdots = 0,$$

where the A, B, ..., are free of the arbitrary constants $e_1, \dots, e_r$.

We briefly [KURTZWEG] call these differential equations the *defining equations of the group, since they completely define the totality of all infinitesimal transformations of this group and therefore, they define the group itself.*

What has been said can now be illustrated precisely by means of a couple of examples.

In the general infinitesimal transformation:

$$\xi_1 \frac{\partial f}{\partial x_1} + \xi_2 \frac{\partial f}{\partial x_2}$$

of the six-term linear group:

$$x_1' = a_1 x_1 + a_2 x_2 + a_3$$
$$x_2' = a_4 x_1 + a_5 x_2 + a_6,$$

ξ_1 and ξ_2 have the form:

$$\xi_1 = e_1 + e_2 x_1 + e_3 x_2, \qquad \xi_2 = e_4 + e_5 x_1 + e_6 x_2.$$

From this, we find that the defining equations of the group are the following:

$$\frac{\partial^2 \xi_1}{\partial x_1^2} = 0, \quad \frac{\partial^2 \xi_1}{\partial x_1 \partial x_2} = 0, \quad \frac{\partial^2 \xi_1}{\partial x_2^2} = 0,$$
$$\frac{\partial^2 \xi_2}{\partial x_1^2} = 0, \quad \frac{\partial^2 \xi_2}{\partial x_1 \partial x_2} = 0, \quad \frac{\partial^2 \xi_2}{\partial x_2^2} = 0.$$

The known eight-term group:

$$x_1' = \frac{a_1 x_1 + a_2 x_2 + a_3}{a_7 x_1 + a_8 x_2 + 1}, \qquad x_2' = \frac{a_4 x_1 + a_5 x_2 + a_6}{a_7 x_1 + a_8 x_2 + 1}$$

of all projective transformations of the plane serves as a second example. Its general infinitesimal transformation:

$$\xi_1 = e_1 + e_2 x_1 + e_3 x_2 + e_7 x_1^2 + e_8 x_1 x_2$$
$$\xi_2 = e_4 + e_5 x_1 + e_6 x_1 + e_7 x_1 x_2 + e_8 x_2^2$$

will be defined by means of relations between the differential quotients of second order of the ξ, namely by means of:

$$\frac{\partial^2 \xi_1}{\partial x_1^2} - 2\frac{\partial^2 \xi_2}{\partial x_1 \partial x_2} = 0, \quad \frac{\partial^2 \xi_2}{\partial x_2^2} - 2\frac{\partial^2 \xi_1}{\partial x_1 \partial x_2} = 0, \quad \frac{\partial^2 \xi_1}{\partial x_2^2} = 0, \quad \frac{\partial^2 \xi_2}{\partial x_1^2} = 0.$$

By renewed differentiation, one finds that all third-order differential quotients of ξ_1 and of ξ_2 vanish.

§ 50. Conversely, when does a system of linear homogeneous differential equations:

$$\sum_{\nu=1}^{n} A_{\mu\nu}(x)\,\xi_\nu + \sum_{\nu,\,\pi}^{1\cdots n} B_{\mu\nu\pi}(x)\,\frac{\partial \xi_\nu}{\partial x_\pi} + \cdots = 0 \qquad (\mu = 1, 2\cdots)$$

define the general infinitesimal transformation of a finite continuous group?

Naturally, the first condition is that the most general solutions $\xi_1, \dots, \xi_n$ of the system depend only on a finite number of arbitrary constants. Let this condition be fulfilled. Then according to Proposition 3, p. 197, of the preceding chapter, it is always possible, by differentiation and elimination, to bring the system to a certain specific form in which all differential quotients of the highest, say the s-th, order can be expressed in terms of those of lower order, and in terms of $x_1,\dots,x_n$, whereas the corresponding property does not, in any case, hold true for all differential quotients of order $(s-1)$; in addition, differentiating once with respect to the x produces no new relation between $x_1,\dots,x_n$, $\xi_1,\dots,\xi_n$ and the differential quotients of first order up to the $(s-1)$-th. We assume that the system is of this form and we imagine that it has been solved in the way indicated on p. 194; then for $k=0,1,\dots,s$, we obtain at each time that a certain number of differential quotients, say ν_k, amongst the ε_k of the k-th order differential quotients of the ξ, are represented as linear homogeneous functions of the remaining differential quotients of k-th order, and of certain differential quotients of $(k-1)$-th, ..., first, zeroth order, with coefficients which depend only upon $x_1,\dots,x_n$, where $\nu_s=\varepsilon_s$, but else, ν_k is always smaller than ε_k.

As was shown in the preceding chapter, p. 195, under the assumptions made, the most general system of solutions $\xi_1,\dots,\xi_n$ of our differential equations comprises precisely:

$$(\varepsilon_0-\nu_0)+(\varepsilon_1-\nu_1)+\cdots+(\varepsilon_{s-1}-\nu_{s-1})=r$$

arbitrary constants; this most general system of solutions can be deduced from r particular systems of solutions $\xi_{1i},\dots,\xi_{ri}$ with the help of r constants of integration as follows:

$$\xi_i=\sum_{k=1}^{r} e_k\,\xi_{ki} \qquad (i=1\cdots n),$$

and here, the particular systems of solutions in question must only be such that the r expressions:

$$\sum_{i=1}^{n}\xi_{1i}\frac{\partial f}{\partial x_i},\quad\dots,\quad\sum_{i=1}^{n}\xi_{ri}\frac{\partial f}{\partial x_i}$$

represent as many independent infinitesimal transformations.

Now according to Theorem 24, p. 172, for $\xi_1\frac{\partial f}{\partial x_1}+\cdots+\xi_n\frac{\partial f}{\partial x_n}$ to be the general infinitesimal transformation of an r-term group, a certain condition is necessary and sufficient, namely: when $\xi_{k1},\dots,\xi_{kn}$ and $\xi_{j1},\dots,\xi_{jn}$ are two particular systems of solutions, then the expression:

$$\sum_{\nu=1}^{n}\left(\xi_{k\nu}\frac{\partial\xi_{ji}}{\partial x_\nu}-\xi_{j\nu}\frac{\partial\xi_{ki}}{\partial x_\nu}\right) \qquad (i=1\cdots n)$$

must always represent a system of solutions. As a result, we have the

Theorem 28. *If $\xi_1,\dots,\xi_n$, as functions of $x_1,\dots,x_n$, are determined by certain linear and homogeneous partial differential equations:*

$$\sum_{\nu=1}^{n}A_{\mu\nu}(x)\,\xi_\nu+\sum_{\nu,\,\pi}^{1\cdots n}B_{\mu\nu\pi}(x)\,\frac{\partial\xi_\nu}{\partial x_\pi}+\cdots=0 \qquad (\mu=1,2\cdots),$$

then the expression $\xi_1\frac{\partial f}{\partial x_1}+\cdots+\xi_n\frac{\partial f}{\partial x_n}$ represents the general infinitesimal transformation of a finite continuous group if and only if: firstly the most general system of *solutions of these differential equations depends only on a finite number of arbitrary constants, and secondly from any two particular systems of solutions $\xi_{k1},\dots,\xi_{kn}$ and $\xi_{j1},\dots,\xi_{jn}$, by the formation of the expression:*

$$\sum_{\nu=1}^{n}\left(\xi_{k\nu}\,\frac{\partial\xi_{ji}}{\partial x_\nu}-\xi_{j\nu}\,\frac{\partial\xi_{ki}}{\partial x_\nu}\right)\qquad (i=1\cdots n),$$

one always obtains a new system of solutions.

If $\xi_1\,\frac{\partial f}{\partial x_1}+\cdots+\xi_n\,\frac{\partial f}{\partial x_n}$ really is the general infinitesimal transformation of a finite group, then naturally, the above differential equations are the defining equations of this group.

If the defining equations of a finite group are given, the numbers $\nu_0,\ \nu_1,\ \dots,\ \nu_{s-1}$ discussed earlier on can be determined by differentiation and by elimination; the number:

$$r=(\varepsilon_0-\nu_0)+(\varepsilon_1-\nu_1)+\cdots+(\varepsilon_{s-1}-\nu_{s-1})$$

indicates how many parameters the group contains.

In the first one of the two former examples, one has:

$$s=2,\quad \nu_0=0,\quad \nu_1=0,\quad \varepsilon_0=2,\quad \varepsilon_1=4,$$

and therefore $r=6$. In the second example, one has:

$$s=3,\quad \nu_0=0,\quad \nu_1=0,\quad \nu_2=4,\quad \varepsilon_0=2,\quad \varepsilon_1=4,\quad \varepsilon_2=6,$$

whence $r=8$.

§ 51. Now, let again:

$$X_kf=\sum_{i=1}^{n}\xi_{ki}(x_1,\dots,x_n)\,\frac{\partial f}{\partial x_i}\qquad (k=1\cdots r)$$

be independent infinitesimal transformations of an r-term group. We imagine that the defining equations of this group are set up and are brought to the form discussed above, hence resolved with respect to ν_k of the ε_k differential quotients of order k of the ξ $(k=0,1,\dots,s)$; here, as said earlier on, ν_k is always $<\varepsilon_k$, except in the case $k=s$, in which $\nu_s=\varepsilon_s$.

The coefficients in the resolved defining equations are visibly *rational* functions of the ξ together with their differential quotients of first order, up to th s-th. Now, since as a matter of principle (cf. p. 187), we restrict ourselves to systems of values $x_1,\dots,x_n$ for which all ξ_{ki} behave regularly, then the coefficients will in general also behave regularly for the systems of values $x_1,\dots,x_n$ coming into consideration, but obviously only in general: there can well exist points $x_1,\dots,x_n$ in which all ξ_{ki} indeed behave regularly, but in which not all coefficients of the solved defining equations do so. In what follows, one must always take heed of this distinction between the different points $x_1, \dots, x_n$.

Let $x_1^0, \dots, x_n^0$ be *a point for which all coefficients in the solved defining equations behave regularly*; then $\xi_1, \dots, \xi_n$ are, in a certain neighborhood of $x_1^0, \dots, x_n^0$, ordinary power series with respect to the $x_k - x_k^0$:

$$\xi_i = g_i^0 + \sum_{\nu=1}^{n} g'_{i\nu}\,(x_\nu - x_\nu^0) + \sum_{\nu,\,\pi}^{1\cdots n} g''_{i\nu\pi}\,(x_\nu - x_\nu^0)(x_\pi - x_\pi^0) + \cdots,$$

where, always, the terms of the same order are thought to be combined together. Here, the coefficients g^0, g', ... must be determined in such a way that the given differential equations are identically satisfied after insertion of the power series expansions for $\xi_1, \dots, \xi_n$.

If we now remember that our group is r-term, then we immediately realize that on the whole, r of the coefficients g^0, g', ... remain undetermined, hence that certain amongst the initial values which the ξ and its differential quotients take for $x_1 = x_1^0$, ..., $x_n = x_n^0$, can be chosen arbitrarily. Now, since our differential equations show that all differential quotients of order s and of higher order can be expressed in terms of the differential quotients of orders zero up to $(s-1)$, and in terms of $x_1, \dots, x_n$, it follows that exactly r amongst the initial values g^0, g', g'', ..., $g^{(s-1)}$ can be chosen arbitrarily, or, what amounts to the same: the $\varepsilon_0 + \varepsilon_1 + \cdots + \varepsilon_{s-1}$ mentioned initial values must be linked together by certain $\sum \varepsilon_k - r$ independent relations. At present, we want to set up these relations.

Our system of differential equations produces all relations which exist, at an *arbitrary* x, between the ξ and their differential quotients of first order up to the $(s-1)$-th. Hence, when we make the substitution $x_i = x_i^0$ in our differential equations, we obtain certain relationships by which the initial values g^0, g', ..., $g^{(s-1)}$ are linked together. In this manner, we obtain ν_0 linear homogeneous relations between the g^0 alone, furthermore ν_1 relations between the g' and certain g^0, and in general ν_k relations between the $g^{(k)}$ and certain $g^{(k-1)}$, ..., g', g^0. The relations in question are resolved, and to be precise, they are resolved with respect to ν_0 of the g^0, with respect to ν_1 of the g', and so on, and in total, there are $\sum \nu_k = \sum \varepsilon_k - r$ independent relations. Now, since according to what has been said, there do not exist more than $\sum \varepsilon_k - r$ independent relations between g^0, g', ..., $g^{(s-1)}$, we have therefore found all relations by means of which these $\sum \varepsilon_k$ initial values are linked together; at the same time, we obtain $\nu_0 + \cdots + \nu_{s-1}$ of the quantities $g^0, \dots, g^{s-1}$ represented as *linear homogeneous* functions of the $\sum(\varepsilon_k - \nu_k) = r$ remaining ones, which stay entirely arbitrary.

We use the preceding observations in order to draw conclusions from them about the constitution of a particular system of solutions to our differential equations.

At first, it can be shown that there does not exist a particular system of solutions $\xi_1, \dots, \xi_n$ whose power series expansion with respect to the $x_i - x_i^0$ begins with terms of order s or yet higher order. Indeed, in the power series expansion of such a system of solutions, the coefficients g^0, g', ..., $g^{(s-1)}$ would all be equal to zero, so all $g^{(s)}$,

$g^{(s+1)}, \dots$ would also vanish and the system of solutions would therefore reduce to: $\xi_1 = 0, \dots, \xi_n = 0$. This system of solutions certainly satisfies the given differential equations, but alone it does not deliver an infinitesimal transformation of our group, and is therefore useless.

However, there is a certain number of particular systems of solutions $\xi_1, \dots, \xi_n$ whose power series expansions begin with terms of order lower than the s-th, let us say with terms of k-th order. The coefficients $g^0, g', \dots, g^{(k-1)}$ can then all be chosen equal to zero; so the existing relations between these are all satisfied, and there remains only ν_k relations between the $g^{(k)}$, ν_{k+1} relations between the $g^{(k+1)}$ and certain $g^{(k)}, \dots$, and lastly, ν_{s-1} relations between the $g^{(s-1)}$ and certain $g^{(s-2)}, \dots, g^{(k)}$. As a result, there are in sum still $(\varepsilon_k - \nu_k) + \cdots + (\varepsilon_{s-1} - \nu_{s-1})$ of the constants $g^{(k)}, \dots, g^{(s-1)}$ which can be chosen arbitrarily, and when one disposes of these constants in such a way that not all the ε_k quantities $g^{(k)}$ vanish, then one always obtains a particular system of solutions $\xi_1, \dots, \xi_n$ whose power series expansions with respect to the $x_i - x_i^0$ contain terms of the k-th order, but no terms of lower order.

If $\xi_1, \dots, \xi_n$ is a particular system of solutions of our differential equations, then the infinitesimal transformation:

$$\xi_1 \frac{\partial f}{\partial x_1} + \cdots + \xi_n \frac{\partial f}{\partial x_n}$$

belongs to our group. From what has been said above, it follows that this group always contains infinitesimal transformations whose power series expansion with respect to the $x_i - x_i^0$ begins with terms of order k, only as soon as k is one of the numbers $0, 1, 2, \dots, s-1$. By contrast, there are no infinitesimal transformations in the group whose power series expansions begin with terms of order s or of higher order. Naturally, all of this is proved only under the assumption that the coefficients of the resolved differential equations behave regularly at the point $x_1^0, \dots, x_n^0$.

In order to meet, at least to some extent, the requirements for conciseness of the expression, we henceforth want to say: *an infinitesimal transformation is of the k-th order in the* $x_i - x_i^0$ *when its power series expansion with respect to the* $x_i - x_i^0$ *begins with terms of order k*. Then we can enunciate the preceding result as follows:

If $x_1^0, \dots, x_n^0$ is a point for which the coefficients in the resolved defining equations of the group $X_1 f, \dots, X_r f$ behave regularly, then the group contains certain infinitesimal transformations of zeroth order in the $x_i - x_i^0$, certain of the first order, and in general, certain of the k-th order, where k means one arbitrary number amongst $0, 1, \dots, s-1$; by contrast, the group contains no infinitesimal transformation of order s or of higher order in the $x_i - x_i^0$.

It is clear that two infinitesimal transformations of different orders in the $x_i - x_i^0$ are always independent of each other. Actually, when determining whether several given infinitesimal transformations are independent of each other or not, the consideration of the terms of lowest order in their power series expansions already settles the question many times; indeed, if the terms of lowest order, taken for themselves, determine independent infinitesimal transformations, then the given infinitesimal transformations are also independent of each other.

The general expression of an infinitesimal transformation which belongs to our group and which is of order k with respect to the $x_i - x_i^0$ contains, as we know, $(\varepsilon_k - \nu_k) + \cdots + (\varepsilon_{s-1} - \nu_{s-1}) = \rho_k$ arbitrary and essential constants, namely the $\varepsilon_k - \nu_k$ which can be chosen arbitrarily amongst the $g^{(k)}$, the $\varepsilon_{k+1} - \nu_{k+1}$ arbitrarily amongst the $g^{(k+1)}$, and so on; however here, the arbitrariness of these $\varepsilon_k - \nu_k$ quantities $g^{(k)}$ is restricted inasmuch as not all $g^{(k)}$ are allowed to vanish simultaneously. From this, it follows that ρ_k independent infinitesimal transformations of our group can be exhibited which are of order k in the $x_i - x_i^0$; but it is easy to see that from these ρ_k infinitesimal transformations, one can derive in total ρ_{k+1} independent transformations which are of the $(k+1)$-th order, or yet of higher order. The general expression of an infinitesimal transformation which is linearly deduced from these ρ_k transformations indeed contains exactly the same arbitrary constants as the general expression of an infinitesimal transformation of order k in the $x_i - x_i^0$, with the only difference that in the first expression, all the $\varepsilon_k - \nu_k$ available $g^{(k)}$ can be set equal to zero, which always gives an infinitesimal transformation of order $(k+1)$ or of higher order. Consequently, amongst these ρ_k infinitesimal transformations of order k, there are only $\rho_{k+1} - \rho_k = \varepsilon_k - \nu_k$ which are independent of each other and out of which no infinitesimal transformation of $(k+1)$-th order or of higher order in the $x_i - x_i^0$ can be linearly deduced.

We recapitulate the present result in the

Theorem 29. *To every r-term group $X_1 f, \dots, X_r f$ in n variables $x_1, \dots, x_n$ is associated a completely determined whole number $s \geqslant 1$ of such a nature that, in the neighborhood of a point x_i^0 for which the coefficients of the resolved defining equations behave regularly, the group contains certain infinitesimal transformations of zeroth, of first, ..., of $(s-1)$-th order in the $x_i - x_i^0$, but none of s-th or of higher order. In particular, one can always select r independent infinitesimal transformations of the group such that, for each one of the s values $0, 1, \dots, s-1$ of the number k, exactly $\varepsilon_k - \nu_k$ mutually independent infinitesimal transformations of order k in the $x_i - x_i^0$ are extant out of which no infinitesimal transformation of order $(k+1)$ or of higher order can be linearly deduced. At the same time, the number ν_k can be determined from the defining equations for the general infinitesimal transformation $\xi_1 \frac{\partial f}{\partial x_1} + \cdots + \xi_n \frac{\partial f}{\partial x_n}$ of the group, and from ε_k, which denotes the number of all differential quotients of order k of the $\xi_1, \dots, \xi_n$ with respect to $x_1, \dots, x_n$ and is always larger than ν_k.*

Example. Earlier on, we have already mentioned the equations:

$$\frac{\partial^2 \xi_1}{\partial x_1^2} = \frac{\partial^2 \xi_1}{\partial x_1 \partial x_2} = \frac{\partial^2 \xi_1}{\partial x_2^2} = \frac{\partial^2 \xi_2}{\partial x_1^2} = \frac{\partial^2 \xi_2}{\partial x_1 \partial x_2} = \frac{\partial^2 \xi_2}{\partial x_2^2} = 0$$

as the defining equations of the six-term linear group:

$$x_1' = a_1 x_1 + a_2 x_2 + a_3, \qquad x_2' = a_4 x_1 + a_5 x_2 + a_6.$$

These defining equations have already been presented in the resolved form; all the appearing coefficients are equal to zero, hence they behave regularly. Amongst the infinitesimal transformations of the group, we can select the following six mutually independent ones:

$$\frac{\partial f}{\partial x_1},\quad \frac{\partial f}{\partial x_2},\quad (x_1-x_1^0)\frac{\partial f}{\partial x_1},\quad (x_2-x_2^0)\frac{\partial f}{\partial x_1},\quad (x_1-x_1^0)\frac{\partial f}{\partial x_2},\quad (x_2-x_2^0)\frac{\partial f}{\partial x_2};$$

the first two of them are of zeroth order, and the last four are of first order in the $x_i - x_i^0$.

For calculations with infinitesimal transformations, the expressions of the form:

$$X(Y(f)) - Y(X(f)) = [X, Y]$$

play an important rôle. So if Xf and Yf are expanded in the neighborhood of the point x_i^0 with respect to powers of $x_i - x_i^0$, the related question is how does the transformation $[X, Y]$ behave at this point?

Let the power series expansion of Xf begin with terms of order μ, let that of Yf begin with terms of order ν, that is to say, let:

$$Xf = \sum_{k=1}^{n} (\xi_k^{(\mu)} + \cdots)\frac{\partial f}{\partial x_k}, \qquad Yf = \sum_{j=1}^{n} (\eta_j^{(\nu)} + \cdots)\frac{\partial f}{\partial x_j},$$

where the $\xi^{(\mu)}$ and the $\eta^{(\nu)}$ denote homogeneous functions of order μ and of order ν, respectively, in the $x_i - x_i^0$, while the terms of higher order in the $x_i - x_i^0$ are left out. Under these assumptions, the power series expansion for $[X, Y]$ is, if one only considers terms of the lowest order, the following:

$$[X,Y] = \sum_{j=1}^{n}\left\{\sum_{k=1}^{n}\left(\xi_k^{(\mu)}\frac{\partial \eta_j^{(\nu)}}{\partial x_k} - \eta_k^{(\nu)}\frac{\partial \xi_j^{(\mu)}}{\partial x_k}\right) + \cdots\right\}\frac{\partial f}{\partial x_j}.$$

So the terms of lowest order in $[X, Y]$ are of order $\mu + \nu - 1$ and they stem solely from the terms of orders μ and ν in Xf and in Yf, respectively.

Theorem 30. *If Xf and Yf are two infinitesimal transformations whose power series expansions with respect to the powers of $x_1 - x_1^0, \dots, x_n - x_n^0$ begin with terms of orders μ and ν, respectively, then the power series expansion of the infinitesimal transformation $XYf - YXf = [X, Y]$ begins with terms of order $(\mu + \nu - 1)$ which are entirely determined by the terms of orders μ and ν in Xf and in Yf, respectively. If these terms of order $(\mu + \nu - 1)$ vanish, then concerning the power series expansion of $[X, Y]$, it can only be said that it starts with terms of order $(\mu + \nu)$, or of higher order.*

If the two numbers μ and ν are greater than one, then the number $\mu + \nu - 1$ is larger than both of them. This remark is often of great utility for calculations with infinitesimal transformations of various orders.

For the derivation of Theorem 30, it is not assumed that the two infinitesimal transformations Xf and Yf belong to a group; the only assumption is that both Xf and Yf can be expanded in powers of $x_k - x_k^0$.

§ 52. Let the defining equations of an r-term group be given in the form described earlier, hence resolved with respect to ν_k of the ε_k differential quotients of order k of $\xi_1, \dots, \xi_n$. Moreover, let x_i^0 be a point in which the coefficients of the resolved defining equations behave regularly.

Under these assumptions, we can expand the infinitesimal transformations of our group in ordinary power series of the $x_i - x_i^0$. We even know that our group contains a completely determined number of independent infinitesimal transformations, namely $\varepsilon_0 - \nu_0$, which are of zeroth order in the $x_i - x_i^0$ and out of which no infinitesimal transformation of first order or of higher order can be linearly deduced; moreover, the group contains a completely determined number of independent infinitesimal transformations, namely $\varepsilon_1 - \nu_1$, of first order in the $x_i - x_i^0$ out of which none of second order or of higher order can be linearly deduced, and so on.

Shortly, our group associates to every point of the indicated nature a series of s whole numbers $\varepsilon_0 - \nu_0$, $\varepsilon_1 - \nu_1$, ..., $\varepsilon_{s-1} - \nu_{s-1}$ and these whole numbers are the same for all points of this kind.

Now, there can also be points $\overline{x}_i$ in special position, hence points in the neighborhood of which the coefficients of the resolved defining equations no longer behave regularly, while by contrast, all infinitesimal transformations of the group can be expanded in ordinary power series in the $x_i - \overline{x}_i$. If $\overline{x}_1, \dots, \overline{x}_n$ is a determined point of this sort, then naturally, there is in our group a completely determined number of infinitesimal transformations of zeroth order in the $x_i - \overline{x}_i$ out of which no infinitesimal transformation of higher order can be linearly deduced, and so on.

Consequently, our group also associates to every point in special position a determined, obviously finite series of whole numbers; frequently, to two different points in special position there will also be associated two different series of whole numbers.

The matter is best clarified with an example.

The defining equations of the two-term group $\frac{\partial f}{\partial x_1}$, $x_2^2 \frac{\partial f}{\partial x_1}$ read in the resolved form are:

$$\xi_2 = 0, \quad \frac{\partial \xi_1}{\partial x_1} = -\frac{\partial \xi_2}{\partial x_1} = \frac{\partial \xi_2}{\partial x_2}, = 0$$
$$\frac{\partial^2 \xi_1}{\partial x_1^2} = \frac{\partial^2 \xi_1}{\partial x_1 \partial x_2} = 0, \qquad \frac{\partial^2 \xi_1}{\partial x_2^2} = \frac{1}{x_2} \frac{\partial \xi_1}{\partial x_2},$$
$$\frac{\partial^2 \xi_2}{\partial x_1^2} = \frac{\partial^2 \xi_2}{\partial x_1 \partial x_2} = \frac{\partial^2 \xi_2}{\partial x_2^2} = 0.$$

The coefficients appearing here behave regularly for all points x_1, x_2 located in the finite, except only for the points of the line $x_2 = 0$.

First, let us consider a point x_1^0, x_2^0 with nonvanishing x_2^0. We have $s = 2$, and moreover $\varepsilon_0 = 2$, $\varepsilon_1 = 4$, $\nu_0 = 1$, $\nu_1 = 3$, hence to the point x_1^0, x_2^0 are associated the two numbers $1, 1$. All infinitesimal transformations of the group can be linearly deduced from the two:

$$\frac{\partial f}{\partial x_1}, \qquad \left(x_2 - x_2^0 + \frac{1}{2x_0}(x_2 - x_2^0)^2\right)\frac{\partial f}{\partial x_1}$$

amongst which the first is of zeroth order in the $x_i - x_i^0$, and the second of first order.

Next, let us consider a point $\overline{x}_1, \overline{x}_2 = 0$.

To such a point, the group associates the three numbers $1, 0, 1$, since amongst its infinitesimal transformations there are none of first order in the $x_i - \overline{x}_i$, but one of second order, hence of s-th order, namely: $x_2^2 \frac{\partial f}{\partial x_1}$. —

If $x_1^0, \dots, x_n^0$ is a point for which the coefficients of the resolved defining equations behave regularly, then according to Theorem 29, the group certainly contains infinitesimal transformations of zeroth, of first, ..., of $(s-1)$-th orders in the $x_i - x_i^0$, but none of s-th or of higher order. Now, our example discussed just now shows that for a point $\overline{x}_i$ in which not all the coefficients in question behave regularly, no general statement of this kind holds anymore: the group can very well contain infinitesimal transformations of s-th order in the $x_i - \overline{x}_i$, and perhaps also some of higher order; on the other hand, it can occur that for one number $k < s$, the group actually contains no infinitesimal transformation of k-th order in the $x_i - \overline{x}_i$.

If $x_1^0, \dots, x_n^0$ denotes an arbitrary point in which all the ξ behave regularly, then as already said, the infinitesimal transformations of our group can be classified according to their orders in the $x_i - x_i^0$. It is of great importance that this classification is maintained when in place of the x, new variables $y_1, \dots, y_n$ are introduced. Of course, the concerned change of variables must, in the neighborhood of the point $x_1^0, \dots, x_n^0$, possess the following properties: $y_1, \dots, y_n$ must firstly be ordinary power series in the $x_i - x_i^0$:

$$(2) \qquad y_k = y_k^0 + \sum_{i=1}^{n} a_{ki}\,(x_i - x_i^0) + \cdots \qquad (k=1\cdots n);$$

and secondly, $x_1, \dots, x_n$ must also be representable as ordinary power series in the $y_k - y_k^0$ and in fact, so that every x_i for $y_1 = y_1^0$, ..., $y_n = y_n^0$ must take the value x_i^0. If the first of these two requirements is satisfied, it is known that the second is then always satisfied, when the determinant $\sum \pm a_{11} \cdots a_{nn}$ is different from zero.

Now, in order to prove that the discussed classification is maintained after the transition to the variables $y_1, \dots, y_n$, we need only show that every infinitesimal transformation of μ-th order in the $x_i - x_i^0$ converts, by the introduction of the new variables $y_k - y_k^0$, into an infinitesimal transformation of μ-th order in the $y_k - y_k^0$. But this is not difficult.

The general form of an infinitesimal transformation of the μ-th order in the $x_i - x_i^0$ is:

$$Xf = \sum_{j=1}^{n} (\xi_j^{(\mu)} + \cdots) \frac{\partial f}{\partial x_j};$$

here, $\xi_1^{(\mu)}, \ldots, \xi_n^{(\mu)}$ denote entire rational functions[1] (Translator's note: GANZE RATIONALE FUNCTIONEN, that is to say, polynomials.) which are homogeneous of order μ and do not all vanish; the terms of orders $(\mu+1)$ and higher are left out.

By the introduction of $y_1, \ldots, y_n$ we obtain:

$$Xf = \sum_{k=1}^{n} Xy_k \frac{\partial f}{\partial y_k};$$

here at first, the Xy_k are ordinary power series in the $x_i - x_i^0$:

$$Xy_k = \sum_{j=1}^{n} a_{kj}\, \xi_j^{(\mu)} + \cdots$$

and they begin with terms of order μ. These terms of order μ do not all vanish, since otherwise one would have:

$$\sum_{j=1}^{n} a_{kj}\, \xi_j^{(\mu)} = 0 \qquad (k=1\cdots n),$$

which is impossible, because the determinant $\sum \pm a_{11} \cdots a_{nn}$ is different from zero, and because $\xi_1^{(\mu)}, \ldots, \xi_n^{(\mu)}$ do not all vanish. Now, if in $Xy_1, \ldots, Xy_n$ we express the x_i in terms of the y_i, we obtain n ordinary power series in the $y_i - y_i^0$. These power series likewise begin with terms of order μ which do not all vanish. Indeed, one obtains the terms of order μ in question by substituting, in the n expressions:

$$\sum_{j=1}^{n} a_{kj}\, \xi_j^{(\mu)} \qquad (k=1\cdots n),$$

the x for the y by means of the equations:

$$y_k = y_k^0 + \sum_{i=1}^{n} a_{ki}\,(x_i - x_i^0) \qquad (k=1\cdots n);$$

but since the n shown expressions do not all vanish, they also do not all vanish after introduction of the y.

Consequently, the infinitesimal transformation Xf is transferred, by the introduction of the y, to an infinitesimal transformation which is of order μ in the $y_i - y_i^0$. But this was to be shown.

As a result, we have the

Proposition 1. *If, in an infinitesimal transformation Xf which is of order μ in $x_1 - x_1^0, \ldots, x_n - x_n^0$, one introduces new variables:*

$$y_k = y_k^0 + \sum_{i=1}^{n} a_{ki}(x_i - x_i^0) + \sum_{i,\,j}^{1\cdots n} a_{kij}\,(x_i - x_i^0)\,(x_j - x_j^0) + \cdots$$
$$(k=1\cdots n),$$

where the determinant $\sum \pm a_{11} \cdots a_{nn}$ is different from zero, then Xf converts into an infinitesimal transformation of order μ in $y_1 - y_1^0, \ldots, y_n - y_n^0$.

From this, we immediately obtain the somewhat more specific

Proposition 2. *If, in the neighborhood of the point $x_1^0, \ldots, x_n^0$, an r-term group contains exactly τ_μ independent infinitesimal transformations of order μ in the $x_i - x_i^0$ out of which none of higher order can be linearly deduced, and if new variables:*

$$y_k = y_k^0 + \sum_{i=1}^{n} a_{ki}(x_i - x_i^0) + \cdots \qquad (k=1\cdots n),$$

are introduced in this group, where the determinant $\sum \pm a_{11} \cdots a_{nn}$ is different from zero, then in turn in the neighborhood of the point y_k^0, the new group which one finds in this way contains exactly τ_μ independent infinitesimal transformations of order μ in the $y_k - y_k^0$ out of which none of higher order can be linearly deduced.

One therefore sees: the series of whole numbers which the initial group associates to the point $x_1^0, \ldots, x_n^0$ is identical to the series of whole numbers which the new groups associates to the point $y_1^0, \ldots, y_n^0$.

§ 53. If one knows the defining equations of an r-term group and if one has resolved them in the way discussed earlier on, then as we have seen, one can immediately identify the numbers $\varepsilon_k - \nu_k$ defined above. For every point $x_1^0, \ldots, x_n^0$ in which the coefficients of the resolved defining equations behave regularly, one therefore knows the number of all independent infinitesimal transformations of the group which are of order k in the $x_i - x_i^0$ and which possess the property that out of them, no infinitesimal transformation of order $(k+1)$ or of higher order can be linearly deduced.

Naturally, one can compute the numbers in question also for the points $x_1^0, \ldots, x_n^0$ in which the coefficients of the resolved defining equations do not behave regularly. For that, the knowledge of the defining equations already suffices, however it is incomparably more convenient when r arbitrary independent infinitesimal transformations are already given, which is what we will assume in the sequel. Then one proceeds as follows.

At first, one determines how many independent infinitesimal transformations of order k or higher in the $x_i - x_i^0$ the group contains. To this end, one expands the general infinitesimal transformation:

$$e_1 X_1 f + \cdots + e_r X_r f$$

with respect to the powers of the $x_i - x_i^0$ and then in the n expressions:

$$e_1\,\xi_{1i} + \cdots + e_r\,\xi_{ri} \qquad\qquad {\scriptstyle (i=1\cdots n)},$$

one sets equal to zero all coefficients of zeroth, of first, ..., of $(k-1)$-th order. In this way, one obtains a certain number of linear homogeneous equations between $e_1, \dots, e_r$; one then easily determines how many independent infinitesimal transformations are extant amongst these equations by calculating certain determinants; if $r - \omega_k$ is the number of independent equations, then it follows that the group contains exactly ω_k independent infinitesimal transformations of order k or higher in the $x_i - x_i^0$. So obviously, $\omega_k - \omega_{k+1}$ is the number of independent infinitesimal transformations of order k out of which no infinitesimal transformation of higher order can be linearly deduced.

It is almost unnecessary to make the observation that the operations just indicated also remain applicable to every point $x_1^0, \dots, x_n^0$ for which the coefficients of the resolved defining equations behave regularly.

Somewhat more precisely, we want to occupy ourselves with the infinitesimal transformations $\sum e_j X_j f$ of the group $X_1 f, \dots, X_r f$ whose power series expansion in the $x_i - x_i^0$ contain only terms of the first and higher orders, but none of the zeroth. At first, we shall examine how many independent infinitesimal transformations of this nature there are and we shall show how one can set them up in a simple manner. Here, by $x_1^0, \dots, x_n^0$, we understand a completely arbitrary, though determined, point.

Evidently, such infinitesimal transformations are characterized by the fact that they, and also the one-term groups generated by them, leave at rest the point $x_i = x_i^0$ (cf. Chap. 7, p. 147), or, what amounts to the same, by the fact that they are the only ones amongst the infinitesimal transformations $\sum e_j X_j f$ which do not attach any direction to the point $x_i = x_i^0$.

Analytically, the most general transformation $\sum e_j X_j f$ of the concerned constitution will be determined by the equations:

$$e_1\,\xi_{1i}(x_1^0, \dots, x_n^0) + \cdots + e_r\,\xi_{ri}(x_1^0, \dots, x_n^0) = 0 \qquad\qquad {\scriptstyle (i=1\cdots n)}.$$

Now, if in the matrix:

$$\text{(3)} \qquad \begin{vmatrix} \xi_{11}(x^0) & \cdot & \cdot & \xi_{1n}(x^0) \\ \cdot & & \cdot\;\cdot & \cdot \\ \xi_{r1}(x^0) & \cdot & \cdot & \xi_{rn}(x^0) \end{vmatrix},$$

all $(r+1)\times(r+1)$ determinants vanish, but not all $h \times h$ determinants, then h of the quantities $e_1, \dots, e_r$ can be represented as linear homogeneous functions of the $r - h$ remaining ones, which are completely arbitrary. As a result, we obtain the following simple but important result:

Proposition 3. *If all* $(h+1)\times(h+1)$ *determinants of the matrix:*

$$\begin{vmatrix} \xi_{11}(x) & \cdot & \cdot & \xi_{1n}(x) \\ \cdot & & \cdot\;\cdot & \cdot \\ \xi_{r1}(x) & \cdot & \cdot & \xi_{rn}(x) \end{vmatrix}$$

vanish for $x_1 = x_1^0, \dots, x_n = x_n^0$, *but not all* $h \times h$ *determinants vanish, then the r-term group:*

$$X_k f = \sum_{i=1}^{n} \xi_{ki}(x_1, \dots, x_n) \frac{\partial f}{\partial x_i} \qquad {\scriptstyle (k=1\cdots r)}$$

contains exactly $r-h$ *independent infinitesimal transformations which, when expanded in power series in* $x_1 - x_1^0, \dots, x_n - x_n^0$, *contain no term of zeroth order — which, in other words, leave at rest the point* $x_1^0, \dots, x_n^0$.

At the same time, we also obtain the following

Proposition 4. *When all* $(h+1) \times (h+1)$ *determinants of the matrix:*

$$\begin{vmatrix} \xi_{11}(x) & \cdot & \cdot & \xi_{1n}(x) \\ \cdot & & \cdot\ \cdot & \cdot \\ \xi_{r1}(x) & \cdot & \cdot & \xi_{rn}(x) \end{vmatrix}$$

are set to zero, then the resulting equations determine the locus of all points $x_1, \dots, x_n$ *which admit*[2] *(Translator's note: — in the sense that the manifold constituted of such a point is left invariant, i.e. at rest —) at least* $r-h$ *independent infinitesimal transformations of the group* $X_1 f, \dots, X_r f$; *amongst the found points, those which do not bring to zero all* $h \times h$ *determinants of the matrix admit exactly* $r-h$ *independent infinitesimal transformations of the group.*

Earlier on (p. 149), we have underlined that the infinitesimal transformations $X_1 f, \dots, X_r f$ associate to a determined point $x_1, \dots, x_n$ precisely h independent directions when all $(h+1) \times (h+1)$ determinants of the matrix (3) vanish, while by contrast not all $h \times h$ determinants do. From this, we see that the found result can also be expressed as follows.

Proposition 5. *If an r-term group* $X_1 f, \dots, X_r f$ *of the space* $x_1, \dots, x_n$ *contains exactly* $r-h$ *independent infinitesimal transformations which leave at rest a determined point* $x_1^0, \dots, x_n^0$, *then the infinitesimal transformations of the group associate to this point exactly h independent directions.*

Now we continue one step further to set up all infinitesimal transformations $\sum e_j X_j f$ which leave at rest a determined point $x_1^0, \dots, x_n^0$.

We assume that in the matrix (3), all $(h+1) \times (h+1)$ determinants vanish, but not all $h \times h$ determinants, and specifically, that in the smaller matrix:

$$\begin{vmatrix} \xi_{11}(x^0) & \cdot & \cdot & \xi_{1n}(x^0) \\ \cdot & & \cdot\ \cdot & \cdot \\ \xi_{h1}(x^0) & \cdot & \cdot & \xi_{hn}(x^0) \end{vmatrix},$$

not all $h \times h$ determinants are equal to zero.

Under these assumptions, there are obviously no infinitesimal transformations of the form $e_1 X_1 f + \cdots + e_h X_h f$ which leave at rest the point $x_1^0, \dots, x_n^0$; by contrast, the $r-h$ infinitesimal transformations:

$$X_{h+k}f + \lambda_{k1}\,X_1 f + \cdots + \lambda_{kh}\,X_h f \qquad {\scriptstyle (k=1\cdots r-h)}$$

do leave it at rest, as soon as one chooses the constants λ in an appropriate way, which visibly is possible only in one way. Given this, $r-h$ independent infinitesimal transformations are found whose power series expansions contain no term of order zero; naturally, out of these $r-h$ transformation, every other transformation of the same constitution can be linearly deduced. From this, it follows that amongst the infinitesimal transformations of our group which are of zeroth order in the $x_i - x_i^0$, there are only h independent transformations out of which no transformation of first order or of higher order can be linearly deduced; of course, $X_1 f, \dots, X_r f$ are transformations of zeroth order of this nature; hence they attach to the point $x_1^0, \dots, x_n^0$ exactly h independent directions.

With these words, we have the

Proposition 6. *If the r infinitesimal transformations:*

$$X_k f = \sum_{i=1}^{n} \xi_{ki}(x_1, \dots, x_n)\, \frac{\partial f}{\partial x_i} \qquad {\scriptstyle (k=1\cdots r)}$$

of an r-term group of the space $x_1, \dots, x_n$ are constituted in such a way that for $x_1 = x_1^0, \dots, x_n = x_n^0$, all $(h+1)\times(h+1)$ determinants, but not all $h \times h$ determinants, of the matrix:

$$\begin{vmatrix} \xi_{11}(x) & \cdot & \cdot & \xi_{1n}(x) \\ \cdot & \cdot & \cdot & \cdot \\ \xi_{r1}(x) & \cdot & \cdot & \xi_{rn}(x) \end{vmatrix}$$

vanish, and especially, if not all $h \times h$ determinants of the matrix:

$$\begin{vmatrix} \xi_{11}(x) & \cdot & \cdot & \xi_{1n}(x) \\ \cdot & \cdot & \cdot & \cdot \\ \xi_{h1}(x) & \cdot & \cdot & \xi_{hn}(x) \end{vmatrix}$$

are zero for $x_i = x_i^0$, then firstly: all infinitesimal transformations:

$$e_1 X_1 f + \cdots + e_h X_h f$$

are of zeroth order in the $x_i - x_i^0$ and they attach to the point $x_1^0, \dots, x_n^0$ exactly h independent directions, and secondly: one can always choose $h(r-h)$ constants λ_{kj}, but only in one way, so that in the $r-h$ independent infinitesimal transformations:

$$X_{h+k}f + \lambda_{k1}\,X_1 f + \cdots + \lambda_{kh}\,X_h f \qquad {\scriptstyle (k=1\cdots r-h)}$$

all terms of zeroth order in the $x_i - x_i^0$ are missing; then out of these $r-h$ infinitesimal transformations, one can linearly deduce all infinitesimal transformations of the group $X_1 f, \dots, X_r f$ which are of the first order in the $x_i - x_i^0$, or of higher order.

For the sequel, it is useful to state this proposition in a somewhat more specific way.

We want to assume that all $(q+1)\times(q+1)$ determinants of the matrix:

$$\begin{vmatrix} \xi_{11}(x) & \cdot & \cdot & \xi_{1n}(x) \\ \cdot & \cdot & \cdot & \cdot \\ \xi_{r1}(x) & \cdot & \cdot & \xi_{rn}(x) \end{vmatrix} \tag{4}$$

vanish *identically*, but that this is not the case for all $q\times q$ determinants and , in particular, that not all $q\times q$ determinants of the matrix:

$$\begin{vmatrix} \xi_{11}(x) & \cdot & \cdot & \xi_{1n}(x) \\ \cdot & \cdot & \cdot & \cdot \\ \xi_{q1}(x) & \cdot & \cdot & \xi_{qn}(x) \end{vmatrix} \tag{5}$$

are identically zero.

Under these assumptions, it is impossible to exhibit q not all vanishing functions $\chi_1(x),\dots,\chi_q(x)$ which make identically equal to zero the expression $\chi_1(x)X_1f+\cdots+\chi_q(x)X_qf$. By contrast, one can determine $q(r-q)$ functions $\varphi_{jk}(x)$ so that the $r-q$ equations:

$$X_{q+j}f=\varphi_{j1}(x_1,\dots,x_n)X_1f+\cdots+\varphi_{jq}(x_1,\dots,x_n)X_qf \qquad (j=1\cdots r-q)$$

are identically satisfied; indeed, every φ_{jk} will be equal to a quotient whose numerator is a certain $q\times q$ determinant of the matrix (4) and whose denominator is a not identically vanishing $q\times q$ determinant of the matrix (5) (cf. the analogous developments in Chap. 7, p. 136).

Now, let $x_1^0,\dots,x_n^0$ be a point in general position, or more precisely, a point for which not all $q\times q$ determinants of (5) vanish. Then the expressions $\varphi_{jk}(x^0)$ are determined, finite constants, and at the same time, the $r-q$ infinitesimal transformations:

$$X_{q+j}f-\varphi_{j1}(x^0)X_1f-\cdots-\varphi_{jq}(x^0)X_qf \qquad (j=1\cdots r-q)$$

belong to our group. These infinitesimal transformations are clearly independent of each other and in addition, they possess the property that their power series expansions with respect to the $x_i-x_i^0$ lack all zeroth order terms. Hence according to Proposition 3, p. 212, every infinitesimal transformation of our group whose power series expansion with respect to the $x_i-x_i^0$ only contains terms of first order or of higher order must be linearly expressible by means of the $r-q$ infinitesimal transformations just found.

Thus, the following holds true.

Proposition 7. *If the first q of the infinitesimal transformations $X_1f,\dots,X_rf$ of an r-term group are not linked by linear relations of the form:*

$$\chi_1(x_1,\dots,x_n)X_1f+\cdots+\chi_q(x_1,\dots,x_n)X_qf=0,$$

while $X_{q+1}f,\dots,X_rf$ can be linearly expressed in terms of $X_1f,\dots,X_qf$:

$$X_{q+j}f \equiv \sum_{k=1}^{q} \varphi_{jk}(x_1,\dots,x_n)\,X_kf \qquad (j=1\cdots r-q),$$

then in the neighborhood of every point $x_1^0,\dots,x_n^0$ in general position, the group contains exactly q independent infinitesimal transformations, for example $X_1f,\dots,X_qf$, which are of zeroth order and out of which no infinitesimal transformation of first order or of higher order in the $x_i-x_i^0$ can be linearly deduced. By contrast, in the neighborhood of x_i^0, the group contains exactly $r-q$ independent infinitesimal transformations, for instance:

$$X_{q+j}f - \sum_{k=1}^{q} \varphi_{jk}(x_1^0,\dots,x_n^0)\,X_kf \qquad (j=1\cdots r-q),$$

which contain no terms of zeroth order in the $x_i-x_i^0$, hence which leave at rest the point $x_1^0,\dots,x_n^0$.

We have pointed out above more precisely what is to be understood, in this proposition, for a point in general position.

Chapter 12
Determination of All Subgroups of an r-term Group

If all the transformations of a ρ-term group are contained in a group with more than ρ parameters, say with r parameters, then the ρ-term group is called a *subgroup* of the r-term group.

The developments of Chap. 4 have already given us examples of subgroups of an r-term group; indeed, every r-term group contains ∞^{r-1} one-term subgroups. In the present chapter, we will first describe a few specific methods which enable us to find subgroups of a given group. Then we consider the question of how one should proceed in order to determine all subgroups of a given group. We obtain the important result that the determination of all continuous subgroups of an r-term group can always be achieved by resolution of *algebraic* equations.

§ 54. In the preceding chapter, we imagined the infinitesimal transformations of a given r-term group expanded with respect to powers of the $x_i - x_i^0$, where it is understood that x_i^0 is a system of values for which all these transformations behave regularly.

For the infinitesimal transformations of the group, there resulted in this way a classification which will now lead us towards the existence of certain subgroups. However, the considerations of this section find an application only to groups which in any case for certain points x_i^0, contain not only infinitesimal transformations of zeroth order, but also some of higher order in the $x_i - x_i^0$. In the neighborhood of the point $x_1^0, \dots, x_n^0$, let an r-term group of the space $x_1, \dots, x_n$ contain exactly ω_k independent infinitesimal transformations:

$$Y_1 f, \dots, Y_{\omega_k} f,$$

whose power series expansions with respect to the $x_i - x_i^0$ start with terms of order k or of higher order.

We want to assume that $k \geqslant 1$. Then if we combine two infinitesimal transformations $Y_i f$ and $Y_j f$, we obtain (Theorem 30, p. 207) an infinitesimal transformation $[Y_i, Y_j]$ of order $(2k-1)$ or higher, hence at least of order k. Consequently, $[Y_i, Y_j]$ must be linearly expressible in terms of the $Y f$:

S. Lie, *Theory of Transformation Groups I*,
DOI 10.1007/978-3-662-46211-9_12

$$[Y_i, Y_j] = \sum_{\nu=1}^{\omega_k} d_{ij\nu} Y_\nu f,$$

or, what is the same: the $Y_1 f, \dots, Y_{\omega_k} f$ generate an ω_k-term subgroup of the given group. Hence the following holds true.

Proposition 1. *If an r-term group of the space $x_1, \dots, x_n$ contains, in the neighborhood of $x_1^0, \dots, x_n^0$, exactly ω_k independent infinitesimal transformations of order k or of higher order and if $k \geqslant 1$ here, then these ω_k infinitesimal transformations generate an ω_k-term subgroup of the group in question.*

If the point $x_1^0, \dots, x_n^0$ is procured so that, for it the coefficients of the resolved defining equations of the group $X_1 f, \dots, X_r f$ behave regularly, then ω_k has the value:

$$(\varepsilon_k - \nu_k) + \cdots + (\varepsilon_{s-1} - \nu_{s-1})$$

(cf. Chap. 11, p. 206).

The case $k = 1$ is particularly important, hence we want to dwell on it.

If:

$$X_j f = \sum_{i=1}^{n} \xi_{ji}(x_1, \dots, x_n) \frac{\partial f}{\partial x_i} \qquad (j=1\cdots r)$$

are independent infinitesimal transformations of the r-term group, then according to Chap. 11, p. 212 sq., one finds the number ω_1 by examining the determinants of the matrix:

$$\begin{vmatrix} \xi_{11}(x^0) & \cdot & \cdot & \xi_{1n}(x^0) \\ \cdot & \cdot & \cdot & \cdot \\ \xi_{r1}(x^0) & \cdot & \cdot & \xi_{rn}(x^0) \end{vmatrix}.$$

Moreover, we recall (cf. p. 212 sq.) that all infinitesimal transformations of the group which contain only terms of first order or of higher order in the $x_i - x_i^0$ are characterized by the fact that they leave at rest the point $x_1^0, \dots, x_n^0$. Hence if $k = 1$, we can also enunciate the above proposition as follows.

Proposition 2. *If, in a group of the space $x_1, \dots, x_n$, there are precisely ω_1 independent infinitesimal transformations which leave invariant a determined point $x_1^0, \dots, x_n^0$, then these transformations generate an ω_1-term subgroup of the concerned group.*

It is clear that in the variables $x_1, \dots, x_n$, there are no more than n infinitesimal transformations which are of zeroth order in the $x_i - x_i^0$ and out of which no infinitesimal transformation of first order or of higher order can be linearly deduced. From this, we conclude that every r-term group in $n < r$ variables contains at least $r - n$ independent infinitesimal transformations which are of first or higher order in the $x_i - x_i^0$. We therefore have the

Proposition 3. *Every r-term group in $n < r$ variables contains subgroups with at least $r - n$ parameters.*

From Proposition 7 of the preceding chapter (p. 215), we finally obtain for the points $x_1^0, \dots, x_n^0$ in general position the

Proposition 4. *If the infinitesimal transformations $X_1 f, \dots, X_q f, \dots, X_r f$ of an r-term group in the space $x_1, \dots, x_n$ are constituted in such a manner that $X_1 f, \dots, X_q f$ are linked by no linear relation of the form:*

$$\chi_1(x_1, \dots, x_n) X_1 f + \cdots + \chi_q(x_1, \dots, x_n) X_q f \equiv 0,$$

while by contrast $X_{q+1} f, \dots, X_r f$ can be expressed linearly in terms of $X_1 f, \dots, X_q f$:

$$X_{q+j} f \equiv \varphi_{j1}(x_1, \dots, x_n) X_1 f + \cdots + \varphi_{jq}(x_1, \dots, x_n) X_q f \qquad (j=1 \cdots r-q),$$

and if in addition $x_1^0, \dots, x_n^0$ is a point in general position, then the $r-q$ infinitesimal transformations:

$$X_{q+j} f - \sum_{\mu=1}^{q} \varphi_{j\mu}(x_1^0, \dots, x_n^0) X_\mu f \qquad (j=1 \cdots r-q)$$

are all of first or higher order in the $x_i - x_i^0$ and they generate an $(r-q)$-term subgroup whose transformations are characterized by the fact that they leave invariant the point $x_1^0, \dots, x_n^0$.

By a point in general position, as on p. 215, we understand here a point which does not bring to zero all $q \times q$ determinants of the matrix:

$$\begin{vmatrix} \xi_{11}(x) & \cdot & \cdot & \xi_{1n}(x) \\ \cdot & \cdot & \cdot & \cdot \\ \xi_{q1}(x) & \cdot & \cdot & \xi_{qn}(x) \end{vmatrix}.$$

§ 55. Here we also want to draw attention to a somewhat more general method which often leads, in a very simple way, to the determination of certain subgroups of a given group. This method is founded on the following

Theorem 31. *If the r-term group $X_1 f, \dots, X_r f$ contains some infinitesimal transformations for which a given system of equations:*

$$\Omega_i(x_1, \dots, x_n) = 0 \qquad (i=1,2\cdots)$$

remains invariant, and if every infinitesimal transformation of this nature can be linearly deduced from the m infinitesimal transformations:

$$Y_k f = \sum_{\nu=1}^{r} h_{k\nu} X_\nu f \qquad (k=1 \cdots m),$$

then $Y_1 f, \dots, Y_m f$ *generate an m-term subgroup of the group* $X_1 f, \dots, X_r f$.

The correctness of this theorem follows almost immediately from Proposition 5 in Chap. 7, p. 134. Indeed, according to this proposition the system $\Omega_i = 0$ admits all infinitesimal transformations of the form $[Y_k, Y_j]$. Since the $[Y_k, Y_j]$ also belong to the group $X_1 f, \dots, X_r f$, then under the assumptions made, none of these infinitesimal transformations can be independent of $Y_1 f, \dots, Y_m f$, hence on the contrary, there must exist relations of the form:

$$[Y_k, Y_j] = \sum_{\mu=1}^{m} h_{kj\mu} Y_\mu f,$$

that is to say, the $Y_k f$ generate a group.

Clearly, we can also state the above theorem as follows.

Proposition 5. *If, amongst the infinitesimal transformations of an r-term group in the variables* $x_1, \dots, x_n$, *there are precisely m independent transformations by which a certain manifold of the space* $x_1, \dots, x_n$ *remains invariant, then these m infinitesimal transformations generate an m-term subgroup of the r-term group.*

The simplest case is the case where the invariant system of equations represents an invariant point, so that it has the form:

$$x_1 = x_1^0, \; \dots, \; x_n = x_n^0.$$

The subgroup which corresponds to this system of equations is of course generated by all infinitesimal transformations whose power series expansions with respect to the $x_i - x_i^0$ start with terms of first order, or of higher order. Thus, we arrive here at one of the subgroups that we have already found in the preceding section.

As a second example, we consider a subgroup of the eight-term general projective group of the plane. The equation of a nondegenerate conic section admits exactly three independent infinitesimal projective transformations of the plane; hence these three infinitesimal transformations generate a three-term subgroup of the general projective group.

Lastly, yet another example, obtained from the ten-term group of all conformal point transformations of R_3. In this group, there are exactly six independent infinitesimal transformations which leave invariant an arbitrarily chosen sphere. These transformations generate a six-term subgroup of the ten-term group.

Theorem 31 is only a special case of the following more general

Theorem 32. *If, in the variables* $x_1, \dots, x_n$, *an arbitrary group is given, finite or infinite, continuous or not continuous, then the totality of all transformations contained in it which leave invariant an arbitrary system of equations in* $x_1, \dots, x_n$, *also forms a group.*

The proof of this theorem is very simple. Any two infinitesimal transformations of the group which, when executed one after the other, leave invariant the system of equations give a transformation which again belongs to the group and which at the same time leaves invariant the system of equations. As a result, the proof that the totality of transformations defined in the theorem effectively forms a group is produced.

Instead of a system of equations, one can naturally also consider a system of differential equations, and then the theorem would still remain valid.

Furthermore, if the given group is continuous, then the subgroup which is defined through the invariant system of equations can very well be discontinuous.

§ 56. Now that we have become acquainted with some methods which enable us to find individual subgroups of a given group, we turn to the more general problem of determining *all* continuous subgroups of a given r-term group $X_1 f, \dots, X_r f$.

Some m arbitrary independent infinitesimal transformations:

$$Y_\mu f = \sum_{\rho=1}^{r} h_{\mu\rho}\, X_\rho f \qquad (\mu = 1 \cdots m)$$

of our group generate an m-term subgroup if and only if all:

$$[Y_\mu, Y_\nu] = \sum_{\rho,\,\sigma}^{1\cdots r} h_{\mu\rho}\, h_{\nu\sigma}\, [X_\rho, X_\sigma]$$

are expressible only by means of $Y_1 f, \dots, Y_m f$. If we insert here the values:

$$[X_\rho, X_\sigma] = \sum_{\tau=1}^{r} c_{\rho\sigma\tau} X_\tau f,$$

then:

$$[Y_\mu, Y_\nu] = \sum_{\rho\sigma\tau}^{1\cdots r} h_{\mu\rho}\, h_{\nu\sigma}\, c_{\rho\sigma\tau} X_\tau f,$$

and it is demanded that these equations take the form:

$$[Y_\mu, Y_\nu] = \sum_{\pi=1}^{m} l_{\mu\nu\pi} Y_\pi f = \sum_{\tau=1}^{r} \sum_{\pi=1}^{m} l_{\mu\nu\pi}\, h_{\pi\tau} X_\tau f.$$

For this to hold, it is necessary and sufficient that the equations:

$$\sum_{\rho,\,\sigma}^{1\cdots r} h_{\mu\rho}\, h_{\nu\sigma}\, c_{\rho\sigma\tau} = \sum_{\pi=1}^{m} l_{\mu\nu\pi}\, h_{\pi\tau} \tag{1}$$

$$(\mu, \nu = 1 \cdots m;\ \tau = 1 \cdots r)$$

can be satisfied; for this to be possible, all the $(m+1) \times (m+1)$ determinants of the matrix:

$$(2) \qquad \begin{vmatrix} \sum_{\rho,\sigma}^{1\cdots r} h_{\mu\rho}h_{\nu\sigma}c_{\rho\sigma 1} & h_{11} & \cdot & h_{m1} \\ \cdot & \cdot & \cdot & \cdot \\ \sum_{\rho,\sigma}^{1\cdots r} h_{\mu\rho}h_{\nu\sigma}c_{\rho\sigma r} & h_{1r} & \cdot & h_{mr} \end{vmatrix}$$

must vanish.

As a result, we have a series of algebraic equations for the determination of the mr unknowns $h_{\pi\rho}$. But because $Y_1 f,\dots,Y_m f$ should be independent infinitesimal transformations, one should from the beginning exclude every system of values $h_{\pi\rho}$ which brings to zero all $m \times m$ determinants of the matrix:

$$\begin{vmatrix} h_{11} & \cdot & \cdot & h_{m1} \\ \cdot & \cdot & \cdot & \cdot \\ h_{1r} & \cdot & \cdot & h_{mr} \end{vmatrix}.$$

If the $h_{\pi\rho}$ are determined so that all these conditions are satisfied, then for every pair of numbers μ, ν, the equations (1) reduce to exactly m equations that determine completely the unknown constants $l_{\mu\nu 1},\dots,l_{\mu\nu m}$. Therefore, every system of solutions $h_{\pi\rho}$ provides an m-term subgroup, and it is clear that in this way, one finds all m-term subgroups.

With this, we have a general method for the determination of all subgroups of a given r-term group; however in general, this method is practically applicable only when the number r is not too large; nonetheless it shows that the problem of determining all these subgroups necessitates only algebraic operations, which is already a very important result.

Theorem 33. *The determination of all continuous subgroups of a given r-term group $X_1 f,\dots,X_r f$ necessitates only algebraic operations; the concerned operations are completely determined in terms of the constants c_{iks} in the relationships:*[1]

$$[X_i, X_k] = \sum_{s=1}^{r} c_{iks} X_s f \qquad (i,k=1\cdots r).$$

In specific cases, the determination of all subgroups of a given group will often be facilitated by the fact that one knows from the beginning certain subgroups and actually also certain properties of the concerned group; naturally, there is also a simplification when one has already settled the corresponding problem for one subgroup of the given group. In addition, we shall see later that the matter is not actually of setting up *all* subgroups, but rather, that it suffices to identify certain of these subgroups (cf. the studies on types of subgroups, Chap. 23).

§ 57. Let the m independent infinitesimal transformations:

$$Y_\mu f = \sum_{k=1}^{r} h_{\mu k} X_k f \qquad (\mu=1\cdots m)$$

[1] LIE, Archiv for Mathematik of Naturv., Vol. 1, Christiania 1876.

generate an m-term subgroup of the r-term group $X_1 f, \dots, X_r f$. Then the general infinitesimal transformation of this subgroup is:

$$\sum_{\mu=1}^{m} \alpha_\mu Y_\mu f = \sum_{\mu=1}^{m} \sum_{k=1}^{r} \alpha_\mu h_{\mu k} X_k f,$$

where the α_μ denote arbitrary parameters.

As a consequence of this, all infinitesimal transformations $e_1 X_1 f + \cdots + e_r X_r f$ of the r-term group which belong to the m-term subgroup $Y_1 f, \dots, Y_m f$ are defined by the equations:

$$e_k = \sum_{\mu=1}^{m} \alpha_\mu h_{\mu k} \qquad (k = 1 \cdots r).$$

Hence if we imagine here that the m arbitrary parameters α_μ are eliminated, we obtain exactly $r - m$ independent linear homogeneous equations between $e_1, \dots, e_r$, so that we can say:

Proposition 6. *If $e_1 X_1 f + \cdots + e_r X_r f$ is the general infinitesimal transformation of an r-term group, then the infinitesimal transformations of an arbitrary m-term subgroup of this group can be defined by means of $r - m$ independent linear homogeneous relations between $e_1, \dots, e_r$.*

The infinitesimal transformations which are common to two distinct subgroups of an r-term group $X_1 f, \dots, X_r f$ generate in turn a subgroup; indeed, according to Chap. 9, Proposition 2, p. 174, the infinitesimal transformations common to the two groups generate a group, which, naturally, is contained in $X_1 f, \dots, X_r f$ as a subgroup.

Now, if we assume that one of the two groups is m-term, and the other μ-term, then their common infinitesimal transformations will be defined by means of $r - m + r - \mu$ linear homogeneous equations between the e, some equations which need not, however, be mutually independent.

From this, we conclude that amongst the common infinitesimal transformations, there are at least $r - (2r - m - \mu) = m + \mu - r$ which are independent. Therefore, we have the statement:

Proposition 7. *If an r-term group contains two subgroups with m and μ parameters, respectively, then these two subgroups have at least $m + \mu - r$ independent infinitesimal transformations in common. The infinitesimal transformations of the r-term group which are actually common to the two subgroups generate in turn a subgroup.*

This proposition can evidently be generalized:

Actually, if amongst the infinitesimal transformations $e_1 X_1 f + \cdots + e_r X_r f$ of an r-term group, two *families* are sorted, the one by means of $r - m$ linear homogeneous equations, the other by means of $r - \mu$ such equations, then there are at least $m + \mu - r$ independent infinitesimal transformations which are common to the two families.

Chapter 13
Transitivity, Invariants, Primitivity

The concepts of transitivity and of primitivity which play such a broad rôle in the theory of substitutions, shall be extended [AUSGEDEHNT] here to finite continuous transformation groups. In passing, let us mention that these concepts can be extended to all groups, namely to finite groups and infinite groups, and to continuous groups and non-continuous groups.[1]

§ 58. A finite continuous group in the variables $x_1,\dots,x_n$ is called *transitive* when, in the space $(x_1,\dots,x_n)$, there is an n-times extended domain inside which each point can be transferred to any other point by means of at least one transformation of the group. One calls *intransitive* every group which is not transitive.

According to this definition, an r-term group:

$$x_i' = f_i(x_1,\dots,x_n,\, a_1,\dots,a_r) \qquad {\scriptstyle (i=1\cdots n)}$$

is transitive when in general,[1] (Translator's note: As usual (cf. Chap. 1), one reasons *generically*, hence the concept of transitivity is essentially considered for (sub)domains and for generic values of the x and of the x'.) to every system of values $x_1,\dots,x_n$, $x_1',\dots,x_n'$ at least one system of values $a_1,\dots,a_r$ can be determined so that the equations $x_i' = f_i(x,a)$ are satisfied by the concerned values of x, a, x'. In other words: *The equations $x_i' = f_i(x,a)$ of a* transitive group *can be resolved with respect to n of the r parameters $a_1,\dots,a_r$. If by contrast such a resolution is impossible, and if rather, from the equations $x_i' = f_i(x,a)$ of the group, one can*

[1] After LIE had integrated a few differential equations with known continuous groups in 1869, in 1871 and in 1872, he stated in conjunction with KLEIN the problem of translating the concepts of the theory of substitutions as far as possible into the theory of continuous transformation groups. LIE settled this problem in detail; on the basis of the presentation and of the concepts exhibited here, as early as 1874, he developed the fundamentals of a general theory of integration of complete systems which admit known infinitesimal transformations (Verh. d. G. d. W. zu Christiania, 1874). He reduced this problem to the case where the known infinitesimal transformations generate a finite continuous group which is imprimitive, since its transformations permute the characteristic manifolds of the complete system. Amongst other things, he determined all cases where the integration of the complete system can be performed by means of quadratures.

S. Lie, *Theory of Transformation Groups I*,
DOI 10.1007/978-3-662-46211-9_13

derive equations which are free of the parameters a and which contain only the variables $x_1,\dots,x_n$, $x_1',\dots,x_n'$, *then the group is not transitive, it is* intransitive.

From this, we see that every transitive group of the space $x_1,\dots,x_n$ contains at least n essential parameters.

If a *transitive* group in n variables has exactly n essential parameters, then in general, it contains one, but only one, transformation which transfers an arbitrary point of the space to another arbitrary point; hence in particular, aside from the identity transformation, it contains no transformation which leaves invariant a point in general position. We call *simply transitive* [EINFACH TRANSITIV] every group of this nature.

The general criterion for transitivity and, respectively, for intransitivity of a group given above is practically applicable only when one knows the finite equations of the group. But should there not be a criterion the application of which would require only the knowledge of the infinitesimal transformations of the group? We will show that such a criterion can indeed be exhibited. At the same time, we will find means in order to recognize how many and which relations there are between the x and the x' only for an intransitive group with known infinitesimal transformations.

Let us be given r independent infinitesimal transformations:

$$X_k f = \sum_{i=1}^{n} \xi_{ki}(x_1,\dots,x_n)\,\frac{\partial f}{\partial x_i} \qquad {\scriptstyle (k=1\cdots r)}$$

which generate an r-term group. This group, the finite equations of which we can imagine written down in the form:

$$\Phi_i = -x_i' + x_i + \sum_{k=1}^{r} e_k\,\xi_{ki}(x) + \cdots \qquad {\scriptstyle (i=1\cdots n),}$$

is, according to the preceding, transitive if and only if the n equations $\Phi_1 = 0$, ..., $\Phi_n = 0$ are solvable with respect to n of the r parameters $e_1,\dots,e_r$. Consequently, for the transitivity of our group, it is necessary and sufficient that not all $n\times n$ determinants of the matrix:

$$(1)\qquad \begin{vmatrix} \frac{\partial \Phi_1}{\partial e_1} & \cdot & \cdot & \frac{\partial \Phi_n}{\partial e_1} \\ \cdot & \cdot & \cdot & \cdot \\ \frac{\partial \Phi_1}{\partial e_r} & \cdot & \cdot & \frac{\partial \Phi_n}{\partial e_r} \end{vmatrix}$$

vanish identically. From this, it follows that the group is certainly transitive when the corresponding determinants do not all vanish for $e_1 = 0,\dots,e_r = 0$, hence when not all $n\times n$ determinants of the matrix:

$$(2)\qquad \begin{vmatrix} \xi_{11} & \cdots & \xi_{1n} \\ \cdot & \cdot\;\cdot & \cdot \\ \xi_{r1} & \cdots & \xi_{rn} \end{vmatrix}$$

are identically zero.

As a result, we have found a sufficient condition for the transitivity of our group. We now want to study what happens when this condition is not satisfied.

So, let all $n \times n$ determinants of the above-written matrix (2) be identically zero; in order to embrace all possibilities in one case, we in addition want to assume that all determinants of sizes $(n-1)$, $(n-2)$, ..., $(q+1)$ vanish identically, whereas not all $q \times q$ determinants do this.

Under these circumstances, it is certain that not all $q \times q$ determinants of the matrix (1) vanish, hence we can conclude that from the n equations $\Phi_1 = 0$, ..., $\Phi_n = 0$, at most $n-q$ equations between the x and x' only that are free of $e_1, \dots, e_r$ and are mutually independent can be deduced.

But now, under the assumptions made, the r equations $X_1 f = 0$, ..., $X_r f = 0$ reduce to $q < n$ independent ones, say to: $X_1 f = 0, \dots, X_q f = 0$, while $X_{q+1} f, \dots, X_r f$ can be expressed as follows:

$$X_{q+\nu} f \equiv \varphi_{\nu 1}(x) X_1 f + \cdots + \varphi_{\nu q}(x) X_q f \qquad (\nu = 1 \cdots r-q).$$

Consequently, for all values of i and k, one will have:

$$[X_i, X_k] = \sum_{j=1}^{q} \left(c_{ikj} + \sum_{\nu=1}^{r-q} c_{ik,\,q+\nu}\, \varphi_{\nu j} \right) X_j f,$$

that is to say: the q equations $X_1 f = 0$, ..., $X_q f = 0$ form a q-term complete system with $n-q$ independent solutions, which can be denoted by $\Omega_1(x), \dots, \Omega_{n-q}(x)$. These solutions admit every infinitesimal transformation of the form $e_1 X_1 f + \cdots + e_r X_r f$, hence every infinitesimal transformation, and in consequence of that, also every finite transformation of our r-term group $X_1 f, \dots, X_r f$ (cf. Chap. 6, p. 113). Analytically, this is expressed by saying that between the variables x and x', which appear in the transformation equations of our group, the following $n-q$ equations free of the e are extant:

$$\Omega_1(x_1', \dots, x_n') = \Omega_1(x_1, \dots, x_n), \quad \dots\dots, \quad \Omega_{n-q}(x_1', \dots, x_n') = \Omega_{n-q}(x_1, \dots, x_n).$$

Above, we said that between the x and the x' alone, there could exist at most $n-q$ independent relations, and therefore we have found all the relations in question.

In particular, we realize that the group $X_1 f, \dots, X_r f$ is intransitive as soon as all $n \times n$ determinants of the matrix (2) vanish identically. With this, it is shown that the sufficient condition found a short while ago for the transitivity of the group $X_1 f, \dots, X_r f$ is not only sufficient, but also necessary.

We formulate the obtained results as propositions. At the head, we state the

Theorem 34. *The r-term group $X_1 f, \dots, X_r f$ of the space $x_1, \dots, x_n$ is transitive when amongst the r equations $X_1 f = 0, \dots, X_r f = 0$, exactly n mutually independent ones are found, and it is intransitive in the opposite case.*

Then we have the following proposition:

Proposition 1. *From the finite equations:*

$$x_i' = f_i(x_1,\dots,x_n;a_1,\dots,a_r) \qquad {\scriptstyle (i=1\cdots n)}$$

of an r-term group with the infinitesimal transformations $X_1f,\dots,X_rf$, one can eliminate the r parameters $a_1,\dots,a_r$ only when the group is intransitive; in this case, one obtains between the x and the x′ a certain number of relations that can be brought to the form:

$$\Omega_k(x_1',\dots,x_n') = \Omega_k(x_1,\dots,x_n) \qquad {\scriptstyle (k=1,2\cdots)};$$

here, $\Omega_1(x)$, $\Omega_2(x)$, … is a system of independent solutions of the complete system which is determined by the r equations $X_1f=0$, …, $X_rf=0$.

In Chap. 6, p. 112, we saw that the solutions of the linear partial differential equations $Xf=0$ are the only invariants of the one-term group Xf; accordingly, the common solutions of the equations $X_1f=0$, …, $X_rf=0$ are the only invariants of the group $X_1f,\dots,X_rf$. We can hence state the proposition:

Proposition 2. *If the r-term group $X_1f,\dots,X_rf$ of the space $x_1,\dots,x_n$ is transitive, then it has no invariant; if it is intransitive, then the common solutions of the equations $X_1f=0$, …, $X_rf=0$ are its only invariants.*

In order to get straight [KLARSTELLEN] the conceptual sense [BEGRIFFLICHEN SINN] of the obtained analytic results, we now want to interpret $x_1,\dots,x_n$ as point coordinates of a space of n dimensions.

Let the complete system mentioned in the theorem be q-term and let the number q be smaller than n, so that the group $X_1f,\dots,X_rf$ is intransitive. Let the functions $\Omega_1(x_1,\dots,x_n)$, …, $\Omega_{n-q}(x_1,\dots,x_n)$ be independent solutions of the complete system in question, and let $C_1,\dots,C_{n-q}$ denote arbitrary constants. *Then the equations:*

$$\Omega_1 = C_1, \ \dots, \ \Omega_{n-q} = C_{n-q}$$

decompose the whole space into ∞^{n-q} different q-times extended subsidiary domains [THEILGEBIETE] *which all remain invariant by the group $X_1f,\dots,X_rf$.* Every point of the space belongs to a completely determined subsidiary domain and can be transferred only to points of the same subsidiary domain by transformations of the group. Still, the points of a subsidiary domain are transformed transitively, that is to say, every point in general position in the concerned subsidiary domain can be transferred to any other such point by means of at least one transformation of the group.

If $x_1,\dots,x_n$ is a point in general position, then we also call the functions $\Omega_1(x),\dots,\Omega_{n-q}(x)$ the invariants of the point $x_1,\dots,x_n$ with respect to the group $X_1f,\dots,X_rf$. The number of these invariants indicates the degree of intransitivity of our group, since the larger the number of invariants is, the smaller is the dimension number of the subsidiary domains, inside which the point $x_1,\dots,x_n$ stays through the transformations of the group.

Relative to a transitive group, a point in general position clearly has no invariant.

The theorem stated above enables us to determine whether a given group $X_1 f, \dots, X_r f$ is transitive or not. At present, one can give various other versions of the criterion contained there for the transitivity or the intransitivity of a group.

First, with a light change of terminology, we can say:

An r-term group $X_1 f, \dots, X_r f$ in $x_1, \dots, x_n$ is transitive if and only if amongst its infinitesimal transformations, there are exactly n — say $X_1 f, \dots, X_n f$ — which are linked by no relation of the form:

$$\chi_1(x_1, \dots, x_n) X_1 f + \cdots + \chi_n(x_1, \dots, x_n) X_n f = 0,$$

whereas $X_{n+1} f, \dots, X_r f$ is expressed as follows in terms of $X_1 f, \dots, X_n f$:

$$X_{n+j} f = \varphi_{j1}(x) X_1 f + \cdots + \varphi_{jn}(x) X_n f$$
$$(j=1 \cdots r-n).$$

If there are no infinitesimal transformations of this constitution, then the group is intransitive.

Now, if we remember that every infinitesimal transformation $X_k f$ attaches a direction to each point of the space $x_1, \dots, x_n$, and if we add what has been said in Chap. 6, p. 117, concerning the independence of such directions which pass through the same point, then we can also state as follows the criterion for the transitivity of a group:

Proposition 4. *A group $X_1 f, \dots, X_r f$ in the variables $x_1, \dots, x_n$ is transitive when it contains n infinitesimal transformations which attach to every point in general position n independent directions; if the group contains no infinitesimal transformation of this constitution, then it is intransitive.*

On the other hand, let us recall the discussions in Chap. 11, p. 215, where we imagined the infinitesimal transformations of the group expanded with respect to the powers of the $x_i - x_i^0$ in the neighborhood of a point x_i^0 in general position. Since a transitive group $X_1 f, \dots, X_r f$ in the variables $x_1, \dots, x_n$ contains n infinitesimal transformations, say: $X_1 f, \dots, X_n f$ which are linked by no relation $\chi_1(x) X_1 f + \cdots + \chi_n(x) X_n f = 0$, we obtain the following proposition:

Proposition 5. *An r-term group $X_1 f, \dots, X_r f$ in the n variables $x_1, \dots, x_n$ is transitive if, in the neighborhood of a point x_i^0 in general position, it contains exactly n independent infinitesimal transformations of zeroth order in the $x_i - x_i^0$ out of which no infinitesimal transformation of first order or of higher order can be linearly deduced. If the number of such infinitesimal transformations of zeroth order is smaller than n, then the group is intransitive.*

From this, one sees *that one need only know the defining equations of the group $X_1 f, \dots, X_r f$ in order to settle its transitivity or its intransitivity.*

Finally, we can also state the first part of Proposition 5 as follows:

The group $X_1 f, \dots, X_r f$ is transitive when in the neighborhood of a point x_i^0 in general position, it contains exactly $r - n$ independent infinitesimal transformations whose power series expansions with respect to the $x_i - x_i^0$ start with terms of first or of higher order, hence when the group contains exactly $r-n$, and not more, independent infinitesimal transformations which leave invariant a point in general position. —

If one only knows the infinitesimal transformations $X_1 f,\dots,X_r f$ of an intransitive group then, as we have seen, one finds the invariants of the group by integrating the complete system $X_1 f = 0,\dots,X_r f = 0$. Now, it is of great importance that this integration is not necessary when the finite equations of the group are known, so that in this case, the invariants of the group are found instead by plain elimination.

In order to prove this, we imagine that the finite equations of an intransitive group are given:

$$x_i' = f_i(x_1,\dots,x_n\, a_1,\dots,a_r) \qquad (i=1\cdots n),$$

and then we eliminate the parameters $a_1,\dots,a_r$ from them. According to the preceding, it must be possible to bring the $n-q$ independent equations obtained in this way:

$$W_\mu(x_1,\dots,x_n,x_1',\dots,x_n') = 0 \qquad (\mu=1\cdots n-q), \tag{3}$$

to the form:

$$\Omega_\mu(x_1',\dots,x_n') = \Omega_\mu(x_1,\dots,x_n) \qquad (\mu=1\cdots n-q), \tag{4}$$

where the $\Omega_\mu(x)$ are the sought invariants. Hence, when we solve the equations (3) with respect to $n-q$ of the variables $x_1',\dots,x_n'$:

$$x_\mu' = \Pi_\mu(x_1,\dots,x_n,x_{n-q+1}',\dots,x_n') \qquad (\mu=1\cdots n-q),$$

which is always possible, then we obtain $n-q$ functions $\Pi_1,\dots,\Pi_{n-q}$ in which the variables $x_1,\dots,x_n$ occur only in the combinations $\Omega_1(x),\dots,\Omega_{n-q}(x)$. Consequently, the $n-q$ expressions:

$$\Pi_\mu(x_1,\dots,x_n,\alpha_{n-q+1},\dots,\alpha_n) \qquad (\mu=1\cdots n-q)$$

in which $\alpha_{n-q+1},\dots,\alpha_n$ denote constants, represent invariants of our group, and in fact clearly, $n-q$ independent invariants. —

The following therefore holds true.

Theorem 35. *If one knows the finite transformations:*

$$x_i' = f_i(x_1,\dots,x_n,a_1,\dots,a_r) \qquad (i=1\cdots n)$$

of an intransitive group, then one can find the invariants of this group by means of elimination.

§ 59. When we studied how a point in general position behaves relative to the transformations of an r-term group, we were naturally led to the concepts of transitivity and of intransitivity [WURDEN WIR MIT NOTHWENDIGKEIT AUF DIE BEGRIFFE TRANSITIVITÄT UND INTRANSITIVITÄT GEFÜHRT]. We obtained in this way a division [EINTHEILUNG] of all r-term groups of a space of n dimensions in two different classes, exactly the same way as in the theory of substitutions; but at the same time, we also obtained a division of the intransitive groups, namely according to the number of the invariants that a point in general position possesses with respect to the concerned group.

Correspondingly to the process of the theory of substitutions, we now can also go further and study the behavior of two or more points in general position relative to an r-term group. This gives us a new classification of the groups of R_n.

Let:

$$y_i = f_i(x_1,\dots,x_n,a_1,\dots,a_r) \qquad {\scriptstyle (i=1\cdots n)}$$

be an r-term group and let $X_1f,\dots,X_rf$ be r independent infinitesimal transformations of it.

At first, we want to consider two points $x_1',\dots,x_n'$ and $x_1'',\dots,x_n''$ and to seek their invariants relative to our group, that is to say: we seek all functions of $x_1',\dots,x_n'$, $x_1'',\dots,x_n''$ which remain invariant by the transformations of our group.

To this end, we write the infinitesimal transformations X_kf once in the x' under the form $X_k'f$ and once in the x'' in the form $X_k''f$; then simply, the sought invariants are the invariants of the r-term group:

$$(5) \qquad X_k'f + X_k''f \qquad {\scriptstyle (k=1\cdots r)}$$

in the variables $x_1',\dots,x_n', x_1'',\dots,x_n''$.

If $J_1(x),\dots,J_{\rho_1}(x)$ are the invariants of the group $X_1f,\dots,X_rf$, then without effort, the $2\rho_1$ functions:

$$J_1(x'),\dots,J_{\rho_1}(x'),\;\; J_1(x''),\dots,J_{\rho_1}(x'')$$

are invariants, and in fact, independent invariants of the group (5); but in addition, there can be a certain number, say ρ_2, of invariants:

$$J_1'(x_1',\dots,x_n',x_1'',\dots,x_n''),\;\;\dots,\;\; J_{\rho_2}'(x_1',\dots,x_n',x_1'',\dots,x_n'')$$

which are mutually independent and are independent of the above $2\rho_1$ invariants. So in this case, two points in general position have $2\rho_1+\rho_2$ independent invariants relative to the group $X_1f,\dots,X_rf$, amongst which however, only ρ_2 have to be considered as essential, because each one of the two points already has ρ_1 invariants for itself. Under these assumptions, from the equations:

$$y_i' = f_i(x_1',\dots,x_n',a_1,\dots,a_r), \qquad y_i'' = f_i(x_1'',\dots,x_n'',a_1,\dots,a_r) \qquad {\scriptstyle (i=1\cdots n)}$$

there result the following relations free of the a:

$$J_k(y') = J_k(x'), \qquad J_k(y'') = J_k(x'') \qquad (k=1\cdots\rho_1)$$
$$J'_j(y', y'') = J'_j(x', x'') \qquad (j=1\cdots\rho_2).$$

Therefore, if we imagine that the quantities $x'_1, \dots, x'_n$ and $y'_1, \dots, y'_n$ are chosen fixed, then the totality of all positions $y''_1, \dots, y''_n$ which the point $x''_1, \dots, x''_n$ can take are determined by the equations:

$$J_k(y'') = J_k(x''), \qquad J'_j(y', y'') = J'_j(x', x'')$$
$$(k=1\cdots\rho_1\,;\; j=1\cdots\rho_2)\,;$$

so there are $\infty^{n-\rho_1-\rho_2}$ distinct positions of this sort.

In a similar way, one can determine the invariants that three, four and more points have relative to the group. *So one finds a series of whole numbers ρ_1, ρ_2, ρ_3, ... which are characteristic of the group and which are independent of the choice of variables.* If one computes these numbers one after the other, one always comes to a number ρ_m which vanishes, while at the same time all numbers ρ_{m+1}, ρ_{m+2}, ... are equal to zero.

We do not want pursue this behavior any further, but it must be observed that analogous results hold for every family of ∞^r transformations, whether or not this family constitutes a group.

§ 60. We saw above that an intransitive group $X_1f, \dots, X_rf$ decomposes the entire space $(x_1, \dots, x_n)$ into a continuous family of q-times extended manifolds of points:

$$\Omega_1(x_1, \dots, x_n) = C_1, \;\; \dots, \;\; \Omega_{n-q}(x_1, \dots, x_n) = C_{n-q}$$

which all remain invariant by the transformations of the group. Here, the Ω denote independent solutions of the q-term complete system which is determined by the equations $X_1f = 0, \dots, X_rf = 0$.

Each point of the space belongs to one and to only one of the ∞^{n-q} manifolds $\Omega_1 = a_1$, ..., $\Omega_{n-q} = a_{n-q}$, so in the sense provided by Chap. 6, p. 116, we are dealing with a decomposition of the space. This decomposition remains invariant by all transformations of the group $X_1f, \dots, X_rf$; and at the same time, each one of the individual subsidiary domains in which the space is decomposed stays invariant.

It can also happen for transitive groups that there exists a decomposition of the space into ∞^{n-q} q-times extended manifolds $\Omega_1 = a_1$, ..., $\Omega_{n-q} = a_{n-q}$ which remain invariant by the group. But naturally, each individual manifold amongst the ∞^{n-q} manifolds need not remain invariant, since otherwise the group would be intransitive; these ∞^{n-q} manifolds must rather be permuted by the group, while the totality of them remains invariant.

Now, a group of R_n is called *imprimitive* [IMPRIMITIV] when it determines at least one invariant decomposition of the space by ∞^{n-q} q-times extended manifolds;

a group for which there appears absolutely no invariant decomposition is called *primitive* [PRIMITIV]. That the values $q=0$ and $q=n$ are excluded hardly requires mention here.

Intransitivity is obviously a special case of imprimitivity: every intransitive group is also imprimitive. On the other hand, every primitive group is necessarily transitive.

Now, in order to obtain an analytic definition of the imprimitivity of an r-term group $X_1 f, \dots, X_r f$, we need only recall that every decomposition of the space into ∞^{n-q} q-times extended manifolds $y_1 = \text{const.}, \dots, y_{n-q} = \text{const.}$ is analytically defined by the q-term complete system $Y_1 f = 0, \dots, Y_q f = 0$, the solutions of which are the y_k. The fact that the concerned decomposition remains invariant by the group $X_1 f, \dots, X_r f$ amounts to the fact that the corresponding q-term complete system admits all transformations of the group.

The q-term complete system $Y_k f = 0$ admits our group as soon as it admits the general one-term group $\lambda_1 X_1 f + \cdots + \lambda_r X_r f$. According to Theorem 20, Chap. 8, p. 155, this is the case when, between the $X_i f$ and the $Y_k f$, there exist relationships of the following form:

$$[X_i, Y_k] = \sum_{\nu=1}^{q} \psi_{ik\nu}(x)\, Y_\nu f.$$

Consequently, it is a necessary and sufficient condition for the imprimitivity of the group $X_1 f, \dots, X_r f$ that there exists a q-term complete system:

$$Y_1 f = 0, \ \dots, \ Y_q f = 0 \qquad (q<n)$$

which stands in such relationships with the $X_k f$.

Furthermore, the group $X_1 f, \dots, X_r f$ can also be imprimitive in several ways, that is to say, there can exist many, and even infinitely many systems that the group admits.

Later, we will develop a method for constructing all complete systems which remain invariant by a given group. Hence in particular, we will also be able to determine whether the group in question is primitive or not. Naturally, the latter question requires a special examination only for transitive groups.

Now, we make a brief remark.

Let the equations:

$$y_1 = C_1, \ \dots, \ y_{n-q} = C_{n-q}$$

represent a decomposition of the space $x_1, \dots, x_n$ which is invariant by the group $X_1 f, \dots, X_r f$. Then if we introduce $y_1, \dots, y_{n-q}$ together with q other appropriate functions $z_1, \dots, z_q$ of $x_1, \dots, x_n$ as new independent variables, the infinitesimal transformations $X_k f$ receive the specific form:

$$X_k f = \sum_{\mu=1}^{n-q} \omega_{k\mu}(y_1, \dots, y_{n-q}) \frac{\partial f}{\partial y_\mu} + \sum_{j=1}^{q} \zeta_{kj}(y_1, \dots, y_{n-q}, z_1, \dots, z_q) \frac{\partial f}{\partial z_j}$$

$$(k=1 \cdots r).$$

Now, according to Chap. 12, p. 220, all infinitesimal transformations $e_1 X_1 f + \cdots + e_r X_r f$ which leave invariant a determined manifold $y_1 = y_1^0, \dots, y_{n-q} = y_{n-q}^0$ generate a subgroup. If we want to find the infinitesimal transformations in question, then we only have to seek all systems of values $e_1, \dots, e_r$ which satisfy the $n-q$ equations:

$$\sum_{k=1}^{r} e_k\, \omega_{k\mu}(y_1^0, \dots, y_{n-q}^0) = 0 \qquad (\mu = 1 \cdots n-q).$$

If $r > n-q$, then there are always systems of values $e_1, \dots, e_r$ of this constitution, and consequently in this case, the group $X_1 f, \dots, X_r f$ certainly contains subgroups with a least $r-n+q$ parameters.

Furthermore, it is clear that the r *reduced* [VERKÜRZTEN] infinitesimal transformations:

$$\overline{X}_k f = \sum_{\mu=1}^{n-q} \omega_{k\mu}(y_1, \dots, y_{n-q}) \frac{\partial f}{\partial y_\mu} \qquad (k = 1 \cdots r)$$

in the $n-q$ variables $y_1, \dots, y_{n-q}$ generate a group; however, this group possibly contains not r, but only a smaller number of essential parameters. Clearly, the calculations just indicated amount to the determination of all infinitesimal transformations $e_1 \overline{X}_1 f + \cdots + e_r \overline{X}_r f$ which leave invariant the point $y_1^0, \dots, y_{n-q}^0$ of the $(n-q)$-times extended space $y_1, \dots, y_{n-q}$.

Chapter 14
Determination of All Systems of Equations Which Admit a Given r-term Group

If a system of equations remains invariant by all transformations of an r-term group $X_1 f, \dots, X_r f$, we say that it admits the group in question. Every system of equations of this form admits all transformations of the general one-term group $\sum e_k X_k f$, and therefore in particular, all ∞^{r-1} *infinitesimal* transformations $\sum e_k X_k f$ of the r-term group.

Now on the other hand, we have shown earlier on that every system of equations which admits the r *infinitesimal* transformations $X_1 f, \dots, X_r f$ and therefore also, all ∞^{r-1} infinitesimal transformations $\sum e_k X_k f$, allows at the same time all *finite* transformations of the one-term group $\sum e_k X_k f$, that is to say, all transformations of the group $X_1 f, \dots, X_r f$ (cf. Theorem 14, p. 127). Hence if all systems of equations which admit the r-term group $X_1 f, \dots, X_r f$ are to be determined, this shall be a problem which is completely settled by the developments of Chap. 7. Indeed, the problem of setting up all systems of equations which admit r given infinitesimal transformations is solved in complete generality there.

However, if the $X_1 f, \dots, X_r f$ which are considered here generate an r-term group then this leads to a major simplification in comparison to the general case. Hence it appears to be completely legitimate to independently settle the special case where the $X_k f$ generate a group.

The treatment of the addressed problem turns out to be essentially the same whether or not one knows the finite equations of the concerned group. In the first case, no integration is required. But in the second case, one generally needs integration; however, some operations which were necessary for the general problem in Chapter 7 can be dropped.

We shall treat these two cases, above all because of the applications, but also in order to afford a deeper insight into the problem; in fact, the concerned developments mutually complement one another.[1]

Lastly, let us mention that from now on, we shall frequently translate the common symbolism of the theory of substitutions into the theory of transformation groups.

[1] LIE, Math. Ann. Vol. XI, pp. 510–512, Vol. XVI, p. 476. Archiv for Math. og Nat., Christiania 1878, 1882, 1883. Math. Ann. Vol. XXIV.

S. Lie, *Theory of Transformation Groups I*,
DOI 10.1007/978-3-662-46211-9_14

So for example, we denote by S, T, ... individual transformations, and by S^{-1}, T^{-1}, ... the corresponding inverse transformations. By ST, we understand the transformation which is obtained when the transformation S is executed first, and then the transformation T. From this, it follows that expressions of the form SS^{-1}, TT^{-1} mean the identity transformation.

§ **61.** We consider an arbitrary point P of the space. The totality of all positions that this point takes by the ∞^r transformations of the group forms a certain manifold M; we shall show that this manifold admits the group, or in other words, that every point of M is transferred to a point again of M by every transformation of the group.

Indeed, let P' be any point of M, and let P' come from P by the transformation S of our group, what we want to express by means of the symbolic equation:

$$(P)\,S = (P').$$

Next, if T is a completely arbitrary transformation of the group, then by the execution of T, P' is transferred to:

$$(P')\,T = (P)\,ST;$$

but since the transformation ST belongs to the group as well, $(P)\,ST$ is also a point of M, and our claim is therefore proved.

Obviously, every manifold invariant by the group which contains the point P must at the same time contain the manifold M. That is why we can also say: M is the *smallest* manifold invariant by the group to which the point P belongs.

But still, there is something more. It can be shown that with the help of transformations of the group, every point of M can be transferred to any other point of this manifold. Indeed, if P' and P'' are any two points of M and if they are obtained from P by means of the transformations S and U, respectively, one has the relations:

$$(P)\,S = (P'), \qquad (P)\,U = (P'');$$

from the first one, it follows:

$$(P')\,S^{-1} = (P)\,SS^{-1} = (P);$$

hence with the help of the second one, we get:

$$(P')\,S^{-1}U = (P''),$$

that is to say, by the transformation $S^{-1}U$ which likewise belongs to the group, the point P' is transferred to the point P''. As a result, the assertion stated above is proved.

From this, we realize that the manifold M can also be defined as the totality of all positions which any of its other points, not just P, take by the ∞^r transformations of the group.

Consequently, the following holds true.

Theorem 36. *If, on a point P of the space $(x_1,\dots,x_n)$, one executes all ∞^r transformations of an r-term group of this space, then the totality of all positions that the point takes in this manner forms a manifold invariant by the group; this manifold contains no smaller subsidiary domain invariant by the group, and it is itself contained in all invariant manifolds in which the point P lies.*

If one assumes that the finite equations $x_i' = f_i(x_1,\dots,x_n,\,a_1,\dots,a_r)$ of the r-term group are known, then without difficulty, one can indicate for every point $x_1^0,\dots,x_n^0$ the smallest invariant manifold to which it belongs. Indeed, by the above, the manifold in question consists of the totality of all positions $x_1,\dots,x_n$ that the point $x_1^0,\dots,x_n^0$ takes by the transformations of the group; but evidently, the totality of these positions is represented by the n equations:

$$x_i = f_i(x_1^0,\dots,x_n^0,\,a_1,\dots,a_r) \qquad (i=1\cdots n)$$

in which the parameters a are to be interpreted as independent variables. If one eliminates the a, one obtains the sought manifold represented by equations between the x alone.

Here, it is to be recalled that in the equations $x_i = f_i(x^0,a)$, the a are not completely arbitrary; indeed, all systems of values $a_1,\dots,a_r$ for which the determinant:

$$\sum \pm \left[\frac{\partial f_1(x,a)}{\partial x_1}\right]_{x=x^0} \cdots \left[\frac{\partial f_n(x,a)}{\partial x_n}\right]_{x=x^0}$$

vanish are excluded from the beginning, because we always use only transformations which are solvable. From this, it follows that in certain circumstances, one obtains, by elimination of the a, a manifold which contains, aside from the points to which $x_1^0,\dots,x_n^0$ is transferred by the *solvable* transformations of the group, yet other points; then as one easily sees, the latter points in turn form an invariant manifold.

Furthermore, it is to be remarked that the discussed elimination can take different shapes for different systems of values x_k^0; indeed, the elimination of the a need not always lead to the same number of relations between the x, which again means that the smallest invariant manifolds in question need not all have the same dimension number.

If, amongst all smallest invariant manifolds of the same dimension number, one takes infinitely many such invariant manifolds according to an arbitrary analytic rule, then their totality also forms an invariant manifold. In this way, all invariant manifolds can obviously be obtained.

So we have the

Theorem 37. *If one knows the finite equations of an r-term group $X_1f,\dots,X_rf$ in the variables $x_1,\dots,x_n$, then without integration, one can find all invariant systems* *of equations invariant by the group, or, what is the same, all manifolds invariant by it.*

§ 62. In Chap. 12, p. 219, we have seen that an r-term group G_r in $x_1, \dots, x_n$ associates to every point of the space a completely determined subgroup, namely the subgroup which consists of all transformations of G_r which leave the point invariant.

Let the point P be invariant by an m-term subgroup of G_r, but by no subgroup with more terms; let S be the general symbol of a transformation of this m-term subgroup, so that one hence has: $(P)\,S = (P)$. Moreover, let T be a transformation which transfers the point P to the new position P':

$$(P)\,T = P'.$$

Now, if T is an arbitrary transformation of G_r which transfers P to P' as well, we have:

$$(P)\,\mathsf{T} = (P') = (P)\,T,$$

hence we have:

$$(P)\,\mathsf{T}\,T^{-1} = (P).$$

From this, it is evident that $\mathsf{T}T^{-1}$ belongs to the transformations S, hence that:

$$\mathsf{T} = S\,T$$

is the general form of a transformation of the same constitution as T. Now, since there are precisely ∞^m transformations S, we see that:

G_r contains exactly ∞^m transformations which transfer P to P'.

On the other hand, if we ask for all transformations S' of G_r which leave invariant the point P', then we have to fulfill the condition $(P')\,S' = (P')$. From it, we see that:

$$(P)\,T\,S' = (P)\,T \qquad \text{and} \qquad (P)\,T\,S'\,T^{-1} = (P),$$

and consequently $T\,S'\,T^{-1}$ is a transformation S, that is to say S' has the form:

$$S' = T^{-1}\,S\,T.$$

One easily sees here that S can be a completely arbitrary transformation of the subgroup associated to the point P; as a result, our group contains exactly ∞^m different transformations S', and in turn now, they obviously form an m-term subgroup of G_r.

The results of this section obtained up to now can be summarized as follows.

Proposition 1. *If an r-term group G_r of R_n contains exactly ∞^m and not more transformations S which leave the point P invariant, and if in addition it contains at least one transformation T which transfers the point P to the point P', then it contains on* *the whole ∞^m different transformations which transfer P to P'; the general form of these transformations is: ST. In addition, G_r contains exactly ∞^m transformations which leave invariant the point P'; their general form is: $T^{-1}ST$.*

From the second part of this proposition, it follows that the points which admit exactly ∞^m transformations of our group are permuted by the transformations of the group, while their totality remains invariant.

Hence the following holds true.

Theorem 38. *The totality of all points which admit the same number, say ∞^m, and not more transformations of an r-term group, remains invariant by all transformations of the group.*

We have proved this theorem by applying considerations which are borrowed from the theory of substitutions. But at the same time, we want to show how one can conduct the proof in case one abstains from such considerations, or from a more exact language.

A point $x_1^0, \dots, x_n^0$ which allows ∞^m transformations of the r-term group:

$$X_k f = \sum_{i=1}^{n} \xi_{ki}(x_1, \dots, x_n) \frac{\partial f}{\partial x_i} \qquad (k=1\cdots r)$$

admits precisely m independent infinitesimal transformations of this group. Hence the group contains, in the neighborhood of $x_1^0, \dots, x_n^0$, exactly m independent infinitesimal transformations whose power series expansions with respect to the $x_i - x_i^0$ begin with terms of first order or of higher order. Now, if we imagine that new variables:

$$\overline{x}_i = \overline{x}_i^0 + \sum_{k=1}^{n} \alpha_{ki}(x_k - x_k^0) + \cdots \qquad (i=1\cdots n)$$

$$\sum \pm \alpha_{11} \cdots \alpha_{nn} \neq 0,$$

are introduced in the group, then according to Chap. 11, p. 210, we obtain a new group in the $\overline{x}_i$ which, in the neighborhood of $\overline{x}_i^0$, contains in the same way exactly m independent infinitesimal transformations of first order or of higher order. In particular, if we imagine that the transition from the x_i to the $\overline{x}_i$ is a transformation of the group $X_1 f, \dots, X_r f$, then the group in the $\overline{x}_i$ is simply identical to the group:

$$\overline{X}_k f = \sum_{i=1}^{n} \xi_{ki}(\overline{x}_1, \dots, \overline{x}_n) \frac{\partial f}{\partial \overline{x}_i} \qquad (k=1\cdots r)$$

(cf. Chap. 3, p. 46). In other words, if, by a transformation of our group, the point x_i^0 is transferred to the point $\overline{x}_i^0$, then this point also admits precisely m independent infinitesimal transformations of the group. But with this, Theorem 38 is visibly proved.

From Theorem 38 we see immediately that the following proposition also holds true:

Proposition 2. *The totality of all points $x_1,\dots,x_n$ which admit m or more independent infinitesimal transformations of the r-term group $X_1f,\dots,X_rf$ remains invariant by this group.*

If we associate this proposition with the developments in Chap. 11, Proposition 4, p. 213, we obtain a new important result. At that time, we indeed saw that the points $x_1,\dots,x_n$ which admit m or more independent infinitesimal transformations $e_1X_1f+\cdots+e_rX_rf$ of the r-term group $X_1f,\dots,X_rf$ are characterized by the fact that all $(r-m+1)\times(r-m+1)$ determinants of a certain matrix are brought to zero. At present, we recognize that the system of equations which is obtained by equating to zero these $(r-m+1)\times(r-m+1)$ determinants admits all transformations of the group $X_1f,\dots,X_rf$. As a result, we have the

Theorem 39. *If r independent infinitesimal transformations:*

$$X_kf=\sum_{i=1}^{n}\xi_{ki}(x_1,\dots,x_n)\,\frac{\partial f}{\partial x_i}\qquad\qquad (k=1\cdots r)$$

generate an r-term group, then by equating to zero all $(r-m+1)\times(r-m+1)$ determinants of the matrix:

$$(1)\qquad\qquad \begin{vmatrix} \xi_{11}(x) & \cdot & \cdot & \xi_{1n} \\ \cdot & \cdot & \cdot & \cdot \\ \xi_{r1}(x) & \cdot & \cdot & \xi_{rn} \end{vmatrix},$$

one always obtains a system of equations which admits all transformations of the group $X_1f,\dots,X_rf$; this holds true for every number $m\leqslant r$, provided only that there actually exist systems of values $x_1,\dots,x_n$ which bring to zero all the $(r-m+1)\times(r-m+1)$ determinants in question.

Later in the course of this chapter (§§ 66 and 67, p. 249 and 251 resp.), we will give two different purely analytic proofs of the above important theorem. Temporarily, we observe only the following:

Theorem 39 shows that there is an essential difference between the problem of Chap. 7, p. 138 up to 147, and the problem of the present chapter.

If $X_1f,\dots,X_rf$ generate an r-term group, then by equating to zero all $(r-m+1)\times(r-m+1)$ determinants of the matrix (1), one always obtains an invariant system of equations, only as soon as all these determinants really can vanish at the same time. But this is no longer true when it is only assumed about the X_kf that, when set to zero, they constitute a complete system consisting of r, or fewer equations. In this case, it is certainly possible that there are invariant systems of equations which embrace the equations obtained by equating to zero the determinants in question, but it is not at all always the case that one obtains an invariant system of equations by equating to zero these determinants, just like that. To get this, further operations are rather necessary in general, as is explained in Chap. 7, p. 139 sq.

In the last section of this chapter, p. 253, we will study this point in more detail.

§ 63. In the preceding section, we saw that, *from the infinitesimal transformations* of an r-term group, one can derive *without integration* certain systems of equations which remain invariant by the concerned group. Now, we will show how one finds *all* systems of equations which admit an r-term group *with given infinitesimal transformations*:

$$X_k f = \sum_{i=1}^{n} \xi_{ki}(x_1,\dots,x_n)\,\frac{\partial f}{\partial x_i} \qquad (k=1\cdots r).$$

We assume that amongst the r equations $X_1 f=0,\dots,X_r f=0$, exactly q mutually independent are extant, hence that in the matrix (1), all $(q+1)\times(q+1)$ determinants vanish identically, but not all $q\times q$ determinants.

Exactly as in Chap. 7, p. 135 and 138, *we can then distribute in $q+1$ different classes the systems of equations which admit the r infinitesimal transformations $X_1 f,\dots,X_r f$. In one and the same class, we include the systems of equations by virtue of which all $(p+1)\times(p+1)$ determinants of the matrix* (1) *vanish, but not all $p\times p$ determinants, where it is understood that p is one of the $q+1$ numbers* q, $q-1$, …, 1, 0. If we prefer to apply the terminology of the theory of manifolds, we must say: to one and the same class belong the invariant manifolds whose points admit the same number, say exactly $r-p$, of infinitesimal transformations $e_1 X_1 f + \cdots + e_r X_r f$. To every point of such a manifold, the infinitesimal transformations $X_1 f,\dots,X_r f$ attach exactly p independent directions, which in turn are in contact with the manifold (cf. Chap. 123, p. 149).

The usefulness of this classification is that it makes it possible to consider each individual class and to determine the systems of equations which belong to it, or, respectively, the manifolds.

If the number p equals q, then the determination of all invariant systems of equations which belong to the concerned class is achieved by Theorem 17 in Chap. 7, p. 137. Every such system of equations can be represented by relations between the common solutions of the equations $X_1 f = 0$, …, $X_r f = 0$. Since these r equations determine a q-term complete system, they will of course possess common solutions only when q is smaller than n.

At present, we can disregard the case $p=q$. Hence we assume from now on that p is one of the numbers 0, 1, …, $q-1$ and we state the problem of determining all manifolds invariant by the group $X_1 f,\dots,X_r f$ which belong to the class defined by the number p.

The first step for solving this problem is the determination of the locus of all points for which all $(p+1)\times(p+1)$ determinants of the matrix (1) vanish, whereas not all $p\times p$ determinants do. Indeed, the corresponding locus clearly contains all manifolds invariant by the group which belong to our class; furthermore, according to Theorem 38, p. 239, this locus itself constitutes an invariant manifold.

In order to find the sought locus, we at first study the totality of all points for which all $(p+1)\times(p+1)$ determinants of the matrix (1) vanish, that is to say, we calculate all the $(p+1)\times(p+1)$ determinants of the matrix in question — they can be denoted $\Delta_1, \Delta_2, \dots, \Delta_\rho$ — and we set them equal to zero:

$$\Delta_1 = 0, \; \dots, \; \Delta_\rho = 0.$$

The equations so obtained then represent a manifold which, according to Theorem 39, p. 240, is invariant by the group and which contains the sought locus.

If there is in fact no system of values $x_1,\dots,x_n$ which brings to zero all the Δ, or if the $(p+1)\times(p+1)$ determinants can vanish only in such a way that all $p\times p$ determinants also vanish at the same time, then it is clear that actually no manifold invariant by the group $X_1f,\dots,X_rf$ belongs to the class which is defined by p. Consequently, we see that not all of our $q+1$ classes need to be represented by manifolds which belong to it.

We assume that for the p chosen by us, neither of the two exceptional cases just discussed occurs, so that there really are systems of values $x_1,\dots,x_n$ for which all $(p+1)\times(p+1)$ determinants of the matrix (1), but not all $p\times p$ determinants, vanish.

As we have already observed, the manifold $\Delta_1 = 0,\dots,\Delta_\rho = 0$ remains invariant by the group $X_1f,\dots,X_rf$. Now, if this manifold is reducible, it therefore consists of a discrete number of finitely many different manifolds, so it decomposes without difficulty into as many different individual invariant manifolds. Indeed, the group $X_1f,\dots,X_rf$ is generated by infinitesimal transformations; hence when it leaves invariant the totality of finitely many manifolds, then each individual manifold must stay at rest.[1] (Translator's note: In fact, the argument is that each stratum is kept invariant because the group acts close to the identity.)

Let $M_1, M_2, \dots$ be the individual irreducible, and so invariant by the group, manifolds into which the manifold $\Delta_1 = 0,\dots,\Delta_\rho = 0$ decomposes. Then amongst these manifolds, there may possibly be some at all points of which all $p\times p$ determinants of the matrix (1) also vanish. When we exclude all manifolds of this special constitution, we still keep certain manifolds M_1, M_2, ..., the totality of which clearly forms the locus of all points for which all $(p+1)\times(p+1)$ determinants of the matrix (1) vanish, but not all $p\times p$ determinants.

With this, we have found the sought locus; at the same time, we see that this locus can consist of a discrete number of individual invariant manifolds M_1, $\mathsf{M}_2, \dots$ which, naturally, all belong to the class defined by p.

Clearly, each manifold invariant by our group which belongs to the class defined by p is contained in one of the manifolds M_1, $\mathsf{M}_2, \dots$ So in order to find all such manifolds, we need only to examine each individual manifold M_1, M_2, ... and to determine the invariant manifolds contained in them which belong to the said class. According to a remark made earlier on (Chap. 7, Proposition 6, p. 149), each invariant manifold belonging to the class p is at least p-times extended.

§ 64. The problem to which we have been led at the end of the preceding section is a special case of the following general problem:

Suppose the equations of an irreducible manifold M which remain invariant by the transformations of the r-term group $X_1 f,\dots,X_r f$ are given. In general, the infinitesimal transformations $X_1 f,\dots,X_r f$ attach exactly p independent directions to the points of the manifold with which they are naturally in contact, and the manifold is at least p-times extended. We seek all invariant subsidiary domains contained in M, to the points of which the transformations $X_1 f,\dots,X_r f$ attach exactly p independent directions.

We now want to solve this problem.

We imagine the equations of M presented in the resolved form:

$$x_{s+i} = \varphi_{s+i}(x_1,\dots,x_s) \qquad (i=1\cdots n-s),$$

but here it should not be forgotten that by the choice of a *determined* resolution, we exclude all systems of values $x_1,\dots,x_n$ for which precisely this resolution is not possible. It is conceivable that we exclude in this manner certain invariant subsidiary domains of M which are captured by another resolution.

The case $p = 0$ requires no special treatment, because obviously, the manifold M then consists only of invariant points.

In order to be able to solve the problem for the remaining values of p, we must begin by mentioning a few remarks that stand in close connection to the analytic developments in Chap. 7, p. 140 and 142, and which in principle already have great importance.

Since the manifold M remains invariant by the transformations of our group, its points are permuted by the transformations of the group. Hence, if we disregard all points lying outside of M, then our group $X_1 f,\dots,X_r f$ determines a certain group of transformations of the points of M. However, this new group need not contain r essential parameters, since it can happen that a subgroup of the group $X_1 f,\dots,X_r f$ leaves all points of M individually fixed.

We first summarize what has been said:

Theorem 40. *The points of a manifold that remains invariant by an r-term group of the space $(x_1,\dots,x_n)$ are in their turn transformed by a continuous group with r or fewer parameters.*

Since we have assumed that the invariant manifold M is irreducible, we can consider it as being a space in itself. The analytic expression of the transformation group by which the points of this space are transformed must therefore be obtained by covering the points of M by means of a related coordinate system and by establishing how these coordinates are transformed by the group $X_1 f,\dots,X_r f$.

Under the assumptions made, the group which transforms the points of M can be immediately indicated. Indeed, we need only to interpret $x_1,\dots,x_s$ as coordinates of the points of M, and in the finite equations $x_i' = f_i(x,a)$ of the group $X_1 f,\dots,X_r f$, to replace the $x_{s+1},\dots,x_n$ by $\varphi_{s+1},\dots,\varphi_n$ and to leave out $x_{s+1}',\dots,x_n'$; then we obtain the equations of the concerned group, which are the following:

$$x_i' = f_i(x_1,\dots,x_s,\, \varphi_{s+1},\dots,\varphi_n;\, a_1,\dots,a_r) \qquad (i=1\cdots n).$$

One could convince oneself directly that one really has to consider a group in the variables $x_1,\dots,x_n$. For that, one would only need to execute two transformations one after another[2] (Translator's note: Recall from Chaps. 4 and 9 that only closure under composition counts for Lie.) in the form just written, and then to take account of two facts: firstly, that the transformations $x_i' = f_i(x_1,\dots,x_n,\,a_1,\dots,a_r)$ form a group, and secondly that the system of equations $x_{s+i} = \varphi_{s+i}$ admits this group.

If on the other hand we wanted to know the *infinitesimal* transformations of the group in $x_1,\dots,x_s$, then we would only need to leave out the terms with $\partial f/\partial x_{s+1}$, ..., $\partial f/\partial x_n$ in the $X_k f$ and to replace $x_{s+1},\dots,x_n$ by the φ in the remaining terms. One finds the r infinitesimal transformations:

$$\overline{X}_k f = \sum_{\nu=1}^{s} \xi_{k\nu}(x_1,\dots,x_s,\,\varphi_{s+1},\dots,\varphi_n)\,\frac{\partial f}{\partial x_\nu}$$

$$(k=1\cdots r),$$

which, however, need not be independent of each other.

We will verify directly that the reduced infinitesimal transformations $\overline{X}_k f$ generate a group. The concerned computation is mostly similar to the one executed in Chap. 7, p. 145.

At that time, we indicated the execution of the substitution $x_{s+i} = \varphi_{s+i}$ by means of the symbol $[\,]$. So we have at first:

$$\overline{X}_k f = \sum_{\nu=1}^{s} [\xi_{k\nu}]\,\frac{\partial f}{\partial x_\nu}.$$

Furthermore as before, we see via Eq. (3) of Chap. 7, p. 126, that:

$$\overline{X}_k[\Omega] \equiv [X_k\Omega],$$

where it is understood that Ω is a completely arbitrary function of $x_1,\dots,x_n$. From this, we get:

$$[\overline{X}_k,\overline{X}_j] = \sum_{\nu=1}^{s} \big\{[X_k\xi_{j\nu}] - [X_j\xi_{k\nu}]\big\}\,\frac{\partial f}{\partial x_\nu};$$

and since relations of the form:

$$[X_k,X_j] = \sum_{\pi=1}^{r} c_{kj\pi}\,X_\pi f,$$

or, what is the same, of the form:

$$X_k\xi_{j\nu} - X_j\xi_{k\nu} = \sum_{\pi=1}^{r} c_{kj\pi}\,\xi_{\pi\nu}$$

hold true, then we obtain simply:

$$[\overline{X}_k, \overline{X}_j] = \sum_{\pi=1}^{r} c_{kj\pi} \overline{X}_\pi f.$$

As a result, it is proved in a purely analytic way that $\overline{X}_1 f, \dots, \overline{X}_r f$ really generate a group.

The infinitesimal transformations $X_1 f, \dots, X_r f$ of the space $x_1, \dots, x_n$ attach exactly p independent directions $\mathrm{d}x_1 : \cdots : \mathrm{d}x_n$ to every point in general position on the manifold M, and these directions, as is known, are in contact with the manifold. It can be foreseen that the infinitesimal transformations $\overline{X}_1 f, \dots, \overline{X}_r f$ of the space $x_1, \dots, x_s$, or what is the same, of the manifold M, also attach to every point $x_1, \dots, x_s$ in general position exactly p independent directions $\mathrm{d}x_1 : \cdots : \mathrm{d}x_s$. We shall verify that this is really so.

Under the assumptions made, after the substitution $x_{s+i} = \varphi_{s+i}$, all $(p+1) \times (p+1)$ determinants of the matrix (1) vanish, but not all $p \times p$ determinants, and therefore, amongst the r equations:

$$(2) \qquad [\xi_{k1}] \frac{\partial f}{\partial x_1} + \cdots + [\xi_{kn}] \frac{\partial f}{\partial x_n} \qquad (k=1\cdots r),$$

exactly p independent ones are extant. From this, it follows that amongst the r equations $\overline{X}_k f = 0$, there are at most p independent ones; our problem is to prove that there are exactly p. This is not difficult.

Since the system of equations $x_{s+i} - \varphi_{s+i} = 0$ admits the infinitesimal transformations $X_k f$, we have identically:

$$[X_k(x_{s+i} - \varphi_{s+i})] \equiv 0,$$

or if written in greater length:

$$[\xi_{k,s+i}] \equiv \sum_{\nu=1}^{s} [\xi_{k\nu}] \frac{\partial \varphi_{s+i}}{\partial x_\nu}.$$

Hence, if by $\chi_1, \dots, \chi_r$ we denote arbitrary functions of $x_1, \dots, x_s$, we then have:

$$\sum_{k=1}^{r} \chi_k [\xi_{k,s+i}] \equiv \sum_{k=1}^{r} \chi_k \overline{X}_k \varphi_{s+i}.$$

Now, if there are r functions $\psi_1, \dots, \psi_r$ of $x_1, \dots, x_n$ not all vanishing such that the equation:

$$\sum_{k=1}^{r} \psi_k(x_1, \dots, x_s) \overline{X}_k f \equiv 0$$

is identically satisfied, then we have:

$$\sum_{k=1}^{r} \psi_k [\xi_{k,s+i}] \equiv \sum_{k=1}^{r} \psi_k \overline{X}_k \varphi_{s+i} \equiv 0,$$

and consequently also:

$$\sum_{k=1}^{r} \psi_k(x_1,\dots,x_s) \sum_{\nu=1}^{n} [\xi_{k\nu}] \frac{\partial f}{\partial x_\nu} \equiv 0.$$

As a result, it is proved that amongst the equations:

$$\overline{X}_1 f = 0, \ \dots, \ \overline{X}_r f = 0,$$

there are exactly as many independent equations as there are amongst the equations (2), that is to say, exactly p independent ones.

At present, we are at last in a position to settle the problem posed at the beginning of the section, on p. 243.

The problem is the determination of certain subsidiary domains invariant by the group $X_1 f,\dots,X_r f$ in the invariant manifold M, namely the subsidiary domains to the points of which the infinitesimal transformations $X_1 f,\dots,X_r f$ attach exactly p independent directions. According to the preceding, these subsidiary domains can be defined as certain manifolds of the space $x_1,\dots,x_s$; as such, they are characterized by the fact that they admit the group $\overline{X}_1 f,\dots,\overline{X}_r f$ and that, to their points, are attached exactly p independent directions by the infinitesimal transformations $\overline{X}_1 f,\dots,\overline{X}_r f$. Consequently, our problem amounts to the following:

In a space M of s dimensions, let the group $\overline{X}_1 f,\dots,\overline{X}_r f$ be given, whose infinitesimal transformations attach, to the points of this space in general position, exactly $p \leqslant s$ independent directions. We seek all invariant manifolds contained in M having the same constitution.

But we have already solved this problem above (p. 241); only then we had the group $X_1 f,\dots,X_r f$ in place of the group $\overline{X}_1 f,\dots,\overline{X}_r f$, the number n in place of the number s, the number q in place of the number p. Thus, the wanted manifolds are represented by means of relations between the solutions of the p-term complete system that the equations $\overline{X}_1 f = 0, \dots, \overline{X}_r f = 0$ determine. If one adds these relations to the equations of M, then one obtains the equations of the invariant subsidiary domains of M in terms of the initial variables $x_1,\dots,x_n$.

Naturally, there are invariant subsidiary domains in M of the demanded sort only when s is larger than p, and there are none, when the numbers s and p are equal to one another.

With this, we therefore have the following important result:

Theorem 41. *If an s-times extended manifold of the space $x_1,\dots,x_n$ admits the r-term group $X_1 f,\dots,X_r f$ and if, to the points of this manifold, the infinitesimal transformations attach exactly p independent directions which then surely fall into the manifold,*[3] *(Translator's note: Act of "intrinsicness": directions attached to M inside the ambient space in fact happen to be intrinsically attached to M.) then $s \geqslant p$; in case $s > p$, the manifold decomposes into ∞^{s-p} p-times extended subsidiary domains, each of which admits the group $X_1 f,\dots,X_r f$.*

At the same time, the problem to which we were led at the end of the preceding section (p. 243) is also completely solved, and with this, the determination of all systems of equations that the group $X_1 f, \dots, X_r f$ admits is achieved. In view of applications, we put together once more the required guidelines [MASSREGELN].

Theorem 42. *If the r independent infinitesimal transformations:*

$$X_k f = \sum_{i=1}^{n} \xi_{ki}(x_1, \dots, x_n) \frac{\partial f}{\partial x_i} \qquad (k=1\cdots r)$$

generate an r-term group and if at the same time, all $(q+1) \times (q+1)$ *determinants of the matrix:*

$$\begin{vmatrix} \xi_{11} & \cdot & \cdot & \cdot & \xi_{1n} \\ \cdot & \cdot & \cdot & \cdot & \cdot \\ \xi_{r1} & \cdot & \cdot & \cdot & \xi_{rn} \end{vmatrix}$$

vanish identically, whereas not all $q \times q$ *determinants do, then one finds as follows all systems of equations, or, what is the same, all manifolds that the group admits.*

One distributes the systems of equations or the manifolds in question in $q+1$ *different classes by including in the same class the systems of equations by virtue of which*[4] *(Translator's note: This just means systems of equations* including *the equations of* $(p+1) \times (p+1)$ *minors.) all* $(p+1) \times (p+1)$ *determinants of the above matrix, but not all* $p \times p$ *determinants, vanish, where it is understood that* p *is one of the numbers* q, $q-1, \dots, 1, 0$.

Then in order to find all invariant systems of equations which belong to a determined class, one forms all $(p+1) \times (p+1)$ *determinants* $\Delta_1, \Delta_2, \dots, \Delta_\rho$ *of the matrix and one sets them equal to zero. If there is no system of values* $x_1, \dots, x_n$ *which brings to zero all the* ρ *determinants* Δ_i *at the same time, then in fact, the class which is defined by the number* p *contains no invariant manifold; and the same evidently also holds for the classes with numbers* $p-1$, $p-2, \dots, 1, 0$. *On the other hand, if all systems of values* $x_1, \dots, x_n$ *which make* $\Delta_1, \dots, \Delta_\rho$ *equal to zero would at the same time bring to zero all* $p \times p$ *determinants of the matrix, then also in this case, the class with the number* p *would not be present as manifolds. If neither of these two cases occurs, then the system of equations:*

$$\Delta_1 = 0, \ \dots, \ \Delta_\rho = 0$$

represents the manifold M *invariant by the group inside which all invariant manifolds with the class number* p *are contained. If* M *decomposes into a discrete number of manifolds* M_1, $\mathsf{M}_2, \dots$, *then these manifolds remain individually invariant, but in each one of them, infinitely many invariant subsidiary domains can yet be contained which belong to the same class as* M. *In order to find these subsidiary domains, one sets the equations of, say* M_1, *in resolved form:*

$$x_{s+i} = \varphi_{s+i}(x_1, \dots, x_s) \qquad (i=1\cdots n-s),$$

where the whole number s is at least equal to p. Lastly, one forms the r equations:

$$\overline{X}_k f = \sum_{\nu=1}^{s} \xi_{k\nu}(x_1,\dots,x_s,\,\varphi_{s+1},\dots,\varphi_n)\,\frac{\partial f}{\partial x_\nu} = 0,$$

and one computes any $s-p$ independent solutions:

$$\omega_1(x_1,\dots,x_s),\ \dots,\ \omega_{s-p}(x_1,\dots,x_s)$$

of the p-term complete system determined by these equations. The general analytic expression for the sought invariant subsidiary domains of M_1 *is then:*

$$x_{s+i} - \varphi_{s+i}(x_1,\dots,x_n) = 0, \qquad \psi_j(\omega_1,\dots,\omega_{s-p}) = 0$$
$$(i = 1 \cdots n-s;\ j = 1 \cdots m),$$

where the $m \leqslant s-p$ relations $\psi_j = 0$ are completely arbitrary. —

Naturally, $\mathsf{M}_2, \dots$ *must also be treated in the same manner as* M_1*. In addition, for p, one has to insert one after the other all the $q+1$ numbers q, $q-1, \dots, 1, 0$.*

§ 65. In order to apply the preceding results to an example, we consider the three-term group:

$$X_1 f = \frac{\partial f}{\partial y} + x\frac{\partial f}{\partial z}, \qquad X_2 f = y\frac{\partial f}{\partial y} + z\frac{\partial f}{\partial z},$$
$$X_3 f = (-z+xy)\frac{\partial f}{\partial x} + y^2\frac{\partial f}{\partial y} + yz\frac{\partial f}{\partial z}$$

of ordinary space. The group is transitive, because the determinant:

$$\Delta = \begin{vmatrix} 0 & 1 & x \\ 0 & y & z \\ -z+xy & y^2 & yz \end{vmatrix} = -(z-xy)^2$$

does not vanish identically. From this, we conclude that the surface of second degree: $z - xy = 0$ remains invariant by the group, and else that no further surface does.

For the points of the surface $z - xy = 0$, and only for these points, all the 2×2 subdeterminants of Δ vanish, while its 1×1 subdeterminants cannot vanish simultaneously. From this, it follows that the invariant surface decomposes into ∞^1 invariant curves, but that apart from these, no other invariant curves exist, while there are actually no invariant points.

In order to find the ∞^1 curves on the surface $z - xy = 0$, we choose x and y as coordinates for the points of the surface and we form, according to the instructions given above, the reduced infinitesimal transformations in x, y:

$$\overline{X}_1 f = \frac{\partial f}{\partial y}, \qquad \overline{X}_2 f = y\frac{\partial f}{\partial y}, \qquad \overline{X}_3 f = y^2\frac{\partial f}{\partial y}.$$

The three equations $\overline{X}_k f = 0$ reduce to a single one whose solution is x. Therefore, the sought curves are represented by the equations:

$$z - xy = 0, \qquad x = \text{const.},$$

that is to say, all individuals of a family of generatrices on the surface of second degree remain invariant.

§ 66. Here, we give one of the two new proofs promised on p. 240 for the important Theorem 39.

As before, we denote by $\Delta_1(x), \dots, \Delta_\rho(x)$ all the $(p+1)\times(p+1)$ determinants of the matrix:

$$\text{(3)} \qquad \begin{vmatrix} \xi_{11}(x) & \cdot & \cdot & \xi_{1n}(x) \\ \cdot & \cdot & \cdot & \cdot \\ \xi_{r1}(x) & \cdot & \cdot & \xi_{rn}(x) \end{vmatrix}.$$

In addition, we assume that there are systems of values $x_1, \dots, x_n$ which bring to zero all the ρ determinants Δ. Then it is to be proved that the system of equations:

$$\Delta_1(x) = 0, \;\dots,\; \Delta_\rho(x) = 0$$

admits all transformations:

$$x_i' = f_i(x_1, \dots, x_n, a_1, \dots, a_r) \qquad {\scriptstyle (i=1\cdots n)}$$

of the r-term group $X_1 f, \dots, X_r f$.

According to Chap. 6, p. 114, this only amounts to proving that the system of equations:

$$\Delta_1(x') = 0, \;\dots,\; \Delta_\rho(x') = 0$$

is equivalent, after the substitution $x_i' = f_i(x, a)$, to the system of equations:

$$\Delta_1(x) = 0, \;\dots,\; \Delta_\rho(x) = 0;$$

here, it is completely indifferent whether or not the equations $\Delta_1 = 0, \dots, \Delta_\rho = 0$ are mutually independent.

In order to prove that our system of equations really possesses the property in question, we proceed as follows:

In Chap. 3, p. 48,[5] (Translator's note: The matrix denoted $\widetilde{\rho}(a)$ is denoted here by $\omega(a)$.) we have seen that by virtue of the equations $x_i' = f_i(x, a)$, a relation of the form:

(4)
$$\sum_{k=1}^{r} e'_k X'_k f = \sum_{k=1}^{r} e_k X_k f$$

holds, in which the e'_k are related to the e_k by the r equations:

$$e_j = -\sum_{\pi,\,k}^{1\cdots r} \vartheta_{kj}(a)\,\alpha_{\pi k}(a)\,e'_\pi = \sum_{\pi=1}^{r} \omega_{j\pi}(a)\,e'_\pi.$$

If we insert the above expression of the e_k in (4) and if we compare the coefficients of the two sides, we obtain r relations:

$$X'_k f = \sum_{j=1}^{r} \omega_{jk}(a)\,X_j f \qquad (k=1\cdots r)$$

which clearly reduce to identities as soon as one expresses the x' in terms of the x by means of the equations $x'_i = f_i(x,a)$.

By inserting the function x'_i in place of f in the equations just found, we obtain the equations:

$$X'_k x'_i = \xi_{ki}(x') = \sum_{j=1}^{r} \omega_{jk}(a)\,X_j x'_i$$
$$= \sum_{j=1}^{r} \omega_{jk}(a) \sum_{\nu=1}^{n} \xi_{j\nu}(x)\,\frac{\partial f_i(x,a)}{\partial x_\nu}$$

which express directly the $\xi_{ki}(x')$ as functions of the x and a. Thanks to this, we are in a position to study the behavior of the equations $\Delta(x') = 0$ after the substitution $x'_i = f_i(x,a)$.

The determinants $\Delta_1(x'), \dots, \Delta_\rho(x')$ are made up from the matrix:

$$\begin{vmatrix} \xi_{11}(x') & \cdot & \cdot & \xi_{1n}(x') \\ \cdot & \cdot & \cdot & \cdot \\ \xi_{r1}(x') & \cdot & \cdot & \xi_{rn}(x') \end{vmatrix}$$

in the same way as the determinants $\Delta_1(x), \dots, \Delta_\rho(x)$ are made up from the matrix (3). Now, if we imagine that the values found a while ago:

$$\xi_{ki}(x') = \sum_{j=1}^{r} \omega_{jk}\,X_j x'_i$$

are inserted in the matrix just written and then that the determinants $\Delta(x')$ are computed, we realize the following: the determinants $\Delta_\sigma(x')$ have the form:

$$\Delta_\sigma(x') = \sum_{\tau=1}^{\rho} \chi_{\sigma\tau}(a)\,D_\tau \qquad (\sigma=1\cdots\rho),$$

where the $\chi_{\sigma\tau}$ are certain determinants formed with the $\omega_{jk}(a)$, while $D_1,\dots,D_\rho$ denote all the $(p+1)\times(p+1)$ determinants of the matrix:

$$\begin{vmatrix} X_1x_1' & \cdot & \cdot & X_1x_n' \\ \cdot & \cdot & \cdot & \cdot \\ X_rx_1' & \cdot & \cdot & X_rx_n' \end{vmatrix}.$$

Lastly, if the replace each X_kx_i' by its value:

$$X_kx_i' = \sum_{\nu=1}^{n} \xi_{k\nu}(x)\,\frac{\partial f_i(x,a)}{\partial x_\nu},$$

we obtain for the determinants D_τ expressions of the form:

$$D_\tau = \sum_{\mu=1}^{\rho} \psi_{\tau\mu}(x,a)\,\Delta_\mu(x),$$

where the $\psi_{\tau\mu}$ are certain determinants formed with the $\frac{\partial f_i(x,a)}{\partial x_\nu}$.

With this, it is proved that, after the substitution $x_i' = f_i(x,a)$, the $\Delta_\sigma(x')$ take the form:

$$\Delta_\sigma(x') = \sum_{\tau,\,\mu}^{1\cdots\rho} \chi_{\sigma\tau}(a)\,\psi_{\tau\mu}(x,a)\,\Delta_\mu(x) \qquad (\sigma=1\cdots\rho).$$

Now, since the functions $\chi_{\sigma\tau}(a)$, $\psi_{\tau\mu}(x,a)$ behave regularly for all systems of values x,a coming into consideration, it is clear that the system of equations $\Delta_\sigma(x') = 0$ is equivalent, after the substitution $x_i' = f_i(x,a)$, to the system of equations $\Delta_\sigma(x) = 0$, hence that the latter system of equations admits all transformations $x_i' = f_i(x,a)$. But this is what was to be proved.

§ 67. Theorem 39 is so important that it appears not to be superfluous to produce yet a third proof of it.

According to Proposition 3 of Chap. 7 (p. 129), the system of equations $\Delta_1 = 0$, $\dots$, $\Delta_\rho = 0$ certainly admits all transformations of the r-term group $X_1f,\dots,X_rf$ when there exist relations of the form:

$$X_k\Delta_\sigma = \sum_{\tau=1}^{\rho} \omega_{\sigma\tau}(x_1,\dots,x_n)\,\Delta_\tau \qquad (k=1\cdots r;\ \sigma=1\cdots\rho)$$

and when in addition the functions $\omega_{\sigma\tau}$ behave regularly for the systems of values which satisfy the equations $\Delta_1 = 0$, $\dots$, $\Delta_\rho = 0$. Now, it is in our case not more difficult to prove that the system of equations $\Delta_\sigma = 0$ satisfies this property. But in order not to be too extensive, we want to execute this proof only in a special case. In this way, one will clearly see how to treat the most general case.

We will firstly assume that our group is simply transitive. So it contains n independent infinitesimal transformations and in addition, the determinant:

$$\Delta = \begin{vmatrix} \xi_{11}(x) & \cdot & \cdot & \cdot & \xi_{1n}(x) \\ \cdot & \cdot & \cdot & \cdot & \cdot \\ \xi_{n1}(x) & \cdot & \cdot & \cdot & \xi_{nn}(x) \end{vmatrix}$$

does not vanish identically.

Furthermore, we will restrict ourselves to establishing that there are n relations of the form:

$$X_i\Delta = \omega_i(x_1,\ldots,x_n)\,\Delta \qquad (i=1\cdots n),$$

so that the equation $\Delta = 0$ admits all transformations of the group. By contrast, we will not consider the invariant systems of equations which are obtained by equating to zero all subdeterminants of the determinant Δ.

If we express the $(n-1)\times(n-1)$ subdeterminants of Δ as partial differential quotients of Δ with respect to the $\xi_{\mu\nu}$, we obtain for $X_i\Delta$ the expression:

$$X_i\Delta = \sum_{\mu,\,\nu}^{1\cdots n} X_i\,\xi_{\mu\nu}\,\frac{\partial\Delta}{\partial\xi_{\mu\nu}}.$$

Now, $X_1f,\ldots,X_nf$ generate an n-term group, so there are relations of the form:

$$[X_i,\,X_\mu] = \sum_{s=1}^{n} c_{i\mu s}X_sf,$$

or, in more detail:

$$X_i\,\xi_{\mu\nu} - X_\mu\,\xi_{i\nu} = \sum_{s=1}^{n} c_{i\mu s}\,\xi_{s\nu}.$$

Consequently, for $X_i\,\xi_{\mu\nu}$, we obtain the following expression:

$$X_i\,\xi_{\mu\nu} = \sum_{s=1}^{n}\left(\xi_{\mu s}\,\frac{\partial\xi_{i\nu}}{\partial x_s} + c_{i\mu s}\,\xi_{s\nu}\right).$$

If, in this expression, we insert the equation above for $X_i\Delta$, then we get:

$$X_i\,\Delta = \sum_{\mu,\,\nu,\,s}^{1\cdots n}\left(\xi_{\mu s}\,\frac{\partial\xi_{i\nu}}{\partial x_s} + c_{i\mu s}\,\xi_{s\nu}\right)\frac{\partial\Delta}{\partial\xi_{\mu\nu}}.$$

Here according to a known proposition about determinants, the coefficients of $\partial\xi_{i\nu}/\partial x_s$ and of $c_{i\mu s}$ can be expressed in terms of Δ. Namely, one has:

$$\sum_{\mu=1}^{n} \xi_{\mu s} \frac{\partial \Delta}{\partial \xi_{\mu\nu}} = \varepsilon_{s\nu} \Delta,$$

$$\sum_{\nu=1}^{n} \xi_{s\nu} \frac{\partial \Delta}{\partial \xi_{\mu\nu}} = \varepsilon_{s\mu} \Delta,$$

where the quantities $\varepsilon_{\pi\rho}$ vanish as soon as π and ρ are different from each other, while $\varepsilon_{\pi\pi}$ always has the value 1. Using these formulas, we obtain:

$$X_i \Delta = \Delta \left\{ \sum_{\nu,\, s}^{1\cdots n} \varepsilon_{s\nu} \frac{\partial \xi_{i\nu}}{\partial x_s} + \sum_{\mu,\, s}^{1\cdots n} \varepsilon_{s\mu}\, c_{i\mu s} \right\},$$

and from this, it follows lastly that:

$$X_i \Delta = \Delta \sum_{s=1}^{n} \left(\frac{\partial \xi_{is}}{\partial x_s} + c_{iss} \right) \qquad (i=1\cdots n). \tag{5}$$

Since, as always, only systems of values $x_1, \dots, x_n$ for which all $\xi_{ki}(x)$ behave regularly are taken into consideration, then clearly, the factor of Δ in the right-hand side behaves regularly for the considered systems of values $x_1, \dots, x_n$. Hence, if the equation $\Delta = 0$ can be satisfied by such systems of values $x_1, \dots, x_n$, then according to Proposition 3, p. 129, it admits all transformations of the group $X_1 f, \dots, X_n f$.

§ 68. As was already underlined on p. 235, the developments of the present chapter have great similarities with those of Chap. 7, p. 135 sq. Therefore, it is important to be conscious of the differences between the two theories.

We have already mentioned the first difference on page 240. It consists in what follows:

When the r independent infinitesimal transformations $X_1 f, \dots, X_r f$ generate an r-term group, then each system of equations $\Delta_1 = 0, \dots, \Delta_\rho = 0$ obtained by forming determinants as mentioned more than enough admits all infinitesimal transformations $X_1 f, \dots, X_r f$. By contrast, when the r infinitesimal transformations are only subjected to the restriction that the independent equations amongst the equations $X_1 f = 0, \dots, X_r f = 0$ form a complete system, then in general, none of the systems of equations $\Delta_1 = 0, \dots, \Delta_\rho = 0$ needs to admit the infinitesimal transformations $X_1 f, \dots, X_r f$.

Amongst certain conditions, one can in fact also be sure from the beginning, in the second one of the two cases just mentioned, that a system of equations $\Delta_1 = 0, \dots, \Delta_\rho = 0$ obtained by forming determinants admits the infinitesimal transformations $X_1 f, \dots, X_r f$.

Let the r independent infinitesimal transformations $X_1 f, \dots, X_r f$ be constituted in such a way that the independent equations amongst the r equations $X_1 f = 0, \dots, X_r f = 0$ form a complete system, so that there are relations of the form:

(6)
$$[X_i, X_k] = \gamma_{ik1}(x_1,\dots,x_n)X_1f + \dots + \gamma_{ikr}(x_1,\dots,x_n)X_rf$$
$$(i,\,k=1\cdots r).$$

Let the $(p+1)\times(p+1)$ determinants of the matrix:

(7)
$$\begin{vmatrix} \xi_{11}(x) & \cdot & \cdot & \xi_{1n}(x) \\ \cdot & \cdot & \cdot & \cdot \\ \xi_{r1}(x) & \cdot & \cdot & \xi_{rn}(x) \end{vmatrix}$$

be denoted by $\Delta_1,\dots,\Delta_\rho$. If there are systems of values $x_1,\dots,x_n$ for which all determinants $\Delta_1,\dots,\Delta_\rho$ vanish and if *all functions* $\gamma_{ikj}(x_1,\dots,x_n)$ *behave regularly for these systems of values*, then it can be shown that the system of equations $\Delta_1=0$, $\dots$, $\Delta_\rho=0$ admits the infinitesimal transformations $X_1f,\dots,X_rf$.

In what follows, this proposition plays no rôle; it will therefore suffice that we prove it only in a special simple case; the proof for the general proposition can be executed in an entirely similar way.

We want to assume that $r=n$ and that the n equations $X_1f=0$, $\dots$, $X_nf=0$ are mutually independent; in addition, let $p=n-1$. The mentioned matrix then reduces to the not identically vanishing determinant:

$$\Delta = \sum \pm \xi_{11}\cdots\xi_{nn},$$

and it contains only a single $(p+1)\times(p+1)$ determinant, namely itself. We will show that the equation $\Delta=0$ then certainly admits the infinitesimal transformations $X_1f,\dots,X_rf$ when the functions γ_{ikj} in the equations:

$$[X_i, X_k] = \sum_{j=1}^{n} \gamma_{ikj}X_jf \qquad (i,\,k=1\cdots n)$$

behave regularly for the systems of values $x_1,\dots,x_n$ which bring Δ to zero.

According to Chap. 7, Proposition 3, p. 129, we need only to show that each $X_k\Delta$ can be represented in the form $\omega_k(x_1,\dots,x_n)\Delta$ and that the ω_k behave regularly for the systems of values of $\Delta=0$. This proof succeeds in the same way as in the preceding section. We simply compute the expressions $X_k\Delta$ and we find in the same way as before:

$$X_k\Delta = \Delta\sum_{\nu=1}^{n}\left\{\frac{\partial\xi_{k\nu}}{\partial x_\nu} + \gamma_{k\nu\nu}(x_1,\dots,x_n)\right\}$$
$$(k=1\cdots n).$$

The computation necessary for that is exactly the previous one, although the constants c_{ikj} are replaced by the functions $\gamma_{ikj}(x)$; but still, that there occurs no difference has its reason in the fact that in the preceding sections, no use was made of the constancy property of the c_{ikj}.

Clearly, the factors of Δ in the right-hand side of the above equations behave regularly for the systems of values of $\Delta=0$, hence we see that the equation $\Delta=0$ really admits the infinitesimal transformations $X_1f,\dots,X_nf$.

A second important difference between the case of an r-term *group* $X_1 f, \dots, X_r f$ and the more general case of Chap. 7 comes out as soon as one already knows a manifold M which admits all infinitesimal transformations $X_1 f, \dots, X_r f$.

We want to assume that $X_1 f, \dots, X_r f$ attach exactly p independent directions to the points of M and especially, that $X_1 f, \dots, X_p f$ determine p independent directions. Under these assumptions, for the points of M, there are relations of the form:

$$X_{p+k} f = \varphi_{k1}(x_1, \dots, x_n)\, X_1 f + \cdots + \varphi_{kp}(x_1, \dots, x_n)\, X_p f$$
$$(k = 1 \cdots r - p),$$

where the φ_{kj} behave regularly; on the other hand, there is no relation of the form:

$$\chi_1(x_1, \dots, x_n)\, X_1 f + \cdots + \chi_p(x_1, \dots, x_n)\, X_p f = 0.$$

Now, if $X_1 f, \dots, X_r f$ generate an r-term group, then for the points of M, all $[X_i, X_k]$ can be represented in the form:

$$[X_i, X_k] = \sum_{j=1}^{p} \psi_{ikj}(x_1, \dots, x_n)\, X_j f \qquad (i, k = 1 \cdots r),$$

and here, *the* ψ_{ikj} *behave regularly*. By contrast, if $X_1 f, \dots, X_r f$ only possess the property that the independent equations amongst the equations $X_1 f = 0, \dots, X_r f = 0$ form a complete system, then such a representation of the $[X_i, X_k]$ for the points of M is not possible in all cases; but it is always possible when the functions γ_{ikj} in the equations (6) behave regularly for the systems of values $x_1, \dots, x_n$ on M. We have already succeeded in making use of this condition in Chap. 7, p. 144 sq.

Chapter 15
Invariant Families of Infinitesimal Transformations

We study in this chapter the general linear combination:

$$e_1 X_1 + \cdots + e_q X_q$$

of $q \geqslant 1$ given arbitrary local infinitesimal transformations:

$$X_k = \sum_{i=1}^{n} \xi_{ki}(x) \frac{\partial}{\partial x_i} \qquad\qquad (k=1\cdots q)$$

having analytic coefficients $\xi_{ki}(x)$ and which are assumed to be independent of each other. When one introduces new variables $x'_i = \varphi_i(x_1, \dots, x_n)$ in place of the x_l, every transformation X_k of this general combination receives another form, but it may sometimes happen under certain circumstances that the complete family in its wholeness remains unchanged, namely that there are functions $e'_k = e'_k(e_1, \dots, e_q)$ such that:

$$\varphi_*\big(e_1 X_1 + \cdots + e_q X_q\big) = e'_1(e) X'_1 + \cdots + e'_q(e) X'_q,$$

where, as in previous circumstances, the $X'_k = \sum_{i=1}^{n} \xi_{ki}(x') \frac{\partial}{\partial x'_i}$ denote the same vector fields, but viewed in the target space $x'_1, \dots, x'_n$.

Definition 15.1. The family $e_1 X_1 + \cdots + e_q X_q$ of infinitesimal transformations is said to *remain invariant after the introduction of the new variables* $x' = \varphi(x)$ if there are functions $e'_k = e'_k(e_1, \dots, e_q)$ depending on φ such that:

$$(1) \qquad \varphi_*\big(e_1 X_1 + \cdots + e_q X_q\big) = e'_1(e) X'_1 + \cdots + e'_q(e) X'_q;$$

alternatively, one says that the family *admits* the transformation which is represented by the concerned change of variables.

Proposition 2. *Then the functions $e'_k(e)$ in question are necessarily linear:*

S. Lie, *Theory of Transformation Groups I*,
DOI 10.1007/978-3-662-46211-9_15

$$e'_k = \sum_{j=1}^{q} \rho_{kj}\, e_j \qquad {\scriptstyle (k=1\cdots q)},$$

with the constant matrix $\big(\rho_{kj}\big)_{1\leqslant k\leqslant q}^{1\leqslant j\leqslant q}$ *being invertible:* $e_k = \sum_{j=1}^{q} \widetilde{\rho}_{kj}\, e'_j$.

Proof. Indeed, through the change of coordinates $x' = \varphi(x)$, if we write that the vector fields X_k are transferred to:

$$\varphi_*(X_k) = \sum_{i=1}^{n} X_k(x'_i)\,\frac{\partial}{\partial x'_i} =: \sum_{i=1}^{n} \eta_{ki}(x'_1,\dots,x'_n)\,\frac{\partial}{\partial x'_i} \qquad {\scriptstyle (k=1\cdots q)},$$

with their coefficients $\eta_{ki} = \eta_{ki}(x')$ being expressed in terms of the target coordinates, and if we substitute the resulting expression into (1), we get the following linear relations:

$$\text{(1')} \qquad \sum_{k=1}^{q} e'_k\, \xi_{ki}(x') = \sum_{k=1}^{q} e_k\, \eta_{ki}(x') \qquad {\scriptstyle (i=1\cdots n)}.$$

The idea is to substitute here for x' exactly the same number q of different systems of fixed values:

$$x_1^{(1)},\dots,x_n^{(1)},x_1^{(2)},\dots,x_n^{(2)},\dots\dots,x_1^{(q)},\dots,x_n^{(q)}$$

that are mutually in general position and considered as constant. In fact, according to the proposition on p. 79, or equivalently, according to the assertion formulated just below the long matrix located on p. 78, the linear independence of $X_1,\dots,X_q$ insures that for most such q points, the long $q \times qn$ matrix in question:

$$\begin{pmatrix} \xi_{11}^{(1)} & \cdots & \xi_{1n}^{(1)} & \xi_{11}^{(2)} & \cdots & \xi_{1n}^{(2)} & \cdots\cdots & \xi_{11}^{(q)} & \cdots & \xi_{1n}^{(q)} \\ \cdots & \cdots & \cdots & \cdots & \cdots & \cdots & \cdots\cdots & \cdots & \cdots & \cdots \\ \xi_{q1}^{(1)} & \cdots & \xi_{qn}^{(1)} & \xi_{q1}^{(2)} & \cdots & \xi_{qn}^{(2)} & \cdots\cdots & \xi_{q1}^{(q)} & \cdots & \xi_{qn}^{(q)} \end{pmatrix}$$

has rank equal to q, where we have set $\xi_{ki}^{(\nu)} := \xi_{xi}\big(x^{(\nu)}\big)$. Consequently, while considering the values of $\xi_{ki}\big(x^{(\nu)}\big)$ and of $\eta_{ki}\big(x^{(\nu)}\big)$ as *constant*, the linear system above is solvable with respect to the unknowns e'_k and we obtain:

$$e'_k = \sum_{j=1}^{q} \rho_{kj}\, e_j \qquad {\scriptstyle (k=1\cdots q)},$$

for some constants ρ_{kj}. In addition, we claim that the determinant of the matrix $\big(\rho_{kj}\big)_{1\leqslant k\leqslant q}^{1\leqslant j\leqslant q}$ is in fact nonzero. Indeed, the linear independence of $X_1,\dots,X_q$ being obviously equivalent to the linear independence of $\varphi_*(X_1),\dots,\varphi_*(X_q)$, the other corresponding long matrix:

$$\begin{pmatrix} \eta_{11}^{(1)} \cdots \eta_{1n}^{(1)} & \eta_{11}^{(2)} \cdots \eta_{1n}^{(2)} & \cdots\cdots & \eta_{11}^{(q)} \cdots \eta_{1n}^{(q)} \\ \cdots \cdots \cdots & \cdots \cdots \cdots & \cdots\cdots & \cdots \cdots \cdots \\ \eta_{q1}^{(1)} \cdots \eta_{qn}^{(1)} & \eta_{q1}^{(2)} \cdots \eta_{qn}^{(2)} & \cdots\cdots & \eta_{q1}^{(q)} \cdots \eta_{qn}^{(q)} \end{pmatrix}$$

then also has rank equal to q and we therefore can also solve symmetrically:

$$e_k = \sum_{j=1}^{q} \widetilde{\rho}_{kj}\, e'_j \qquad {\scriptstyle (k=1\cdots q)},$$

with coefficients $\widetilde{\rho}_{kj}$ which necessarily coincide with the elements of the inverse matrix. □

As usual, we understand by:

$$X_k f = \sum_{i=1}^{n} \xi_{ki}\, \frac{\partial f}{\partial x_i} \qquad {\scriptstyle (k=1\cdots q)}$$

independent infinitesimal transformations; temporarily, this shall be the only assumption which we make about the $X_k f$.

We consider the family of ∞^{q-1} infinitesimal transformations which is represented by the expression:

$$e_1 X_1 f + \cdots + e_q X_q f$$

with the q arbitrary parameters $e_1, \dots, e_q$. When we introduce, in this expression, new independent variables x' in place of the x, then each infinitesimal transformations in our family takes another form; evidently, we then obtain in general a completely new family of ∞^{q-1} infinitesimal transformations. However, in certain circumstances, it can happen that the new family does not essentially differ in its form from the original family, when for arbitrary values of the e, there is a relation of the form:

$$\text{(1)} \qquad \sum_{k=1}^{q} e_k \sum_{i=1}^{n} \xi_{ki}(x_1, \dots, x_n)\, \frac{\partial f}{\partial x_i} = \sum_{k=1}^{q} e'_k \sum_{i=1}^{n} \xi_{ki}(x'_1, \dots, x'_n)\, \frac{\partial f}{\partial x'_i},$$

in which the e'_k do not depend upon the x, but only upon $e_1, \dots, e_q$.

If there is such a relation, which we can also write briefly as:

$$\text{(2)} \qquad \sum_{k=1}^{q} e_k X_k f = \sum_{k=1}^{q} e'_k X'_k f,$$

then we say: *the family of infinitesimal transformations* $\sum e_k X_k f$ *remains invariant after the introduction of the new variables* x', or: *it admits the transformation which is represented by the concerned change of variables.*

§ 69. Let the family of the ∞^{q-1} infinitesimal transformations $\sum e_k X_k f$ remain invariant through the transition to the variables x', so that there is a relation of the form:

$$\sum_{k=1}^{q} e_k X_k f = \sum_{k=1}^{q} e'_k X'_k f, \tag{2}$$

in which the e' are certain functions of only the e. To begin with, we study this relationship of dependence between the e and the e'; in this way, we reach the starting point for the more precise study of such families of infinitesimal transformations.

The expressions $X_k f$ can be written as:

$$X_k f = \sum_{i=1}^{n} X_k x'_i \frac{\partial f}{\partial x'_i},$$

or, when one expresses the $X_k x'_i$ in terms of the x', as:

$$X_k f = \sum_{i=1}^{n} \eta_{ki}(x'_1, \dots, x'_n) \frac{\partial f}{\partial x'_i}.$$

If we insert these values in the equation (2), we can equate the coefficients of the $\partial f / \partial x'_i$ in the two sides, and so we obtain the following linear relations between the e and the e':

$$\sum_{k=1}^{q} e'_k \xi_{ki}(x') = \sum_{k=1}^{q} e_k \eta_{ki}(x') \qquad (i=1\cdots n). \tag{2'}$$

According to our assumption, it is possible to enter for the e' functions of the e alone so that the equations (2') are satisfied for all values of the x'. It can be shown that the concerned functions of the e are completely determined.

Since the equations (2') are supposed to hold true for all values of the x', then they must also be satisfied when we replace $x'_1, \dots, x'_n$ by any other system of variables. We want to do this, and to write down the equations (2') in exactly q different systems of variables $x'_1, \dots, x'_n$, $x''_1, \dots, x''_n$, $\cdots$, $x^{(q)}_1, \dots, x^{(q)}_n$:

$$\sum_{k=1}^{q} e'_k \xi_{ki}(x^{(\nu)}) = \sum_{k=1}^{q} e_k \eta_{ki}(x^{(\nu)}) \qquad (i=1\cdots n)$$
$$(\nu=1\cdots q).$$

The equations so obtained are solvable with respect to $e'_1, \dots, e'_q$, because under the assumptions made, according to the developments of Chap. 4, p. 78, not all $q \times q$ determinants of the matrix:

$$\begin{vmatrix} \xi'_{11} & \cdot & \cdot & \xi'_{1n} & \xi''_{11} & \cdot & \cdot & \xi''_{1n} & \cdot & \cdot & \cdot & \xi^{(q)}_{1n} \\ \cdot & \cdot & \cdot & \cdot & \cdot & \cdot & \cdot & \cdot & \cdot & \cdot & \cdot & \cdot \\ \xi'_{q1} & \cdot & \cdot & \xi'_{qn} & \xi''_{q1} & \cdot & \cdot & \xi''_{qn} & \cdot & \cdot & \cdot & \xi^{(q)}_{qn} \end{vmatrix}$$

vanish.

In addition, since the mentioned equations are certainly compatible with each other,[1] (Translator's note: — and since, furthermore, the lemma on p. 78 insures that, with a suitable choice of generic fixed points $x'_1, \dots, x'_n, x''_1, \dots, x''_n, \cdots, x^{(q)}_1, \dots, x^{(q)}_n$, the rank of the considered matrix of ξ's is maximal equal to q —) we obtain the e' represented as linear homogeneous functions of the e:

$$e'_k = \sum_{j=1}^{q} \rho_{kj}\, e_j \qquad (k=1\cdots q).$$

Here naturally, the ρ_{kj} are independent of the $x', x'', \dots, x^{(q)}$ and hence are absolute constants; the determinant of the ρ_{kj} is different from zero, because visibly the e_k can, in exactly the same way, be represented as linear homogeneous functions of the e'.

Although, under the assumptions made, the *family* of infinitesimal transformations $\sum e_k X_k f$ remains invariant after the introduction of the x', in general, its individual transformations are permuted. However, there always exists at least one infinitesimal transformation $\sum e^0_k X_k f$ which remains invariant, since the condition which the coefficients e^0_k of such an infinitesimal transformation:

$$\sum_{k=1}^{q} e^0_k X_k f = \omega \sum_{k=1}^{q} e^0_k X'_k f$$

must satisfy can be replaced by the q equations:

$$\omega\, e^0_k = \sum_{j=1}^{q} \rho_{kj}\, e^0_j \qquad (k=1\cdots q)$$

and these last equations can always be satisfied without all the e^0_k being zero.

For a closer illustration of what has been said, an *example* is more suitable.

In the family of the ∞^3 transformations:

$$e_1 \frac{\partial f}{\partial x_1} + e_2 \frac{\partial f}{\partial x_2} + e_3 \left(x_1^2 \frac{\partial f}{\partial x_1} + x_1 x_2 \frac{\partial f}{\partial x_2} \right) + e_4 \left(x_1 x_2 \frac{\partial f}{\partial x_1} + x_2^2 \frac{\partial f}{\partial x_2} \right),$$

we introduce new variables by setting:

$$x'_1 = a_1 x_1 + a_2 x_2, \qquad x'_2 = a_3 x_1 + a_4 x_2.$$

At the same time, the family receives the new form:

$$e_1'\frac{\partial f}{\partial x_1'}+e_2'\frac{\partial f}{\partial x_2'}+e_3'\left(x_1'^2\frac{\partial f}{\partial x_1'}+x_1'x_2'\frac{\partial f}{\partial x_2'}\right)+e_4'\left(x_1'x_2'\frac{\partial f}{\partial x_1'}+x_2'^2\frac{\partial f}{\partial x_2'}\right),$$

where $e_1',\dots,e_4'$ are expressed as follows:

$$e_1'=a_1\,e_1+a_2\,e_2,\quad e_2'=a_3\,e_1+a_4\,e_2,\quad e_3'=\frac{a_4\,e_3-a_3\,e_4}{a_1a_4-a_2a_3},\quad e_4'=\frac{a_1\,e_4-a_2\,e_3}{a_1a_4-a_2a_3}.$$

Consequently, the family remains invariant in the above sense. If one wanted to know which individual infinitesimal transformations of the family remain invariant, one would only have to determine ω from the equation:

$$\begin{vmatrix} a_1-\omega & a_2 \\ a_3 & a_4-\omega \end{vmatrix}\cdot\begin{vmatrix} a_1\omega-1 & a_2 \\ a_3 & a_4\omega-1 \end{vmatrix}=0,$$

and to choose $e_1,\dots,e_4$ so that $e_k'=\omega\,e_k$; the concerned systems of values of the e_k provide the invariant infinitesimal transformations.

Now, by coming back to the general case, we want to specialize in a certain direction the assumptions made above. Namely, we assume that the transition from the x to the x' is a completely arbitrary transformation of a determined group. Correspondingly, we state the following question:

Under which conditions does the family $\sum e_k\,X_kf$ remain invariant through every transformation $x_i'=f_i(x_1,\dots,x_n,t)$ of the one-term group Yf, that is to say, under which conditions does a relation:

$$\sum_{k=1}^{q}e_k\,X_kf=\sum_{k=1}^{q}e_k'\,X_k'f,$$

hold for all systems of values $e_1,\dots e_q,t$, in which the e_k', aside from upon the e_j, only depend upon t?

When, in order to introduce new variables in X_kf, we apply the general transformation:

$$x_i'=x_i+t\,Y\,x_i+\cdots\qquad {\scriptstyle (i=1\cdots n)}$$

of the one-term group Yf, we obtain according to Chap. 4, p. 73:

$$X_kf=X_k'f+t\big(X_k'Y'f-Y'X_k'f\big)+\cdots;$$

hence also inversely:

$$X_k'f=X_kf+t[Y,X_k]+\cdots, \tag{3}$$

which is more convenient for what follows.

Now, if every infinitesimal transformation $X_kf+t[Y,X_k]+\cdots$ should belong to the family $e_1X_1f+\cdots+e_qX_qf$, and in fact so for every value of t, then obviously every infinitesimal transformation $[Y,X_k]$ would also be contained in this family. As a result, certain necessary conditions for the invariance of our family would be found, some conditions which amount to the fact that q relations of the form:

$$[Y, X_k] = \sum_{j=1}^{q} g_{kj} X_j f \qquad (k=1\cdots q) \tag{4}$$

should hold, in which the g_{kj} denote absolute constants.

If the family of infinitesimal transformations:

$$e_1 X_1 f + \cdots + e_q X_q f$$

is constituted so that for every k, a relation of the form (4) holds true, then we want to say that *the family admits the infinitesimal transformation* Yf. By adopting this terminology, we can state the result just obtained as follows:

If the family of infinitesimal transformations:

$$e_1 X_1 f + \cdots + e_q X_q f$$

admits all transformations of the one-term group Yf, *then it also admits the infinitesimal transformation* Yf.

But the converse also holds true, as we will now show.

We assume that the family of transformations $\sum e_k X_k f$ admits the infinitesimal transformation Yf, hence that relations of the form (4) hold true. If now the family $\sum e_k X_k f$ is supposed to simultaneously admit all finite transformations of the one-term group Yf, then it must be possible to determine $e_1', \dots, e_q'$ as functions of $e_1, \dots, e_q$ in such a way that the equation:

$$\sum_{k=1}^{q} e_k' X_k' f = \sum_{k=1}^{q} e_k X_k f$$

is identically satisfied, as soon as one introduces the variable x in place of x' in the $X_k' f$. Consequently, if $X_k' f$ takes the form:

$$X_k' f = \sum_{i=1}^{n} \zeta_{ki}(x_1, \dots, x_n, t) \frac{\partial}{\partial x_i}$$

after the introduction of the x, then one must be able to determine the e_k' so that the expression:

$$\sum_{k=1}^{q} e_k' X_k' f = \sum_{k=1}^{q} \sum_{i=1}^{n} e_k' \zeta_{ki}(x_1, \dots, x_n, t) \frac{\partial f}{\partial x_i}$$

is free of t, hence so that the differential quotient:

$$\frac{\partial}{\partial t} \sum_{k=1}^{q} e_k' X_k' f = \sum_{i=1}^{n} \frac{\partial f}{\partial x_i} \frac{\partial}{\partial t} \sum_{k=1}^{q} e_k' \zeta_{ki}(x_1, \dots, x_n, t)$$

vanishes;[2] (Translator's note: Indeed, differentiation with respect to t of $e_1 X_1 + \cdots + e_r X_r$ yields: $0 \equiv \frac{\partial}{\partial t} \sum_{k=1}^{q} e_k X_k$.) but at the same time, the e must still also satisfy the initial condition: $e'_k = e_k$ for $t = 0$.

In order to be able to show that under the assumptions made there really are functions e' of the required constitution, we must first calculate the differential quotient:

$$\frac{\partial}{\partial t} X'_k f = \sum_{i=1}^{n} \frac{\partial \zeta_{ki}(x_1, \ldots, x_n, t)}{\partial t} \frac{\partial f}{\partial x_i};$$

for this, we shall take an indirect route.

Above, we saw that $X'_k f$ can be expressed in the following way in terms of $x_1, \ldots, x_n$ and t:

$$X'_k f = X_k f + t[Y, X_k] + \cdots,$$

when the independent variables x' entering the X'_k are determined by the equations $x'_i = f_i(x_1, \ldots, x_n, t)$ of the one-term group Yf. So the desired differential quotient can be obtained by differentiation of the infinite power series in t lying in the right-hand side, or differently enunciated: it is the coefficient of τ^1 in the expansion of the expression:

$$X_k f + (t + \tau)[Y, X_k] + \cdots = \sum_{i=1}^{n} \xi_{ki}(x''_1, \ldots, x''_n) \frac{\partial f}{\partial x''_i} = X''_k f$$

with respect to powers of τ. Here, the x'' mean the quantities:

$$x''_i = f_i(x_1, \ldots, x_n, t + \tau).$$

However, the expansion coefficient [ENTWICKELUNGSCOEFFICIENT] discussed just above appears at first as an infinite series of powers of t; nevertheless, there is no difficulty in finding a finite closed expression for it.

As we know, the transition from the variables x to the variables $x'_i = f_i(x_1, \ldots, x_n, t)$ occurs through a transformation of the one-term *group* Yf, and to be precise, through a transformation with the parameter t. One comes from the x to the $x''_i = f_i(x_1, \ldots, x_n, t + \tau)$ through a transformation of *the same* group, namely through the transformation with the parameter $t + \tau$. But this transformation can be substituted for the succession of two transformations, of which the first one possesses the parameter t, and the second one the parameter τ; consequently, the transition from the x' to the x'' is likewise obtained through a transformation of the one-term group Yf, namely through the transformation whose parameter is τ:

$$x''_i = f_i(x'_1, \ldots, x'_n, \tau).$$

From this, we conclude that the series expansion of $X''_k f$ with respect to powers of τ reads:

$$X''_k f = X'_k f + \tau[Y', X'_k] + \cdots.$$

As a result, we have then found a finite closed expression for the expansion coefficient mentioned a short while ago; the sought differential quotient $\frac{\partial(X_k' f)}{\partial t}$ is hence:

$$\frac{\partial}{\partial t} X_k' f = [Y', X_k'] = Y' X_k' f - X_k' Y' f. \tag{5}$$

Naturally, this formula holds generally, and one can also choose as the two infinitesimal transformations $X_k f$ and Yf. However, in our specific case, $X_1 f, \dots, X_q f, Yf$ are not absolutely arbitrary, but they are linked together through the relations (4). So under the assumptions made above, we obtain:

$$\frac{\partial(X_k' f)}{\partial t} = \sum_{\nu=1}^{q} g_{k\nu} X_\nu' f \qquad (k=1\cdots q). \tag{6}$$

Now, if we form the differential quotient of $\sum e_k' X_k' f$ with respect to t, we obtain:

$$\begin{aligned} \frac{\partial}{\partial t} \sum_{k=1}^{q} e_k' X_k' f &= \sum_{k=1}^{q} \frac{\mathrm{d}e_k'}{\mathrm{d}t} X_k' f + \sum_{k=1}^{q} e_k' \sum_{\nu=1}^{q} g_{k\nu} X_\nu' f \\ &= \sum_{k=1}^{q} \left\{ \frac{\mathrm{d}e_k'}{\mathrm{d}t} + \sum_{\nu=1}^{q} g_{\nu k} e_\nu' \right\} X_k' f. \end{aligned}$$

Obviously, this expression vanishes only when the e_k' satisfy the differential equations:

$$\frac{\mathrm{d}e_k'}{\mathrm{d}t} + \sum_{\nu=1}^{q} g_{\nu k} e_\nu' = 0 \qquad (k=1\cdots q). \tag{7}$$

But from this the e_k' can be determined as functions of t in such a way that for $t=0$, each e_k' converts into the corresponding e_k; in addition, the e' are linear homogeneous functions of the e.

If one puts the values of the e' in question in the expression $\sum e_k' X_k' f$ and returns afterwards from the x' to the initial variables $x_1, \dots, x_n$, then $\sum e_k' X_k'$ will be independent of t, that is to say, it will be equal to $\sum e_k X_k f$. Consequently, the family of infinitesimal transformations $\sum e_k X_k f$ effectively remains invariant by the change of variables in question.

As a result, we can state the following theorem:

Theorem 43. *A family of ∞^{q-1} infinitesimal transformations $e_1 X_1 f + \cdots + e_q X_q f$ remains invariant, through the introduction of new variables x' which are defined by the equations of a one-term group:*

$$x_i' = x_i + t\, Y x_i + \cdots \qquad (i=1\cdots n),$$

if and only if, between Yf and the $X_k f$ there are q relations of the form:

$$(4)\qquad [Y, X_k] = \sum_{\nu=1}^{q} g_{k\nu} X_\nu f \qquad (k=1\cdots q),$$

in which the $g_{k\nu}$ denote constants. If these conditions are satisfied, then by the concerned change of variables, $\sum e_k X_k f$ receives the form $\sum e'_k X'_k f$, where $e'_1, \dots, e'_q$ determine themselves through the differential equations:

$$\frac{\mathrm{d}e'_k}{\mathrm{d}t} + \sum_{\nu=1}^{q} g_{\nu k}\, e'_\nu = 0 \qquad (k=1\cdots q),$$

by taking account of the initial conditions: $e'_k = e_k$ for $t = 0$.[1]

If one performs the integration of which the preceding theorem speaks, hence determines $e'_1, \dots, e'_r$ from the differential equations:

$$\frac{\mathrm{d}e'_k}{\mathrm{d}t} = -\sum_{\nu=1}^{q} g_{\nu k}\, e'_\nu \qquad (k=1\cdots q)$$

taking as a basis the initial conditions: $e'_k = e_k$ for $t = 0$, then one obtains equations of the form:

$$e'_k = \sum_{j=1}^{q} d_{kj}(t)\, e_j \qquad (k=1\cdots q).$$

It is clear that these equations represent the finite transformations of a certain one-term group, namely the one which is generated by the infinitesimal transformation:

$$\sum_{k=1}^{q} \left\{ \sum_{\nu=1}^{q} g_{\nu k}\, e_\nu \right\} \frac{\partial f}{\partial e_k}$$

(cf. Chap. 4).

From the theorem just stated, we want to derive an important proposition which is certainly closely suggested.

If the family of ∞^{q-1} infinitesimal transformations $e_1 X_1 f + \cdots + e_q X_q$ admits the two infinitesimal transformations $y_1 f$ and $Y_2 f$, then it also admits at the same time each transformation $c_1 Y_1 f + c_2 Y_2 f$ which is linearly deduced from $Y_1 f$ and $Y_2 f$; this follows immediately from the fact that the infinitesimal transformation:

$$\big[c_1 Y_1 f + c_2 Y_2 f,\, X_k f\big] = c_1 [Y_1, X_k] + c_2 [Y_2, X_k]$$

can, in our case, be linearly expressed in terms of the $X_i f$. But our family $e_1 X_1 f + \cdots + e_q X_q f$ also admits the infinitesimal transformation $[Y_1, Y_2]$. Indeed, one forms the Jacobi identity:

$$\big[[Y_1, Y_2], X_k\big] + \big[[Y_2, X_k], Y_1\big] + \big[[X_k, Y_1], Y_2\big] = 0,$$

[1] LIE, Archiv for Mathematik og Naturvidenskab Vol. 3, Christiania 1878.

and one takes into account that $[Y_1, X_k]$ and $[Y_2, X_k]$ can be linearly expressed in terms of the $X_i f$, so one realizes that this is also the case for $\big[[Y_1, Y_2], X_k\big]$.

By combining these two observations, one obtains the announced

Proposition 1. *If the most general infinitesimal transformation which leaves invariant a family of ∞^{q-1} infinitesimal transformations:*

$$e_1 X_1 f + \cdots + e_q X_q f$$

can be linearly deduced from a bounded number of infinitesimal transformations, say from $Y_1 f, \dots, Y_m f$, then the $Y_k f$ generate an m-term group.

This proposition can be further generalized; indeed, it is evident that the totality of *all finite* transformations which leave invariant the family $\sum e_k X_k f$ always forms a group.

§ 70. Let the conditions of the latter theorem be satisfied, namely let the family of the ∞^{q-1} infinitesimal transformations $\sum e_k X_k f$ be invariant by all transformations of the one-term group Y_f.

Now, according to Chap. 4, p. 70, the following holds true: if, after the introduction of new variables, the infinitesimal transformation Xf is transferred to Zf, then at the same time, the transformations of the one-term group Xf are transferred to the transformations of the one-term group Zf. So we deduce that under the assumptions of Theorem 43, not only the family of the ∞^{q-1} infinitesimal transformations $e_1 X_1 f + \cdots + e_r X_r f$ remains invariant, but also the family of the ∞^{q-1} one-term groups generated by these infinitesimal transformations, and naturally also, the totality of the ∞^q finite transformations which belong to these one-term groups.

But we want to go a step further: we want to study how the analytic expression of the individual finite transformations of the one-term groups $e_1 X_1 f + \cdots + e_q X_q f$ behave, when the new variables:

$$x_i' = x_i + t\,Y x_i + \cdots$$

are introduced in place of the x.

The answer to this question is given by Proposition 3 in Chap. 4, p. 71. Indeed, from this proposition, it easily results that, after the introduction of the new variables x', every finite transformation:

$$\overline{x}_i = x_i + \sum_k^{1\cdots q} e_k X_k x_i + \sum_{k,\,j}^{1\cdots q} \frac{e_k e_j}{1 \cdot 2} X_k X_j x_i + \cdots \qquad (i = 1 \cdots n) \tag{8}$$

receives the form:

$$\overline{x}_i' = x_i' + \sum_k^{1\cdots q} e_k' X_k' x_i' + \sum_{k,\,j}^{1\cdots q} \frac{e_k' e_j'}{1\cdot 2} X_k' X_j' x_i' + \cdots, \tag{9}$$

where the connection between the e_k and the e_k' is prescribed through the relation:

$$\sum_{k=1}^{q} e_k X_k f + \sum_{k=1}^{q} e_k' X_k' f. \tag{10}$$

Consequently, we see directly that the concerned family of ∞^q finite transformations in the new variables x' are of exactly the same form as in the initial variables x. But in addition, we remark that a finite transformation which has the parameters $e_1, \dots, e_q$ in the x possesses, after the introduction of the new variables x', the parameters $e_1', \dots, e_q'$.

Now, as just said, the connection between the e and the e' through the identity (10) is completely prescribed; hence this identity is absolutely sufficient when the question is to determine the new form which an arbitrary finite transformation (8) takes after the transition to the x'.

In order to be as clear as possible, we shall *interpret* $\sum e_k X_k f$ *virtually as the symbol of the finite transformation:*

$$x_i' = x_i + \sum_{k=1}^{q} e_k X_k x_i + \cdots \qquad (i=1\cdots n),$$

where the absolute[3] *(Translator's note: —namely the values themselves, but not the 'absolute values'* $|e_k|$ *in the modern sense—) values of the* e_k *then come into consideration, not only their ratio.* Then we can simply say:

After the introduction of the new variables x'*, the finite transformation* $\sum e_k X_k f$ *is transferred to the finite transformation* $\sum e_k' X_k' f$.

In the later studies of this chapter, the symbol $\sum e_k X_k f$ will sometimes be employed as the symbol of a finite transformation, and sometimes as the symbol of an infinitesimal transformation. Hence in each individual case, we shall underline which one of the two interpretations of the symbol is meant.

§ 71. Let the family of the ∞^q transformations $e_1 X_1 f + \cdots + e_q X_q f$ remain invariant by all transformations of the one-term group Yf. It can happen that Yf itself is an infinitesimal transformation of the family $\sum e_k X_k f$; indeed, the case where Yf is an arbitrary one of the ∞^{q-1} infinitesimal transformations $\sum e_k X_k f$ is of special interest. This will occur if and only if between the $X_k f$, there are relations of the form:

$$[X_i, X_k] = \sum_{s=1}^{q} g_{iks} X_s f.$$

Hence from the theorem of the preceding section, we obtain the following more special theorem in which we permit ourselves to write r instead of q and c_{iks} instead of g_{iks}.

Theorem 44. *For a family of ∞^r finite transformations:*

$$e_1 X_1 f + \cdots + e_r X_r f \qquad \text{or:} \qquad \overline{x}_i = \mathfrak{P}_i(x_1, \dots, x_n, e_1, \dots, e_r)$$

to remain invariant by every transformation which belongs to it — so that, after the introduction of the new variables:

$$x_i' = \mathfrak{P}_i(x_1, \dots, x_n, h_1, \dots, h_r), \qquad \overline{x}_i' = \mathfrak{P}_i(\overline{x}_1, \dots, \overline{x}_n, h_1, \dots, h_r)$$

in place of the x and the $\overline{x}$, it takes the form:

$$\overline{x}_i' = \mathfrak{P}_i(x_1', \dots, x_n', l_1, \dots, l_r)$$

where the l only depend upon $e_1, \dots, e_r$ and $h_1, \dots, h_r$ — it is necessary and sufficient that the Xf stand pairwise in the relationships:

$$[X_i, X_k] = \sum_{s=1}^{r} c_{iks} X_s f,$$

where the c_{iks} are absolute constants.

This theorem states an important property that the family of the ∞^r finite transformations $\sum e_k X_k f$ possesses as soon as relations of the form $[X_i, X_k]$ exist. It is noteworthy that, for the proof of this theorem, we have used the results of the preceding chapter only for the smallest part; moreover, we have used no more than a few developments of Chaps. 2, 3, 4 and 8. Namely, one should observe that we have made no use of Theorem 24, Chap. 9, p. 172.

If one assumes that the latter theorem is known, then one can shorten the proof of Theorem 44 as follows: one shows at first, as above, that the relations $[X_i, X_k] = \sum c_{iks} X_s f$ are necessary; then from Theorem 24, p. 172, it follows that the ∞^{r-1} infinitesimal transformations $\sum e_k X_k f$ generate an r-term group. If $x_i' = \mathfrak{P}_i(x_1, \dots, x_n, h_1, \dots, h_r)$ are the finite equations of this group, then according to Theorem 5, p. 45, there is an identity of the form:

$$\sum_{k=1}^{r} e_k X_k f = \sum_{k=1}^{r} e_k' X_k' f;$$

with this, the proof of Theorem 44 is produced.

One does not even need to refer to Theorem 5, p. 45, but one can conclude in the following way:

The equations $x_i' = \mathfrak{P}_i(x, h)$ of our group, when resolved, give a transformation of the form:

$$x_i = \mathfrak{P}_i\big(x_1', \dots, x_n', \chi_1(h), \dots, \chi_r(h)\big),$$

that is to say, the transformation which is inverse to the transformation with the parameters $h_1,\dots,h_r$. Now, if one imagines that these values of the x_i are inserted into the equations $\overline{x}_i = \mathfrak{P}_i(x,e)$ and if one takes into consideration that one has to deal with a group, then one realizes that there exist certain equations of the form:

$$\overline{x}_i = \mathfrak{P}_i\big(x_1',\dots,x_n',\, \psi_1(h,e),\dots,\, \psi_r(h,e)\big).$$

Lastly, if one inserts these expressions for the $\overline{x}_i$ in the equations $\overline{x}_i' = \mathfrak{P}_i(\overline{x},h)$, then one obtains:

$$\overline{x}_i' = \mathfrak{P}_i(x_1',\dots,x_n',\, l_1,\dots,l_r).$$

This is the new form that the transformations $\overline{x}_i = \mathfrak{P}_i(x,e)$ take after the introduction of the new variables x'. Here evidently, the l are functions of only the e and the h, exactly as is claimed in Theorem 44, p. 269.

As a result, the connection which exists between Theorem 24, p. 172, and Theorem 44 of the present chapter is clarified.

§ 72. In order to be able to state the gained results more briefly, or, if one wants, more clearly, we will, as earlier on, translate the symbolism of the theory of substitutions into the theory of transformation groups.

We shall denote all finite transformations $e_1 X_1 f + \cdots + e_r X_r f$ by the common symbol T, and the individual transformations by marking an appended index, so that for instance the symbol $T_{(a)}$ denotes the finite transformation:

$$a_1 X_1 f + \cdots + a_r X_r f.$$

Using this terminology, we can state Theorem 24, p. 172, as follows:

Proposition 2. *If r independent infinitesimal transformations $X_1 f,\dots,X_r f$ stand pairwise in relationships of the form:*

$$[X_i,\, X_k] = \sum_{s=1}^{r} c_{iks}\, X_s f,$$

then the family of all finite transformations $\sum e_k X_k f$, or $T_{(e)}$, also contains, simultaneously with the two transformations $T_{(a)}$ and $T_{(b)}$, the transformation $T_{(a)}\, T_{(b)}$; hence there is a symbolic equation of the form:

$$T_{(a)}\, T_{(b)} = T_{(c)},$$

in which the parameters c are functions of the a and of the b.

Correspondingly, from Theorem 44 of the present chapter, we obtain the following proposition, which, however, does not exhaust the complete content of the theorem:

Proposition 3. *If r independent infinitesimal transformations $X_1 f, \dots, X_r f$ stand pairwise in the relationships $[X_i, X_k]$, then the family of the ∞^r finite transformations $\sum e_k X_k f$, or $T_{(e)}$, also contains, simultaneously with the transformations $T_{(a)}$ and $T_{(b)}$, the transformation $T_{(a)}^{-1}\, T_{(b)}\, T_{(a)}$, whence there exists an equation of the form:*

$$T_{(a)}^{-1}\, T_{(b)}\, T_{(a)} = T_{(c')}$$

in which the parameters c' are functions of the a and of the b.

Obviously, the existence of the symbolic relation: $T_{(a)}^{-1}\, T_{(b)}\, T_{(a)} = T_{(c')}$ is a consequence of the former relation: $T_{(a)}\, T_{(b)} = T_{(c)}$, hence the latter proposition is also a consequence of the preceding one, as we already have seen in the previous sections.

Finally, by combining the two Theorems 44 and 24, p. 172, we obtain the following curious result.

Theorem 45. *If a family of ∞^r finite transformations: $a_1 X_1 f + \cdots + a_r X_r f$, or shortly $T_{(a)}$, possesses the property that the transformation $T_{(a)}^{-1}\, T_{(b)}\, T_{(a)}$ always belongs to the family, whatever values the parameters $a_1, \dots, a_r$, $b_1, \dots, b_r$ can have, then the family of ∞^r transformations in question forms an r-term group, that is to say: $T_{(a)}\, T_{(b)}$ is always a transformation which also belongs to the family.*

§ 73. If the transformation $T_{(a)}^{-1}\, T_{(b)}\, T_{(a)}$ coincides with the transformation $T_{(b)}$, a fact that we express by means of the symbolic equation:

$$T_{(a)}^{-1}\, T_{(b)}\, T_{(a)} = T_{(b)},$$

then we say: *the transformation $T_{(b)}$ remains invariant by the transformation $T_{(a)}$.*

But in this case, we also have:

$$T_{(b)}^{-1}\, T_{(a)}\, T_{(b)} = T_{(a)},$$

that is to say, the transformation $T_{(a)}$ remains invariant by the transformation $T_{(b)}$; on the other hand:

$$T_{(a)}\, T_{(b)} = T_{(b)}\, T_{(a)}$$

is an equation which expresses that the two transformations $T_{(a)}$ and $T_{(b)}$ are interchangeable with one another.

We already remarked in Theorem 6, p. 65, that the transformations of an arbitrary one-term group are interchangeable by pairs. At present, we can also settle the more general question of when the transformations of two *different* one-term groups Xf and Yf are interchangeable one with another.

In the general finite transformation $e Xf$ of the one-term group Xf, we introduce the new variables x_i' which are defined by the finite equations $x_i' = x_i + t\, Y x_i + \cdots$ of the one-term group Yf. The transformations of our two one-term groups will then be interchangeable if and only if every transformation of the form eXf remains invariant after the introduction of the x', whence eXf is equal to $eX'f$.

According to Theorem 43, p. 265, for the existence of an equation of the form:

$$eXf = e'Xf,$$

it is necessary and sufficient that the infinitesimal transformations Xf and Yf satisfy a relation:

$$[Y, X] = gXf;$$

at the same time, the e' determines itself through the differential equation:

$$\frac{\mathrm{d}e'}{\mathrm{d}t} + ge' = 0.$$

But in our case, e' is supposed, for every value of t, to be equal to e, hence e' absolutely does not depend on t, that is to say, the differential quotient $\mathrm{d}e'/\mathrm{d}t$ vanishes, and with it, the quantity g too. At the same time, this condition is evidently necessary and sufficient. Consequently, the following holds true:

Proposition 4. *The finite transformations of two one-term groups Xf and Yf are interchangeable with one another if and only if the expression $[X, Y]$ vanishes identically.*

It stands to reason to call *interchangeable* two infinitesimal transformations Xf and Yf which stand in the relationship $[X, Y] \equiv 0$.

If we introduce this terminology, we can say that the finite transformations of two one-term groups are interchangeable by pairs if and only if the infinitesimal transformations of the two groups are so.

Moreover, from the latter proposition follows the

Theorem 46. *The finite transformations of an r-term group $X_1f, \dots, X_rf$ are pairwise interchangeable if and only if all expressions $[X_i, X_k]$ vanish identically, or stated differently, if and only if the infinitesimal transformations $X_1f, \dots, X_rf$ are interchangeable by pairs*[2].

§ 74. From the general developments of §§ 69 and 70, we will now draw a few further consequences that are of importance.

We again assume that the r independent infinitesimal transformations $X_1f, \dots, X_rf$ stand pairwise in the relationships:

$$[X_i, X_k] = \sum_{s=1}^{r} c_{iks} X_s f,$$

[2] LIE, Gesellschaft der Wissenschaften zu Christiania 1872; Archiv for Mathematik og Naturvidenskab Vol. 8, p. 180, 1882; Math. Annalen Vol. 24, p. 557, 1884.

whence according to Theorem 24 of Chap. 9, p. 172, the ∞^r finite transformations $\sum e_k X_k f$ form an r-term group. Now, if the family of transformations $\sum e_k X_k f$ remains invariant by all transformations of a one-term group Yf, then according to Theorem 43, p. 265, there exist r equations of the form:

$$[Y, X_k] = \sum_{s=1}^{r} g_{ks} X_s f. \tag{4}$$

These equations show that the $r+1$ infinitesimal transformations $X_1 f, \dots, X_r f$, Yf generate an $(r+1)$-term group to which $X_1 f, \dots, X_r f$ belongs as an r-term subgroup.

In addition, it is clear that the relations (4) do not essentially change their form when one inserts in place of Yf a completely arbitrary infinitesimal transformation of the $(r+1)$-term group.

Hence, if $x_i' = \psi_i(x_1, \dots, x_n, a_1, \dots, a_{r+1})$ are the finite equations of the $(r+1)$-term group, then the family of finite transformations $\sum e_k X_k f$ remains invariant when one introduces in the same way the new variables x' in place of the x, that is to say, only the parameters vary in the analytic expression of the r-term group.

A similar property would hold true if, instead of the single transformation Yf, one had several, say m, such transformations all of which satisfied relations of the form (4); in addition, we want to add the assumption that these m infinitesimal transformations $Y_1 f, \dots, Y_m f$, together with $X_1 f, \dots, X_r f$, generate an $(r+m)$-term group. Then if we introduce new variables in the group $X_1 f, \dots, X_r f$ by means of an arbitrary transformation of the $(r+m)$-term group, the family of finite transformations of our r-term group remains invariant, although the parameters are changed in their analytic representation.

We want to express this relationship between the two groups by saying shortly: *if the r-term group $X_1 f, \dots, X_r f$ remains invariant by all transformations of the $(r+m)$-term group, it is an invariant subgroup of it.*

If we translate the terminology commonly used in the theory of substitutions, we can also interpret the definition of invariants subgroups as follows:

If T is the symbol of an arbitrary transformation of the $(r+m)$-term group G, and if S is an arbitrary transformation of a subgroup of G, then this subgroup is invariant in G when the transformation $T^{-1}ST$ also always belongs to the subgroup in question.

In what has been said above, the analytic conditions for the invariance of a subgroup are completely exhibited; so we need only to summarize them once again:

Theorem 47. *If the r-term group $X_1 f, \dots, X_r f$ is contained in an $(r+m)$-term group $X_1 f, \dots, X_r f, Y_1 f, \dots, Y_m f$, then it is an invariant subgroup of it when every $[Y_i, X_k]$ is expressible in terms of $X_1 f, \dots, X_r f$ linearly with constant coefficients.*

On can state Theorem 47 in the following more general version — though only in a formal sense:

Proposition 5. *If the totality of all infinitesimal transformations* $e_1 X_1 f + \cdots + e_m X_m f$ *forms an invariant family in the r-term group* $X_1 f, \ldots, X_m f, \ldots, X_r f$, *then* $X_1 f, \ldots, X_m f$ *generate an m-term invariant subgroup of the r-term group.*

Indeed, since all $[X_i, X_k]$, in which $i \leqslant m$, can be linearly deduced from $X_1 f, \ldots, X_m f$, then in particular the same holds true of all $[X_i, X_k]$ in which both i and k are $\leqslant m$. As a result, $X_1 f, \ldots, X_m f$ generate an m-term subgroup to which Theorem 47 can immediately be applied.

First, a few examples of invariant subgroups.

Proposition 6. [3] *If the r independent infinitesimal transformations* $X_1 f, \ldots, X_r f$ *generate an r-term group, then the totality of all infinitesimal transformations* $[X_i, X_k]$ *also generates a group; if the latter group contains r parameters, then it is identical to the group* $X_1 f, \ldots, X_r f$; *if it contains fewer than r parameters, then it is an invariant subgroup of the group* $X_1 f, \ldots, X_r f$; *if one adds to the* $[X_i, X_k]$ *arbitrarily many mutually independent infinitesimal transformations* $e_1 X_1 f + \cdots + e_r X_r f$ *that are also independent of the* $[X_i, X_k]$, *then one always obtains again an invariant subgroup of the group* $X_1 f, \ldots, X_r f$.

It is clear that the $[X_i, X_k]$ can at most generate an r-term group, since they all belong to the group $X_1 f, \ldots, X_r f$; the fact that they effectively generate a group comes immediately from the relations:

$$[X_i, X_k] = \sum_{s=1}^{r} c_{iks} X_s f,$$

for one indeed has:

$$\big[[X_i, X_k], [X_j, X_l]\big] = \sum_{s,\sigma}^{1\cdots r} c_{iks}\, c_{jl\sigma}\, [X_s, X_\sigma].$$

The claim that the group generated by the $[X_i, X_k]$ in the group $X_1 f, \ldots, X_r f$ is invariant becomes evident from the equations:

$$\big[X_j, [X_i, X_k]\big] = \sum_{s=1}^{r} c_{iks}\, [X_j, X_s].$$

The last part of the proposition does not require any further explanation.

The following generalization of the proposition just proved is noteworthy:

Proposition 7. *If m infinitesimal transformations* $Z_1 f, \ldots, Z_m f$ *of the r-term group* $X_1 f, \ldots, X_r f$ *generate an m-term subgroup of this group, and if the subgroup in question is invariant in the r-term group, then the subgroup which is generated by all the infinitesimal transformations* $[Z_\mu, Z_\nu]$ *is also invariant in the r-term group.*

[3] In the Archiv for Mathematik og Naturvidenskab Vol. 8, p. 390, Christiania 1883, LIE observed that the $[X_i, X_k]$ form an invariant subgroup. KILLING had realized this independently in the year 1886.

The proof of this is very simple. Under the assumptions of the proposition, there are relations of the form:

$$[X_k, Z_\mu] = \sum_{\lambda=1}^{m} h_{k\mu\lambda} Z_\lambda f \qquad (k=1\cdots r;\ \mu=1\cdots m),$$

where the $h_{k\mu\lambda}$ are constants. Next, if we form the Jacobi identity (cf. Chap. 5, p. 109):

$$[X_k, [Z_\mu, Z_\nu]] + [Z_\mu, [Z_\nu, X_k]] + [Z_\nu, [X_k, Z_\mu]] = 0,$$

and if we insert in it the expressions written above for $[Z_\mu, X_k]$ and $[Z_\nu, X_k]$, we obtain the equations:

$$[X_k, [Z_\mu, Z_\nu]] = \sum_{\lambda=1}^{m} \{h_{k\nu\lambda}\,[Z_\mu, Z_\lambda] - h_{k\mu\lambda}\,[Z_\nu, Z_\lambda]\},$$

from which it follows that the subgroup of the group $X_1 f, \dots, X_r f$ generated by the $[Z_\mu, Z_\nu]$ is invariant in the former group. But this is what was to be proved. —

Let the r-term group $X_1 f, \dots, X_r f$, or shortly G_r, contain an $(r-1)$-term invariant subgroup and let $Y_1 f, \dots, Y_r f$ be r independent infinitesimal transformations of G_r selected in a such a way that $Y_1 f, \dots, Y_{r-1} f$ is this invariant subgroup. Then there exist relations of the form[4] (Translator's note: Indeed, the invariance yields such relations for all $i = 1, \dots, r-1$ and all $k = 1, \dots, r$, and also by skew-symmetry for all $i = 1, \dots, r$ and all $k = 1, \dots, r-1$; then only $[Y_n, Y_n]$ remains, which, anyway, is zero.):

$$[Y_i, Y_k] = c_{ik1} Y_1 f + \cdots + c_{ik,r-1} Y_{r-1} f \qquad (i, k=1\cdots r),$$

whence all $[Y_i, Y_k]$ and also all $[X_i, X_k]$ can be linearly deduced from only $Y_1 f, \dots, Y_{r-1} f$. From this, we conclude that every $(r-1)$-term invariant subgroup of G_r contains all the infinitesimal transformations $[X_i, X_k]$ and so, we realize that the following proposition holds true:

Proposition 8. *In the r-term group $X_1 f, \dots, X_r f$, there is an $(r-1)$-term invariant subgroup if and only if the infinitesimal transformations $[X_i, X_k]$ generate a group with fewer than r parameters; if there are, amongst the $[X_i, X_k]$ exactly $r_1 < r$ mutually independent infinitesimal transformations, then one obtains all $(r-1)$-term invariant subgroups of the group $X_1 f, \dots, X_r f$ by adding to the $[X_i, X_k]$ in the most general way $r - r_1 - 1$ infinitesimal transformations $e_1 X_1 f + \cdots + e_r X_r f$ which are mutually independent and are independent of the $[X_i, X_k]$.*

Proposition 1 in Chap. 12, p. 218, provides us with another example of an invariant subgroup.

Namely, if, in the neighborhood of a point $x_1^0, \dots, x_n^0$, a group contains infinitesimal transformations of first or of higher order in the $x_i - x_i^0$, then each time, all infinitesimal transformations of order k ($k > 0$) and higher generate a subgroup.

Now, by bracketing [KLAMMEROPERATION], two infinitesimal transformations of respective orders k and $k+\nu$ produce a transformation $[X, Y]$ of the order $2k+\nu-1$. Because k is >0, one has $2k+\nu-1 \geqslant k+\nu$, hence $[X, Y]$ must be linearly expressible in terms of the infinitesimal transformations of orders $k+\nu$ and higher. In other words: all the infinitesimal transformations of orders $k+\nu$ and higher generate a group which is invariant in the group generated by the infinitesimal transformations of orders k and higher. And as said, all of that holds true under the only assumption that the number k is larger than zero.

In particular, if the numbers k and $k+\nu$ can be chosen in such a way that the group contains no infinitesimal transformation of orders $(2k+\nu-1)$ and higher in the neighborhood of $x_1^0, \dots, x_n^0$, then the expression $[X, Y]$ must vanish identically. Hence the following holds true:

Proposition 9. *If, in the neighborhood of a point $x_1^0, \dots, x_n^0$, a group contains no infinitesimal transformation of $(s+1)$-th order, or of higher order, and if, by contrast, it contains transformations of k-th order, where k satisfies the condition $2k-1>s$, then all infinitesimal transformations of the group which are of the k-th order and of higher order generate a group with pairwise interchangeable transformations.*

Here, the point $x_1^0, \dots, x_n^0$ absolutely need not be such that the coefficients of the resolved defining equations of the group (cf. Chap. 11, p. 204 sq.) behave regularly. —

Consequently [CONSEQUENTERWEISE] one must say that each finite continuous group contains two invariant subgroups, namely firstly itself and secondly the identity transformation. One realizes this by setting m, in Theorem 47, firstly equal to r, and secondly equal to zero; in the two cases the condition for the invariance of the subgroup $X_1 f, \dots, X_r f$ is satisfied by itself, only as soon as $X_1 f, \dots, X_r f$, $Y_1 f, \dots, Y_m f$ is an $(r+m)$-term group.

The groups which contain absolutely no invariant subgroup, disregarding the two which are always present, are of special importance. That is why these groups are also supposed to have a special name, and they should be called *simple* [EINFACH]. In contrast to this, a group is called *compound* [ZUSAMMENGESETZT] when, aside from the two invariant subgroups indicated above, it also contains other invariant subgroups.

In conclusion, here are two further propositions concerning invariant subgroups:

Proposition 10. *The transformations which are common to two invariant subgroups of a group G form in the same way a subgroup which is invariant in G.*

The transformations in question certainly form a subgroup of G (cf. Chap. 9, Proposition 2, p. 174); this subgroup must be invariant in G, since by all transformations of G, it is transferred to a group which belongs to the two invariant subgroups, that is to say, to itself.

Proposition 11. *If two invariant subgroups $Y_1 f, \dots, Y_m f$ and $Z_1 f, \dots, Z_p f$ of a group G have no infinitesimal transformations in common, then all expressions $[Y_i, Z_k]$ vanish identically, that is to say, every transformation of one subgroup is interchangeable with every other transformation of the other subgroup.*

Indeed, under the assumptions made, every expression $[Y_i, Z_k]$ must be expressible with constant coefficients both in terms of $Y_1 f, \dots, Y_m f$ and in terms of $Z_1 f, \dots, Z_p f$; but since the two subgroups have no infinitesimal transformations in common, then the conclusion is nothing else but that all expressions $[Y_i, Z_k]$ vanish identically. The rest follows from Proposition 4, p. 263.

§ 75. There are r-term groups for which one can select r independent infinitesimal transformations: $Y_1 f, \dots, Y_r f$ so that for every $i < r$, the i independent infinitesimal transformations $Y_1 f, \dots, Y_i f$ generate an i-term group which is invariant in the $(i+1)$-term group $Y_1 f, \dots, Y_{i+1} f$. Then between $Y_1 f, \dots, Y_r f$, there are relations of the form:

$$(11)\qquad [Y_i, Y_{i+k}] = \overline{c}_{i,i+k,1}\, Y_1 f + \cdots + \overline{c}_{i,i+k,i+k-1}\, Y_{i+k-1} f \qquad (i=1 \cdots r-1;\ k=1 \cdots r-i).$$

In the integration theory of those systems of differential equations which admit finite groups, it follows that the groups of the specific constitution just defined occupy a certain outstanding position in comparison to all other groups.[4]

Later, in the chapter on linear homogeneous groups, we will occupy ourselves more precisely with this special category of groups; at present, we want only to show in which way one can determine whether a given r-term group $X_1 f, \dots, X_r f$ belongs, or does not belong, to the category in question.

In order that it be possible to choose, amongst the infinitesimal transformations $e_1 X_1 f + \cdots + e_r X_r f$, r mutually independent ones: $Y_1 f, \dots, Y_r f$ which stand in relationships of the form (11), there must above all exist an $(r-1)$-term invariant subgroup in G_r: $X_1 f, \dots, X_r f$. Thanks to Proposition 8, p. 275, we are in a position to determine whether this is the case: according to this proposition, the group $X_1 f, \dots, X_r f$ contains an $(r-1)$-term invariant subgroup only when the group generated by all $[X_i, X_k]$ contains fewer than r parameters, say r_1; if this condition is satisfied, then one obtains all $(r-1)$-term invariant subgroups of G_r by adding to the $[X_i, X_k]$ in the most general way $r - r_1 - 1$ infinitesimal transformations $e_1 X_1 f + \cdots + e_r X_r f$ that are mutually independent and are independent of the $[X_i, X_k]$.

However, not every r-term group which contains an $(r-1)$-term invariant subgroup belongs to the specific category defined above; for it to belong to this category, it must contain an $(r-1)$-term invariant subgroup, which in turn must contain an $(r-2)$-term invariant subgroup, and again the latter subgroup must contain an $(r-3)$-term invariant subgroup, and so on.

From this, we see how we have to proceed with the group $X_1 f, \dots, X_r f$: amongst all $(r-1)$-term invariant subgroups of G_r, we must select those which contain at least an $(r-2)$-term invariant subgroup and we must determine all their $(r-2)$-term invariant subgroups; according to what precedes, this presents no difficulty.

[4] LIE, Ges. der Wiss. zu Christiania, 1874, p. 273. Math. Ann. Vol. XI, p. 517 and 518. Archiv for Math. og Nat. Vol. 3, 1878, p. 105 sq., Vol. 8, 1883.

Afterwards, amongst the found $(r-2)$-term subgroups, we must select those which contain $(r-3)$-term invariant subgroups, and so forth.

It is clear that in this manner, we arrive at the answer to the question which we have asked about the group $X_1 f, \dots, X_r f$. Either we realize that this group does not belong to the discussed specific category, or we find r independent infinitesimal transformations $Y_1 f, \dots, Y_r f$ of our group which are linked by relations of the form (11).

In certain circumstances, the computations just indicated will be made more difficult by the fact that the subgroups to be studied contain arbitrary parameters which specialize themselves in the course of the computations. So we will indicate yet another process which also leads to an answer to our question, but which avoids all computations with arbitrary parameters.

We assume that amongst the $[X_i, X_k]$, one finds exactly $r_1 \leqslant r$ which are independent and that all $[X_i, X_k]$ can be linearly deduced from $X_1' f, \dots, X_{r_1}' f$, and furthermore correspondingly, that all $[X_i', X_k']$ can be linearly deduced from the $r_2 \leqslant r_1$ independent transformations $X_1'' f, \dots, X_{r_2}'' f$, all $[X_i'', X_k'']$ from the $r_3 \leqslant r_2$ independent $X_1''' f, \dots, X_{r_3}''' f$, and so on. Then according to Proposition 6, p. 274, $X_1' f, \dots, X_{r_1}' f$ generate an r_1-term invariant subgroup G_{r_1} of G_r: $X_1 f, \dots, X_r f$, and furthermore $X_1'' f, \dots, X_{r_2}'' f$ generate an r_2-term invariant subgroup G_{r_2} of G_{r_1}, and so forth; briefly, we obtain a series of subgroups G_{r_1}, G_{r_2}, G_{r_3}, ..., of G_r in which each subgroup is contained in all the preceding ones and is invariant in the immediately preceding one, in all cases. But now, according to Proposition 7, p. 274, G_{r_2} is at first invariant not only in G_{r_1}, but also in G_r, and furthermore, according to the same proposition, G_{r_3} is not only invariant in G_{r_2}, but also in G_{r_1} and even in G_r itself, and so forth. One sees that in the series of groups: G_r, G_{r_1}, G_{r_2}, ..., each individual group is contained in and is invariant in all the preceding groups.

In the series of whole numbers r, r_1, r_2, ..., there is none which is larger than the preceding one, and on the other hand, none which is smaller than zero. Consequently, there must exist a positive number q of such a nature that r_{q+1} is equal to r_q, while for $j < q$, it always holds true that: $r_{j+1} < r_j$. Evidently, one then has:

$$r_q = r_{q+1} = r_{q+2} = \cdots,$$

so actually:

$$r_{q+k} = r_q \qquad (k=1,2\cdots).$$

Now, there are two cases to be distinguished, according to whether the number r_q has the value zero, or is larger than zero.[5] (Translator's note: In this case, $r_q \geqslant 2$ in fact, for a one-term group is always solvable.)

In the case $r_q = 0$, it is always possible, as we will show, to select r independent infinitesimal transformations $Y_1 f, \dots, Y_r f$ of G_r which stand in relationships of the form (11).

Indeed, we choose as $Y_r f$, $Y_{r-1} f$, ..., $Y_{r_1+1} f$ any $r - r_1$ infinitesimal transformations $e_1 X_1 f + \cdots + e_r X_r f$ of G_r that are mutually independent and are independent of $X'_1 f, \dots, X'_{r_1} f$; as $Y_{r_1} f$, $Y_{r_1-1} f$, …, $Y_{r_2+1} f$, we choose any $r_1 - r_2$ infinitesimal transformations $e'_1 X'_1 f + \cdots + e'_{r_1} X'_{r_1} f$ of G_{r_1} that are mutually independent and are independent of $X''_1 f, \dots, X''_{r_2} f$, etc.; lastly, as $Y_{r_{q-1}} f$, $Y_{r_{q-1}-1} f$, …, $Y_1 f$, we choose any r_{q-1} independent infinitesimal transformations $e_1^{(q-1)} X_1^{(q-1)} f + \cdots + e_{r_{q-1}}^{(q-1)} X_{r_{q-1}}^{(q-1)} f$ of $G_{r_{q-1}}$. In this way, we obviously obtain r mutually independent infinitesimal transformations $Y_1 f, \dots, Y_r f$ of our G_r; about them, we claim that for every $i < r$, the i transformations $Y_1 f, \dots, Y_i f$ generate an i-term group which is invariant in the $(i+1)$-term group: $Y_1 f, \dots, Y_{i+1} f$. If we succeed in proving this claim, then we will also have proved that $Y_1 f, \dots, Y_r f$ stand in the relationships (11).

Let j be any of the numbers $1, 2, \dots, q$. Then it is clear that the r_j mutually independent infinitesimal transformations $Y_1 f, \dots, Y_{r_j} f$ generate an r_j-term group, namely the group G_{r_j} defined above; certainly, it must be remarked that in the case $j = q$, G_{r_j} reduces to the identity transformation.

As we have already remarked earlier, according to Proposition 6, p. 274, G_{r_j} is invariant in the group $G_{r_{j-1}}$; but from this proposition, one can conclude even more, namely one can conclude that we always obtain an invariant subgroup of $G_{r_{j-1}}$ when, to the infinitesimal transformations $Y_1 f, \dots, Y_{r_j} f$ of G_{r_j}, we add any infinitesimal transformations of the group $G_{r_{j-1}}$ that are mutually independent and are independent of $Y_1 f, \dots, Y_{r_j} f$. Now, since $Y_{r_j+1} f, \dots, Y_{r_{j-1}} f$ belong to $G_{r_{j-1}}$, and furthermore, since they are mutually independent and independent of $Y_1 f, \dots, Y_{r_j} f$, then it follows that each of the following systems of infinitesimal transformations:

$$Y_1 f, \dots, Y_{r_j} f,\ Y_{r_j+1} f, \dots, Y_{r_{j-1}-2} f, Y_{r_{j-1}-1} f$$
$$Y_1 f, \dots, Y_{r_j} f,\ Y_{r_j+1} f, \dots, Y_{r_{j-1}-2} f$$
$$\dots\dots\dots\dots\dots\dots\dots\dots\dots\dots$$
$$Y_1 f, \dots, Y_{r_j} f,\ Y_{r_j+1} f,\ Y_{r_j+2} f$$
$$Y_1 f, \dots, Y_{r_j} f,\ Y_{r_j+1} f$$

generates an invariant subgroup of $G_{r_{j-1}}$.

In this way, between the two groups $G_{r_{j-1}}$ and G_{r_j}, there are certain groups — let us call them $\Gamma_{r_{j-1}-1}, \Gamma_{r_{j-1}-2}, \dots, \Gamma_{r_j+1}$ — which interpolate them and which possess the following properties: each one of them has a number of terms exactly one less than that of the group just preceding it in the series, each one of them is contained in $G_{r_{j-1}}$ and in all the groups preceding it in the series, and each one of them is invariant in $G_{r_{j-1}}$, hence also invariant in all the groups preceding it in the series, and in particular, invariant in the group immediately preceding it. It then follows that G_{r_j} is contained in the Γ_{r_j+1} as an invariant subgroup.

What has been said holds for all values: $1, 2, \dots, q$ of the number j, and consequently, we have effectively proved that $Y_1 f, \dots, Y_r f$ are independent infinitesimal transformations of the group $X_1 f, \dots, X_r f$ such that, for any $i < r$, $Y_1 f, \dots, Y_i f$ always generate an i-term group which is invariant in the $(i+1)$-term group $Y_1 f, \dots, Y_{i+1} f$. But this is what we wanted to prove.

The case where the whole number r_q defined above vanishes is now settled, so it still remains to consider the case where r_q is larger than zero. We shall show that it is impossible in this case to select r infinitesimal transformations $Y_1 f, \dots, Y_r f$ in the group $X_1 f, \dots, X_r f$ which stand in relationships of the form (11).

If $r_q > 0$, then the r_q-term group G_{r_q} which is generated by the r_q independent infinitesimal transformations $X_1^{(q)} f, \dots, X_{r_q}^{(q)} f$ certainly contains no $(r_q - 1)$-term invariant subgroup. Indeed, since $r_{q+1} = r_q$, then amongst the infinitesimal transformations $[X_i^{(q)}, X_k^{(q)}]$, one finds exactly r_q that are mutually independent, whence by taking account of Proposition 8, p. 275, the property of the group G_{r_q} just stated follows.

At present, we assume[6] (Translator's note: (reasoning by contradiction)) that in the group $X_1 f, \dots, X_r f$, there are r independent infinitesimal transformations $Y_1 f, \dots, Y_r f$ which are linked by relations of the form (11), and we denote by $\mathfrak{G}_i$ the i-term group which, under this assumption, is generated by $Y_1 f, \dots, Y_i f$.

According to p. 279, the group G_{r_q} is invariant in the group $X_1 f, \dots, X_r f$, but as we saw just now, it contains no $(r_q - 1)$-term invariant subgroup. Now, since each $\mathfrak{G}_i$ contains an $(i-1)$-term invariant subgroup, namely the group $\mathfrak{G}_{i-1}$, it then follows immediately that G_{r_q} cannot coincide with the group $\mathfrak{G}_{r_q}$, and at the same time, it also follows that there exists an integer number m which is at least equal to r_q and is smaller than r, such that G_{r_q} is contained in none of the groups $\mathfrak{G}_{r_q}, \mathfrak{G}_{r_q+1}, \dots, \mathfrak{G}_m$, while by contrast, it is contained in all groups: $\mathfrak{G}_{m+1}, \mathfrak{G}_{m+2}, \dots, \mathfrak{G}_r$.

As a result, we have an m-term group $\mathfrak{G}_m$ and an r_q-term group G_{r_q} which are both contained in the $(m+1)$-term group $\mathfrak{G}_{m+1}$ as subgroups, and to be precise, which are both evidently contained in it as invariant[7] (Translator's note: Since, as we saw, G_{r_q} is invariant in $G_r = \mathfrak{G}_r$, it is then trivially invariant in $\mathfrak{G}_{m+1} \subset \mathfrak{G}_r$.) subgroups. According to Chap. 12, Proposition 7, p. 223, the transformations common to $\mathfrak{G}_m$ and to G_{r_q} form a group Γ which has at least $r_q - 1$ parameters, and which, according to Chap. 15, Proposition 10, p. 276, is invariant in $\mathfrak{G}_{m+1}$. Now, since under the assumptions made, G_{r_q} is not contained in $\mathfrak{G}_m$, it follows that Γ is exactly $(r_q - 1)$-term and is at the same time invariant in G_{r_q}.

This is a contradiction, since according to the preceding, G_{r_q} contains absolutely no invariant $(r_q - 1)$-term subgroup. Consequently, the assumption which we took as a starting point is false, namely the assumption that in the group $X_1 f, \dots, X_r f$, one can indicate r independent infinitesimal transformations $Y_1 f, \dots, Y_r f$ which stand mutually in relationships of the form (11). From this, we see that in the case $r_q > 0$, there are no infinitesimal transformations $Y_1 f, \dots, Y_r f$ of this constitution.

In the preceding developments, a simple process has been provided by means of which one can realize whether a given r-term group $X_1 f, \dots, X_r f$ belongs, or does not belong, to the specific category defined on page 277.

Proposition 12. [5] *If $X_1 f, \dots, X_r f$ are independent infinitesimal transformations of an r-term group, if $X_1' f, \dots, X_{r_1}' f$ ($r_1 \leqslant r$) are independent infinitesimal transforma-*

[5] In this proposition, r_1 naturally has the same meaning as on p. 278, and likewise r_2, r_3, etc.

tions from which all $[X_i, X_k]$ *can be linearly deduced, if furthermore* $X_1''f, \dots, X_{r_2}''f$ ($r_2 \leqslant r_1$) *are independent infinitesimal transformations from which all* $[X_i', X_k']$ *can be linearly deduced, and if one defines in a corresponding way* $r_3 \leqslant r_2$ *mutually independent infinitesimal transformations* $X_1'''f, \dots, X_{r_3}'''f$, *and so on, then the* $X'f$ *generate an* r_1*-term group, the* $X''f$ *an* r_2*-term group, the* $X'''f$ *an* r_3*-term group, and so on, and to be precise, each one of these groups is invariant in all the preceding groups, and also in the group* $X_1f, \dots, X_rf$. — *In the series of numbers* r, r_1, r_2, ... *there is a number, say* r_q, *which is equal to all the subsequent numbers:* r_{q+1}, r_{q+2}, ..., *while by contrast the numbers* r, r_1, ..., r_q *are all distinct from one another. Now, if* $r_q = 0$, *then it is always possible to indicate, in the group* $X_1f, \dots, X_rf$, r *mutually independent infinitesimal transformations* $Y_1f, \dots, Y_rf$ *such that for every* $i < r$, *the transformations* $Y_1f, \dots, Y_if$ *generate an* i*-term group which is invariant in the* $(i+1)$*-term group:* $Y_1f, \dots, Y_{i+1}f$, *so that there exist relations of the specific form:*

$$[Y_i, Y_{i+k}] = c_{i,i+k,1}\, Y_1 f + \cdots + c_{i,i+k,i+k-1}\, Y_{i+k-1} f$$

$$(i=1 \cdots r-1;\; k=1 \cdots r-i).$$

However, if $r_q > 0$, *it is not possible to determine* r *independent infinitesimal transformations* $Y_1f, \dots, Y_rf$ *in the group* $X_1f, \dots, X_rf$ *having the constitution defined above.*

Chapter 16
The Adjoint Group

Let $x_i' = f_i(x_1,\dots,x_n,a_1,\dots,a_r)$ be an r-term group with the r infinitesimal transformations:

$$X_k f = \sum_{i=1}^{n} \xi_{ki}(x)\,\frac{\partial f}{\partial x_i} \qquad {\scriptstyle (k=1\cdots r)}.$$

If one introduces the x_i' as new variables in the expression $\sum e_k X_k f$, then as has been already shown in Chap. 3, Proposition 4, p. 48, one gets for all values of the e_k an equation of the form:

$$\sum_{k=1}^{r} e_k X_k f = \sum_{k=1}^{r} e_k' X_k' f.$$

Here, the e_k' are certain linear, homogeneous functions of the e_k with coefficients that depend upon $a_1,\dots,a_r$:

$$e_k' = \sum_{j=1}^{r} \rho_{kj}(a_1,\dots,a_r)\, e_j. \tag{1}$$

If one again introduces in $\sum e_k' X_k' f$ the new variables $x_i'' = f_i(x,b)$, then one obtains:

$$\sum_{k=1}^{r} e_k' X_k' f = \sum_{k=1}^{r} e_k'' X_k'' f,$$

where:

$$e_k'' = \sum_{j=1}^{r} \rho_{kj}(b_1,\dots,b_r)\, e_j'. \tag{1'}$$

But now, because the equations $x_i' = f_i(x,a)$ represent a group, the x'' are consequently linked with the x through relations of the form $x_i'' = f_i(x,c)$ in which the c depend only upon a and b:

$$c_k = \varphi_k(a_1,\dots,a_r,b_1,\dots,b_r).$$

S. Lie, *Theory of Transformation Groups I*,
DOI 10.1007/978-3-662-46211-9_16

Hence if one passes directly from the x to the x'', one finds:

$$\sum_{k=1}^{r} e_k X_k f = \sum_{k=1}^{r} e''_k X''_k f,$$

and to be precise, one has:

$$(1'') \qquad e''_k = \sum_{j=1}^{r} \rho_{kj}(c_1,\dots,c_r)\, e_j = \sum_{j=1}^{r} \rho_{kj}\big(\varphi_1(a,b),\dots,\varphi_r(a,b)\big)\, e_j.$$

From this, it can be deduced that the totality of all transformations $e'_k = \sum \rho_{kj}(a)\, e_j$ forms a group. Indeed, by combination of the equations (1) and (1') we get:

$$e''_k = \sum_{j,\,\nu}^{1\cdots r} \rho_{kj}(b_1,\dots,b_r)\, \rho_{j\nu}(a_1,\dots,a_r)\, e_\nu,$$

which must naturally coincide with the equations (1") and in fact, for all values of the e, of the a and of the b. Consequently, there are the r^2 identities:

$$\rho_{k\nu}\big(\varphi_1(a,b),\dots,\varphi_r(a,b)\big) \equiv \sum_{j=1}^{r} \rho_{j\nu}(a_1,\dots,a_r)\, \rho_{kj}(b_1,\dots,b_r),$$

from which it results that the family of the transformations $e'_k = \sum \rho_{kj}(a)\, e_j$ effectively forms a group.

To every r-term group $x'_i = f_i(x,\, a)$ therefore belongs a fully determined linear homogeneous group:

$$(1) \qquad e'_k = \sum_{j=1}^{r} \rho_{kj}(a_1,\dots,a_r)\, e_j \qquad (k=1\cdots r),$$

which we want to call the *adjoint group*[1] [ADJUNGIRTE GRUPPE] of the group $x'_i = f_i(x,\, a)$.

We consider *for example* the two-term group $x' = ax+b$ with the two independent infinitesimal transformations: $\frac{\mathrm{d}f}{\mathrm{d}x}$, $x\,\frac{\mathrm{d}f}{\mathrm{d}x}$. We find:

$$e_1\,\frac{\mathrm{d}f}{\mathrm{d}x} + e_2\, x\,\frac{\mathrm{d}f}{\mathrm{d}x} = e_1\, a\,\frac{\mathrm{d}f}{\mathrm{d}x'} + e_2\,(x'-b)\,\frac{\mathrm{d}f}{\mathrm{d}x'} = e'_1\,\frac{\mathrm{d}f}{\mathrm{d}x'} + e'_2\, x'\,\frac{\mathrm{d}f}{\mathrm{d}x'},$$

whence we obtain for the adjoint group of the group $x' = ax+b$ the following equations:

$$e'_1 = a\, e_1 - b\, e_2, \qquad e'_2 = e_2,$$

which visibly represent a group.

The adjoint group of the group $x'_i = f_i(x,\, a)$ contains, in the form found above, precisely r arbitrary parameters: $a_1,\dots,a_r$. But for every individual group

[1] LIE, Archiv for Math., Vol. 1, Christiania 1876.

$x_i' = f_i(x, a)$, special work is required to investigate whether the parameters $a_1, \dots, a_r$ are all essential in the adjoint group. Actually, we shall shortly see that there are r-term groups whose adjoint group does not contain r essential parameters.

Furthermore, in all circumstances, one special transformation appears in the adjoint group of the group $x_i' = f_i(x, a)$, namely the identity transformation; for if one sets for $a_1, \dots, a_r$ in the equations (1) the system of values which produces the identity transformation $x_i' = x_i$ in the group $x_i' = f_i(x, a)$, then one obtains the transformation: $e_1' = e_1, \dots, e_r' = e_r$, which hence is always present in the adjoint group. However, as we shall see, it can happen that the adjoint group consists only of the identity transformation: $e_1' = e_1, \dots, e_r' = e_r$.

§ 76. In order to make accessible the study of the adjoint group, we must above all determine its infinitesimal transformations. We easily reach this end by an application of Theorem 43, Chap. 15, p. 265; but we must in the process replace the equations $x_i' = f_i(x, a)$ of our group by the equivalent *canonical* equations:

$$x_i' = x_i + \frac{t}{1} \sum_{k=1}^{r} \lambda_k X_k x_i + \cdots \qquad (i=1\cdots n), \tag{2}$$

which represent the ∞^{r-1} one-term subgroups of the group $x_i' = f_i(x, a)$. According to Chap. 4, p. 81, the a_k are defined here as functions of t and $\lambda_1, \dots, \lambda_r$ by the simultaneous system:

$$\frac{\mathrm{d}a_k}{\mathrm{d}t} = \sum_{j=1}^{r} \lambda_j \, \alpha_{jk}(a_1, \dots, a_r) \qquad (k=1\cdots r). \tag{3}$$

By means of the equations (2), we therefore have to introduce the new variables x_i' in $\sum e_k X_k f$ and we must then obtain a relation of the form:

$$\sum_{k=1}^{r} e_k X_k f = \sum_{k=1}^{r} e_k' X_k' f.$$

The infinitesimal transformation denoted by Yf in Theorem 43 on p. 265 now reads as: $\lambda_1 X_1 f + \cdots + \lambda_r X_r f$; we therefore obtain in our case:

$$\begin{aligned} Y\big(X_k(f)\big) - X_k\big(Y(f)\big) &= \sum_{\nu=1}^{r} \lambda_\nu \, [X_\nu, X_k] \\ &= \sum_{s=1}^{r} \Big\{ \sum_{\nu=1}^{r} \lambda_\nu \, c_{\nu k s} \Big\} X_s f. \end{aligned}$$

Consequently, we obtain the following differential equations for $e_1', \dots, e_r'$:

$$\frac{\mathrm{d}e_s'}{\mathrm{d}t} + \sum_{\nu=1}^{r} \lambda_\nu \sum_{k=1}^{r} c_{\nu k s} \, e_k' = 0 \qquad (s=1\cdots r). \tag{4}$$

We consider the integration of these differential equations as an executable operation, for it is known that it requires only the resolution of an algebraic equation of r-th degree. So if we perform the integration on the basis of the initial condition: $e'_k = e_k$ for $t = 0$, we obtain r equations of the form:

$$e'_k = \sum_{j=1}^{r} \psi_{kj}(\lambda_1 t, \dots, \lambda_r t)\, e_j \qquad (k=1\cdots r), \tag{5}$$

which are equivalent to the equations (1), as soon as the a_k are expressed as functions of $\lambda_1 t, \dots, \lambda_r t$ in the latter.

From this, it follows that the equations (5) represent the adjoint group too. But now we have derived the equations (5) in exactly the same way as if we would have wanted to determine all finite transformations which are generated by the infinitesimal transformations:

$$\sum_{\nu=1}^{r} \lambda_\nu \sum_{k,\,s}^{1\cdots r} c_{k\nu s}\, e_k \frac{\partial f}{\partial e_s} = \sum_{\nu=1}^{r} \lambda_\nu\, E_\nu f$$

(cf. p. 51 above). *Consequently, we conclude that the adjoint group* (1) *consists of the totality of all one-term groups of the form* $\lambda_1 E_1 f + \cdots + \lambda_r E_r f$.

If amongst the family of all infinitesimal transformations $\lambda_1 E_1 f + \cdots + \lambda_r E_r f$ there are exactly ρ transformations and no more which are independent, say $E_1 f, \dots, E_\rho f$, then all the finite transformations of the one-term groups $\lambda_1 E_1 f + \cdots + \lambda_r E_r f$ are already contained in the totality of all finite transformations of the $\infty^{\rho-1}$ groups $\lambda_1 E_1 f + \cdots + \lambda_\rho E_\rho f$. The totality of these ∞^ρ finite transformations forms the adjoint group: $e'_k = \sum \rho_{kj}(a)\, e_j$, which therefore contains only ρ essential parameters (Chap. 4, Theorem 8, p. 75).

According to the preceding, it is to be supposed that $E_1 f, \dots, E_\rho f$ are linked together by relations of the form:

$$[E_\mu, E_\nu] = \sum_{s=1}^{\rho} g_{\mu\nu s}\, E_s f;$$

we can also confirm this by a computation. By a direct calculation, we obtain:

$$E_\mu\big(E_\nu(f)\big) - E_\nu\big(E_\mu(f)\big) = \sum_{\sigma,\,k,\,\pi}^{1\cdots r} \big(c_{\pi\mu k}\, c_{k\nu\sigma} - c_{\pi\nu k}\, c_{k\mu\sigma}\big)\, e_\pi \frac{\partial f}{\partial e_\sigma}.$$

But between the c_{iks}, there exist the relations:

$$\sum_{k=1}^{r} \big(c_{\pi\mu k}\, c_{k\nu\sigma} + c_{\mu\nu k}\, c_{k\pi\sigma} + c_{\nu\pi k}\, c_{k\mu\sigma}\big) = 0,$$

which we have deduced from the Jacobi identity some time ago (cf. Chap. 9, Theorem 27, p. 185). If we then use for this that $c_{\nu\pi k} = -c_{\pi\nu k}$ and $c_{k\pi\sigma} = -c_{\pi k\sigma}$, we can bring the right-hand side of our equation for $[E_\mu, E_\nu]$ to the form:

$$\sum_{k=1}^{r} c_{\mu\nu k} \sum_{\sigma,\,\pi}^{1\cdots r} c_{\pi k\sigma}\, e_\pi \frac{\partial f}{\partial e_\sigma},$$

whence we get:

$$[E_\mu, E_\nu] = \sum_{k=1}^{r} c_{\mu\nu k} E_k f.$$

Lastly, under the assumptions made above, the right-hand side can be expressed by means of $E_1 f, \dots, E_\rho f$ alone, so that relations of the form:

$$[E_\mu, E_\nu] = \sum_{s=1}^{\rho} g_{\mu\nu s} E_s f$$

really hold, in which the $g_{\mu\nu s}$ denote constants.

Before we continue, we want to recapitulate in cohesion [IM ZUSAMMENHANGE WIEDERHOLEN] the results of the chapter obtained up to now.

Theorem 48. *If, in the general infinitesimal transformation $e_1 X_1 f + \cdots + e_r X_r f$ of the r-term group $x_i' = f_i(x, a)$, one introduces the new variables x' in place of the x, then one obtains an expression of the form:*

$$e_1' X_1' f + \cdots + e_r' X_r' f;$$

in the process, the e' are linked to the e by equations of the form:

$$e_k' = \sum_{j=1}^{r} \rho_{kj}(a_1, \dots, a_r)\, e_j \qquad (k=1\cdots r),$$

which represent a group in the variables e, the so-called adjoint group of the group $x_i' = f_i(x, a)$. This adjoint group contains the identity transformation and is generated by certain infinitesimal transformations; if, between $X_1 f, \dots, X_r f$, there exist the relations:

$$[X_i, X_k] = \sum_{s=1}^{r} c_{iks} X_s f \qquad (i,k=1\cdots r),$$

and if one sets:

$$E_\mu f = \sum_{k,\,j}^{1\cdots r} c_{j\mu k}\, e_j \frac{\partial f}{\partial e_k} \qquad (\mu=1\cdots r),$$

then $\lambda_1 E_1 f + \cdots + \lambda_r E_r f$ is the general infinitesimal transformation of the adjoint group and between $E_1 f, \dots, E_r f$, there are at the same time the relations:

$$[E_i, E_k] = \sum_{s=1}^{r} c_{iks} E_s f \qquad (i,k=1\cdots r).$$

If two r-term groups $X_1 f, \ldots, X_r f$ and $Y_1 f, \ldots, Y_r f$ are constituted in such a way that one simultaneously has:

$$[X_i, X_k] = \sum_{s=1}^{r} c_{iks} X_s, \qquad\qquad [Y_i, Y_k] = \sum_{s=1}^{r} c_{iks} Y_s f,$$

with the same constants c_{iks} in the two cases, then both groups obviously have the same adjoint group. Later, we will also see that in certain circumstances, certain groups which do not possess an equal number of terms can nonetheless have the same adjoint group.

§ 77. Now, by what can one recognize how many independent infinitesimal transformation there are amongst $E_1 f, \ldots, E_r f$?

If $E_1 f, \ldots, E_r f$ are not all independent of each other, then there is at least one infinitesimal transformation $\sum g_\mu E_\mu f$ which vanishes identically. From the identity:

$$\sum_{\mu=1}^{r} g_\mu \sum_{k,\,j}^{1\cdots r} c_{j\mu k}\, e_j \frac{\partial f}{\partial e_k} \equiv 0,$$

we get:

$$\sum_{\mu=1}^{r} g_\mu\, c_{j\mu k} = 0$$

for all values of j and k, and consequently the expression:

$$\left[X_j, \sum_{\mu=1}^{r} g_\mu X_\mu f\right] = \sum_{k=1}^{r} \left\{ \sum_{\mu=1}^{r} g_\mu\, c_{j\mu k} \right\} X_k f$$

vanishes, that is to say: the infinitesimal transformation $\sum g_\mu X_\mu f$ is interchangeable with all the r infinitesimal transformations $X_j f$. Conversely, if the group $X_1 f, \ldots, X_r f$ contains an infinitesimal transformation $\sum g_\mu X_\mu f$ which is interchangeable with all the $X_k f$, then it follows in the same way that the infinitesimal transformation $\sum g_\mu E_\mu f$ vanishes identically.

In order to express this relationship in a manner which is as brief as possible, we introduce the following terminology:

An infinitesimal transformation $\sum g_\mu X_\mu f$ of the r-term group $X_1 f, \ldots, X_r f$ is called an excellent infinitesimal transformation of this group if it is interchangeable with all the $X_k f$.

Incidentally, the excellent infinitesimal transformations of the group $X_1 f, \ldots, X_r f$ are also characterized by the fact that they keep their form through the introduction of the new variables $x_i' = f_i(x, a)$, whichever values the parameters $a_1, \ldots, a_r$ take. Indeed, if the infinitesimal transformation $\sum g_\mu X_\mu f$ is excellent, then according to Chap. 15, p. 259, there is a relation of the form:

$$\sum g_\mu X_\mu f = \sum g_\mu X'_\mu f.$$

In addition, the cited developments show that each finite transformation of the one-term group $\sum g_\mu X_\mu f$ is interchangeable with every finite transformation of the group $X_1 f, \dots, X_r f$.

According to what has been said above, to every excellent infinitesimal transformation of the group $X_1 f, \dots, X_r f$, there corresponds a linear relation between $E_1 f, \dots, E_r f$. If $\sum g_\mu X_\mu f$ is an excellent infinitesimal transformation, then there exists between the Ef simply the relation: $\sum g_\mu E_\mu f = 0$. Consequently, between $E_1 f, \dots, E_r f$ there are exactly as many independent relations of this sort as there are independent excellent infinitesimal transformations in the group $X_1 f, \dots, X_r f$. If there are exactly m and not more such independent transformations, then amongst the infinitesimal transformations $E_1 f, \dots, E_r f$ there are exactly $r - m$ and not more which are independent, and likewise, the adjoint group contains the same number of essential parameters.

We therefore have the

Theorem 49. *The adjoint group $e'_k = \sum \rho_{kj}(a)\, e_j$ of an r-term group $X_1 f, \dots, X_r f$ contains r essential parameters if and only if none of the ∞^{r-1} infinitesimal transformations $\sum g_\mu X_\mu f$ are excellent inside the group $X_1 f, \dots, X_r f$; by contrast, the adjoint group has fewer than r essential parameters, namely $r - m$, when the group $X_1 f, \dots, X_r f$ contains exactly m and not more independent excellent infinitesimal transformations.*[2]

Let us take for example the group:

$$\frac{\partial f}{\partial x_1}, \quad \dots, \quad \frac{\partial f}{\partial x_r} \qquad (r \leqslant n).$$

Its infinitesimal transformations are all excellent, since all c_{iks} are zero. So all r expressions $E_1 f, \dots, E_r f$ vanish identically, and the adjoint group reduces to the identity transformation.

On the other hand, let us take the four-term group:

$$x\frac{\partial f}{\partial x}, \quad y\frac{\partial f}{\partial x}, \quad x\frac{\partial f}{\partial y}, \quad y\frac{\partial f}{\partial y}.$$

This group contains a single excellent infinitesimal transformation, namely:

$$x\frac{\partial f}{\partial x} + y\frac{\partial f}{\partial y};$$

its adjoint group is hence only three-term.

If the group $X_1 f, \dots, X_r f$ contains the excellent infinitesimal transformation $\sum g_\mu X_\mu f$, then $\sum g_\mu X_\mu f$ is a linear partial differential equation which remains

[2] LIE, Math. Ann. Vol. XXV, p. 94.

invariant by the group; in consequence of that, the group is imprimitive (cf. p. 233). From this, we conclude that the following proposition holds:

Proposition 1. *If the group $X_1 f, \dots, X_r f$ in the variables $x_1, \dots, x_n$ is primitive, then it contains no excellent infinitesimal transformation.*

If we combine this proposition with the latter theorem, we obtain the

Proposition 2. *The adjoint group of an r-term primitive group contains r essential parameters.*

Later, we shall give a somewhat more general version of the last two propositions (cf. Chapter 24 on systatic groups).

According to the preceding, the infinitesimal transformations of an r-term group $X_1 f, \dots, X_r f$ and those of the adjoint group $E_1 f, \dots, E_r f$ can be mutually ordered in such a way that, to every infinitesimal transformation $\sum e_k X_k$ which is not excellent there corresponds a nonvanishing infinitesimal transformation $\sum e_k E_k f$, while to every excellent infinitesimal transformation $\sum e_k X_k f$ is associated the *identity* transformation in the adjoint group. In addition, if one takes account of the fact that the two systems of equations:

$$[X_i, X_k] = \sum_{s=1}^{r} c_{iks} X_s f, \qquad [E_i, E_k] = \sum_{s=1}^{r} c_{iks} E_s f$$

have exactly the same form, then one could be led to the presumption that the adjoint group cannot contain excellent infinitesimal transformations. Nevertheless, this presumption would be false; which is shown by the group $\partial f / \partial x_2$, $x_1 \partial f / \partial x_2$, $\partial f / \partial x_1$ whose adjoint group consists of two interchangeable infinitesimal transformations and therefore contains two independent excellent infinitesimal transformations.

§ 78. The starting point of our study was the remark that the expression $\sum e_k X_k f$ takes the similar form $\sum e_k' X_k' f$ after the introduction of the new variables $x_i' = f_i(x, a)$. But now, according to Chap. 15, p. 268, the expression $\sum e_k X_k f$ can be interpreted as the symbol of the general finite transformation of the group $X_1 f, \dots, X_r f$; here, the quantities $e_1, \dots, e_r$ are to be considered as the parameters of the finite transformations of the group $X_1 f, \dots, X_r f$. Consequently, we can also say: after the transition to the variables x', the finite transformations of the group $X_1 f, \dots, X_r f$ are permuted with each other, while their totality remains invariant. Thanks to the developments of the preceding section, we can add that the concerned permutation [VERTAUSCHUNG] is achieved by a transformation of the adjoint group.

But when a "*permutation of the finite transformations* $\sum e_k X_k f$" is spoken of, the interpretation fundamentally lies in the fact that one imagines these transformations as *individuals* [INDIVIDUEN]; we now want to pursue this interpretation in some detail.

Every individual in the family $\sum e_k X_k f$ is determined by the associated values of $e_1, \dots, e_r$; for the sake of graphic clarity, we hence imagine the e_k as right-angled point coordinates of an r-times extended manifold. The points of this manifold then represent all finite transformations $\sum e_k X_k f$; hence, they are transformed by our known linear homogeneous group:

$$e'_k = \sum_{j=1}^{r} \rho_{kj}(a)\, e_j.$$

At the same time, the origin of coordinates $e_1 = 0, \dots, e_r = 0$, i.e. the image-point [BILDPUNKT] of the identity transformation $x'_i = x_i$, obviously remains invariant.

Every finite transformation e^0_k belongs to a completely determined one-term group, whose transformations are defined by the equations:

$$\frac{e_1}{e^0_1} = \cdots = \frac{e_r}{e^0_r};$$

but according to our interpretation, these equations represent a straight line through $e_k = 0$, that is to say:

Every one-term subgroup of the group $X_1 f, \dots, X_r f$ is represented, in the space $e_1, \dots, e_r$, by a straight line which passes through the origin of coordinates $e_k = 0$; conversely, every straight line through the origin of coordinates represents such a one-term group.

Every other subgroup of the r-term group $X_1 f, \dots, X_r f$ consists of one-term subgroups and these one-term groups are determined by the infinitesimal transformations that the subgroup in question contains. But according to Chap. 11, Proposition 6, p. 223, the infinitesimal transformations of a subgroup of the group $X_1 f, \dots, X_r f$ can be defined by means of *linear homogeneous* equations between $e_1, \dots, e_r$; these equations naturally also define the one-term groups which belong to the subgroup, hence they actually define the finite transformations of the subgroup in question. Expressed differently:

Every m-term subgroup of the group $X_1 f, \dots, X_r f$ is represented, in the space $e_1, \dots, e_r$, by a straight[2] (Translator's note: The "straight" character just means that the manifold in question is a linear subspace of the space $e_1, \dots, e_r$.) [EBEN] *m-times extended manifold which passes through the origin of coordinates: $e_1 = 0$, $\dots$, $e_r = 0$.*

Of course, the converse does not hold true in general; it occurs only very exceptionally that *every* straight manifold through the origin of coordinates represents a subgroup.

The adjoint group $e'_k = \sum \rho_{kj}(a)\, e_j$ now transforms linearly the points e_k. If an arbitrary point $\overline{e}_k$ remains invariant by all transformations of the group, and in consequence of that, also every point $m\overline{e}_k = \text{Const.}\, \overline{e}_k$, then according to the preceding, this means that the transformations of the one-term group $\sum \overline{e}_k X_k f$ are interchangeable with all transformations of the group $X_1 f, \dots, X_r f$.

Every *invariant* subgroup of the group $X_1 f, \dots, X_r f$ is represented by a straight manifold containing the origin of coordinates which keeps its position through all transformations $e'_k = \sum \rho_{kj}(a)\, e_j$. On the other hand, every manifold through the origin of coordinates which remains invariant by the adjoint group represents an invariant subgroup of $X_1 f, \dots, X_r f$ (cf. Chap. 15, Proposition 5, p. 273).

Now, let M be any straight manifold through the origin of coordinates which represents a subgroup of G_r: $X_1 f, \dots, X_r f$. If an arbitrary transformation of the adjoint group is executed on M, this gives a new straight manifold which, likewise, represents a subgroup of G_r. Obviously, this new subgroup is equivalent [ÄHNLICH], through a transformation of G_r, to the initial one, and if we follow the process of the theory of substitutions, we can express this as follows: *the two subgroups just discussed are conjugate* [GLEICHBERECHTIGT][3] *(Translator's note: Literally in German: they are equal, they are considered on the same basis, or they have the same rights.) inside the group* $X_1 f, \dots, X_r f$.

The totality of all subgroups which, inside the group G_r, are conjugate to the subgroup which is represented by M is represented by a family of straight manifolds, namely by the family that one obtains as soon as one executes on M all transformations of the adjoint group. Two different manifolds of this family naturally represent conjugate subgroups.

From this we conclude that every invariant subgroup of G_r is conjugate only to itself inside G_r.

Since the family of all subgroups g which are conjugate to a given one results from the latter by the execution of all transformations $e'_k = \sum \rho_{kj}(a)\, e_j$, then this family itself must be reproduced by all transformations $e''_k = \sum \rho_{kj}(b)\, e'_j$. This is because if one executes the transformations $e'_k = \sum \rho_{kj}(a)\, e_j$ and $e''_k = \sum \rho_{kj}(a)\, e'_j$ one after the other, then one obtains the same result as if one would have applied all transformations:

$$e''_k = \sum \rho_{kj}(c)\, e_j \qquad (c_k = \varphi_k(a, b))$$

to the initially given subgroup; in the two cases, one obtains the said family.

Now, if all conjugate subgroups g have a continuous number of mutually common transformations, then the totality of all these transformations are represented by a straight manifold in the space $e_1, \dots, e_r$. Naturally, this straight manifold remains invariant by all transformations of the adjoint group and hence, according to what was said above, it represents an invariant subgroup of G_r. Consequently, the following holds true.

Theorem 50. *If the subgroups of a group G_r which are conjugate inside G_r to a determined subgroup have a family of transformations in common, then the totality of these transformations forms a subgroup invariant in G_r.*

§ 79. We distribute [EINTHEILEN] the subgroups of an arbitrary r-term group G_r into different classes which we call *types* [TYPEN] *of subgroups of G_r*. We include in the *same* type the groups which are mutually conjugate inside G_r; groups which are not conjugate inside G_r belong to *different* types.

If one knows any subgroup of G_r, then at once, one can determine all subgroups which belong to the same type. Thanks to this fact, the enumeration [AUFZÄHLUNG] of the subgroups of a given group is essentially facilitated, since one clearly does not have to write down all subgroups, but rather, one needs only to enumerate the different types of subgroups by indicating a representative for every type, hence a subgroup which belongs to the type in question.

We also speak of different types for the finite transformations of a group. We include in the same type two finite transformations: $e_1^0 X_1 f + \cdots + e_r^0 X_r f$ and $\overline{e}_1 X_1 f + \cdots + \overline{e}_r X_r f$ of the r-term group $X_1 f, \dots, X_r f$ if and only if they are conjugate inside G_r, that is to say: when the adjoint group contains a transformation which transfers the point $e_1^0, \dots, e_r^0$ to the point $\overline{e}_1, \dots, \overline{e}_r$. It is clear, should it be mentioned, that in the concerned transformation $e_k' = \sum \gamma_{kj} e_j$ of the adjoint group the determinant of the coefficients γ_{kj} should not vanish.

At present, we want to take up again the question concerning the types of subgroups of a given group, though not in complete generality; instead, we want at least to show how one has to proceed in order to find the extant types of one-term groups and of finite transformations.

At first, we ask for all types of finite transformations.

Let $e_1^0 X_1 f + \cdots + e_r^0 X_r f$ be any finite transformation of the group $X_1 f, \dots, X_r f$. If, on the point $e_1^0, \dots, e_r^0$, we execute all transformations of the adjoint group, we obtain the image-points [BILDPUNKTE] of all the finite transformations of our group which are conjugate to $\sum e_k^0 X_k f$ and thus, belong to the same type as $\sum e_k^0 X_k f$. According to Chap. 14, p. 237, the totality of all these points forms a manifold invariant by the adjoint group and to be precise, a so-called *smallest* invariant manifold, as was said at that time.

We can consider the finite equations of the adjoint group as known; consequently, we are in a position to indicate without integration the equations of the just mentioned smallest invariant manifold (cf. Chap. 14, Theorem 37, p. 237). Now, since every such *smallest* invariant manifold represents the totality of all finite transformations which belong to a certain type, then with this, all types of finite transformations of our group are found. As a result, the following holds true.

Proposition 3. *If an r-term group $x_i' = f_i(x_1, \dots, x_n, a_1, \dots, a_r)$ with the r independent infinitesimal transformations $X_1 f, \dots, X_r f$ is presented, then one finds in the following way all types of finite transformations $e_1 X_1 f + \cdots + e_r X_r f$ of this group: one sets up the finite equations $e_k' = \sum \rho_{kj}(a)\, e_j$ of the adjoint group of the group $X_1 f, \dots, X_r f$ and one then determines, in the space of the e, the smallest manifolds which remain invariant by the adjoint group; these manifolds represent the demanded types.*

With all of that, one does not forget that only the transformations of the adjoint group for which the determinant of the coefficients does not vanish are permitted. Concerning this point, one may compare with what was said in Chap. 14, p. 237.

Next, we seek all types of one-term subgroups, or, what is the same: all *types of infinitesimal transformations* of our group.

Two one-term groups $\sum e_k^0 X_k f$ and $\sum \overline{e}_k X_k f$ are conjugate inside G_r when there is, in the adjoint group, a transformation which transfers the straight line:

$$\frac{e_1}{e_1^0} = \cdots = \frac{e_r}{e_r^0}$$

to the straight line:

$$\frac{e_1}{\overline{e}_1} = \cdots = \frac{e_r}{\overline{e}_r};$$

these two straight lines passing through the origin of coordinates are indeed the images of the one-term subgroups in question. Hence if we imagine that all transformations of the adjoint group are executed on the first one of these two straight lines, we obtain all straight lines passing through the origin of coordinates that represent one-term subgroups which are conjugate to the group $\sum e_k^0 X_k f$. Naturally, the totality of all these straight lines forms a manifold invariant by the adjoint group, namely the smallest invariant manifold to which the straight line: $e_1/e_1^0 = \cdots = e_r/e_r^0$ belongs. It is clear that this manifold is represented by a system of equations which is *homogeneous* in the variables $e_1, \dots, e_r$.

Conversely, it stands to reason that every system of equations homogeneous in the e which admits the adjoint group represents an invariant family of one-term groups. Now, since every system of equations homogeneous in the e is characterized to be homogeneous by the fact that it admits all transformations of the form:

$$e_1' = \lambda\, e_1, \ \dots, \ e_r' = \lambda\, e_r, \tag{6}$$

it follows that we obtain all invariant families of one-term groups by looking up at all manifolds of the space $e_1, \dots, e_r$ which, aside from the transformations of the adjoint group, also admit all transformations of the form (6). In particular, if we seek all smallest invariant manifolds of this constitution, we clearly obtain all existing types of one-term subgroups.

The transformations (6) form a one-term group whose infinitesimal transformations read:

$$Ef = \sum_{k=1}^{r} e_k \frac{\partial f}{\partial e_k}.$$

If we add Ef to the infinitesimal transformations $E_1 f, \dots, E_r f$ of the adjoint group, we again obtain the infinitesimal transformations of a group; indeed, the expressions $[E_k, E]$ all vanish identically, as their computation shows. Visibly, everything amounts to the determination of the smallest manifolds which remain invariant by the group $E_1 f, \dots, E_r f, Ef$. But according to the instructions in Chap 14, p. 237, this determination can be accomplished, since together with the finite equations of the adjoint group, the finite equations of the group just mentioned are also known without effort. Hence we have the

Proposition 4. *If an r-term group $x_i' = f_i(x_1, \dots, x_n, a_1, \dots, a_r)$ with the r independent infinitesimal transformations $X_1 f, \dots, X_r f$ is presented, then one finds as follows all types of one-term subgroups of this group, or what is the same, all types*

of infinitesimal transformations $E_1 f, \dots, E_r f$*: one sets up the infinitesimal transformations of the adjoint group of the group* $X_1 f, \dots, X_r f$*, then one computes the finite equations of the group which is generated by the* $r+1$ *infinitesimal transformations* $E_1 f, \dots, E_r f$ *and:*

$$Ef = \sum_{k=1}^{r} e_k \frac{\partial f}{\partial e_k}$$

and lastly, one determines, in the space $e_1, \dots, e_r$*, the smallest manifolds invariant by the group just defined; these manifolds represent the demanded types.*

In the preceding, it is shown how one can determine all types of finite transformations and all types of one-term subgroups of a given r-term group. Now in a few words, we will consider somewhat more precisely the connection which exists between these two problems; we will see that the settlement of one of these two problems facilitates the settlement of the other.

On the first hand, we assume that we already know all types of *finite transformations* of the group $X_1 f, \dots, X_r f$, so that all smallest manifolds of the space $e_1, \dots, e_r$ which admit the adjoint group $E_1 f, \dots, E_r f$ are known to us. Then how must one proceed in order to find the smallest manifolds invariant by the group Ef, $E_1 f, \dots, E_r f$?

It is clear that every manifold invariant by the group $Ef, E_1 f, \dots, E_r f$ also admits the adjoint group; consequently, every sought manifold must either be one of the already known manifolds, or it must contain at least one of the known manifolds. Hence in order to find all the sought manifolds, we only need to take the known manifolds one after the other and for each of them, to find the smallest manifold invariant by the group $Ef, E_1 f, \dots, E_r f$ in which it is contained.

Let:

$$W_1(e_1, \dots, e_r) = 0, \quad \dots, \quad W_m(e_1, \dots, e_r) = 0, \tag{7}$$

or shortly M, be one of the known manifolds which admits the adjoint group $E_1 f, \dots, E_r f$. Now, how can one find the smallest manifold which admits the group $Ef, E_1 f, \dots, E_r f$ and which in addition comprises the manifold M?

The sought manifold necessarily contains the origin of coordinates $e_1 = 0, \dots, e_r = 0$ and moreover, it consists in nothing but straight lines passing through it, hence it certainly contains the manifold M' which is formed of the straight lines between the points of M and the origin of coordinates. Now, if we can prove that M' admits the infinitesimal transformations $Ef, E_1 f, \dots, E_r f$, then at the same time, we have proved that M' is the sought manifold.

Visibly, the equations of M' are obtained by eliminating the parameter τ from the equations:

$$W_1(e_1 \tau, \dots, e_r \tau) = 0, \quad \dots, \quad W_m(e_1 \tau, \dots, e_r \tau) = 0. \tag{8}$$

Consequently, M' can be interpreted as the totality of all the ∞^1 manifolds that are represented by the equations (8) with the arbitrary parameter τ. But the totality of the manifolds (8) obviously admits the infinitesimal transformation Ef, since the ∞^1

systems of equations (8) are permuted with each other by the finite transformations:

$$e_1' = \lambda\, e_1, \;\dots,\; e_r' = \lambda\, e_r$$

of the one-term group Ef. Furthermore, it is easy to see that each of the individual systems of equations (8) allows the infinitesimal transformations $E_1f,\dots,E_rf$. Indeed, because the system of equations (7) allows the infinitesimal transformations:

$$E_\mu f = \sum_{k,\,j}^{1\cdots r} c_{j\mu k}\, e_j \frac{\partial f}{\partial e_k} \qquad (\mu = 1\cdots r)$$

then the system (8) with the parameter τ admits the transformations:

$$\sum_{k,\,j}^{1\cdots r} c_{j\mu k}\, e_j\, \tau \frac{\partial f}{\partial (e_k \tau)} = E_\mu f,$$

that is to say, it admits $E_1f,\dots,E_rf$ themselves, whichever value τ takes.

From this, we conclude that the totality of the ∞^1 manifolds (8) admits the infinitesimal transformations Ef, $E_1f,\dots,E_rf$, hence that the manifold M' which coincides with this totality really is the sought manifold; as said, this manifold is analytically represented by the equations which are obtained by elimination of the parameter τ.

If we imagine that for *every* manifold M, the accompanying manifold M' is formed, then according to the above, we obtain all types of one-term subgroups of the group $X_1f,\dots,X_rf$.—

On the other hand, we assume that we know all types of *one-term subgroups* of the group $X_1f,\dots,X_rf$, and we then seek to determine from them all types of finite transformations of these groups.

All smallest manifolds invariant by the group Ef, $E_1f,\dots,E_rf$ are known to us and we must seek all smallest manifolds invariant by the group $E_1f,\dots,E_rf$. But now, since every sought manifold is contained in one of the known manifolds, we only have to consider for itself each individual known manifold and to find the manifolds of the demanded constitution that are located in each such manifold.

Let the q-times extended manifold M be one of the smallest manifolds invariant by the group Ef, $E_1f,\dots,E_rf$. Then for the points of M (cf. Chap. 14, p. 247), all the $(q+1)\times(q+1)$ determinants, but not all $q\times q$ determinants, of the matrix:

$$\text{(9)} \qquad \begin{vmatrix} e_1 & \cdot\;\cdot\;\cdot & e_r \\ \sum_{k=1}^r c_{k11}\, e_k & \cdot\;\cdot\;\cdot & \sum_{k=1}^r c_{k1r}\, e_k \\ \cdot & \cdot\;\cdot\;\cdot & \cdot \\ \sum_{k=1}^r c_{kr1}\, e_k & \cdot\;\cdot\;\cdot & \sum_{k=1}^r c_{krr}\, e_k \end{vmatrix}$$

vanish.

Now, the question whether M decomposes in subsidiary domains which remain invariant by the group $E_1f,\dots,E_rf$ is settled by the behavior of the $q\times q$ subdeterminants of the determinant:

$$\Delta = \begin{vmatrix} \sum_{k=1}^{r} c_{k11}\, e_k & \cdot & \cdot & \cdot & \sum_{k=1}^{r} c_{k1r}\, e_k \\ \cdot & & \cdot & & \cdot \\ \sum_{k=1}^{r} c_{kr1}\, e_k & \cdot & \cdot & \cdot & \sum_{k=1}^{r} c_{krr}\, e_k \end{vmatrix}.$$

If, for the points of M, not all the subdeterminants in question vanish, then no decomposition of M in smaller manifolds invariant by the adjoint group takes place.

By contrast, if the concerned subdeterminants all vanish for the points of M, then M decomposes into ∞^1 $(q-1)$-times extended subsidiary domains invariant by the group $E_1 f, \dots, E_r f$; that there are exactly ∞^1 such subsidiary domains follows from the fact that surely for the points of M, not all $(q-1)\times(q-1)$ subdeterminants of Δ vanish, for otherwise, all the $q \times q$ determinants of the matrix (9) would vanish simultaneously. Naturally, one can set up without integration the equations of the discussed subsidiary domains, since the finite equations of the group $E_1 f, \dots, E_r f$ are known.

In the present chapter, up to now we have always considered the finite transformations of the group $X_1 f, \dots, X_r f$ only as individuals and we have interpreted them as points of an r-times extended space.

There is another standpoint which is equally legitimate. We can also consider as individuals the one-term subgroups, or, what amounts to the same, the infinitesimal transformations of our group, and interpret them as points of a now $(r-1)$-times extended space. Then clearly, we must understand the quantities $e_1, \dots, e_r$ as *homogeneous* coordinates in this space.

Our intention is not to pursue these views; above all, by considering what was said earlier on, it appears for instance evident that every m-term subgroup of the group $X_1 f, \dots, X_r f$ is represented, in the $(r-1)$-times extended space, by a smooth[4] (Translator's note: Namely, the projectivization of a linear subspace (Lie subalgebra) or "straight" [EBEN] manifold (cf. 291).) [EBEN] manifold of $m-1$ dimensions. We only want to derive a simple proposition which follows by taking as a basis the new interpretation and which can often be useful for conceptual research on transformation groups.

We imagine that the general finite transformation:

$$x_i' = x_i + \frac{t}{1} \sum_{k=1}^{r} \lambda_k^0\, X_k x_i + \cdots \qquad (i=1\cdots n)$$

of the one-term group $\lambda_1^0 X_1 f + \cdots + \lambda_r^0 X_r f$ is executed on the infinitesimal transformation: $e_1^0 X_1 f + \cdots + e_r^0 X_r f$ of our r-term group, that is to say, we imagine that, in place of $x_1, \dots, x_n$, the new variables $x_1', \dots, x_n'$ are introduced in the expression $e_1^0 X_1 f + \cdots + e_r^0 X_r f$. According to p. 285, the infinitesimal transformation: $e_1^0 X_1 f + \cdots + e_r^0 X_r f$ is transferred at the same time to the ∞^1 infinitesimal transformations: $e_1 X_1 f + \cdots + e_r X_r f$ of our group, where $e_1, \dots, e_r$ are certain functions of $e_1^0, \dots, e_r^0$ and t that determine themselves through the differential equations:

$$(4') \qquad \frac{\mathrm{d}e_s}{\mathrm{d}t} = -\sum_{\nu=1}^{r} \lambda_\nu^0 \sum_{k=1}^{r} c_{\nu k s}\, e_k \qquad (s=1\cdots r),$$

with the initial conditions: $e_1 = e_1^0, \ldots, e_r = e_r^0$ for $t = 0$.

Since every infinitesimal transformation of our group is represented by a point of the $(r-1)$-times extended space mentioned a short while ago, we can also obviously say: If all ∞^1 transformations of the one-term group: $\lambda_1^0 X_1 f + \cdots + \lambda_r^0 X_r f$ are executed on the infinitesimal transformation: $e_1^0 X_1 f + \cdots + e_r^0 X_r f$, then the image-point of this infinitesimal transformation moves on a certain curve of the space $e_1, \ldots, e_r$.

Now, there is a very simple definition for the tangent to this curve at the point $e_1^0 : \cdots : e_r^0$. Namely, the equations of the tangent in question have the form:

$$e_s \sum_{\nu,\,k}^{1\cdots r} c_{\nu k \tau}\, \lambda_\nu^0\, e_k^0 - e_\tau \sum_{\nu,\,k}^{1\cdots r} c_{\nu k s}\, \lambda_\nu^0\, e_k^0 = 0$$

$$(s,\, \tau = 1\cdots r),$$

whence the point whose homogeneous coordinates have the values:

$$e_s = \sum_{\nu,\,k}^{1\cdots r} c_{\nu k s}\, \lambda_\nu^0\, e_k^0 \qquad (s=1\cdots r)$$

obviously lie on the tangent. But this point is visibly the image-point of the infinitesimal transformation:

$$\left[\sum_{\nu=1}^{r} \lambda_\nu^0 X_\nu f,\ \sum_{k=1}^{r} e_k^0 X_k f\right] = \sum_{s=1}^{r} \left\{ \sum_{\nu,\,k}^{1\cdots r} c_{\nu k s}\, \lambda_\nu^0\, e_k^0 \right\} X_s f$$

which is obtained by *combination* of the two infinitesimal transformations $\sum \lambda_\nu^0 X_\nu f$ and $\sum e_k^0 X_k f$. Consequently, we have the

Proposition 5. *If one interprets the ∞^{r-1} infinitesimal transformations $e_1 X_1 f + \cdots + e_r X_r f$ of an r-term group $X_1 f, \ldots, X_r f$ as points of an $(r-1)$-times extended space by considering $e_1, \ldots, e_r$ as homogeneous coordinates in this space, then the following happens: If all transformations of a determined one term group: $\lambda_1^0 X_1 f + \cdots + \lambda_r^0 X_r f$ are executed on a determined infinitesimal transformation $e_1^0 X_1 f + \cdots + e_r^0 X_r f$, then the image-point of the transformation $e_1^0 X_1 f + \cdots + e_r^0 X_r f$ describes a curve, the tangent of which at the point $e_1^0 : \cdots : e_r^0$ may be obtained by connecting, through a straight line, this point to the image-point of the infinitesimal transformation $[\sum \lambda_\nu^0 X_\nu f,\, \sum e_k^0 X_k f]$; on the other hand, if all transformations of the one-term group $\sum e_k^0 X_k f$ are executed on the infinitesimal transformation $\sum \lambda_\nu^0 X_\nu f$, then the image-point of the transformation $\sum \lambda_\nu^0 X_\nu f$ describes a curve, the tangent of which one obtains by connecting, through a straight line, this point to the image-point of the infinitesimal transformation $[\sum e_k^0 X_k f,\, \sum \lambda_\nu^0 X_\nu f]$.*

Chapter 17
Composition and Isomorphism

Several problems that one can raise concerning an r-term group $X_1 f, \dots, X_r f$ require, for their solution, only the knowledge of the constants c_{iks} in the relations:

$$[X_i, X_k] = \sum_{s=1}^{r} c_{iks} X_s f.$$

For instance, we have seen that the determination of all subgroups of the group $X_1 f, \dots, X_r f$ depends only on the constants c_{iks} and that exactly the same also holds true for the determination of all types of subgroups (cf. Theorem 33, p. 222 and Chap. 16, p. 291 and 292).

It is evident that the constants c_{iks} actually play the rôle of certain properties of the group $X_1 f, \dots, X_r f$. For the totality of these properties, we introduce a specific terminology, we call them the *composition* [ZUSAMMENSETZUNG] of the group, and we thus say that *the constants c_{iks} in the relations:*

$$[X_i, X_k] = \sum_{s=1}^{r} c_{iks} X_s f \tag{1}$$

determine the composition of the r-term group $X_1 f, \dots, X_r f$.

§ 80. The system of the c_{iks} which determines the composition of the r-term group $X_1 f, \dots, X_r f$ is in turn not completely determined. Indeed, the individual c_{iks} receive in general other numerical values when one chooses, in place of $X_1 f, \dots, X_r f$, any other r independent infinitesimal transformations $e_1 X_1 f + \cdots + e_r X_r f$.

From this, it follows that two different systems of c_{iks} can represent, in certain circumstances, the composition of one and the same group. But how can one recognize that this is the case?

We start from the relations:

S. Lie, *Theory of Transformation Groups I*,
DOI 10.1007/978-3-662-46211-9_17

(1) $$[X_i, X_k] = \sum_{s=1}^{r} c_{iks} X_s f \qquad (i, k = 1 \cdots r)$$

which exist between r determined independent infinitesimal transformations $X_1 f, \dots, X_r f$ of our group. We seek the general form of the relations by which r arbitrary independent infinitesimal transformations $\mathscr{X}_1 f, \dots, \mathscr{X}_r f$ of the group $X_1 f, \dots, X_r f$ are linked.

If the concerned relations have the form:

(2) $$[\mathscr{X}_i, \mathscr{X}_k] = \sum_{s=1}^{r} c'_{iks} \mathscr{X}_s f,$$

then the system of the constants c'_{iks} is the most general one which represents the composition of the group $X_1 f, \dots, X_r f$. Hence we are concerned only with the computation of the c'_{iks}.

Since $\mathscr{X}_1 f, \dots, \mathscr{X}_r f$ are supposed to be arbitrary independent infinitesimal transformations of the group $X_1 f, \dots, X_r f$, we have:

$$\mathscr{X}_k f = \sum_{j=1}^{r} h_{kj} X_j f \qquad (k = 1 \cdots r),$$

where the constants h_{kj} can take all the possible values which do not bring to zero the determinant:

$$D = \sum \pm h_{11} \cdots h_{rr}.$$

By a calculation, we obtain:

$$[\mathscr{X}_i, \mathscr{X}_k] = \sum_{j,\,\pi}^{1\cdots r} h_{ij} h_{k\pi} [X_j, X_\pi] = \sum_{j,\,\pi,\,s}^{1\cdots r} h_{ij} h_{k\pi} c_{j\pi s} X_s f;$$

on the other hand, it follows from (2) that:

$$[\mathscr{X}_i, \mathscr{X}_k] = \sum_{\pi,\,s}^{1\cdots r} h_{\pi s} c'_{ik\pi} X_s f.$$

If we compare these two expressions with each other for $[\mathscr{X}_i, \mathscr{X}_k]$ and if we take account of the fact that the $X_1 f, \dots, X_r f$ are independent infinitesimal transformations, we obtain the relations:

(3) $$\sum_{\pi=1}^{r} h_{\pi s} c'_{ik\pi} = \sum_{j,\,\pi}^{1\cdots r} h_{ij} h_{k\pi} c_{j\pi s} \qquad (s = 1 \cdots r).$$

Under the assumptions made, these equations can be solved with respect to the $c'_{ik\pi}$, hence one has:[1] (Translator's note: Implicitly here, one sees a standard way to represent the matrix which is the inverse of (h_{kj}).)

$$c'_{ik\rho} = \frac{1}{D}\sum_{s=1}^{r}\left\{\frac{\partial D}{\partial h_{\rho s}}\sum_{j,\,\pi}^{1\cdots r} h_{ij}\,h_{k\pi}\right\}c_{j\pi s} \tag{4}$$

$$(i,\,k,\,\rho = 1\cdots r).$$

With these words and according to the above, we have found the general form of all systems of constants which determine the composition of the group $X_1f,\dots,X_rf$. At the same time, we have at least theoretical means to decide whether a given system of constants $\overline{c}_{iks}$ determines the composition of the group $X_1f,\dots,X_rf$; namely, such a system obviously possesses this property if and only if one can choose the parameters h_{kj} in such a way that $c'_{iks} = \overline{c}_{iks}$.—

If *two* r-term groups are given, we can compare their compositions. Clearly, thanks to the above developments, we are in a position to decide whether the two groups have one and the same composition, or have different compositions. Here, we do not need to pay heed to the number of variables.

We say that two r-term groups which have one and the same composition are equally composed [GLEICHZUSAMMENGESETZT].

If:

$$X_kf = \sum_{i=1}^{n}\xi_{ki}(x_1,\dots,x_n)\,\frac{\partial f}{\partial x_i} \qquad (k=1\cdots r)$$

are independent infinitesimal transformations of an r-term group and if:

$$Y_kf = \sum_{\mu=1}^{m}\eta_{k\mu}(y_1,\dots,y_m)\,\frac{\partial f}{\partial y_\mu} \qquad (k=1\cdots r)$$

are independent infinitesimal transformations of a second r-term group, and in addition, if there are relations:

$$[X_i,\,X_k] = \sum_{s=1}^{r} c_{iks}\,X_sf,$$

then obviously, these two groups are equally composed when, and only when, amongst the infinitesimal transformations $e_1\,Y_1f+\cdots+e_r\,Y_rf$ of the second group, one can indicate r mutually independent transformations $\mathcal{Y}_1f,\dots,\mathcal{Y}_rf$ such that the relations:

$$[\mathcal{Y}_i,\,\mathcal{Y}_k] = \sum_{s=1}^{r} c_{iks}\,\mathcal{Y}_sf$$

hold identically.

If the relations holding between $Y_1f,\dots,Y_rf$ have the form:

$$[Y_i,\,Y_k] = \sum_{s=1}^{r} \overline{c}_{iks}\,Y_sf,$$

then we can say: the two groups are equally composed when and only when it is possible to choose the parameters h_{kj} in the equations (4) in such a way that every c'_{iks} is equal to the corresponding $\overline{c}_{iks}$.

One can also compare the compositions of two groups which do not have the same number of parameters. This is made possible by the introduction of the general concept: *isomorphism* [ISOMORPHISMUS].

The r-term group $X_1 f, \dots, X_r f$*:*

$$[X_i, X_k] = \sum_{s=1}^{r} c_{iks} X_s f$$

is said to be isomorphic [ISOMORPH] *to the* $(r-q)$*-term group:* $Y_1 f, \dots, Y_{r-q} f$ *when it is possible to choose r infinitesimal transformations:*

$$\mathcal{Y}_k f = h_{k1} Y_1 f + \cdots + h_{k,r-q} Y_{r-q} f$$
$$(k=1 \cdots r)$$

in the $(r-q)$ *so that not all* $(r-q) \times (r-q)$ *determinants of the matrix:*

$$\begin{vmatrix} h_{11} & \cdot & \cdot & h_{1,r-q} \\ \cdot & \cdot & \cdot & \cdot \\ \cdot & \cdot & \cdot & \cdot \\ h_{r1} & \cdot & \cdot & h_{r,r-q} \end{vmatrix}$$

vanish, and so that at the same time, the relations:

$$[\mathcal{Y}_i, \mathcal{Y}_k] = \sum_{s=1}^{r} c_{iks} \mathcal{Y}_s f$$

hold identically.[1]

Let there be an isomorphism in this sense and let $\mathcal{Y}_1 f, \dots, \mathcal{Y}_r f$ be already chosen in the indicated way. Then if we always associate to the infinitesimal transformation $e_1 X_1 f + \cdots + e_r X_r f$ of the r-term group the infinitesimal transformation:

$$e_1 \mathcal{Y}_1 f + \cdots + e_r \mathcal{Y}_r f$$

of the $(r-q)$-term group, whichever values the constants $e_1, \dots, e_r$ may have, then the following clearly holds true: when $\mathrm{Y}_1 f$ is the transformation of the $(r-q)$-term group which is associated to the transformation $\Xi_1 f$ of the other group, and when, correspondingly, $\mathrm{Y}_2 f$ is associated to the transformation $\Xi_2 f$, then the transformation $[\Xi_1, \Xi_2]$ always corresponds to the transformation $[\mathrm{Y}_1, \mathrm{Y}_2]$. We express this more briefly as follows: through the indicated correspondence of the infinitesimal transformations of the two groups, the groups are *isomorphically related one to an-*

[1] Cf. Volume III, [1], p. 701, remarks.

other. Visibly, this isomorphic condition is completely determined when one knows that, to $X_1f,\dots,X_rf$, are associated the transformations $\mathscr{Y}_1f,\dots,\mathscr{Y}_rf$, respectively.

One makes a distinction between *holoedric* and *meroedric isomorphisms*. The *holoedric* case occurs when the number q, which appears in the definition of the isomorphism, has the value zero; the *meroedric* case when q is larger than zero. Correspondingly, one says that the two groups are holoedrically, or meroedrically, isomorphic one to another.

Visibly, the property of being equally composed [DIE EIGENSCHAFT DES GLEICHZUSAMMENGESETZTSEINS] of two groups is a special case of isomorphism; namely, two equally composed groups are always holoedrically isomorphic, and conversely.

Two meroedrically isomorphic groups are, for example, the two:

$$\frac{\partial f}{\partial x_1}, \quad x_1\frac{\partial f}{\partial x_1}, \quad x_1^2\frac{\partial f}{\partial x_1}, \quad \frac{\partial f}{\partial x_2}$$

and:

$$\frac{\partial f}{\partial y_1}, \quad y_1\frac{\partial f}{\partial y_1}, \quad y_1^2\frac{\partial f}{\partial y_1},$$

with, respectively, four and three parameters. We obtain that these two groups are meroedrically isomorphic when, to the four infinitesimal transformations:

$$X_1f=\frac{\partial f}{\partial x_1}, \quad X_2f=x_1\frac{\partial f}{\partial x_1}, \quad X_3f=x_1^2\frac{\partial f}{\partial x_1}, \quad X_4f=\frac{\partial f}{\partial x_1}+\frac{\partial f}{\partial x_2}$$

of the first, the following four, say, are associated:

$$\mathscr{Y}_1f=\frac{\partial f}{\partial y_1}, \quad \mathscr{Y}_2f=y_1\frac{\partial f}{\partial y_1}, \quad \mathscr{Y}_3f=y_1^2\frac{\partial f}{\partial y_1}, \quad \mathscr{Y}_4f=\frac{\partial f}{\partial y_1}.$$

In the theory of substitutions, one also speaks of isomorphic groups, although the notion of isomorphism happens to be apparently different from the one here.[2] Later (cf. Chap. 21: *The Group of Parameters*) we will convince ourselves that nevertheless, the concept of isomorphism following from our definition corresponds perfectly to the concept which one obtains as soon as one translates the definition of the theory of substitution directly into the theory of finite continuous groups.

At present, we shall at first derive a few simple consequences from our definition of isomorphism.

In the preceding chapter, we have seen that to every r-term group $X_1f,\dots,X_rf$ is associated a certain linear homogeneous group, the adjoint group, as we have named it. From the relations which exist between the infinitesimal transformations of the adjoint group (cf. Chap. 16, p. 287), it becomes immediately evident that the group $X_1f,\dots,X_rf$ is isomorphic to its adjoint group; however, the two groups are holoedrically isomorphic only when the group $X_1f,\dots,X_rf$ contains no excellent

[2] Camille JORDAN, Traité des substitutions, Paris 1870.

infinitesimal transformation, because the adjoint group is r-term only in this case, whereas it always contains less than r parameters in the contrary case. Thus:

Theorem 55. *To every r-term group $X_1 f, \dots, X_r f$ is associated an isomorphic linear homogeneous group, namely the adjoint group; this group is holoedrically isomorphic to the group $X_1 f, \dots, X_r f$ only when the latter contains no excellent infinitesimal transformation.*

We do not consider here the question of whether to every r-term group which contains excellent infinitesimal transformations one can also associate an holoedrically isomorphic linear homogeneous group. Yet by an example, we want to show that this is in any case possible in many circumstances, also when the given r-term group contains excellent infinitesimal transformations.

Let the r-term group $X_1 f, \dots, X_r f$ contain precisely $r-m$ independent excellent infinitesimal transformations, and let $X_1 f, \dots, X_r f$ be chosen in such a way that $X_{m+1} f, \dots, X_r f$ are excellent infinitesimal transformations; then between $X_1 f, \dots, X_r f$, there exist relations of the form:

$$[X_\mu, X_\nu] = c_{\mu\nu 1} X_1 f + \dots + c_{\mu\nu r} X_r f$$
$$[X_\mu, X_{m+k}] = [X_{m+k}, X_{m+j}] = 0$$
$$(\mu, \nu = 1 \cdots m;\ k, j = 1 \cdots r-m).$$

In the associated adjoint group $E_1 f, \dots, E_r f$, there are only m independent infinitesimal transformations: $E_1 f, \dots, E_m f$, while $E_{m+1} f, \dots, E_r f$ vanish identically, and hence $E_1 f, \dots, E_m f$ are linked together by the relations:

$$[E_\mu, E_\nu] = c_{\mu\nu 1} E_1 f + \dots + c_{\mu\nu m} E_m f.$$

In particular, if all $c_{\mu,\nu,m+1}, \dots, c_{\mu\nu r}$ vanish, one can always indicate an r-term linear homogeneous group which is holoedrically isomorphic to the group $X_1 f, \dots, X_r f$. In this case namely, $X_1 f, \dots, X_m f$ actually generate an m-term group to which the group $E_1 f, \dots, E_m f$ is holoedrically isomorphic. Hence if we set:

$$\mathsf{E}_{m+1} f = e_{r+1} \frac{\partial f}{\partial e_{r+1}}, \ \dots, \ \mathsf{E}_r f = e_{2r-m} \frac{\partial f}{\partial e_{2r-m}},$$

then the r independent infinitesimal transformations:

$$E_1 f, \dots, E_m f, \ \mathsf{E}_{m+1} f, \dots, \mathsf{E}_r f$$

obviously generate a linear homogeneous group which is holoedrically isomorphic to the group $X_1 f, \dots, X_r f$.

But also in the cases where not all $c_{\mu,\nu,m+1}, \dots, c_{\mu\nu r}$ vanish, one can often easily indicate a holoedrically isomorphic linear homogeneous group. As an example, we use the three-term group $X_1 f, X_2 f, X_3 f$:

$$[X_1, X_2] = X_3 f, \quad [X_1, X_3] = [X_2, X_3] = 0,$$

which contains an excellent infinitesimal transformation, namely $X_3 f$; it is holoedrically isomorphic to the linear homogeneous group:

$$\mathsf{E}_1 f = \alpha_3 \frac{\partial f}{\partial \alpha_1}, \quad \mathsf{E}_2 f = \alpha_1 \frac{\partial f}{\partial \alpha_2}, \quad \mathsf{E}_3 f = \alpha_3 \frac{\partial f}{\partial \alpha_2}.$$

As we have seen in Chap. 9, p. 185, the constants c_{iks} in the equations:

$$(1) \qquad [X_i, X_k] = \sum_{s=1}^{r} c_{iks} X_s f$$

satisfy the relations:

$$(5) \qquad \begin{cases} c_{iks} + c_{kis} = 0 \\ \sum_{\nu=1}^{r} \left\{ c_{ik\nu}\, c_{\nu js} + c_{kj\nu}\, c_{\nu is} + c_{ji\nu}\, c_{\nu ks} \right\} = 0 \\ (i, k, j, s = 1 \cdots r). \end{cases}$$

With the help of these relations, in Chap. 16, p. 287, we succeeded in proving that the r infinitesimal transformations:

$$(6) \qquad E_\mu f = \sum_{k,\, j}^{1 \cdots r} c_{j\mu k}\, e_j \frac{\partial f}{\partial e_k} \qquad (\mu = 1 \cdots r)$$

of the group adjoint to the group $X_1 f, \dots, X_r f$ stand pairwise in the relationships:

$$(7) \qquad [E_\mu, E_\nu] = \sum_{s=1}^{r} c_{\mu\nu s} E_s f.$$

But for this proof of the relations (7), we have used no more than the fact that the c_{iks} satisfied the equations (5), namely we have made no use of the fact that we knew r infinitesimal transformations $X_1 f, \dots, X_r f$ which were linked together by the relations (1). Thanks to the cited developments, it is hence established that the r infinitesimal transformations (6) always stand in the relationships (7) when the c_{iks} satisfy the equations (5).

Consequently, if we know a system of c_{iks} which satisfies the relations (5), we can immediately indicate r linear homogeneous infinitesimal transformations $E_1 f, \dots, E_r f$, namely the transformations (6), which stand pairwise in the relationships:

$$[E_i, E_k] = \sum_{s=1}^{r} c_{iks} E_s f.$$

It is evident that the so obtained infinitesimal transformations $E_1 f, \dots, E_r f$ generate a group, and to be precise, a group with r or fewer parameters; clearly, they generate a group with exactly r parameters only when they are mutually independent, hence when it is impossible that the equations:

$$g_1 E_1 f + \cdots + g_r E_r f = 0,$$

or the r^2 equivalent equations:

$$g_1 c_{j1k} + g_2 c_{j2k} + \cdots + g_r c_{jrk} \qquad (j, k = 1 \cdots r)$$

are satisfied by not all vanishing quantities $g_1, \dots, g_r$.

As a result, we have the

Theorem 52. [3] *When the constants c_{iks} $(i, k, s = 1 \cdots r)$ possess values such that all relations of the form:*

$$(5) \qquad \begin{cases} c_{iks} + c_{kis} = 0 \\ \sum_{\nu=1}^{r} \left\{ c_{ik\nu}\, c_{\nu js} + c_{kj\nu}\, c_{\nu is} + c_{ji\nu}\, c_{\nu ks} \right\} = 0 \\ (i, k, j, s = 1 \cdots r) \end{cases}$$

are satisfied, then the r linear homogeneous infinitesimal transformations:

$$E_\mu f = \sum_{k,\, j}^{1 \cdots r} c_{j\mu k}\, e_j \frac{\partial f}{\partial e_k} \qquad (\mu = 1 \cdots r)$$

stand pairwise in the relationships:

$$[E_i, E_k] = \sum_{s=1}^{r} c_{iks}\, E_s f \qquad (i, k = 1 \cdots r),$$

and hence, they generate a linear homogeneous group. In particular, if the c_{iks} are constituted so that not all $r \times r$ determinants, the horizontal series of which have the form:

$$\left| c_{j1k}\ c_{j2k}\ \cdots\ c_{jrk} \right| \qquad (j, k = 1 \cdots r),$$

vanish, then $E_1 f, \dots, E_r f$ are independent infinitesimal transformations and they generate an r-term group whose composition is determined by the system of the c_{iks}, and which contains no excellent infinitesimal transformation. In all other cases, the group generated by $E_1 f, \dots, E_r f$ has fewer than r parameters.

§ 81. The results of the preceding section suggest the presumption that actually, every system of c_{iks} which satisfies the relations (5) represents the composition of a certain r-term group. This presumption corresponds to the truth [DIESE VERMUTHUNG ENTSPRICHT DER WAHRHEIT], for the following holds.[2] (Translator's note: This is the so-called *Third Fundamental Theorem* of Lie's theory, cf. Vol. III.)

---

[3] LIE, Archiv for Math. og Nat. Vol. 1, p. 192, Christiania 1876.

Proposition 1. *If the constants* c_{iks} $(i,k,s\!=\!1\cdots r)$ *possess values such that the relations:*

$$(5)\qquad \begin{cases} c_{iks}+c_{kis}=0 \\ \sum_{\nu=1}^{r}\left\{c_{ik\nu}\,c_{\nu js}+c_{kj\nu}\,c_{\nu is}+c_{ji\nu}\,c_{\nu ks}\right\}=0 \\ (i,\,k,\,j,\,s\!=\!1\cdots r) \end{cases}$$

are satisfied, then there are always, in a space of the appropriate number of dimensions, r independent infinitesimal transformations $X_1f,\dots,X_rf$ *which stand pairwise in the relationships:*

$$[X_i,X_k]=\sum_{s=1}^{r}c_{iks}\,X_sf$$

and hence generate an r-term group of the composition c_{iks}.

For the time being, we suppress the proof of this important proposition, in order not to be forced to digress for a long while, and we will perform this proof only in the next volume. Of course, until then, we will make the least possible use of this proposition.

From Proposition 1, it results that the totality of all possible compositions of r-term groups is represented by the totality of all systems of c_{iks} which satisfy the equations (5). If one knows all such systems of c_{iks}, then with this, one knows at the same time all compositions of r-term groups.

But now, as we have seen on p. 303, there are in general infinitely many systems of c_{iks} which represent one and the same composition; if a system of c_{iks} is given which represents a composition, then one finds all systems of c'_{iks} which represent the same composition by means of the equations (4), in which it is understood that the h_{kj} are arbitrary parameters. Hence, when one knows all systems of c_{iks} which satisfy the equations (5), more work required in order to find out which of these systems represent different compositions. In order to be able to execute this research, we must at first consider more closely the equations (4).

For the moment, we disregard the fact that the c_{iks} are linked together by some relations; rather, we consider the c_{iks}, and likewise the c'_{iks}, as variables independent of each other. On the basis of this conception, it will be shown that the equations (4) represent a continuous transformation group in the variables c_{iks}.

In order to prove the claimed property of the equations (4), we will directly execute one after the other two transformations (4), or, what is the same, two transformations:

$$(3)\qquad \sum_{\pi=1}^{r}h_{\pi s}\,c'_{ik\pi}=\sum_{j,\,\pi}^{1\cdots r}h_{ij}\,h_{k\pi}\,c_{j\pi s}$$

$$(i,\,k,\,s\!=\!1\cdots r).$$

To begin with, we therefore transfer the c_{iks} to the c'_{iks} by means of the transformation (3) and then the c'_{iks} to the c''_{iks} by means of the transformation:

$$(3') \qquad \sum_{\pi=1}^{r} h'_{\pi s}\, c''_{ik\pi} = \sum_{j,\,\pi}^{1\cdots r} h'_{ij}\, h'_{k\pi}\, c'_{j\pi s}.$$

In this way, we obtain a new transformation, the equations of which are obtained when the c'_{iks} are taken away from (3) and (3'). It is to be proved that this new transformation has the form:

$$(3'') \qquad \sum_{\pi=1}^{r} h''_{\pi s}\, c''_{ik\pi} = \sum_{j,\,\pi}^{1\cdots r} h''_{ij}\, h''_{k\pi}\, c'_{j\pi s},$$

where the h'' are functions of only the h and the h'.

We multiply (3') by $h_{s\sigma}$ and we sum with respect to s; then we receive:

$$\sum_{\pi,\,s}^{1\cdots r} h'_{\pi s}\, h_{s\sigma}\, c''_{ik\pi} = \sum_{j,\,\pi,\,s}^{1\cdots r} h'_{ij}\, h'_{k\pi}\, h_{s\sigma}\, c'_{j\pi s},$$

or, because of (3):

$$\sum_{\pi,\,s}^{1\cdots r} h'_{\pi s}\, h_{s\sigma}\, c''_{ik\pi} = \sum_{j,\,\pi,\,\tau,\,\rho}^{1\cdots r} h'_{ij}\, h'_{k\pi}\, h_{j\tau}\, h_{\pi\rho}\, c_{\tau\rho\sigma}.$$

This is the discussed new transformation; it converts into (3'') when one sets:

$$h''_{\pi\sigma} = \sum_{s=1}^{r} h'_{\pi s}\, h_{s\sigma}.$$

As a result, it is proved that the transformations (4) effectively form a group.

Now, we claim that the transformations of this group leave invariant the equations (5).

Let c_{iks} be a system of constants which satisfies the relations (5), hence according to Proposition 1, which represents the composition of a certain r-term group. Then as we know, the system of the c'_{iks} which is determined by the relations (4) represents in the same way a composition, namely the same composition as the one of the system of the c_{iks}; consequently, the c'_{iks} also satisfy relations of the form (5). Thus, the transformations (4) transfers every system c_{iks} which satisfies (5) to a system c'_{iks} having the same constitution, that is to say, they leave invariant the system of equations (5), which is what we claimed.

At present, we interpret the r^3 variables c_{iks} as point coordinates in a space of r^3 dimensions.

In this space, a certain manifold M which is invariant by the transformations of the group (4) is sorted by the equations (5). Every point of M — as we can say — represents a composition of r-term groups, and conversely, every possible composition of r-term groups is represented by certain points of M. Two different points of M represent one and the same composition when there is, in the group (4), a transformation which transfers the first point to the other.

Hence, if P is an arbitrary point of M, then the totality of all positions that the point P takes by the transformations of the group (4) coincides with the totality of all points which represent the same composition as P. We know from before (Chap. 14, p. 237) that this totality of points forms a manifold invariant by the group (4), and to be precise, a so-called smallest invariant manifold.

From this, it follows that one has to proceed as follows in order to find all different types of r-term groups:

One determines all smallest manifolds located in M that remain invariant by the group (4); on *each* such manifold, one chooses an arbitrary point c_{iks}: the systems of values c_{iks} which belong to the chosen points then represent all types of different compositions of r-term groups.

Since the finite equations of the group (4) are here, the determination of the smallest invariant manifolds has to be considered as an executable operation; it requires only the resolution of algebraic equations.

We therefore have the

Theorem 53. *The determination of all essentially different compositions of r-term groups requires only algebraic operations.*

§ **82.** Now, let $X_1 f, \dots, X_r f$ be an r-term group G_r of the composition:

$$[X_i,\, X_k] = \sum_{s=1}^{r} c_{iks}\, X_s f.$$

Furthermore, let $Y_1 f, \dots, Y_{r-q} f$ be an $(r-q)$-term group isomorphic to G_r, and to be precise, meroedrically isomorphic, so that q is therefore larger than zero. We want to denote this second group shortly by G_{r-q}.

Let the two groups be, in the way indicated on p. 304 and 305, isomorphically related to one another; so in G_{r-q}, let r infinitesimal transformations $\mathscr{Y}_1 f, \dots, \mathscr{Y}_r f$ be chosen which stand in the relationships:

$$[\mathscr{Y}_i,\, \mathscr{Y}_k] = \sum_{s=1}^{r} c_{iks}\, \mathscr{Y}_s f,$$

where $\mathscr{Y}_1 f, \dots, \mathscr{Y}_{r-q} f$ are mutually independent, whereas $\mathscr{Y}_{r-q+1} f, \dots, \mathscr{Y}_r f$ are defined by the identities:

(8)
$$\mathscr{Y}_{r-q+k} f \equiv d_{k1}\, \mathscr{Y}_1 f + \cdots + d_{k,\,r-q}\, \mathscr{Y}_{r-q} f \qquad (k = 1 \cdots q).$$

Under the assumptions made, one obviously has:

$$\left[\mathscr{Y}_j f,\, \mathscr{Y}_{r-q+k} f - \sum_{\mu=1}^{r-q} d_{k\mu}\, \mathscr{Y}_\mu f\right] \equiv 0$$

$$(j=1\cdots r;\; k=1\cdots q),$$

or:

$$\sum_{s=1}^{r} \left\{ c_{j,r-q+k,s} - \sum_{\mu=1}^{r-q} d_{k\mu}\, c_{j\mu s} \right\} \mathscr{Y}_s f \equiv 0.$$

If we replace here $\mathscr{Y}_{r-q+1} f, \dots, \mathscr{Y}_r f$ by their values (8), we obtain linear relations between $\mathscr{Y}_1 f, \dots, \mathscr{Y}_{r-q} f$; but obviously, such linear relations can hold only when the coefficients of every individual $\mathscr{Y}_1 f, \dots, \mathscr{Y}_{r-q} f$ take the value zero.

From the vanishing of these coefficients, it follows that the rq expressions:

$$\left[X_j f,\, X_{r-q+k} f - \sum_{\mu=1}^{r-q} d_{k\mu}\, X_\mu f\right] = \sum_{s=1}^{r} \left\{ c_{j,r-q+k,s} - \sum_{\mu=1}^{r-q} d_{k\mu}\, c_{j\mu s} \right\} X_s f$$

can be linearly deduced from the q infinitesimal transformations:

$$\text{(9)} \qquad X_{r-q+k} f - \sum_{\mu=1}^{r-q} d_{k\mu}\, X_\mu f \qquad (k=1\cdots q).$$

Expressed differently: the q independent infinitesimal transformations (9) generate a q-term *invariant* subgroup of the group $X_1 f, \dots, X_r f$.

Theorem 54. *If the $(r-q)$-term group $Y_1 f, \dots, Y_{r-q} f$ is isomorphic to the r-term group: $X_1 f, \dots, X_r f$, and if $\mathscr{Y}_1 f, \dots, \mathscr{Y}_r f$ are infinitesimal transformations of the $(r-q)$-term group such that firstly $\mathscr{Y}_1 f, \dots, \mathscr{Y}_{r-q} f$ are mutually independent while by contrast, $\mathscr{Y}_{r-q+1}, \dots, \mathscr{Y}_r f$ can be linearly deduced from $\mathscr{Y}_1 f, \dots, \mathscr{Y}_{r-q} f$:*

$$\mathscr{Y}_{r-q+k} f \equiv d_{k1}\, \mathscr{Y}_1 f + \cdots + d_{k,r-q}\, \mathscr{Y}_{r-q} f \qquad (k=1\cdots q),$$

and secondly such that, simultaneously with the relations:

$$[X_i,\, X_k] = \sum_{s=1}^{r} c_{iks}\, X_s f,$$

the analogous relations:

$$[\mathscr{Y}_i,\, \mathscr{Y}_k] = \sum_{s=1}^{r} c_{iks}\, \mathscr{Y}_s f$$

hold, then the q infinitesimal transformations:

$$X_{r-q+k} f - d_{k1}\, X_1 f - \cdots - d_{k,r-q}\, X_{r-q} f \qquad (k=1\cdots q)$$

generate a q-term invariant subgroup of the group $X_1 f, \dots, X_r f$.

G_r: $X_1 f, \dots, X_r f$ and G_{r-q}: $Y_1 f, \dots, Y_{r-q} f$ are, as we know, isomorphically related when, to every infinitesimal transformation of the form: $e_1 X_1 f + \cdots + e_r X_r f$, we associate the infinitesimal transformation:

$$\sum_{k=1}^{r} e_k \mathscr{Y}_k f = \sum_{k=1}^{r-q} \left\{ e_k + \sum_{j=1}^{q} e_{r-q+j} d_{jk} \right\} \mathscr{Y}_k f.$$

Through this correspondence, the transformations of G_r which belong to the q-term group (9) are the only ones which correspond to the identically vanishing infinitesimal transformation of G_{r-q}. Consequently, in general, to every one-term subgroup of G_r, there corresponds a completely determined one-term subgroup of G_{r-q}, and it is only to the one-term subgroups of the group (9) that there correspond no one-term subgroups of G_{r-q}, for the associated one-term groups indeed reduce to the identity transformation. Conversely, to one and the same one-term subgroup $h_1 \mathscr{Y}_1 f + \cdots + h_{r-q} \mathscr{Y}_{r-q} f$ of G_{r-q}, there correspond in total ∞^q different one-term subgroups of G_r, namely all the groups of the form:

$$\sum_{k=1}^{r-q} h_k X_k f + \sum_{j=1}^{q} \lambda_j \left\{ X_{r-q+j} f - \sum_{\mu=1}^{r-q} d_{j\mu} X_\mu f \right\},$$

where $\lambda_1, \dots, \lambda_q$ denote arbitrary constants.

Now, a certain correspondence between the subgroups of G_r and the subgroups of G_{r-q} actually takes place.

If m arbitrary mutually independent infinitesimal transformations:

$$l_{\mu 1} X_1 f + \cdots + l_{\mu r} X_r f \qquad (\mu = 1 \cdots m)$$

generate an m-term subgroup of G_r, then the m infinitesimal transformations:

$$\sum_{k=1}^{r} l_{\mu k} \mathscr{Y}_k f = \sum_{k=1}^{r-q} \left\{ l_{\mu k} + \sum_{j=1}^{q} l_{\mu, r-q+j} d_{jk} \right\} \mathscr{Y}_k f$$

$$(\mu = 1 \cdots m)$$

obviously generate a subgroup of G_{r-q}. This subgroup is at most m-term, and in particular, it is 0-term, that is to say, it consists of only the identity transformation when the m-term subgroup of G_r is contained in the q-term group (9), and in fact clearly, only in this case.

Conversely, if m' arbitrary infinitesimal transformations:

$$l_{\mu 1} \mathscr{Y}_1 f + \cdots + l_{\mu, r-q} \mathscr{Y}_{r-q} f \qquad (\mu = 1 \cdots m')$$

generate an m'-term subgroup of G_{r-q}: $Y_1 f, \dots, Y_{r-q} f$, then the m' infinitesimal transformations:

$$l_{\mu 1} X_1 f + \cdots + l_{\mu, r-q} X_{r-q} f \qquad (\mu = 1 \cdots m'),$$

together with the q transformations:

$$X_{r-q+k}f - d_{k1}X_1f - \cdots - d_{k,r-q}X_{r-q}f \qquad (k=1\cdots q)$$

always generate an $(m'+q)$-term subgroup of G_r.

From this, we see that to every subgroup of G_r corresponds a completely determined subgroup of G_{r-q} which, in certain circumstances, consists of only the identity transformation; and moreover, we see that to every subgroup of G_{r-q} corresponds at least one subgroup of G_r. If we know all subgroups of G_r and if we determine all the subgroups of G_{r-q} corresponding to them, we then obtain all subgroups of G_{r-q}. Thus, the following holds.

Proposition 2. *If one has isomorphically related an r-term group of which one knows all subgroups of an $(r-q)$-term group, then one can also indicate straightaway all subgroups of the $(r-q)$-term group.*

Let $X_1f,\dots,X_rf$, or shortly G_r, be an r-term group of the composition:

$$[X_i, X_k] = \sum_{s=1}^{r} c_{iks}X_sf.$$

We want to study what different compositions that a group meroedrically isomorphic to G_r can have.

Up to now, we know only the following: If G_r can be isomorphically related to an $(r-q)$-term group, then there is in G_r a completely determined q-term invariant subgroup which corresponds to the identity transformation in the $(r-q)$-term group. Now, we claim that this proposition can be reversed in the following way: If G_r contains a q-term invariant subgroup, then there is always an $(r-q)$-term group G_{r-q} which is isomorphic to G_r and which can be isomorphically related to G_r in such a way that the q-term invariant group inside G_r corresponds to the identity transformation[3] (Translator's note: The notion of *quotient group* will in fact not come up here; instead, Lie uses Proposition 1 above.) in G_{r-q}.

For reasons of convenience, we imagine that the r independent infinitesimal transformations $X_1f,\dots,X_rf$ are chosen so that $X_{r-q+1}f,\dots,X_rf$ generate the said invariant subgroup. Then our claim clearly amounts to the fact that, in arbitrary variables $y_1,y_2,\dots$, there are r infinitesimal transformations $\mathcal{Y}_1f,\dots,\mathcal{Y}_rf$ which satisfy the following two conditions: firstly, $\mathcal{Y}_{r-q+1}f,\dots,\mathcal{Y}_rf$ vanish identically, while $\mathcal{Y}_1f,\dots,\mathcal{Y}_{r-q}f$ are mutually independent, and secondly, the relations:

$$(10) \qquad [\mathcal{Y}_i, \mathcal{Y}_k] = c_{ik1}\mathcal{Y}_1f + \cdots + c_{ikr}\mathcal{Y}_rf \qquad (i,k=1\cdots r)$$

must hold identically.

Since $X_{r-q+1}f,\dots,X_rf$ generate a subgroup invariant in G_r, all c_{ik1}, c_{ik2}, $\dots$, $c_{ik,r-q}$ in which at least one of the two indices i and k is larger than $r-q$, are equal to zero. Thus, if in the relations (10), we set equal to zero all the infinitesimal transformations $\mathcal{Y}_{r-q+1}f,\dots,\mathcal{Y}_rf$, then all relations for which not both i and k are smaller than $r-q+1$ will be identically satisfied, and we keep only the following relations between $\mathscr{Y}_1 f, \dots, \mathscr{Y}_{r-q} f$:

$$(11) \qquad [\mathscr{Y}_i, \mathscr{Y}_k] = c_{ik1}\, \mathscr{Y}_1 f + \cdots + c_{ik,r-q}\, \mathscr{Y}_{r-q} f \qquad (i, k = 1 \cdots r-q).$$

So, we need only to prove that there are $r-q$ independent infinitesimal transformations $\mathscr{Y}_1 f, \dots, \mathscr{Y}_{r-q} f$ which are linked together by the relations (11), or, what is the same: that there is an $(r-q)$-term group, the composition of which is represented by the system of constants c_{iks} $(i, k, s = 1 \cdots r-q)$.

In order to prove this, we start from the fact that, for $i, k, j = 1, \dots, r-q$, the Jacobi identity:

$$\big[[X_i, X_k], X_j\big] + \big[[X_k, X_j], X_i\big] + \big[[X_j, X_i], X_k\big] = 0,$$

holds, and it can also be written as:

$$\sum_{\mu=1}^{r-q} \Big\{ c_{ik\mu} [X_\mu, X_j] + c_{kj\mu} [X_\mu, X_i] + c_{ji\mu} [X_\mu, X_k] \Big\}$$
$$+ \sum_{\pi=1}^{q} \Big[X_{r-q+\pi} f,\; c_{ik,r-q+\pi} X_j f + c_{kj,r-q+\pi} X_i f + c_{ji,r-q+\pi} X_k f \Big] = 0.$$

If we develop here the left-hand side and if we take into consideration that the coefficients of $X_1 f, \dots, X_{r-q} f$ must vanish, we obtain, between the constants c_{iks} which appear in (11), the following relations:

$$\sum_{\mu=1}^{r-q} \Big\{ c_{ik\mu}\, c_{\mu j\nu} + c_{kj\mu}\, c_{\mu i\nu} + c_{ji\mu}\, c_{\mu k\nu} \Big\} = 0$$
$$(i, k, j, \nu = 1 \cdots r-q).$$

These relations show that the c_{iks} $(i, k, s = 1 \cdots r-q)$ determine a composition in the sense defined earlier on. As was observed on p. 309, there surely are $(r-q)$-term groups the composition of which is determined by the c_{iks} $(i, k, s = 1 \cdots r-q)$.

As a result, the claim stated above is proved. If we then add what we had already known for a while, we obtain the[4]

Proposition 3. *If one knows all invariant subgroups of the r-term group* $X_1 f, \dots, X_r f$:

$$[X_i, X_k] = \sum_{s=1}^{r} c_{iks} X_s f,$$

then one can indicate all compositions which a group isomorphic to the group $X_1 f, \dots, X_r f$ *can have.*

In order to set up the concerned compositions, one must proceed as follows:

[4] We will show later that the cited Proposition 1, p. 309 is not at all indispensable for the developments of the text. Cf. LIE, Archiv for Math. og Nat. Vol. 10, p. 357, Christiania 1885.

If the $q > 0$ independent infinitesimal transformations:

$$g_{\mu 1} X_1 f + \cdots + g_{\mu r} X_r f \qquad (\mu = 1 \cdots q)$$

generate a q-term invariant subgroup of the group: $X_1 f, \ldots, X_r f$, or shortly G_r, then one sets:

$$g_{\mu 1} \mathscr{Y}_1 f + \cdots + g_{\mu r} \mathscr{Y}_r f = 0 \qquad (\mu = 1 \cdots q),$$

one solves with respect to q of the r expressions $\mathscr{Y}_1 f, \ldots, \mathscr{Y}_r f$, and one eliminates them from the relations:

$$[\mathscr{Y}_i, \mathscr{Y}_k] = \sum_{s=1}^{r} c_{iks} \mathscr{Y}_s f.$$

Between the $r - q$ expressions remaining amongst the expressions $\mathscr{Y}_1 f, \ldots, \mathscr{Y}_r f$, one then obtains relations which define the composition of an $(r-q)$-term group isomorphic to G_r. If one proceeds in this way for every individual subgroup of G_r, one obtains all the desired compositions.

Since every group is its own invariant subgroup, it follows that to every r-term group G_r is associated a meroedrically isomorphic group, namely the group which is formed by the identity transformation. If G_r is simple (cf. Chap. 15, p. 276), then the identity transformation is evidently the only group which is meroedrically isomorphic to it.

§ 83. In this section, we consider an important case in which groups occur that are isomorphic to a given group.

Let the r-term group:

$$X_k f = \sum_{i=1}^{n} \xi_{ki}(x_1, \ldots, x_n) \frac{\partial f}{\partial x_i} \qquad (k = 1 \cdots r)$$

of the space $x_1, \ldots, x_n$ be imprimitive, and let:

$$u_1(x_1, \ldots, x_n) = \text{const.}, \quad \ldots, \quad u_{n-q}(x_1, \ldots, x_n) = \text{const.}$$

be a decomposition of the space into ∞^{n-q} q-times extended manifolds invariant by the group.

According to Chap. 7 and to Chap. 13, p. 233, the $n - q$ mutually independent functions $u_1, \ldots, u_{n-q}$ satisfy relations of the form:

$$X_k u_\nu = \omega_{k\nu}(u_1, \ldots, u_{n-q}) \qquad (k = 1 \cdots r;\ \nu = 1 \cdots n-q),$$

hence the r expressions:

$$\sum_{\nu=1}^{n-q} X_k u_\nu \frac{\partial f}{\partial u_\nu} = \sum_{\nu=1}^{n-q} \omega_{k\nu}(u_1, \ldots, u_{n-q}) \frac{\partial f}{\partial u_\nu} = \overline{X}_k f$$

$$(k = 1 \cdots r)$$

represent as many infinitesimal transformations in the variables $u_1, \dots, u_{n-q}$. We claim that $\overline{X}_1 f, \dots, \overline{X}_r f$ generate a group isomorphic[4] (Translator's note: This notion of *reduced* group $\overline{X}_1 f, \dots, \overline{X}_r f$ appearing here is central in the general algorithm, devised by LIE and developed later in Vol. III, towards the classification of all imprimitive transformation groups, and it also appropriately unveils the true mathematical causality of the nowadays seemingly unusual notion of isomorphism which was introduced earlier on, p. 304.) to the group $X_1 f, \dots, X_r f$.

For the proof, we form:

$$\begin{aligned}\overline{X}_i\big(\overline{X}_k(f)\big) - \overline{X}_k\big(\overline{X}_i(f)\big) &= \sum_{\nu=1}^{n-q} \big(\overline{X}_i\, \omega_{k\nu} - \overline{X}_k\, \omega_{i\nu}\big) \frac{\partial f}{\partial u_\nu} \\ &= \sum_{\nu=1}^{n-q} \big(X_i\, \omega_{k\nu} - X_k\, \omega_{i\nu}\big) \frac{\partial f}{\partial u_\nu}\,;\end{aligned}$$

now, we have:

$$X_i\big(X_k(f)\big) - X_k\big(X_i(f)\big) = \sum_{s=1}^{r} c_{iks}\, X_s f,$$

or, when we insert u_ν in place of f:

$$X_i\, \omega_{k\nu} - X_k\, \omega_{i\nu} = \sum_{s=1}^{r} c_{iks}\, X_s\, u_\nu = \sum_{s=1}^{r} c_{iks}\, \omega_{s\nu},$$

hence, after inserting the found values of $X_i\, \omega_{k\nu} - X_k\, \omega_{i\nu}$, we obtain:

$$\overline{X}_i\big(\overline{X}_k(f)\big) - \overline{X}_k\big(\overline{X}_i(f)\big) = \sum_{s=1}^{r} \sum_{\nu=1}^{n-q} c_{iks}\, \omega_{s\nu} \frac{\partial f}{\partial u_\nu},$$

or, what amounts to the same:

$$\big[\overline{X}_i, \overline{X}_k\big] = \sum_{s=1}^{r} c_{iks}\, \overline{X}_s f.$$

But this is what was to be proved.

The group $\overline{X}_1 f, \dots, \overline{X}_r f$ has a very simple conceptual meaning.

From the definition of imprimitivity, it follows that the totality of the ∞^{n-q} manifolds $u_1 = \text{const.}, \dots, u_{n-q} = \text{const.}$ remains invariant by the group $X_1 f, \dots, X_r f$, hence that the ∞^{n-q} manifolds are permuted by every transformation of this group. Consequently, to every transformation of the group $X_1 f, \dots, X_r f$, there corresponds a certain permutation of our ∞^{n-q} manifolds, or, what is the same, a transformation in the $n-q$ variables $u_1, \dots, u_{n-q}$. It is clear that the totality of all the so obtained transformations in the variables u form a group, namely just the group which is generated by $\overline{X}_1 f, \dots, \overline{X}_r f$.

Naturally, the group $\overline{X}_1 f, \dots, \overline{X}_r f$ need not be holoedrically isomorphic to the initial group, but evidently, it is only meroedrically isomorphic to it when, amongst the infinitesimal transformations $e_1 \overline{X}_1 f + \cdots + e_r \overline{X}_r f$, there is at least one which vanishes identically without $e_1, \dots, e_r$ being all zero, hence when at least one amongst the infinitesimal transformations $e_1 X_1 f + \cdots + e_r X_r f$ leaves individually invariant each one of our ∞^{n-q} manifolds.

Thus, we have the

Proposition 4. *If the r-term group $X_1 f, \dots, X_r f$ of the space $x_1, \dots, x_n$ is imprimitive and if the equations:*

$$u_1(x_1, \dots, x_n) = \text{const.}, \quad \dots, \quad u_{n-q}(x_1, \dots, x_n) = \text{const.}$$

represent a decomposition of the space into ∞^{n-q} q-times extended manifolds, then the infinitesimal transformations:

$$\sum_{\nu=1}^{n-q} X_k u_\nu \frac{\partial f}{\partial u_\nu} = \sum_{\nu=1}^{n-q} \omega_{k\nu}(u_1, \dots, u_{n-q}) \frac{\partial f}{\partial u_\nu} = \overline{X}_k f$$
$$(k=1 \cdots r)$$

in the variables $u_1, \dots, u_{n-q}$, generate a group isomorphic to the group $X_1 f, \dots, X_r f$ which indicates in which way the ∞^{n-q} manifolds are permuted by the transformations of the group $X_1 f, \dots, X_r f$. If, amongst the infinitesimal transformations $e_1 X_1 f + \cdots + e_r X_r f$, there are exactly $r - \rho$ independent ones which leave individually invariant each one of the ∞^{n-q} manifolds, then the group $\overline{X}_1 f, \dots, \overline{X}_r f$ is just ρ-term.

Using Theorem 54, p. 312, we obtain the

Proposition 5. *If the group $X_1 f, \dots, X_r f$ of the space $x_1, \dots, x_n$ is imprimitive and if:*

$$u_1(x_1, \dots, x_n) = \text{const.}, \quad \dots, \quad u_{n-q}(x_1, \dots, x_n) = \text{const.}$$

is a decomposition of the space in ∞^{n-q} q-times extended manifolds, then the totality of all infinitesimal transformations $e_1 X_1 f + \cdots + e_r X_r f$ which leave individually invariant each one of these manifolds generates an invariant subgroup of the group $X_1 f, \dots, X_r f$.

At present, we consider an arbitrary r-term group $X_1 f, \dots, X_r f$ which leaves invariant a manifold of the space $x_1, \dots, x_n$.

According to Chap. 14, p. 243, the points of this manifold are in turn transformed by a group, the infinitesimal transformations of which can be immediately indicated when the equations of the manifold are in resolved form. Indeed, if the equations of the manifold read in the following way:

$$x_1 = \varphi_1(x_{n-m+1}, \dots, x_n), \quad \dots, \quad x_{n-m} = \varphi_{n-m}(x_{n-m+1}, \dots, x_n),$$

and if $x_{n-m+1}, \dots, x_n$ are chosen as coordinates for the points of the manifold, then the infinitesimal transformations of the group in question are of the form (cf. Chap. 14, p. 244):

$$\overline{X}_k f = \sum_{\mu=1}^{m} \xi_{k,n-m+\mu}(\varphi_1,\dots,\varphi_{n-m},x_{n-m+1},\dots,x_n)\,\frac{\partial f}{\partial x_{n-m+\mu}}$$

$$(k=1\cdots r).$$

What is more, as we have already proved at that time, the following relations hold:

$$[\overline{X}_i, \overline{X}_k] = \sum_{s=1}^{r} c_{iks}\,\overline{X}_s f.$$

From this, we see that the group $\overline{X}_1 f,\dots,\overline{X}_r f$ is isomorphic to the group $X_1 f,\dots,X_r f$; in particular, it is meroedrically isomorphic to it when, amongst the infinitesimal transformations:

$$e_1 \overline{X}_1 f + \cdots + e_r \overline{X}_r f,$$

there is at least one transformation which vanishes identically, so that at least one amongst the infinitesimal transformations $e_1 X_1 f + \cdots + e_r X_r f$ leaves fixed every individual point of the manifold.

We can therefore complement Theorem 40 in Chap. 14, p. 243, as follows:

Proposition 6. *If the r-term group $X_1 f,\dots,X_r f$ in the variables $x_1,\dots,x_n$ leaves invariant the manifold:*

$$x_1 = \varphi_1(x_{n-m+1},\dots,x_n), \;\; \dots, \;\; x_{n-m} = \varphi_{n-m}(x_{n-m+1},\dots,x_n),$$

then the reduced infinitesimal transformations:

$$\overline{X}_k f = \sum_{\mu=1}^{m} \xi_{k,n-m+\mu}(\varphi_1,\dots,\varphi_{n-m},x_{n-m+1},\dots,x_n)\,\frac{\partial f}{\partial x_{n-m+\mu}}$$

$$(k=1\cdots r)$$

in the variables $x_{n-m+1},\dots,x_n$ generate a group isomorphic to the r-term group which indicates in which way the points of the manifold are permuted by the transformations of the group $X_1 f,\dots,X_r f$. If, amongst the infinitesimal transformations $e_1 X_1 f + \cdots + e_r X_r f$, there are exactly $r-\rho$ independent transformations which leave invariant every individual point of the manifold, then the group $\overline{X}_1 f,\dots,\overline{X}_r f$ is just ρ-term.

In addition, the following also holds true.

Proposition 7. *If the r-term group $X_1 f,\dots,X_r f$ of the space $x_1,\dots,x_n$ leaves invariant a manifold, then the totality of all infinitesimal transformations $e_1 X_1 f + \cdots + e_r X_r f$ which leave fixed every individual point of this manifold generate an invariant subgroup of the r-term group.*

Let:

$$X_k f = \sum_{i=1}^{n} \xi_{ki}(x_1,\dots,x_n)\,\frac{\partial f}{\partial x_i} \qquad (k=1\cdots r)$$

be an r-term *intransitive* group of the R_n, and one which contains only a *discrete* number of invariant subgroups.

The complete system which is determined by the equations:

$$X_1 f = 0, \ \dots, \ X_r f = 0$$

is q-term under these assumptions, where $q < n$, and it possesses $n-q$ independent solutions. Hence we can imagine that the variables $x_1,\dots,x_n$ are chosen from the beginning in such a way that $x_{q+1},\dots,x_n$ are solutions of the complete system; if we do this, then because $\xi_{k,q+j} = X_k x_{q+j}$, all $\xi_{k,q+1},\dots,\xi_{kr}$ vanish.

Evidently, each one of the ∞^{n-q} q-times extended manifolds:

$$x_{q+1} = a_{q+1}, \ \dots, \ x_n = a_n$$

now remains invariant by the group $X_1 f,\dots,X_r f$; but their points are permuted and according to Chap. 14, p. 243, by a group. If we choose $x_1,\dots,x_q$ as coordinates for the points of the manifold, then the infinitesimal transformations of the group in question will be:

$$\overline{X}_k f = \sum_{i=1}^{q} \xi_{ki}(x_1,\dots,x_q,\, a_{q+1},\dots,a_n)\,\frac{\partial f}{\partial x_i}$$

$$(k=1\cdots r)$$

Whether this group is r-term or not remains temporarily uncertain, and in any case, it is isomorphic to the group $X_1 f,\dots,X_r f$.

If the group $\overline{X}_1 f,\dots,\overline{X}_r f$ is ρ-term ($\rho \leqslant r$), then the group $X_1 f,\dots,X_r f$ contains an $(r-\rho)$-term invariant subgroup which leaves fixed every individual point of the manifold $x_{q+1} = a_{q+1},\dots,x_n = a_n$ (cf. Propositions 6 and 7). This invariant subgroup cannot change with the values of $a_{q+1},\dots,a_n$, since otherwise, there would be in the group $X_1 f,\dots,X_r f$ a continuous series of invariant subgroups, and this would contradict our assumption. Consequently, there exists in the group $X_1 f,\dots,X_r f$ an $(r-\rho)$-term invariant subgroup which leaves fixed all points of an arbitrary manifold amongst the ∞^{n-q} manifolds:

$$x_{q+1} = a_{q+1}, \ \dots, \ x_n = a_n,$$

hence which actually leaves fixed all points of the space $x_1,\dots,x_n$. But now, the identity transformation is the only one which leaves at rest all points of the space $x_1,\dots,x_n$, so one has $r-\rho = 0$ and $\rho = r$, that is to say: the group $\overline{X}_1 f,\dots,\overline{X}_r f$ is holoedrically isomorphic to the group $X_1 f,\dots,X_r f$.

As a result, the following holds true.

Proposition 8. *If $X_1 f, \dots, X_r f$ are independent infinitesimal transformations of an r-term group with the absolute invariants $\Omega_1(x_1, \dots, x_n)$, ..., $\Omega_{n-q}(x_1, \dots, x_n)$, and if there is only, in this group, a discrete number of invariant subgroups, then the points of an arbitrary invariant domain: $\Omega_1 = a_1$, ..., $\Omega_{n-q} = a_{n-q}$ are transformed by a group* holoedrically *isomorphic to the group $X_1 f, \dots, X_r f$.*

References

1. Engel, F.; Lie, S.: *Theorie der transformationsgruppen. Dritter und letzter Abschnitt.* Unter Mitwirkung von Dr. Friedrich Engel, bearbeitet von Sophus Lie, Verlag und Druck von B.G. Teubner, Leipzig und Berlin, xxix+836 pp. (1890). Reprinted by Chelsea Publishing Co., New York, N.Y. (1970)

Chapter 18
Finite Groups, the Transformations of Which Form Discrete Continuous Families

So far, we have only occupied ourselves with continuous transformation groups, hence with groups which are represented by *one* system of equations of the form:

$$x_i' = f_i(x_1, \dots, x_n, a_1, \dots, a_r) \qquad (i=1\cdots n).$$

In this chapter, we shall also briefly treat the finite groups which cannot be represented by a single system of equations, but which can only be represented by several such systems; these are the groups about which we have already mentioned in the Introduction, p. 11.[1]

Thus, we imagine that a series of systems of equations of the form:

$$x_i' = f_i^{(k)}\big(x_1, \dots, x_n, a_1^{(k)}, \dots, a_{r_k}^{(k)}\big) \qquad (i=1\cdots n) \tag{1}$$
$$(k=1,2\cdots)$$

is presented, in which each system contains a finite number r_k of arbitrary parameters $a_1^{(k)}, \dots, a_{r_k}^{(k)}$, and we assume that the totality of all transformations which are represented by these systems of equations forms a group.

Since each of the systems of equations (1) represents a continuous family of transformations, our group consists of a discrete number of continuous transformations. It is clear that every continuous family of transformations of our group either coincides with one of the families (1), or must be contained in one of these families. Of course, we assume that none of the families (1) is contained in one of the remaining families.

It is our intention to develop the foundations of a general theory of the sorts of groups just defined, but for reasons of simplicity we will introduce a few restrictions, which, incidentally, are not to be considered as essential.

Firstly, we make the assumption that the transformations of the group (1) are ordered as inverses by pairs. Hence, although the transformations of the family:

[1] LIE, Verhandlungen der Gesellschaft der Wissenchaften zu Christiania, Nr. 12, p. 1, 1883.

S. Lie, *Theory of Transformation Groups I*,
DOI 10.1007/978-3-662-46211-9_18

$$x_i' = f_i^{(k)}\big(x_1,\dots,x_n,\, a_1^{(k)},\dots,a_{r_k}^{(k)}\big) \qquad (i=1\cdots n)$$

are not already ordered as inverses by pairs, the totality of the associated inverse transformations is supposed to form a family which belongs to the group, hence which is contained amongst the families (1).

Secondly, we assume that the number of the families (1) is finite, say equal to m. However, when the propositions we derive are also true for infinitely many families (1), we will occasionally point out that this is the case.

§ 84. To the two assumptions about the group (1) which we have made in the introduction of the chapter, we want to temporarily add the third assumption that all families of the group should contain the same number, say r, of essential parameters. In the next section, we show that this third assumption follows from the first two, and hence is superfluous.

Let:

$$x_i' = f_i^{(k)}(x_1,\dots,x_n,\, a_1,\dots,a_r) \qquad (i=1\cdots n)$$

be one of the m families of ∞^r transformations of which our group consists, so k is any of the numbers $1,\dots,m$.

By resolution of the equations written above, we obtain a family of transformations:

$$x_i = F_i^{(k)}(x_1',\dots,x_n',\, a_1,\dots,a_r) \qquad (i=1\cdots n)$$

which, under the assumptions made, equally belongs to the group.

Therefore, when the two transformations:

$$\begin{aligned} x_i &= F_i^{(k)}(x_1',\dots,x_n',\, a_1,\dots,a_r) \\ x_i'' &= f_i^{(k)}(x_1,\dots,x_n,\, a_1+h_1,\dots,a_r+h_r) \end{aligned} \qquad (i=1\cdots n)$$

are executed one after the other, we again obtain a transformation of our group, namely the following one:

$$\text{(2)} \qquad x_i'' = f_i^{(k)}\big(F_1^{(k)}(x',a),\dots,F_n^{(k)}(x',a),\, a_1+h_1,\dots,a_r+h_r\big) \qquad (i=1\cdots n).$$

Here, we expand the right-hand side with respect to powers of $h_1,\dots,h_r$ and we find:

$$x_i'' = f_i^{(k)}\big(F^{(k)}(x',a),a\big) + \sum_{j=1}^{r} h_j \left[\frac{\partial f_i^{(k)}(x,a)}{\partial a_j}\right]_{x=F^{(k)}(x',a)} + \cdots,$$

where all the omitted terms in $h_1,\dots,h_r$ are of second and higher order. But if we take into account that the two transformations:

$$x_i = F_i^{(k)}(x',a), \qquad x_i' = f_i^{(k)}(x,a) \qquad (i=1\cdots n)$$

are inverse to each other, and if in addition, we adopt the abbreviation:

$$\left[\frac{\partial f_i^{(k)}}{\partial a_j}\right]_{x=F^{(k)}(x',a)} = \eta_{ji}^{(k)}(x_1',\dots,x_n',\,a_1,\dots,a_r), \tag{3}$$

then we see that the transformation just found has the form:

$$x_i'' = x_i' + \sum_{j=1}^{r} h_j\,\eta_{ji}^{(k)}(x',a) + \cdots \qquad (i=1\cdots n). \tag{4}$$

It is easy to see that there are no functions $\chi_1(a),\dots,\chi_r(a)$ independent of $x_1',\dots,x_n'$ which satisfy the n equations:

$$\sum_{j=1}^{r} \chi_j(a_1,\dots,a_r)\,\eta_{ji}^{(k)}(x',a) = 0 \qquad (i=1\cdots n)$$

identically, without all vanishing. Indeed, if one makes the substitution $x_\nu' = f_\nu^{(k)}(x,a)$ in these equations, one obtains the equations:

$$\sum_{j=1}^{r} \chi_j(a)\,\frac{\partial f_i^{(k)}(x,a)}{\partial a_j} = 0 \qquad (i=1\cdots n),$$

which must also be satisfied identically; but according to Chap. 2, Theorem 2.2, p. 15, this is impossible, because the parameters $a_1,\dots,a_r$ in the transformation equations $x_i' = f_i^{(k)}(x,a)$ are essential.

From this, we conclude that the r infinitesimal transformations:

$$\sum_{i=1}^{n} \eta_{ji}^{(k)}(x_1',\dots,x_n',\,a_1,\dots,a_r)\,\frac{\partial f}{\partial x_i'} \qquad (j=1\cdots r) \tag{5}$$

are always mutually independent when $a_1,\dots,a_r$ is a system of values in general position.

By $a_1^0,\dots,a_r^0$, we want to understand a system of values in general position and we want to set:

$$\eta_{ji}^{(1)}(x',a^0) = \xi_{ji}(x_1',\dots,x_n'),$$

by conferring the special value 1 to the number k. We will show that all infinitesimal transformations (5) can be linearly deduced from the r mutually independent infinitesimal transformations:

$$X_j'f = \sum_{i=1}^{n} \xi_{ji}(x_1',\dots,x_n')\,\frac{\partial f}{\partial x_i'} \qquad (j=1\cdots r),$$

whichever value k has as one of the integers 1, 2, ..., m and whichever value $a_1,\dots,a_r$ has.

The proof here has great similarities with the developments in Chap. 4, p. 87 sq.

We execute two transformations of our group one after the other, namely first the transformation:

$$x_i'' = x_i' + \sum_{j=1}^{r} h_j\, \xi_{ji}(x') + \cdots \qquad (i=1\cdots n) \tag{6}$$

which results from the transformation (4) when one sets $k = 1$ and $a_1 = a_1^0$, ..., $a_r = a_r^0$, and then, the transformation:

$$x_i''' = x_i'' + \sum_{j=1}^{r} \rho_j\, \eta_{ij}^{(k)}(x'', a) + \cdots \qquad (i=1\cdots n)$$

which, likewise, is of the form (4). In this way, we obtain the following transformation belonging to our group:

$$x_i''' = x_i' + \sum_{j=1}^{r} h_j\, \xi_{ji}(x') + \sum_{j=1}^{r} \rho_j\, \eta_{ji}^{(k)}(x', a) + \cdots$$
$$(i=1\cdots n),$$

where the omitted terms are of second and higher order in the $2r$ quantities $h_1, \dots, h_r, \rho_1, \dots, \rho_r$.

Disregarding the a, the $2r$ arbitrary parameters $h_1, \dots, h_r, \rho_1, \dots, \rho_r$ appear in the latter transformation, whereas our group can contain only transformations with r essential parameters. From Proposition 4 of Chap. 4, p. 80, it therefore follows that, amongst the $2r$ infinitesimal transformations: (5) and $X_1'f, \dots, X_r'f$, only r can be in existence that are mutually independent. But because $X_1'f, \dots, X_r'f$ are mutually independent, the infinitesimal transformations (5) must be linearly expressible in terms of $X_1'f, \dots, X_r'f$, for all values 1, 2, ..., m of k and for all values of the a.

At present, thanks to considerations analogous to those in Chap. 3, we realize that identities of the form:

$$\sum_{i=1}^{n} \eta_{ji}^{(k)}(x', a)\, \frac{\partial f}{\partial x_i'} \equiv \sum_{\pi=1}^{r} \psi_{j\pi}^{(k)}(a_1, \dots, a_r)\, X_\pi' f$$
$$(k=1\cdots m;\ j=1\cdots r)$$

hold, where the $\psi_{j\pi}^{(k)}$ are completely determined analytic functions of $a_1, \dots, a_r$.

Lastly, if we remember the equations (3) which can evidently also be written as:

$$\eta_{ji}^{(k)}\big(f_1^{(k)}(x,a), \dots, f_n^{(k)}(x,a), a_1, \dots, a_r\big) \equiv \frac{\partial f_i^{(k)}(x,a)}{\partial a_j},$$

and if we compare these equations to the identities:

$$\eta_{ji}^{(k)}(x', a) \equiv \sum_{\pi=1}^{r} \psi_{j\pi}^{(k)}(a)\, \xi_{\pi i}(x'),$$

we then obtain the identities:

$$\frac{\partial f_i^{(k)}(x,a)}{\partial a_j} \equiv \sum_{\pi=1}^{r} \psi_{j\pi}^{(k)}(a_1,\dots,a_r)\,\xi_{\pi i}\big(f_1^{(k)}(x,a),\dots,f_n^{(k)}(x,a)\big) \tag{7}$$

The functions $\xi_{\pi i}$ here are independent of the index k, but by contrast, the $\psi_{j\pi}^{(k)}$ are not.

Thus, we have the

Theorem 55. *If the m systems of equations:*

$$x_i' = f_i^{(k)}(x_1,\dots,x_n,\,a_1,\dots,a_r) \qquad (i=1\cdots n)$$
$$(k=1\cdots m),$$

in each of which the r parameters $a_1,\dots,a_r$ are essential, represent all transformations of a group (and if at the same time, all these transformations can be ordered as inverses by pairs),[2] *then there are r independent infinitesimal transformations:*

$$X_j f = \sum_{i=1}^{n} \xi_{ji}(x_1,\dots,x_n)\,\frac{\partial f}{\partial x_i} \qquad (j=1\cdots r)$$

which stand in such a relationship to the group that each family:

$$x_i' = f_i^{(k)}(x_1,\dots,x_n,\,a_1,\dots,a_r) \qquad (i=1\cdots n)$$

satisfies differential equations of the form:

$$\frac{\partial f_i^{(k)}}{\partial a_j} = \sum_{\pi=1}^{r} \psi_{j\pi}^{(k)}(a_1,\dots,a_r)\,\xi_{\pi i}(x_1',\dots,x_n') \tag{7'}$$
$$(i=1\cdots n;\; j=1\cdots r).$$

From this, it follows at first that, according to Theorems 21, p. 164 and 24, p. 172, the r infinitesimal transformations $X_1 f,\dots,X_r f$ generate an r-term group.

Furthermore, it is clear that Theorem 25 in Chap. 9, p. 174, finds an application to each one of the families $x_i' = f_i^{(k)}(x,a)$: every transformation $x_i' = f_i^{(k)}(x,a)$ whose parameters $a_1,\dots,a_r$ lie in a certain neighborhood of $\overline{a}_1,\dots,\overline{a}_r$ can be obtained by executing firstly the transformation $\overline{x}_i = f_i^{(k)}(x,\overline{a})$ and then, a certain transformation of the r-term group $X_1 f,\dots,X_r f$.

Now, since our group contains all transformations of the form (6):

[2] As one easily realizes, Theorem 55 still remains correct when the words in brackets are deleted. (Compare with the developments of Chaps. 3, 4 and 9.)

$$x_i'' = x_i' + \sum_{j=1}^{r} h_j\,\xi_{ji}(x') + \cdots \qquad (i=1\cdots n)$$

and since one of these transformations is the identity transformation, namely the one with the parameters $h_1 = 0, \dots, h_r = 0$, then the identity transformation must occur in one of the families $x_i' = f_i^{(k)}(x,a)$. Consequently, one amongst these families is just the r-term group $X_1 f, \dots, X_r f$. Of course, this is the only r-term group generated by infinitesimal transformations which is contained in the group $x_i' = f_i^{(k)}(x,a)$.

As a result, we have established the following theorem:

Theorem 56. *Every group:*

$$x_i' = f_i^{(k)}(x_1,\dots,x_n,\,a_1,\dots,a_r) \qquad (i=1\cdots n)$$
$$(k=1\cdots m)$$

having the constitution indicated in the preceding theorem contains one, and only one, r-term group with transformations inverse by pairs. This r-term group is generated by the infinitesimal transformations $X_1 f, \dots, X_r f$ defined in the previous theorem, its ∞^r transformations form one of the m families $x_i' = f_i^{(k)}(x,a)$ and in addition, they stand in the following relationship with respect to each one of the remaining $m-1$ families: If $\overline{x}_i = f_i^{(k)}(x,\overline{a})$ is an arbitrary transformation of the family $x_i' = f_i^{(k)}(x,a)$, then every other transformation of this family whose parameters $a_1,\dots,a_r$ lie in a certain neighborhood of $\overline{a}_1,\dots,\overline{a}_r$ can be obtained by executing firstly the transformation $\overline{x}_i = f_i^{(k)}(x,\overline{a})$, and then, a certain transformation $x_i' = \omega_i(\overline{x}_1,\dots,\overline{x}_n)$ of the r-term group $X_1 f, \dots, X_r f$.

For example, we consider the group consisting of the two families of ∞^1 transformations that are represented by the two systems of equations:

$$x' = x\cos a - y\sin a, \qquad y' = x\sin a + y\cos a$$
$$x' = x\cos a + y\sin a, \qquad y' = x\sin a - y\cos a.$$

For the two families, one obtains by differentiation with respect to a:

$$\frac{\mathrm{d}x'}{\mathrm{d}a} = -y', \qquad \frac{\mathrm{d}y'}{\mathrm{d}a} = x',$$

so that in the present case, the functions $\psi_{j\pi}^{(k)}$ mentioned in Theorem 55 are independent of the index k.

The first one of the above two families is a one-term group which is generated by the infinitesimal transformation $y\partial f/\partial x - x\partial f/\partial y$. The general transformation of the second family will be obtained when one executes firstly the transformation:

$$\overline{x} = x, \qquad \overline{y} = -y,$$

and then the transformation:

$$x' = \overline{x}\cos a - \overline{y}\sin a, \qquad y' = \overline{x}\sin a + \overline{y}\cos a$$

of the one-term group $y\partial f/\partial x - x\partial f/\partial y$, hence the general transformation of the first family. —

It should not go unmentioned that the two Theorems 55 and 56 also remain valid when the concerned group consists of an *infinite* number of discrete continuous families which all contain the same number of essential parameters. —

Before we go further, we make a few important remarks.

Let:

$$x_i' = f_i(x_1,\dots,x_n,\, a_1,\dots,a_r) \qquad (i=1\cdots n)$$

be an r-term continuous group which does not contain the identity transformation. Then according to Theorem 25, Chap. 9, p. 174, there is an r-term group $X_1 f,\dots,X_r f$ with the identity transformation and with pairwise inverse transformations which stands in the following relationship to the group $x_i' = f_i(x,a)$, or shortly $\mathfrak{G}$: If one executes firstly a transformation $\overline{x}_i = f_i(x,\overline{a})$ of $\mathfrak{G}$ and then a transformation:

$$x_i' = \overline{x}_i + \sum_{k=1}^{r} e_k \overline{X}_k \overline{x}_i + \cdots \qquad (i=1\cdots n)$$

of the group $X_1 f,\dots,X_r f$, then one always obtains a transformation of $\mathfrak{G}$.

Now, one can easily prove that one then also always obtains a transformation of $\mathfrak{G}$ when one firstly executes a transformation of the group $X_1 f,\dots,X_r f$ and afterwards a transformation of the group $\mathfrak{G}$. We do not want to spend time in order to produce this proof in detail, and we only want to remark that for this proof, one may employ considerations completely similar to those of Chap. 4 (Cf. Theorem 58).

Now, if we take these two relationships between $\mathfrak{G}$ and the group $X_1 f,\dots,X_r f$ and in addition, if we take into account that we have to deal with two groups, then we realize immediately that the transformations of the group $\mathfrak{G}$ and of the group $X_1 f,\dots,X_r f$, when combined, again form a group, but to be precise, a group the transformations of which cannot be ordered as inverses by pairs.

At present, we also attract into the circle of our considerations the family of transformations:

$$x_i' = F_i(x_1,\dots,x_n,\, a_1,\dots,a_r) \qquad (i=1\cdots n)$$

which are inverses of the transformations $x_i' = f_i(x,a)$. According to Theorem 2[1] (Translator's note: This theorem simply states that if transformation equations $x_i' = f(x,a)$ are stable by composition, the same holds for the inverse transformations $x_i = F_i(x',a)$ modulo possible shrinkings of domains.) ([1], p. 19), this family of transformations also forms a group, that may be called $\mathfrak{G}'$. We shall show that the transformations of the three groups $\mathfrak{G}$, $\mathfrak{G}'$, $X_1 f,\dots,X_r f$, when taken together, again form a group, and naturally, a group with pairwise inverse transformations.

Let T be the general symbol of a transformation of $\mathfrak{G}$, whence T^{-1} is the general symbol of a transformation of $\mathfrak{G}'$; by S, it will always be understood a transformation of the group $X_1 f, \dots, X_r f$.

We already know that all transformations T and S taken together form a group and that the transformations T^{-1} taken for themselves do the same. From this, we realize the existence of relations which have the following form:

$$T_\alpha T_\beta = T_\gamma, \qquad S_\lambda S_\mu = S_\nu, \qquad T_\beta^{-1} T_\alpha^{-1} = T_\gamma^{-1}$$
$$T_\alpha S_\lambda = T_\pi, \qquad S_\lambda T_\alpha = T_\rho.$$

We can also write the second series of these relations as follows:

$$S_\lambda^{-1} T_\alpha^{-1} = T_\pi^{-1}, \qquad T_\alpha^{-1} S_\lambda^{-1} = T_\rho^{-1}.$$

Now, since the group of the S consists of pairwise inverse transformations, it is immediately clear that the T^{-1} together with the S form a group. Moreover, we have:

$$T_\alpha^{-1} T_\pi = T_\alpha^{-1} T_\alpha S_\lambda = S_\lambda,$$
$$T_\alpha T_\rho^{-1} = T_\alpha T_\alpha^{-1} S_\lambda^{-1} = S_\lambda^{-1},$$

and therefore, the totality of all S, T, T^{-1} also forms a group.

With these words, we bring the promised proof: the transformations of the three groups $\mathfrak{G}$, $\mathfrak{G}'$, $X_1 f, \dots, X_r f$ form a group together, and naturally, a group with pairwise inverse transformations.

§ 85. At present, we take up a standpoint more general than in the previous section. We drop the special assumption[2] (Translator's note: (i.e. the third one)) made there and we only maintain the two assumptions which were set out in the Introduction.

Thus, we consider a group G which consists of m discrete families with, respectively, r_1, r_2, ..., r_m essential parameters and which, for each of its transformations, also contains the inverse transformation. We will prove that the numbers r_1, r_2, ..., r_m are all mutually equal. Then from this, it follows that the assumption: $r_1 = r_2 = \cdots = r_m$ made in the previous section was not a restriction.

We execute two transformations of the group one after the other, firstly a transformation:

$$x_i' = f_i^{(k)}(x_1, \dots, x_n, a_1, \dots, a_{r_k}) \qquad (i=1\cdots n)$$

of a family with r_k parameters, and secondly a transformation:

$$x_i'' = f_i^{(j)}(x_1', \dots, x_n', b_1, \dots, b_{r_j}) \qquad (i=1\cdots n)$$

of a family with r_j parameters.

In this way, we find a transformation:

$$x_i'' = f_i^{(j)}\big(f_1^{(k)}(x,a), \dots, f_n^{(k)}(x,a), b_1, \dots, b_{r_j}\big)$$

which belongs to our group and which formally contains $r_k + r_j$ arbitrary parameters. Now, if the largest number amongst the numbers $r_1, r_2, \dots, r_m$ has the value r, then amongst these $r_k + r_j$ arbitrary parameters, there are no more than r which are essential, but also no fewer essential ones than indicated by the largest of the two numbers r_k and r_j. So in particular, if both numbers r_k and r_j are equal to r, then the last written transformation contains exactly r essential parameters. Consequently, all families of our group which contain exactly r essential parameters already form a group Γ when taken together.

Now, we can immediately apply Theorem 56 of the previous section to the group Γ. From this, we see that Γ and hence also G contains an r-term group:

$$x_i' = f_i(x_1, \dots, x_n, a_1, \dots, a_r) \qquad (i=1\cdots n)$$

which is generated by r independent infinitesimal transformations. Thus, if we first execute the transformation $x_i' = f_i(x,a)$ and then the transformation:

$$\text{(8)} \qquad x_i'' = f_i^{(j)}(x_1', \dots, x_n', b_1, \dots, b_{r_j}),$$

we again obtain a transformation of the group G, namely the following one:

$$\text{(9)} \qquad x_i'' = f_i^{(j)}\big(f_1(x,a), \dots, f_n(x,a), b_1, \dots, b_{r_j}\big) \qquad (i=1\cdots n).$$

Of the $r + r_j$ parameters of this transformation, exactly r are essential, so the equations just written represent a continuous family of ∞^r transformations of the group G. But since the group $x_i' = f_i(x,a)$ contains the identity transformation, there are special values of the parameters $a_1, \dots, a_r$ for which the functions $f_1(x,a), \dots, f_n(x,a)$ reduce to $x_1, \dots, x_n$, respectively. Consequently, the family of the ∞^{r_j} transformations (8) is contained in the family of the ∞^r transformations (9). According to the remarks made in the introduction of the chapter, this is possible only when the two families coincide, hence when r_j is equal to r.

As a result, it is proved that the numbers $r_1, r_2, \dots, r_m$ are all equal to one another. Consequently, we have the

Theorem 57. *If a group whose transformations are pairwise inverse to one another consists of m continuous families of transformations and if each one of these families contains only a finite number of arbitrary parameters, then the families all have the same number of essential parameters.*

Furthermore, this theorem still remains valid when the number of the families of which the group consists is infinitely large, when each of these infinitely many families contains just a finite number ρ_k of arbitrary parameters and when at the same time, amongst all the numbers ρ_k, a largest one is extant.

§ 86. As up to now, let G consist of m discrete, continuous families of ∞^r transformations; in addition, we assume that the transformations of G are mutually inverse by pairs.

According to Theorem 57 in the preceding section, there is in the group G one and only one r-term group generated by r independent infinitesimal transformations. Now, if S is the symbol of the general transformation of this r-term group and T is the symbol of an arbitrary transformation of G, then in the same way:

$$T^{-1}ST$$

is the symbol of the general infinitesimal transformation of an r-term group generated by infinitesimal transformations. Since this new group is contained in G, it must coincide with the group of all S; according to the terminology [TERMINOLOGIE] introduced in Chap. 15, p. 273, we can also express this as: the discussed r-term group remains invariant by every transformation T. Thus the following holds true.

Theorem 58. *If a group G with pairwise inverse transformations consists of several families of transformations, then the largest group generated by infinitesimal transformations which is contained in G remains invariant by every transformation of G.*

From this theorem, it follows how one can construct groups which consist of several continuous families of ∞^r transformations.

Let $X_1 f, \dots, X_r f$ be an r-term group in the variables $x_1, \dots, x_n$ and again let S be the symbol of the general transformation of this group.

Now, when a group with pairwise inverse transformations contains all transformations of the r-term group $X_1 f, \dots, X_r f$, and in addition contains a finite number, say $m-1$, of discrete families of ∞^r transformations, then in consequence of Theorem 56, p. 328, it possesses the form:

$$T_0 S, \;\; T_1 S, \;\; \dots, \;\; T_{m-1} S. \tag{10}$$

Here, T_0 means the identity transformation and $T_1, \dots, T_{m-1}$ are, according to the last theorem, constituted in such a way that their totality leaves invariant all transformations S. This property of the T_ν is expressed analytically by the fact that for every transformation S_k of the group $X_1 f, \dots, X_r f$, a relation of the form:

$$T_\nu^{-1} S_k T_\nu = S_j$$

exists, where the transformation S_j again belongs to the group $X_1 f, \dots, X_r f$. Incidentally, such a relation also holds true when ν is equal to zero, for indeed one then has $S_k = S_j$.

Since $T_1, \dots, T_{m-1}$ themselves belong to the transformations (10), the totality of all transformations (10) can then be a group only when all transformations $T_\mu T_\nu$ also belong to this totality. Hence, aside from the above relations, the T_i must also satisfy relations of the form:

$$T_\mu\, T_\nu = T_\pi\, S_\tau,$$

where μ and ν denote arbitrary numbers amongst the numbers $1, 2, \dots, m-1$, while π runs through the values $0, 1, \dots, m-1$. If one of the numbers μ, ν, say μ, is equal to zero, then actually, there is already a relation of the form indicated, since indeed $T_\pi = T_\nu$ and S_τ is, like T_0, the identity transformation.

On the other hand, if the transformations T_μ possess the properties indicated, then for all values $0, 1, \dots, m-1$ of the two numbers μ and ν, there exist relations of the form:

$$T_\mu\, S_k\, T_\nu\, S_l = T_\mu\, T_\nu\, S_j\, S_l = T_\pi\, S_\tau\, S_j\, S_l = T_\pi\, S_\rho,$$

and therefore, the totality of all transformations (10) forms a group. It is easy to see that in any case and in general, the transformations of such a group order as inverses by pairs.

At present, it is yet to be asked how the transformations T_μ must be constituted in order that the m families (10) are all distinct from each other.

Obviously, all the families (10) are distinct from each other when no transformation of the group belongs simultaneously to two of these families. On the other hand, if any two of these families, say: $T_\mu\, S$ and $T_\nu\, S$, have a transformation in common, then they are identical, because from the existence of a relation of the form:

$$T_\mu\, S_k = T_\nu\, S_l,$$

it immediately follows that:

$$T_\nu = T_\mu\, S_k\, S_l^{-1},$$

hence the family of the transformations $T_\nu\, S$ has the form:

$$T_\mu\, S_k\, S_l^{-1}\, S,$$

that is to say, it is identical to the family: $T_\mu\, S$.

Thus, for the m families (10) to be distinct from each other, it is necessary and sufficient that no two of the transformations $T_0, T_1, \dots, T_{m-1}$ be linked by a relation of the form:

$$T_\nu = T_\mu\, S_j \qquad (\nu \neq \mu).$$

We summarize the result obtained in the

Theorem 59. *If S is the symbol of the general transformation of the r-term group $X_1 f, \dots, X_r f$, and if moreover, $T_1, \dots, T_{m-1}$ are transformations which leave invariant the group $X_1 f, \dots, X_r f$ and which, in addition, are linked together and jointly with the identity transformation T_0 by relations of the form:*

$$T_\mu\, T_\nu = T_\pi\, S,$$

but not by relations of the form:

$$T_\nu = T_\mu\, S_j \qquad (\nu \neq \mu),$$

then the totality of all transformations:

$$T_0 S, \ T_1 S, \ \dots, \ T_{m-1} S$$

form a group with pairwise inverse transformations which consists of m discrete continuous families of ∞^r transformations and which, at the same time, contains all transformations of the group $X_1 f, \dots, X_r f$. If one chooses the transformations $T_1, \dots, T_{m-1}$ in all possible ways, then one obtains all groups having the constitution indicated.

In the next chapter, we give a general method for the determination of all transformations which leave invariant a given group $X_1 f, \dots, X_r f$.

If one has two different systems of transformations $T_1, \dots, T_{m-1}$, say $T_1, \dots, T_{m-1}$ and $T_1', \dots, T_{m-1}'$, then obviously, the two groups:

$$\begin{aligned} &T_0 S, \ T_1 S, \ \dots, \ T_{m-1} S \\ &T_0 S, \ T_1', \ \ \dots, \ T_{m-1}' S \end{aligned}$$

are always distinct when and only when it is not possible to represent each one of the transformations $T_1', \dots, T_{m-1}'$ in the form:

$$T_\mu' = T_{i_\mu} S_{k_\mu}.$$

Needless to say, one can frequently arrange that the m transformations $T_0, T_1, \dots, T_{m-1}$ already form a discontinuous group for themselves.

Example. The n infinitesimal transformations:

$$\frac{\partial f}{\partial x_1}, \ \dots, \ \frac{\partial f}{\partial x_n}$$

generate an r-term group. The totality of all transformations that leave this group invariant forms a finite continuous group which is generated by the $n+n^2$ infinitesimal transformations:

$$\frac{\partial f}{\partial x_i}, \quad x_i \frac{\partial f}{\partial x_k} \qquad (i,\, k = 1 \cdots n)$$

Now, if amongst the ∞^{nn} transformations:

$$x_i' = a_{i1} x_1 + \cdots + a_{in} x_n \qquad (i = 1 \cdots n)$$

of the group:

$$x_i \frac{\partial f}{\partial x_k} \qquad (i,\, k = 1 \cdots n),$$

one chooses m arbitrary transformations that form a discontinuous group as transformations $T_0, T_1, \dots, T_{m-1}$, and if one sets for S the general transformation:

$$x_1' = x_1 + a_1, \ \ldots, \ x_n' = x_n + a_n$$

of the group $\partial f / \partial x_1, \ldots, \partial f / \partial x_n$, then one always obtains a group which consists of m discrete families and which comprises all ∞^n transformations of the group $\partial f / \partial x_1, \ldots, \partial f / \partial x_n$.

Naturally, Theorem 58, p. 332, also holds true when the group G consists of infinitely many families of ∞^r transformations. Hence if one wants to construct such a group, one only has to seek infinitely many discrete transformations:

$$T_1, \, T_2, \, \ldots$$

which leave invariant the group $X_1 f, \ldots, X_r f$ and which, in addition, satisfy pairwise relations of the form:

$$T_\mu \, T_\nu = T_\pi \, S_\tau,$$

but by contrast, which are neither mutually, nor together with the identity transformation, linked by relations of the form:

$$T_\mu = T_\nu \, S_j.$$

The totality of all transformations:

$$T_0 \, S, \ T_1 \, S, \ T_2 \, S, \ \ldots$$

then forms a group G which comprises the group $X_1 f, \ldots, X_r f$ and which consists of infinitely many different families of ∞^r transformations.

But the transformations of the group G found in this way are in general not ordered as inverses by pairs; in order that they enjoy this property, each one of the transformations $T_1, T_2, \ldots$, must also satisfy, aside from the relations indicated above, relations of the form:

$$T_\mu^{-1} = T_{k_\mu} \, S_{j_\mu}.$$

§ 87. Now, we make a few observations about the *invariant* subgroups as we have considered them in the previous three sections.

Let G be a group which consists of the m discrete families:

$$x_i^{(k)} = f_i^{(k)}\big(x_1, \ldots, x_n, a_1^{(k)}, \ldots, a_r^{(k)}\big) \qquad {\scriptstyle (i=1\cdots n)} \atop {\scriptstyle (k=1\cdots m)}$$

of ∞^r transformations and which, for each of its transformations, also contains the inverse transformation. In particular, let:

$$x_i' = f_i^{(1)}(x_1, \ldots, x_n, a_1, \ldots, a_r) \qquad {\scriptstyle (i=1\cdots n)}$$

be the r-term group generated by r infinitesimal transformations: $X_1 f, \dots, X_r f$ which is contained in G.

In accordance with Chap. 6, p. 111, we say that every function which admits all transformations of G, hence which satisfies the m equations:

$$\mathfrak{U}\big(x_1^{(k)}, \dots, x_n^{(k)}\big) = \mathfrak{U}(x_1, \dots, x_n) \qquad (k=1 \cdots m),$$

is an *invariant* of G. We want to show how one can find the invariants of G.

Evidently, every invariant of G is at the same time an invariant of the r-term group $X_1 f, \dots, X_r f$ and therefore, it is a solution of the complete system which is determined by the equations:

$$X_1 f = 0, \;\dots,\; X_r f = 0.$$

If this complete system is n-term, then the group $X_1 f, \dots, X_r f$ and therefore also the group G possess in fact no invariant. So, we assume that the said complete system is $(n-q)$-term and we denote by $u_1, \dots, u_q$ any q of its solutions that are independent.

According to Theorem 58, the group $X_1 f, \dots, X_r f$ remains invariant by all transformations of G; consequently, the $(n-q)$-term complete system determined by the equations:

$$X_1 f = 0, \;\dots,\; X_r f = 0$$

also admits all transformations of G. Consequently (cf. Chap. 8, p. 153), the solutions $u_1, \dots, u_q$ of this complete system satisfy relations of the form:

$$u_j\big(x_1^{(k)}, \dots, x_n^{(k)}\big) = \omega_j^{(k)}\big(u_1(x), \dots, u_q(x), a_1^{(k)}, \dots, a_r^{(k)}\big) \tag{11}$$

$$(j=1 \cdots q;\; k=1 \cdots m).$$

It can be shown here that the functions $\omega_j^{(k)}$ are all free of the parameters $a_1^{(k)}, \dots, a_r^{(k)}$.

By $\overline{a}_1^{(k)}, \dots, \overline{a}_r^{(k)}$, we want to denote an arbitrary fixed system of values. If the system of values $a_1^{(k)}, \dots, a_r^{(k)}$ lies in a certain neighborhood of $\overline{a}_1^{(k)}, \dots, \overline{a}_r^{(k)}$, then according to Theorem 56, p. 328, the transformation:

$$x_i^{(k)} = f_i^{(k)}\big(x_1, \dots, x_n, a_1^{(k)}, \dots, a_r^{(k)}\big)$$

can be obtained by executing firstly the transformation:

$$\overline{x}_i^{(k)} = f_i^{(k)}\big(x_1, \dots, x_n, \overline{a}_1^{(k)}, \dots, \overline{a}_r^{(k)}\big)$$

and then, a certain transformation:

$$x_i^{(k)} = f_i^{(1)}\big(\overline{x}_1^{(k)}, \dots, \overline{x}_n^{(k)}, \alpha_1, \dots, \alpha_r\big)$$

of the group $X_1 f, \ldots, X_r f$. So we have:

$$x_i^{(k)} = f_i^{(1)}\big(f_1^{(k)}(x, \overline{a}^{(k)}),\; \ldots,\; f_n^{(k)}(x, \overline{a}^{(k)}),\; \alpha_1, \ldots, \alpha_r\big).$$

Now, $u_1, \ldots, u_q$ are invariants of the group $X_1 f, \ldots, X_r f$ and therefore, they satisfy relations of the form:

$$u_j\big(x_1^{(k)}, \ldots, x_n^{(k)}\big) = u_j\big(\overline{x}_1^{(k)}, \ldots, \overline{x}_n^{(k)}\big).$$

On the other hand, we have:

$$u_j\big(\overline{x}_1^{(k)}, \ldots, \overline{x}_n^{(k)}\big) = \overline{\omega}_j^{(k)}\big(u_1(x), \ldots, u_q(x)\big),$$

where the $\overline{\omega}_j^{(k)}$ depend only upon $u_1, \ldots, u_q$ and contain no arbitrary parameters, since $\overline{a}_1^{(k)}, \ldots, \overline{a}_r^{(k)}$ are numerical constants, indeed. Thus we obtain:

$$u_j\big(x_1^{(k)}, \ldots, x_n^{(k)}\big) = \overline{\omega}_j^{(k)}\big(u_1(x), \ldots, u_q(x)\big), \tag{11'}$$

and with this, it is proved that the functions $\omega_j^{(k)}$ in the equations (11) are effectively free of the parameters $a_1^{(k)}, \ldots, a_r^{(k)}$.

According to what has bee said above, every invariant $\mathfrak{U}(x_1, \ldots, x_n)$ of the group G satisfies m equations of the form:

$$\mathfrak{U}\big(x_1^{(k)}, \ldots, x_n^{(k)}\big) = \mathfrak{U}(x_1, \ldots, x_n) \qquad (k=1 \cdots m);$$

but since it is in addition a function of only $u_1, \ldots, u_q$, say:

$$\mathfrak{U}(x_1, \ldots, x_n) = J(u_1, \ldots, u_q),$$

then at the same time, it satisfies the m relations:

$$J\big(\omega_1^{(k)}(u_1, \ldots, u_q),\; \ldots,\; \omega_q^{(k)}(u_1, \ldots, u_q)\big) = J(u_1, \ldots, u_q) \qquad (k=1 \cdots m). \tag{12}$$

Conversely, every function $J(u_1, \ldots, u_q)$ that satisfies the m functional equations just written is obviously an invariant of the group G. Consequently, in order to find all invariants of G, we only need to fulfill these functional equations in the most general way.

The problem of determining all solutions of the functional equations (12) is visibly identical to the problem of determining all functions of $u_1, \ldots, u_q$ which admit the m transformations:

$$u_j' = \omega_j^{(k)}(u_1, \ldots, u_q) \qquad (j=1 \cdots q) \quad (k=1 \cdots m). \tag{13}$$

But these m transformations form a discontinuous group, as one realizes without difficulty from the group property [GRUPPENEIGENSCHAFT] of G. Thus, the problem stated at the outset of the section leads back to a problem of the theory of discontinuous groups.

We summarize the obtained result in the

Theorem 60. *If a group G (whose transformations are pairwise inverse) consists of several discrete families of ∞^r transformations:*

$$x_i^{(k)} = f_i^{(k)}\big(x_1,\dots,x_n,\, a_1^{(k)},\, \dots,\, a_r^{(k)}\big) \qquad {\scriptstyle (i=1\cdots n)} \qquad {\scriptstyle (k=1,2\cdots)},$$

then all invariants of G are at the same time also invariants of the r-term continuous group: $X_1 f,\dots,X_r f$ determined by G. If one knows the invariants of the latter group, hence if one knows any q arbitrary independent solutions $u_1,\dots,u_q$ of the $(n-q)$-term complete system which is determined by the equations:

$$X_1 f = 0,\ \dots,\ X_r f = 0,$$

then one finds the invariants of G in the following way: one first forms the relations:

$$u_j\big(x_1^{(k)},\dots,x_n^{(k)}\big) = \omega_j^{(k)}\big(u_1(x),\dots,u_q(x)\big) \qquad {\scriptstyle (j=1\cdots q)} \qquad {\scriptstyle (k=1,2\cdots)},$$

which, under the assumptions made, exist, and in which the $\omega_j^{(k)}$ depend only upon the two indices j and k; afterwards, one determines all functions of $u_1,\dots,u_q$ which admit the discontinuous group formed by the transformations:

$$u_j' = \omega_j^{(k)}(u_1,\dots,u_q) \qquad {\scriptstyle (j=1\cdots q)} \qquad {\scriptstyle (k=1,2\cdots)}.$$

The concerned functions are the invariants of the group G.

A similar theorem clearly holds true when the group G consists of infinitely many continuous families of transformations.

One can propose to oneself the problem of finding all invariants that a given family of transformations:

$$x_i' = \varphi_i(x_1,\dots,x_n,\, a_1,\dots,a_r) \qquad {\scriptstyle (i=1\cdots n)}$$

possesses, or all invariants that are common to several such families.[3] The family, respectively the families, can here be completely arbitrary and they need not belong to a finite group.

---

[3] Cf. LIE, Berichte der K. Sächs. Ges. d. W., 1. August 1887.

We do not intend to treat this problem exhaustively; let it only be remarked that the concerned invariants are solutions, though not arbitrary solutions, of a certain complete system that can easily be indicated. Indeed, since the sought invariants, aside from the given transformations, also obviously admit the associated inverse transformations, then one can very easily set up certain infinitesimal transformations by which they also remain invariant. In general, these infinitesimal transformations contain arbitrary elements, and in particular, certain parameters; when set equal to zero, these infinitesimal transformations provide linear partial differential equations which must be satisfied by the sought invariants. Now, it is always possible to set up the smallest complete system which embraces all these differential equations. If one knows a system of solutions $u_1, u_2, \dots$ of this complete system, then one forms an arbitrary function $\Omega(u_1, u_2, \dots)$ of them, one executes on it the general transformation of the given family and one determines Ω in the most general way in order that Ω behaves invariantly.

References

1. Engel, F.; Lie, S.: *Theorie der transformationsgruppen. Erster Abschnitt.* Unter Mitwirkung von Dr. Friedrich Engel, bearbeitet von Sophus Lie, Verlag und Druck von B.G. Teubner, Leipzig und Berlin, xii+638 pp. (1888). Reprinted by Chelsea Publishing Co., New York, N.Y. (1970)

Chapter 19
Theory of the Similarity [AEHNLICHKEIT] of r-term Groups

It is often of the utmost importance to answer the question whether a given r-term group $x_i' = f_i(x_1, \dots, x_s, a_1, \dots, a_r)$ of s-times extended space is *similar* [ÄHNLICH] to another given r-term group $y_i' = F_i(y_1, \dots, y_s, b_1, \dots, b_r)$ of the same space, hence whether one can introduce, in place of the x and of the a, new *variables*: $y_1, \dots, y_s$ and new *parameters*: $b_1, \dots, b_r$ so that the first group converts into the second group (Chap. 3, p. 29). If one knows, in a given case, that such a transfer of one of the groups to the other is possible, then a second question raises itself: how can one accomplish the concerned transfer in the most general way?

In the present chapter, we provide means for answering the two questions.

To begin with, we show that the first of the two questions can be replaced by the following more simple question: under which conditions does there exist a transformation:

$$y_i = \Phi(x_1, \dots, x_s) \qquad (i=1\cdots s)$$

of such a nature that r arbitrary independent infinitesimal transformations of the group $x_i' = f_i(x,a)$ are transferred to infinitesimal transformations of the group $y_i' = F_i(y,b)$ by the introduction of the variables $y_1, \dots, y_s$? We settle this simpler question by setting up certain conditions which are necessary for the existence of a transformation $y_i = \Phi_i(x)$ of the demanded constitution, and which also prove to be sufficient. At the same time, we shall see that all possibly existing transformations $y_i = \Phi_i(x)$ of the demanded constitution can be determined by integrating complete systems. With that, the second of the two questions stated above will then also be answered.

§ 88. Let the two r-term groups: $x_i' = f_i(x,a)$ and $y_i' = F_i(y,b)$ be similar to each other and, to be precise, let the first be transferred to the second when the new variables $y_i = \Phi_i(x_1, \dots, x_s)$ are introduced in place of $x_1, \dots, x_s$, and when the new parameters $b_k = \beta_k(a_1, \dots, a_r)$ are introduced in place of $a_1, \dots, a_r$. Furthermore, let:

S. Lie, *Theory of Transformation Groups I*,
DOI 10.1007/978-3-662-46211-9_19

$$X_k f = \sum_{i=1}^{s} \xi_{ki}(x_1,\dots,x_s)\,\frac{\partial f}{\partial x_i} \qquad (k=1\cdots r)$$

be r arbitrary independent infinitesimal transformations of the group $x_i' = f_i(x,a)$; by the introduction of the new variables $y_1,\dots,y_s$, let them take the form:

$$X_k f = \sum_{i=1}^{s} X_k y_i\,\frac{\partial f}{\partial y_i} = \sum_{i=1}^{s} \mathfrak{y}_{ki}(y_1,\dots,y_s)\,\frac{\partial f}{\partial y_i} = \mathfrak{Y}_k f$$
$$(k=1\cdots r).$$

We decompose the transition from the group $x_i' = f_i(x,a)$ to the group $y_i' = F_i(y,b)$ in a series of steps.

First, we bring the group $x_i' = f_i(x,a)$ to the canonical form:

$$\text{(1)} \qquad x_i' = x_i + \sum_{k=1}^{r} e_k\,\xi_{ki}(x_1,\dots,x_s) + \cdots \qquad (i=1\cdots s)$$

by introducing for $a_1,\dots,a_r$ certain (needless to say) perfectly determined functions of $e_1,\dots,e_r$. Then evidently, we must obtain from (1) the equations $y_i' = F_i(y,b)$ when we introduce the new variables $y_1,\dots,y_s$ in place of the x, and when in addition, we insert for $e_1,\dots,e_r$ some completely determined functions of $b_1,\dots,b_r$.

But now, according to Chap. 4, p. 71, after the introduction of the variables $y_1,\dots,y_s$, the equations (1) take the form:

$$\text{(1')} \qquad y_i' = y_i + \sum_{k=1}^{r} e_k\,\mathfrak{y}_{ki}(y_1,\dots,y_s) + \cdots \qquad (i=1\cdots s),$$

hence the equations (1') must become identical to the equations $y_i' = F_i(y,b)$ when one expresses $e_1,\dots,e_r$ in terms of $b_1,\dots,b_r$ in the indicated way. Consequently, the equations (1') are a form of the group $y_i' = F_i(y,b)$, and to be precise, a canonical form, as an examination teaches us [WIE DER AUGENSCHEIN LEHRT]. In other words: $\mathfrak{Y}_1 f,\dots,\mathfrak{Y}_r f$ are independent infinitesimal transformations of this group.

Thus, if after the introduction of the new variables $y_1,\dots,y_s$ and of the new parameters $b_1,\dots,b_r$, the r-term group $x_i' = f_i(x_1,\dots,x_s,\,a_1,\dots,a_r)$ converts into the group $y_i' = F_i(y_1,\dots,y_s,\,b_1,\dots,b_r)$, then the infinitesimal transformations of the first group convert into the infinitesimal transformations of the second group, also after the introduction of the variables $y_1,\dots,y_s$.

Obviously, the converse also holds true: if the two groups $x_i' = f_i(x,a)$ and $y_i' = F_i(y,b)$ stand in the mutual relationship that the infinitesimal transformations of the first are transferred to the infinitesimal transformations of the second after the introduction of the new variables $y_1,\dots,y_s$, then one can always transfer the group $x_i' = f_i(x,a)$ to the group $y_i' = F_i(y,b)$ by means of appropriate choices of the variables and of the parameters. Indeed, by the introduction of the y, the canonical form (1) of the group $x_i' = f_i(x,a)$ is transferred to the canonical form (1') of the group $y_i' = F_i(y,b)$.

As a result, we have the

Theorem 61. *Two r-term groups in the same number of variables are similar to each other if and only if it is possible to transfer any r arbitrary independent infinitesimal transformations of the first to infinitesimal transformations of the second by the introduction of new variables.*

When the question is to examine whether the two r-term groups $x_i' = f_i(x,a)$ and $y_i = F_i(y,b)$ are similar to each other, then one needs only to consider [INS AUGE FASSEN] the infinitesimal transformations of the two groups and to ask whether they can be transferred to each other.

From the above, one can immediately draw another more important conclusion.

We know that between the infinitesimal transformations $X_1 f, \dots, X_r f$ of the group $x_i' = f_i(x,a)$, there exist relations of the form:

$$X_i\big(X_k(f)\big) - X_k\big(X_i(f)\big) = [X_i, X_k] = \sum_{\sigma=1}^{r} c_{ik\sigma} X_\sigma f.$$

According to Chap. 5, Proposition 2, p. 100, after the introduction of the new variables $y_1, \dots, y_s$, these relations receive the form:

$$\mathfrak{Y}_i\big(\mathfrak{Y}_k(f)\big) - \mathfrak{Y}_k\big(\mathfrak{Y}_i(f)\big) = [\mathfrak{Y}_i, \mathfrak{Y}_k] = \sum_{\sigma=1}^{r} c_{ik\sigma} \mathfrak{Y}_\sigma f,$$

so the r independent infinitesimal relations $\mathfrak{Y}_1 f, \dots, \mathfrak{Y}_r f$ of the group $y_i' = F_i(y,b)$ are linked together by exactly the same relations as the infinitesimal transformations $X_1 f, \dots, X_r f$ of the group $x_i' = f_i(x,a)$.

According to the terminology introduced in Chap. 17, p. 303 and 305, we can therefore say:

Theorem 62. *If two r-term groups in the same number of variables are similar to each other, then they are also equally composed, or, what is the same, holoedrically isomorphic.*

At the same time, it is clear that the transformation $y_i = \Phi_i(x_1, \dots, x_s)$ establishes a holoedrically isomorphic relationship between the two groups: $x_i' = f_i(x,a)$ and $y_i' = F_i(y,b)$ since it associates to the r independent infinitesimal transformations $X_1 f, \dots, X_r f$ of the first group the r independent transformations $\mathfrak{Y}_1 f, \dots, \mathfrak{Y}_r f$ of the other, and since through this correspondence, the two groups are obviously related to each other in a holoedrically isomorphic way.

§ 89. At present, we imagine that two arbitrary r-term groups in the same number of variables are presented; let their infinitesimal transformations be:

$$X_k f = \sum_{i=1}^{s} \xi_{ki}(x_1, \dots, x_s) \frac{\partial f}{\partial x_i} \qquad (k=1\cdots r)$$

and:

$$Z_k f = \sum_{i=1}^{s} \zeta_{ki}(x_1,\ldots,x_s)\,\frac{\partial f}{\partial y_i} \qquad (k=1\cdots r).$$

We ask: are these two groups similar to each other, or not?

Our answer to this question must obviously fall in the negative sense as soon as the two groups are not equally composed; indeed, according to Theorem 62, only equally composed groups can be similar to each other. Consequently, we need only to occupy ourselves with the case when the two groups are equally composed; the question of whether this case really happens may always be settled thanks to an algebraic discussion [ALGEBRAISCHE DISCUSSION].

Accordingly, we assume from now on that the two presented groups are *equally composed*.

According to Theorem 61, the two equally composed groups $X_1 f,\ldots,X_r f$ and $Z_1 f,\ldots,Z_r f$ are similar to each other if and only if there is a transformation $y_i = \Phi_i(x_1,\ldots,x_s)$ of such a constitution that $X_1 f,\ldots,X_r f$ convert into infinitesimal transformations of the group $Z_1 f,\ldots,Z_r f$ after the introduction of the new variables $y_1,\ldots,y_s$. If we combine this with the remark at the end of the previous section, we realize that the two groups are similar to each other if and only if a holoedrically isomorphic relation can be established between them so that it is possible, by the introduction of appropriate new variables: $y_i = \Phi_i(x_1,\ldots,x_s)$ to transfer the r infinitesimal transformations $X_1 f,\ldots,X_r f$ precisely to the infinitesimal transformations $\mathfrak{Y}_1 f,\ldots,\mathfrak{Y}_r f$ of the group $Z_1 f,\ldots,Z_r f$ that are associated through the isomorphic relation.

The next step that we must climb towards the answer to the stated question is therefore to relate the two groups in the most general holoedrically isomorphic way.

In the group $Z_1 f,\ldots,Z_r f$, we choose in the most general way r independent infinitesimal transformations:

$$Y_k f = \sum_{j=1}^{r} g_{kj} Z_j f \qquad (k=1\cdots r)$$

such that together with the relations:

$$[X_i,\,X_k] = \sum_{\sigma=1}^{r} c_{ik\sigma} X_\sigma f, \tag{2}$$

there are at the same time the relations:

$$[Y_i,\,Y_k] = \sum_{\sigma=1}^{r} c_{ik\sigma} Y_\sigma f. \tag{2'}$$

If this occurs — and only algebraic operations are required for it —, then we associate the infinitesimal transformations $Y_1 f,\ldots,Y_r f$ to $X_1 f,\ldots,X_r f$, respectively, and we obtain that the two groups are holoedrically isomorphically related to each other in the most general way.

In the present chapter, by the g_{kj}, we understand everywhere *the most general system of constants which satisfies the requirement just stated.*

Here, it is to be remarked that in general, the g_{kj} depend upon arbitrary elements, once upon arbitrary parameters and then upon certain arbitrarinesses [WILLKÜRLICHKEITEN] which are caused by the algebraic operations that are necessary for the determination of the g_{kj}; indeed, it is conceivable that there are several discrete families of systems of values g_{kj} which are of the demanded constitution.

At present, the question is whether, amongst the reciprocal isomorphic relationships between the two groups found in this way, there is one which has the property indicated above. In other words: is it possible to specialize the arbitrary elements which occur in the coefficients g_{kj} in such a way that $X_1f,\dots,X_rf$ can be transferred to $Y_1f,\dots,Y_rf$, respectively, by means of the introduction of appropriate new variables: $y_i = \Phi_i(x_1,\dots,x_s)$?

When, but only when, this question has been answered, one may conclude that the two groups $X_1f,\dots,X_rf$ and $Z_1f,\dots,Z_rf$ are similar to each other.

Let $X_1f,\dots,X_nf$ ($n \leqslant r$) be linked together by no linear relation of the form:

$$\chi_1(x_1,\dots,x_s)\,X_1f+\cdots+\chi_n(x_1,\dots,x_s)\,X_nf=0,$$

while by contrast, $X_{n+1}f,\dots,X_rf$ can be linearly expressed in terms of $X_1f,\dots,X_nf$:

$$X_{n+k}f \equiv \varphi_{k1}(x_1,\dots,x_s)\,X_1f+\cdots+\varphi_{kn}(x_1,\dots,x_s)\,X_nf \qquad (3)$$
$$(k=1\cdots r-n).$$

Now, if the transformation $y_i = \Phi_i(x_1,\dots,x_s)$ is constituted in such a way that, after introducing the y, $X_1f,\dots,X_rf$ are transferred to certain infinitesimal transformations $\mathfrak{Y}_1f,\dots,\mathfrak{Y}_rf$ which belong to the group $Z_1f,\dots,Z_rf$, then naturally, $\mathfrak{Y}_1f,\dots,\mathfrak{Y}_nf$ are not linked together by a linear relation of the form:

$$\psi_1(y_1,\dots,y_s)\,\mathfrak{Y}_1f+\cdots+\psi_n(y_1,\dots,y_s)\,\mathfrak{Y}_nf=0;$$

by contrast, we visibly obtain for $\mathfrak{Y}_{n+1}f,\dots,\mathfrak{Y}_rf$ expressions of the form:

$$\mathfrak{Y}_{n+k}f \equiv \sum_{\nu=1}^{n} \overline{\varphi}_{k\nu}(y_1,\dots,y_s)\,\mathfrak{Y}_\nu f \qquad (k=1\cdots r-n),$$

in which the $\overline{\varphi}_{k\nu}(y)$ come into existence after the introduction of the variables y in place of the x, so that the $n(r-n)$ equations:

$$\overline{\varphi}_{k\nu}(y_1,\dots,y_s)=\varphi_{k\nu}(x_1,\dots,x_s)$$

are hence identities after the substitution: $y_i = \Phi_i(x_1,\dots,x_s)$.

From this, we conclude that:

If there is no linear relation between $X_1f,\dots,X_nf$, while $X_{n+1}f,\dots,X_rf$ can be expressed linearly in terms of $X_1f,\dots,X_nf$ by virtue of the relations (3), *then the two equally composed groups $X_1f,\dots,X_rf$ and $Z_1f,\dots,Z_rf$ can be similar only when* *the arbitrary elements in the coefficients g_{kj} defined above can be chosen in such a way that the infinitesimal transformations:*

$$Y_k f = \sum_{j=1}^{r} g_{kj} Z_j f \qquad (k=1\cdots r)$$

possess the following properties: firstly, $Y_1 f, \dots, Y_n f$ are linked by no linear relation, while by contrast, $Y_{n+1} f, \dots, Y_r f$ are linearly expressible in terms of $Y_1 f, \dots, Y_n f$:

$$\text{(3')} \qquad Y_{n+k} f \equiv \sum_{\nu=1}^{n} \psi_{k\nu}(y_1, \dots, y_s)\, Y_\nu f \qquad (k=1\cdots r-n),$$

and secondly, the $n(r-n)$ equations:

$$\text{(4)} \qquad \varphi_{k\nu}(x_1, \dots, x_s) - \psi_{k\nu}(y_1, \dots, y_s) = 0 \qquad (k=1\cdots r-n;\ \nu=1\cdots n),$$

are compatible with each other and they give relations neither between the x alone nor between the y alone.

These conditions are *necessary* for the similarity between the two equally composed groups $X_1 f, \dots, X_r f$ and $Z_1 f, \dots, Z_r f$. We claim that the same conditions are *sufficient*. Put more precisely, we claim: *when the said conditions are satisfied, then there is always a transformation $y_i = \Phi_i(x_1, \dots, x_s)$ which transfers the infinitesimal transformations $X_1 f, \dots, X_r f$ to, respectively, $Y_1 f, \dots, Y_r f$, so the two groups $X_1 f, \dots, X_r f$ and $Z_1 f, \dots, Z_r f$ are similar to each other.*

The proof of this claim will be produced as we develop a method which leads to the determination of a transformation having the indicated constitution.

Our more present standpoint is therefore the following:

In the s variables $x_1, \dots, x_s$, let an r-term group:

$$X_k f = \sum_{i=1}^{s} \xi_{ki}(x_1, \dots, x_s) \frac{\partial f}{\partial x_i} \qquad (k=1\cdots r)$$

be presented, the composition of which is determined by the relations:

$$[X_i, X_k] = \sum_{\sigma=1}^{r} c_{ik\sigma} X_\sigma f.$$

Between $X_1 f, \dots, X_n f$, there is at the same time no linear relation of the form:

$$\chi_1(x_1, \dots, x_s) X_1 f + \cdots + \chi_n(x_1, \dots, x_s) X_n f = 0,$$

while by contrast $X_{n+1} f, \dots, X_r f$ express themselves in terms of $X_1 f, \dots, X_n f$:

$$\text{(3)} \qquad X_{n+k} f \equiv \sum_{\nu=1}^{n} \varphi_{k\nu}(x_1, \dots, x_s)\, X_\nu f \qquad (k=1\cdots n-r).$$

Furthermore, let us be given an r-term group:

$$Z_k f = \sum_{i=1}^{s} \zeta_{ki}(y_1, \dots, y_s) \frac{\partial f}{\partial y_i} \qquad (k=1 \cdots r)$$

which is equally composed with the group $X_1 f, \dots, X_r f$, and let r independent infinitesimal transformations:

$$Y_k f = \sum_{j=1}^{r} \overline{g}_{kj} Z_j f = \sum_{i=1}^{s} \eta_{ki}(y_1, \dots, y_s) \frac{\partial f}{\partial y_i} \qquad (k=1 \cdots r)$$

in this group be chosen in such a way that firstly, the relations:

$$[Y_i, Y_k] = \sum_{\sigma=1}^{r} c_{ik\sigma} Y_\sigma f$$

are identically satisfied, and such that secondly, $Y_1 f, \dots, Y_n f$ are linked together by no relation of the form:

$$\psi_1(y_1, \dots, y_s) Y_1 f + \cdots + \psi_n(y_1, \dots, y_s) Y_n f = 0,$$

while by contrast $Y_{n+1} f, \dots, Y_r f$ express themselves in the following way:

$$\text{(3')} \qquad Y_{n+k} f \equiv \sum_{\nu=1}^{n} \psi_{k\nu}(y_1, \dots, y_s) Y_\nu f \qquad (k=1 \cdots n-r),$$

and such that, lastly, the $n(r-n)$ equations:

$$\text{(4)} \qquad \varphi_{k\nu}(x_1, \dots, x_s) - \psi_{k\nu}(y_1, \dots, y_s) = 0 \qquad (k=1 \cdots r-n;\ \nu=1 \cdots n)$$

are compatible with each other and give relations neither between the x alone, nor between the y alone.

To seek a transformation:

$$\text{(5)} \qquad y_i = \Phi_i(x_1, \dots, x_s) \qquad (i=1 \cdots s)$$

which transfers $X_1 f, \dots, X_r f$ to $Y_1 f, \dots, Y_r f$, respectively.

We can add:

The sought transformation is constituted in such a way that the equations (4) *are identities after the substitution:* $y_1 = \Phi_1(x), \dots, y_s = \Phi_s(x)$.

With these words, the problem which is to be settled is enunciated.

§ 90. Before we attack in its complete generality the problem stated at the end of the preceding section, we want to consider a special case, the settlement of which turns out to be substantially simpler; we mean the case $n = r$ which, obviously, can occur only when s is at least equal to r.

Thus, let the whole number n defined above be equal to r.

It is clear that in this case, neither between $X_1 f, \dots, X_r f$, nor between $Z_1 f, \dots, Z_r f$ does there exists a linear relation. From this, it follows that the r infinitesimal transformations:

$$Y_k f = \sum_{j=1}^{r} g_{kj} Z_j f \qquad (k = 1 \cdots r)$$

satisfy by themselves the conditions indicated on p. 346, without it being necessary to specialize further the arbitrary elements contained in the g_{kj}. Indeed, firstly there is no linear relation between $Y_1 f, \dots, Y_r f$, and secondly, the equations (4) reduce to the identity $0 = 0$, hence they are certainly compatible with each other and they produce relations neither between the x alone, nor between the y alone.

Thus, if the claim stated on p. 346 is correct, our two r-term groups must be similar, and to be precise, there must exist a transformation $y_i = \Phi_i(x_1, \dots, x_s)$ which transfers $X_1 f, \dots, X_r f$ to $Y_1 f, \dots, Y_r f$, respectively. We seek to determine such a transformation.

When we introduce the variables $y_1, \dots, y_s$ in $X_k f$ by means of the transformation:

$$\text{(5)} \qquad y_i = \Phi_i(x_1, \dots, x_s) \qquad (i = 1 \cdots s),$$

we obtain:

$$X_k f = \sum_{i=1}^{s} X_k y_i \frac{\partial f}{\partial y_i} = \sum_{i=1}^{s} X_k \Phi_i \frac{\partial f}{\partial y_i};$$

so by comparing to:

$$Y_k f = \sum_{i=1}^{s} \eta_{ki}(y_1, \dots, y_s) \frac{\partial f}{\partial y_i} = \sum_{i=1}^{s} Y_k y_i \frac{\partial f}{\partial y_i},$$

we realize that the transformation (5) transfers $X_1 f, \dots, X_r f$ to $Y_1 f, \dots, Y_r f$, respectively, if and only if the rs equations:

$$\text{(6)} \qquad Y_k y_i - X_k \Phi_i = 0 \qquad (k = 1 \cdots r;\ i = 1 \cdots s)$$

become identities after the substitution: $y_1 = \Phi_1(x), \dots, y_s = \Phi_s(x)$.

Now, if we set:

$$X_k f + Y_k f = \Omega_k f \qquad (k = 1 \cdots r),$$

we have:

$$\Omega_k(y_i - \Phi_i) = Y_k y_i - X_k \Phi_i;$$

thus, when the equations (5) represent a transformation having the constitution demanded, the rs expressions $\Omega_k(y_i - \Phi_i)$ all vanish by means of (5), or, what amounts to the same: the system of equations (5) admits the r infinitesimal transformations $\Omega_1 f,\dots,\Omega_r f$ (cf. Chap. 7, p. 124 sq.).

On the other hand, the following obviously holds true: every system of equations solvable with respect to $x_1,\dots,x_s$ of the form (5) which admits the r infinitesimal transformations $\Omega_1 f,\dots,\Omega_r f$ represents a transformation which transfers $X_1 f,\dots,X_r f$ to $Y_1 f,\dots,Y_r f$, respectively.

At present, if we take into account that the r infinitesimal transformations $\Omega_1 f,\dots,\Omega_r f$ stand pairwise in the relationships:

$$[\Omega_i,\,\Omega_k] = [X_i,\,X_k] + [Y_i,\,Y_k] = \sum_{\sigma=1}^{r} c_{ik\sigma}\,\Omega_\sigma f,$$

and therefore generate an r-term group in the $2s$ variables $x_1,\dots,x_s$, $y_1,\dots,y_s$, we can thus say:

The totality of all transformations (5) *which transfer $X_1 f,\dots,X_r f$ to $Y_1 f,\dots,Y_r f$, respectively, is identical to the totality of all systems of equations solvable with respect to $x_1,\dots,x_n$ of the form* (5) *which admit the r-term group $\Omega_1 f,\dots,\Omega_r f$.*

The determination of a transformation of the discussed constitution therefore leads back to the determination of a certain system of equations in the $2s$ variables $x_1,\dots,x_s$, $y_1,\dots,y_s$; this system of equations must possess the following properties: *it must consist of s independent equations, it must be solvable both with respect to $x_1,\dots,x_s$ and with respect to $y_1,\dots,y_s$, and lastly, it must admit the r-term group $\Omega_1 f,\dots,\Omega_r f$.*

For the resolution of this new problem, we can base ourselves on the developments of Chap. 14.

Under the assumptions made, the $r\times r$ determinants of the matrix:

$$\begin{vmatrix} \xi_{11}(x) & \cdot & \cdot & \xi_{1s}(x) \\ \cdot & \cdot & \cdot & \cdot \\ \xi_{r1}(x) & \cdot & \cdot & \xi_{rs}(x) \end{vmatrix}$$

do not all vanish identically and even less all $r\times r$ determinants of the matrix:

$$\tag{7} \begin{vmatrix} \xi_{11}(x) & \cdot & \cdot & \xi_{1s}(x) & \eta_{11}(y) & \cdot & \cdot & \eta_{1s}(y) \\ \cdot & \cdot & \cdot & \cdot & \cdot & \cdot & \cdot & \cdot \\ \xi_{r1}(x) & \cdot & \cdot & \xi_{rs}(x) & \eta_{r1}(y) & \cdot & \cdot & \eta_{rs}(y) \end{vmatrix}.$$

Hence when a system of equations brings to zero all $r\times r$ determinants of the matrix (7), it must necessarily contain relations between the x alone.

Our problem is the determination of a system of equations of the form:

$$\tag{8} y_1 - \Phi_1(x_1,\dots,x_s) = 0, \ \dots, \ y_s - \Phi_s(x_1,\dots,x_s) = 0$$

which admits the group $X_k f + Y_k f$, and which is at the same time solvable with respect to $x_1,\dots,x_s$.

A system of equations of this nature certainly contains no relation between $x_1,\dots,x_s$ only, so it does not bring to zero all $r\times r$ determinants of the matrix (7) and according to Theorem 17 in Chap. 7, p. 137, it can be brought to a such a form that it contains only relations between the solutions of the complete system:

$$\Omega_k f = X_k f + Y_k f = 0 \qquad (k=1\cdots r). \tag{9}$$

Consequently, our problem will be solved when we succeed in determining s independent relations between the solutions of this complete system that are solvable with respect to $x_1,\dots,x_s$ and with respect to $y_1,\dots,y_s$.

The complete system (9) possesses $2s-r$ independent solutions which we can evidently choose in such a way that $s-r$ of them, say:

$$u_1(x_1,\dots,x_s),\ \dots,\ u_{s-r}(x_1,\dots,x_s)$$

depend only upon the x, so that $s-r$ other solutions:

$$v_1(y_1,\dots,y_s),\ \dots,\ v_{s-r}(y_1,\dots,y_s)$$

depend only upon the y, while by contrast, the r remaining ones:

$$w_1(x_1,\dots,x_s,y_1,\dots,y_s),\ \dots,\ w_r(x_1,\dots,x_s,y_1,\dots,y_s)$$

must contain certain x and certain y as well. Here, the s functions $u_1,\dots,u_{s-r}$, $w_1,\dots,w_r$ are mutually independent as far as $x_1,\dots,x_s$ are concerned, and the s functions $v_1,\dots,v_{s-r}$, $w_1,\dots,w_r$ are so too, as far as $y_1,\dots,y_s$ are concerned; from this, it follows that the r equations of the complete system are solvable with respect to r of the differential quotients $\partial f/\partial y_1,\ \dots,\ \partial f/\partial y_s$, and with respect to r of the differential quotients $\partial f/\partial x_1,\ \dots,\ \partial f/\partial x_s$ as well (cf. Chap. 5, Theorem 12, p. 105).

Now, when are s mutually independent relations between the u, v, w solvable both with respect to $x_1,\dots,x_s$ and with respect to $y_1,\dots,y_s$? Clearly, when and only when they can be solved both with respect to $u_1,\dots,u_{s-r}$, $w_1,\dots,w_r$ and with respect to $v_1,\dots,v_{s-r}$, $w_1,\dots,w_r$, hence when they can be brought to the form:

$$\begin{aligned} v_1 &= \mathfrak{F}_1(u_1,\dots,u_{s-r}),\ \dots,\ v_{s-r} = \mathfrak{F}_{s-r}(u_1,\dots,u_{s-r}),\\ w_1 &= \mathfrak{G}_1(u_1,\dots,u_{s-r}),\ \dots,\ \ w_r = \mathfrak{G}_r(u_1,\dots,u_{s-r}), \end{aligned} \tag{10}$$

where $\mathfrak{F}_1,\dots,\mathfrak{F}_{s-r}$ denote arbitrary *mutually independent* functions of their arguments, while the functions $\mathfrak{G}_1,\dots,\mathfrak{G}_r$ are submitted to absolutely no restriction.

The equations (10) represent the most general system of equations which consists of s independent equations, which admits the group $\Omega_1 f,\dots,\Omega_r f$ and which can be solved both with respect to $x_1,\dots,x_s$ and with respect to $y_1,\dots,y_s$; at the same time, they represent the most general transformation between $x_1,\dots,x_s$ and $y_1,\dots,y_s$ which transfers $X_1 f,\dots,X_r f$ to $Y_1 f,\dots,Y_r f$, respectively. It is therefore proved that there exist transformations which achieve this transfer, hence that our two groups are effectively similar to each other.

But even more: the equations (10) actually represent the most general transformation which transfers the group $X_1 f, \dots, X_r f$ to the group $Z_1 f, \dots, Z_r f$.

In fact, if $y_i = \Psi_i(x_1, \dots, x_s)$ is an arbitrary transformation which transfers one group to the other, then it converts $X_1 f, \dots, X_r f$ into certain infinitesimal transformations $\mathfrak{Y}_1 f, \dots, \mathfrak{Y}_r f$ of the group $Z_1 f, \dots, Z_r f$ which stand pairwise in the relationships:

$$[\mathfrak{Y}_i, \mathfrak{Y}_k] = \sum_{\sigma=1}^{r} c_{ik\sigma} \mathfrak{Y}_\sigma f,$$

and which can hence be obtained from the r infinitesimal transformations:

$$Y_k f = \sum_{j=1}^{r} g_{kj} Z_j f \qquad (k=1\cdots r)$$

when one specializes in an appropriate way the arbitrary elements appearing in the g_{kj}. But now, all transformations which convert $X_1 f, \dots, X_r f$ into $Y_1 f, \dots, Y_r f$, respectively, are contained in the form (10), so in particular, the transformation $y_i = \Psi_i(x_1, \dots, x_n)$ is also contained in this form.

Now, by summarizing the obtained result, we can say:

Theorem 63. *If the two r-term groups:*

$$X_k f = \sum_{i=1}^{s} \xi_{ki}(x_1, \dots, x_s) \frac{\partial f}{\partial x_i} \qquad (k=1\cdots r)$$

and:

$$Z_k f = \sum_{i=1}^{s} \zeta_{ki}(y_1, \dots, y_s) \frac{\partial f}{\partial y_i} \qquad (k=1\cdots r)$$

are equally composed and if neither $X_1 f, \dots, X_r f$ nor $Z_1 f, \dots, Z_r f$ are linked together by linear relations, then the two groups are also similar to each other. One obtains the most general transformation which transfers one group to the other in the following way: One chooses the r^2 constants g_{kj} in the most general way so that the r infinitesimal transformations:

$$Y_k f = \sum_{j=1}^{r} g_{kj} Z_j f \qquad (k=1\cdots r)$$

are mutually independent and so that, together with the relations:

$$[X_i, X_k] = \sum_{\sigma=1}^{r} c_{ik\sigma} X_\sigma f,$$

there are at the same time the relations:

$$[Y_i, Y_k] = \sum_{\sigma=1}^{r} c_{ik\sigma} Y_\sigma f;$$

then, one forms the r-term complete system:

$$X_k f + Y_k f = 0 \qquad (k=1\cdots r)$$

in the $2s$ *variables* $x_1, \dots, x_s$, $y_1, \dots, y_s$ *and one determines* $2s - r$ *independent solutions of it, namely* $s - r$ *independent solutions:*

$$u_1(x_1, \dots, x_s), \; \dots, \; u_{s-r}(x_1, \dots, x_s)$$

which contain only the x, plus $s - r$ *independent solutions:*

$$v_1(y_1, \dots, y_s), \; \dots, \; v_{s-r}(y_1, \dots, y_s)$$

which contain only the y, and r solutions:

$$w_1(x_1, \dots, x_s, y_1, \dots, y_s), \; \dots, \; w_r(x_1, \dots, x_s, y_1, \dots, y_s)$$

which are mutually independent and which are independent of $u_1, \dots, u_{s-r}$, $v_1, \dots, v_{s-r}$*; if this takes place, then the system of equations:*

$$v_1 = \mathfrak{F}_1(u_1, \dots, u_{s-r}), \; \dots, \; v_{s-r} = \mathfrak{F}_{s-r}(u_1, \dots, u_{s-r}),$$
$$w_1 = \mathfrak{G}_1(u_1, \dots, u_{s-r}), \; \dots, \; w_r = \mathfrak{G}_r(u_1, \dots, u_{s-r}),$$

represents the demanded transformation; here, $\mathfrak{G}_1, \dots, \mathfrak{G}_r$ *are perfectly arbitrary functions of their arguments; by contrast,* $\mathfrak{F}_1, \dots, \mathfrak{F}_{s-r}$ *are subjected to the restriction that they must be mutually independent.*

From this, in particular, we obtain the

Proposition 1. *If the* $r \leqslant s$ *independent infinitesimal transformations:*

$$X_k f = \sum_{i=1}^{s} \xi_{ki}(x_1, \dots, x_s) \frac{\partial f}{\partial x_i} \qquad (k=1\cdots r)$$

stand pairwise in the relationships:

$$[X_i, X_k] = 0 \qquad (i, k=1\cdots r),$$

without being, however, linked together by a linear relation of the form:

$$\sum_{k=1}^{r} \chi_k(x_1, \dots, x_s) X_k f = 0,$$

then they generate an r-term group which is similar to the group of transformations:

$$Y_1 f = \frac{\partial f}{\partial y_1}, \; \dots, \; Y_r f = \frac{\partial f}{\partial y_r}.$$

A case of the utmost importance is when the two numbers s and r are equal to each other, so that the two groups $X_1 f, \dots, X_r f$ and $Z_1 f, \dots, Z_r f$ are transitive, or more precisely: simply transitive (cf. Chap. 13, p. 226). We want to enunciate Theorem 63 for this special case:

Theorem 64. *Two simply transitive equally composed groups in the same number of variables are always also similar to each other. If:*

$$X_k f = \sum_{i=1}^{r} \xi_{ki}(x_1, \dots, x_r) \frac{\partial f}{\partial x_i} \qquad (k=1\cdots r)$$

and:

$$Z_k f = \sum_{i=1}^{r} \zeta_{ki}(y_1, \dots, y_r) \frac{\partial f}{\partial y_i} \qquad (k=1\cdots r)$$

are the infinitesimal transformations of the two groups, then one finds in the following way the most general transformation which transfers the one group to the other: One chooses the r^2 constants g_{kj} in the most general way so that the r infinitesimal transformations:

$$Y_k f = \sum_{j=1}^{r} g_{kj} Z_j f \qquad (k=1\cdots r)$$

are mutually independent and so that, together with the relations:

$$[X_i, X_k] = \sum_{\sigma=1}^{r} c_{ik\sigma} X_\sigma f,$$

there are at the same time the relations:

$$[Y_i, Y_k] = \sum_{\sigma=1}^{r} c_{ik\sigma} Y_\sigma f;$$

moreover, one forms the r-term complete system:

$$X_k f + Y_k f = 0 \qquad (k=1\cdots r)$$

in the $2r$ variables $x_1, \dots, x_r$, $y_1, \dots, y_r$ and one determines r arbitrary independent solutions:

$$w_1(x_1, \dots, x_r, y_1, \dots, y_r), \; \dots, \; w_r(x_1, \dots, x_r, y_1, \dots, y_r)$$

of it; then the r equations:

$$w_1 = a_1, \; \dots, \; w_r = a_r$$

with the r arbitrary constants $a_1, \dots, a_r$ represent the most general transformation having the constitution demanded.

§ 91. Now, we turn to the treatment of the general problem that we stated at the end of § 89 (p. 347).

First, we can prove exactly as in the previous section that every transformation: $y_i = \Phi_i(x_1, \dots, x_s)$ which transfers $X_1 f, \dots, X_r f$ to $Y_1 f, \dots, Y_r f$, respectively, represents a system of equations which admits the r-term group $\Omega_k f = X_k f + Y_k f$ and that, on the other hand, every system of equations solvable with respect to $x_1, \dots, x_s$:

$$y_i = \Phi_i(x_1, \dots, x_s)$$

which admits the group $\Omega_1 f, \dots, \Omega_r f$ represents a transformation which transfers $X_1 f, \dots, X_r f$ to $Y_1 f, \dots, Y_r f$, respectively.

Now, according to p. 346, every transformation $y_i = \Phi_i(x_1, \dots, x_s)$ which transfers $X_1 f, \dots, X_r f$ to $Y_1 f, \dots, Y_r f$, respectively, is constituted in such a way that the equations (4) become identities after the substitution: $y_1 = \Phi_1(x), \dots, y_s = \Phi_s(x)$. Consequently, we can also enunciate as follows the problem formulated on p. 347:

To seek, in the $2s$ *variables $x_1, \dots, x_s$, $y_1, \dots, y_s$, a system of equations which admits the r-term group $\Omega_1 f, \dots, \Omega_r f$, which consists of exactly s independent equations that are solvable both with respect to $x_1, \dots, x_s$ and to $y_1, \dots, y_s$, and lastly, which embraces*[1] *(Translator's note: See the footnote on p. 132.) the $n(r-n)$ equations:*

$$\varphi_{k\nu}(x_1, \dots, x_s) - \psi_{k\nu}(x_1, \dots, x_s) = 0 \qquad (k=1 \cdots r-n;\ \nu=1 \cdots n). \tag{4}$$

For the solution of this problem, it is of great importance that *the system of equations* (4) *admits in turn the r-term group $\Omega_1 f, \dots, \Omega_r f$.*

In order to prove this, we imagine that the matrix which is associated to the infinitesimal transformations $\Omega_1 f, \dots, \Omega_r f$:

$$\begin{vmatrix} \xi_{11}(x) & \cdot & \cdot & \xi_{1s}(x) & \eta_{11}(y) & \cdot & \cdot & \eta_{1s}(y) \\ \cdot & \cdot & \cdot & \cdot & \cdot & \cdot & \cdot & \cdot \\ \xi_{r1}(x) & \cdot & \cdot & \xi_{rs}(x) & \eta_{r1}(y) & \cdot & \cdot & \eta_{rs}(y) \end{vmatrix} \tag{11}$$

is written down. We will show that the equations (4) are a system of equations which is obtained by setting equal to zero all $(n+1) \times (n+1)$ determinants of the matrix (11). With this, according to Theorem 39, Chap. 14, p. 240, it will be proved that the system of equations (4) admits the group $\Omega_1 f, \dots, \Omega_r f$.

Amongst the $(n+1) \times (n+1)$ determinants of the matrix (11), there are in particular those of the form:

$$\Delta = \begin{vmatrix} \xi_{1k_1}(x) & \cdot\cdot & \xi_{1k_n}(x) & \eta_{1\sigma}(y) \\ \cdot & \cdot\cdot & \cdot & \cdot \\ \xi_{nk_1}(x) & \cdot\cdot & \xi_{nk_n}(x) & \eta_{n\sigma}(y) \\ \xi_{n+j,k_1}(x) & \cdot\cdot & \xi_{n+j,k_n}(x) & \eta_{n+j,\sigma}(y) \end{vmatrix}.$$

If we replace in Δ the members of the last horizontal row by their values from (3) and (3'), namely by the following values:

$$\xi_{n+j,k_\mu}(x) \equiv \sum_{\nu=1}^{n} \varphi_{j\nu}(x)\,\xi_{\nu,k_\mu}(x), \qquad \eta_{n+j,\sigma}(y) \equiv \sum_{\nu=1}^{n} \psi_{j\nu}(y)\,\eta_{\nu\sigma}(y),$$

and if we subtract from the last horizontal row the first n rows, after we have multiplied them before by $\varphi_{j1}(x), \dots, \varphi_{jn}(x)$, respectively, then we obtain:

$$\Delta = \sum \pm \xi_{1k_1}(x) \cdots \xi_{nk_n}(x) \sum_{\nu=1}^{n} \eta_{\nu\sigma}(y)\{\psi_{j\nu}(y) - \varphi_{j\nu}(x)\}.$$

Here, under the assumptions made earlier on, the determinants of the form:

$$D = \sum \pm \xi_{1k_1}(x) \cdots \xi_{nk_n}(x)$$

do not all vanish identically, and likewise not all determinants:

$$\mathfrak{D} = \sum \pm \eta_{1k_1}(y) \cdots \eta_{nk_n}(y)$$

vanish identically.

Obviously, a system of equations that brings to zero all determinants Δ must either contain all equations of the form $D = 0$, or it must contain the $s(r-n)$ equations:

$$\sum_{\nu=1}^{n} \eta_{\nu\sigma}(y)\{\psi_{j\nu}(y) - \varphi_{j\nu}(x)\} = 0 \qquad (\sigma = 1 \cdots s;\; j = 1 \cdots r-n);$$

in the latter case, it embraces either all equations of the form $\mathfrak{D} = 0$, or the $n(r-n)$ equations:

(4) $$\varphi_{j\nu}(x) - \psi_{j\nu}(y) = 0 \qquad (j = 1 \cdots r-n;\; \nu = 1 \cdots n).$$

In the latter of these three cases, we now observe the following:

The system of equation (4) brings to zero not only all determinants Δ, but actually also all $(n+1)\times(n+1)$ determinants of the matrix (11). One realizes this immediately when one writes the matrix in the form:

$$\begin{vmatrix} \xi_{11}(x) & \cdot\,\cdot & \xi_{1s}(x) & \eta_{11}(y) & \cdot\,\cdot & \eta_{1s}(y) \\ \cdot & \cdot\,\cdot & \cdot & \cdot & \cdot\,\cdot & \cdot \\ \sum_\nu^{1\cdots n} \varphi_{1\nu}\,\xi_{\nu 1} & \cdot\,\cdot & \sum_\nu^{1\cdots n} \varphi_{1\nu}\,\xi_{\nu s} & \sum_\nu^{1\cdots n} \psi_{1\nu}\,\eta_{\nu 1} & \cdot\,\cdot & \sum_\nu^{1\cdots n} \psi_{1\nu}\,\eta_{\nu s} \\ \cdot & \cdot\,\cdot & \cdot & \cdot & \cdot\,\cdot & \cdot \\ \sum_\nu^{1\cdots n} \varphi_{r-n,\nu}\,\xi_{\nu 1} & \cdot\,\cdot & \sum_\nu^{1\cdots n} \varphi_{r-n,\nu}\,\xi_{\nu s} & \sum_\nu^{1\cdots n} \psi_{r-n,\nu}\,\eta_{\nu 1} & \cdot\,\cdot & \sum_\nu^{1\cdots n} \psi_{r-n,\nu}\,\eta_{\nu s} \end{vmatrix},$$

and when one makes afterwards the substitution: $\psi_{k\nu}(y) = \varphi_{k\nu}(x)$; the $(n+1)\times(n+1)$ determinants of the obtained matrix are all identically zero. Consequently, (4) belongs to the systems of equations that one obtains by setting equal to zero all $(n+1)\times(n+1)$ determinants of the matrix (11), and therefore, according to the theorem cited above, it admits the r-term group $\Omega_1 f,\dots,\Omega_r f$.

The important property of the system of equations (4) just proved can also be realized in another, somewhat more direct way.

According to p. 124 and to p. 235, the system of equations (4) admits in any case the r-term group $\Omega_1 f,\dots,\Omega_r f$ only when all equations of the form:

$$\Omega_j\big(\varphi_{k\nu}(x) - \psi_{k\nu}(y)\big) = X_j\,\varphi_{k\nu}(x) - Y_j\,\psi_{k\nu}(y) = 0$$

are a consequence of (4). That this condition is satisfied in the present case can be easily verified.

For $j = 1,\dots,r$ and $k = 1,\dots,r-n$, we have:

$$\begin{aligned}[X_j, X_{n+k}] &= \Big[X_j f,\ \sum_{\nu=1}^{n} \varphi_{k\nu}\,X_\nu f\Big] \\ &= \sum_{\nu=1}^{n} X_j\,\varphi_{k\nu}\,X_\nu f + \sum_{\nu=1}^{n} \varphi_{k\nu}\,[X_j, X_\nu].\end{aligned}$$

Moreover, we have in general:

$$[X_j, X_\mu] = \sum_{\pi=1}^{r} c_{j\mu\pi}\,X_\pi f = \sum_{\nu=1}^{n} \Big\{c_{j\mu\nu} + \sum_{\tau=1}^{r-n} c_{j\mu,n+\tau}\,\varphi_{\tau\nu}\Big\}\,X_\nu f.$$

If we insert these values into the preceding equation and if in addition, we take into account that $X_1 f,\dots,X_n f$ are not linked together by a linear relation, we then find:

$$(12)\qquad \begin{cases} X_j\,\varphi_{k\nu} = c_{j,n+k,\nu} + \sum_{\tau=1}^{r-n} c_{j,n+k,n+\tau}\,\varphi_{\tau\nu} \\ \qquad\qquad - \sum_{\mu=1}^{n} \varphi_{k\mu}\Big\{c_{j\mu\nu} + \sum_{\tau=1}^{r-n} c_{j,\mu,n+\tau}\,\varphi_{\tau\nu}\Big\}. \end{cases}$$

Consequently, we have:

$$X_j\,\varphi_{k\nu} = \Pi_{jk\nu}(\varphi_{11},\,\varphi_{12},\,\dots) \qquad (j=1\cdots r;\ k=1\cdots r-n;\ \nu=1\cdots r),$$

and a completely similar computation gives:

$$Y_j\,\psi_{k\nu} = \Pi_{jk\nu}(\psi_{11},\psi_{12},\dots),$$

where in the two cases $\Pi_{jk\nu}$ denote the same functions of their arguments.
At present, we obtain:

$$\Omega_j(\varphi_{k\nu}-\psi_{k\nu}) = \Pi_{jk\nu}(\varphi_{11},\varphi_{12},\dots) - \Pi_{jk\nu}(\psi_{11},\psi_{12},\dots),$$

from which it is to be seen that the expressions $\Omega_j(\varphi_{k\nu}-\psi_{k\nu})$ effectively vanish by virtue of (4).

In general, the $n(r-n)$ equations $\varphi_{k\nu}(x) = \psi_{k\nu}(x)$ are not mutually independent, but rather, they can be replaced by a smaller number of mutually independent equations, say by the following $s-\rho \leqslant s$ ones:

(13) $$\varphi_k(x_1,\dots,x_s) = \psi_k(y_1,\dots,y_s) \qquad (k=1\cdots s-\rho).$$

Naturally, each $X_j\,\varphi_k$ will then be a function of $\varphi_1,\dots,\varphi_{s-\rho}$ alone:

(14) $$X_j\,\varphi_k = \pi_{jk}(\varphi_1,\dots,\varphi_{s-\rho}) \qquad (j=1\cdots r;\ k=1\cdots s-\rho),$$

and every $Y_j\,\psi_k$ will be the same function of $\psi_1,\dots,\psi_{s-\rho}$:

(14') $$Y_j\,\psi_k = \pi_{jk}(\psi_1,\dots,\psi_{s-\rho}) \qquad (j=1\cdots r;\ k=1\cdots s-\rho),$$

hence all $\Omega_j(\varphi_k-\psi_k)$ vanish by virtue of the system of equations: $\varphi_1=\psi_1$, ..., $\varphi_{s-\rho}=\psi_{s-\rho}$. But since this system of equations is presented in a form which satisfies the requirement given on p. 123 sq., then according to p. 127 and to p. 235, we conclude that it admits the r-term group $\Omega_1 f,\dots,\Omega_r f$.

Now, if $s-\rho=s$, so $\rho=0$, then the s equations $\varphi_1=\psi_1$, ..., $\varphi_s=\psi_s$ actually already represent a transformation in the variables $x_1,\dots,x_s$, $y_1,\dots,y_s$. This transformation obviously transfers $X_1 f,\dots,X_r f$ to $Y_1 f,\dots,Y_r f$, respectively, and at the same time, it is the most general transformation that does this.

Thus, the case $s-\rho=s$ is settled from the beginning, but by contrast, the case $s-\rho<s$ requires a closer examination.

In order to simplify the additional considerations, we first want to introduce appropriate new variables in place of the variables x and y.

The n mutually independent equations: $X_1 f=0$, ..., $X_n f=0$ form an n-term complete system in the s variables $x_1,\dots,x_s$, hence they have $s-n$ independent solutions in common, and likewise, the n equations: $Y_1 f=0$, ..., $Y_n f=0$ in the variables $y_1,\dots,y_s$ have exactly $s-n$ independent solutions in common.

It stands to reason to simplify the infinitesimal transformations $X_1 f,\dots,X_r f$ and $Y_1 f,\dots,Y_r f$ by introducing, in place of the x, new independent variables of which $s-n$ are independent solutions of the complete system: $X_1 f=0$, ..., $X_n f=0$, and by introducing, in place of the y, new independent variables of which $s-n$ are independent solutions of the complete system: $Y_1 f=0,\dots,Y_n f=0$.

On the other hand, it stands to reason to simplify the equations (13) by introducing the functions $\varphi_1(x),\dots,\varphi_{s-\rho}(x)$ and $\psi_1(y),\dots,\psi_{s-\rho}(y)$ as new variables.

We want to attempt to combine the two simplifications as far as possible.

We start by introducing appropriate new variables in place of the x.

Amongst the solutions of the complete system $X_1 f = 0, \dots, X_n f = 0$, there may be some which can be expressed in terms of the functions $\varphi_1(x), \dots, \varphi_{s-\rho}(x)$ alone; all solutions $F(\varphi_1,\dots,\varphi_{s-\rho})$ of this nature determine themselves from the n differential equations:

$$(15) \qquad \sum_{\nu=1}^{s-\rho} X_k\,\varphi_\nu\,\frac{\partial F}{\partial \varphi_\nu} = \sum_{\nu=1}^{s-\rho} \pi_{k\nu}(\varphi_1,\dots,\varphi_{s-\rho})\,\frac{\partial F}{\partial \varphi_\nu} = 0 \qquad (k=1\cdots n)$$

in the $s-\rho$ variables $\varphi_1,\dots,\varphi_{s-\rho}$. We assume that these equations possess exactly $s-q \leqslant s-\rho$ independent solutions, say the following ones:

$$\mathfrak{U}_1(\varphi_1,\dots,\varphi_{s-\rho}), \ \dots, \ \mathfrak{U}_{s-q}(\varphi_1,\dots,\varphi_{s-\rho}).$$

Under these assumptions:

$$\mathfrak{U}_1\big(\varphi_1(x),\dots,\varphi_{s-\rho}(x)\big) = u_1(x), \ \dots, \ \mathfrak{U}_{s-q}\big(\varphi_1(x),\dots,\varphi_{s-\rho}(x)\big) = u_{s-q}(x)$$

are obviously independent solutions of the complete system: $X_1 f = 0, \dots, X_n f = 0$ which can be expressed in terms of $\varphi_1(x),\dots,\varphi_{s-\rho}(x)$ alone, and such that every other solution of the same constitution is a function of $u_1(x),\dots,u_{s-q}(x)$ only. Naturally, we also have at the same time the inequality: $s-q \leqslant s-n$, hence neither of the two numbers ρ and n is larger than q.

Now, let:

$$u_{s-q+1} = u_{s-q+1}(x), \ \dots, \ u_{s-n} = u_{s-n}(x)$$

be $q-n$ arbitrary mutually independent solutions of the complete system $X_1 f = 0$, $\dots$, $X_n f = 0$ that are also independent of $u_1(x),\dots,u_{s-q}(x)$. We will show that the $s-\rho+q-n$ functions: $u_{s-q+1}(x)$, $\dots$, $u_{s-n}(x)$, $\varphi_1(x),\dots,\varphi_{s-\rho}(x)$ are mutually independent.

Since $\varphi_1(x),\dots,\varphi_{s-\rho}(x)$ are mutually independent, there are, amongst the functions $\varphi_1(x)$, $\dots$, $\varphi_{s-\rho}(x)$, $u_{s-q+1}(x)$, $\dots$, $u_{s-n}(x)$ at least $s-\rho$, say exactly $s-\rho+q-n-h$ that are mutually independent, where $0 \leqslant h \leqslant q-n$. We assume that precisely $\varphi_1(x)$, $\dots$, $\varphi_{s-\rho}(x)$, $u_{s-q+h+1}(x)$, $\dots$, $u_{s-n}(x)$ are mutually independent, while $u_{s-q+1}(x)$, $\dots$, $u_{s-q+h}(x)$ can be expressed in terms of $\varphi_1(x)$, $\dots$, $\varphi_{s-\rho}(x)$, $u_{s-q+h+1}(x)$, $\dots$, u_{s-n} only. Thus, there must exist, between the quantities $u_{s-q+1},\dots,u_{s-n}$, $\varphi_1,\dots,\varphi_{s-\rho}$ relations of the form:

$$(16) \qquad u_{s-q+j} = \chi_j\big(u_{s-q+h+1},\dots,u_{s-n},\,\varphi_1,\dots,\varphi_{s-\rho}\big) \qquad (j=1\cdots h)$$

which reduce to identities after the substitution:

$$(17) \qquad \begin{cases} u_{s-q+1} = u_{s-q+1}(x), \ \dots, \ u_{s-n} = u_{s-n}(x) \\ \varphi_1 = \varphi_1(x), \ \dots, \ \varphi_{s-\rho} = \varphi_{s-\rho}(x). \end{cases}$$

If we interpret the substitution (17) by the symbol [ ], we then have:

$$[u_{s-q+j} - \chi_j] \equiv 0 \qquad (j=1\cdots h),$$

whence we get:

$$X_k\,[u_{s-q+j} - \chi_j] = -\sum_{\nu=1}^{s-\rho} [\pi_{k\nu}(\varphi_1,\dots,\varphi_{s-\rho})] \left[\frac{\partial \chi_j}{\partial \varphi_\nu}\right] \equiv 0 \qquad (k=1\cdots n;\ j=1\cdots h),$$

that is to say: all the expressions:

$$\text{(18)} \qquad \sum_{\nu=1}^{s-\rho} \pi_{k\nu}(\varphi_1,\dots,\varphi_{s-\rho}) \frac{\partial \chi_j}{\partial \varphi_\nu} \qquad (k=1\cdots n;\ j=1\cdots h)$$

vanish identically after the substitution (17). But now, these expressions are all free of $u_{s-q+1},\dots,u_{s-q+h}$, so if they were not identically zero and they vanished identically after the substitution (17), then the functions $u_{s-q+h+1}(x)$, ..., $u_{s-n}(x)$, $\varphi_1(x)$, ..., $\varphi_{s-\rho}(x)$ would not be mutually independent, but this contradicts the assumption. Consequently, the expressions (18) are in fact identically zero, or, what amounts to the same, the functions $\chi_1,\dots,\chi_h$ are solutions of the n differential equations (15) in the $s-\rho$ variables $\varphi_1,\dots,\varphi_{s-\rho}$. From this, it results that $\varphi_1,\dots,\varphi_{s-\rho}$ appear in the χ_j only in the combination: $\mathfrak{U}_1(\varphi_1,\dots,\varphi_{s-\rho})$, ..., $\mathfrak{U}_{s-q}(\varphi_1,\dots,\varphi_{s-\rho})$, so that the h equations (16) can be replaced by h relations of the form:

$$\text{(19)} \qquad u_{s-q+j} = \overline{\chi}_j(u_{s-q+h+1},\dots,u_{s-n},\,u_1,\dots,u_{s-q}) \qquad (j=1\cdots h),$$

which in turn reduce to identities after the substitution:

$$u_1 = u_1(x), \ \dots, \ u_{s-n} = u_{s-n}(x).$$

Obviously, relations of the form (19) cannot exist, for $u_1,\dots,u_{s-n}$ are independent solutions of the complete system $X_1 f = 0,\dots,X_n f = 0$; consequently, h is equal to zero. As a result, it is proved that the $s-\rho+q-n$ functions $\varphi_1(x),\dots,\varphi_{s-\rho}(x)$, $u_{s-q+1}(x)$, ..., $u_{s-n}(x)$ really are independent of each other.

We therefore see that, between the $s-n+s-\rho$ quantities $u_1,\dots,u_{s-n}$, $\varphi_1,\dots,\varphi_{s-\rho}$, no other relations exist that are independent from the $s-q$ relations:

$$\text{(20)} \qquad u_1 = \mathfrak{U}_1(\varphi_1,\dots,\varphi_{s-\rho}), \ \dots, \ u_{s-q} = \mathfrak{U}_{s-q}(\varphi_1,\dots,\varphi_{s-\rho})$$

which reduce to identities after the substitution:

$$u_1 = u_1(x), \ \dots, \ u_{s-q} = u_{s-q}(x), \quad \varphi_1 = \varphi_1(x), \ \dots, \ \varphi_{s-\rho} = \varphi_{s-\rho}(x).$$

From what has just been said, it follows that, amongst the $s-n+s-\rho$ functions $u_1(x)$, ..., $u_{s-n}(x)$, $\varphi_1(x)$, ..., $\varphi_{s-\rho}(x)$, there exist exactly $s-n+s-\rho-(s-q) = s-n+q-\rho$ that are mutually independent, namely for instance the $s-n$ functions: $u_1(x),\dots,u_{s-n}(x)$ to which $q-\rho \geqslant 0$ of the functions $\varphi_1(x),\dots,\varphi_{s-\rho}(x)$ are added. If we agree that the equations (20) can be resolved precisely with respect to $\varphi_{q-\rho+1}, \dots, \varphi_{s-\rho}$, we can conclude that precisely the $s-n+q-\rho$ functions: $u_1(x)$, $\dots$, $u_{s-n}(x)$, $\varphi_1(x)$, $\dots$, $\varphi_{q-\rho}(x)$ are mutually independent. Here, the number $q-\rho$, or shortly m, is certainly not larger than n, since the sum $s-n+q-\rho$ can naturally not exceed the number s of the variables x.

At present, we have gone so far that we can introduce new independent variables $x_1', \dots, x_s'$ in place of $x_1, \dots, x_s$; we choose them in the following way:

We set simply:

$$x_{q+1}' = u_1(x_1, \dots, x_s), \quad \dots, \quad x_s' = u_{s-q}(x_1, \dots, x_s),$$

furthermore:

$$x_{n+1}' = u_{s-q+1}(x_1, \dots, x_s), \quad \dots, \quad x_q' = u_{s-n}(x_1, \dots, x_s),$$

and:

$$x_1' = \varphi_1(x_1, \dots, x_s), \quad \dots, \quad x_m' = \varphi_m(x_1, \dots, x_s),$$

where $m = q-\rho$ is not larger than n; in addition, we set:

$$x_{m+1}' = \lambda_1(x_1, \dots, x_s), \quad \dots, \quad x_n' = \lambda_{n-m}(x_1, \dots, x_s),$$

where $\lambda_1(x), \dots, \lambda_{n-m}(x)$ denote arbitrary mutually independent functions that are also independent of $u_1(x)$, $\dots$, $u_{s-n}(x)$, $\varphi_1(x)$, $\dots$, $\varphi_m(x)$.

In a completely similar way, we introduce new independent variables in place of $y_1, \dots, y_s$.

We form the $s-q$ functions:

$$\mathfrak{U}_1\big(\psi_1(y), \dots, \psi_{s-\rho}(y)\big) = v_1(y), \quad \dots, \quad \mathfrak{U}_{s-q}\big(\psi_1(y), \dots, \psi_{s-\rho}(y)\big) = v_{s-q}(y)$$

that are evidently independent solutions of the n-term complete system: $Y_1 f = 0$, $\dots$, $Y_n f = 0$; moreover, we determine any $q-n$ arbitrary mutually independent solutions: $v_{s-q+1}(y)$, $\dots$, $v_{s-n}(y)$ of the same complete system that are also independent of $v_1(y)$, $\dots$, $v_{s-q}(y)$. Then it is clear that the $s-n+q-\rho$ functions:

$$v_1(y), \dots, v_{s-n}(y), \quad \psi_1(y), \dots, \psi_{q-\rho}(y)$$

are mutually independent.

We now choose the new variables $y_1', \dots, y_s'$ in exactly the same way as the variables x' a short while ago.

We set simply:

$$y_{q+1}' = v_1(y_1, \dots, y_s), \quad \dots, \quad y_s' = v_{s-q}(y_1, \dots, y_s),$$

furthermore:

$$y_{n+1}' = v_{s-q+1}(y_1, \dots, y_s), \quad \dots, \quad y_q' = v_{s-n}(y_1, \dots, y_s),$$

and:

$$y_1' = \psi_1(y_1,\dots,y_s),\ \dots,\ y_m' = \psi_m(y_1,\dots,y_s);$$

in addition, we set:

$$y_{m+1}' = \Lambda_1(y_1,\dots,y_s),\ \dots,\ y_n' = \Lambda_{n-m}(y_1,\dots,y_s),$$

where $\Lambda_1,\dots,\Lambda_{n-m}$ are arbitrary mutually independent functions that are also independent of: $v_1(y),\ \dots,\ v_{s-n}(y),\ \psi_1(y),\ \dots,\ \psi_m(y)$.

Now, we introduce the new variables x' and y' in the infinitesimal transformations X_kf, Y_kf and in the equations (13).

Since all $X_k x_{n+1}',\dots,X_k x_s'$ vanish identically and since all $X_k\,\varphi_1,\ \dots,\ X_k\,\varphi_m$ depend only on $\varphi_1,\dots,\varphi_{s-\rho}$, that is to say, only on $x_1',\dots,x_m',\, x_{q+1}',\dots,x_s'$, the X_kf are of the following form in the variables $x_1',\dots,x_s'$:

$$\begin{aligned} X_kf = \sum_{\mu=1}^{m} \mathfrak{x}_{k\mu}(x_1',\dots,x_m',\ x_{q+1}',\dots,x_s')\,\frac{\partial f}{\partial x_\mu'} + \\ + \sum_{j=1}^{n-m} \mathfrak{x}_{k,m+j}(x_1',\dots,x_m',\dots,x_n',\dots,x_q',\dots,x_s')\,\frac{\partial f}{\partial x_{m+j}'} = \Xi_kf. \end{aligned}$$

In the same way, we have:

$$\begin{aligned} Y_kf = \sum_{\mu=1}^{m} \mathfrak{x}_{k\mu}(y_1',\dots,y_m',\ y_{q+1}',\dots,y_s')\,\frac{\partial f}{\partial y_\mu'} + \\ + \sum_{j=1}^{n-m} \mathfrak{y}_{k,m+j}(y_1',\dots,y_m',\dots,y_n',\dots,y_q',\dots,y_s')\,\frac{\partial f}{\partial y_{m+j}'} = H_kf. \end{aligned}$$

Here, the $\mathfrak{x}_{k\mu}(y_1',\dots,y_m',y_{q+1}',\dots,y_s')$ denote the same functions of their arguments as the $\mathfrak{x}_{k\mu}(x_1',\dots,x_m',x_{q+1}',\dots,x_s')$. Indeed, it results from the equations (14) and (14') that $Y_k\,y_\mu'$ is the same function of $y_1',\dots,y_m',\ y_{q+1}',\dots,y_s'$ as $X_k\,x_\mu'$ is of $x_1',\dots,x_m',\ x_{q+1}',\dots,x_s'$, where it is understood that μ is an arbitrary number amongst $1,2,\dots,m$.

On the other hand, we must determine which form the system of equations (13) receives in the new variables.

The system of equations can evidently be replaced by the following one:

$$\begin{gathered} \varphi_1 - \psi_1 = 0,\ \dots,\ \varphi_m - \psi_m = 0, \\ \mathfrak{U}_1(\varphi_1,\dots,\varphi_{s-\rho}) - \mathfrak{U}_1(\psi_1,\dots,\psi_{s-\rho}) = 0,\ \dots, \\ \mathfrak{U}_{s-q}(\varphi_1,\dots,\varphi_{s-\rho}) - \mathfrak{U}_{s-q}(\psi_1,\dots,\psi_{s-\rho}) = 0. \end{gathered}$$

If we introduce our new variables in this system, we obviously obtain the simple system:

$$(21) \qquad \begin{cases} x_1' - y_1' = 0, \ \dots, \ x_m' - y_m' = 0, \\ x_{q+1}' - y_{q+1}' = 0, \ \dots, \ x_s' - y_s' = 0; \end{cases}$$

so the system of equations (13), or, what is the same, the system of equations (4), can be brought to this form after the introduction of the new variables.

Finally, if we remember that neither $X_1 f, \dots, X_n f$ nor $Y_1 f, \dots, Y_n f$ are linked by linear relations and that all $(n+1) \times (n+1)$ determinants of the matrix (11) vanish by means of (4) while not all $n \times n$ determinants do, we then realize that in the same way, neither $\Xi_1 f, \dots, \Xi_n f$, nor $H_1 f, \dots, H_n f$ are linked by linear relations, and that all $(n+1) \times (n+1)$ determinants, but not all $n \times n$ determinants, of the matrix which can be formed with the coefficients of the differential quotients of f in the r infinitesimal transformations $\Omega_k f = \Xi_k f + H_k f$ vanish by means of (21).

Thanks to the preceding developments, the problem stated at the outset of the section, p. 354, leads back to the following simpler problem:

To seek, in the 2s variables $x_1', \dots, x_s'$, $y_1', \dots, y_s'$, a system of equations which admits the r-term group:

$$\Omega_k f = \Xi_k f + H_k f \qquad (k=1\cdots r)$$

and in addition which consists of s independent equations that are solvable with respect to $x_1', \dots, x_s'$ and to $y_1', \dots, y_s'$, and lastly, which comprises the $s - q + m$ equations (21).

In order to solve this new problem, we remember Chap. 14, p. 243 sq.; from what was said at that time, we deduce that every system of equations which admits the group $\Xi_k f + H_k f = \Omega_k f$ and which comprises at the same time the equations (21) can be obtained by adding to the equations (21) a system of equations in the $s+q-m$ variables $x_1', \dots, x_m', \dots, x_n', \dots, x_q', \dots, x_s', \dots, y_{m+1}', \dots, y_q'$ that admits the r-term group:

$$\begin{aligned} \overline{\Omega}_k f = & \sum_{\mu=1}^{m} \mathfrak{x}_{k\mu}(x_1', \dots, x_m', x_{q+1}', \dots, x_s') \frac{\partial f}{\partial x_\mu'} + \\ & + \sum_{j=1}^{n-m} \mathfrak{x}_{k,m+j}(x_1', \dots, x_m', \dots, x_n', \dots, x_q', \dots, x_s') \frac{\partial f}{\partial x_{m+j}'} + \\ & + \sum_{j=1}^{n-m} \mathfrak{y}_{k,m+j}(x_1', \dots, x_m', y_{m+1}', \dots, y_n', \dots, y_q', x_{q+1}', \dots, x_s') \frac{\partial f}{\partial y_{m+j}'}. \end{aligned}$$

Here, $\overline{\Omega}_k f$ is obtained by leaving out all terms with $\partial f / \partial y_1', \dots, \partial f / \partial y_m'$ in $\Omega_k f = \Xi_k f + H_k f$ and by making everywhere the substitution:

$$y_1' = x_1', \ \dots, \ y_m' = x_m', \ \ y_{q+1}' = x_{q+1}', \ \dots, \ y_s' = x_s'$$

by means of (21) in the remaining terms. This formation of $\overline{\Omega}_k f$ shows that in the matrix formed with $\overline{\Omega}_1 f, \dots, \overline{\Omega}_r f$:

$$(22)\qquad \begin{vmatrix} \mathfrak{x}_{11}(x') & \cdot & \cdot & \mathfrak{x}_{1n}(x') & \mathfrak{y}_{1,m+1}(x',y') & \cdot & \cdot & \mathfrak{y}_{1n}(x',y') \\ \cdot & & \cdot & \cdot & \cdot & & \cdot & \cdot \\ \mathfrak{x}_{r1}(x') & \cdot & \cdot & \mathfrak{x}_{rn}(x') & \mathfrak{y}_{r,m+1}(x',y') & \cdot & \cdot & \mathfrak{y}_{rn}(x',y') \end{vmatrix},$$

all $(n+1)\times(n+1)$ determinants vanish identically, but not all $n\times n$ determinants. If we add that neither $\Xi_1 f,\dots,\Xi_n f$ nor $H_1 f,\dots,H_n f$ are linked by linear relations, we realize immediately that in particular the two $n\times n$ determinants:

$$(23)\qquad \sum \pm \mathfrak{x}_{11}(x')\cdots \mathfrak{x}_{nn}(x')$$

and:

$$(24)\qquad \begin{vmatrix} \mathfrak{x}_{11}(x') & \cdot & \cdot & \mathfrak{x}_{1m}(x') & \mathfrak{y}_{1,m+1}(x',y') & \cdot & \cdot & \mathfrak{y}_{1n}(x',y') \\ \cdot & \cdot & \cdot & \cdot & \cdot & \cdot & \cdot & \cdot \\ \mathfrak{x}_{n1}(x') & \cdot & \cdot & \mathfrak{x}_{nm}(x') & \mathfrak{y}_{n,m+1}(x',y') & \cdot & \cdot & \mathfrak{y}_{nn}(x',y') \end{vmatrix}$$

are not identically zero.

But now, the matter for us is not to find *all* systems of equations which comprise (21) and which admit the group $\Omega_1 f,\dots,\Omega_r f$, but only to determine systems of equations of this sort which consist of exactly s independent equations that are solvable both with respect to $x_1',\dots,x_s'$ and to $y_1',\dots,y_s'$. Consequently, we do not have to set up *all* systems of equations in $x_1',\dots,x_s'$, $y_{m+1}',\dots,y_q'$ which admit $\overline{\Omega}_1 f,\dots,\overline{\Omega}_r f$, but only the systems which consist of exactly $s-(m+s-q)=q-m$ independent equations and which in addition are solvable both with respect to $x_{m+1}',\dots,x_n',\dots,x_q'$ and with respect to $y_{m+1}',\dots,y_n',\dots,y_q'$.

If a system of equations in $x_1',\dots,x_s'$, $y_{m+1}',\dots,y_q'$ satisfies the requirement just stated, then it can be brought to the form:

$$(25)\qquad y_{m+j}'=\Phi_{m+j}(x_1',\dots,x_m',\dots,x_n',\dots,x_q',\dots,x_s') \qquad {\scriptstyle (j=1\cdots q-m)},$$

where the functions $\Phi_{m+1},\dots,\Phi_q$ are in turn independent relative to $x_{m+1}',\dots,x_q'$. Now, it is clear that a system of equations of the form (25) cannot bring to zero all $n\times n$ determinants of the matrix (22), since in any case, the determinant (23) cannot be equal to zero by virtue of (25). On the other hand, since all $(n+1)\times(n+1)$ determinants of (22) vanish identically, then according to Chap. 14, p. 241, it follows that every system of equations of the form (25) which admits the group $\overline{\Omega}_1 f,\dots,\overline{\Omega}_r f$ is represented by relations between the solutions of the equations: $\overline{\Omega}_1 f=0$, …, $\overline{\Omega}_r f=0$, or, what is the same, by relations between the solutions of the n-term complete system: $\overline{\Omega}_1 f=0$, …, $\overline{\Omega}_n f=0$.

The n-term complete system $\overline{\Omega}_1 f=0$, …, $\overline{\Omega}_n f=0$ contains $s+q-m$ independent variables and therefore possesses $s-n+q-m$ independent solutions; one can immediately indicate $s-n+q-n$ independent solutions, namely: x_{n+1}', …, x_q', …, x_s', y_{n+1}', …, y_q', while the $n-m$ remaining ones must be determined by integration and can obviously be brought to the form:

$$\omega_1(x_1',\dots,x_m',\dots,x_s',y_{m+1}',\dots,y_q'),\ \dots,\ \omega_{n-m}(x_1',\dots,x_m',\dots,x_s',y_{m+1}',\dots,y_q').$$

Since the two determinants (23) and (24) do not vanish identically, the equations of our complete system are solvable both with respect to $\partial f/\partial x_1', \dots, \partial f/\partial x_m', \dots, \partial f/\partial x_n'$ and with respect to $\partial f/\partial x_1', \dots, \partial f/\partial x_m', \partial f/\partial y_{m+1}', \dots, \partial f/\partial y_n'$ and therefore, its $s-n+q-m$ independent solutions $x_{n+1}',\dots,x_s', y_{n+1}',\dots,y_q', \omega_1,\dots,\omega_{n-m}$ are mutually independent both relative to $x_{n+1}', \dots, x_q', \dots, x_s', y_{m+1}',\dots,y_q'$ and to $x_{m+1}', \dots, x_n', \dots, x_q', \dots, x_s', y_{n+1}',\dots,y_q'$ (cf. Theorem 12, p. 105). Consequently, the functions $\omega_1,\dots,\omega_{n-m}$ are mutually independent both relative to $y_{m+1}',\dots,y_n'$ and to $x_{m+1}',\dots,x_n'$.

Every system of equations (25) which satisfies the stated requirements is represented by relations between the solutions $x_{n+1}',\dots,x_s', y_{n+1}',\dots,y_q', \omega_1,\dots,\omega_{n-m}$ and, to be precise, by $q-m$ relations that are solvable both with respect to $y_{m+1}',\dots,y_q'$ and to $x_{m+1}',\dots,x_q'$. Visibly, these relations must be solvable both with respect to $\omega_1,\dots,\omega_{n-m}, x_{n+1}',\dots,x_q'$ and to $\omega_1,\dots,\omega_{n-m}, y_{n+1}',\dots,y_q'$, so they must have the form:

$$(26)\quad \begin{cases} \omega_\mu(x_1',\dots,x_m', x_{m+1}',\dots,x_s', y_{m+1}',\dots,y_q') = \chi_\mu(x_{n+1}',\dots,x_q',\dots,x_s') \\ \qquad\qquad (\mu = 1 \cdots n-m) \\ y_{n+1}' = \Pi_1(x_{n+1}',\dots,x_q',\dots,x_s'), \;\dots,\; y_q' = \Pi_{q-n}(x_{n+1}',\dots,x_q',\dots,x_s'), \end{cases}$$

where $\Pi_1,\dots,\Pi_{q-n}$ are mutually independent relative to $x_{n+1}',\dots,x_q'$.

Conversely, every system of equations of the form (26) in which $\Pi_1,\dots,\Pi_{q-n}$ are mutually independent relative to $x_{n+1}',\dots,x_q'$ is solvable both with respect to $y_{m+1}',\dots,y_q'$ and with respect to $x_{m+1}',\dots,x_q'$ and since in addition, it admits the group $\overline{\Omega}_1 f,\dots,\overline{\Omega}_r f$, then it possesses all properties which the sought system of equations in the variables $x_1',\dots,x_s', y_{m+1}',\dots,y_q'$ should have.

From this, we conclude that the equations (26) represent the most general system of equations which admits the group $\overline{\Omega}_1 f,\dots,\overline{\Omega}_r f$, which consists of exactly $q-m$ independent equations, and which is solvable both with respect to $y_{m+1}',\dots,y_q'$ and to $x_{m+1}',\dots,x_q'$; here, $\chi_1,\dots,\chi_{n-m}$ are absolutely arbitrary functions of their arguments and $\Pi_1,\dots,\Pi_{q-n}$, though with the restriction for the latter that they must be mutually independent relative to $x_{n+1}',\dots,x_q'$.

Thus, if we add the system of equation (26) to the equations (21), we obtain the most general system of equations in $x_1',\dots,x_s', y_1',\dots,y_s'$ which admits the group $\Omega_k = \Xi_k f + H_k f$, which comprises the equations (21), which consists of s independent equations and which is solvable both with respect to $x_1',\dots,x_s'$ and to $y_1',\dots,y_s'$. Finally, if, in this system of equations, we express the variables x' and y' in terms of the initial variables x and y, we obtain the most general system of equations which admits the group $\Omega_k f = X_k f + Y_k f$, which comprises the equations (4), which consists of s independent equations and which is solvable both with respect to $x_1,\dots,x_s$ and to $y_1,\dots,y_s$. In other words: we obtain the most general transformation $y_i = \Phi_i(x_1,\dots,x_s)$ which transfers $X_1 f,\dots,X_r f$ to $Y_1 f,\dots,Y_r f$, respectively.

As a result, the problem stated on p. 347 is settled, and we have achieved even more than what was actually required, because we not only know a transformation of the demanded constitution, we know all these transformations. At the same time, the claim stated on p. 346 is proved, namely it is proved that the two groups $X_1f,\dots,X_rf$ and $Z_1f,\dots,Z_rf$ are similar to each other, under the assumptions made there.

Lastly, on the basis of the preceding developments, we can determine the most general transformation which transfers the group $X_1f,\dots,X_rf$ to the group $Z_1f,\dots,Z_rf$; indeed, if we remember the considerations of p. 344 sq., and if we combine them with the result just obtained, we then see immediately that the transformation in question can be found in the following way: In the infinitesimal transformations:

$$Y_kf = \sum_{j=1}^{r} \overline{g}_{kj} Z_jf \qquad (k=1\cdots r)$$

defined on p. 347, one chooses the constants $\overline{g}_{kj}$ in the most general way and afterwards, following the method given by us, one determines the most general transformation which transfers $X_1f,\dots,X_rf$ to $Y_1f,\dots,Y_rf$, respectively; this then is at the same time the most general transformation which actually transfers the group $X_1f,\dots,X_rf$ to the group $Z_1f,\dots,Z_rf$.

As the equations (26) show, the transformation found in this way contains $n-m+q-n=q-m$ arbitrary functions of $s-n$ arguments, and in addition, certain arbitrary elements which come from the $\overline{g}_{kj}$; these are firstly certain arbitrary parameters, and secondly, certain arbitrarinesses which come from the fact that the $\overline{g}_{kj}$ are determined by algebraic operations. From this, it follows that the said transformation cannot in all cases be represented by a single system of equations. —

Now we summarize our results.

First, we have the

Theorem 65. *Two r-term groups:*

$$X_kf = \sum_{i=1}^{s} \xi_{ki}(x_1,\dots,x_s)\frac{\partial f}{\partial x_i} \qquad (k=1\cdots r)$$

and:

$$Z_kf = \sum_{i=1}^{s} \zeta_{ki}(y_1,\dots,y_s)\frac{\partial f}{\partial y_i} \qquad (k=1\cdots r)$$

in the same number of variables are similar to each other if and only if the following conditions are satisfied:

Firstly: the two groups must be equally composed; so, if the relations:

$$[X_i, X_k] = \sum_{\sigma=1}^{r} c_{ik\sigma} X_\sigma f$$

hold, it must be possible to determine r^2 constants g_{kj} such that the r infinitesimal transformations:

$$\mathfrak{Y}_kf = \sum_{j=1}^{r} g_{kj} Z_jf \qquad (k=1\cdots r)$$

are mutually independent and such that the relations:

$$[\mathfrak{Y}_i, \mathfrak{Y}_k] = \sum_{\sigma=1}^{r} c_{ik\sigma} \mathfrak{Y}_\sigma f$$

are identically satisfied.

Secondly: if $X_1 f, \dots, X_r f$ *are constituted in such a way that* $X_1 f, \dots, X_n f$ *(say) are linked together by no relation of the form:*

$$\chi_1(x_1, \dots, x_s) X_1 f + \cdots + \chi_n(x_1, \dots, x_s) X_n f = 0,$$

while by contrast $X_{n+1} f, \dots, X_r f$ *express themselves linearly in terms of* $X_1 f, \dots, X_n f$*:*

$$X_{n+k} f \equiv \sum_{\nu=1}^{n} \varphi_{k\nu}(x_1, \dots, x_s) X_\nu f \qquad (k=1 \cdots r-n),$$

them amongst the systems of g_{kj} *which satisfy the above requirements, there must exist at least one, say the system:* $g_{kj} = \overline{g}_{kj}$*, which is constituted in such a way that, of the r infinitesimal transformations:*

$$Y_k f = \sum_{j=1}^{r} \overline{g}_{kj} Z_j f \qquad (k=1 \cdots r),$$

the first n are linked together by no linear relation of the form:

$$\psi_1(y_1, \dots, y_s) Y_1 f + \cdots + \psi_n(y_1, \dots, y_s) Y_n f = 0,$$

while $Y_{n+1} f, \dots, Y_r f$ *express themselves in terms of* $Y_1 f, \dots, Y_n f$*:*

$$Y_{n+k} f \equiv \sum_{\nu=1}^{n} \psi_{k\nu}(y_1, \dots, y_s) Y_\nu f \qquad (k=1 \cdots r-n),$$

and such that in addition, the $n(r-n)$ *equations:*

$$\varphi_{k\nu}(x_1, \dots, x_s) - \psi_{k\nu}(y_1, \dots, y_s) = 0 \qquad (k=1 \cdots r-n;\ \nu=1 \cdots n)$$

neither contradict themselves mutually, nor produce relations between the x alone or between the y alone.[1]

Moreover, we have the

Proposition 2. *If the two r-term groups* $X_1 f, \dots, X_r f$ *and* $Z_1 f, \dots, Z_r f$ *are similar to each other and if the r infinitesimal transformations:*

$$Y_k f = \sum_{j=1}^{r} \overline{g}_{kj} Z_j f \qquad (k=1 \cdots r)$$

[1] LIE, Archiv for Math. og Naturv. Vols. 3 and 4, Christiania 1878 and 1879; Math. Ann. Vol. XXV, pp. 96–107.

are chosen as is indicated by Theorem 65, then there always exists at least one transformation:

$$y_1 = \Phi_1(x_1,\dots,x_s),\ \dots,\ y_s = \Phi_s(x_1,\dots,x_s)$$

which transfers $X_1 f,\dots,X_r f$ to $Y_1 f,\dots,Y_r f$, respectively. One can set up each transformation of this nature as soon as one has integrated certain complete systems. One obtains the most general transformation which actually transfers the group $X_1 f,\dots,X_r f$ to the group $Z_1 f,\dots,Z_r f$ by choosing the constants $\overline{g}_{kj}$ in the $Y_k f$ in the most general way, and afterwards, by seeking the most general transformation which transfers $X_1 f,\dots,X_r f$ to $Y_1 f,\dots,Y_r f$, respectively.

The complete systems which are spoken of in this proposition do not appear at all in one case, namely in the case where the number ρ defined earlier on is equal to zero, hence when amongst the $n(r-n)$ functions $\varphi_{k\nu}(x_1,\dots,x_s)$, there are exactly s which are mutually independent. Indeed, we have already remarked on p. 357 that in this case, the equations:

$$\varphi_{k\nu}(x_1,\dots,x_s) - \psi_{k\nu}(y_1,\dots,y_s) = 0 \qquad (k=1\cdots r-n;\ \nu=1\cdots n)$$

are solvable both with respect to $y_1,\dots,y_s$ and with respect to $x_1,\dots,x_s$, and that they represent the most general transformation which transfers $X_1 f,\dots,X_r f$ to $Y_1 f,\dots,Y_r f$, respectively.

On the other hand, there are cases in which the integration of the mentioned complete systems is executable, for instance, this is always so when the finite equations of both groups $X_1 f,\dots,X_r f$ and $Y_1 f,\dots,Y_r f$ are known; however, we cannot get involved in this sort of question.

§ **92.** Let the two r-term groups $X_1 f,\dots,X_r f$ and $Z_1 f,\dots,Z_r f$ be similar to each other, so that the transformations $X_1 f,\dots,X_r f$ convert into the infinitesimal transformations of the other group.

If we now choose r independent infinitesimal transformations:

$$Y_k f = \sum_{j=1}^{r} g_{kj} Z_j f \qquad (k=1\cdots r)$$

in the group $Z_1 f,\dots,Z_r f$, then according to the preceding, there exists a transformation which transfers $X_1 f,\dots,X_r f$ to $Y_1 f,\dots,Y_r f$, respectively, if and only if $Y_1 f,\dots,Y_r f$ possess the properties indicated in Theorem 65.

We assume that $Y_1 f,\dots,Y_r f$ satisfy this requirement and that $y_i = \Phi_i(x_1,\dots,x_s)$ is a transformation which transfers $X_1 f,\dots,X_r f$ to $Y_1 f,\dots,Y_r f$, respectively. Then by the transformation $y_i = \Phi_i(x)$, the general infinitesimal transformation: $e_1 X_1 f + \cdots + e_r X_r f$ is of the form: $e_1 Y_1 f + \cdots + e_r Y_r f$, hence our transformation associates to every infinitesimal transformation of the group $X_1 f,\dots,X_r f$ a completely determined infinitesimal transformation of the group $Z_1 f,\dots,Z_r f$, and conversely. The univalent and invertible relationships which is established in this way between the infinitesimal transformations of the two groups is, according to p. 343, a holoedrically isomorphic one.

Now, let $x_1^0, \dots, x_s^0$ be a point in general position, i.e. a point for which the transformations $X_1 f, \dots, X_n f$ produce n independent directions, so that the functions $\varphi_{k\nu}(x)$ in the identities:

$$(3) \qquad X_{n+k} f \equiv \sum_{\nu=1}^{n} \varphi_{k\nu}(x_1, \dots, x_n)\, X_\nu f \qquad (k=1\cdots r-n)$$

behave regularly. According to Chap. 11, p. 216, the infinitesimal transformations $e_1 X_1 f + \cdots + e_r X_r f$ that leave invariant the point $x_1^0, \dots, x_s^0$ have the form:

$$(27) \qquad \sum_{k=1}^{r-n} \varepsilon_k \left\{ X_{q+k} f - \sum_{\nu=1}^{n} \varphi_{k\nu}(x_1^0, \dots, x_s^0)\, X_\nu f \right\},$$

where it is understood that $\varepsilon_1, \dots, \varepsilon_{r-n}$ are arbitrary parameters, and all these infinitesimal transformations generate an $(r-n)$-term subgroup of the group $X_1 f, \dots, X_r f$.

If, in the identities (3), we introduce the new variables $y_1, \dots, y_s$ by means of the transformation: $y_i = \Phi_i(x_1, \dots, x_s)$, then we obtain between $Y_1 f, \dots, Y_r f$ the identities:

$$(3') \qquad Y_{n+k} f \equiv \sum_{\nu=1}^{n} \psi_{k\nu}(y_1, \dots, y_s)\, Y_\nu f \qquad (k=1\cdots r-n).$$

Hence with $y_i^0 = \Phi_i(x_1^0, \dots, x_s^0)$, after the introduction of the variables y, the infinitesimal transformation (27) is transferred to:

$$(27') \qquad \sum_{k=1}^{r-n} \varepsilon_k \left\{ Y_{q+k} - \sum_{\nu=1}^{n} \psi_{k\nu}(y_1^0, \dots y_s^0)\, Y_\nu f \right\},$$

that is to say, to the most general infinitesimal transformation $e_1 Y_1 f + \cdots + e_r Y_r f$ which leaves fixed the point: $y_1^0, \dots, y_s^0$ in general position. At the same time, all infinitesimal transformations of the form (27') naturally generate an $(r-n)$-term subgroup of the group $Z_1 f, \dots, Z_r f$.

In that, we have an important property of the holoedrically isomorphic relationship which is established between the two groups by the transformation: $y_i = \Phi_i(x_1, \dots, x_s)$. Indeed, to the most general subgroup of $X_1 f, \dots, X_r f$ which leaves fixed an arbitrarily chosen point: $x_1^0, \dots, x_s^0$ in general position, this holoedrically isomorphic relationship always associates the most general subgroup of $Y_1 f, \dots, Y_r f$ which leaves fixed a point: $y_1^0, \dots, y_s^0$ in general position. Exactly the same association is found in the reverse direction; in other words: when the point: $x_1^0, \dots, x_s^0$ runs through all possible positions, then the point: $y_1^0, \dots, y_s^0$ also runs through all possible positions.

Now conversely, for two r-term groups in the same number of variables to be similar to each other, then obviously, one must be able to produce between them a holoedrically isomorphic relationship of the constitution just described. We claim that this necessary condition is at the same time also sufficient; we will show that the two groups are really similar, when such a holoedrically isomorphic relationship can be produced between them.

In fact, let $Z_1 f, \dots, Z_r f$ be any r-term group which can be related in a holoedrically isomorphic way to the group $X_1 f, \dots, X_r f$ in the said manner; let $e_1 Y_1 f + \cdots + e_r Y_r f$ be the infinitesimal transformation of the group $Z_1 f, \dots, Z_r f$ which is associated to the general infinitesimal transformation $e_1 X_1 f + \cdots + e_r X_r f$ of the group $X_1 f, \dots, X_r f$ through the concerned holoedrically isomorphic relationship.

If $X_1 f, \dots, X_n f$ are linked together by no linear relations, while $X_{n+1} f, \dots, X_r f$ can be expressed by means of $X_1 f, \dots, X_n f$ in the known way, then the most general infinitesimal transformation contained in the group $X_1 f, \dots, X_r f$ which leaves invariant the point: $x_1^0, \dots, x_s^0$ in general position reads as follows:

$$\sum_{k=1}^{r-n} \varepsilon_k \left\{ X_{n+k} f - \sum_{\nu=1}^{n} \varphi_{k\nu}(x_1^0, \dots, x_s^0) \, X_\nu f \right\}. \tag{27}$$

Under the assumptions made, to it corresponds, in the group $Z_1 f, \dots, Z_r f$, the infinitesimal transformation:

$$\sum_{k=1}^{r-n} \varepsilon_k \left\{ Y_{n+k} f - \sum_{\nu=1}^{n} \varphi_{k\nu}(x_1^0, \dots, x_s^0) \, Y_\nu f \right\}, \tag{28}$$

which now is in turn the most general transformation of the group $Z_1 f, \dots, Z_r f$ which leaves at rest a certain point: $y_1^0, \dots, y_s^0$ in general position. Here, if the point $x_1^0, \dots, x_s^0$ runs through all possible positions, then the point $y_1^0, \dots, y_s^0$ does the same.

Since (28) is the most general infinitesimal transformation $e_1 Y_1 f + \cdots + e_r Y_r f$ which leaves invariant the point: $y_1^0, \dots, y_s^0$ in general position, $Y_1 f, \dots, Y_n f$ can be linked together by no linear relation, and by contrast, $Y_{n+1} f, \dots, Y_r f$ must be expressible by means of $Y_1 f, \dots, Y_n f$ in the known way. From this, we deduce that the most general infinitesimal transformation $e_1 Y_1 f + \cdots + e_r Y_r f$ which leaves fixed the point: $y_1^0, \dots, y_s^0$ can also be represented by the following expression:

$$\sum_{k=1}^{r-n} \varepsilon_k' \left\{ Y_{n+k} f - \sum_{\nu=1}^{n} \psi_{k\nu}(y_1^0, \dots, y_s^0) \, Y_\nu f \right\}. \tag{28'}$$

Evidently, every infinitesimal transformation contained in the expression (28) is identical to one of the infinitesimal transformations (28'), so for arbitrarily chosen $\varepsilon_1, \dots, \varepsilon_{r-n}$, it must always be possible to determine $\varepsilon_1', \dots, \varepsilon_{r-n}'$ so that the equation:

$$\sum_{k=1}^{r-n} (\varepsilon_k - \varepsilon_k') \, Y_{n+k} f - \sum_{\nu=1}^{n} \left\{ \sum_{k=1}^{r-n} \left(\varepsilon_k \, \varphi_{k\nu}(x^0) - \varepsilon_k' \, \psi_{k\nu}(y^0) \right) \right\} Y_\nu f = 0$$

is identically satisfied.

Because $Y_1 f, \dots, Y_r f$ are independent infinitesimal transformations, the equation just written decomposes into the following ones:

$$\varepsilon_k - \varepsilon_k' = 0, \qquad \sum_{j=1}^{r-n} \big(\varepsilon_j\, \varphi_{j\nu}(x^0) - \varepsilon_j'\, \psi_{j\nu}(y^0)\big) = 0 \tag{29}$$

$$(k=1\cdots r-n;\ \nu=1\cdots n).$$

From this, it follows immediately that:

$$\varepsilon_1' = \varepsilon_1, \ \dots, \ \varepsilon_{r-n}' = \varepsilon_{r-n},$$

and in addition, thanks to the arbitrariness of the ε, we obtain the equations:

$$\varphi_{k\nu}(x_1^0, \dots, x_s^0) - \psi_{k\nu}(y_1^0, \dots, y_s^0) = 0 \qquad (k=1\cdots r-n;\ \nu=1\cdots n),$$

which must therefore, under the assumptions made, hold for the two points $x_1^0, \dots, x_s^0$ and $y_1^0, \dots, y_s^0$. At present, if we recall that the point $y_1^0, \dots, y_s^0$ runs through all possible positions, as soon as $x_1^0, \dots, x_s^0$ does this, then we realize immediately that, under the assumptions made, the $n(r-n)$ equations:

$$\varphi_{k\nu}(x_1, \dots, x_s) - \psi_{k\nu}(y_1, \dots, y_s) = 0 \qquad (k=1\cdots r-n;\ \nu=1\cdots n) \tag{4}$$

are compatible with each other and produce relations neither between the x alone, nor between the y alone.

According to Theorem 65, p. 365, it is thus proved that the two groups $X_1 f, \dots, X_r f$ and $Z_1 f, \dots, Z_r f$ are similar to each other, and this is just what we wanted to prove.

We therefore have the

Proposition 3. *Two r-term groups G and Γ in the same number of variables are similar to each other if and only if it is possible to relate them in a holoedrically isomorphic way so that the most general subgroup of G which leaves invariant a determined point in general position always corresponds, in whichever way the point may be chosen, to the most general subgroup of Γ which leaves invariant a certain point in general position, and so that the same correspondence also holds in the reverse direction.*

But at the same time, it is also proved that, under the present assumption, there exists a transformation: $y_i = \Phi_i(x_1, \dots, x_s)$ which transfers $X_1 f, \dots, X_r f$ to $Y_1 f, \dots, Y_r f$, respectively. We can therefore also state the following somewhat more specific proposition:

Proposition 4. *If the r independent infinitesimal transformations:*

$$X_k f = \sum_{i=1}^{s} \xi_{ki}(x_1, \dots, x_s)\, \frac{\partial f}{\partial x_i} \qquad (k=1\cdots r)$$

generate an r-term group G, and if the r independent infinitesimal transformations:

$$Y_k f = \sum_{i=1}^{s} \eta_{ki}(y_1,\dots,y_s)\,\frac{\partial f}{\partial y_i} \qquad {\scriptstyle (k=1\cdots r)}$$

generate an r-term group Γ, then there is a transformation: $y_i = \Phi_i(x_1,\dots,x_s)$ which transfers $X_1 f,\dots,X_r f$ to $Y_1 f,\dots,Y_r f$, respectively, if and only if the following conditions are satisfied:

Firstly: if G contains exactly $r-n$ independent infinitesimal transformations which leave invariant an arbitrarily chosen point in general position, then Γ must also contain exactly $r-n$ independent infinitesimal transformations having this constitution.

Secondly: if one associates to every infinitesimal transformation $e_1 X_1 f + \cdots + e_r X_r f$ of G the infinitesimal transformation $e_1 Y_1 f + \cdots + e_r Y_r f$ of Γ, then the two groups must be related to each other in a holoedrically isomorphic way, in the manner indicated by the previous proposition.

Because the most general subgroup of G which leaves invariant a determined point in general position is completely defined by this point, and moreover, because the holoedrically isomorphic relationship between G and Γ mentioned several times associates to every subgroup of G of this kind a subgroup of Γ constituted in the same way, it follows that this relationship between G and Γ also establishes a correspondence between the points $x_1,\dots,x_s$ and the points $y_1,\dots,y_s$; however, this correspondence is in general infinitely multivalent, for to every point $x_1,\dots,x_s$ there obviously correspond all points $y_1,\dots,y_s$ which satisfy the equations (4), and conversely. As a result, this agrees with the fact that there are in general infinitely many transformations which transfer $X_1 f,\dots,X_r f$ to $Y_1 f,\dots,Y_r f$, respectively.

For *transitive* groups, the criterion of similarity enunciated in Proposition 3 turns out to be particularly simple. We shall see later (Chap. 21) that two r-term transitive groups G and Γ in the same number, say s, of variables are already similar to each other when it is possible to relate them in a holoedrically isomorphic way so that, to a single $(r-s)$-term subgroup of G which leaves fixed one point in general position there corresponds an $(r-s)$-term subgroup of Γ having the same constitution.

If an r-term group:

$$X_k f = \sum_{i=1}^{s} \xi_{ki}(x_1,\dots,x_s)\,\frac{\partial f}{\partial x_i} \qquad {\scriptstyle (k=1\cdots r)}$$

having the composition:

$$[X_i, X_k] = \sum_{\sigma=1}^{r} c_{ik\sigma} X_\sigma f$$

is presented, then one can ask for all transformations:

$$x_i' = \Phi_i(x_1,\dots,x_s) \qquad {\scriptstyle (i=1\cdots s)}$$

that leave it invariant, namely one can ask for all transformations through which the group is similar to itself.

Thanks to the developments of the previous sections, we are in a position to determine all the transformations in question.

To begin with, we relate in the most general holoedrically isomorphic way the group $X_1 f, \dots, X_r f$ to itself, hence we choose in the most general way r independent infinitesimal transformations:

$$\Xi_k f = \sum_{j=1}^{r} g_{kj} X_j f \qquad (k=1\cdots r)$$

that stand pairwise in the relationships:

$$[\Xi_i, \Xi_k] = \sum_{\sigma=1}^{r} c_{ik\sigma} \Xi_\sigma f.$$

Afterwards, we specialize the arbitrary elements contained in the g_{kj} in such a way that the following conditions are satisfied: when there are no relations between $X_1 f, \dots, X_n f$, while $X_{n+1} f, \dots, X_r f$ express themselves by means of $X_1 f, \dots, X_n f$:

$$X_{n+k} f \equiv \sum_{\nu=1}^{n} \varphi_{k\nu}(x_1, \dots, x_s) X_\nu f \qquad (k=1\cdots r-n),$$

then firstly, $\Xi_1 f, \dots, \Xi_n f$ should also be linked together by no linear relation, while by contrast $\Xi_{n+1} f, \dots, \Xi_r f$ also express themselves by means of $\Xi_1 f, \dots, \Xi_n f$:

$$\Xi_{n+k} f \equiv \sum_{\nu=1}^{n} \psi_{k\nu}(x_1, \dots, x_s) \Xi_\nu f \qquad (k=1\cdots r-n),$$

and secondly, the $n(r-n)$ equations:

$$\varphi_{k\nu}(x_1', \dots, x_s') = \psi_{k\nu}(x_1, \dots, x_s) \qquad (k=1\cdots r-n;\ \nu=1\cdots n)$$

should be mutually compatible, and they should produce relations neither between the x alone, nor between the x' alone.

If all of this is realized, then following the introduction of § 91, we determine the most general transformation $x_i' = \Phi_i(x_1, \dots, x_s)$ which transfers $\Xi_1 f, \dots, \Xi_r f$ to $X_1' f, \dots, X_r' f$, respectively, where:

$$X_k' f = \sum_{i=1}^{r} \xi_{ki}(x_1', \dots, x_s') \frac{\partial f}{\partial x_i'}.$$

The transformation in question is then the most general one by which the group $X_1 f, \dots, X_r f$ is transferred to itself.

It is clear that the totality of all transformations which leave invariant the group $X_1 f, \dots, X_r f$ forms a group by itself, namely *the largest group in which* $X_1 f, \dots, X_r f$

is contained as an invariant subgroup. This group can be finite or infinite, continuous or not continuous, but in all circumstances its transformations order as inverses by pairs, because when a transformation $x_i' = \Phi_i(x_1,\dots,x_s)$ leaves the group $X_1f,\dots,X_rf$ invariant, then the associated inverse transformation also does this.

If by chance the group just defined consists of a finite number of different families of transformations and if at the same time each of these families contains only a finite number of arbitrary parameters, then according to Chap. 18, p. 327 sq., there is in the concerned group a family of transformations which constitutes a finite continuous group; then this family is the largest continuous group in which $X_1f,\dots,X_rf$ is contained as an invariant subgroup.

Up to now, we have spoken of similarity only for groups which contain the same number of variables. However, it was not excluded that certain of the variables were not transformed by the concerned groups, so that they did not appear in the infinitesimal transformations.

Now, for the groups which do not contain the same number of variables, one can also speak of similarity; indeed, one can always complete the number of variables in one of the groups so that one adds a necessary number of variables that are not transformed by the concerned group. Then one has two groups in the same number of variables and one can examine whether they are similar to each other, or not.

In the sequel, unless the contrary is expressly stressed, we shall actually interpret the concept of similarity in the original, narrower sense.

§ 93. In order to illustrate the general theory of similarity by an example, we will examine whether the two three-term groups in two independent variables:

$$X_1f = \frac{\partial f}{\partial x_1}, \qquad X_2f = x_1\frac{\partial f}{\partial x_1} + x_2\frac{\partial f}{\partial x_2},$$
$$X_3f = x_1^2\frac{\partial f}{\partial x_1} + (2x_1x_2 + Cx_2^2)\frac{\partial f}{\partial x_2}$$

and:

$$Y_1f = \frac{\partial f}{\partial y_1} + \frac{\partial f}{\partial y_2}, \qquad Y_2f = y_1\frac{\partial f}{\partial y_1} + y_2\frac{\partial f}{\partial y_2},$$
$$Y_3f = y_1^2\frac{\partial f}{\partial y_1} + y_2^2\frac{\partial f}{\partial y_2}$$

are similar to each other.

As we remarked, the Y_kf here are already chosen in such a way that one has at the same time:

$$[X_1, X_2] = X_1f, \qquad [X_1, X_3] = 2X_2f, \qquad [X_2, X_3] = X_3f$$

and:

$$[Y_1, Y_2] = Y_1f, \qquad [Y_1, Y_3] = 2Y_2f, \qquad [Y_2, Y_3] = Y_3f$$

Although the $Y_k f$ are not chosen in the most general way so that the shown relations hold, it is nevertheless not necessary to do this, because the result would not be modified by this.

We find:

$$X_3 f \equiv -(x_1^2 + C x_1 x_2) X_1 f + (2x_1 + C x_2) X_2 f$$

and:

$$Y_3 f \equiv -y_1 y_2 Y_1 f + (y_1 + y_2) Y_2 f,$$

whence one must have:

$$y_1 y_2 = x_1^2 + C x_1 x_2, \qquad y_1 + y_2 = 2x_1 + C x_2.$$

As long as the constant C does not vanish, these equations determine a transformation, and consequently, according to our general theory, the two groups are similar to each other in the case $C \neq 0$. But if $C = 0$, we obtain a relation between y_1 and y_2 alone, hence in any case, there exists no transformation which transfers $X_1 f$, $X_2 f$, $X_3 f$ to $Y_1 f$, $Y_2 f$, $Y_3 f$, respectively; one can easily convince oneself that in this case, the two groups are actually not similar to each other.

§ **94.** Subsequently to the theory of the similarity of r-term groups, we want to briefly treat a somewhat more general question and to indicate its solution.

We imagine, in the variables $x_1, \dots, x_s$, that any p infinitesimal transformations: $X_1 f, \dots, X_p f$ are presented, hence not necessarily some which generate a finite group, and likewise, we imagine in $y_1, \dots, y_s$ that any p infinitesimal transformations $Y_1 f, \dots, Y_p f$ are presented. We ask under which conditions there is a transformation $y_i = \Phi_i(x_1, \dots, x_s)$ which transfers $X_1 f, \dots, X_p f$ to $Y_1 f, \dots, Y_p f$, respectively. Here, we do not demand that $X_1 f, \dots, X_p f$ should be independent infinitesimal transformations, since this assumption would in fact not interfere with the generality of the following considerations, but it would complicate the presentation, because it would always have to be taken into account.

First, we can lead the above general problem back to the special case where the independent equations amongst the equations: $X_1 f = 0, \dots, X_p f = 0$ form a complete system.

Indeed, if there is a transformation which transfers $X_1 f, \dots, X_p f$ to $Y_1 f, \dots, Y_p f$, respectively, then according to Chap. 4, p. 100, every expression:

$$X_k\big(X_j(f)\big) - X_j\big(X_k(f)\big) = [X_k, X_j]$$

also converts into the corresponding expression:

$$Y_k\big(Y_j(f)\big) - Y_j\big(Y_k(f)\big) = [Y_k, Y_j].$$

Hence, if the independent equations amongst the equations $X_1 f = 0, \dots, X_p f = 0$ do not already form a complete system, then to $X_1 f, \dots, X_p f$ we can add all the expressions $[X_k, X_j]$, and also, we must only add to $Y_1 f, \dots, Y_p f$ all the expressions $[Y_k, Y_j]$.

The question of whether there exists a transformation of the demanded constitution then amounts to the question of whether there exists a transformation which transfers $X_1 f, \dots, X_p f$, $[X_k, X_j]$ to $Y_1 f, \dots, Y_p f$, $[Y_k, Y_j]$, respectively, where one has to set for k and j the numbers $1, 2, \dots, p$ one after the other.

Now, if the independent equations amongst the equations: $X_1 f = 0$, ..., $X_p f = 0$, $[X_k, X_j] = 0$ also do not form a complete system, then we add the expressions $\big[X_i, [X_k, X_j]\big]$ and the $\big[[X_i, X_k], [X_j, X_l]\big]$, and the corresponding expressions in the $Y f$ as well. If we continue in this way, we obtain at the end that our initial problem leads back to the following one:

In the variables $x_1, \dots, x_s$, let r infinitesimal transformations: $X_1 f, \dots, X_r f$ be presented, of which $n \leqslant r$, say $X_1 f, \dots, X_n f$, are linked together by no linear relation, while $X_{n+1} f, \dots, X_r f$ can be linearly expressed in terms of $X_1 f, \dots, X_n f$:

$$X_{n+k} f \equiv \sum_{\nu=1}^{n} \varphi_{k\nu}(x_1, \dots, x_s)\, X_\nu f \qquad (k = 1 \cdots n - r);$$

in addition, let relations of the form:

$$[X_k, X_j] = \sum_{\nu=1}^{n} \varphi_{kj\nu}(x_1, \dots, x_s)\, X_\nu f \qquad (k, j = 1 \cdots r)$$

hold, so that the independent equations amongst the equations $X_1 f = 0$, ..., $X_r f = 0$ form an n-term complete system. Moreover, in the variables $y_1, \dots, y_s$, let r infinitesimal transformations $Y_1 f, \dots, Y_r f$ be presented. To determine whether there is a transformation $y_i = \Phi_i(x_1, \dots, x_s)$ which transfers $X_1 f, \dots, X_r f$ to $Y_1 f, \dots, Y_r f$, respectively.

For a transformation of the constitution demanded here to exist, $Y_1 f, \dots, Y_r f$ should naturally not be linked together by a linear relation, while $Y_{n+1} f, \dots, Y_r f$ should be linearly expressible in terms of $Y_1 f, \dots, Y_n f$:

$$Y_{n+k} f \equiv \sum_{\nu=1}^{n} \psi_{k\nu}(y_1, \dots, y_s)\, Y_\nu f \qquad (k = 1 \cdots r - n),$$

and there should be relations of the form:

$$[Y_k, Y_j] = \sum_{\nu=1}^{n} \psi_{kj\nu}(y_1, \dots, y_s)\, Y_\nu f \qquad (k, j = 1 \cdots r).$$

In addition, the equations:

$$(30) \qquad \begin{cases} \varphi_{k\nu}(x_1, \dots, x_s) - \psi_{k\nu}(y_1, \dots, y_s) = 0 & (k = 1 \cdots r - n;\ \nu = 1 \cdots n) \\ \varphi_{kj\nu}(x_1, \dots, x_s) - \psi_{kj\nu}(y_1, \dots, y_s) = 0 & (k, j = 1 \cdots r;\ \nu = 1 \cdots n) \end{cases}$$

should neither mutually contradict, nor lead to relations between the x alone or the y alone, since these equations will obviously reduce to identities after the substitution:

$y_i = \Phi_i(x_1,\dots,x_s)$, when the transformation: $y_i = \Phi_i(x)$ converts $X_1 f,\dots,X_r f$ into $Y_1 f,\dots,Y_r f$, respectively.

We want to assume that all these conditions are satisfied and that all the equations (30) reduce to the ρ mutually independent equations:

$$\varphi_1(x) - \psi_1(x) = 0, \ \dots, \ \varphi_\rho(x) - \psi_\rho(x) = 0. \tag{31}$$

Thanks to considerations completely similar to those of p. 348 sq., we realize that the determination of a transformation which transfers $X_1 f,\dots,X_r f$ to $Y_1 f,\dots,Y_r f$, respectively, amounts to determining a system of equations in the $2s$ variables $x_1,\dots,x_s$, $y_1,\dots,y_s$, and to be precise, a system of equations having the following constitution: it must admit the r infinitesimal transformations: $\Omega_k f = X_k f + Y_k f$, it must consist of exactly s independent equations, it must be solvable both with respect to $x_1,\dots,x_s$ and with respect to $y_1,\dots,y_s$, and lastly, it must comprise the ρ equations (31).

A system of equations which comprises the ρ equations (31) and which admits the r infinitesimal transformations $\Omega_k f$ embraces at the same time all the $r\rho$ equations:

$$\Omega_k\big(\varphi_j(x) - \psi_j(x)\big) = X_k\,\varphi_j(x) - Y_k\,\psi_j(y) = 0$$
$$(k=1\cdots r;\ j=1\cdots \rho).$$

If these equations are not a consequence of (31), we can again deduce from them new equations which must be contained in the sought system of equations, and so on. If we proceed in this way, then at the end, we must come either to relations which contradict each other, or to relations between the x alone, or to relations between the y alone, or lastly, to a system of $\sigma \leqslant s$ independent equations:

$$\varphi_1(x) - \psi_1(y) = 0, \ \dots, \ \varphi_\sigma(x) - \psi_\sigma(y) = 0 \qquad (\sigma \geqslant \rho) \tag{32}$$

which possesses the following two properties: it produces no relation between the x or the y alone, and it admits the r infinitesimal transformations $\Omega_k f$, so that each one of the $r\sigma$ equations:

$$\Omega_k\big(\varphi_j(x) - \psi_j(y)\big) = 0 \qquad (k=1\cdots r;\ j=1\cdots \sigma).$$

is a consequence of (32).

Evidently, there can be a transformation which transfers $X_1 f,\dots,X_r f$ to $Y_1 f,\dots,Y_r f$, respectively, only when we are led to a system of equations (32) having the constitution just defined by means of the indicated operations. So we need to consider only this case.

If the whole number σ is precisely equal to s, then the system of equations (32) taken for itself represents a transformation which achieves the demanded transfer and to be precise, it is obviously the only transformation which does this. We will also show that in the case $\sigma < s$, a transformation exists which transfers $X_1 f,\dots,X_r f$ to $Y_1 f,\dots,Y_r f$, respectively; here, we produce the proof of this by indicating a method which leads us to the determination of a transformation having the demanded constitution.

It is clear that the equations (32) neither cancel the independence of the equations: $X_1 f = 0, \dots, X_n f = 0$, nor do they cancel the independence of the equations: $Y_1 f = 0, \dots, Y_n f = 0$, for they neither produce relations between the x alone, nor between the y alone.

Furthermore, it is to be observed that relations of the form:

$$[\Omega_k, \Omega_j] = \sum_{\nu=1}^{n} \varphi_{kj\nu}(x)\, \Omega_\nu f = \sum_{\nu=1}^{n} \psi_{kj\nu}(y)\, \Omega_\nu f$$

$$(k, j = 1, 2 \cdots n)$$

hold, in which the coefficients $\varphi_{kj\nu}(x) = \psi_{kj\nu}(y)$ behave regularly in general for the systems of values of the system of equations: $\varphi_1 - \psi_1 = 0, \dots, \varphi_\sigma - \psi_\sigma = 0$. Thus, the case settled in Theorem 19, p. 146 is present here.

As in p. 357 sq., we introduce the solutions of the n-term complete system $X_1 f = 0, \dots, X_n f = 0$ as new x and those of the complete system $Y_1 f = 0, \dots, Y_n f = 0$ as new y, and on the occasion, exactly as we did at that time, we have to make a distinction between the solutions which can be expressed in terms of $\varphi_1(x), \dots, \varphi_\sigma(x)$ or, respectively, in terms of $\psi_1(y), \dots, \psi_\sigma(y)$, and the solutions which are independent of the φ, or respectively, of the ψ. In this way, we simplify the form of the equations (32), and then, exactly as on p. 361 sq., we can determine a system of equations which admits $\Omega_1 f, \dots, \Omega_r f$, which contains exactly s independent equations, which is solvable both with respect to $x_1, \dots, x_s$ and with respect to $y_1, \dots, y_s$, and lastly, which embraces the equations (32). Obviously, the obtained system of equations represents a transformation which transfers $X_1 f, \dots, X_r f$ to $Y_1 f, \dots, Y_r f$, respectively.

Thus, we have the following Theorem.

Theorem 66. *If, in the variables $x_1, \dots, x_s$, p infinitesimal transformations $X_1 f, \dots, X_p f$ are presented and if, in the variables $y_1, \dots, y_s$, p infinitesimal transformations $Y_1 f, \dots, Y_p f$ are also presented, then one can always decide, by means of differentiations and of eliminations, whether there is a transformation $y_i = \Phi_i(x_1, \dots, x_s)$ which transfers $X_1 f, \dots, X_r f$ to $Y_1 f, \dots, Y_r f$, respectively; if there is such a transformation, then one can determine the most general transformation which accomplishes the concerned transfer, as soon as one has integrated certain complete systems.*[2]

---

[2] LIE, Archiv for Math. og Naturv., Vol. 3, p. 125, Christiania 1878.

Chapter 20
Groups, the Transformations of Which Are Interchangeable With All Transformations of a Given Group

Thanks to the developments of §§ 89, 90, 91, pp. 343 up to 367, we are in the position to determine all transformations which leave invariant a given r-term group:

$$X_k f = \sum_{i=1}^{s} \xi_{ki}(x_1,\dots,x_s)\,\frac{\partial f}{\partial x_i} \qquad {\scriptstyle (k=1\cdots r)}.$$

Amongst all transformations of this nature, we now want to pick out those which in addition possess the property *of leaving invariant every individual transformation of the group* $X_1 f,\dots,X_r f$, and we want to occupy ourselves more closely with these transformations.

When a transformation T leaves invariant every individual transformation S of the group $X_1 f,\dots,X_r f$, then according to p. 271, it stands with respect to S in the relationship:

$$T^{-1}\,ST = S,$$

or, what is the same, in the relationship:

$$ST = TS,$$

hence it is interchangeable with T. Thus, we can also characterize the transformations just defined in the following way: *they are the transformations which are interchangeable with all transformations of the group* $X_1 f,\dots,X_r f$.

§ 95. According to Chap. 15, p. 268, the expression $e_1 X_1 f + \cdots + e_r X_r f$ can be regarded as the general symbol of a transformation of the group $X_1 f,\dots,X_r f$. Consequently, a transformation $x_i' = \Phi_i(x_1,\dots,x_s)$ will leave invariant every individual transformation of the group $X_1 f,\dots,X_r f$ when, for arbitrary choice of the e, it leaves invariant the expression $e_1 X_1 f + \cdots + e_r X_r f$, hence when the expression $e_1 X_1 f + \cdots + e_r X_r f$ takes the form: $e_1 X_1' f + \cdots + e_r X_r' f$ after the introduction of the new variables $x_i' = \Phi_i(x)$, where:

S. Lie, *Theory of Transformation Groups I*,
DOI 10.1007/978-3-662-46211-9_20

$$X_k' f = \sum_{i=1}^{s} \xi_{ki}(x_1', \dots, x_s') \frac{\partial f}{\partial x_i'}.$$

Here, for this to hold, it is necessary and sufficient that the r infinitesimal transformations: $X_1 f, \dots, X_r f$ are of the form $X_1' f, \dots, X_r' f$, respectively, after the introduction of the new variables x_i'.

It is clear that there are transformations $x_i' = \Phi_i(x)$ having the constitution demanded; indeed, the identity transformation $x_i' = x_i$ is such a transformation; besides, this results from Proposition 2 of the preceding chapter (p. 366), because this proposition shows that there are transformations which transfer $X_1 f, \dots, X_r f$ to $X_1' f, \dots, X_r' f$, respectively. In addition, it follows from the results found at that time that the most general transformation of the demanded constitution is represented by the most general system of equations:

$$x_1' = \Phi_1(x_1, \dots, x_s), \;\; \dots, \;\; x_s' = \Phi_s(x_1, \dots, x_s)$$

which admits the r-term group: $X_1 f + X_1' f, \;\dots,\; X_r f + X_r' f$ in the $2s$ variables $x_1, \dots, x_s, x_1', \dots, x_s'$.

If one executes, one after the other, two transformations which leave invariant every individual transformation of the group: $X_1 f, \dots, X_r f$, then obviously, one always obtains a transformation which does the same; consequently, the totality of all transformations of this constitution forms a group G. This group can be discontinuous, and it can even reduce to the identity transformation; it can consist of several discrete families of which each one contains only a finite number of arbitrary parameters, it can be infinite; but its transformations are always ordered as inverses by pairs, since if a transformation leaves invariant all transformations of the group: $X_1 f, \dots, X_r f$, then the associated inverse transformation naturally possesses the same property.

If the group G just defined contains only a finite number of arbitrary parameters, then it belongs to the category of groups which was discussed in Chap. 18, and according to Theorem 56, p. 328, it certainly comprises a finite number of continuous subgroups generated by infinitesimal transformations. On the other hand, if the group G is infinite, then thanks to considerations similar to those of Chap. 18, it can be proved that it comprises one-term groups, and in fact, infinitely many such groups that are generated by infinitely many independent infinitesimal transformations.

Now, we directly take up the problem of determining all one-term groups which are contained in the group G.

According to Chap. 15, p. 271 and 271, the r expressions $X_1 f, \dots, X_r f$ remain invariant by all transformations of the one-term group Zf when the r relations:

$$[X_k, Z] = X_k\big(Z(f)\big) - Z\big(X_k(f)\big) = 0 \qquad (k=1\cdots r)$$

are identically satisfied, hence when the infinitesimal transformation Zf is interchangeable with all transformations of the group $X_1 f, \dots, X_r f$. Thus, the determination of all one-term groups having the constitution defined a short while ago amounts to the determination of the most general infinitesimal transformation Zf which is interchangeable with all $X_k f$.

If one has two infinitesimal transformations $Z_1 f$ and $Z_2 f$ which are interchangeable with all $X_k f$, then all expressions $[Z_1, X_k]$ and $[Z_2, X_k]$ vanish identically, whence the Jacobi identity:

$$\big[[Z_1, Z_2], X_k\big] + \big[[Z_2, X_k], Z_1\big] + \big[[X_k, Z_1], Z_2\big] \equiv 0$$

reduces to:

$$\big[[Z_1, Z_2], X_k\big] \equiv 0.$$

Thus, the following holds:

Proposition 1. *If the two infinitesimal transformations $Z_1 f$ and $Z_2 f$ are interchangeable with all infinitesimal transformations of the r-term group $X_1 f, \dots, X_r f$, then so is the transformation $[Z_1, Z_2]$.*

On the other hand, every infinitesimal transformation $aZ_1 f + bZ_2 f$ is at the same time interchangeable with $X_1 f, \dots, X_r f$, whichever values the constants a and b may have. Hence if by chance there is only a finite number, say q, of independent infinitesimal transformations $Z_1 f, \dots, Z_q f$ which are interchangeable with $X_1 f, \dots, X_r f$, then the most general infinitesimal transformation having the same constitution has the form: $\lambda_1 Z_1 f + \cdots + \lambda_q Z_q f$, where it is understood that $\lambda_1, \dots, \lambda_q$ are arbitrary parameters. Then because of Proposition 1, there must exist relations of the form:

$$[Z_i, Z_k] = \sum_{\sigma=1}^{q} c'_{ik\sigma} Z_\sigma f,$$

so that *$Z_1 f, \dots, Z_q f$ generate a q-term group.*

At present, we seek to determine directly the most general infinitesimal transformation:

$$Zf = \sum_{i=1}^{s} \zeta_i(x_1, \dots, x_s) \frac{\partial f}{\partial x_i}$$

which is interchangeable with all infinitesimal transformations of the group $X_1 f, \dots, X_r f$.

The r condition-equations [BEDINGUNGSGLEICHUNGEN]:

$$[X_1, Z] = 0, \; \dots, \; [X_r, Z] = 0$$

decompose immediately into the following rs equations:

$$X_k \zeta_i = Z \xi_{ki} \qquad (k=1 \cdots r;\ i=1 \cdots s),$$

or, if written at more length:

$$\text{(1)} \qquad \sum_{\nu=1}^{s} \xi_{k\nu}(x) \frac{\partial \zeta_i}{\partial x_\nu} = \sum_{\nu=1}^{s} \frac{\partial \xi_{ki}}{\partial x_\nu} \zeta_\nu \qquad (k=1 \cdots r;\ i=1 \cdots s).$$

The question is to determine the most general solutions $\zeta_1, \dots, \zeta_s$ to these differential equations.

Let:

$$\zeta_i = \omega_i(x_1, \dots, x_s) \qquad (i=1\cdots n) \tag{2}$$

be any system of solutions of (1), whence all the expressions:

$$X_k\,\omega_i - \sum_{\nu=1}^{s} \frac{\partial \xi_{ki}}{\partial x_\nu}\,\zeta_\nu$$

vanish identically after the substitution: $\zeta_1 = \omega_1(x), \dots, \zeta_s = \omega_s(x)$; in other words: the system of equations (2) in the $2s$ variables $x_1, \dots, x_s$, $\zeta_1, \dots, \zeta_s$ admits the r infinitesimal transformations:

$$W_k\,F = X_k f + \sum_{i=1}^{s} \left\{ \sum_{\nu=1}^{s} \frac{\partial \xi_{ki}}{\partial x_\nu}\,\zeta_\nu \right\} \frac{\partial f}{\partial \zeta_i} \qquad (k=1\cdots r).$$

Conversely, if a system of equations of the form (2) admits the infinitesimal transformations $W_1 f, \dots, W_r f$, then $\omega_1(x), \dots, \omega_s(x)$ are obviously solutions of the differential equations (1). Consequently, the integration of the differential equations (1) is equivalent to the determination of the most general system of equations (2) which admits the infinitesimal transformations $W_1 f, \dots, W_r f$.

Every system of equations which admits $W_1 f, \dots, W_r f$ also allows the infinitesimal transformation $W_k(W_j(f)) - W_j(W_k(f)) = [W_k, W_j]$; we compute it.

We have:

$$\begin{aligned}
[W_k, W_j] = [X_k, X_j] &+ \sum_{i,\,\mu,\,\nu}^{1\cdots s} \left\{ \xi_{k\nu}\,\frac{\partial^2 \xi_{ji}}{\partial x_\mu \partial x_\nu} - \xi_{j\nu}\,\frac{\partial^2 \xi_{ki}}{\partial x_\mu \partial x_\nu} \right\} \zeta_\mu\,\frac{\partial f}{\partial \zeta_i} \\
&+ \sum_{i,\,\mu,\,\nu}^{1\cdots s} \left\{ \frac{\partial \xi_{k\nu}}{\partial x_\mu}\,\frac{\partial \xi_{ji}}{\partial x_\nu} - \frac{\partial \xi_{j\nu}}{\partial x_\mu}\,\frac{\partial \xi_{ki}}{\partial x_\nu} \right\} \zeta_\mu\,\frac{\partial f}{\partial \zeta_i},
\end{aligned}$$

and here, the right-hand side can be written:

$$[X_k, X_j] + \sum_{i,\,\mu}^{1\cdots s} \frac{\partial}{\partial x_\mu} \sum_{\nu=1}^{s} \left\{ \xi_{k\nu}\,\frac{\partial \xi_{ji}}{\partial x_\nu} - \xi_{j\nu}\,\frac{\partial \xi_{ki}}{\partial x_\nu} \right\} \zeta_\mu\,\frac{\partial f}{\partial \zeta_i}.$$

But now, there are relations of the form:

$$[X_k, X_j] = \sum_{\sigma=1}^{r} c_{kj\sigma}\,X_\sigma f,$$

from which it follows:

$$\sum_{\nu=1}^{s} \left\{ \xi_{k\nu}\,\frac{\partial \xi_{ji}}{\partial x_\nu} - \xi_{j\nu}\,\frac{\partial \xi_{ki}}{\partial x_\nu} \right\} = \sum_{\sigma=1}^{r} c_{kj\sigma}\,\xi_{\sigma i},$$

and therefore, we get simply:

$$[W_k, W_j] = \sum_{\sigma=1}^{r} c_{kj\sigma} W_\sigma f.$$

From this, we see that $W_1 f, \dots, W_r f$ generate an r-term group in the $2s$ variables $x_1, \dots, x_s$, $\zeta_1, \dots, \zeta_s$; but this was closely presumed.

There can be n, amongst the infinitesimal transformations $X_1 f, \dots, X_r f$, say $X_1 f, \dots, X_n f$ that are linked together by no linear relation of the form:

$$\chi_1(x_1, \dots, x_s) X_1 f + \dots + \chi_n(x_1, \dots, x_s) X_n f = 0,$$

while $X_{n+1} f, \dots, X_r f$ can be expressed in the following way:

$$X_{n+k} f \equiv \sum_{\nu=1}^{n} \varphi_{k\nu}(x_1, \dots, x_s) X_\nu f \qquad (k=1 \cdots r-n). \tag{3}$$

Under these assumptions, the differential equations (1) can also be written as:

$$X_\nu \zeta_i - Z \xi_{\nu i} = 0 \qquad (\nu=1 \cdots n;\ i=1 \cdots s)$$

$$\sum_{\nu=1}^{n} \varphi_{k\nu}(x) X_\nu \zeta_i - Z \xi_{n+k,i} = 0 \qquad (k=1 \cdots r-n;\ i=1 \cdots s).$$

Hence if the expressions $X_1 \zeta_1, \dots, X_n \zeta_i$ are taken away, we obtain between $\zeta_1, \dots, \zeta_s$ the finite equations:

$$\sum_{\pi=1}^{s} \left\{ \frac{\partial \xi_{n+k,i}}{\partial x_\pi} - \sum_{\nu=1}^{n} \varphi_{k\nu} \frac{\partial \xi_{\nu i}}{\partial x_\pi} \right\} \zeta_\pi = 0 \tag{4}$$

$$(k=1 \cdots r-n;\ i=1 \cdots s).$$

It stands to reason that every system of equations of the form (2) which admits the group $W_1 f, \dots, W_r f$ must comprise the equations (4).

The equations (4) are linear and homogeneous in $\zeta_1, \dots, \zeta_s$; so if amongst them, one finds exactly s that are mutually independent, then $\zeta_1 = 0$, ..., $\zeta_s = 0$ is the only system of solutions which satisfies the equations. In this case, there is only one system of equations of the form (2) which admits the group $W_1 f, \dots, W_r f$, namely the system of equations: $\zeta_1 = 0, \dots, \zeta_s = 0$, hence there is no infinitesimal transformation Zf which is interchangeable with all $X_k f$.

Alternatively, assume that the equations (4) reduce to fewer than s, say to $m < s$ independent equations. We will see that in this case, aside from the useless system $\zeta_1 = 0$, ..., $\zeta_s = 0$, there are also other systems of the form (2) which admit the group $W_1 f, \dots, W_r f$.

Above all, we observe that the system of equations (4) admits the group $W_1 f, \dots, W_r f$.

In order to prove this, we write down the matrix which is associated to the infinitesimal transformations $W_1 f, \dots, W_r f$:

$$(5)\qquad \begin{vmatrix} \xi_{11} & \cdot\cdot & \xi_{1s} & \sum_{\nu=1}^{s}\frac{\partial\xi_{11}}{\partial x_\nu}\zeta_\nu & \cdot\cdot & \sum_{\nu=1}^{s}\frac{\partial\xi_{1s}}{\partial x_\nu}\zeta_\nu \\ \cdot & \cdot\cdot & \cdot & \cdot & \cdot\cdot & \cdot \\ \xi_{r1} & \cdot\cdot & \xi_{rs} & \sum_{\nu=1}^{s}\frac{\partial\xi_{r1}}{\partial x_\nu}\zeta_\nu & \cdot\cdot & \sum_{\nu=1}^{s}\frac{\partial\xi_{rs}}{\partial x_\nu}\zeta_\nu \end{vmatrix}.$$

We will show that (4) belongs to the system of equations that one obtains by setting equal to zero all $(n+1)\times(n+1)$ determinants of this matrix. As a result, according to Chap. 14, Theorem 39, p. 240, it will be proved that (4) admits the group $W_1 f,\dots,W_r f$.

Amongst the $(n+1)\times(n+1)$ determinants of the matrix (5), there are those of the form:

$$D = \begin{vmatrix} \xi_{1k_1} & \cdot\cdot & \xi_{1k_n} & \sum_{\nu=1}^{s}\frac{\partial\xi_{1\sigma}}{\partial x_\nu}\zeta_\nu \\ \cdot & \cdot\cdot & \cdot & \cdot \\ \xi_{nk_1} & \cdot\cdot & \xi_{nk_n} & \sum_{\nu=1}^{s}\frac{\partial\xi_{n\sigma}}{\partial x_\nu}\zeta_\nu \\ \cdot & \cdot\cdot & \cdot & \cdot \\ \xi_{n+j,k_1} & \cdot\cdot & \xi_{n+j,k_n} & \sum_{\nu=1}^{s}\frac{\partial\xi_{n+j,\sigma}}{\partial x_\nu}\zeta_\nu \end{vmatrix}.$$

If we use the identities following from (3):

$$\xi_{n+j,\pi} \equiv \sum_{\nu=1}^{n}\varphi_{j\nu}(x)\,\xi_{\nu\pi} \qquad (\pi=1\cdots s;\ j=1\cdots r-n),$$

then we can obviously write D as:

$$D=\sum \pm\xi_{1k_1}\cdots\xi_{nk_n}\sum_{\nu=1}^{s}\left\{\frac{\partial\xi_{n+j,\sigma}}{\partial x_\nu}-\sum_{\tau=1}^{n}\varphi_{j\tau}\frac{\partial\xi_{\tau\sigma}}{\partial x_\nu}\right\}\zeta_\nu.$$

Now, under the assumptions made, not all determinants of the form:

$$\sum \pm\xi_{1k_1}\cdots\xi_{nk_n}$$

vanish identically; so if we set all determinants of the form D equal to zero, we obtain either relations between the $x_1,\dots,x_s$ alone, or we obtain the system of equations (4). But as is easy to see, this system of equations actually brings to zero all $(n+1)\times(n+1)$ determinants of the matrix (5), hence it admits the group $W_1 f,\dots,W_r f$.

The determination of the most general system of equations (2) which admits the group $W_1 f,\dots,W_r f$ and which comprises the equations (4) can now be executed on the basis of Chap. 14, p. 246 up to p. 248.

The system of equations (4) brings to zero all $(n+1)\times(n+1)$ determinants of the matrix (5), but not all $n\times n$ determinants; likewise, a system of equations of the form (2) cannot make zero all $n\times n$ determinants of the matrix (5). Thus, we proceed in the following way: We solve the equations (4) with respect to m of the quantities $\zeta_1,\dots,\zeta_s$, say with respect to $\zeta_1,\dots,\zeta_m$, then, following the introduction of the cited developments, we form the reduced infinitesimal transformations $\overline{W}_1 f,\dots,\overline{W}_r f$ in the $2s-m$ variables $x_1,\dots,x_s$, $\zeta_{m+1},\dots,\zeta_s$, and lastly, we determine $2s-m-n$

arbitrary independent solutions of the n-term complete system which is formed by the n equations: $\overline{W}_1 f = 0, \ldots, \overline{W}_n f = 0$.

The n equations $\overline{W}_1 f = 0, \ldots, \overline{W}_n f = 0$ are solvable with respect to n of the differential quotients $\partial f / \partial x_1, \ldots, \partial f / \partial x_s$, hence its $2s - m - n$ independent solutions are mutually independent relatively to $s - n$ of the x and to the variables $\zeta_{m+1}, \ldots, \zeta_s$ (cf. Chap. 15, Theorem 12, p. 105). Now, amongst the solutions of the complete system $\overline{W}_k f = 0$, there are exactly $s - n$ independent ones which satisfy at the same time the n-term complete system: $X_1 f = 0, \ldots, X_n f = 0$, and hence depend only upon the x, and they can be called:

$$\mathfrak{u}_1(x_1, \ldots, x_s), \ \ldots, \ \mathfrak{u}_{s-n}(x_1, \ldots, x_s).$$

So, if:

$$\mathfrak{B}_1(\zeta_{m+1}, \ldots, \zeta_s, x_1, \ldots, x_s), \ \ldots, \ \mathfrak{B}_{s-m}(\zeta_{m+1}, \ldots, \zeta_s, x_1, \ldots, x_s)$$

are $s - m$ arbitrary mutually independent, and independent of the $\mathfrak{u}$, solutions of the complete system $\overline{W}_k f = 0$, then necessarily, these solutions are mutually independent relative to $\zeta_{m+1}, \ldots, \zeta_s$.

Now, we obtain the most general system of equations (2) which admits the group $W_1 f, \ldots, W_r f$ by adding to the equations (4), in the most general way, $s - m$ mutually independent relations between $\mathfrak{u}_1, \ldots, \mathfrak{u}_{s-n}, \mathfrak{B}_1, \ldots, \mathfrak{B}_{s-m}$ that are solvable with respect to $\zeta_{m+1}, \ldots, \zeta_s$. It is clear that these relations must be solvable with respect to $\mathfrak{B}_1, \ldots, \mathfrak{B}_{s-m}$, so that they can be brought to the form:

$$\text{(6)} \quad \mathfrak{B}_\mu(\zeta_{m+1}, \ldots, \zeta_s, x_1, \ldots, x_s) = \Omega_\mu\big(\mathfrak{u}_1(x), \ldots, \mathfrak{u}_{s-n}(x)\big) \qquad (\mu = 1 \cdots s - m).$$

Here, the Ω_μ are subject to no restriction at all, and they are absolutely arbitrary functions of their arguments.

Thus, if we add the equations (6) to (4) and, what is always possible, if we solve all of them with respect to $\zeta_1, \ldots, \zeta_s$, then we obtain the most general system of equations (2) which admits the group $W_1 f, \ldots, W_r f$ and as a result, the most general system of solutions to the differential equations (1). Visibly, this most general system of solutions contains $s - m$ arbitrary functions of $\mathfrak{u}_1, \ldots, \mathfrak{u}_{s-n}$.

But now, the differential equations (1) are linear and homogeneous in the unknowns $\zeta_1, \ldots, \zeta_s$; thus, it can be concluded that its most general system of solutions $\zeta_1, \ldots, \zeta_s$ can be deduced from $s - m$ particular systems of solutions:

$$\zeta_{\mu 1}(x), \ \ldots, \ \zeta_{\mu s}(x) \qquad (\mu = 1 \cdots s - m)$$

in the following way:

$$\zeta_i = \chi_1(\mathfrak{u}_1, \ldots, \mathfrak{u}_{s-n})\, \zeta_{1i} + \cdots + \chi_{s-m}(\mathfrak{u}_1, \ldots, \mathfrak{u}_{s-n})\, \zeta_{s-m,\,i}$$
$$(i = 1 \cdots s),$$

where the χ are completely arbitrary functions of the u; naturally, the particular system of solutions in question must be constituted in such a way that there are no $s-m$ functions $\psi_1(\mathfrak{u}_1,\dots,\mathfrak{u}_{s-n})$, ..., $\psi_{s-m}(\mathfrak{u}_1,\dots,\mathfrak{u}_{s-n})$ which satisfy identically the s equations:

$$\psi_1(\mathfrak{u})\,\zeta_{1i}+\cdots+\psi_{s-m}(\mathfrak{u})\,\zeta_{s-m,\,i}=0 \qquad (i=1\cdots s).$$

It can even be proved that actually, there are no $s-m$ functions $\Psi_1(x_1,\dots,x_s)$, ..., $\Psi_{s-m}(x_1,\dots,x_s)$ which satisfy identically the $s-m$ equations:

$$\Psi_1(x)\,\zeta_{1i}+\cdots+\Psi_{s-m}\,\zeta_{s-m,\,i}=0 \qquad (i=1\cdots s-m).$$

Indeed, the $s-m$ equations:

$$\zeta_{m+\sigma}=\sum_{\mu=1}^{s-m}\chi_\mu(\mathfrak{u})\,\zeta_{\mu,\,m+\sigma} \qquad (\sigma=1\cdots s-m)$$

are obviously equivalent to the equations (6), hence they must be solvable with respect to $\chi_1,\dots,\chi_{s-m}$ and the determinant:

$$\sum\pm\zeta_{1,\,m+1}\cdots\zeta_{s-m,\,m+s-m}$$

should not vanish identically.

According to these preparations, we can finally determine the form that the most general infinitesimal transformation Zf interchangeable with $X_1f,\dots,X_rf$ possesses. The concerned transformation reads:

$$Zf=\sum_{\mu=1}^{s-m}\chi_\mu(\mathfrak{u}_1,\dots,\mathfrak{u}_{s-m})\,Z_\mu f,$$

where the $s-m$ infinitesimal transformations:

$$Z_\mu f=\sum_{i=1}^{s}\zeta_{\mu i}(x_1,\dots,x_s)\,\frac{\partial f}{\partial x_i} \qquad (\mu=1\cdots s-m)$$

are linked together by no linear relation of the form:

$$\Psi_1(x_1,\dots,x_s)\,Z_1f+\cdots+\Psi_{s-m}(x_1,\dots,x_s)\,Z_{s-m}f=0.$$

In addition, since according to Proposition 1, p. 381, every infinitesimal transformation $[Z_\mu,Z_\nu]$ is also interchangeable with $X_1f,\dots,X_rf$, there are relations of the specific form:

$$[Z_\mu,Z_\nu]=\sum_{\pi=1}^{s-m}\omega_{\mu\nu\pi}(\mathfrak{u}_1,\dots,\mathfrak{u}_{s-m})\,Z_\pi f \qquad (\mu,\,\nu=1\cdots s-m).$$

We thus see: There always exists a transformation which is interchangeable with $X_1 f,\dots,X_r f$ when and only when the number m defined on p. 383 is smaller than the number s of the variables x. If $m < s$ and if at the same time, $n < s$, then the group $X_1 f,\dots,X_r f$ is intransitive, so the most general infinitesimal transformation Zf interchangeable with $X_1 f,\dots,X_r f$ depends upon arbitrary functions. If by contrast $n = s$, then the group $X_1 f,\dots,X_r f$ is transitive, so the most general transformation Zf interchangeable with $X_1 f,\dots,X_r f$ can be linearly deduced from $s - m$ independent infinitesimal transformations: $Z_1 f,\dots,Z_{s-m} f$; according to a remark made earlier on (p. 381), the concerned infinitesimal transformations then generate an $(s-m)$-term group. —

We know that an infinitesimal transformation interchangeable with $X_1 f,\dots,X_r f$ exists only when the equations (4) reduce to less than s independent equations. We can give a somewhat more lucid interpretation of this condition by remembering the identities (3), or the equivalent identities:

$$\xi_{n+k,i} - \sum_{\nu=1}^{n} \varphi_{k\nu}(x)\, \xi_{\nu i} \equiv 0 \qquad (k=1\cdots r-n;\ i=1\cdots s),$$

that define the functions $\varphi_{k\nu}$. Indeed, if we differentiate the identities just written with respect to x_j, we obtain the following identities:

$$\frac{\partial \xi_{n+k,i}}{\partial x_j} - \sum_{\nu=1}^{n} \left\{ \varphi_{k\nu} \frac{\partial \xi_{\nu i}}{\partial x_j} + \xi_{\nu i} \frac{\partial \varphi_{k\nu}}{\partial x_j} \right\} \equiv 0$$
$$(k=1\cdots r-n;\ i=1,\ j\cdots s),$$

by virtue of which the equations (4) can be replaced by the equivalent equations:

$$\sum_{\nu=1}^{n} \xi_{\nu i} \sum_{j=1}^{s} \zeta_j \frac{\partial \varphi_{k\nu}}{\partial x_j} = 0 \qquad (i=1\cdots s;\ k=1\cdots r-n).$$

But since not all determinants of the form $\sum \pm \xi_{k_1} \cdots \xi_{k_n}$ vanish identically, then in turn, the latter equations are equivalent to:

$$\text{(4')} \qquad \sum_{j=1}^{s} \zeta_j \frac{\partial \varphi_{k\nu}}{\partial x_j} = 0 \qquad (k=1\cdots r-n;\ \nu=1\cdots n).$$

Thus, there is a transformation Zf interchangeable with $X_1 f,\dots,X_r f$ only when the linear equations (4') in the ζ_j reduce to fewer than s independent equations.

The equations (4') are more clearly arranged than the equations (4), and they have in addition a simple meaning, for they express that each one of the $n(r-n)$ functions $\varphi_{k\nu}(x_1,\dots,x_s)$ admits all the infinitesimal transformations Zf.

If, amongst the equations (4), there are exactly m that are mutually independent, then naturally, amongst the equations (4'), there are also exactly m that are mutually independent, and therefore, m is nothing but the number of independent functions amongst the $n(r-n)$ functions $\varphi_{k\nu}(x)$.

We recapitulate the obtained result:

Theorem 67. *If, amongst the r infinitesimal transformations:*

$$X_k f = \sum_{i=1}^{s} \xi_{ki}(x_1, \dots, x_s) \frac{\partial f}{\partial x_i} \qquad (k=1\cdots r),$$

of an r-term group, $X_1 f, \dots, X_n f$, *say, are linked together by no linear relation, while* $X_{n+1} f, \dots, X_r f$ *can be expressed linearly in terms of* $X_1 f, \dots, X_n f$*:*

$$X_{n+j} f \equiv \sum_{\nu=1}^{n} \varphi_{j\nu}(x_1, \dots, x_s) X_\nu f \qquad (j=1\cdots r-n),$$

and if amongst the $n(r-n)$ *functions* $\varphi_{k\nu}$*, there exist exactly s that are mutually independent, then there is no infinitesimal transformation which is interchangeable with all* $X_k f$*; by contrast, if amongst the functions* $\varphi_{k\nu}$*, there exist fewer than s, say only m, that are mutually independent, then there are infinitesimal transformations which are interchangeable with* $X_1 f, \dots, X_r f$*, and to be precise, the most general infinitesimal transformations* Zf *of this nature are of the form:*

$$Zf = \psi_1(\mathfrak{u}_1, \dots, \mathfrak{u}_{s-n}) Z_1 f + \cdots + \psi_{s-m}(\mathfrak{u}_1, \dots, \mathfrak{u}_{s-n}) Z_{s-m} f,$$

where $\mathfrak{u}_1, \dots, \mathfrak{u}_{s-n}$ *denote independent solutions of the n-term complete system:* $X_1 f = 0, \dots, X_n f = 0$*, where furthermore* $\psi_1, \dots, \psi_{s-m}$ *mean arbitrary functions of their arguments, and lastly, where the infinitesimal transformations:*

$$Z_\mu f = \sum_{i=1}^{s} \zeta_{\mu i}(x_1, \dots, x_s) \frac{\partial f}{\partial x_i} \qquad (\mu=1\cdots s-m)$$

interchangeable with $X_1 f, \dots, X_r f$ *and linked together by no linear relation stand pairwise in relationships of the form:*[1]

$$[Z_\mu, Z_\nu] = \sum_{\pi=1}^{s-m} \omega_{\mu\nu\pi}(\mathfrak{u}_1, \dots, \mathfrak{u}_{s-m}) Z_\pi f.$$

Furthermore, a part of the result stated in this proposition follows immediately from the developments of the previous chapter. Indeed, if the number of independent functions amongst the functions $\varphi_{k\nu}(x_1, \dots, x_s)$ is exactly equal to s, then according to p. 357, the equations:

$$\varphi_{k\nu}(x_1', \dots, x_s') = \varphi_{k\nu}(x_1, \dots, x_s) \qquad (k=1\cdots r-n;\ \nu=1\cdots n)$$

represent the most general transformation which transfers $X_1 f, \dots, X_r f$ to $X_1' f, \dots, X_r' f$, respectively. By contrast, if amongst the $\varphi_{k\nu}$, there are fewer than s that are mutually independent, then according to Theorem 65, p. 365, there is a

[1] LIE suggested this general theorem in Math. Ann. Vol. XXV.

continuous set [MENGE] of transformations which leaves invariant every $X_k f$; we have already remarked on p. 380 that the totality of all such transformations forms a group, about which it can be directly seen that it comprises one-term groups.

It seems to be adequate to expressly state the following proposition:

Proposition 2. *If there exists an infinitesimal transformation which is interchangeable with all infinitesimal transformations $X_1 f, \dots, X_r f$ of an r-term group of the space $x_1, \dots, x_s$, then the most general infinitesimal transformation interchangeable with $X_1 f, \dots, X_r f$ contains arbitrary functions, as soon as the group $X_1 f, \dots, X_r f$ is intransitive, but by contrast, it contains only arbitrary parameters, as soon as the group is transitive.*

§ 96. Of outstanding significance is the case where the group $X_k f$ is simply transitive, thus the case $s = r = n$.

If one would desire, in this case, to know the most general transformation $x_i' = \Phi_i(x_1, \dots, x_n)$ by virtue of which every $X_k f$ takes the form $X_k' f$, then one would only have to seek n independent solutions $\Omega_1, \dots, \Omega_n$ of the n-term complete system:

$$X_k f + X_k' f = 0 \qquad {\scriptstyle (k=1\cdots n)}$$

and to set these solutions equal to arbitrary constants $a_1, \dots, a_n$. The equations:

$$\Omega_k(x_1, \dots, x_n, x_1', \dots, x_n') = a_k \qquad {\scriptstyle (k=1\cdots n)}$$

are then solvable both with respect to the x and with respect to the x', and they represent the transformation demanded.

From the beginning, we know that the totality of all transformations $\Omega_k = a_k$ forms a group. At present, we see that this group is n-term and simply transitive; this follows immediately from the form in which the group is presented, resolved with respect to its n parameters.

The group:

$$\Omega_1(x, x') = a_1, \ \dots, \ \Omega_n(x, x') = a_n$$

contains the identity transformation and n independent infinitesimal transformations:

$$Z_k f = \sum_{i=1}^{n} \zeta_{ki}(x_1, \dots, x_n) \frac{\partial f}{\partial x_i} \qquad {\scriptstyle (k=1\cdots n)},$$

and this results from the developments of the preceding section; but this is also clear in itself, since the transformations of the group are ordered as inverses by pairs, and hence Theorem 56 in Chap. 18, p. 328, finds an application. Besides, it follows from the transitivity of the group $Z_1 f, \dots, Z_n f$ that $Z_1 f, \dots, Z_n f$ are linked together by no linear relation of the form $\sum \chi_i(x_1, \dots, x_s) Z_i f = 0$.

Between the two *simply transitive* groups $X_k f$ and $Z_i f$, there is a full relationship of reciprocity [RECIPROCITÄTSVERHÄLTNISS]. If the $X_k f$ are given, then the general infinitesimal transformation $Zf = \sum e_i Z_i f$ is completely defined by the n equations:

$$X_k\big(Z(f)\big) - Z\big(X_k(f)\big) = 0 \qquad (k=1\cdots n);$$

on the other hand, if the $Z_i f$ are given then the n equations:

$$X\big(Z_i(f)\big) - Z_i\big(X(f)\big) = 0 \qquad (i=1\cdots n)$$

determine in the same way the general infinitesimal transformation:

$$Xf = \sum e_k X_k f.$$

However, the peculiar relationship in which the two groups stand has not yet been written down exhaustively. Indeed, as we will now show, the two groups are also equally composed; since both are simply transitive, it then follows immediately that they are similar to each other (cf. Chap. 19, Theorem 64, p. 353).

In the infinitesimal transformations $X_k f$ and $Z_i f$, we imagine that the $\xi_{k\nu}$ and the $\zeta_{i\nu}$ are expanded in powers of $x_1, \dots, x_n$, and we assume at the same time that $x_1 = 0$, $\dots$, $x_n = 0$ is a point in general position. According to Chap. 13, p. 229, since the group $X_k f$ is simply transitive, it contains exactly n infinitesimal transformations of zeroth order in $x_1, \dots, x_n$ out of which no infinitesimal transformation of first order, or of higher order, can be linearly deduced.

We can therefore imagine that $X_1 f, \dots, X_n f$ are replaced by n other independent infinitesimal transformations $\mathfrak{X}_1 f, \dots, \mathfrak{X}_n f$ which have the form:

$$\mathfrak{X}_k f = \frac{\partial f}{\partial x_k} + \sum_{\mu,\,\nu}^{1\cdots n} h_{k\mu\nu}\, x_\mu \frac{\partial f}{\partial x_\nu} + \cdots$$
$$(k=1\cdots n),$$

after leaving out the terms of second order and of higher order. In the same way, we can replace $Z_1 f, \dots, Z_n f$ by n independent infinitesimal transformations of the form:

$$\mathfrak{Z}_i f = -\frac{\partial f}{\partial x_i} + \sum_{\mu,\,\nu}^{1\cdots n} l_{i\mu\nu}\, x_\mu \frac{\partial f}{\partial x_\nu} + \cdots$$
$$(i=1\cdots n).$$

After these preparations, we form the equations $[\mathfrak{X}_k, \mathfrak{Z}_i] = 0$; the same equation takes the form:

$$\sum_{\nu=1}^{n} (l_{ik\nu} + h_{ki\nu}) \frac{\partial f}{\partial x_\nu} + \cdots = 0,$$

from which it follows that:

$$l_{ik\nu} = -h_{ki\nu}.$$

Computations of the same sort yield:

$$[\mathfrak{X}_k, \mathfrak{X}_j] = \sum_{\nu=1}^{n} (h_{jk\nu} - h_{kj\nu}) \frac{\partial f}{\partial x_\nu} + \cdots$$

and:

$$\begin{aligned}[\mathfrak{Z}_k, \mathfrak{Z}_j] &= \sum_{\nu=1}^{n} (-l_{jk\nu} + l_{kj\nu}) \frac{\partial f}{\partial x_\nu} + \cdots \\ &= \sum_{\nu=1}^{n} (h_{kj\nu} - h_{jk\nu}) \frac{\partial f}{\partial x_\nu} + \cdots .\end{aligned}$$

On the other hand, we have:

$$[\mathfrak{X}_k, \mathfrak{X}_j] = \sum_{\nu=1}^{n} \mathfrak{c}_{kj\nu} \mathfrak{X}_\nu f, \qquad [\mathfrak{Z}_k, \mathfrak{Z}_i] = \sum_{\nu=1}^{n} \mathfrak{c}'_{kj\nu} \mathfrak{Z}_\nu f;$$

here, we insert the expressions just found for: $\mathfrak{X}_\nu f$, $\mathfrak{Z}_\nu f$, $[\mathfrak{X}_k, \mathfrak{X}_j]$, $[\mathfrak{Z}_k, \mathfrak{Z}_j]$ and afterwards, we make the substitution: $x_1 = 0, \dots, x_n = 0$; we obtain:

$$\mathfrak{c}_{kj\nu} = h_{jk\nu} - h_{kj\nu} = \mathfrak{c}'_{kj\nu}.$$

Thus, our two groups are effectively equally composed, and in consequence of that, as already remarked above, they are similar to each other. As a result, the following holds:

Theorem 68. *If $X_1 f, \dots, X_n f$ are independent infinitesimal transformations of a simply transitive group in the variables $x_1, \dots, x_n$, then the n equations $[X_k, Z] = 0$ define the general infinitesimal transformation Zf of a second simply transitive group $Z_1 f, \dots, Z_n f$ which has the same composition as the group $X_1 f, \dots, X_n f$ and which is at the same time similar to it. The relationship between these two simply transitive groups is a reciprocal relationship: each one of the two groups consists of the totality of all one-term groups whose transformations are interchangeable with all transformations of the other group.*[2]

It is convenient to call the groups $X_k f$ and $Z_i f$ *reciprocal* [RECIPROKE] transformation groups, or always, the one as the reciprocal group of the other.

If we recall from Chap. 16, p. 288, that the excellent infinitesimal transformations $e_1 X_1 f + \cdots + e_r X_r f$ of the group $X_1 f, \dots, X_n f$ are defined by the n equations $[X_k, Y] = 0$, then we can state the following proposition:

Proposition 3. *The common infinitesimal transformations of two reciprocal simply transitive groups are at the same time the excellent infinitesimal transformations of both groups.*

Let the two n-term groups $X_1 f, \dots, X_n f$ and $Z_1 f, \dots, Z_n f$ in the n variables $x_1, \dots, x_n$ be simply transitive and reciprocal to each other. Then, if after the introduction of new variables $x'_1, \dots, x'_n$, the $X_k f$ are transferred to $X'_k f$ and the

---

[2] LIE communicated Theorem 68 in the Gesellschaft der Wissenschaften zu Christiania in Nov. 1882 and in May 1883; cf. also Math. Ann. Vol. XXV, p. 107 sq.

$Z_k f$ to $Z'_k f$, the two simply transitive groups $X'_1 f, \dots, X'_n f$ and $Z'_1 f, \dots, Z'_n f$ are also reciprocal to each other; this follows immediately from the relations: $[X'_k, Z'_j] = [X_k, Z_j] \equiv 0$.

From this, in particular, we get the following

Proposition 4. *If an n-term simply transitive group in n variables remains invariant by a transformation, then at the same time, its reciprocal simply transitive group remains invariant.*

Lastly, from this proposition we obtain the next one:

Proposition 5. *The largest group of the space $x_1, \dots, x_n$ in which an n-term simply transitive group of this space is contained as an invariant subgroup coincides with the largest subgroup in which the reciprocal group of this simply transitive group is contained as an invariant subgroup.*

We want to give a couple of simple examples of the preceding general developments concerning simply transitive groups.

The group:

$$\frac{\partial f}{\partial x}, \qquad x\frac{\partial f}{\partial x} + y\frac{\partial f}{\partial y}$$

of x, y-space is simply transitive. The finite equations of the reciprocal group are obtained by integration of the complete system:

$$\frac{\partial f}{\partial x} + \frac{\partial f}{\partial x'} = 0, \qquad x\frac{\partial f}{\partial x} + y\frac{\partial f}{\partial y} + x'\frac{\partial f}{\partial x'} + y'\frac{\partial f}{\partial y'} = 0$$

in the following form:

$$\frac{x' - x}{y} = a, \qquad \frac{y'}{y} = b,$$

hence they are solved with respect to x' and y':

$$x' = x + ay, \qquad y' = by.$$

The infinitesimal transformations of the reciprocal group are therefore:

$$y\frac{\partial f}{\partial x}, \qquad y\frac{\partial f}{\partial y}.$$

An interesting example is provided by the six-term projective group of a non-degenerate surface of second order in ordinary space. Indeed, this group contains two three-term simply transitive groups which are reciprocal to each other, of which the first group leaves fixed all generatrices of the one family, while the other group leaves fixed all generatrices of the other family.

If $z - xy = 0$ is the equation of the surface, then:

$$X_1 f = \frac{\partial f}{\partial x} + y\frac{\partial f}{\partial z}, \qquad X_2 f = x\frac{\partial f}{\partial x} + z\frac{\partial f}{\partial z},$$
$$X_3 f = x^2\frac{\partial f}{\partial x} + (xy - z)\frac{\partial f}{\partial y} + xz\frac{\partial f}{\partial z}$$

is one of the two simply transitive groups and:

$$Z_1 f = \frac{\partial f}{\partial y} + x\frac{\partial f}{\partial z}, \qquad Z_2 f = y\frac{\partial f}{\partial y} + z\frac{\partial f}{\partial z}$$
$$Z_3 f = (xy - z)\frac{\partial f}{\partial x} + y^2\frac{\partial f}{\partial y} + yz\frac{\partial f}{\partial z}$$

is the other.

The two reciprocal groups are naturally similar to each other; but it is to be observed that they are similar through a *projective* transformation, namely through every projective transformation which interchanges the two families of generatrices of the surface.

§ 97. We proceed to a general study of reciprocal simply transitive groups.

Let $X_1 f, \dots, X_n f$ and $Z_1 f, \dots, Z_n f$ be two simply transitive and reciprocal groups in the variables $x_1, \dots, x_n$.

If n were equal to 1, then the two groups would be identical to each other, as one easily convinces oneself; we therefore assume that n is larger than 1. Then the group $Z_1 f, \dots, Z_n f$ certainly contains subgroups. If $Z_1 f, \dots, Z_m f$ $(m < n)$ is such a subgroup, then the m equations:

$$Z_1 f = 0, \ \dots, \ Z_m f = 0$$

form an m-term complete system which, as follows from the identities:

$$[X_i, Z_1] \equiv 0, \ \dots, \ [X_i, Z_m] \equiv 0 \qquad (i = 1 \cdots n),$$

admits the group $X_1 f, \dots, X_n f$ (cf. Chap. 8, Theorem 20, p. 155). Consequently, the group $X_1 f, \dots, X_n f$ is *imprimitive*.

If $u_1, \dots, u_{n-m}$ are independent solutions of the complete system $Z_1 f = 0, \dots, Z_m f = 0$, then according to Chap. 8, Proposition 1, p. 153, there are relations of the form:

$$X_i u_\nu = \omega_{i\nu}(u_1, \dots, u_{n-m}) \qquad (i = 1 \cdots n; \ \nu = 1 \cdots n - m),$$

and hence (cf. p. 157) the ∞^{n-m} m-times extended manifolds:

$$u_1 = \text{const.}, \ \dots, \ u_{n-m} = \text{const.}$$

are mutually permuted by the group $X_1 f, \dots, X_r f$.

We therefore have the:

Proposition 6. *Every simply transitive group: $X_1f,\dots,X_nf$ in $n>1$ variables: $x_1,\dots,x_n$ is imprimitive; if $Z_1f,\dots,Z_nf$ is the associated reciprocal group and if $Z_1f,\dots,Z_mf$ is an arbitrary subgroup of it with the invariants $u_1,\dots,u_{n-m}$, then the m-term complete system $Z_1f=0,\ \dots,\ Z_mf=0$ admits the group $X_1f,\dots,X_nf$ and the ∞^{n-m} m-times extended manifolds: $u_1=\text{const.},\ \dots,\ u_{n-m}=\text{const.}$ are mutually permuted by this group.*

The preceding proposition shows that every m-term subgroup of the group $Z_1f,\dots,Z_nf$ provides a completely determined decomposition of the space $x_1,\dots,x_n$ in ∞^{n-m} m-times extended manifolds invariant by the group $X_1f,\dots,X_nf$. Hence, if we imagine that all subgroups of the group $Z_1f,\dots,Z_nf$ are determined, and that all the associated invariants are computed, then we obtain infinitely many decompositions of the space invariant by the group $X_1f,\dots,X_nf$. It can be proved that all existing invariant decompositions are found in this way, so that Proposition 6 can be reversed.

Indeed, let $Y_1f=0,\ \dots,\ Y_mf=0$ be any m-term complete system which admits the group $X_1f,\dots,X_nf$, and let $u_1,\dots,u_{n-m}$ be independent solutions of this complete system, so that the family of the ∞^{n-m} m-times extended manifolds:

$$u_1=\text{const.},\ \dots,\ u_{n-m}=\text{const.}$$

represent a decomposition of the space $x_1,\dots,x_n$ invariant by the group $X_1f,\dots,X_nf$. We claim that the group $Z_1f,\dots,Z_nf$ contains a completely determined m-term subgroup which leaves individually invariant each one of these ∞^{n-m} manifolds; this is just the reverse of Proposition 6.

First, we introduce the functions $u_1,\dots,u_{n-m}$ and m arbitrary mutually independent functions: $v_1,\dots,v_m$ of the x that are also independent of $u_1,\dots,u_{n-m}$ as variables in our reciprocal groups, and we get:

$$X_kf=\sum_{\nu=1}^{n-m}\omega_{k\nu}(u_1,\dots,u_{n-m})\frac{\partial f}{\partial u_\nu}+\sum_{\mu=1}^{m}X_k\,v_\mu\,\frac{\partial f}{\partial v_\mu}\qquad(k=1\cdots n)$$

and:

$$Z_kf=\sum_{\nu=1}^{n-m}Z_k\,u_\nu\,\frac{\partial f}{\partial u_\nu}+\sum_{\mu=1}^{m}Z_k\,v_\mu\,\frac{\partial f}{\partial v_\mu}\qquad(k=1\cdots n),$$

where the $X_k\,v_\mu$, $Z_k\,u_\nu$ and $Z_k\,v_\mu$ are certain functions of the u and the v. Our claim now obviously amounts to the fact that m independent infinitesimal transformations should be linearly deducible from $Z_1f,\dots,Z_nf$ which should absolutely not transform $u_1,\dots,u_{n-m}$, hence in which the coefficients of $\partial f/\partial u_1,\ \dots,\ \partial f/\partial u_{n-m}$ should all be zero.

In order to be able to prove this, we must compute the coefficients of $\partial f/\partial u_1,\dots,$ $\partial f/\partial u_{n-m}$ in the general infinitesimal transformation $Zf=e_1Z_1f+\cdots+e_nZ_nf$ of the group $Z_1f,\dots,Z_nf$.

The infinitesimal transformation Zf is completely determined by the relations:

$$X_1\big(Z(f)\big)-Z\big(X_1(f)\big)=0,\ \dots,\ X_n\big(Z(f)\big)-Z\big(X_n(f)\big)=0.$$

If we replace f in these equations by u_ν, we obtain the relations:

$$X_1(Z\,u_\nu)-Z\,\omega_{1\nu}=0,\ \dots,\ X_n(Z\,u_\nu)-Z\,\omega_{n\nu}=0 \qquad (\nu=1\cdots n-m).$$

Consequently, the functions $Z\,u_1,\dots,Z\,u_{n-m}$ are solutions of the differential equations:

$$(7)\qquad X_k\,\rho_\nu-\sum_{\mu=1}^{n-m}\frac{\partial\omega_{k\nu}}{\partial u_\mu}\,\rho_\mu=0 \qquad (k=1\cdots n;\ \nu=1\cdots n-m),$$

in which $\rho_1,\dots,\rho_{n-m}$ are to be seen as the unknown functions.

If:

(8)

$$\rho_1=\psi_1(u_1,\dots,u_{n-m},v_1,\dots,v_m),\ \dots,\ \rho_{n-m}=\psi_{n-m}(u_1,\dots,u_{n-m},v_1,\dots,v_m)$$

is an arbitrary system of solutions to the differential equations (7), then all the $n(n-m)$ expressions:

$$X_k\,\psi_\nu-\sum_{\mu=1}^{n-m}\frac{\partial\omega_{k\nu}}{\partial u_\mu}\,\rho_\mu$$

vanish identically after the substitution: $\rho_1=\psi_1,\ \dots,\ \rho_{n-m}=\psi_{n-m}$. Hence if we interpret the equations (8) as a system of equations in the $n+n-m$ variables: $u_1,\dots,u_{n-m},\ v_1,\dots,v_m,\ \rho_1,\dots,\rho_{n-m}$, we realize immediately that this system of equations admits the n infinitesimal transformations:

$$(9)\qquad U_kf=X_kf+\sum_{\nu=1}^{n-m}\left\{\sum_{\mu=1}^{n-m}\frac{\partial\omega_{k\nu}}{\partial u_\mu}\,\rho_\mu\right\}\frac{\partial f}{\partial\rho_\nu} \qquad (k=1\cdots n).$$

The converse is also clear: if one known an arbitrary system of equations of the form (8) which admits the infinitesimal transformations $U_1f,\dots,U_nf$, then one also knows a system of solutions of the differential equations (7), since the functions $\psi_1,\dots,\psi_{n-m}$ are such a system.

From this, it follows that the determination of the most general system of solutions to the differential equations (7) amounts to determining, in the $2n-m$ variables u, v, ρ, the most general system of equations (8) that admits the infinitesimal transformations $U_1f,\dots,U_nf$.

Every system of equations which admits $U_1f,\dots,U_nf$ also allows all the infinitesimal transformations $[U_i,\,U_k]$. By a calculation, we find:

$$\begin{aligned}[U_i,\,U_k]=[X_i,\,X_k]&+\sum_{\mu,\,\nu,\,\pi}^{1\cdots n-m}\left(\omega_{i\pi}\,\frac{\partial^2\omega_{k\nu}}{\partial u_\mu\partial u_\pi}-\omega_{k\pi}\,\frac{\partial^2\omega_{i\nu}}{\partial u_\mu\partial u_\pi}\right)\rho_\mu\,\frac{\partial f}{\partial\rho_\nu}\\&+\sum_{\mu,\,\nu,\,\pi}^{1\cdots n-m}\left(\frac{\partial\omega_{i\nu}}{\partial u_\mu}\,\frac{\partial\omega_{k\pi}}{\partial u_\nu}-\frac{\partial\omega_{k\nu}}{\partial u_\mu}\,\frac{\partial\omega_{i\pi}}{\partial u_\nu}\right)\rho_\mu\,\frac{\partial f}{\partial\rho_\pi},\end{aligned}$$

or else, if written differently:

$$[U_i, U_k] = [X_i, X_k] + \sum_{\mu,\, \nu}^{1\cdots n-m} \frac{\partial}{\partial u_\mu} \left(X_i\, \omega_{k\nu} - X_k\, \omega_{i\nu}\right) \rho_\mu \frac{\partial f}{\partial \rho_\nu}.$$

Now, since $X_1 f, \dots, X_n f$ generate a group, there exist relations of the form:

$$[X_i, X_k] = X_i\big(X_k(f)\big) - X_k\big(X_i(f)\big) = \sum_{\sigma=1}^{n} c_{ik\sigma} X_\sigma f,$$

hence in particular, we have:

$$\begin{aligned} X_i\left(X_k\, u_\nu\right) - X_k\left(X_i\, u_\nu\right) &= X_i\, \omega_{k\nu} - X_k\, \omega_{i\nu} \\ &= \sum_{\sigma=1}^{n} c_{ik\sigma}\, \omega_{\sigma\nu}, \end{aligned}$$

and we get:

$$[U_i, U_k] = \sum_{\sigma=1}^{n} c_{ik\sigma} U_\sigma f.$$

Consequently, the infinitesimal transformations $U_1 f, \dots, U_n f$ generate an n-term group in the $2n-m$ variables $u_1, \dots, u_{n-m}, v_1, \dots, v_m, \rho_1, \dots, \rho_{n-m}$.

Thus, the question at present is to determine the most general system of equations of the form (8) that admits the group $U_1 f, \dots, U_n f$. Since $X_1 f, \dots, X_n f$ are linked together by no linear relation, not all $n \times n$ determinants of the matrix which is associated to $U_1 f, \dots, U_n f$ vanish, and likewise, not all these determinants vanish by virtue of a system of equations of the form (8). It therefore follows from Chap. 14, Theorem 42, p. 247, that every system of equations of the form (8) which admits the group $U_1 f, \dots, U_n f$ is represented by relations between the solutions of the n-term complete system $U_1 f = 0, \dots, U_n f = 0$. Now, this complete system possesses exactly $n-m$ independent solutions, say:

$$\Psi_\mu\left(u_1, \dots, u_{n-m}, v_1, \dots, v_m, \rho_1, \dots, \rho_{n-m}\right) \qquad {\scriptstyle (k=1\cdots n-m)}$$

and to be precise, $\Psi_1, \dots, \Psi_{n-m}$ are mutually independent relative to $\rho_1, \dots, \rho_{n-m}$, because the complete system is solvable with respect to the differential quotients $\partial f/\partial u_1, \dots, \partial f/\partial u_{n-m}, \partial f/\partial v_1, \dots, \partial f/\partial v_m$ (cf. Chap. 5, Theorem 12, p. 105). From this, we conclude that the most general system of equations (8) that admits the group $U_1 f, \dots, U_n f$ can be given the form:

$$\Psi_1 = C_1, \; \dots, \; \Psi_{n-m} = C_{n-m},$$

where $C_1, \dots, C_{n-m}$ denote arbitrary constants.

If we solve the system of equations just found with respect to $\rho_1, \dots, \rho_{n-m}$, which is always possible, then we obtain the most general system of solutions to the differential equations (7), hence we see that this most general system of solutions contains exactly $n-m$ arbitrary, essential constants. Now, since the differential equations (7) are linear and homogeneous in the unknowns $\rho_1, \dots, \rho_{n-m}$, the said most general system of solutions must be of the form:

$$\rho_\mu = C'_1\,\psi_\mu^{(1)} + C'_2\,\psi_\mu^{(2)} + \cdots + C'_{n-m}\,\psi_\mu^{(n-m)} \qquad (\mu = 1 \cdots n-m), \tag{10}$$

where the $n-m$ systems of functions:

$$\psi_1^{(\nu)},\ \psi_2^{(\nu)},\ \dots,\ \psi_{n-m}^{(\nu)} \qquad (\nu = 1 \cdots n-m)$$

represent the same number $n-m$ of linearly independent systems of solutions to the differential equations (7) that are free of arbitrary constants, and where the C' are arbitrary constants. Of course, the determinant of the ψ does not vanish, for the equations (10) must be solvable with respect to $C'_1, \dots, C'_{n-m}$.

The functions: $Zu_1, \dots, Zu_{n-m}$ are, according to p. 395, solutions of the differential equations (7), hence they have the form:

$$Zu_\mu = \overline{C}_1\psi_\mu^{(1)} + \cdots + \overline{C}_{n-m}\psi_\mu^{(n-m)} \qquad (\mu = 1 \cdots n-m).$$

Here, the $\overline{C}_\nu$ are constants about which we temporarily do not know anything more precise; it could still be conceivable that they are linked together by linear relations.

From the values of the Zu_μ, it follows that the general infinitesimal transformation Zf of the group: $Z_1f, \dots, Z_nf$ can be given the following representation:

$$Zf = \sum_{\mu,\,\nu}^{1\cdots n-m} \overline{C}_\mu\,\psi_\nu^{(\mu)}\,\frac{\partial f}{\partial u_\nu} + \sum_{\mu=1}^{m} Zv_\mu\,\frac{\partial f}{\partial v_\mu}.$$

Now, since the $n-m$ expressions:

$$\sum_{\nu=1}^{n-m} \psi_\nu^{(\mu)}\,\frac{\partial f}{\partial u_\nu} \qquad (\mu = 1 \cdots n-m)$$

represent independent infinitesimal transformations, then from $Z_1f, \dots, Z_nf$, one can obviously deduce linearly at least m, hence say exactly $m+\varepsilon$, independent infinitesimal transformations in which the coefficients of $\partial f/\partial u_1, \dots, \partial f/\partial u_{n-m}$ are equal to zero. These $m+\varepsilon$ infinitesimal transformations naturally generate an $(m+\varepsilon)$-term subgroup of the group: $Z_1f, \dots, Z_nf$, and in fact, a subgroup which leaves fixed each one of the ∞^{n-m} m-times extended manifolds $u_1 = \text{const.}, \dots, u_{n-m} = \text{const.}$ But this is only possible when the whole number ε is equal to zero, since if ε were > 0, the group $Z_1f, \dots, Z_nf$ could not be simply transitive.

As a result, the claim made on p. 394 is proved and we can therefore state the following proposition:

Proposition 7. *If $X_1f, \dots, X_nf$ and $Z_1f, \dots, Z_nf$ are two reciprocal simply transitive groups in the n variables $x_1, \dots, x_n$, and if:*

$$u_1(x_1,\dots,x_n)=\text{const.},\ \dots,\ u_{n-m}(x_1,\dots,x_n)=\text{const.}$$

is an arbitrary decomposition of the space $x_1,\dots,x_n$ *into* ∞^{n-m} *m-times extended manifolds that is invariant by the group* $X_1f,\dots,X_nf$*, then the group* $Z_1f,\dots,Z_nf$ *always contains an m-term subgroup which leaves individually fixed each one of these* ∞^{n-m} *manifolds.*

By combining this proposition with Proposition 6, p. 394, we obtain the:

Theorem 69. *If the n-term group* $X_1f,\dots,X_nf$ *in the n variables* $x_1,\dots,x_n$ *is simply transitive, then one finds all m-term complete systems that this groups admits, or, what is the same, all invariant decompositions of the space* $x_1,\dots,x_n$ *in* ∞^{n-m} *m-times extended manifolds, in the following way: One first determines the simply transitive group:* $Z_1f,\dots,Z_nf$ *which is reciprocal to* $X_1f,\dots,X_nf$ *and one sets up all m-term subgroups of the former; if, say:*

$$\mathsf{Z}_\mu f=g_{\mu 1}Z_1f+\cdots+g_{\mu n}Z_nf \qquad (\mu=1\cdots m)$$

is one of the found subgroups, then the equations $\mathsf{Z}_1f=0,\ \dots,\ \mathsf{Z}_mf=0$ *represent one of the sought complete systems and they determine a decomposition of the space* $x_1,\dots,x_n$ *into* ∞^{n-m} *m-times extended manifolds that is invariant by the group* $X_1f,\dots,X_nf$*; if, for each one of the found subgroups, one forms the m-term complete system which the subgroup provides, then one obtains all m-term complete systems that the group* $X_1f,\dots,X_nf$ *admits. If one undertakes the indicated study for each of the numbers* $m=1,2,\dots,n-1$*, then one actually obtains all complete systems that the group* $X_1f,\dots,X_nf$ *admits, and therefore at the same time, all decompositions of the space* $x_1,\dots,x_n$ *that are invariant by this group.*

The above theorem contains a solution to the problem of determining all possible ways in which a given *simply transitive* group can be imprimitive.

Let the equations:

$$u_1=\text{const.},\ \dots,\ u_{n-m}=\text{const.}$$

again represent an arbitrary decomposition of the space $x_1,\dots,x_n$ into ∞^{n-m} m-times extended manifolds invariant by the group $X_1f,\dots,X_nf$, so that hence, when $u_1,\dots,u_{n-m}$, together with appropriate functions $v_1,\dots,v_m$, are introduced as new variables, $X_1f,\dots,X_nf$ receive the form:

$$X_kf=\sum_{\nu=1}^{n-m}\omega_{k\nu}(u_1,\dots,u_{n-m})\frac{\partial f}{\partial u_\nu}+\sum_{\mu=1}^{m}\overline{\xi}_{k\mu}(u_1,\dots,u_{n-m},v_1,\dots,v_m)\frac{\partial f}{\partial v_\mu}.$$

Here, not all $(n-m)\times(n-m)$ determinants of the matrix:

$$\begin{vmatrix}\omega_{11} & \cdot & \cdot & \omega_{1,n-m}\\ \cdot & \cdot & \cdot & \cdot\\ \omega_{n1} & \cdot & \cdot & \omega_{n,n-m}\end{vmatrix}$$

can vanish identically, since otherwise, $X_1 f, \dots, X_n f$ would be linked together by a linear relation, which is contrary to the assumption. So, if by $u_1^0, \dots, u_{n-m}^0$ we understand a general system of values, then according to Chap. 13, p. 234, the group $X_1 f, \dots, X_n f$ contains exactly ∞^{m-1} different infinitesimal transformations which leave invariant the system of equations:

$$u_1 = u_1^0, \; \dots, \; u_{n-m} = u_{n-m}^0;$$

naturally, these infinitesimal transformations then generate an m-term subgroup, the most general subgroup of the group $X_1 f, \dots, X_n f$ which leaves fixed the m-times extended manifold $u_1 = u_1^0, \dots, u_{n-m} = u_{n-m}^0$, or shortly M.

On the other hand, the said m-times extended manifold M now also admits m independent infinitesimal transformations of the reciprocal group $Z_1 f, \dots, Z_n f$, for according to Proposition 7, p. 397, this group contains an m-term subgroup which leaves individually fixed each of the ∞^{n-m} manifolds:

$$u_1 = \text{const.}, \; \dots, \; u_{n-m} = \text{const.}$$

In addition, it is clear that M cannot admit a larger subgroup of the group $Z_1 f, \dots, Z_n f$, for $u_1^0, \dots, u_{n-m}^0$ is supposed to be a general system of values.

From this, we see that M allows exactly m infinitesimal transformations both from the two reciprocal groups $X_1 f, \dots, X_n f$ and $Z_1 f, \dots, Z_n f$, hence from each one, a completely determined m-term subgroup.

We can make these two m-term subgroups visible by choosing an arbitrary system of values: $v_1^0, \dots, v_m^0$ in general position and by expanding the infinitesimal transformations of our two reciprocal groups in powers of: $u_1 - u_1^0$, $\dots$, $u_{n-m} - u_{n-m}^0$, $v_1 - v_1^0$, $\dots$, $v_m - v_m^0$. Indeed, if we disregard the terms of first order and of higher order, then similarly as on p. 390, we can replace the infinitesimal transformations: $X_1 f, \dots, X_n f$ by n transformations of the form:

$$\mathfrak{X}_1 f = \frac{\partial f}{\partial u_1} + \cdots, \; \dots, \; \mathfrak{X}_{n-m} f = \frac{\partial f}{\partial u_{n-m}} + \cdots,$$
$$\mathfrak{X}_{n-m+1} f = \frac{\partial f}{\partial v_1} + \cdots, \; \dots, \; \mathfrak{X}_n f = \frac{\partial f}{\partial v_m} + \cdots,$$

and we can also replace $Z_1 f, \dots, Z_n f$ by n transformations of the form:

$$\mathfrak{Z}_1 f = -\frac{\partial f}{\partial u_1} + \cdots, \; \dots, \; \mathfrak{Z}_{n-m} f = -\frac{\partial f}{\partial u_{n-m}} + \cdots,$$
$$\mathfrak{Z}_{n-m+1} f = -\frac{\partial f}{\partial v_1} + \cdots, \; \dots, \; \mathfrak{Z}_n f = -\frac{\partial f}{\partial v_m} + \cdots.$$

Here, $\mathfrak{X}_{n-m+1} f, \dots, \mathfrak{X}_n f$ are obviously independent infinitesimal transformations which leave invariant the manifold: $u_1 = u_1^0$, $\dots$, $u_{n-m} = u_{n-m}^0$, and $\mathfrak{Z}_{n-m+1} f, \dots, \mathfrak{Z}_n f$ are independent infinitesimal transformations which do the same; thus, $\mathfrak{X}_{n-m+1}f,\dots,\mathfrak{X}_nf$ and $\mathfrak{Z}_{n-m+1}f,\dots,\mathfrak{Z}_nf$ are the two m-term subgroups about which we have just spoken.

From this representation of the two subgroups, one can derive a few notable conclusions.

Between $\mathfrak{X}_1f,\dots,\mathfrak{X}_nf$, there are relations of the form:

$$[\mathfrak{X}_i,\mathfrak{X}_k]=\sum_{\nu=1}^{n}\mathfrak{c}_{ik\nu}\,\mathfrak{X}_\nu f,$$

and according to p. 390 sq., between $\mathfrak{Z}_1f,\dots,\mathfrak{Z}_nf$, there are the same relations:

$$[\mathfrak{Z}_i,\mathfrak{Z}_k]=\sum_{\nu=1}^{n}\mathfrak{c}_{ik\nu}\,\mathfrak{Z}_\nu f,$$

with the same constants $\mathfrak{c}_{ik\nu}$. From this, it follows that the two subgroups: $\mathfrak{X}_{n-m+1}f,\dots,\mathfrak{X}_nf$ and $\mathfrak{Z}_{n-m+1}f,\dots,\mathfrak{Z}_nf$ are equally composed. But we obtain even more. Indeed, since the two groups: $X_1f,\dots,X_nf$ and: $Z_1f,\dots,Z_nf$ are related to each other in a holoedrically isomorphic way, when to every infinitesimal transformation: $e_1\,\mathfrak{X}_1f+\cdots+e_r\mathfrak{X}_rf$ is associated the infinitesimal transformation: $e_1\,\mathfrak{Z}_1f+\cdots+e_r\mathfrak{Z}_rf$, and since through this association, the subgroup: $\mathfrak{X}_{n-m+1}f,\dots,\mathfrak{X}_rf$ obviously corresponds to the subgroup: $\mathfrak{Z}_{n-m+1}f,\dots,\mathfrak{Z}_rf$, then it becomes evident that the two reciprocal groups can be related to each other in a holoedrically isomorphic way so that the two m-term subgroups which leave M invariant correspond to each other.

Summarizing the results of pp. 398 sq., we have the:

Proposition 8. *If an m-times extended manifold of the space $x_1,\dots,x_n$ admits exactly m independent infinitesimal transformations, and hence an m-term subgroup, of a simply transitive group $X_1f,\dots,X_nf$ of this space, then it admits at the same time exactly m independent infinitesimal transformations, and hence an m-term subgroup, of the simply transitive group: $Z_1f,\dots,Z_nf$ which is reciprocal to $X_1f,\dots,X_nf$. The two m-term subgroups defined in this way are equally composed and it is possible to relate the two simply transitive reciprocal groups in a holoedrically isomorphic way so that these m-term subgroups correspond to each other.*

§ 98. The largest portion of the results of the preceding section can be derived by means of simple conceptual considerations. We shall now undertake this, and at the same time, we shall obtain a few further results.

As up to now, let $X_1f,\dots,X_nf$ and $Z_1f,\dots,Z_nf$ be two reciprocal simply transitive groups in the variables $x_1,\dots,x_n$; moreover, let $Z_1f,\dots,Z_mf$ again be an arbitrary m-term subgroup of the group $Z_1f,\dots,Z_nf$ and let $u_1,\dots,u_{n-m}$ be its invariants. Let the letter S be the general symbol of a transformation of the group

$Z_1 f, \dots, Z_m f$, and lastly, let T be an arbitrarily chosen transformation of the group $X_1 f, \dots, X_n f$.

Now, if P is any point of the space $x_1, \dots, x_n$, then:

$$(P') = (P)\,S$$

is the general symbol of a point on the manifold:

$$u_1 = \text{const.}, \;\dots,\; u_{n-m} = \text{const.}$$

which passes through the point P. Furthermore, since the transformations S and T are mutually interchangeable, we have:

$$(P')\,T = (P)\,S\,T = (P)\,T\,S,$$

hence, if we denote by Π the point $(P)\,T$, we have:

$$(P')\,T = (\Pi)\,S. \tag{11}$$

Here, $(\Pi)\,S$ is the general symbol of a point on the manifold: $u_\nu = \text{const.}$ passing though Π. Consequently, our symbolic equation (11) says that the transformation T, hence actually every transformation of the group $X_1 f, \dots, X_n f$, permutes the ∞^{n-m} manifolds: $u_1 = \text{const.}, \dots, u_{n-m} = \text{const.}$, by transferring each one of these manifolds to a manifold of the same family.

As a result, Proposition 6, p. 394, is derived.

But we can also prove the converse of this proposition by means of such conceptual considerations.

Let us imagine that an arbitrary decomposition of the space $x_1, \dots, x_n$ into ∞^{n-m} m-times extended manifolds invariant by the group $X_1 f, \dots, X_n f$ is given, and let us assume that M is one of these ∞^{n-m} manifolds. By P and P', let us understand two arbitrary points of M, and by T, an arbitrary transformation of the group $X_1 f, \dots, X_n f$.

By the execution of T, let the point P be transferred to Π, so we have:

$$(\Pi) = (P)\,T\,;$$

on the other hand, there is always in the reciprocal group $Z_1 f, \dots, Z_n f$ one, and only one, transformation which transfers P to P':

$$(P') = (P)\,S.$$

Because of:

$$(P)\,S\,T = (P)\,T\,S,$$

we also have:

$$(P')\,T = (\Pi)\,S. \tag{12}$$

We must attempt to interpret this equation.

To begin with, we assume that the point Π also belongs to the manifold M. In this case, the transformation T has the property of leaving M invariant. Indeed, T permutes the mentioned ∞^{n-m} manifolds, but on the other hand, T transfers a point of M, namely P, to a point of M, namely the point Π; thus, T must transfer all points of M to points of M, hence it must leave M invariant. Now, since P' lies on M, then the point $(P')T$ also belongs to the manifold M, and because of (12), the point $(\Pi)S$ too; but by appropriate choice of T, Π can be an arbitrary point of M, hence the transformation S also transfers every point of M to a point of M, so that it also leaves invariant the manifold M.

On the other hand, we can suppose that Π is any point of another arbitrary manifold amongst the ∞^{n-m} in question; if we make this assumption, then we immediately realize from (12) that S leaves at rest the concerned manifold.

With these words, it is proved that the group $Z_1 f, \dots, Z_n f$ contains a transformation, namely the transformation S, which leaves individually fixed each one of our ∞^{n-m} manifolds. But there are evidently ∞^m different such transformations, since after fixing P, the point P' can be chosen inside M in ∞^m different ways. However, there are no more than ∞^m transformations of this sort in the group $Z_1 f, \dots, Z_n f$, because this group is simply transitive; consequently, the ∞^m existing transformations generate an m-term subgroup of the group $Z_1 f, \dots, Z_n f$.

As a result, Proposition 7, p. 397, is proved.

Obviously, the manifold M also admits, aside from the ∞^m transformations of the group $Z_1 f, \dots, Z_n f$, ∞^m transformations of the group $X_1 f, \dots, X_n f$, which in turn form an m-term subgroup of this group. This is a result which is stated in Proposition 8, p. 400.

Something essentially new arises when the number m in the above developments is chosen to be equal to n. Up to now, this case has not come into consideration, because to it, there corresponds no decomposition of the space $x_1, \dots, x_n$.

If m is equal to n, then the manifold M coincides with the space $x_1, \dots, x_n$ itself; hence P and P' are arbitrary points of the space, and by appropriate choice of P and P', S can be any transformation of the reciprocal group $Z_1 f, \dots, Z_n f$.

If we choose P and P' fixed, the transformation S is completely determined; next, if T is an arbitrary transformation of the group $X_1 f, \dots, X_n f$, we have:

$$(P)\,S\,T = (P)\,T\,S,$$

or, because $(P') = (P)\,S$:

$$(P')\,T = (P)\,T\,S. \tag{13}$$

Here, by an appropriate choice of the transformation T, the point $(P)\,T$ can be brought to coincidence with any arbitrary point of the space $x_1, \dots, x_n$; the same holds for the point $(P')\,T$. Consequently, thanks to the equation (13), we are in a position to indicate, for every point $\mathfrak{P}$ of the space, the new position $\mathfrak{P}'$ that it reaches by the transformation S; we need only to determine the transformation T of the group $X_1 f, \dots, X_n f$ which transfers P to $\mathfrak{P}$, hence which satisfies the symbolic equation:

$$(P)\,\mathsf{T} = (\mathfrak{P}).$$

Then we have:

$$(\mathfrak{P}') = (\mathfrak{P})\,S = (P)\,\mathsf{T}\,S,$$

whence:

$$(\mathfrak{P}') = (P')\,\mathsf{T}.$$

Now, if we let the point $\mathfrak{P}$ take all possible positions, or, what is the same, if we set for T one after the other all transformations of the group $X_1 f, \dots, X_n f$, then we obtain that, to every point of the space $x_1, \dots, x_n$ is associated another completely determined point, hence we obtain a transformation of the space $x_1, \dots, x_n$, namely the transformation S. Lastly, if we choose the points P and P' in all possible ways, then we obviously obtain all transformations of the group $Z_1 f, \dots, Z_n f$.

Thus, we can say:

If two points P and P' of the space $x_1, \dots, x_n$ are transformed in a cogredient way [IN COGREDIENTER WEISE] *by means of each one of the ∞^n transformations of a simply transitive group $X_1 f, \dots, X_n f$, then the transformation which transfers each of the ∞^n positions taken by P to the corresponding position of the point P' belongs to the group $Z_1 f, \dots, Z_n f$ reciprocal to the group $X_1 f, \dots, X_n f$. If one chooses the points P and P' in all possible ways, then one obtains all transformations of the group $Z_1 f, \dots, Z_n f$.*

It goes without saying that here, the points P and P' are always to be understood as points in general position, or put more precisely, points which lie on no manifold invariant by the group $X_1 f, \dots, X_n f$.

If one knows the finite equations of the group $X_1 f, \dots, X_n f$, then one can use the construction just found for the transformations of the group $Z_1 f, \dots, Z_n f$ in order to set up the finite equations of this group.

Let:

$$x_i' = f_i(x_1, \dots, x_n, a_1, \dots, a_n) \qquad (i=1 \cdots n)$$

be the finite equations of the group $X_1 f, \dots, X_n f$. If one calls the coordinates of the point say $x_1^0, \dots, x_n^0$, and those of the point P' say $u_1^0, \dots, u_n^0$, then by the ∞^n transformations of the group $X_1 f, \dots, X_n f$, P receives the ∞^n different positions:

$$y_i = f_i(x_1^0, \dots, x_n^0, a_1, \dots, a_n) \qquad (i=1 \cdots n)$$

and P' receives the ∞^n positions:

$$y_i' = f_i(u_1^0, \dots, u_n^0, a_1, \dots, a_n) \qquad (i=1 \cdots n).$$

Every system of values of the a provides positions for P and P' which correspond to each other; hence if we eliminate from the equations $y_i = f_i(x^0, a)$ and $y_i' = f_i(u^0, a)$ the parameters a, we obtain the equations:

$$(14) \qquad y_i' = \mathfrak{F}_i(y_1, \dots, y_n, x_1^0, \dots, x_n^0, u_1^0, \dots, u_n^0) \qquad (i=1 \cdots n)$$

of a transformation of the group $Z_1 f, \dots, Z_n f$, namely the transformation which transfers the point P to P'. Lastly, if we let $x_1^0, \dots, x_n^0$ and $u_1^0, \dots, u_n^0$ take all possible values, we obtain all transformations of the group $Z_1 f, \dots, Z_n f$.

In the equations (14) of the group $Z_1 f, \dots, Z_n f$ just found, there appear $2n$ arbitrary parameters; however, this is only fictitious, only n of these parameters are essential. Indeed, one can obtain every individual transformation of the group $Z_1 f, \dots, Z_n f$ in ∞^n different ways, since one can always arbitrarily choose the point P, whereas the point P' is determined by the concerned transformation after the fixed choice of P.

From this, it follows that one can also derive in this way all transformations of the group $Z_1 f, \dots, Z_n f$, by choosing the point P fixed once for all, and only by letting the point P' take all possible positions; that is to say, one can insert determined numbers for the quantities $x_1^0, \dots, x_n^0$ and one only needs to interpret $u_1^0, \dots, u_n^0$ as arbitrary parameters.

Thus, the following holds:

Proposition 9. *If the finite equations:*

$$x_i' = f_i(x_1, \dots, x_n, a_1, \dots, a_n) \qquad (i=1 \cdots n)$$

of a simply transitive n-term group of the space $x_1, \dots, x_n$ are presented, then one finds the equations of the reciprocal simply transitive group in the following way:

In the equations:

$$y_i = f_i(x_1^0, \dots, x_n^0, a_1, \dots, a_n) \qquad (i=1 \cdots n),$$

one confers a fixed value to the x^0 and afterwards, one eliminates the n quantities $a_1, \dots, a_n$ from these equations and from the equations:

$$y_i' = f_i(u_1^0, \dots, u_n^0, a_1, \dots, a_n) \qquad (i=1 \cdots n);$$

the resulting equations:

$$y_i' = \mathfrak{F}_i(y_1, \dots, y_n, x_1^0, \dots, x_n^0, u_1^0, \dots, u_n^0) \qquad (i=1 \cdots n)$$

with the n arbitrary parameters are the equations of the reciprocal group. The assumption is that the x_k^0 are chosen so that the point $x_1^0, \dots, x_n^0$ lies on no manifold which remains invariant by the group: $x_i' = f_i(x, a)$.

§ 99. Proposition 7, p. 397, is a special case of a general proposition that also holds true for certain groups which are not simply transitive. We now want to derive this general proposition; on the occasion, we obtain at the same time a new proof of Proposition 7.

Let $X_1 f, \dots, X_n f$ be an n-term group in the variables $x_1, \dots, x_n$ and let the number n be no larger than s. In addition, we make the assumption that $X_1 f, \dots, X_n f$ are linked together by no linear relation of the form:

$$\chi_1(x_1,\dots,x_s)\,X_1f+\cdots+\chi_n(x_1,\dots,x_s)\,X_nf=0.$$

The proposition to be proved amounts to the following: if one knows an arbitrary m-term complete system: $\mathfrak{Y}_1f=0,\,\dots,\,\mathfrak{Y}_mf=0$ that the group $X_1f,\dots,X_nf$ admits, then one can always bring the complete system to a form: $Y_1f=0,\dots,Y_mf=0$ such that *the infinitesimal transformations* $Y_1f,\dots,Y_mf$ *are all interchangeable with* $X_1f,\dots,X_nf$, and that in the relations:

$$[Y_i,\,Y_k]=\sum_{\nu=1}^{m}\tau_{ik\nu}(x_1,\dots,x_s)\,Y_\nu f$$

which hold between $Y_1f,\dots,Y_mf$, the $\tau_{ik\nu}$ are all solutions of the n-term complete system: $X_1f=0,\,\dots,\,X_nf=0$.

In the special case $s=n$, where the group $X_1f,\dots,X_nf$ is simply transitive, the infinitesimal transformations $Y_1f,\dots,Y_mf$ obviously belong to the simply transitive group $Z_1f,\dots,Z_nf$ reciprocal to $X_1f,\dots,X_nf$; moreover, since the n-term complete system $X_1f=0,\,\dots,\,X_nf=0$ possesses in this case no other solutions than $f=$ const., the functions $\tau_{ik\nu}$ are then plain constants so that $Y_1f,\dots,Y_mf$ generate an m-term subgroup of the group $Z_1f,\dots,Z_nf$. We therefore have Proposition 7, p. 397.

However, we deal at present with the general case.

We therefore imagine that an m-term complete system:

$$\mathfrak{Y}_1f=0,\;\dots,\;\mathfrak{Y}_mf=0$$

is presented which admits the group $X_1f,\dots,X_nf$, so that the Xf and $\mathfrak{Y}f$ are linked together by relations of the form:

$$[\mathfrak{Y}_\mu,\,X_k]=\sum_{\nu=1}^{m}\alpha_{\mu k\nu}(x_1,\dots,x_s)\,\mathfrak{Y}_\nu f\qquad(k=1\cdots n;\;\mu=1\cdots m)\tag{15}$$

(cf. Chap. 13, p. 233).

To begin with, we now attempt to determine m functions $\rho_1,\dots,\rho_m$ of the x so that the infinitesimal transformation:

$$Yf=\sum_{\mu=1}^{m}\rho_\mu(x_1,\dots,x_s)\,\mathfrak{Y}_\mu f$$

is interchangeable with all the n infinitesimal transformations X_kf. We therefore have to satisfy the n equations:

$$\begin{aligned}[X_k,\,Y]&=\sum_{\mu=1}^{m}X_k\rho_\mu\,\mathfrak{Y}_\mu f+\sum_{\mu=1}^{m}\rho_\mu\,[X_k,\,\mathfrak{Y}_\mu]\\&=\sum_{\mu=1}^{m}\Big\{X_k\rho_\mu-\sum_{\nu=1}^{m}\alpha_{\nu k\mu}\,\rho_\nu\Big\}\,\mathfrak{Y}_\mu f=0,\end{aligned}$$

or, because $\mathfrak{Y}_1 f, \dots, \mathfrak{Y}_m f$ cannot be linked together by linear relations, the following mn relations:

$$X_k \rho_\mu - \sum_{\nu=1}^{m} \alpha_{\nu k \mu} \rho_\nu = 0 \qquad (k=1\cdots n;\ \mu=1\cdots m). \tag{16}$$

These are differential equations by means of which the ρ are to be determined.

If:

$$\rho_1 = P_1(x_1, \dots, x_s), \ \dots, \ \rho_m = P_m(x_1, \dots, x_s) \tag{17}$$

is a system of solutions of the differential equations (16), then the equations:

$$X_k P_\mu - \sum_{\nu=1}^{m} \alpha_{\nu k \mu} \rho_\nu = 0$$

are satisfied identically after the substitution: $\rho_1 = P_1, \dots, \rho_m = P_m$. Expressed differently: the system of equations (17) in the $s+m$ variables $x_1, \dots, x_s$, $\rho_1, \dots, \rho_m$ admits the infinitesimal transformations:

$$\mathfrak{U}_k f = X_k f + \sum_{\mu=1}^{m} \left\{ \sum_{\nu=1}^{m} \alpha_{\nu k \mu} \rho_\nu \right\} \frac{\partial f}{\partial \rho_\mu} \qquad (k=1\cdots n).$$

Conversely, if a system of equations of the form (17) admits the infinitesimal transformations $\mathfrak{U}_1 f, \dots, \mathfrak{U}_n f$, then the functions $P_1, \dots, P_m$ are obviously solutions of the differential equations (16).

From this, we see that the integration of the differential equations (16) amounts to determining, in the $s+m$ variables x, ρ, the most general system of equations (17), which admits the infinitesimal transformations $\mathfrak{U}_1 f, \dots, \mathfrak{U}_n f$.

The system of equations to be determined also admits the infinitesimal transformations: $[\mathfrak{U}_i, \mathfrak{U}_k]$.

By calculation, we find:

$$\begin{aligned}
[\mathfrak{U}_i, \mathfrak{U}_k] = [X_i, X_k] &+ \sum_{\mu, \nu}^{1\cdots m} \left\{ X_i \alpha_{\nu k \mu} - X_k \alpha_{\nu i \mu} \right\} \rho_\nu \frac{\partial f}{\partial \rho_\mu} \\
&+ \sum_{\mu, \nu, \pi}^{1\cdots m} \left\{ \alpha_{\nu i \pi} \alpha_{\pi k \mu} - \alpha_{\nu k \pi} \alpha_{\pi i \mu} \right\} \rho_\nu \frac{\partial f}{\partial \rho_\mu}.
\end{aligned}$$

Here, in order to simplify the right-hand side, we form the Jacobi identity (cf. Chap. 5, p. 109):

$$\big[\mathfrak{Y}_\nu, [X_i, X_k]\big] + \big[X_i, [X_k, \mathfrak{Y}_\nu]+\big] \big[X_k, [\mathfrak{Y}_\nu, X_i]\big] = 0,$$

which, using (15), can be written as:

$$[\mathfrak{Y}_\nu, [X_i, X_k]] = \sum_{\mu=1}^{m} \{[X_i f, \alpha_{\nu k\mu}\, \mathfrak{Y}_\mu f] - [X_k f, \alpha_{\nu i\mu}\, \mathfrak{Y}_\mu f]\},$$

or:

$$[\mathfrak{Y}_\nu, [X_i, X_k]] = \sum_{\mu=1}^{m} \{X_i\, \alpha_{\nu k\mu} - X_k\, \alpha_{\nu i\mu}\}\, \mathfrak{Y}_\mu f$$
$$+ \sum_{\mu,\,\pi}^{1\cdots m} \{\alpha_{\nu i\mu}\, \alpha_{\mu k\pi} - \alpha_{\nu k\mu}\, \alpha_{\mu i\pi}\}\, \mathfrak{Y}_\pi f.$$

But now, $X_1 f, \dots, X_n f$ generate an n-term group, whence:

$$[X_i, X_k] = \sum_{\sigma=1}^{n} c_{ik\sigma}\, X_\sigma f,$$

from which it follows:

$$[\mathfrak{Y}_\nu, [X_i, X_k]] = \sum_{\mu=1}^{m} \left\{ \sum_{\sigma=1}^{n} c_{ik\sigma}\, \alpha_{\nu\sigma\mu} \right\} \mathfrak{Y}_\mu f.$$

If we still take into account that $\mathfrak{Y}_1 f, \dots, \mathfrak{Y}_m f$ are linked together by no linear relations, we obtain:

$$X_i\, \alpha_{\nu k\mu} - X_k\, \alpha_{\nu i\mu} + \sum_{\pi=1}^{m} \{\alpha_{\nu i\pi}\, \alpha_{\pi k\mu} - \alpha_{\nu k\pi}\, \alpha_{\pi i\mu}\}$$
$$= \sum_{\sigma=1}^{n} c_{ik\sigma}\, \alpha_{\nu\sigma\mu}.$$

By inserting these values into the expression found above for $[\mathfrak{U}_i, \mathfrak{U}_k]$, we get:

$$[\mathfrak{U}_i, \mathfrak{U}_k] = \sum_{\sigma=1}^{n} c_{ik\sigma}\, \mathfrak{U}_\sigma f,$$

so $\mathfrak{U}_1 f, \dots, \mathfrak{U}_n f$ generate an n-term group in the $s+m$ variables x, ρ.

At present, the question is to determine the most general system of equations (17) which admits the n-term group $\mathfrak{U}_1 f, \dots, \mathfrak{U}_n f$.

Since $X_1 f, \dots, X_n f$ are linked together by no linear relation, not all $n \times n$ determinants vanish identically in the matrix associated to $\mathfrak{U}_1 f, \dots, \mathfrak{U}_n f$; but obviously, these $n \times n$ determinants cannot all be equal to zero, even by virtue of a system of equations of the form (17). Consequently, every system of equations of the form (17) which admits the group $\mathfrak{U}_1 f, \dots, \mathfrak{U}_n f$ is represented by relations between the solutions of the n-term complete system: $\mathfrak{U}_1 f = 0, \dots, \mathfrak{U}_n f = 0$ (cf. Chap. 14, Theorem 42, p. 247).

The n-term complete system $\mathfrak{U}_1 f = 0, \dots, \mathfrak{U}_n f = 0$ possesses $s+m-n$ independent solutions; $s-n$ of these solutions can be chosen in such a way that they are free of the ρ and depend only upon the x, they are the independent solutions of the

n-term complete system: $X_1 f = 0, \dots, X_n f = 0$, and we can call them:

$$\mathfrak{v}_1(x_1,\dots,x_s), \ \dots, \ \mathfrak{v}_{s-n}(x_1,\dots,x_s).$$

Furthermore, let:

$$\Omega_1(\rho_1,\dots,\rho_m,x_1,\dots,x_s), \ \dots, \ \Omega_m(\rho_1,\dots,\rho_m,x_1,\dots,x_s)$$

be m arbitrary mutually independent solutions of the complete system: $\mathfrak{U}_1 f = 0, \dots, \mathfrak{U}_n f = 0$ that are also independent of the $\mathfrak{v}$.

Since the equations: $\mathfrak{U}_1 f = 0, \dots, \mathfrak{U}_n f = 0$ are solvable with respect to n of the differential quotients $\partial f/\partial x_1, \dots, \partial f/\partial x_s$, the $s-n+m$ functions $\mathfrak{v}_1,\dots,\mathfrak{v}_{s-n}$, $\Omega_1,\dots,\Omega_m$ must be mutually independent relative to $s-n$ of the variables $x_1,\dots,x_s$, and relative to $\rho_1,\dots,\rho_m$. Consequently, $\Omega_1,\dots,\Omega_m$ are mutually independent relative to $\rho_1,\dots,\rho_m$.

After these preparations, we can identify the most general system of equations which admits the group $\mathfrak{U}_1 f,\dots,\mathfrak{U}_n f$ and which can be given the form (17).

The concerned system of equations consists of m relations between $\mathfrak{v}_1,\dots,\mathfrak{v}_{s-n}$, $\Omega_1,\dots,\Omega_m$ and is solvable with respect to $\rho_1,\dots,\rho_m$; consequently, it is solvable with respect to $\Omega_1,\dots,\Omega_m$ and has the form:

$$\text{(18)} \qquad \Omega_\mu(\rho_1,\dots,\rho_m,x_1,\dots,x_s) = \Psi_\mu\big(\mathfrak{v}_1(x),\dots,\mathfrak{v}_{s-n}(x)\big) \qquad (\mu = 1\cdots m),$$

where the Ψ are absolutely arbitrary functions of their arguments. If we solve this system of equations with respect to $\rho_1,\dots,\rho_m$, then for $\rho_1,\dots,\rho_m$, we receive expressions that represent the most general system of solutions to the differential equations (16).

Thus, the most general system of solutions (16) contains m arbitrary functions of $\mathfrak{v}_1,\dots,\mathfrak{v}_{s-n}$. Now, since the equations (16) in the unknowns $\rho_1,\dots,\rho_m$ are linear and homogeneous, one can conclude that the most general system of solutions $\rho_1,\dots,\rho_m$ to it can be deduced from m particular systems of solutions:

$$P_1^{(\mu)}(x_1,\dots,x_s), \ \dots, \ P_m^{(\mu)}(x_1,\dots,x_s) \qquad (\mu = 1\cdots m)$$

in the following way:

$$\text{(19)} \qquad \rho_\nu = \chi_1(\mathfrak{v}_1,\dots,\mathfrak{v}_{s-n})\,P_\nu^{(1)} + \cdots + \chi_m(\mathfrak{v}_1,\dots,\mathfrak{v}_{s-n})\,P_\nu^{(m)} \qquad (\mu = 1\cdots m),$$

where it is understood that the χ are arbitrary functions of their arguments. Naturally, the m particular systems of solutions must be constituted in such a way that it is not possible to determine m functions $\psi_1,\dots,\psi_m$ of $\mathfrak{v}_1,\dots,\mathfrak{v}_{s-n}$ such that the m equations:

$$\sum_{\mu=1}^{m} \psi_\mu(\mathfrak{v}_1,\dots,\mathfrak{v}_{s-n})\,P_\nu^{(\mu)} = 0 \qquad (\nu = 1\cdots m)$$

are satisfied simultaneously.

Lastly, it is still to be observed that the equations (19) are solvable with respect to $\chi_1, \dots, \chi_m$, so that the determinant of the $P_\nu^{(\mu)}$ does not vanish identically. Indeed, according to the preceding, one must be able to bring the equations (19) to the form (18), and this is obviously impossible when the determinant of the $P_\nu^{(\mu)}$ vanishes.

At present, we can write down the most general transformation:

$$Yf = \sum_{\mu=1}^{m} \rho_\mu \, \mathfrak{Y}_\mu f$$

which is interchangeable with $X_1 f, \dots, X_n f$ (cf. above, p. 405). Its form is:

$$Yf = \sum_{\nu=1}^{m} \chi_\nu(\mathfrak{v}_1, \dots, \mathfrak{v}_{s-n}) \sum_{\mu=1}^{m} P_\mu^{(\nu)}(x_1, \dots, x_s) \, \mathfrak{Y}_\mu f,$$

or:

$$Yf = \sum_{\nu=1}^{m} \chi_\nu(\mathfrak{v}_1, \dots, \mathfrak{v}_{s-n}) \, Y_\nu f,$$

if we set:

$$\sum_{\mu=1}^{m} P_\mu^{(\nu)} \, \mathfrak{Y}_\mu f = Y_\nu f \qquad (\nu = 1 \cdots m).$$

Here, obviously, $Y_1 f, \dots, Y_m f$ are all interchangeable with $X_1 f, \dots, X_n f$; furthermore, the equations: $Y_1 f = 0$, ..., $Y_m f = 0$ are equivalent to the equations: $\mathfrak{Y}_1 f = 0$, ..., $\mathfrak{Y}_m f = 0$, and hence they in turn form an m-term complete system that admits the group $X_1 f, \dots, X_n f$.

As a consequence of this, between $Y_1 f, \dots, Y_m f$, there are relations of the form:

$$[Y_\mu, Y_\nu] = \sum_{\pi=1}^{m} \tau_{\mu\nu\pi}(x_1, \dots, x_s) \, Y_\pi f.$$

But from the Jacobi identity:

$$\big[X_k, [Y_\mu, Y_\nu]\big] + \big[Y_\mu, [Y_\nu, X_k]\big] + \big[Y_\nu, [X_k, Y_\mu]\big] \equiv 0,$$

in which the last two terms are identically zero, it immediately follows that:

$$\big[X_k, [Y_\mu, Y_\nu]\big] \equiv 0.$$

We must therefore have:

$$\sum_{\pi=1}^{m} X_k \, \tau_{\mu\nu\pi} \, Y_\pi f \equiv 0,$$

or, because $Y_1 f, \dots, Y_m f$ are linked together by no linear relation:

$$X_k\,\tau_{\mu\nu\pi} \equiv 0 \qquad\qquad (k=1\cdots n),$$

that is to say: the $\tau_{\mu\nu\pi}$ are all solutions of the complete system: $X_1 f = 0, \ldots, X_n f = 0$, they are functions of $\mathfrak{v}_1, \ldots, \mathfrak{v}_{s-n}$ alone.

The complete system: $Y_1 f = 0, \ldots, Y_m f = 0$ possesses all the properties indicated on p 405; we can therefore enunciate the statement:

Theorem 70. *If an m-term complete system:* $\mathfrak{Y}_1 f = 0, \ldots, \mathfrak{Y}_m f = 0$ *in the s variables* $x_1, \ldots, x_s$ *admits the n-term group* $X_1 f, \ldots, X_n f$ *and if this group is constituted in such a way that between* $X_1 f, \ldots, X_n f$*, there is no linear relation of the form:*

$$\chi_1(x_1,\ldots,x_s)\,X_1 f + \cdots + \chi_n(x_1,\ldots,x_s)\,X_n f = 0,$$

then it is possible to determine m^2 *functions* $P_\mu^{(\nu)}(x_1,\ldots,x_s)$ *with not identically vanishing determinant such that the m infinitesimal transformations:*

$$Y_\nu f = \sum_{\mu=1}^{m} P_\mu^{(\nu)}(x_1,\ldots,x_s)\,\mathfrak{Y}_\mu f \qquad\qquad (\nu=1\cdots m)$$

are all interchangeable with $X_1 f, \ldots, X_n f$*. In turn, the equations:* $\mathfrak{Y}_1 f = 0, \ldots, \mathfrak{Y}_m f = 0$ *equivalent to* $Y_1 f = 0, \ldots, Y_m f = 0$ *form an m-term complete system that admits the group* $X_1 f, \ldots, X_n f$*. Lastly, between* $Y_1 f, \ldots, Y_m f$*, there are relations of the specific form:*

$$[Y_\mu, Y_\nu] = \sum_{\pi=1}^{m} \vartheta_{\mu\nu\pi}(\mathfrak{v}_1,\ldots,\mathfrak{v}_{s-n})\,Y_\pi f,$$

where $\mathfrak{v}_1, \ldots, \mathfrak{v}_{s-n}$ *are independent solutions of the n-term complete system:* $X_1 f = 0, \ldots, X_n f = 0$.[3]

---

[3] During the summer semester 1887, in a lecture about the general theory of integration of the differential equations that admit a finite continuous group, LIE developed Theorem 70, which has Theorem 69 as a special case.

Chapter 21
The Group of Parameters

If one executes three transformations of the space $x_1,\dots,x_n$ subsequently, say the following ones:

$$\text{(1)}\qquad \begin{cases} x_i' = f_i(x_1,\dots,x_n) & (i=1\cdots n)\\ x_i'' = g_i(x_1',\dots,x_n') & (i=1\cdots n)\\ x_i''' = h_i(x_1'',\dots,x_n'') & (i=1\cdots n), \end{cases}$$

the one gets a new transformation:

$$x_i''' = \omega_i(x_1,\dots,x_n) \qquad (i=1\cdots n)$$

of the space $x_1,\dots,x_n$.

The equations of the new transformation are obtained when the $2n$ variables: $x_1',\dots,x_n', x_1'',\dots,x_n''$ are eliminated from the $3n$ equations (1). Clearly, one can execute this elimination in two different ways, since one can begin either by taking away the x', or by taking away the x''. In the first case, one obtains firstly between the x and the x'' the relations:

$$x_i'' = g_i\big(f_1(x),\,\dots,\,f_n(x)\big) \qquad (i=1\cdots n),$$

and afterwards, one has to insert these values of $x_1'',\dots,x_n''$ into the equations:

$$x_i''' = h_i(x_1'',\dots,x_n'') \qquad (i=1\cdots n).$$

In the second case, one obtains firstly between the x' and the x''' the relations:

$$x_i''' = h_i\big(g_1(x'),\,\dots,\,g_n(x')\big) \qquad (i=1\cdots n),$$

and then one has to replace the x' by their values:

$$x_i' = f_i(x_1,\dots,x_n) \qquad (i=1\cdots n).$$

S. Lie, *Theory of Transformation Groups I*,
DOI 10.1007/978-3-662-46211-9_21

The obvious observation that in both ways indicated above, one obtains as the final result the same transformation has content [INHALT] when one interprets the transformation as an operation and when one applies the symbolism of substitution theory.

For the three substitutions (1), we want to introduce the symbols: S, T, U one after the other, so that the transformation:

$$x_i'' = g_i\big(f_1(x), \dots, f_n(x)\big) \qquad (i=1 \cdots n)$$

has the symbol: ST, the transformation:

$$x_i''' = h_i\big(g_1(x'), \dots, g_n(x')\big) \qquad (i=1 \cdots n),$$

the symbol: TU and lastly, the transformation: $x_i''' = \omega_i(x_1, \dots, x_n)$, the symbol: STU. We can then express the two ways of forming the transformation: $x_i''' = \omega_i(x_1, \dots, x_n)$ discussed above by saying that this transformation is obtained both when one first executes the transformation: ST and then the transformation U, and when one first executes the transformation S, and then the transformation TU.

Symbolically, these facts can be expressed by the equations:

$$(ST)U = S(TU) = STU, \tag{2}$$

and it is known that they say that the operations S, T, U satisfy the so-called *associative rule* [ASSOCIATIVE GESETZ].

We can therefore say:

The transformations of an n-times extended space are operations for which the law of associativity holds true.

At present, let us consider the specific case where S, T, U are arbitrary transformations of an r-term group:

$$x_i' = f_i(x_1, \dots, x_n, a_1, \dots, a_r) \qquad (i=1 \cdots n).$$

In this case, the transformations ST, TU and STU naturally also belong to the group in question. We now want to see what can be concluded from the validity of the associative law.

§ 100. Let the transformations S, T, U of our group be:

$$\begin{cases} S: & x_i' = f_i(x_1, \dots, x_n, a_1, \dots, a_r) \\ T: & x_i'' = f_i(x_1', \dots, x_n', b_1, \dots, b_r) \\ U: & x_i''' = f_i(x_1'', \dots, x_n'', c_1, \dots, c_r). \end{cases} \tag{3}$$

For the transformation ST, there result from this in the known way equations of the form:

$$x_i'' = f_i\big(x_1,\dots,x_n,\ \varphi_1(a,b),\ \dots,\ \varphi_r(a,b)\big),$$

where the functions $\varphi_1(a,b), \dots, \varphi_r(a,b)$ are, according to Chap. 3, p. 53, mutually independent both relative to $a_1,\dots,a_r$ and relative to $b_1,\dots,b_r$.

Furthermore, the equations of the transformation $(ST)U$ are:

$$x_i''' = f_i\big(x_1,\dots,x_n,\ \varphi_1(\varphi(a,b),c),\ \dots,\ \varphi_r(\varphi(a,b),c)\big). \tag{4}$$

On the other hand, we have for the transformation TU the equations:

$$x_i''' = f_i\big(x_1',\dots,x_n',\ \varphi_1(b,c),\ \dots,\ \varphi_r(b,c)\big),$$

hence for $S(TU)$ the following:

$$x_i''' = f_i\big(x_1,\dots,x_n,\ \varphi_1(a,\ \varphi(b,c)),\ \dots,\ \varphi_r(a,\ \varphi(b,c))\big). \tag{4'}$$

Now, we have: $(ST)U = S(TU)$, whence the two transformations (4) and (4') must be identical to each other; by comparison of the two parameters in the two transformations, we therefore obtain the following relations:

$$\varphi_k\big(\varphi_1(a,b),\ \dots,\ \varphi_r(a,b),\ c_1,\dots,c_r\big) = \varphi_k\big(a_1,\dots,a_r,\ \varphi_1(b,c),\ \dots,\ \varphi_r(b,c)\big)$$
$$(k=1\cdots r),$$

or more shortly:

$$\varphi_k\big(\varphi(a,b),\ c\big) = \varphi_k\big(a,\ \varphi(b,c)\big) \qquad (k=1\cdots r). \tag{5}$$

Thus, the functions $\varphi_1,\dots,\varphi_r$ must identically satisfy these relations for all values of the a, b, c.

The equations (5) express the law of associativity for three arbitrary transformations of the group $x_i' = f_i(x,a)$. But they can also be interpreted in another way; namely, they say that the equations:

$$a_k' = \varphi_k(a_1,\dots,a_r,b_1,\dots,b_r) \qquad (k=1\cdots r) \tag{6}$$

in the variables $a_1,\dots,a_r$ represent a group, and to be precise, a group with the r parameters $b_1,\dots,b_r$.

Indeed, if we execute two transformations (6), say:

$$a_k' = \varphi_k(a_1,\dots,a_r,b_1,\dots,b_r)$$

and:

$$a_k'' = \varphi_k(a_1',\dots,a_r',c_1,\dots,c_r)$$

one after the other, we obtain the transformation:

$$a_k'' = \varphi_k\big(\varphi_1(a,b),\ \dots,\ \varphi_r(a,b),\ c_1,\dots,c_r\big)$$

which, by virtue of (5), takes the form:

$$a_k'' = \varphi_k\big(a_1,\dots,a_r,\ \varphi_1(b,c),\ \dots,\ \varphi_r(b,c)\big),$$

and hence belongs likewise to the transformations (6). As a result, our assertion is proved.

We want to call the group:

$$(6) \qquad a_k' = \varphi_k(a_1,\dots,a_r,b_1,\dots,b_r) \qquad (k=1\cdots r)$$

the parameter group [PARAMETERGRUPPE] *of the group* $x_i' = f_i(x_1,\dots,x_n,\,a_1,\dots,a_r)$.

According to the already cited page 53, the equations $a_k' = \varphi_k(a,b)$ are solvable with respect to the r parameters $b_1,\dots,b_r$:

$$b_k = \psi_k(a_1,\dots,a_r,a_1',\dots,a_r') \qquad (k=1\cdots r).$$

From this, we conclude that the parameter group is r-term, is transitive, and in fact, simply transitive. In addition, the equations (5) show that the parameter group is its own parameter group.

We can therefore say in summary:

Theorem 71. *If the functions* $f_i(x_1,\dots,x_n,\,a_1,\dots,a_r)$ *in the equations* $x_i' = f_i(x,a)$ *of an r-term group satisfy the functional equations:*

$$f_i\big(f_1(x,a),\ \dots,\ f_n(x,a),b_1,\dots,b_r\big) = f_i\big(x_1,\dots,x_n,\ \varphi_1(a,b),\ \dots,\ \varphi_r(a,b)\big) \qquad (i=1\cdots n),$$

then the r relations:

$$a_i' = \varphi_i(a_1,\dots,a_r,b_1,\dots,b_r) \qquad (i=1\cdots r)$$

determine an r-term group between the $2r$ *variables a and* a'*: the parameter group of the original group. This parameter group is simply transitive and is its own parameter group.*

§ 101. If the r-term group: $x_i' = f_i(x_1,\dots,x_n,a_1,\dots,a_r)$ is generated by the r independent infinitesimal transformations:

$$X_k f = \sum_{i=1}^{n} \xi_{ki}(x_1,\dots,x_n)\,\frac{\partial f}{\partial x_i} \qquad (k=1\cdots r),$$

then according to Chap. 9, p. 172, its transformations are ordered together as inverses by pairs. Evidently, the transformations of the associated parameter group: $a_k' = \varphi_k(a,b)$ are then also ordered together as inverses by pairs, whence according to Chap. 9, p. 184 above, the parameter group contains exactly r independent infinitesimal transformations and is generated by them.

We can also establish this important property of the parameter group in the following way, where we find at the same time the infinitesimal transformations of this group.

In the equations: $x_i' = f_i(x_1,\dots,x_n,a_1,\dots,a_r)$, if the x_i' are considered as functions of the x and of the a, then according to Theorem 3, p. 40, there are differential equations of the form:

$$\frac{\partial x_i'}{\partial a_k} = \sum_{j=1}^{r} \psi_{kj}(a_1,\dots,a_r)\,\xi_{ji}(x_1',\dots,x_n') \qquad (i=1\cdots n;\, k=1\cdots r), \tag{7}$$

and they can also be written:

$$\xi_{ji}(x_1',\dots,x_n') = \sum_{k=1}^{r} \alpha_{jk}(a_1,\dots,a_r)\,\frac{\partial x_i'}{\partial a_k} \qquad (j=1\cdots r;\, i=1\cdots n). \tag{7'}$$

Here, according to Chap. 3, Proposition 3.4, p. 54, the ξ_{ji} and the α_{jk} are determined by the equations:

$$\xi_{ji}(x_1',\dots,x_n') = \left[\frac{\partial x_i'}{\partial b_j}\right]_{b=b^0}, \qquad \alpha_{jk}(a_1,\dots,a_r) = \left[\frac{\partial a_k}{\partial b_j}\right]_{b=b^0},$$

whose meaning has been explained in the indicated locations.

For the group $a_k' = \varphi_k(a,b)$, one naturally obtains analogous differential equations:

$$\frac{\partial a_i'}{\partial b_k} = \sum_{j=1}^{r} \psi_{kj}(b_1,\dots,b_r)\,\overline{\psi}_{ji}(a_1',\dots,a_r') \qquad (i,\, k=1\cdots r)$$

in which the ψ_{kj} have the same signification as in (7), while the $\overline{\psi}_{ji}(a)$ are determined by:

$$\overline{\psi}_{ji}(a_1,\dots,a_r) = \left[\frac{\partial a_i}{\partial b_j}\right]_{b=b^0}.$$

In consequence, the $\overline{\psi}_{ji}$ are the same functions of their arguments as the α_{ji}; in correspondence to the formulas (7) and (7'). We therefore obtain the two following ones:

$$\frac{\partial a_i'}{\partial b_k} = \sum_{j=1}^{r} \psi_{kj}(b_1,\dots,b_r)\,\alpha_{ji}(a_1',\dots,a_r') \qquad (i,\, k=1\cdots r) \tag{8}$$

and:

$$\alpha_{ji}(a_1',\dots,a_r') = \sum_{k=1}^{r} \alpha_{jk}(b_1,\dots,b_r)\,\frac{\partial a_i'}{\partial b_k} \qquad (j,\, i=1\cdots r). \tag{8'}$$

Now, the group: $x_i' = f_i(x_1,\dots,x_n,a_1,\dots,a_r)$ contains the identity transformation and in fact, we can always suppose that the parameters $a_1^0,\dots,a_r^0$ of the identity transformation lie in the region $((a))$ defined on p. 26. According to p. 40,[1] (Translator's note: According to the footnote, this determinant is equal to $(-1)^r$.) the determinant:

$$\sum \pm \psi_{11}(a^0)\cdots\psi_{rr}(a^0)$$

is then certainly different from zero.

On the other hand, it is clear that, also in the family of transformations: $a_k' = \varphi_k(a,b)$, the identity transformation: $a_1' = a_1,\ \dots,\ a_r' = a_r$ appears, and even, that the transformation:

$$a_k' = \varphi_k(a_1,\dots,a_r,a_1^0,\dots,a_r^0) \qquad (k=1\cdots r)$$

is the identity transformation. Thus, if we take into account the existence of the differential equations (8) and if we apply Theorem 9, p. 82, we realize immediately that the family of the ∞^r transformations: $a_k' = \varphi_k(a,b)$ coincides with the family of the ∞^{r-1} one-term groups:

$$\sum_{k=1}^{r} \lambda_k \sum_{i=1}^{r} \alpha_{ki}(a_1,\dots,a_r)\,\frac{\partial f}{\partial a_i}.$$

Consequently, the group: $a_k' = \varphi_k(a,b)$ is generated by the r infinitesimal transformations:

$$A_k f = \sum_{j=1}^{r} \alpha_{kj}(a_1,\dots,a_r)\,\frac{\partial f}{\partial a_j} \qquad (k=1\cdots r).$$

Naturally, these infinitesimal transformations are independent of each other, since there exists between them absolutely no relation of the form:

$$\chi_1(a_1,\dots,a_r)A_1 f + \cdots + \chi_r(a_1,\dots,a_r)A_r f = 0,$$

as we have already realized in Chap. 3, Theorem 3, p. 40; on the other hand, it can also be concluded from this that the parameter group is simply transitive (Theorem 71).

In the preceding, we have not only proved that the group: $a_k' = \varphi_k(a,b)$ is generated by infinitesimal transformations, but we have also found these infinitesimal transformations themselves. From this, we can immediately deduce a new important property of the group: $a_k' = \varphi_k(a,b)$.

The r infinitesimal transformations: $X_1 f,\dots,X_r f$ of the group: $x_i' = f_i(x,a)$ are linked together by relations of the form:

$$[X_i, X_k] = \sum_{s=1}^{r} c_{iks} X_s f,$$

and according to Theorem 21, p. 164, there are, between $A_1 f, \dots, A_r f$, relations of the same form:

$$[A_i, A_k] = \sum_{s=1}^{r} c_{iks} A_s f,$$

with the same constants c_{iks}. By applying the terminology introduced in Chap. 17, p. 303 and p. 305, we can hence state the following theorem:

Theorem 72. *Every r-term group:*

$$x_i' = f_i(x_1, \dots, x_n, a_1, \dots, a_r) \qquad (i=1\cdots n)$$

is equally composed with its parameter group:

$$a_k' = \varphi_k(a_1, \dots, a_r, b_1, \dots, b_r) \qquad (k=1\cdots r),$$

or, what is the same, it is holoedrically isomorphic to it. If the infinitesimal transformations of the group: $x_i' = f_i(x, a)$ *satisfy the relations:*

$$\xi_{ji}(x_1', \dots, x_n') = \sum_{k=1}^{r} \alpha_{jk}(a_1, \dots, a_r) \frac{\partial x_i'}{\partial a_k} \qquad (i=1\cdots n;\ j=1\cdots r),$$

then the r expressions:

$$A_j f = \sum_{k=1}^{r} \alpha_{jk}(a_1, \dots, a_r) \frac{\partial f}{\partial a_k} \qquad (j=1\cdots r)$$

represent r independent infinitesimal transformations of the parameter group.[1]

§ 102. As usual, let:

$$X_k f = \sum_{i=1}^{n} \xi_{ki}(x_1, \dots, x_n) \frac{\partial f}{\partial x_i} \qquad (k=1\cdots r)$$

be independent infinitesimal transformations of an r-term group in the variables $x_1, \dots, x_n$.

Under the guidance of Chap. 9, p. 171 and p. 172, one determines r independent infinitesimal transformations:

$$\overline{A}_k f = \sum_{j=1}^{r} \overline{\alpha}_{kj}(a_1, \dots, a_r) \frac{\partial f}{\partial a_j} \qquad (k=1\cdots r)$$

which are linked together by no linear relation of the form:

$$\chi_1(a_1, \dots, a_r) \overline{A}_1 f + \cdots + \chi_r(a_1, \dots, a_r) \overline{A}_r f = 0,$$

---

[1] LIE, Gesellschaft d. W. zu Christiania 1884, no. 15.

but which stand pairwise in the relationships:

$$[\overline{A}_i, \overline{A}_k] = \sum_{s=1}^{r} c_{iks} \overline{A}_s f\,;$$

in other words, one determines r independent infinitesimal transformations of any simply transitive r-term group which has the same composition as the group $X_1 f, \dots, X_r f$.

Afterwards, one selects any system of values: $\overline{a}_1, \dots, \overline{a}_r$ in the neighborhood of which the $\overline{\alpha}_{ki}(a_1, \dots, a_r)$ behave regularly and for which the determinant $\sum \pm \overline{\alpha}_{11} \cdots \overline{\alpha}_{rr}$ does not vanish, and then, relative to this system of values, one determines the general solutions of the r-term complete system:

$$\sum_{i=1}^{n} \xi_{ki}(x') \frac{\partial f}{\partial x'_i} + \sum_{j=1}^{r} \overline{\alpha}_{kj}(a) \frac{\partial f}{\partial a_j} = X'_k f + \overline{A}_k f = 0 \qquad (k=1 \cdots r) \tag{9}$$

in the $n+r$ variables: $x'_1, \dots, x'_n, a_1, \dots, a_r$.

Now, if $\overline{F}_1(x'_1, \dots, x'_n, a_1, \dots, a_r), \dots, \overline{F}_n(x', a)$ are the general solutions in question, one sets:

$$x_i = \overline{F}_i(x'_1, \dots, x'_n, a_1, \dots, a_r) \qquad (i=1 \cdots n)\,;$$

by resolution with respect to $x'_1, \dots, x'_n$, one obtains equations of the form:

$$x'_i = \overline{f}_i(x_1, \dots, x_n, a_1, \dots, a_r) \qquad (i=1 \cdots n)$$

which, according to Theorem 23, p. 169, represent the finite transformations of an r-term group, namely of the group: $X_1 f, \dots, X_r f$ itself.

In the mentioned theorem, we already observed that the equations: $x'_i = \overline{f}_i(x, a)$ can be brought to the form:

$$x'_i = x_i + \sum_{k=1}^{r} e_k X_k x_i + \sum_{k,\,j}^{1 \dots r} \frac{e_k e_j}{1 \cdot 2} X_k X_j x_i + \cdots \qquad (i=1 \cdots n)$$

after the introduction of new parameters, so that these equations represent the finite equations of the r-term group which is generated by the infinitesimal transformations: $X_1 f, \dots, X_r f$.

It has no influence whether we choose the group $\overline{A}_1 f, \dots, \overline{A}_r f$ or any other group amongst the infinitely many simply transitive groups which have the same composition as the group $X_1 f, \dots, X_r f$, and it is completely indifferent whether we choose the system of values $\overline{a}_1, \dots, \overline{a}_r$ or any other system of values: in the way indicated, we always obtain an analytic representation for the finite transformations of the group: $X_1 f, \dots, X_r f$.

We have already made this remark in Chap. 9, though naturally not in the same words. But at present, we have reached a point where we can see directly why one always obtains the equations of the same group for various choices of the group: $\overline{A}_1 f, \dots, \overline{A}_r f$ and of the systems of values: $\overline{a}_1, \dots, \overline{a}_r$.

Indeed, let:

$$\mathfrak{A}_k f = \sum_{j=1}^{r} \beta_{kj}(\mathfrak{a}_1, \dots, \mathfrak{a}_r) \frac{\partial f}{\partial \mathfrak{a}_j} \qquad (k=1 \cdots r)$$

be any other simply transitive group equally composed with the group: $X_1 f, \dots, X_r f$ for which the relations:

$$[\mathfrak{A}_i, \mathfrak{A}_k] = \sum_{s=1}^{r} c_{iks} \mathfrak{A}_s f$$

are identically satisfied, and moreover, let: $\mathfrak{a}_1^0, \dots, \mathfrak{a}_r^0$ be any system of values in the neighborhood of which all $\beta_{kj}(\mathfrak{a})$ behave regularly and for which the determinant $\sum \pm \beta_{11} \cdots \beta_{rr}$ does not vanish.

Because the two groups: $\overline{A}_1 f, \dots, \overline{A}_r f$ and $\mathfrak{A}_1 f, \dots, \mathfrak{A}_r f$ are equally composed and are both simply transitive, then (Theorem 64, p. 353) there are ∞^r different transformations:

$$a_k = \lambda_k(\mathfrak{a}_1, \dots, \mathfrak{a}_r, C_1, \dots, C_r) \qquad (k=1 \cdots r)$$

which transfer $\mathfrak{A}_1 f, \dots, \mathfrak{A}_r f$ to: $\overline{A}_1 f, \dots, \overline{A}_r f$, respectively. The equations of these transformations are solvable with respect to the arbitrary parameters: $C_1, \dots, C_r$, so in particular, one can choose $C_1, \dots, C_r$ in such a way that the equations:

$$\overline{a}_k = \lambda_k(\mathfrak{a}_1^0, \dots, \mathfrak{a}_r^0, C_1, \dots, C_r) \qquad (k=1 \cdots r)$$

are satisfied.

If: $C_1^0, \dots, C_r^0$ are the values of $C_1, \dots, C_r$ obtained in this way, and if one sets:

$$\lambda_k(\mathfrak{a}_1, \dots, \mathfrak{a}_r, C_1^0, \dots, C_r^0) = \pi_k(\mathfrak{a}_1, \dots, \mathfrak{a}_r) \qquad (k=1 \cdots r),$$

then the equations:

$$a_k = \pi_k(\mathfrak{a}_1, \dots, \mathfrak{a}_r) \qquad (k=1 \cdots r) \tag{10}$$

represent a transformation which transfers $\mathfrak{A}_1 f, \dots, \mathfrak{A}_r f$ to: $\overline{A}_1 f, \dots, \overline{A}_r f$, respectively, and which in addition transfers the system of values: $\mathfrak{a}_1^0, \dots, \mathfrak{a}_r^0$ to the system of values: $\overline{a}_1, \dots, \overline{a}_r$.

From this, it follows that we obtain the general solutions of the complete system:

$$X_k' f + \mathfrak{A}_k f = 0 \qquad (k=1 \cdots r) \tag{9'}$$

relative to: $\mathfrak{a}_1 = \mathfrak{a}_1^0$, ..., $\mathfrak{a}_r = \mathfrak{a}_r^0$ when we make the substitution: $a_1 = \pi_1(\mathfrak{a})$, ..., $a_r = \pi_r(\mathfrak{a})$ in the general solutions:

$$\overline{F}_i(x_1', \dots, x_n', a_1, \dots, a_r) \qquad (i=1 \cdots n)$$

of the complete system (9). Therefore, if we had used, instead of the simply transitive group: $\overline{A}_1 f, \dots, \overline{A}_r f$, the group: $\mathfrak{A}_1 f, \dots, \mathfrak{A}_r f$, and if we had used, instead of the system of values: $\overline{a}_1, \dots, \overline{a}_r$, the system of values: $\mathfrak{a}_1^0, \dots, \mathfrak{a}_r^0$, then instead of the equations: $x_i' = \overline{f}_i(x_1, \dots, x_n, a_1, \dots, a_r)$, we would have received the equations:

$$x_i' = \overline{f}_i\big(x_1, \dots, x_n, \pi_1(\mathfrak{a}), \dots, \pi_r(\mathfrak{a})\big) \qquad (i=1\cdots n).$$

But evidently, these equations are transferred to the former ones when the new parameters $a_1, \dots, a_r$ are introduced in place of $\mathfrak{a}_1, \dots, \mathfrak{a}_r$ by virtue of the equations (10).

As in sections 100 and 101, p. 412 sq., we again want to start with a determined form:

$$x_i' = f_i(x_1, \dots, x_n, a_1, \dots, a_r) \qquad (i=1\cdots n)$$

of the group: $X_1 f, \dots, X_r f$, and as before, we let $a_1^0, \dots, a_r^0$ denote the parameters attached to the identity transformation: $x_i' = x_i$.

Then it is easy to identify the complete system, the integration of which leads precisely to the equations: $x_i' = f_i(x, a)$.

The complete system in question simply has the form:

$$X_k' + A_k f = 0 \qquad (k=1\cdots r),$$

where, in the infinitesimal transformations:

$$A_k f = \sum_{j=1}^{r} \alpha_{kj}(a_1, \dots, a_r)\,\frac{\partial f}{\partial a_j} \qquad (k=1\cdots r),$$

the functions $\alpha_{kj}(a)$ are the same as in the differential equations (7'). If:

$$F_i(x_1', \dots, x_n', a_1, \dots, a_r) \qquad (i=1\cdots n)$$

denote the general solutions to the complete system: $X_k' f + A_k f = 0$ relative to: $a_1 = a_1^0, \dots, a_r = a_r^0$, then by resolution with respect to $x_1', \dots, x_n'$, the equations:

$$x_i = F_i(x_1', \dots, x_n', a_1, \dots, a_r) \qquad (i=1\cdots n)$$

give the equations: $x_i' = f_i(x, a)$ exactly. All of this follows from the developments of Chaps. 3 and 9.

We now apply these considerations to the parameter group associated to the group: $x_i' = f_i(x, a)$, whose equations, according to p. 413, have the form:

$$a_k' = \varphi_k(a_1, \dots, a_r, b_1, \dots, b_r) \qquad (k=1\cdots r)$$

and whose identity transformation possesses the parameters: $b_1 = a_1^0, \dots, b_r = a_r^0$.

The complete system, through the introduction of which the equations: $a_k' = \varphi_k(a, b)$ of the parameter group can be found, visibly reads:

$$\sum_{j=1}^{r} \alpha_{kj}(a'_1,\ldots,a'_r)\frac{\partial f}{\partial a'_j} + \sum_{j=1}^{r} \alpha_{kj}(b_1,\ldots,b_r)\frac{\partial f}{\partial b_j} = 0 \qquad (k=1\cdots r).$$

If we determine the general solutions:

$$H_j(a'_1,\ldots,a'_r,b_1,\ldots,b_r) \qquad (j=1\cdots r)$$

of this complete system relative to: $b_k = a_k^0$ and if we then solve the equations:

$$a_j = H_j(a'_1,\ldots,a'_r,b_1,\ldots,b_r) \qquad (j=1\cdots r)$$

with respect to $a'_1,\ldots,a'_r$, we obtain: $a'_k = \varphi_k(a,b)$.

Thus, the following holds:

Theorem 73. *If one knows the infinitesimal transformations:*

$$A_k f = \sum_{j=1}^{r} \alpha_{kj}(a_1,\ldots,a_r)\frac{\partial f}{\partial a_j} \qquad (k=1\cdots r)$$

of the parameter group of an r-term group and if one knows that the identity transformation of this group has the parameters: $a_1^0,\ldots,a_r^0$, whence the identity transformation of the parameter group also has the parameters: $a_1^0,\ldots,a_r^0$, then one finds the finite equations of the parameter group in the following way: One determines the general solutions of the complete system:

$$\sum_{j=1}^{r} \alpha_{kj}(a')\frac{\partial f}{\partial a'_j} + \sum_{j=1}^{r} \alpha_{kj}(b)\frac{\partial f}{\partial b_j} = 0 \qquad (k=1\cdots r)$$

relative to: $b_1 = a_1^0$, ..., $b_r = a_r^0$; if:

$$H_j(a'_1,\ldots,a'_r,b_1,\ldots,b_r) \qquad (j=1\cdots r)$$

are these general solutions, then one obtains the sought equations of the parameter group by solving the r equations:

$$a_j = H_j(a'_1,\ldots,a'_r,b_1,\ldots,b_r) \qquad (j=1\cdots r)$$

with respect to $a'_1,\ldots,a'_r$.

If, in an r-term group $x'_i = f_i(x,a)$, one introduces new parameters $\mathfrak{a}_k$ in place of the a, then the group receives a new form to which a new parameter group is also naturally associated.

The connection between the new and the old parameter groups is very simple. Indeed, if the new parameters are determined by the equations:

$$a_j = \lambda_j(\mathfrak{a}_1,\ldots,\mathfrak{a}_r) \qquad (j=1\cdots r),$$

then the new form of the group: $x'_i = f_i(x,a)$ is the following:

$$x_i' = f_i\big(x_1,\dots,x_n,\,\lambda_1(\mathfrak{a}),\,\dots,\,\lambda_r(\mathfrak{a})\big) \qquad {\scriptstyle (i=1\cdots n)},$$

and the new parameter group reads:

$$\lambda_k(\mathfrak{a}_1',\dots,\mathfrak{a}_r') = \varphi_k\big(\lambda_1(\mathfrak{a}),\,\dots,\,\lambda_r(\mathfrak{a}),\,\lambda_1(\mathfrak{b}),\,\dots,\,\lambda_r(\mathfrak{b})\big)$$
$$ {\scriptstyle (k=1\cdots r)}$$

while the old one was $a_k' = \varphi_k(a,b)$. Consequently, the new one comes from the old one when by executing the substitution $a_j = \lambda_j(\mathfrak{a})$ both on the a and on the b, that is to say, by inserting, for the a', a, b, the following values:

$$a_j' = \lambda_j(\mathfrak{a}'), \quad a_j = \lambda_j(\mathfrak{a}), \quad b_j = \lambda_j(\mathfrak{b}) \qquad {\scriptstyle (j=1\cdots r)}$$

in the equations: $a_k' = \varphi_k(a,b)$.

§ 103. According to Theorem 72, p. 417, every r-term group has the same composition as its parameter group, and consequently, the r-term groups which have the same parameter group are equally composed with each other.

Now, we claim that conversely, two equally composed r-term groups can always be brought, after the introduction of new parameters, to a form such that they both have the same parameter group.

In order to prove this claim, we imagine that the infinitesimal transformations of the two groups are given; let the transformations of the first group be:

$$X_k f = \sum_{i=1}^{n} \xi_{ki}(x_1,\dots,x_n)\,\frac{\partial f}{\partial x_i} \qquad {\scriptstyle (k=1\cdots r)},$$

and the transformations of the other group be:

$$Y_k f = \sum_{\mu=1}^{m} \eta_{k\mu}(y_1,\dots,y_m)\,\frac{\partial f}{\partial y_\mu} \qquad {\scriptstyle (k=1\cdots r)}.$$

Since the two groups are equally composed, we can assume that the $X_k f$ and the $Y_k f$ are already chosen in such a way that, simultaneously with the relations:

$$[X_i,\, X_k] = \sum_{s=1}^{r} c_{iks}\, X_s f,$$

the relations:

$$[Y_i,\, Y_k] = \sum_{s=1}^{r} c_{iks}\, Y_s f$$

hold.

Lastly, we imagine that r independent infinitesimal transformations:

$$A_k f = \sum_{j=1}^{r} \alpha_{kj}(a_1,\dots,a_r)\frac{\partial f}{\partial a_j} \qquad {\scriptstyle (k=1\cdots r)}$$

are given which generate a simply transitive r-term group equally composed with these groups and which stand pairwise in the relationships:

$$[A_i, A_k] = \sum_{s=1}^{r} c_{iks} A_s f.$$

At present, we form the complete system:

$$X_k' f + A_k f = 0 \qquad {\scriptstyle (k=1\cdots r)}$$

and we determine its general solutions:

$$F_i(x_1',\dots,x_n',a_1,\dots,a_r) \qquad {\scriptstyle (i=1\cdots n)}$$

relative to an arbitrary system of values: $a_1 = a_1^0,\dots,a_r = a_r^0$; furthermore, we form the complete system:

$$Y_k' f + A_k f = 0 \qquad {\scriptstyle (k=1\cdots r)}$$

and we determine its general solutions:

$$\mathfrak{F}_\mu(y_1',\dots,y_m',a_1,\dots,a_r) \qquad {\scriptstyle (i=1\cdots n)}$$

relative to the same system of values: $a_k = a_k^0$.

In addition, we solve the n equations: $x_i = F_i(x',a)$ with respect to $x_1',\dots,x_n'$:

$$x_i' = f_i(x_1,\dots,x_n,a_1,\dots,a_r) \qquad {\scriptstyle (i=1\cdots n)}, \tag{11}$$

and likewise, the m equations: $y_\mu = \mathfrak{F}_\mu(y',a)$ with respect to $y_1',\dots,y_m'$:

$$y_\mu' = \mathfrak{f}_\mu(y_1,\dots,y_m,a_1,\dots,a_r) \qquad {\scriptstyle (\mu=1\cdots n)}. \tag{11'}$$

In whichever form the finite equations of the group: $X_1 f,\dots,X_r f$ are present, then obviously, they can be brought to the form (11) by introducing new parameters, and likewise, in whichever form the finite equations of the group: $Y_1 f,\dots,Y_r f$ are present, they can always be given the form (11') by introducing new parameters.

But Theorem 73, p. 421, can be applied to the two groups (11) and (11'). Indeed, in the two groups, the identity transformation has the parameters: $a_1^0,\dots,a_r^0$ and for the two groups, the associated parameter groups contain the r independent infinitesimal transformations: $A_1 f,\dots,A_r f$, and consequently, according to the mentioned theorem, one can indicate for each one of the two groups the associated parameter group, and these parameter groups are the same for both of them.

As a result, the claim stated above is proved. —

Thus, it is at present established that two groups which have the same parameter group are equally composed, and on the other hand, that two groups which are equally composed can be brought, by introducing new parameters, to a form in which they both have the same parameter group. Thus, we can say:

Theorem 74. *Two r-term groups:*

$$x_i' = f_i(x_1, \dots, x_n, a_1, \dots, a_r) \qquad (i=1\cdots n)$$

and:

$$y_\mu' = g_\mu(y_1, \dots, y_m, \mathfrak{a}_1, \dots, \mathfrak{a}_r) \qquad (\mu=1\cdots m)$$

are equally composed if and only if it is possible to represent the parameters: $\mathfrak{a}_1, \dots, \mathfrak{a}_r$ *as independent functions of the a:*

$$\mathfrak{a}_k = \chi_k(a_1, \dots, a_r) \qquad (k=1\cdots r)$$

in such a way that the parameter group of the group:

$$y_\mu' = g_\mu\big(y_1, \dots, y_m, \chi_1(a), \dots, \chi_r(a)\big) \qquad (\mu=1\cdots m)$$

coincides with the parameter group of the group: $x_i' = f_i(x,a)$.

It is particularly noticeable that our two equally composed groups: $X_1 f, \dots, X_r f$ and: $Y_1 f, \dots, Y_r f$ then have the same parameter group when one writes their finite equations in the canonical form:

$$x_i' = x_i + \sum_{k=1}^{r} e_k X_k x_i + \sum_{k,\,j}^{1\cdots r} \frac{e_k e_j}{1\cdot 2} X_k X_j x_i + \cdots \qquad (i=1\cdots n) \tag{12}$$

and, respectively:

$$y_\mu' = y_\mu + \sum_{k=1}^{r} e_k Y_k y_\mu + \sum_{k,\,j}^{1\cdots r} \frac{e_k e_j}{1\cdot 2} Y_k Y_j y_\mu + \cdots \qquad (\mu=1\cdots m). \tag{12'}$$

In fact, from the developments in Chap. 4, Sect. 4.5, it follows that the equations (11) receive the form (12) after the substitution:

$$a_\nu = a_\nu^0 + \sum_{k=1}^{r} e_k \big[A_k a_\nu\big]_{a=a^0} + \sum_{k,\,j}^{1\cdots r} \frac{e_k e_j}{1\cdot 2} \big[A_k A_j a_\nu\big]_{a=a^0} + \cdots \tag{13}$$
$$(\nu=1\cdots r)$$

and that the equations (11') are transferred to (12') by the same substitution. Now, since the equations (13) represent a transformation between the parameters: $a_1, \dots, a_r$ and $e_1, \dots, e_r$ and since the two groups (11) and (11') have their parameter group in common, it follows that one and the same parameter group is associated to the two groups (12) and (12').

Thus, the following holds:

Proposition 1. *If the r independent infinitesimal transformations:*

$$X_k f = \sum_{i=1}^{n} \xi_{ki}(x_1, \dots, x_n) \frac{\partial f}{\partial x_i} \qquad (k=1\cdots r)$$

stand pairwise in the relationships:

$$[X_i, X_k] = \sum_{s=1}^{r} c_{iks} X_s f$$

and if, on the other hand, the r independent infinitesimal transformations:

$$Y_k f = \sum_{\mu=1}^{m} \eta_{k\mu}(y_1, \dots, y_m) \frac{\partial f}{\partial y_\mu} \qquad (k=1\cdots r)$$

stand pairwise in the same relationships:

$$[Y_i, Y_k] = \sum_{s=1}^{r} c_{iks} Y_s f,$$

then the two equally composed groups:

$$x_i' = x_i + \sum_{k=1}^{r} e_k X_k x_i + \sum_{k,\,j}^{1\cdots r} \frac{e_k e_j}{1\cdot 2} X_k X_j x_i + \cdots \qquad (i=1\cdots n)$$

and:

$$y_\mu' = y_\mu + \sum_{k=1}^{r} e_k Y_k y_\mu + \sum_{k,\,j}^{1\cdots r} \frac{e_k e_j}{1\cdot 2} Y_k Y_j y_\mu + \cdots \qquad (\mu=1\cdots m)$$

have one and the same parameter group.

Now, we imagine that two arbitrary equally composed r-term groups:

$$x_i' = f_i(x_1, \dots, x_n, a_1, \dots, a_r) \qquad (i=1\cdots n)$$

and:

$$y_\mu' = \mathfrak{f}_\mu(y_1, \dots, y_m, a_1, \dots, a_r) \qquad (\mu=1\cdots m)$$

are presented which already have a form such that the parameter group for both is the same.

Let the infinitesimal transformations of the parameter group in question be:

$$A_k f = \sum_{j=1}^{r} \alpha_{kj}(a_1, \dots, a_r) \frac{\partial f}{\partial a_j} \qquad (k=1\cdots r)$$

and let them be linked together by the relations:

$$[A_i, A_k] = \sum_{s=1}^{r} c_{iks} A_s f.$$

According to p. 415, $x_1', \dots, x_n'$, when considered as functions of $x_1, \dots, x_n$, $a_1, \dots, a_r$, satisfy certain differential equations of the form:

$$\xi_{ji}(x_1', \dots, x_n') = \sum_{k=1}^{r} \alpha_{jk}(a_1, \dots, a_r) \frac{\partial x_i'}{\partial a_k} \qquad (j=1 \cdots r;\; i=1 \cdots n).$$

Here, the $\xi_{ji}(x')$ are completely determined functions, because by resolution of the equations: $x_i' = f_i(x,a)$, we get, say: $x_i = F_i(x',a)$, whence it holds identically:

$$\xi_{ji}(x_1', \dots, x_n') = \sum_{k=1}^{r} \alpha_{jk}(a_1, \dots, a_r) \left[\frac{\partial f_i(x,a)}{\partial a_k} \right]_{x=F(x',a)},$$

so that we can therefore find the $\xi_{ji}(x')$ without difficulty, when we want.

According to Theorem 21, p. 164, the r expressions:

$$X_k f = \sum_{i=1}^{n} \xi_{ki}(x_1, \dots, x_n) \frac{\partial f}{\partial x_i} \qquad (k=1 \cdots r)$$

are independent infinitesimal transformations and they are linked together by the relations:

$$[X_i, X_k] = \sum_{s=1}^{r} c_{iks} X_s f;$$

naturally, they generate the group: $x_i' = f_i(x,a)$.

For the group: $y_\mu' = \mathfrak{f}_\mu(y,a)$, there are in a corresponding way differential equations of the form:

$$\eta_{j\mu}(y_1', \dots, y_m') = \sum_{k=1}^{r} \alpha_{jk}(a_1, \dots, a_r) \frac{\partial y_\mu'}{\partial a_k} \qquad (j=1 \cdots r;\; \mu=1 \cdots m),$$

where the $\eta_{j\mu}(y')$ are completely determined functions. The r expressions:

$$Y_k f = \sum_{\mu=1}^{m} \eta_{k\mu}(y_1, \dots, y_m) \frac{\partial f}{\partial y_\mu} \qquad (k=1 \cdots r)$$

are independent infinitesimal transformations and they are linked together by the relations:

$$[Y_i, Y_k] = \sum_{s=1}^{r} c_{iks} Y_s f;$$

naturally, they generate the group: $y_\mu' = \mathfrak{f}_\mu(y,a)$.

Now, exactly as on p. 424, we recognize that, after the substitution:

$$a_\nu = a_\nu^0 + \sum_{k=1}^{r} e_k \left[A_k a_\nu\right]_{a=a^0} + \cdots \qquad (\nu=1\cdots r),$$

the groups: $x_i' = f_i(x,a)$ and: $y_\mu' = \mathfrak{f}_\mu(y,a)$ are of the forms:

$$x_i' = x_i + \sum_{k=1}^{r} e_k X_k x_i + \cdots \qquad (i=1\cdots n)$$

and, respectively:

$$y_\mu' = y_\mu + \sum_{k=1}^{r} e_k Y_k y_\mu + \cdots \qquad (\mu=1\cdots m).$$

Therefore, we get the

Proposition 2. *If the two r-term groups:*

$$x_i' = f_i(x_1,\dots,x_n,\,a_1,\dots,a_r) \qquad (i=1\cdots n)$$

and:

$$y_\mu' = \mathfrak{f}_\mu(y_1,\dots,y_m,\,a_1,\dots,a_r) \qquad (\mu=1\cdots m)$$

have the same parameter group, then it is possible to introduce, in place of the a, new parameters: $e_1,\dots,e_r$ so that the two groups receive the forms:

$$x_i' = x_i + \sum_{k=1}^{r} e_k X_k x_i + \cdots \qquad (i=1\cdots n)$$

and, respectively:

$$y_\mu' = y_\mu + \sum_{k=1}^{r} e_k Y_k y_\mu + \cdots \qquad (\mu=1\cdots m)\,;$$

here, $X_1 f,\dots,X_r f$ and $Y_1 f,\dots,Y_r f$ are independent infinitesimal transformations of the two groups such that, simultaneously with the relations:

$$[X_i,\,X_k] = \sum_{s=1}^{r} c_{iks} X_s f$$

there hold the relations:

$$[Y_i,\,Y_k] = \sum_{s=1}^{r} c_{iks} Y_s f,$$

so that the two groups are related to each other in a holoedrically isomorphic way when the infinitesimal transformation $e_1 Y_1 f + \cdots + e_r Y_r f$ is associated to every infinitesimal transformation: $e_1 X_1 f + \cdots + e_r X_r f$.

Now, concerning the two groups: $x_i' = f_i(x,a)$ and: $y_\mu' = \mathfrak{f}_\mu(y,a)$, we make the same assumptions as in Proposition 2; in addition, we assume that: $a_k' = \varphi_k(a,b)$ are the finite equations of their common parameter group.

Under this assumption, the following obviously holds true of each of the two groups: If two transformations of the group which have the parameters: $a_1,\dots,a_r$ and: $b_1,\dots,b_r$, respectively, are executed one after the other, then the resulting transformation belongs to the group and it possesses the parameters: $\varphi_1(a,b)$, ..., $\varphi_r(a,b)$.

We can express this fact somewhat differently if we mutually associate the transformations of the two groups in such a way that every transformation of the one group corresponds to the transformation of the other group which has the same parameters. Indeed, we can then say: if S is a transformation of the one group and if $\mathfrak{S}$ is the corresponding transformation of the other group, and moreover, if T is a second transformation of the one group and if $\mathfrak{T}$ the corresponding transformation of the other group, then the transformation ST of the one group corresponds to the transformation $\mathfrak{S}\mathfrak{T}$ in the other group.

Such a mutual association of the transformations of both groups is then, according to Theorem 74, p. 424, always possible when and only when the two groups are equally composed. So we have the:

Theorem 75. *Two r-term groups are equally composed if and only if it is possible to relate the transformations of the one group to the transformations of the other group in a univalent invertible way so that the following holds true: If, in the one group, one executes two transformations one after the other and if, in the other group, one executes one after the other and in the same order the corresponding transformations, then the transformation that one obtains in the one group corresponds to the transformation that one obtains in the other group.*[2]

The above considerations provide a new important result when they are applied to Proposition 1, p. 424. In order to be able to state this result under the most simple form, we remember two different things: firstly, that the two groups: $X_1f,\dots,X_rf$ and: $Y_1f,\dots,Y_rf$ are related to each other in a holoedrically isomorphic way when to every infinitesimal transformation: $e_1X_1f+\cdots+e_rX_rf$ is associated the infinitesimal transformation: $e_1Y_1f+\cdots+e_rY_rf$, and secondly, that the expression: $e_1X_1f+\cdots+e_rX_rf$ can also be interpreted as the symbol of a finite transformation of the group: $X_1f,\dots,X_rf$ (cf. Chap. 17, p. 305 and Chap. 15, p. 268). By taking this into account, we can say:

Proposition 3. *Let the two equally composed r-term groups:*

$$X_kf = \sum_{i=1}^{n} \xi_{ki}(x_1,\dots,x_n)\frac{\partial f}{\partial x_i} \qquad (k=1\cdots r)$$

and:

[2] Cf. LIE, Archiv for Mathematik og Naturvidenskab 1876; Math. Ann. Vol. XXV, p. 77; G. d. W. zu Christiania, 1884, no. 15.

$$Y_k f = \sum_{\mu=1}^{m} \eta_{k\mu}(y_1,\dots,y_m) \frac{\partial f}{\partial y_\mu} \qquad (k=1\cdots r)$$

be related to each other in a holoedrically isomorphic way when one associates to every infinitesimal transformation: $e_1 X_1 f + \cdots + e_r X_r f$ *the infinitesimal transformation:* $e_1 Y_1 f + \cdots + e_r Y_r f$ *so that, simultaneously with the relations:*

$$[X_i, X_k] = \sum_{s=1}^{r} c_{iks} X_s f$$

there hold the relations:

$$[Y_i, Y_k] = \sum_{s=1}^{r} c_{iks} Y_s f.$$

Then, if one interprets the expressions: $\sum e_k X_k f$ *and* $\sum e_k Y_k f$ *as the general symbols of the finite transformations of the two groups:* $X_1 f,\dots,X_r f$ *and* $Y_1 f,\dots,Y_r f$*, the following holds true: If the two transformations:* $\sum e_k X_k f$ *and* $\sum e_k' X_k f$ *of the group:* $X_1 f,\dots,X_r f$ *give, when executed one after the other, the transformation:* $\sum e_k'' X_k f$*, then the two transformations:* $\sum e_k Y_k f$ *and* $\sum e_k' Y_k f$ *give the transformation:* $\sum e_k'' Y_k f$ *when executed one after the other.*

Thus, if one has related two equally composed r-term groups in a holoedrically isomorphic way in the sense of Chap. 17, p. 305, then at the same time, thanks to this, one has established a univalent invertible relationship between the finite transformations of the two groups, as is described in Theorem 75.

But the converse also holds true: If one has produced a univalent invertible relationship between the transformations of two r-term equally composed groups which has the property described in Theorem 75, then at the same time, thanks to this, the two groups are related to each other in a holoedrically isomorphic way.

In fact, let: $x_i' = f_i(x_1,\dots,x_n,a_1,\dots,a_r)$ be the one group and let: $y_\mu' = \mathfrak{f}_\mu(y_1,\dots,y_m,a_1,\dots,a_r)$ be the transformation of the other group that corresponds to the transformation: $x_i' = f_i(x,a)$. Then the two groups: $x_i' = f_i(x,a)$ and $y_\mu' = \mathfrak{f}_\mu(y,a)$ obviously have one and the same parameter group and hence, according to Proposition 2, p. 427, by introducing appropriate new parameters: $e_1,\dots,e_r$, they can be given the following forms:

$$x_i' = x_i + \sum_{k=1}^{r} e_k X_k x_i + \cdots \qquad (i=1\cdots n)$$

and:

$$y_\mu' = y_\mu + \sum_{k=1}^{r} e_k Y_k y_\mu + \cdots \qquad (\mu=1\cdots m).$$

From this, it results that the said univalent invertible relationship between the transformations of the two groups amounts to the fact that, to every finite transformation: $e_1 X_1 f + \cdots + e_r X_r f$ of the one group is associated the finite transformation: $e_1 Y_1 f + \cdots + e_r Y_r f$ of the other group. Therefore at the same time, to every infinitesimal transformation: $\sum e_k X_k f$ is associated the infinitesimal transformation:

$\sum e_k Y_k f$, and as a result, according to Proposition 2, a holoedrically isomorphic relationship between the two groups is effectively established.

In the theory of substitutions, one defines the holoedric isomorphism of two groups and the holoedrically isomorphic relationship between two groups differently than what we have done in Chap. 17. There, one says that two groups with the same number of substitutions are "equally composed" or "holoedrically isomorphic" when one can produce, between the transformations of the two groups, a univalent invertible relationship which has the property described in Theorem 75, p. 428; if such a relationship between two holoedrically isomorphic groups is really established, then one says that the two groups are "related to each other in a holoedrically isomorphic way".

But already on p. 305, we observed that materially [MATERIELL], our definition of the concept in question corresponds precisely to the one which is usual in the theory of substitutions, as far as such a correspondence may in any case be possible between domains so different as the theory of substitutions and the theory of transformation groups.

Our last developments show that the claim made on p. 305 is correct. From Theorem 75, p. 428, it follows that two r-term groups which are holoedrically isomorphic in the sense of Chap. 17, p. 305, must also be called holoedrically isomorphic in the sense of the theory of substitutions, and conversely. From Proposition 3, p. 428, and from the remarks following, it is evident that two r-term groups which are related to each other in a holoedrically isomorphic way in the sense of Chap. 17, p. 305, are also holoedrically isomorphic in the sense of substitution theory.

It still remains to prove that our definition of the meroedric isomorphism (cf. Chap. 17, p. 305) also corresponds to the definition of meroedric isomorphism that is given by the theory of substitutions.

Let $X_1 f, \dots, X_{r-q} f, \dots, X_r f$ and $Y_1 f, \dots, Y_{r-q} f$ be two meroedrically isomorphic groups and let $X_{r-q+1} f, \dots, X_r f$ be precisely the invariant subgroup of the r-term group which corresponds to the identity transformation of the $(r-q)$-term group. Lastly, let $A_1 f, \dots, A_r f$ be a simply transitive group in the variables $a_1, \dots, a_r$ which has the same composition as $X_1 f, \dots, X_r f$. Here, we want to assume that $a_1, \dots, a_{r-q}$ are solutions of the complete system:

$$A_{r-q+1} f = 0, \;\dots,\; A_r f = 0.$$

After this choice of the variables, $A_1 f, \dots, A_{r-q} f$ has the form:

$$A_k f = \sum_{i=1}^{r-q} \alpha_{ki}(a_1, \dots, a_{r-q}) \frac{\partial f}{\partial a_i} + \sum_{j=r-q+1}^{r} \beta_{kj}(a_1, \dots, a_r) \frac{\partial f}{\partial a_j}$$

$$(k = 1 \cdots r-q),$$

while $A_{r-q+1} f, \dots, A_r f$ have the form:

$$A_k f = \sum_{j=r-q+1}^{r} \beta_{kj}(a_1,\dots,a_r)\frac{\partial f}{\partial a_j} \qquad {\scriptstyle (k=1\cdots r-q+1)}.$$

At the same time, it is clear that the reduced infinitesimal transformations:

$$\overline{A}_k f = \sum_{i=1}^{r-q} \alpha_{ki}(a_1,\dots,a_{r-q})\frac{\partial f}{\partial a_i} \qquad {\scriptstyle (k=1\cdots r-q)}$$

generate a simply transitive group equally composed with the $(r-q)$-term group $Y_1 f,\dots,Y_{r-q} f$.

If one denotes by $B_k f$, $\overline{B}_k f$ the infinitesimal transformations $A_k f$, $\overline{A}_k f$ written down in the variables b instead of the variables a, then according to earlier discussions, one finds the parameter group of the $(r-q)$-term group $Y_1 f,\dots,Y_{r-q} f$ by integrating the complete system:

$$\overline{A}_k f + \overline{B}_k f = 0 \qquad {\scriptstyle (k=1\cdots r-q)},$$

and also the parameter group of the r-term group $X_1 f,\dots,X_r f$ by integrating the complete system:

$$A_k f + B_k f = 0 \qquad {\scriptstyle (k=1\cdots r)}.$$

If the equations of the parameter group associated with the $(r-q)$-term group obtained in this way are, say:

$$a'_k = \varphi_k(a_1,\dots,a_{r-q},\, b_1,\dots,b_{r-q}) \qquad {\scriptstyle (k=1\cdots r-q)},$$

then it is clear that the parameter group associated with the r-term group is representable by equations of the form:

$$a'_k = \varphi_k(a_1,\dots,a_{r-q},\, b_1,\dots,b_{r-q}) \qquad {\scriptstyle (k=1\cdots r-q)}$$
$$a'_i = \psi_i(a_1,\dots,a_r,\, b_1,\dots,b_r) \qquad {\scriptstyle (i=r-q+1\cdots r)}.$$

From this, it follows that the groups $X_1 f,\dots,X_r f$ and $Y_1 f,\dots,Y_{r-q} f$ are meroedrically isomorphic in the sense of substitution theory.

At present, we want to derive the result just obtained in another way. However, we do not consider it to be necessary to describe this second method in detail, although it is in itself notable, because it is very similar to the preceding developments.

We imagine that the transformation equations:

$$x'_i = f_i(x_1,\dots,x_n,\, a_1,\dots,a_r) \qquad {\scriptstyle (i=1\cdots n)}$$

are presented, but we leave undecided whether the a are essential parameters or not. Now, if the f_i satisfy differential equations of the form:

$$\frac{\partial f_i}{\partial a_k} = \sum_{j=1}^{r} \psi_{kj}(a_1,\dots,a_r)\,\xi_{ji}(f_1,\dots,f_n)$$

which can be resolved with respect to the ξ_{ji}:

$$\xi_{ji}(f) = \sum_{k=1}^{r} \alpha_{jk}(a_1,\dots,a_r)\,\frac{\partial f_i}{\partial a_k},$$

then we realize easily (cf. Chap. 4, Sect. 4.6) that every transformation: $x_i' = f_i(x,a)$ whose parameters a lie in a certain neighborhood of $\overline{a}_1,\dots,\overline{a}_r$ can be obtained by executing firstly the transformation: $\overline{x}_i = f_i(x,\overline{a})$, and then a certain transformation:

$$x_i' = w_i(\overline{x}_1,\dots,\overline{x}_n,\, \lambda_1,\dots,\lambda_r)$$

of a one-term group:

$$\lambda_1 X_1 f + \cdots + \lambda_r X_r f = \sum_{k=1}^{r} \lambda_k \sum_{i=1}^{n} \xi_{ki}(x)\,\frac{\partial f}{\partial x_i}.$$

In addition, we find that the concerned values of the parameters λ only depend upon the form of the functions $\alpha_{jk}(a)$, and on the two systems of values $\overline{a}_k$ and a_k as well.

Under the assumptions made, when one sets:

$$\sum_{k=1}^{r} \alpha_{jk}(a)\,\frac{\partial f}{\partial a_k} = A_k f, \qquad \sum_{i=1}^{n} \xi_{ki}(x')\,\frac{\partial f}{\partial x_i'} = X_k' f,$$

one now obtains (compare with pages 161 sq.) that relations of the form:

$$[X_i',\, X_k'] = \sum_{s=1}^{r} \vartheta_{iks}(x_1',\dots,x_n',\, a_1,\dots,a_r)\, X_s' f$$

$$[A_i,\, A_k] = \sum_{s=1}^{r} \vartheta_{iks}(x_1',\dots,x_n',\, a_1,\dots,a_r)\, A_s f$$

hold, in which the ϑ_{iks} are even independent of the x'. However, contrary to the earlier analogous developments, it cannot be proved now that in the *two* latter equations, the ϑ can be set equal to absolute constants. But when we consider only the equations:

$$[X_i',\, X_k'] = \sum_{s=1}^{r} \vartheta_{iks}(a)\, X_s' f,$$

it is clear that by particularizing the a, they provide relations of the form:

$$[X_i',\, X_k'] = \sum_{s=1}^{r} c_{iks}\, X_s' f$$

in which the c_{iks} denote constants. Hence it is also true under our present assumptions that all finite transformations $\lambda_1 X_1 f + \cdots + \lambda_r X_r f$ form a group which possesses the same number of essential parameters as the family $x_i' = f_i(x,a)$.

On the other hand, let $X_1'f,\dots,X_r'f$ denote r infinitesimal transformation that are not necessarily independent and which stand in the relationships:

$$[X_i',X_k'] = \sum_{s=1}^{r} c_{iks}X_s'f;$$

furthermore, let:

$$A_kf = \sum_{i=1}^{r} \alpha_{ki}(a_1,\dots,a_r)\frac{\partial f}{\partial a_i} \qquad {\scriptstyle (k=1\cdots r)}$$

be r independent infinitesimal transformations of a simply transitive group whose composition is given by the equations:

$$[A_i,A_k] = \sum_{s=1}^{r} c_{iks}A_sf.$$

Now, we form the r-term complete system:

$$X_k'f + A_kf = 0 \qquad {\scriptstyle (k=1\cdots r)},$$

we compute its general solutions $F_1,\dots,F_n$ relative to a suitable system of values $a_1 = a_1^0, \dots, a_r = a_r^0$ and we set: $x_1 = F_1, \dots, x_n = F_n$. Then, the equations resulting by resolution:

$$x_k' = f_k(x_1,\dots,x_n,a_1,\dots,a_r)$$

determine a family of at most ∞^r transformations which obviously comprises the identity transformation $x_k' = x_k$.

Now, by proceeding exactly as on p. 167, we first obtain the system of equations:

$$\sum_{\mu=1}^{r} \alpha_{j\mu}(a)\frac{\partial x_\nu'}{\partial a_\mu} = \xi_{j\nu}(x'),$$

and then from it, by resolution:

$$\frac{\partial x_\nu'}{\partial a_\mu} = \sum_{j=1}^{r} \psi_{\mu j}(a)\,\xi_{j\nu}(x').$$

From this, we conclude: *firstly*, that the family $x_i' = f_i(x,a)$ consists of the transformations of all one-term groups $\sum \lambda_k X_kf$, *secondly*, that all these transformations form a *group* with at most r essential parameters, and *thirdly* that from the two transformations:

$$x_i' = f_i(x,a), \qquad x_i'' = f_i(x',b),$$

a third transformation $x_i'' = f_i(x,c)$ comes into existence whose parameters: $c_1 = \varphi_1(a,b), \dots, c_r = \varphi_r(a,b)$ are determined by these two systems of values a_k, b_i and by the form of the functions $\alpha_{kj}(a)$.

But with this, the result obtained earlier on is derived in a new manner, without the need to enter into further details.

If a given r-term group $X_1 f, \dots, X_{r-q} f, \dots, X_r f$ contains a known invariant subgroup, say $X_{r-q+1} f, \dots, X_r f$, then at present, we can easily indicate a meroedrically isomorphic $(r-q)$-term group, the identity transformation of which corresponds to the said invariant subgroup (cf. p. 315, footnote). In fact, one forms a simply transitive group $A_1 f, \dots, A_{r-q} f, \dots, A_r f$ in the variables $a_1, \dots, a_r$ which is equally composed with the r-term group $X_1 f, \dots, X_r f$. At the same time, we can assume, as earlier on, that $a_1, \dots, a_{r-q}$ are invariants of the q-term group $A_{r-q+1} f, \dots, A_r f$. Then if we again set:

$$A_k f = \sum_{i=1}^{r-q} \alpha_{ki}(a_1, \dots, a_{r-q}) \frac{\partial f}{\partial a_i} + \sum_{j=r-q+1}^{r} \beta_{kj}(a) \frac{\partial f}{\partial a_j}$$

$$(k=1 \cdots r-q),$$

then the reduced infinitesimal transformations:

$$\overline{A}_k f = \sum_{i=1}^{r-q} \alpha_{ki}(a_1, \dots, a_{r-q}) \frac{\partial f}{\partial a_i} \qquad (k=1 \cdots r-q)$$

obviously generate an $(r-q)$-term group having the constitution demanded.

In addition, it results from these developments that *every proposition about the composition of $(r-q)$-term groups produces without effort a proposition about the composition of r-term groups with a q-term invariant subgroup*. This general principle, which has its analogue in the theory of substitutions, will be exploited in the third volume.[3]

§ **104.** In Chap. 19, Proposition 3, p. 370, we gave a remarkable form to the criterion for the similarity of equally composed groups; namely, we showed that two equally composed groups in the same number of variables are similar to each other if and only if they can be related to each other in a holoedrically isomorphic, completely specific way.

At that time, we had already announced that the criterion in question could be simplified in an essential way as soon as the two concerned groups are *transitive*; we announced that the following theorem holds:

Theorem 76. *Two equally composed transitive groups:*

$$X_k f = \sum_{i=1}^{n} \xi_{ki}(x_1, \dots, x_n) \frac{\partial f}{\partial x_i} \qquad (k=1 \cdots r)$$

and:

[3] Cf. LIE, Math. Ann. Vol. XXV, p. 137.

$$Z_k f = \sum_{i=1}^{n} \zeta_{ki}(y_1, \dots, y_n) \frac{\partial f}{\partial y_i} \qquad {\scriptstyle (k=1 \cdots r)}$$

in the same number of variables are similar to each other if and only if the following condition is satisfied: If one chooses a determined point $x_1^0, \dots, x_n^0$ which lies on no manifold invariant by the group: $X_1 f, \dots, X_r f$, then it must be possible to relate the two groups to one another in a holoedrically isomorphic way so that the largest subgroup contained in the group: $X_1 f, \dots, X_r f$ which leaves invariant the point: $x_1^0, \dots, x_n^0$ corresponds to the largest subgroup of the group: $Z_1 f, \dots, Z_r f$ which leaves at rest a certain point: $y_1^0, \dots, y_n^0$.[4]

From Proposition 3, p. 370, it is clear that this condition for the similarity of the two groups is necessary; hence we need only to prove that it is also sufficient.

So, we imagine that the assumptions of the theorem are satisfied, namely we imagine that in the group: $Z_1 f, \dots, Z_r f$, independent infinitesimal transformations: $Y_1 f, \dots, Y_r f$ are chosen so that our two groups are related to each other in the holoedrically isomorphic way described in Theorem 76, when to every infinitesimal transformation: $e_1 X_1 f + \cdots + e_r X_r f$ is associated the infinitesimal transformation: $e_1 Y_1 f + \cdots + e_r Y_r f$.

According to the developments of the preceding section, by means of the concerned holoedrically isomorphic relationship, a univalent invertible relationship between the finite transformations of the two groups is also produced. We can describe the latter relationship simply by interpreting, in the known way, $\sum e_k X_k f$ and $\sum e_k Y_k f$ as symbols of the finite transformations of our groups and in addition, by denoting for convenience the finite transformation $\sum e_k X_k f$ shortly by: $T_{(e)}$, and also, the transformation: $\sum e_k Y_k f$ shortly by: $\mathfrak{T}_{(e)}$.

Indeed, under these assumptions, the holoedrically isomorphic relationship between our two groups associates the transformation $\mathfrak{T}_{(e)}$ to the transformation $T_{(e)}$, and conversely, and this association is constituted in such a way that the two transformations: $T_{(e)} T_{(e')}$ and $\mathfrak{T}_{(e)} \mathfrak{T}_{(e')}$ are related to each other, where $e_1, \dots, e_r$ and $e_1', \dots, e_r'$ denote completely arbitrary systems of values.

The point: $x_1^0, \dots, x_n^0$ remains invariant by exactly $r-n$ independent infinitesimal transformations of the group $X_1 f, \dots, X_r f$, and consequently, it admits exactly ∞^{r-n} different finite transformations of this group, transformations which form, as is known, an $(r-n)$-term subgroup. For the most general transformation $T_{(e)}$ which leaves at rest the point: $x_1^0, \dots, x_n^0$, we want to introduce the symbol: $S_{(a_1,\dots,a_{r-n})}$, or shortly: $S_{(a)}$; here, by $a_1, \dots, a_{r-n}$ we understand arbitrary parameters.

We denote by $\mathfrak{S}_{(a_1,\dots,a_{r-n})}$, or shortly by $\mathfrak{S}_{(a)}$, the transformations $\mathfrak{T}_{(e)}$ of the group: $Z_1 f, \dots, Z_r f$ which are associated to the transformations $S_{(a)}$ of the group: $X_1 f, \dots, X_r f$; under the assumptions made, the ∞^{r-n} transformations $\mathfrak{S}_{(a)}$ then form the largest subgroup contained in the group: $Z_1 f, \dots, Z_r f$ which leaves invariant the point: $y_1^0, \dots, y_n^0$. This is the reason why the point: $y_1^0, \dots, y_n^0$ belongs to no manifold which remains invariant by the group: $Z_1 f, \dots, Z_r f$.

After these preparations, we conduct the following reflections.

---

[4] Cf. LIE, Archiv for Math. og Naturv. Christiania 1885, p. 388 and p. 389.

By execution of the transformation $T_{(e)}$ the point x_i^0 is transferred to the point: $(x_i^0)\,T_{(e)}$, whose position naturally depends on the values of the parameters $e_1,\dots,e_r$. According to Chap. 14, Proposition 1, p. 238, this new point in turn admits exactly ∞^{r-n} transformations of the group: $X_1f,\dots,X_rf$, namely all transformations of the form:

$$T_{(e)}^{-1}\,S_{(a)}\,T_{(e)}$$

with the $r-n$ arbitrary parameters: $a_1,\dots,a_{r-n}$.

On the other hand, the point y_i^0 is transferred by the transformation $\mathfrak{T}_{(e)}$ to the point: $(y_i^0)\,\mathfrak{T}_{(e)}$ which, naturally, admits exactly ∞^{r-n} transformations of the group: $Z_1f,\dots,Z_rf$, namely all transformations of the form:

$$\mathfrak{T}_{(e)}^{-1}\,\mathfrak{S}_{(a)}\,\mathfrak{T}_{(e)}.$$

Now, this is visibly the transformations of the group: $Z_1f,\dots,Z_rf$ which is associated to the transformation: $T_{(e)}^{-1}\,S_{(a)}\,T_{(e)}$ of the group: $X_1f,\dots,X_rf$; so, we see that, by means of our holoedrically isomorphic relationship, exactly the same holds true for the points: $(x_i^0)\,T_{(e)}$ and $(y_i^0)\,\mathfrak{T}_{(e)}$ as for the points: x_i^0 and y_i^0; namely, to the largest subgroup contained in the group $X_1f,\dots,X_rf$ which leaves invariant the point: $(x_i^0)\,T_{(e)}$ there corresponds the largest subgroup of the group: $Z_1f,\dots,Z_rf$ by which the point: $(y_i^0)\,\mathfrak{T}_{(e)}$ remains fixed.

Now, we give to the parameters $e_1,\dots,e_r$ in the transformations $T_{(e)}$ and $\mathfrak{T}_{(e)}$ all possible values, gradually. Then because of the transitivity of the group: $X_1f,\dots,X_rf$, the point: $(x_i^0)\,T_{(e)}$ is transferred gradually to all points of the space $x_1,\dots,x_n$ that lie on no invariant manifold, hence to all points in general position. At the same time, the point: $(y_i^0)\,\mathfrak{T}_{(e)}$ is transferred to all points in general position in the space $y_1,\dots,y_n$.

With these words, it is proved that the holoedrically isomorphic relationship between our two groups possesses the properties put together in Proposition 3, p. 370; consequently, according to that proposition, the group: $X_1f,\dots,X_rf$ is similar to the group: $Z_1f,\dots,Z_rf$ and the correctness of Theorem 76, p. 434, is established.

Moreover, for the proof of Theorem 76, one does not need to go back to the cited proposition. Rather, it can be directly concluded from the developments above that a transformation exists which transfers the infinitesimal transformations: $X_1f,\dots,X_rf$ to $Y_1f,\dots,Y_rf$, respectively.

We have seen that through our holoedrically isomorphic relationship, the point: $(y_i^0)\,\mathfrak{T}_{(e)}$ of the space $y_1,\dots,y_n$ is associated to the point: $(x_i^0)\,T_{(e)}$ of the space $x_1,\dots,x_n$, and to be precise, in this way, a univalent invertible relationship is produced between the points of the two spaces in general position, inside a certain region. Now, if we interpret the x_i and the y_i as point coordinates of one and the same n-times extended space: R_n, and both with respect to the same system of coordinates, then there is a completely determined transformation T of the space R_n which always transfers the points with the coordinates: $(x_i^0)\,T_{(e)}$ to the point with the coordinates: $(y_i^0)\,\mathfrak{T}_{(e)}$. We will prove that this transformation T transfers the infinitesimal transformations: $X_1 f, \dots, X_r f$ to: $Y_1 f, \dots, Y_r f$, respectively, so that our two r-term groups are similar to each other precisely thanks to the transformation T.

The transformation T satisfies all the infinitely many symbolic equations:

$$(x_i^0)\, T_{(e')}\, \mathsf{T} = (y_i^0)\, \mathfrak{T}_{(e')}, \tag{14}$$

with the r arbitrary parameters $e_1', \dots, e_r'$, and it is even defined by these equations. From this, it results that T satisfies at the same time the symbolic equations:

$$(x_i^0)\, T_{(e')}\, T_{(e)}\, \mathsf{T} = (y_i^0)\, \mathfrak{T}_{(e')}\, \mathfrak{T}_{(e)},$$

whichever values the parameters e and e' may have.

If we compare these equations with the equations (14) which we can obviously also write as:

$$(x_i^0)\, T_{(e')} = (y_i^0)\, \mathfrak{T}_{(e')}\, \mathsf{T}^{-1},$$

then we obtain:

$$(y_i^0)\, \mathfrak{T}_{(e')}\, \mathsf{T}^{-1}\, T_{(e)}\, \mathsf{T} = (y_i^0)\, \mathfrak{T}_{(e')}\, \mathfrak{T}_{(e)},$$

or, what is the same:

$$(y_i^0)\, \mathfrak{T}_{(e')}\, \mathsf{T}^{-1}\, T_{(e)}\, \mathsf{T}\, \mathfrak{T}_{(e)}^{-1} = (y_i^0)\, \mathfrak{T}_{(e')}. \tag{15}$$

By appropriate choice of $e_1', \dots, e_r'$, the point: $(y_i^0)\, \mathfrak{T}_{(e')}$ can be brought into coincidence with every point in general position in R_m, hence the equations (15) express that every transformation:

$$\mathsf{T}^{-1}\, T_{(e)}\, \mathsf{T}\, \mathfrak{T}_{(e)}^{-1} \tag{16}$$

leaves fixed all points in general position in R_n. But obviously, this is possible only when all transformations (16) coincide with the identity transformation.

With these words, we have proved that the following ∞^r symbolic equations hold:

$$\mathsf{T}^{-1}\, T_{(e)}\, \mathsf{T} = \mathfrak{T}_{(e)}, \tag{17}$$

hence that the transformation T transfers every transformation $T_{(e)}$ of the group: $X_1 f, \dots, X_r f$ to the corresponding transformation $\mathfrak{T}_{(e)}$ of the group: $Z_1 f, \dots, Z_r f$. In other words, the expression: $e_1 X_1 f + \cdots + e_r X_r f$ converts into the expression: $e_1 Y_1 f + \cdots + e_r Y_r f$ after the execution of the transformation T, or, what is the same, the r infinitesimal transformations: $X_1 f, \dots, X_r f$ are transferred, by the execution of T, to $Y_1 f, \dots, Y_r f$, respectively.

Consequently, the two groups: $X_1 f, \dots, X_r f$ and $Z_1 f, \dots, Z_r f$ are similar to each other and Theorem 76, p. 434, is now proved, independently of Proposition 3, p. 370.

§ 105. In § 100, we saw that the same equations defined there: $a_k' = \varphi_k(a,b)$ represented a group, and in fact, we realized that between the r functions $\varphi_k(a,b)$, the r identities:

$$(5)\quad \varphi_k\big(\varphi_1(a,b),\,\dots,\,\varphi_r(a,b),\,c_1,\dots,c_r\big) \equiv \varphi_k\big(a_1,\dots,a_r,\,\varphi_1(b,c),\dots,\varphi_r(b,c)\big)$$

held true.

Now, one convinces oneself in the same way that the equations:

$$(18)\qquad a_k' = \varphi_k(b_1,\dots,b_r,\,a_1,\dots,a_r) \qquad {\scriptstyle (k=1\cdots r)}$$

in the variables a also represent a group with the parameters b and to be precise, a simply transitive r-term group.

We now make a few remarks about this new group.

This group possesses the remarkable property that each of its transformations is interchangeable with every transformation of the parameter group: $a_k' = \varphi_k(a,b)$. Indeed, if one first executes the transformation (18) and then any transformation:

$$a_k'' = \varphi_k(a_1',\dots,a_r',\,c_1,\dots,c_r) \qquad {\scriptstyle (k=1\cdots r)}$$

of the parameter group: $a_k' = \varphi_k(a,b)$, then one obtains the transformation:

$$a_k'' = \varphi_k\big(\varphi_1(b,a),\,\dots,\varphi_r(b,a),\,c_1,\dots,c_r\big) \qquad {\scriptstyle (k=1\cdots r)}$$

which, because of the identities (5), can be brought to the form:

$$a_k'' = \varphi_k\big(b_1,\dots,b_r,\,\varphi_1(a,c),\,\dots,\,\varphi_r(a,c)\big) \qquad {\scriptstyle (k=1\cdots r)}.$$

But this transformation can be obtained by first executing the transformation:

$$a_k' = \varphi_k(a_1,\dots,a_r,\,c_1,\dots,c_r)$$

of the parameter group and then the transformation:

$$a_k'' = \varphi_k(b_1,\dots,b_r,\,a_1',\dots,a_r')$$

of the group (18).

Thus, the group (18) is nothing else than the reciprocal simply transitive group associated to the parameter group;[5] according to Theorem 68, p. 391, it is equally composed with the parameter group, and even similar to it; besides, it is naturally also equally composed with the group: $x_i' = f_i(x,a)$ itself.

---

[5] The observation that the two groups (6) and (18) discussed in the text stand in a relationship such that the transformations of the one are interchangeable with the transformations of the other, is due to ENGEL.

Chapter 22
The Determination of All r-term Groups

we have already emphasized in Chap. 17, p. 309, that every system of constants c_{iks} which satisfies the relations:

$$
(1)\qquad \begin{cases} c_{iks} + c_{kis} = 0 \\ \sum_{\nu=1}^{r} \left(c_{ik\nu}\, c_{\nu js} + c_{kj\nu}\, c_{\nu is} + c_{ji\nu}\, c_{\nu ks} \right) = 0 \\ (i,k,j,s = 1 \cdots r). \end{cases}
$$

represents a possible composition of r-term groups and that there are always r-term groups whose composition is determined just by this system of c_{iks}. As we remarked at the time, the proof for this will be first provided in full generality in the second volume; there, we will imagine that an arbitrary system of c_{iks} having the said constitution is presented and we will prove that in order to find the infinitesimal transformations of an r-term group having the composition c_{iks}, only the integration of simultaneous systems of ordinary differential equations is required in any case. The finite equations of this group can likewise be obtained by integrating ordinary differential equations, according to Chaps. 4 and 9.

Now in the present chapter, we show two kinds of things:

Firstly, we imagine that the finite equations:

$$
x_i' = f_i(x_1, \dots, x_n, a_1, \dots, a_r) \qquad (i = 1 \cdots n)
$$

of an r-term group are presented and we show that by integrating simultaneous systems of ordinary differential equations, one can in every case find all r-term *transitive* groups which are equally composed with the group: $x_i' = f_i(x,a)$.

Secondly, we show that one can determine all *intransitive* r-term groups without integration as soon as one knows all *transitive* groups with r or fewer parameters.

If we combine these results with what was said above and if we add that according to Theorem 53, p. 311, the determination of all possible compositions of groups with a given number of parameters requires only algebraic operations, then we immediately realize the following:

S. Lie, *Theory of Transformation Groups I*,
DOI 10.1007/978-3-662-46211-9_22

If the number r is given, then, aside from executable operations, the determination of all r-term groups requires at most the integration of simultaneous systems of ordinary differential equations.

Naturally, the question of whether the integration of the appearing differential equations is executable or not is of great importance. However, this question will neither be addressed in this chapter, nor in the second volume, since the answer presupposes the theory of integration, which cannot be developed wholly in a work on transformation groups, and which would rather demand a separate treatment.

§ 106. In this section, we put together various methods for the determination of *simply transitive* groups, partly because these method will find applications in the sequel, partly because they are notable in themselves.

First, let the equations:

$$x_i' = f_i(x_1,\dots,x_n,a_1,\dots,a_r) \qquad (i=1\cdots n) \tag{2}$$

of an r-term group be presented. We seek the finite equations of a simply transitive group which is equally composed with the presented group.

According to Chap. 21, p. 414, the parameter group of the group (2) is simply transitive and is equally composed with the group (2). Now, the finite equations:

$$a_k' = \varphi_k(a_1,\dots,a_r,b_1,\dots,b_r) \qquad (k=1\cdots r)$$

of this group can be found by means of executable operations, namely by resolving finite equations. Thus, we can set up the finite equations of *one* simply transitive group having the constitution demanded, namely just the said parameter group.

We therefore have the:

Theorem 77. *If the finite equations of an r-term group are presented, then one can find the equations of an equally composed simply transitive group by means of executable operations.*

Obviously, together with this one simply transitive group, all other simply transitive groups which are equally composed with the group: $x_i' = f_i(x,a)$ are also given, for as a consequence of Theorem 64, p. 353, all these groups are similar to each other.

On the other hand, we assume that not the finite equations, but instead only the infinitesimal transformations:

$$X_k f = \sum_{i=1}^{n} \xi_{ki}(x_1,\dots,x_n)\,\frac{\partial f}{\partial x_i} \qquad (k=1\cdots r)$$

of an r-term group are presented, and we seek to determine the infinitesimal transformations of an equally composed simply transitive group.

Visibly, we have already solved this problem in Chap. 9, p. 171 and 172, though it was interpreted differently there, because at that time, we did not have the concepts of simply transitive group and of being equally composed [GLEICHZUSAMMENGESETZTSEINS]. Now we can state our previous solution to the problem as follows:

Proposition 1. *If the infinitesimal transformations:*

$$X_k f = \sum_{i=1}^{n} \xi_{ki}(x_1,\dots,x_n) \frac{\partial f}{\partial x_i} \qquad (k=1\cdots r)$$

of an r-term group are presented, then one finds the infinitesimal transformations of an equally composed simply transitive group in the following way: One sets:

$$X_k^{(\mu)} = \sum_{i=1}^{n} \xi_{ki}\big(x_1^{(\mu)},\dots,x_n^{(\mu)}\big) \frac{\partial f}{\partial x_i^{(\mu)}} \qquad (k=1\cdots r)$$

$$(\mu=1,2\cdots r-1),$$

one forms the r infinitesimal transformations:

$$W_k f = X_k f + X_k^{(1)} f + \cdots + X_k^{(r-1)} f \qquad (k=1\cdots r)$$

and one determines $rn-r$ arbitrary independent solutions: $u_1,\dots,u_{nr-r}$ of the r-term complete system:

$$W_1 f = 0, \ \dots, \ W_r f = 0;$$

then one introduces $u_1,\dots,u_{nr-r}$ together with r appropriate functions: $y_1,\dots,y_r$ of the $n\,r$ variables $x_i^{(\mu)}$ as new independent variables and in this way, one obtains the infinitesimal transformations:

$$W_k f = \sum_{j=1}^{r} \eta_{kj}(y_1,\dots,y_r,\, u_1,\dots,u_{nr-r}) \frac{\partial f}{\partial y_j} \qquad (k=1\cdots r),$$

and if in addition one understands by $u_1^0,\dots,u_{nr-r}^0$ numerical constants, then the r independent infinitesimal transformations:

$$\mathfrak{W}_k f = \sum_{j=1}^{r} \eta_{kj}(y_1,\dots,y_r,\, u_1^0,\dots,u_{nr-r}^0) \frac{\partial f}{\partial y_j} \qquad (k=1\cdots r)$$

generate a simply transitive group equally composed with the group: $X_1 f,\dots,X_r f$. If: $X_1 f,\dots,X_r f$ are linked together by the relations:

$$[X_i,\, X_k] = \sum_{s=1}^{r} c_{iks} X_s f,$$

then the same relations hold between $\mathfrak{W}_1 f,\dots,\mathfrak{W}_r f$:

$$[\mathfrak{W}_i,\,\mathfrak{W}_k] = \sum_{s=1}^{r} c_{iks}\,\mathfrak{W}_s f.$$

The process which is described in the above proposition requires the integration of ordinary differential equations, and naturally, this integration is not always executable. Theoretically, this is completely indifferent, since in the present chapter, we do not manage without integration. Nevertheless, in the developments of the next section, where the question is to set up simply transitive groups, we will nowhere make use of the described process, and we will apply it only in one place, where it holds, in order to prove the existence of a simply transitive group with certain properties.

Lastly, we still want to take up the standpoint where we imagine only that a composition of an r-term group is given, hence a system of c_{iks} which satisfies the equations (1). We will show that it is at least in very many cases possible, by means of executable operations, to set up the finite equations of a simply transitive group of the composition c_{iks}.

For this, we use Theorem 52, p. 308. According to it, the r infinitesimal transformations:

$$E_\mu f = \sum_{k,\,j}^{1\cdots r} c_{j\mu k}\, e_j \frac{\partial f}{\partial e_k} \qquad (k=1\cdots r)$$

stand pairwise in the relationships:

$$[E_i,\,E_k] = \sum_{s=1}^{r} c_{iks}\,E_s f,$$

hence they generate a linear homogeneous group in the variables: $e_1,\dots,e_r$. This group is in particular r-term when not all $r\times r$ determinants vanish whose horizontal rows are of the form:

$$\left| c_{j1k}\;\; c_{j2k}\;\; \cdots\;\; c_{jrk} \right| \qquad (j,\,k=1\cdots r)$$

and it then evidently has the composition c_{iks}.

Now, the finite equations of the group: $E_1 f,\dots,E_r f$ can be set up by means of executable operations (cf. p. 286); hence when these $r\times r$ determinants are not all equal to zero, we can indicate the finite equations of an r-term group having the composition c_{iks}. But from this, it immediately follows (Theorem 77, p. 440) that we can also indicate the finite equations of a simply transitive group having the composition c_{iks}.

Thus we have proved the following proposition:

Proposition 2. *If a system of constants c_{iks} is presented which satisfies the equations:*

$$
(1)\qquad \begin{cases} c_{iks}+c_{kis}=0 \\ \sum_{\nu=1}^{r}\left(c_{ik\nu}\,c_{\nu js}+c_{kj\nu}\,c_{\nu is}+c_{ji\nu}\,c_{\nu ks}\right)=0 \\ \qquad (i,k,j,s=1\cdots r) \end{cases}
$$

and which is constituted in such a way that not all $r \times r$ determinants vanish whose horizontal rows have the form:

$$
\left|c_{j1k}\;c_{j2k}\;\cdots\;c_{jrk}\right| \qquad (j,k=1\cdots r),
$$

then by means of executable operations, one can always find the finite equations of a simply transitive group having the composition c_{iks}.

Furthermore, the same can also be proved in a completely analogous way for other systems of c_{iks}, but we do not want to go further in that direction. We make only a few more observations.

If r infinitesimal transformations: $X_1f,\dots,X_rf$ are presented which generate an r-term group, then the system of c_{iks} which is associated to the group in question is given at the same time. Now, if this system of c_{iks} is constituted in such a way that not all determinants considered in Proposition 2 vanish, then visibly, the group: $X_1f,\dots,X_rf$ contains no excellent infinitesimal transformation (cf. p. 288), hence we see that the following holds:

If r independent infinitesimal transformations: $X_1f,\dots,X_rf$ are presented which generate an r-term group, then when this group contains no excellent infinitesimal transformation, one can set up in every case, by means of executable operations, the finite equations of a simply transitive group which is equally composed with the group: $X_1f,\dots,X_rf$.

§ 107. We now tackle the first of the two problems, the solution of which we promised in the introduction of the chapter.

So, we imagine that the finite equations:

$$
(2)\qquad x_i'=f_i(x_1,\dots,x_n,a_1,\dots,a_r) \qquad (i=1\cdots n)
$$

of an r-term group are presented and we take up the problem of determining all transitive r-term groups which are equally composed with the presented group.

When an r-term group Γ of any arbitrary space is transitive and has the same composition as the group: $x_i'=f_i(x,a)$, the same two properties evidently hold true for all groups of the same space that are similar to Γ. Hence, we classify the sought groups into classes by including in the same class all groups of the demanded constitution which contain an equal number of parameters and which in addition are similar to each other. We call the totality of all groups which belong to such a class a *type* of transitive group of the given composition.

This classification of the sought groups has the advantage that it is not necessary for us to write down all the sought groups. Indeed, if *one* group of the constitution demanded is known, then at the same time, all groups similar to it are known, hence all groups which belong to the same type. Consequently, it completely suffices that we enumerate how many different types there are of the sought groups and that we indicate for every individual type a representative, hence a group which belongs to the concerned type.

In order to settle our problem, we take the following route:

First, we indicate a process which provides the transitive groups equally composed with the group: $x_i' = f_i(x,a)$. Then, we conduct the proof that, by means of this process, one can obtain every group having the concerned constitution, so that notably, one finds at least one representative group for every type. Lastly, we give criteria to decide whether or not two different groups obtained by our process are similar to each other, hence to decide whether or not they belong to the same type. Thanks to this, we are in a position to have a view of the existing mutually distinct types and at the same time, for each one of these types, to have a representative.

This is the program that will be carried out in the sequel.

Let:

$$Y_k f = \sum_{j=1}^{r} \eta_{kj}(y_1,\dots,y_r)\,\frac{\partial f}{\partial y_j} \qquad {\scriptstyle (k=1\cdots r)}$$

be independent infinitesimal transformations of a simply transitive group which is equally composed with the group: $x_i' = f_i(x,a)$. Moreover, let:

$$Z_k f = \sum_{j=1}^{r} \zeta_{kj}(y_1,\dots,y_r)\,\frac{\partial f}{\partial y_j} \qquad {\scriptstyle (k=1\cdots r)}$$

be independent infinitesimal transformations of the associated reciprocal group which, according to Theorem 68, p. 391, is simply transitive as well, and is equally composed with the group: $x_i' = f_i(x,a)$.

Evidently, all these assumptions can be satisfied. For instance, two simply transitive groups having the constitution indicated are the parameter group of the group: $x_i' = f_i(x,a)$ defined in the previous chapter:

$$a_k' = \varphi_k(a_1,\dots,a_r,\,b_1,\dots,b_r) \qquad {\scriptstyle (k=1\cdots r)}$$

and its reciprocal group:

$$a_k' = \varphi_k(b_1,\dots,b_r,\,a_1,\dots,a_r) \qquad {\scriptstyle (k=1\cdots r)}$$

and the finite equations of both may even be set up by means of executable operations.

After these preparations, we turn to the explanation of the process announced above which produces transitive groups that are equally composed with the group: $x_i' = f_i(x,a)$.

In Chap. 17, p. 316, we showed how every decomposition of the space invariant by a group can be used in order to set up an isomorphic group. We want to apply this to the group: $Y_1f,\dots,Y_rf$.

On the basis of Theorem 69, p. 398, we seek an arbitrary decomposition of the space $y_1,\dots,y_r$ invariant by the group $Y_1f,\dots,Y_rf$. By means of algebraic operations, we determine an m-term subgroup of the group: $Z_1f,\dots,Z_rf$. If:

$$\mathsf{Z}_\mu f = \varepsilon_{\mu 1} Z_1 f + \cdots + \varepsilon_{\mu r} Z_r f \qquad (\mu = 1 \cdots m) \tag{3}$$

are independent infinitesimal transformations of this subgroup, we form the m-term complete system:

$$\mathsf{Z}_1 f = 0, \ \dots, \ \mathsf{Z}_m f = 0$$

and we compute, by integrating it, $r-m$ arbitrary invariants:

$$u_1(y_1,\dots,y_r), \ \dots, \ u_{r-m}(y_1,\dots,y_r)$$

of the group: $\mathsf{Z}_1f,\dots,\mathsf{Z}_mf$. Then the equations:

$$u_1(y_1,\dots,y_r) = \text{const.}, \ \dots, \ u_{r-m}(y_1,\dots,y_r) = \text{const.} \tag{4}$$

determine a decomposition of the space $y_1,\dots,y_r$ into ∞^{r-m} m-times extended manifolds invariant by the group: $Y_1f,\dots,Y_rf$.

Now, we introduce $u_1(y),\dots,u_{r-m}(y)$ together with m other appropriate functions: $v_1,\dots,v_m$ of the y as new variables in $Y_1f,\dots,Y_rf$ and we obtain:

$$Y_k f = \sum_{\nu=1}^{r-m} \omega_{k\nu}(u_1,\dots,u_{r-m}) \frac{\partial f}{\partial u_\nu} + \sum_{\mu=1}^{m} w_{k\mu}(u_1,\dots,u_{r-m}, v_1,\dots,v_m) \frac{\partial f}{\partial v_\mu}$$
$$(k = 1 \cdots r).$$

From this, we lastly form the reduced infinitesimal transformations:

$$U_k f = \sum_{\nu=1}^{r-m} \omega_{k\nu}(u_1,\dots,u_{r-m}) \frac{\partial f}{\partial u_\nu} \qquad (k = 1 \cdots r).$$

According to Chap. 17, Proposition 4, p. 318, they generate a group in the variables: $u_1,\dots,u_{r-m}$ which is isomorphic with the group: $Y_1f,\dots,Y_rf$.

If we had chosen, instead of $u_1(y),\dots,u_{r-m}(y)$, any other independent invariants: $u_1'(y),\dots,u_{r-m}'(y)$ of the group: $\mathsf{Z}_1f,\dots,\mathsf{Z}_mf$, then in place of the group: $U_1f,\dots,U_rf$, we would have obtained another group in the variables $u_1',\dots,u_{r-m}'$, but this group would visibly be similar to the group: $U_1f,\dots,U_rf$, because $u_1'(y),\dots,u_{r-m}'(y)$ are independent functions of: $u_1(y),\dots,u_{r-m}(y)$. If we replaced $u_1(y),\dots,u_{r-m}(y)$ by the most general system of $r-m$ independent invariants of the group: $\mathsf{Z}_1f,\dots,\mathsf{Z}_mf$, then we would obtain the most general group in $r-m$ variables which is similar to the group: $U_1f,\dots,U_rf$.

From the fact that the group: $Y_1 f, \dots, Y_r f$ is simply transitive, it follows, as we have observed already in Chap. 20, p. 398, that not all $(r-m) \times (r-m)$ determinants of the matrix:

$$\begin{vmatrix} \omega_{11}(u) & \cdot & \cdot & \omega_{1,r-m}(u) \\ \cdot & \cdot & \cdot & \cdot \\ \cdot & \cdot & \cdot & \cdot \\ \omega_{r1}(u) & \cdot & \cdot & \omega_{r,r-m}(u) \end{vmatrix}$$

vanish identically; expressed differently: it follows that the group: $U_1 f, \dots, U_{r-m} f$ in the $r-m$ variables u is transitive.

We therefore have a method for setting up transitive groups which are isomorphic with the group: $Y_1 f, \dots, Y_r f$. However, this does not suffice, for we demand transitive groups that are equally composed with the group: $Y_1 f, \dots, Y_r f$, hence are holoedrically isomorphic to it. Thus, the group: $U_1 f, \dots, U_r f$ is useful for us only when it is r-term. Under which conditions is it so?

Since the group: $U_1 f, \dots, U_r f$ is transitive, it contains at least $r-m$ essential parameters, hence it will in general contain exactly $r-l$ essential parameters, where $0 \leqslant l \leqslant m$. Then according to p. 318, in the group: $Y_1 f, \dots, Y_r f$, there are exactly l independent infinitesimal transformations which leave individually fixed each one of the ∞^{r-m} manifolds (4), and these infinitesimal transformations generate an l-term invariant subgroup of the group: $Y_1 f, \dots, Y_r f$. For reasons of brevity, we want to denote the concerned subgroup by g.

If M is an arbitrary manifold amongst the generally located manifolds (4), then M admits exactly m independent infinitesimal transformations: $e_1 Y_1 f + \cdots + e_r Y_r f$ which generate an m-term subgroup γ of the group: $Y_1 f, \dots, Y_r f$ (cf. page 399). Naturally, the invariant subgroup g is contained in this group γ. On the other hand, M admits exactly m independent infinitesimal transformations of the reciprocal group: $Z_1 f, \dots, Z_r f$, namely: $\mathsf{Z}_1 f, \dots, \mathsf{Z}_m f$, which also generate an m-term group. Therefore, according to Chap. 20, p. 400, one can relate the two simply transitive groups: $Y_1 f, \dots, Y_r f$ and $Z_1 f, \dots, Z_r f$ to each other in a holoedrically isomorphic way so that the subgroup γ corresponds to the subgroup: $\mathsf{Z}_1 f, \dots, \mathsf{Z}_m f$. On the occasion, to the invariant subgroup g, there visibly corresponds an l-term invariant subgroup g' of the group: $Z_1 f, \dots, Z_r f$, and in fact, g' is contained in the subgroup: $\mathsf{Z}_1 f, \dots, \mathsf{Z}_m f$.

We therefore see: when the group: $U_1 f, \dots, U_r f$ is exactly $(r-l)$-term, then the group: $\mathsf{Z}_1 f, \dots, \mathsf{Z}_m f$ contains an l-term subgroup g' which is invariant in the group: $Z_1 f, \dots, Z_r f$.

Conversely: when the group: $\mathsf{Z}_1 f, \dots, \mathsf{Z}_m f$ contains an l-term subgroup:

$$\mathsf{Z}'_\lambda f = h_{\lambda 1} \mathsf{Z}_1 f + \cdots + h_{\lambda m} \mathsf{Z}_m f \qquad (\lambda = 1 \cdots l)$$

which is invariant in the group: $Z_1 f, \dots, Z_r f$, then the group: $U_1 f, \dots, U_r f$ can at most be $(r-l)$-term.

Indeed, under the assumptions just made, the l mutually independent equations:

$$\mathsf{Z}'_1 f = 0, \ \dots, \ \mathsf{Z}'_l f = 0 \tag{5}$$

form an l-term complete system with $r-l$ independent solutions:

$$\psi_1(y_1,\dots,y_r), \;\dots,\; \psi_{r-l}(y_1,\dots,y_r).$$

Furthermore, there are relations of the form:

$$[\mathsf{Z}_k,\,\mathsf{Z}'_\lambda] = h_{k\lambda 1}\,\mathsf{Z}'_1 f + \cdots + h_{k\lambda l}\,\mathsf{Z}'_l f$$
$$(k=1\cdots r;\; \lambda=1\cdots l)$$

which express that the complete system (5) admits the group: $Z_1 f,\dots,Z_r f$. Consequently, the equations:

$$\psi_1(y_1,\dots,y_r) = \text{const.},\;\dots,\; \psi_{r-l}(y_1,\dots,y_r) = \text{const.} \tag{6}$$

represent a decomposition of the space $y_1,\dots,y_r$ invariant by the group $Z_1 f,\dots,Z_r f$, and to be precise, a decomposition into ∞^{r-l} l-times extended manifolds.

These ∞^{r-l} manifolds stand in a very simple relationship with respect to the ∞^{r-m} manifolds:

$$u_1(y_1,\dots,y_r) = \text{const.},\;\dots,\; u_{r-m}(y_1,\dots,y_r) = \text{const.}, \tag{4}$$

namely each one of the manifolds (4) consists of ∞^{m-l} different manifolds (6). This follows without effort from the fact that $u_1(y),\dots,u_{r-m}(y)$, as solutions of the complete system:

$$\mathsf{Z}_1 f = 0,\;\dots,\; \mathsf{Z}_m f = 0,$$

simultaneously satisfy the complete system (5), and can therefore be represented as functions of: $\psi_1(y),\dots,\psi_{r-l}(y)$.

Now, according to Chap. 20, p. 397, the reciprocal group: $Y_1 f,\dots,Y_r f$ contains exactly l independent infinitesimal transformations which leave individually fixed each one of the ∞^{r-l} manifolds (6), hence it obviously contains at least l independent infinitesimal transformations which leave individually fixed each one of the ∞^{r-m} manifolds (4); but from this, it immediately follows that, under the assumption made, the group: $U_1 f,\dots,U_r f$ can be at most $(r-l)$-term, as we claimed above.

Thanks to the preceding developments, it is proved that the group: $U_1 f,\dots,U_r f$ is $(r-l)$-term if and only if the group: $\mathsf{Z}_1 f,\dots,\mathsf{Z}_m f$ contains an l-term subgroup invariant in the group: $\mathsf{Z}_1 f,\dots,\mathsf{Z}_r f$, but no larger subgroup of the same nature. In particular, the group: $U_1 f,\dots,U_r f$ is r-term if and only if the group: $\mathsf{Z}_1 f,\dots,\mathsf{Z}_m f$ contains, aside from the identity transformation, no subgroup invariant in the group $\mathsf{Z}_1 f,\dots,\mathsf{Z}_r f$. Here, the word 'subgroup' is to be understood in its widest sense, so that one then also has to consider the group: $\mathsf{Z}_1 f,\dots,\mathsf{Z}_m f$ itself as a subgroup contained in it.

By summarizing the obtained result, we can now say:

Theorem 78. *If the two r-term groups:*

$$Y_k f = \sum_{j=1}^{r} \eta_{kj}(y_1, \dots, y_r) \frac{\partial f}{\partial y_j} \qquad (k=1 \cdots r)$$

and:

$$Z_k f = \sum_{j=1}^{r} \zeta_{kj}(y_1, \dots, y_r) \frac{\partial f}{\partial y_j} \qquad (k=1 \cdots r)$$

are simply transitive and reciprocal to each other, if, moreover:

$$\mathsf{Z}_\mu f = \varepsilon_{\mu_1} Z_1 f + \cdots + \varepsilon_{\mu r} Z_r f \qquad (\mu=1 \cdots m)$$

is an m-term subgroup of the group: $Z_1 f, \dots, Z_r f$*, and if:*

$$u_1(y_1, \dots, y_r), \;\dots,\; u_{r-m}(y_1, \dots, y_r)$$

are independent invariants of this subgroup, then the r infinitesimal transformations:

$$\sum_{\nu=1}^{r-m} Y_k u_\nu \frac{\partial f}{\partial u_\nu} = \sum_{\nu=1}^{r-m} \omega_{k\nu}(u_1, \dots, u_{r-m}) \frac{\partial f}{\partial u_\nu} = U_k f$$
$$(k=1 \cdots r)$$

in the $r-m$ *variables:* $u_1, \dots, u_{r-m}$ *generate a transitive group, isomorphic with the group:* $Y_1 f, \dots, Y_r f$*. This group is* $(r-l)$*-term when there is in the group:* $\mathsf{Z}_1 f, \dots, \mathsf{Z}_m f$ *an l-term subgroup, but no larger subgroup, which is invariant in the group:* $Z_1 f, \dots, Z_r f$*. In particular, it is r-term and equally composed with the group:* $Y_1 f, \dots, Y_r f$ *if and only if the group:* $\mathsf{Z}_1 f, \dots, \mathsf{Z}_m f$ *is neither invariant itself in the group:* $Z_1 f, \dots, Z_r f$*, nor contains, aside from the identity transformation, another subgroup invariant in the group:* $Z_1 f, \dots, Z_r f$*.*

If one replaces $u_1(y), \dots, u_{r-m}(y)$ *by the most general system:* $u_1'(y), \dots, u_{r-m}'(y)$ *of* $r-m$ *independent invariants of the group:* $\mathsf{Z}_1 f, \dots, \mathsf{Z}_m f$*, then in place of the group:* $U_1 f, \dots, U_r f$*, one obtains the most general group in* $r-m$ *variables similar to it. In particular, if the group:* $U_1 f, \dots, U_r f$ *is r-term, then one obtains in this way the most general transitive group equally composed with the group:* $Y_1 f, \dots, Y_r f$ *that belongs to the same type as the group:* $U_1 f, \dots, U_r f$*.*

In addition, from the developments used for the proof of this theorem, we have the following

Proposition 3. *If* $Y_1 f, \dots, Y_r f$ *and* $Z_1 f, \dots, Z_r f$ *are two reciprocal simply transitive groups and if:* $Z_1 f, \dots, Z_l f$ *is an invariant l-term subgroup of the second group, then the invariants of this l-term group can also be defined as the invariants of a certain l-term group which is contained as an invariant subgroup in the r-term group:* $Y_1 f, \dots, Y_r f$*.*

Thanks to Theorem 78, the first part of the program stated on p. 444 is settled, and we are in possession of a process which provides transitive groups equally composed with the group: $x_i' = f_i(x, a)$. We now come to the second part, namely to the proof that every group of this sort can be found thanks to our process.

In $r-m$ variables, let an arbitrary transitive group be presented which is equally composed with the group: $x_i' = f_i(x,a)$; let its infinitesimal transformations be:

$$\mathfrak{X}_k f = \sum_{\nu=1}^{r-m} \mathfrak{x}_{k\nu}(z_1,\dots,z_{r-m})\frac{\partial f}{\partial z_\nu} \qquad (k=1\cdots r)$$

At first, we prove that amongst the simply transitive groups of the same composition, there is in any case one which can be obtained from the group: $\mathfrak{X}_1 f,\dots,\mathfrak{X}_r f$ in a way completely analogous to the way in which the group: $U_1 f,\dots,U_r f$ was obtained from the simply transitive group: $Y_1 f,\dots,Y_r f$.

To this end, under the guidance of Proposition 1, p. 441, we form, in the $r(r-m)$ variables: z, $z^{(1)}$, ..., $z^{(r-1)}$, the r infinitesimal transformations:

$$W_k f = \mathfrak{X}_k f + \mathfrak{X}_k^{(1)} f + \cdots + \mathfrak{X}_k^{(r-1)} f \qquad (k=1\cdots r).$$

Now, if $\varphi_1,\dots,\varphi_R$ are $(r-m)r-r=R$ independent solutions of the r-term complete system:

$$W_1 f = 0, \; \dots, \; W_r f = 0, \tag{7}$$

then we introduce them, together with $z_1,\dots,z_{r-m}$ and with m functions $\overline{z}_1,\dots,\overline{z}_m$ of the z, $z^{(1)}$, ..., $z^{(r-1)}$, as new variables. This is possible, since the equations (7) are solvable with respect to $\partial f/\partial z_1$, ..., $\partial f/\partial z_{r-m}$ because of the transitivity of the group: $\mathfrak{X}_1 f,\dots,\mathfrak{X}_r f$, whence $z_1,\dots,z_{r-m}$ are independent of the functions: $\varphi_1,\dots,\varphi_R$. In the new variables, $W_1 f,\dots,W_r f$ receive the form:

$$W_k f = \mathfrak{X}_k f + \sum_{\mu=1}^{m} w_{k\mu}(z_1,\dots,z_{r-m},\overline{z}_1,\dots,\overline{z}_m,\,\varphi_1,\dots,\varphi_R)\frac{\partial f}{\partial \overline{z}_\mu}$$

$$(k=1\cdots r),$$

and here, when we confer to the φ suitable fixed values $\varphi_1^0,\dots,\varphi_R^0$ and when we set:

$$w_k(z_1,\dots,z_{r-m},\overline{z}_1,\dots,\overline{z}_m,\,\varphi_1^0,\dots,\varphi_R^0) = w_k^0(z_1,\dots,z_{r-m},\overline{z}_1,\dots,\overline{z}_m),$$

then:

$$\mathfrak{W}_k f = \mathfrak{X}_k f + \sum_{\mu=1}^{m} w_{k\mu}^0(z_1,\dots,z_{r-m},\overline{z}_1,\dots,\overline{z}_m)\frac{\partial f}{\partial \overline{z}_\mu} \qquad (k=1\cdots r)$$

are independent infinitesimal transformations of a simply transitive group which is equally composed with the group: $\mathfrak{X}_1 f,\dots,\mathfrak{X}_r f$ and therefore, also equally composed with the group: $x_i' = f_i(x,a)$.

With this, we have found a simply transitive group having the constitution indicated earlier on.

Indeed, the equations:

$$z_1 = \text{const.}, \; \dots, \; z_{r-m} = \text{const.}$$

obviously determine a decomposition of the space: $z_1, \dots, z_{r-m}, \overline{z}_1, \dots, \overline{z}_m$ into ∞^{r-m} m-times extended manifolds invariant by the group: $\mathfrak{W}_1 f, \dots, \mathfrak{W}_r f$. The group: $\mathfrak{X}_1 f, \dots, \mathfrak{X}_r f$ indicates in which way these ∞^{r-m} manifolds are permuted by the group: $\mathfrak{W}_1 f, \dots, \mathfrak{W}_r f$. So, between the two groups: $\mathfrak{X}_1 f, \dots, \mathfrak{X}_r f$ and $\mathfrak{W}_1 f, \dots, \mathfrak{W}_r f$, there is a relationship completely analogous to the one above between the two groups: $U_1 f, \dots, U_r f$ and $Y_1 f, \dots, Y_r f$.

Now, there is no difficulty in proving that the group: $\mathfrak{X}_1 f, \dots, \mathfrak{X}_r f$ is similar to one of the groups that we obtain when we apply the process described in Theorem 78, p. 447, to two determined simply transitive reciprocal groups: $Y_1 f, \dots, Y_r f$ and $Z_1 f, \dots, Z_r f$ having the concerned composition.

Let: $\mathsf{B}_1 f, \dots, \mathsf{B}_r f$ be the simply transitive group reciprocal to: $\mathfrak{W}_1 f, \dots, \mathfrak{W}_r f$. Naturally, this group is equally composed with the group: $Z_1 f, \dots, Z_r f$ and hence, also similar to it (cf. Chap. 19, Theorem 64, p. 353).

We want to assume that the transformation:

$$(8) \qquad \begin{cases} z_1 = z_1(y_1, \dots, y_r), \; \dots, \; z_{r-m} = z_{r-m}(y_1, \dots, y_r) \\ \overline{z}_1 = \overline{z}_1(y_1, \dots, y_r), \; \dots, \; \overline{z}_m = \overline{z}_m(y_1, \dots, y_r) \end{cases}$$

transfers the group: $Z_1 f, \dots, Z_r f$ to the group: $\mathsf{B}_1 f, \dots, \mathsf{B}_r f$. Then according to Chap. 20, p. 391, through the same transformation, the group: $Y_1 f, \dots, Y_r f$ is transferred at the same time to the group: $\mathfrak{W}_1 f, \dots, \mathfrak{W}_r f$, hence by virtue of (8), there are relations of the form:

$$Y_k f = \sum_{j=1}^{r} \mathfrak{h}_{kj} \, \mathfrak{W}_j f \qquad {\scriptstyle (k=1 \cdots r)},$$

where the determinant of the constants $\mathfrak{h}_{kj}$ does not vanish. But from these relations, the following relations immediately follow:

$$(9) \qquad \sum_{\nu=1}^{r-m} Y_k z_\nu \frac{\partial f}{\partial z_\nu} = \sum_{j=1}^{r} \mathfrak{h}_{kj} \, \mathfrak{X}_j f \qquad {\scriptstyle (k=1 \cdots r)},$$

and likewise, they hold identically by virtue of (8).

This is the reason why the equations:

$$z_1(y_1, \dots, y_r) = \text{const.}, \; \dots, \; z_{r-m}(y_1, \dots, y_r) = \text{const.}$$

represent a decomposition of the space $y_1, \dots, y_r$ invariant by the group: $Y_1 f, \dots, Y_r f$, or, what amount to the same, the reason why $z_1(y), \dots, z_{r-m}(y)$ are independent invariants of a completely determined m-term subgroup $\mathfrak{g}$ of the group $Z_1 f, \dots, Z_r f$. Consequently, we obtain the group: $\mathfrak{X}_1 f, \dots, \mathfrak{X}_r f$ thanks to the process described in Theorem 78, p. 447, when we arrange ourselves as follows: As the group: $\mathsf{Z}_1 f, \dots, \mathsf{Z}_m f$, we choose the m-term subgroup $\mathfrak{g}$ of the group: $Z_1 f, \dots, Z_r f$ just defined, and as the functions: $u_1(y), \dots, u_{r-m}(y)$, we choose the just mentioned invariants: $z_1(y), \dots, z_{r-m}(y)$ of the group $\mathfrak{g}$. Indeed, under these assumptions, the relations (9) hold true, in which the right-hand side expressions are independent infinitesimal transformations of the group: $\mathfrak{X}_1 f, \dots, \mathfrak{X}_r f$.

As a result, it is proved that, thanks to the process which is described in Theorem 78, p. 447, one can find every transitive group isomorphic with the group: $x_i' = f_i(x,a)$. We can therefore enunciate the following theorem:

Theorem 79. *If the finite equations:*

$$x_i' = f_i(x_1, \dots, x_n, a_1, \dots, a_r) \qquad (i=1\cdots n)$$

of an r-term group are presented, then one finds in the following way all transitive groups that are equally composed with this group:

To begin with, one determines, which requires only executable operations, two r-term simply transitive groups:

$$Y_k f = \sum_{j=1}^{r} \eta_{kj}(y_1, \dots, y_r) \frac{\partial f}{\partial y_j} \qquad (k=1\cdots r)$$

and:

$$Z_k f = \sum_{j=1}^{r} \zeta_{kj}(y_1, \dots, y_r) \frac{\partial f}{\partial y_j} \qquad (k=1\cdots r)$$

that are reciprocal to each other and are equally composed with the group presented. Afterwards, by means of algebraic operations, one sets up all subgroups of the group: $Z_1 f, \dots, Z_r f$ which are neither invariant in this group, nor contain, aside from the identity transformation, a subgroup invariant in the group: $Z_1 f, \dots, Z_r f$. Under the guidance of Theorem 78, each of the found subgroups produces all transitive groups equally composed with the group: $x_i' = f_i(x,a)$ that belong to a certain type. If one determines these groups for each of the found subgroups, then one obtains all transitive groups that are equally composed with the group: $x_i' = f_i(x,a)$.[1]

The second part of our program is now carried out, and we are in a position to identify all types of transitive groups equally composed with the group: $x_i' = f_i(x,a)$. Every subgroup of the group: $Z_1 f, \dots, Z_r f$ which has the property mentioned in Theorem 79 provides us with such a type. It remains to decide when two different subgroups of the group: $Z_1 f, \dots, Z_r f$ produce different types, and when they produce the same type.

Let the m independent infinitesimal transformations:

$$\mathsf{Z}_\mu f = \sum_{k=1}^{r} \varepsilon_{\mu k} Z_k f \qquad (\mu=1\cdots m)$$

---

[1] LIE, Archiv for Math., Vol. 10, Christiania 1885, and Verh. der Gesellsch. d. W. zu Chr. a. 1884; Berichte der Kgl. Sächs. G. d. W., 1888.

generate an m-term group which is neither invariant in the group: $Z_1 f, \dots, Z_r f$, nor contains, aside from the identity transformation, another subgroup invariant in the group: $Z_1 f, \dots, Z_r f$.

Let the functions:

$$u_1(y_1, \dots, y_r), \; \dots, \; u_{r-m}(y_1, \dots, y_r)$$

be independent invariants of the group: $\mathsf{Z}_1 f, \dots, \mathsf{Z}_m f$. Under these assumptions, according to Theorem 79, the group:

$$U_k f = \sum_{\nu=1}^{r-m} Y_k u_\nu \frac{\partial f}{\partial u_\nu} = \sum_{\nu=1}^{r-m} \omega_{k\nu}(u_1, \dots, u_{r-m}) \frac{\partial f}{\partial u_\nu} \qquad (k=1\cdots r)$$

in the $r-m$ variables: $u_1, \dots, u_{r-m}$ has the same composition as the group: $x_i' = f_i(x,a)$, and is in addition transitive, hence it is a representative of the type of groups which corresponds to the subgroup: $\mathsf{Z}_1 f, \dots, \mathsf{Z}_m f$.

If another subgroup of the group: $Z_1 f, \dots, Z_r f$ is the same type of group, then this subgroup must evidently be m-term, for it is only in this case that it can provide transitive groups in $r-m$ variables that are equally composed with the group: $x_i' = f_i(x,a)$.

So, we assume that:

$$\mathfrak{Z}_\mu f = \sum_{k=1}^{r} \mathfrak{e}_{\mu k} Z_k f \qquad (\mu=1\cdots m)$$

is another m-term subgroup of the group: $Z_1 f, \dots, Z_r f$, and that this subgroup too is neither invariant in the group: $Z_1 f, \dots, Z_r f$, nor contains, aside from the identity transformation, another subgroup invariant in the group: $Z_1 f, \dots, Z_r f$. Let independent invariants of the group: $\mathfrak{Z}_1 f, \dots, \mathfrak{Z}_m f$ be:

$$\mathfrak{u}_1(y_1, \dots, y_r), \; \dots, \; \mathfrak{u}_{r-m}(y_1, \dots, y_r).$$

Under these assumptions, the group:

$$\mathfrak{U}_k f = \sum_{\nu=1}^{r-m} Y_k \mathfrak{u}_\nu \frac{\partial f}{\partial \mathfrak{u}_\nu} = \sum_{\nu=1}^{r-m} \mathfrak{o}_{k\nu}(\mathfrak{u}_1, \dots, \mathfrak{u}_{r-m}) \frac{\partial f}{\partial \mathfrak{u}_\nu} \qquad (k=1\cdots r)$$

in the $r-m$ variables: $\mathfrak{u}_1, \dots, \mathfrak{u}_{r-m}$ is transitive and has the same composition as the group: $x_i' = f_i(x,a)$.

The question of whether the two subgroups: $\mathsf{Z}_1 f, \dots, \mathsf{Z}_m f$ and $\mathfrak{Z}_1 f, \dots, \mathfrak{Z}_m f$ of the group: $Z_1 f, \dots, Z_r f$ provide the same type of group, or not, can now obviously also be expressed as follows: Are the two groups: $U_1 f, \dots, U_r f$ and: $\mathfrak{U}_1 f, \dots, \mathfrak{U}_r f$ similar to each other, or not?

As the two groups we are presently considering are transitive; the question of whether they are similar, or not similar, can be decided on the basis of Theorem 76, p. 434.

From this theorem, we see that the groups: $U_1 f, \dots, U_r f$ and: $\mathfrak{U}_1 f, \dots, \mathfrak{U}_r f$ are similar to each other if and only if it is possible to relate them to each other in a holoedrically isomorphic way so that the following condition is satisfied: The most general subgroup of the group: $U_1 f, \dots, U_r f$ which leaves invariant an arbitrarily chosen, but determined, system of values: $u_1 = u_1^0, \dots, u_{r-m} = u_{r-m}^0$ in general position must correspond to the most general subgroup of the group: $\mathfrak{U}_1 f, \dots, \mathfrak{U}_r f$ which leaves invariant a certain system of values: $\mathfrak{u}_1 = \mathfrak{u}_1^0, \dots, \mathfrak{u}_{r-m} = \mathfrak{u}_{r-m}^0$ in general position.

Now, the groups: $U_1 f, \dots, U_r f$ and: $\mathfrak{U}_1 f, \dots, \mathfrak{U}_r f$ are, because of their derivation, both related to the group: $Y_1 f, \dots, Y_r f$ in a holoedrically isomorphic way, hence we can also obviously express the necessary and sufficient criterion for their similarity as follows: *Similarity happens to hold when and only when it is possible to relate the group: $Y_1 f, \dots, Y_r f$ to itself in a holoedrically isomorphic way so that its largest subgroup G which fixes the manifold:*

$$u_1(y_1, \dots, y_r) = u_1^0, \ \dots, \ u_{r-m}(y_1, \dots, y_r) = u_{r-m}^0 \tag{10}$$

corresponds to its largest subgroup $\mathfrak{G}$ which fixes the manifold:

$$\mathfrak{u}_1(y_1, \dots, y_r) = \mathfrak{u}_1^0, \ \dots, \ \mathfrak{u}_{r-m}(y_1, \dots, y_r) = \mathfrak{u}_{r-m}^0. \tag{11}$$

The criterion found with these words can be brought to another remarkable form. In fact, we consider the group: $Z_1 f, \dots, Z_r f$ reciprocal to the group: $Y_1 f, \dots, Y_r f$. In it, $\mathrm{Z}_1 f, \dots, \mathrm{Z}_m f$ is the largest subgroup which leaves invariant the manifold (10), and $\mathfrak{Z}_1 f, \dots, \mathfrak{Z}_m f$ is the largest subgroup which leaves invariant the manifold (11). According to Chap. 20, Proposition 8, p. 400, one can then relate the two groups: $Y_1 f, \dots, Y_r f$ and: $Z_1 f, \dots, Z_r f$ in a holoedrically isomorphic way so that the subgroup G corresponds to the subgroup: $\mathrm{Z}_1 f, \dots, \mathrm{Z}_m f$; but one can also relate them together in ą holoedrically isomorphic way so that the subgroup $\mathfrak{G}$ corresponds to the subgroup: $\mathfrak{Z}_1 f, \dots, \mathfrak{Z}_m f$.

From this, it results that the group: $Y_1 f, \dots, Y_r f$ can be related to itself in a holoedrically isomorphic way as described just above if and only if it is possible to relate the group: $Z_1 f, \dots, Z_r f$ to itself in a holoedrically isomorphic way so that the subgroup: $\mathrm{Z}_1 f, \dots, \mathrm{Z}_m f$ corresponds to the subgroup: $\mathfrak{Z}_1 f, \dots, \mathfrak{Z}_m f$.

We therefore have the:

Theorem 80. *If the two r-term groups:*

$$Y_k f = \sum_{j=1}^{r} \eta_{kj}(y_1, \dots, y_r) \frac{\partial f}{\partial y_j} \qquad (k=1 \cdots r)$$

and:

$$Z_k f = \sum_{j=1}^{r} \zeta_{kj}(y_1, \dots, y_r) \frac{\partial f}{\partial y_j} \qquad (k=1 \cdots r)$$

are simply transitive and reciprocal to each other, if moreover:

$$\mathsf{Z}_\mu f = \sum_{k=1}^{r} \varepsilon_{\mu k} Z_k f \qquad (\mu = 1 \cdots m)$$

and:

$$\mathfrak{Z}_\mu f = \sum_{k=1}^{r} \mathfrak{e}_{\mu k} Z_k f \qquad (\mu = 1 \cdots m)$$

are two m-term subgroups of the group: $Z_1 f, \dots, Z_r f$ *both of which are neither invariant in this group, nor contain, aside from the identity transformation, another subgroup invariant in this group, and lastly, if:* $u_1(y), \dots, u_{r-m}(y)$ *and:* $\mathfrak{u}_1(y), \dots, \mathfrak{u}_{r-m}(y)$ *are independent invariants of these two m-term subgroups, respectively, then the two transitive groups, both equally composed with the group* $Y_1 f, \dots, Y_r f$*:*

$$U_k f = \sum_{\nu=1}^{r-m} Y_k u_\nu \frac{\partial f}{\partial u_\nu} = \sum_{\nu=1}^{r-m} \omega_{k\nu}(u_1, \dots, u_{r-m}) \frac{\partial f}{\partial u_\nu} \qquad (k = 1 \cdots r)$$

and:

$$\mathfrak{U}_k f = \sum_{\nu=1}^{r-m} Y_k \mathfrak{u}_\nu \frac{\partial f}{\partial \mathfrak{u}_\nu} = \sum_{\nu=1}^{r-m} \mathfrak{o}_{k\nu}(\mathfrak{u}_1, \dots, \mathfrak{u}_{r-m}) \frac{\partial f}{\partial \mathfrak{u}_\nu} \qquad (k = 1 \cdots r)$$

are similar to each other if and only if it is possible to relate the group: $Z_1 f, \dots, Z_r f$ *to itself in a holoedrically isomorphic way so that the subgroup:* $\mathsf{Z}_1 f, \dots, \mathsf{Z}_m f$ *corresponds to the subgroup:* $\mathfrak{Z}_1 f, \dots, \mathfrak{Z}_m f$*.*

Thanks to this theorem, the last part of the program stated on p. 444 is now also settled. At present, we can decide whether two different subgroups of the group: $Z_1 f, \dots, Z_r f$ provide, or do not provide, different types of transitive groups equally composed with the group: $x_i' = f_i(x, a)$. Clearly, for that, only a study of the subgroups of the group $Z_1 f, \dots, Z_r f$ is required, or, what is the same, of the subgroups of the group: $x_i' = f_i(x, a)$.

We recapitulate the necessary operations in a theorem:

Theorem 81. *If the finite equations, or the infinitesimal transformations of an r-term group* Γ *are presented, one can determine in the following way how many different types of transitive groups having the same composition as* Γ *there are: One determines all m-term subgroups of* Γ*, but one excludes those which are either invariant in* Γ *or which contain a subgroup invariant in* Γ *different from the identity transformation. One distributes the found m-term subgroups in classes by including two subgroups always in the same class when it is possible to relate* Γ *to itself in a holoedrically isomorphic way so that the two subgroups correspond to each other. To each class of m-term subgroups obtained in this way there corresponds a completely determined type of transitive group in* $r - m$ *variables equally composed with* Γ*; to different classes there correspond different types. If one undertakes this study for each one of the numbers:* $m = 0, 1, 2, \dots, r-1$*, then one can have a view* [ÜBERSEHEN] *of all existing types.*

We observe the following: The operations required in Theorem 81 are all executable, even when the group Γ is not given, and when only its composition is given. Also in this case, only executable operations are then necessary. Since the number of the subgroups of Γ only depends upon arbitrary parameters, the number of existing types depends at most upon arbitrary parameters. In particular, there is only one type of simply transitive group which is equally composed with the group Γ. But this already results from the developments of the previous section.

By combining the two Theorems 81 and 78, we obtain the following:

Theorem 82. *If the finite equations:*

$$x_i' = f_i(x_1,\dots,x_n,\,a_1,\dots,a_r) \qquad (i=1\cdots n)$$

of an r-term group are presented, then the determination of all equally composed transitive groups requires in all cases, while disregarding executable operations, only the integration of simultaneous systems of ordinary differential equations.[2]

If one wants to list all r-term transitive groups in n variables, then one determines all compositions of r-term groups and one then seeks, for each composition, the associated types of transitive groups in n variables. For given r (and n), all these types decompose into a bounded number of kinds [GATTUNG] so that the types of a kind have the same analytic representation. However, the analytic expressions for all types of a kind contain certain parameters, the number of which we can always imagine to be lowered down to a minimum. That such parameters occur stems from two different facts: firstly, from the fact that, to a given $r > 2$, an unbounded number of different compositions is associated; secondly from the fact that in general, a given r-term group contains an unbounded number of subgroups which produce nothing but different types. We do not consider it to be appropriate here to pursue these considerations further.

§ 108. Once again, we move back to the standpoint we took on p. 451 sq.

We had used two m-term subgroups: $Z_1f,\dots,Z_mf$ and: $\mathfrak{Z}_1f,\dots,\mathfrak{Z}_mf$ of the group: $Z_1f,\dots,Z_rf$ in order to produce transitive groups equally composed with the group: $Y_1f,\dots,Y_rf$, and we have found the two groups: $U_1f,\dots,U_rf$ and: $\mathfrak{U}_1f,\dots,\mathfrak{U}_rf$. At present, the question arises: under which conditions are these two group similar to each other?

We answered this question earlier, when we based ourselves on Theorem 76, p. 434, and in this way, we obtained Theorem 80, p. 453. Now we want to take up this question again and attempt to answer it without using Theorem 76.

Evidently, we are close to presuming that the groups: $U_1f,\dots,U_rf$ and: $\mathfrak{U}_1f,\dots,\mathfrak{U}_rf$ are in any case similar to each other when there is a transformation:

$$\bar{y}_k = \Omega_k(y_1,\dots,y_r) \qquad (k=1\cdots r) \tag{12}$$

[2] Cf. LIE, Math. Annalen Vol. XVI, p. 528.

which converts the subgroup: $\mathsf{Z}_1 f, \dots, \mathsf{Z}_m f$ into the subgroup: $\mathfrak{Z}_1 f, \dots, \mathfrak{Z}_m f$ and which transfers at the same time the group: $Z_1 f, \dots, Z_r f$ into itself. We will show that this presumption corresponds to the truth.

Thus, let (12) be a transformation which possesses the indicated properties. Through this transformation, the invariants of the group: $\mathsf{Z}_1 f, \dots, \mathsf{Z}_m f$ are obviously transferred to the invariants of the group: $\mathfrak{Z}_1 f, \dots, \mathfrak{Z}_m f$, hence we have by virtue of (12):

$$\mathfrak{u}_\nu(\overline{y}_1, \dots, \overline{y}_r) = \chi_\nu\big(u_1(y), \dots, u_{r-m}(y)\big) \qquad (\nu = 1 \cdots r-m),$$

where the functions: $\chi_1, \dots, \chi_{r-m}$ are absolutely determined and are mutually independent relative to: $u_1(y), \dots, u_{r-m}(y)$. On the other hand, according to Chap. 20, p. 391, through the transformation (12), not only the group: $Z_1 f, \dots, Z_r f$ is transferred into itself, but also its reciprocal group: $Y_1 f, \dots, Y_r f$, so we have:

$$\sum_{j=1}^{r} \eta_{kj}(\overline{y}_1, \dots, \overline{y}_r) \frac{\partial f}{\partial \overline{y}_j} = \overline{Y}_k f = \sum_{j=1}^{r} h_{kj} Y_j f \qquad (k = 1 \cdots r),$$

where the h_{kj} are constants, the determinant of which does not vanish.

Now, if we set for f, in the equations just written, an arbitrary function F of: $u_1(y), \dots, u_{r-m}(y)$, or, what is the same, a function of: $\mathfrak{u}_1(\overline{y}), \dots, \mathfrak{u}_{r-m}(\overline{y})$, we get:

$$\overline{Y}_k F = \sum_{\nu=1}^{r-m} \mathfrak{o}_{k\nu}(\mathfrak{u}_1, \dots, \mathfrak{u}_{r-m}) \frac{\partial F}{\partial \mathfrak{u}_\nu} = \sum_{j=1}^{r} h_{kj} \sum_{\nu=1}^{r-m} \omega_{j\nu}(u_1, \dots, u_{r-m}) \frac{\partial F}{\partial u_\nu}.$$

In other words, the two groups: $U_1 f, \dots, U_r f$ and: $\mathfrak{U}_1 f, \dots, \mathfrak{U}_r f$ are similar to each other: obviously:

$$\mathfrak{u}_\nu = \chi_\nu(u_1, \dots, u_{r-m}) \qquad (\nu = 1 \cdots r-m) \tag{13}$$

is a transformation which transfers the one group to the other.

As a result, the presumption enunciated above is proved.

It can be proved that the converse also holds true: When the two groups: $U_1 f, \dots, U_r f$ and: $\mathfrak{U}_1 f, \dots, \mathfrak{U}_r f$ are similar to each other, then there is always a transformation which transfers the subgroup: $\mathsf{Z}_1 f, \dots, \mathsf{Z}_m f$ to the subgroup: $\mathfrak{Z}_1 f, \dots, \mathfrak{Z}_m f$ and which transfers at the same time the group: $Z_1 f, \dots, Z_r f$ to itself.

Thus, let the two groups: $U_k f$ and: $\mathfrak{U}_k f$ be similar to each other and let:

$$\mathfrak{u}_\nu = \psi_\nu(u_1, \dots, u_{r-m}) \qquad (\nu = 1 \cdots r-m) \tag{14}$$

be a transformation which transfers the one group to the other, so that we have:

$$\mathfrak{U}_k f = \sum_{j=1}^{r} \delta_{kj} U_j f = U'_k f \qquad (k = 1 \cdots r). \tag{15}$$

Here, by δ_{kj}, it is to be understood constants, the determinant of which does not vanish. If $\mathfrak{U}_1 f, \dots, \mathfrak{U}_r f$ are linked together by relations of the form:

$$[\mathfrak{U}_i, \mathfrak{U}_k] = \sum_{s=1}^{r} c_{iks}\, \mathfrak{U}_s f,$$

then naturally, $U_1' f, \dots, U_r' f$ are linked together by the relations:

$$[U_i', U_k'] = \sum_{s=1}^{r} c_{iks}\, U_s' f.$$

Above, we have seen that every transformation (12) which leaves invariant the group: $Z_1 f, \dots, Z_r f$ and which transfers the subgroup of the $\mathsf{Z}_\mu f$ to the subgroup of the $\mathfrak{Z}_\mu f$ provides a completely determined transformation (13) which transfers the group of the $U_k f$ to the group of the $\mathfrak{U}_k f$. Under the present assumptions, we already know a transformation which accomplishes the latter transfer, namely the transformation (14). Therefore, we attempt to determine a transformation:

$$\text{(16)} \qquad \overline{y}_k = O_k(y_1, \dots, y_r) \qquad {\scriptstyle (k=1\cdots r)}$$

which leaves invariant the group: $Z_1 f, \dots, Z_r f$, which converts the subgroup: $\mathsf{Z}_1 f, \dots, \mathsf{Z}_m f$ into the subgroup: $\mathfrak{Z}_1 f, \dots, \mathfrak{Z}_m f$ and lastly, which provides exactly the transformation (14).

It is clear that every transformation (16) of the kind demanded must be constituted in such a way that its equations embrace the $r - m$ mutually independent equations:

$$\text{(14')} \qquad \mathfrak{u}_\nu(\overline{y}_1, \dots, \overline{y}_r) = \psi_\nu\big(u_1(y), \dots, u_{r-m}(y)\big) \qquad {\scriptstyle (\nu=1\cdots r-m).}$$

If it satisfies this condition, and in addition, if it leaves invariant the group: $Z_1 f, \dots, Z_r f$, then it satisfies all our demands. Indeed, on the one hand, it transfers the invariants of the group: $\mathsf{Z}_1 f, \dots, \mathsf{Z}_m f$ to the invariants of the group: $\mathfrak{Z}_1 f, \dots, \mathfrak{Z}_m f$, whence it converts the first one of these two groups into the second one, and on the other hand, it visibly produces the transformation (14) by virtue of which the two groups: $U_1 f, \dots, U_r f$ and $\mathfrak{U}_1 f, \dots, \mathfrak{U}_r f$ are similar to each other.

But now, whether we require of the transformation (16) that it leaves invariant the group: $Z_1 f, \dots, Z_r f$, or whether we require that it transfers the group: $Y_1 f, \dots, Y_r f$ into itself, is obviously completely indifferent. We can therefore interpret our problem as follows:

To seek a transformation (16) *which leaves invariant the group: $Y_1 f, \dots, Y_r f$ and which is constituted in such a way that its equations embrace the equations* (14').

For the sake of simplification, we introduce new variables.

We choose m arbitrary mutually independent functions: $v_1(y), \dots, v_m(y)$ that are also independent of: $u_1(y), \dots, u_{r-m}(y)$, and moreover, we choose m arbitrary mutually independent functions: $\mathfrak{v}_1(\overline{y}), \dots, \mathfrak{v}_m(\overline{y})$ that are also independent of

$\mathfrak{u}_1(\overline{y}),\dots,\mathfrak{u}_{r-m}(\overline{y})$. We introduce the functions: $u_1(y),\dots,u_{r-m}(y)$, $v_1(y),\dots,v_m(y)$ as new variables in place of: $y_1,\dots,y_r$ and the functions: $\mathfrak{u}_1(\overline{y}),\dots,\mathfrak{u}_{r-m}(\overline{y})$, $\mathfrak{v}_1(\overline{y}),\dots,\mathfrak{v}_m(\overline{y})$ in place of: $\overline{y}_1,\dots,\overline{y}_r$.

In the new variables, the sought transformation (16) necessarily receives the form:

$$(16') \qquad \begin{cases} \mathfrak{u}_\nu = \psi_\nu(u_1,\dots,u_{r-m}) & (\nu=1\cdots r-m) \\ \mathfrak{v}_\mu = \Psi_\mu(u_1,\dots,u_{r-m},\, v_1,\dots,v_m) & (\mu=1\cdots m), \end{cases}$$

where the fact that the present equations must comprise the equations (14) is already taken account of.

But what do we have in place of the requirement that the transformation (16) should leave invariant the group: $Y_1 f,\dots,Y_r f$?

Clearly, in the new variables: $u_1,\dots,u_{r-m}$, $v_1,\dots,v_m$, the group: $Y_1 f,\dots,Y_r f$ receives the form:

$$U_k f + \sum_{\mu=1}^{m} w_{k\mu}(u_1,\dots,u_{r-m},\, v_1,\dots,v_m)\,\frac{\partial f}{\partial v_\mu} = U_k f + V_k f \qquad (k=1\cdots r).$$

On the other hand, after the introduction of $\mathfrak{u}_1,\dots,\mathfrak{u}_{r-m}$, $\mathfrak{v}_1,\dots,\mathfrak{v}_m$, the infinitesimal transformations:

$$\overline{Y}_k f = \sum_{j=1}^{r} \eta_{kj}(\overline{y}_1,\dots,\overline{y}_r)\,\frac{\partial f}{\partial \overline{y}_j} \qquad (k=1\cdots r)$$

are transferred to:

$$\mathfrak{U}_k f + \sum_{\mu=1}^{m} \mathfrak{w}_{k\mu}(\mathfrak{u}_1,\dots,\mathfrak{u}_{r-m},\, \mathfrak{v}_1,\dots,\mathfrak{v}_m)\,\frac{\partial f}{\partial \mathfrak{v}_\mu} = \mathfrak{U}_k f + \mathfrak{V}_k f \qquad (k=1\cdots r).$$

Consequently, we must require of the transformation (16') that it transfers the group: $U_1 f + V_1 f,\dots,U_r f + V_r f$ to the group: $\mathfrak{U}_1 f + \mathfrak{V}_1 f,\dots,\mathfrak{U}_r f + \mathfrak{V}_r f$; we must attempt to determine the functions: $\Psi_1,\dots,\Psi_m$ accordingly.

As the equation (15) shows, the r independent infinitesimal transformations:

$$\sum_{j=1}^{r} \delta_{1j}\, U_j f,\ \dots,\ \sum_{j=1}^{r} \delta_{rj}\, U_j f$$

are transferred by the transformation:

$$(14) \qquad \mathfrak{u}_\nu = \psi_\nu(u_1,\dots,u_{r-m}) \qquad (\nu=1\cdots r-m)$$

to the transformations:

$$\mathfrak{U}_1 f,\ \dots,\ \mathfrak{U}_r f,$$

respectively. Hence, if a transformation of the form (16') is supposed to convert the group of the $U_k f + V_k f$ to the group of the $\mathfrak{U}_k f + \mathfrak{V}_k f$, then through it, the r

independent infinitesimal transformations:

$$\sum_{j=1}^{r} \delta_{1j}(U_jf+V_jf), \ \dots, \ \sum_{j=1}^{r} \delta_{rj}(U_jf+V_jf)$$

are transferred to:

$$\mathfrak{U}_1f+\mathfrak{V}_1f, \ \dots, \ \mathfrak{U}_rf+\mathfrak{V}_rf.$$

This condition is necessary, and simultaneously also sufficient.

Thanks to the same considerations as in Chap. 19, p. 348 sq., we now recognize that every transformation (16') having the constitution demanded can also be defined as a system of equations in the $2r$ variables u, v, $\mathfrak{u}$, $\mathfrak{v}$ which possesses the form:

$$\text{(16')} \qquad \begin{cases} \mathfrak{u}_\nu = \psi_\nu(u_1,\dots,u_{r-m}) & (\nu=1\cdots r-m) \\ \mathfrak{v}_\mu = \Psi_\mu(u_1,\dots,u_{r-m},v_1,\dots,v_m) & (\mu=1\cdots m), \end{cases}$$

which admits the r-term group:

$$W_kf = \mathfrak{U}_kf+\mathfrak{V}_kf+\sum_{j=1}^{r} \delta_{kj}(U_jf+V_jf) \qquad (k=1\cdots r)$$

and which is solvable with respect to $u_1,\dots,u_{r-m},v_1,\dots,v_m$.

According to our assumption, the system of equations:

$$\text{(14)} \qquad \mathfrak{u}_\nu = \psi_\nu(u_1,\dots,u_{r-m}) \qquad (\nu=1\cdots r-m)$$

represents a transformation which transfers the r independent infinitesimal transformations:

$$\sum_{j=1}^{r} \delta_{kj}U_jf \qquad (k=1\cdots r)$$

to: $\mathfrak{U}_1f,\dots,\mathfrak{U}_rf$, respectively, whence it admits the r-term group:

$$\mathfrak{U}_kf+\sum_{j=1}^{r} \delta_{kj}U_jf \qquad (k=1\cdots r),$$

and consequently, also the group: $W_1f,\dots,W_rf$.

We can therefore use the developments of Chap. 14, pp. 243–246, in order to find a system of equations (16') having the constitution demanded.

To begin with, from $W_1f,\dots,W_rf$, we form certain reduced infinitesimal transformations: $\mathfrak{W}_1f,\dots,\mathfrak{W}_rf$ by leaving out all terms with the differential quotients: $\partial f/\partial\mathfrak{u}_1$, $\dots$, $\partial f/\partial\mathfrak{u}_{r-m}$ and by making the substitution: $\mathfrak{u}_1=\psi_1(u)$, $\dots$, $\mathfrak{u}_{r-m}=\psi_{r-m}(u)$ in all the terms remaining. If this substitution is indicated by the symbol: [ ], then the $\mathfrak{W}_kf$ read as follows:

$$\mathfrak{W}_k f = \sum_{\mu=1}^{m} \big[\mathfrak{w}_{k\mu}(\mathfrak{u}_1,\dots,\mathfrak{u}_{r-m},\mathfrak{v}_1,\dots,\mathfrak{v}_m\big]\,\frac{\partial f}{\partial \mathfrak{v}_\mu} + \sum_{j=1}^{r} \delta_{kj}\,(U_j f + V_j f),$$

or, more shortly:

$$\mathfrak{W}_k f = \big[\mathfrak{V}_k f\big] + \sum_{j=1}^{r} \delta_{kj}\,(U_j f + V_j f) \qquad (k=1\cdots r).$$

Naturally, $\mathfrak{W}_1 f,\dots,\mathfrak{W}_r f$ generate a group in the $r+m$ variables $\mathfrak{v}_1,\dots,\mathfrak{v}_m$, $u_1,\dots,u_{r-m}$, $v_1,\dots,v_m$ and in our case, a group which is evidently r-term.

Now, we determine in the $\mathfrak{v}$, u, v a system of equations of the form:

$$\text{(17)} \qquad \mathfrak{v}_\mu = \Psi_\mu(u_1,\dots,u_{r-m},v_1,\dots,v_m) \qquad (\mu=1\cdots m)$$

which admits the group: $\mathfrak{W}_1 f,\dots,\mathfrak{W}_r f$ and which is solvable with respect to $v_1,\dots,v_m$. Lastly, when we add this system to the equations (14), we then obtain a system of equations of the form (16') which possesses the properties explained above.

Since the r-term group:

$$\sum_{j=1}^{r} \delta_{kj}\,(U_j f + V_j f) \qquad (k=1\cdots r)$$

is simply transitive, then in the matrix which can be formed with the coefficients of $\mathfrak{W}_1 f,\dots,\mathfrak{W}_r f$, it is certain that not all $r\times r$ determinants vanish identically, and they can even less all vanish by virtue of a system of equations of the form (17). Consequently, every system of equations of the form (17) which admits the group: $\mathfrak{W}_1 f,\dots,\mathfrak{W}_r f$ can be represented by m independent relations between m arbitrary independent solutions of the r-term complete system:

$$\text{(18)} \qquad \mathfrak{W}_1 f = 0,\ \dots,\ \mathfrak{W}_r f = 0$$

in the $r+m$ variables $\mathfrak{v}_1,\dots,\mathfrak{v}_m$, $u_1,\dots,u_{r-m}$, $v_1,\dots,v_m$.

The equations (18) are obviously solvable with respect to: $\partial f/\partial u_1$, $\dots$, $\partial f/\partial u_{r-m}$, $\partial f/\partial v_1$, $\dots$, $\partial f/\partial v_m$; on the other hand, they are also solvable with respect to: $\partial f/\partial u_1,\dots,\partial f/\partial u_{r-m}$, $\partial f/\partial \mathfrak{v}_1,\dots,\partial f/\partial \mathfrak{v}_m$, because if we introduce the new variables: $u_1,\dots,u_{r-m}$, $\mathfrak{v}_1,\dots,\mathfrak{v}_m$ in place of $\mathfrak{u}_1,\dots,\mathfrak{u}_{r-m}$, $\mathfrak{v}_1,\dots,\mathfrak{v}_m$ by means of the equations:

$$\text{(14)} \qquad \mathfrak{u}_\mu = \psi_\mu(u_1,\dots,u_{r-m}) \qquad (\mu=1\cdots r-m)$$

in the r infinitesimal transformations:

$$\mathfrak{U}_1 f + \mathfrak{V}_1 f,\ \dots,\ \mathfrak{U}_r f + \mathfrak{V}_r f,$$

then we obtain the infinitesimal transformations:

$$\sum_{j=1}^{r} \delta_{kj} U_j f + [\mathfrak{V}_k f] \qquad (k=1\cdots r)$$

which in turn therefore generate a simply transitive group in the variables: $u_1, \ldots, u_{r-m}, \mathfrak{v}_1, \ldots, \mathfrak{v}_m$.

Consequently, if:

$$P_\mu\big(\mathfrak{v}_1, \ldots, \mathfrak{v}_m, u_1, \ldots, u_{r-m}, v_1, \ldots, v_m\big) \qquad (\mu=1\cdots m)$$

are m arbitrary independent solutions of the complete system (18), then these solutions are mutually independent both relative to $\mathfrak{v}_1, \ldots, \mathfrak{v}_m$ and relative to $v_1, \ldots, v_m$ (cf. Theorem 12, p. 105).

From this, it results that the most general system of equations of the form (17) which can be resolved with respect to $v_1, \ldots, v_m$ and which admits the group: $\mathfrak{W}_1 f, \ldots, \mathfrak{W}_r f$ can be obtained by solving the m equations:

$$(19) \qquad P_\mu(\mathfrak{v}_1, \ldots, \mathfrak{v}_m, u_1, \ldots, u_{r-m}, v_1, \ldots, v_m) = \text{const.} \qquad (\mu=1\cdots m)$$

with respect to $\mathfrak{v}_1, \ldots, \mathfrak{v}_m$.

Now, we can immediately indicate a transformation, and in fact the most general transformation (16'), which transfers the infinitesimal transformations:

$$\sum_{j=1}^{r} \delta_{1j}(U_j + V_j f), \; \ldots, \; \sum_{j=1}^{r} \delta_{rj}(U_j f + V_j f)$$

to the infinitesimal transformations:

$$\mathfrak{U}_1 f + \mathfrak{V}_1 f, \; \ldots, \; \mathfrak{U}_r f + \mathfrak{V}_r f,$$

respectively; this transformation is simply represented by the equations (14) and (19) together. Lastly, if we again introduce the variables: $y_1, \ldots, y_r, \overline{y}_1, \ldots, \overline{y}_r$ in (14) and in (19), we obtain a transformation which leaves invariant the group: $Y_1 f, \ldots, Y_r f$ and whose equations embrace the equations (14'); in other words, we obtain a transformation which leaves the group: $Z_1 f, \ldots, Z_r f$ invariant and which transfers the subgroup: $\mathrm{Z}_1 f, \ldots, \mathrm{Z}_m f$ to the subgroup: $\mathfrak{Z}_1 f, \ldots, \mathfrak{Z}_m f$.

With these words, we have proved that there always exists a transformation having the constitution just described, as soon as the two groups: $U_1 f, \ldots, U_r f$ and $\mathfrak{U}_1 f, \ldots, \mathfrak{U}_r f$ are similar to each other. But since the similarity of the two groups follows from the existence of such a transformation, according to p. 455 sq., we can say:

The two groups: $U_1 f, \ldots, U_r f$ and $\mathfrak{U}_1 f, \ldots, \mathfrak{U}_r f$ are similar to each other if and only if there is a transformation which leaves invariant the group: $Z_1 f, \ldots, Z_r f$ and which transfers the subgroup: $\mathrm{Z}_1 f, \ldots, \mathrm{Z}_m f$ to the subgroup: $\mathfrak{Z}_1 f, \ldots, \mathfrak{Z}_m f$.

Now, the group: $Z_1 f, \ldots, Z_r f$ is simply transitive, hence it is clear that there is always a transformation of this sort when and only when the group: $Z_1 f, \ldots, Z_r f$

can be related to itself in a holoedrically isomorphic way so that the two subgroups: $\mathsf{Z}_1 f, \dots, \mathsf{Z}_m f$ and $\mathfrak{Z}_1 f, \dots, \mathfrak{Z}_m f$ correspond to each other. Consequently, for the similarity of the groups: $U_1 f, \dots, U_r f$ and $\mathfrak{U}_1 f, \dots, \mathfrak{U}_r f$, we find exactly the same criterion as we had expressed in Theorem 80, p. 453.

Furthermore, the preceding developments can also be used to derive a new proof of Theorem 76 in Chap. 21, p. 434.

First, thanks to considerations completely similar to the ones on p. 452, it can be proved that the transitive groups: $U_1 f, \dots, U_r f$ and $\mathfrak{U}_1 f, \dots, \mathfrak{U}_r f$ can be related to each other in the holoedrically isomorphic way described in Theorem 76 if and only if it is possible to relate the group: $Z_1 f, \dots, Z_r f$ to itself in a holoedrically isomorphic way in such a way that the subgroups: $\mathsf{Z}_1 f, \dots, \mathsf{Z}_m f$ and $\mathfrak{Z}_1 f, \dots, \mathfrak{Z}_m f$ correspond to each other. Then, it follows from the preceding results that the conditions of Theorem 76 for the similarity of the two groups: $U_1 f, \dots, U_r f$ and $\mathfrak{U}_1 f, \dots, \mathfrak{U}_r f$ are necessary and sufficient.

§ 109. Now, we turn to the second of the two problems, the settlement of which was announced in the introduction to the chapter (on p. 439): *to determine all r-term intransitive groups*; as was already said at that time, we want to undertake the determination in question under the assumption that all transitive groups with r or fewer parameters are given. Since all transitive groups with an equal number of parameters can be ordered in classes according to their composition and moreover, since all transitive groups having one and the same composition decompose into a series of types (cf. p. 443 sq.), we can express our assumption more precisely by supposing *firstly* that all possible compositions of a group with r or fewer parameters are known and *secondly* by imagining that for each one of these compositions, all possible types of transitive groups having the concerned composition are given.

To begin with, we consider an arbitrary r-term intransitive group.

If $X_1 f, \dots, X_r f$ are independent infinitesimal transformations of an r-term group of the space $x_1, \dots, x_n$, the r equations:

$$X_1 f = 0, \quad \dots, \quad X_r f = 0 \tag{20}$$

have a certain number, say exactly $n - l > 0$, of independent solutions in common. Hence we can imagine that the variables $x_1, \dots, x_n$ are chosen from the beginning in such a way that $x_{l+1}, \dots, x_n$ are such independent solutions. Then $X_1 f, \dots, X_r f$ will receive the form:

$$X_k f = \sum_{\lambda=1}^{l} \xi_{k\lambda}(x_1, \dots, x_l, x_{l+1}, \dots, x_n) \frac{\partial f}{\partial x_\lambda} \qquad (k=1 \cdots r)$$

where now naturally, not all $l \times l$ determinants of the matrix:

$$\begin{vmatrix} \xi_{11}(x) & \cdot & \cdot & \xi_{1l}(x) \\ \cdot & \cdot & \cdot & \cdot \\ \cdot & \cdot & \cdot & \cdot \\ \xi_{r1}(x) & \cdot & \cdot & \xi_{rl}(x) \end{vmatrix}$$

vanish identically, since otherwise, the equations (20) would have more than $n-l$ independent solutions in common.

If the number r, which is at least equal to l, was exactly equal to l, then $X_1f,\dots,X_rf$ would be linked together by no relation of the form:

$$\chi_1(x_{l+1},\dots,x_n)X_1f+\cdots+\chi_r(x_{l+1},\dots,x_n)X_rf=0;$$

but now, r need not be equal to l, hence relations of the form just described can also very well exist without all the functions $\chi_1,\dots,\chi_r$ vanishing. So we want to assume that $X_1f,\dots,X_mf$, say, are linked together by no such relation, while by contrast, $X_{m+1}f,\dots,X_rf$ may be expressed as follows in terms of $X_1f,\dots,X_mf$:

$$\text{(21)}\qquad X_{m+\nu}f \equiv \sum_{\mu=1}^{m} \vartheta_{\nu\mu}(x_{l+1},\dots,x_n)\,X_\mu f \qquad (\nu=1\cdots r-m).$$

Here of course, m satisfies the inequalities: $l \leqslant m \leqslant r$.

By combining the equation (21) with the relations:

$$[X_i, X_k] = \sum_{s=1}^{r} c_{iks}\,X_sf \qquad (i,k=1\cdots r)$$

which hold true in all circumstances, we see that $X_1f,\dots,X_mf$ stand pairwise in the relationships:

$$[X_\lambda, X_\mu] = \sum_{\pi=1}^{m}\left\{c_{\lambda\mu\pi}+\sum_{\nu=1}^{r-m} c_{\lambda,\mu,m+\nu}\,\vartheta_{\nu\pi}(x_{l+1},\dots,x_n)\right\}X_\pi f$$
$$(\lambda,\mu=1\cdots m).$$

Now, if the variables $x_{l+1},\dots,x_n$ are replaced by arbitrary constants: $a_{l+1},\dots,a_n$ and if $x_1,\dots,x_l$ only are still considered as variables, then it is clear that the r infinitesimal transformations:

$$\overline{X}_kf=\sum_{\lambda=1}^{l}\xi_{k\lambda}(x_1,\dots,x_l,a_{l+1},\dots,a_n)\,\frac{\partial f}{\partial x_\lambda} \qquad (k=1\cdots r)$$

in the l variables $x_1,\dots,x_l$ are no longer independent of each other, but can be linearly deduced from the m independent infinitesimal transformations:

$$\overline{X}_\mu f=\sum_{\lambda=1}^{l}\xi_{\mu\lambda}(x_1,\dots,x_l,a_{l+1},\dots,a_n)\,\frac{\partial f}{\partial x_\lambda} \qquad (\mu=1\cdots m).$$

The m infinitesimal transformations $\overline{X}_1 f, \dots, \overline{X}_m f$ are in turn obviously linked together by the relations:

$$(22) \qquad [\overline{X}_\lambda, \overline{X}_\mu] = \sum_{\pi=1}^{m} \left\{ c_{\lambda\mu\pi} + \sum_{\nu=1}^{r-m} c_{\lambda,\mu,m+\nu}\, \vartheta_{\nu\pi}(a_{l+1}, \dots, a_n) \right\} \overline{X}_\pi f,$$

and consequently, whichever values one may confer to the parameters $a_{l+1}, \dots, a_n$, they always generate an m-term group in the variables: $x_1, \dots, x_l$, and of course, a transitive group.

Now, by conferring to the parameters $a_{l+1}, \dots, a_n$ all possible values gradually, one obtains ∞^{n-l} m-term groups in l variables. On the occasion, it is conceivable, though not necessary, that these ∞^{n-l} groups are similar to each other. If this is not the case, then these groups order themselves in $\infty^{n-l-\sigma}$ families, each one consisting of ∞^{σ} groups, and to be precise, in such a way that two groups in the same family are similar, while by contrast, two groups belonging to two different families are not similar.

In all circumstances, our ∞^{n-l} groups belong to the same kind of type [TYPENGATTUNG]*; in the latter case, this kind depends upon essential parameters (cf. p. 455 sq.).*

Now we can summarize how one can find all intransitive r-term groups in n variables. One chooses two numbers l and m so that $l \leqslant m \leqslant r$ and so that in addition $l < n$, and one then forms all kinds [GATTUNGEN] of transitive m-term groups in l variables. If $Y_1 f, \dots, Y_m f$:

$$Y_k f = \sum_{i=1}^{l} \eta_{ki}(x_1, \dots, x_l, \alpha_1, \alpha_2, \dots) \frac{\partial f}{\partial x_i} \qquad (k=1\cdots m)$$

is such a kind with the essential parameters α_1, α_2, $\dots$, then one interprets these parameters as unknown functions of $x_{l+1}, \dots, x_n$, one sets:

$$X_k f = \sum_{i=1}^{m} \beta_{ki}(x_{l+1}, \dots, x_n)\, Y_i f \qquad (k=1\cdots r),$$

and lastly, one attempts to choose the undetermined functions α_j, β_{ki} of $x_{l+1}, \dots, x_n$ in the most general way so that $X_1 f, \dots, X_r f$ become independent infinitesimal transformations of an r-term group. This requirement leads to certain *finite relations* between the α, the β and the c_{iks} of the sought r-term group which must be satisfied in the most general way. The expressions of the infinitesimal transformations determined in this way contain, apart from certain arbitrary constants c_{iks}, certain arbitrary functions of the invariants of the group.

We therefore have the

Theorem 83. *The determination of all intransitive r-term groups:* $X_1 f, \ldots, X_r f$ *in n variables requires, as soon as all transitive groups with r or fewer parameters are found, no integration, but only executable operations.*[3]

It is to be observed that the computations necessary for the determination of all intransitive groups depend only upon the numbers r, l and m. By contrast, the number n plays no direct rôle. Consequently:

Theorem 84. *The determination of all r-term groups in an arbitrary number of variables can be led back to the determination of all r-term groups in r or fewer variables.*

We make a brief remark about the determination of all intransitive r-term groups of the given composition.

If a group of the concerned composition contains only a finite number of invariant subgroups, then m must be equal to r (Proposition 8, p. 320), and consequently, the settlement of the problem just formulated amounts without effort to the determination of all transitive groups of the concerned composition. By contrast, if there occur infinitely many invariant subgroups, m can be smaller than r; then according to the developments just mentioned, the m-term group $\overline{X}_1 f, \ldots, \overline{X}_m f$ discussed above must be meroedrically isomorphic to the sought r-term group. We do not want to undertake to show more closely how the problem can be settled in this case.

---

[3] LIE, Archiv for Math. og Naturv. Vol. 10, Christiania 1885; Math. Ann. Vol. XVI, p. 528, 1880.

Chapter 23
Invariant Families of Manifolds

If $x_1, \dots, x_n$ are point coordinates of an n-times extended space, then a family of manifolds of this space is represented by equations of the form:

$$(1) \qquad \Omega_1(x_1,\dots,x_n,\, l_1,\dots,l_m) = 0, \;\; \dots, \;\; \Omega_{n-q}(x_1,\dots,x_n,\, l_1,\dots,l_m) = 0,$$

in which, aside from the variables $x_1, \dots, x_n$, certain parameters: $l_1, \dots, l_m$ are present.

If one executes an arbitrary transformation:

$$x_i' = f_i(x_1,\dots,x_n) \qquad (i=1\cdots n)$$

of the space $x_1, \dots, x_n$, then each one of the manifolds (1) is transferred to a new manifold, hence the whole family (1) converts into a new family of manifolds. One obtains the equations of this new family when one takes away $x_1, \dots, x_n$ from (1) with the help of the n equations: $x_i' = f_i(x)$. Now in particular, if the new family of manifolds coincides with the original family (1), whence every manifold of the family (1) is transferred by the transformation: $x_i' = f_i(x)$ to a manifold of the same family, then we say *that the family* (1) *admits the transformation in question, or that it remains invariant by it.*

If a family of manifolds in the space $x_1, \dots, x_n$ admits all transformations of an r-term group, then we say *that it admits the group in question.*

Examples of invariant families of manifolds in the space $x_1, \dots, x_n$ have already appeared to us several times; every intransitive group decomposes the space into an invariant family of individually invariant manifolds (Chap. 13, pp. 227–228); every imprimitive group decomposes the space into an invariant family of manifolds that it permutes (p. 232 sq.); also, every manifold individually invariant by a group may be interpreted as an invariant family of manifolds, namely as a family parametrized by a point.

In what follows, we now consider a completely arbitrary family of manifolds. First, we study under which conditions this family admits a single given transformation, or a given r-term group. Then, we imagine that a group is given by which the family remains invariant and we determine the law according to which the manifolds

S. Lie, *Theory of Transformation Groups I*,
DOI 10.1007/978-3-662-46211-9_23

of the family are permuted with each other by the transformations of this group. In this way, we obtain a new process to set up the groups which are isomorphic with a given group. Finally, we give a method for finding all families of manifolds invariant by a given group.

§ 110. Let the equations:

$$\Omega_k(x_1,\dots,x_n,l_1,\dots,l_m)=0 \qquad (k=1\cdots n-q) \tag{1}$$

with the m arbitrary parameters $l_1,\dots,l_m$ represent an arbitrary family of manifolds.

If $l_1,\dots,l_m$ are absolutely arbitrary parameters, then naturally, one should not be able to eliminate all the x from (1); therefore, the equations (1) must be solvable with respect to $n-q$ of the variables $x_1,\dots,x_n$.

By contrast, it is not excluded that the l can be eliminated from (1), and this says nothing but that some relations between the x alone can be derived from (1). Only the number of independent equations free of the l which follow from (1) must be smaller than $n-q$, since otherwise, the parameters $l_1,\dots,l_r$ would be only apparent and the equations (1) would therefore represent not a family of manifolds, but a single manifold.

Before we study how the family of manifolds (1) behaves relative to transformations of the x, we must first make a few remarks about the nature of the equations (1).

The equations (1) contain m parameters $l_1,\dots,l_m$; if we let these parameters take all possible values, then we obtain ∞^m different systems of values: $l_1,\dots,l_m$, but not necessarily ∞^m different manifolds. It must therefore be determined under which conditions the given system of equations represents exactly ∞^m different manifolds, or in other words: one must give a criterion to determine whether the parameters: $l_1,\dots,l_m$ in the equations (1) are *essential*, or not.

In order to find such a criterion, we imagine that the equations: $\Omega_k=0$ are solved with respect to $n-q$ of the x, say with respect to $x_{q+1},\dots,x_n$:

$$x_{q+k}=\psi_{q+k}(x_1,\dots,x_q,l_1,\dots,l_m) \tag{2}$$
$$(k=1\cdots n-q)$$

and we imagine moreover that the functions ψ_{q+k} are expanded with respect to the powers of: $x_1-x_1^0,\dots,x_q-x_q^0$ in the neighborhood of an arbitrary system of values: $x_1^0,\dots,x_q^0$. The coefficients of the expansion, whose number is naturally infinitely large in general, will be analytic functions of $l_1,\dots,l_m$ and we may call them:

$$\Lambda_j(l_1,\dots,l_m) \qquad (j=1,2\cdots).$$

The question amounts just to how many independent functions are extant amongst all the functions $\Lambda_1,\Lambda_2,\dots$.

Indeed, if amongst the Λ, there are exactly l that are mutually independent functions — there are anyway surely no more than l —, then to the ∞^m different systems of values: $l_1, \dots, l_m$, there obviously also correspond ∞^m different systems of values: $\Lambda_1, \Lambda_2, \dots$, and therefore ∞^m different manifolds (2), that is to say, the parameters are essential in the equations (2), and hence also in the equations (1).

Otherwise, assume that amongst the functions $\Lambda_1, \Lambda_2, \dots$, there are not m but fewer functions, say only $m-h$, that are mutually independent. In this case, all the Λ_j can be expressed in terms of $m-h$ of them, say in terms of: $\Lambda'_1, \dots, \Lambda'_{m-h}$, which naturally must be mutually independent. To the ∞^m different systems of values $l_1, \dots, l_m$, there correspond therefore ∞^{m-h} different systems of values $\Lambda'_1, \dots, \Lambda'_{m-h}$, and ∞^{m-h} different systems of values $\Lambda_1, \Lambda_2, \dots$, so that the equations (2), and thus also the equations (1), represent only ∞^{m-h} different manifolds. We can express this in the clearest way by observing that the functions $\psi_{q+k}(x,l)$ contain the parameters $l_1, \dots, l_m$ only in the combinations: $\Lambda'_1, \dots, \Lambda'_{m-h}$, so that the equations (2) are of the form:

$$x_{q+k} = \overline{\Psi}_{q+k}(x_1, \dots, x_q, \Lambda'_1, \dots, \Lambda'_{m-h}) \qquad (k=1\cdots n-q). \tag{2'}$$

From this, it indeed results that we can introduce precisely $\Lambda'_1, \dots, \Lambda'_{m-h}$ as new parameters in place of $l_1, \dots, l_n$, a process by which the number of arbitrary parameters appearing in (2) is lowered to $m-h$.

Thus, we can say:

The equations:

$$\Omega_k(x_1, \dots, x_n, l_1, \dots, l_m) = 0 \qquad (k=1\cdots n-q)$$

represent ∞^m different manifolds only when it is not possible to indicate $m-h<m$ functions $\pi_1, \dots, \pi_{m-h}$ of $l_1, \dots, l_m$ such that, in the resolved equations:

$$x_{q+k} = \psi_{q+k}(x_1, \dots, x_q, l_1, \dots, l_m) \qquad (k=1\cdots n-q),$$

the functions $\psi_{q+1}, \dots, \psi_n$ can be expressed in terms of $x_1, \dots, x_n$ and of $\pi_1, \dots, \pi_{m-h}$ only. By contrast, if it is possible to indicate $m-h<m$ functions π_μ having this constitution, then the equations $\Omega_k = 0$ represent at most ∞^{m-h} manifolds and the parameters $l_1, \dots, l_m$ are hence not essential.

Here, we make a brief remark.

If the functions: $\Lambda_1, \Lambda_2, \dots$ discussed above take the values: Λ_1^0, Λ_2^0 after the substitution: $l_1 = l_1^0, \dots, l_m = l_m^0$, then the equations:

$$\Lambda_j(l_1, \dots, l_m) = \Lambda_j^0 \qquad (j=1,2\cdots)$$

define the totality of all systems of values $l_1, \dots, l_m$ which, when inserted in (2) or in (1), provide the same manifold as the system of values: $l_1^0, \dots, l_m^0$. Now, if amongst the functions $\Lambda_1, \Lambda_2, \dots$, there are m functions that are mutually independent, then the parameters $l_1, \dots, l_m$ are all essential, whence the following obviously holds: Around every system of values $l_1^0, \dots, l_m^0$ in general position, one can delimit a region such that two distinct systems of values $l_1, \dots, l_m$ of the concerned region always provide two manifolds that are also distinct.

The definition for essentiality [WESENTLICHKEIT] and for inessentiality [NICHTWESENTLICHKEIT] of the parameters $l_1, \dots, l_m$ (respectively) given above is satisfied only when the equations: $\Omega_1 = 0, \dots, \Omega_q = 0$ are already solved with respect to $n-q$ of the x. However, it is desirable, in principle and also for what follows, to reshape this definition so that it also applies to a non-resolved system of equations $\Omega_k = 0$.

There is no difficulty in doing that.

Let the parameters $l_1, \dots, l_m$ in the equations:

$$x_{q+k} = \psi_{q+k}(x_1, \dots, x_q, l_1, \dots, l_m) \qquad (k=1 \cdots n-q) \tag{2}$$

be not essential, whence one can indicate $m-h < m$ functions: $\pi_1(l), \dots, \pi_{m-h}(l)$ so that $\psi_{q+1}(x,l), \dots, \psi_n(x,l)$ can be expressed in terms of $x_1, \dots, x_q$ and of $\pi_1(l), \dots, \pi_{m-h}(l)$ alone. Evidently, there is then at least one linear partial differential equation:

$$Lf = \sum_{\mu=1}^{m} \lambda_\mu(l_1, \dots, l_m) \frac{\partial f}{\partial l_\mu} = 0$$

with the coefficients: $\lambda_1(l), \dots, \lambda_\mu(l)$ free of the x which is satisfied identically by all functions: $\pi_1(l), \dots, \pi_{m-h}(l)$, and hence also by all the functions: $x_{q+i} - \psi_{q+i}(x,l)$. Thus, we can also say (cf. Chap. 7, Theorem 15, p. 132):

When the parameters $l_1, \dots, l_m$ are not essential in (2), then the system of equations (2), interpreted as a system of equations in the variables: $x_1, \dots, x_n, l_1, \dots, l_m$, admits an infinitesimal transformation Lf in the variables $l_1, \dots, l_m$ alone.

But the converse also holds true: When the system of equations (2), interpreted as a system of equations in the variables $x_1, \dots, x_n, l_1, \dots, l_m$, admits an infinitesimal transformation Lf in the l alone, then the parameters $l_1, \dots, l_m$ are not essential in the system of equations; it is indeed immediately clear that, under the assumption made, $\psi_{q+1}(x,l), \dots, \psi_n(x,l)$ are solutions of the partial differential equation: $Lf = 0$, whence it is possible to lower the number of arbitrary parameters appearing in (2).

If we bear in mind that the system of equations (2) is only another form of the system of equations (1), then we realize without effort (cf. p. 126) that the following proposition holds:

Proposition 1. *A system of equations:*

$$\Omega_k(x_1, \dots, x_n, l_1, \dots, l_m) = 0 \qquad (k=1 \cdots n-q) \tag{1}$$

with the $l_1, \dots, l_m$ parameters which is solvable with respect to $n-q$ of the variables $x_1, \dots, x_n$ represents ∞^m different manifolds of the space $x_1, \dots, x_n$ if and only if, when it is regarded as a system of equations in the $n+m$ variables: $x_1, \dots, x_n$, $l_1, \dots, l_m$, it admits no infinitesimal transformation:

$$Lf = \sum_{\mu=1}^{m} \lambda_\mu(l_1,\dots,l_m)\frac{\partial f}{\partial l_\mu}$$

in the variables l alone.

§ 111. In the space $x_1,\dots,x_n$, let now a family of ∞^m different manifolds be determined by the $n-q$ equations: $\Omega_k(x,l)=0$, or by the equally good [GLEICHWERTIG] equations:

$$\text{(2)} \qquad x_{q+k} = \psi_{q+k}(x_1,\dots,x_n,\, l_1,\dots,l_m) \qquad (k=1\cdots n-q).$$

If this family is supposed to remain invariant by the transformation: $x_i' = f_i(x_1,\dots,x_n)$, then every manifold of the family must again be transferred by the concerned transformation into a manifold of the family. Hence, if by $l_1',\dots,l_m'$ we understand the parameters of the manifold of the family into which the manifold with the parameters $l_1,\dots,l_m$ is transferred by the transformation: $x_i' = f_i(x)$, then after the introduction of the new variables: $x_1' = f_1(x)$, ..., $x_n' = f_n(x)$, the equations (2) must receive the form:

$$x_{q+k}' = \psi_{q+k}(x_1',\dots,x_q',\, l_1',\dots,l_m') \qquad (k=1\cdots n-q),$$

where the parameters $l_1',\dots,l_m'$ depend naturally only upon the l.

But now, after the introduction of the x', the equation (2) evidently take up the form:

$$x_{q+k}' = \Psi_{q+k}(x_1',\dots,x_q',\, l_1,\dots,l_m) \qquad (k=1\cdots n-q);$$

thus, if the family (2) is supposed to remain invariant by the transformation: $x_i' = f_i(x)$, then it must be possible to satisfy the $n-q$ equations:

$$\text{(3)} \quad \psi_{q+k}(x_1',\dots,x_q',\, l_1',\dots,l_m') = \Psi_{q+k}(x_1',\dots,x_q',\, l_1,\dots,l_m) \qquad (k=1\cdots n-q)$$

independently of the values of the variables $x_1',\dots,x_q'$.

If one expands the two sides of (3), in the neighborhood of an arbitrary system of values $x_1'^0,\dots,x_n'^0$, with respect to the powers of: $x_1'-x_1'^0$, ..., $x_n'-x_n'^0$, if one compares the coefficients, and if one takes into account that $l_1',\dots,l_m'$ are essential parameters, then one realizes that $l_1',\dots,l_m'$ must be entirely determined functions of $l_1,\dots,l_m$:

$$l_\mu' = \chi_\mu(l_1,\dots,l_m) \qquad (\mu=1\cdots m).$$

Conversely, the l must also naturally be representable as functions of the l', because through the transition from the x' to the x, the family of our manifolds must also remain unchanged.

We therefore see that the equations:

$$x_i' = f_i(x_1,\dots,x_n) \quad (i=1\cdots n), \qquad l_\mu' = \chi_\mu(l_1,\dots,l_m) \quad (\mu=1\cdots m)$$

taken together represent a transformation in the $n+m$ variables: $x_1,\dots,x_n, l_1,\dots,l_m$, that leaves invariant the system of equations: $x_{q+k} = \psi_{q+k}(x,l$ in these $n+m$ variables. Consequently, we can say:

A family of ∞^m manifolds:

$$\Omega_k(x_1,\dots,x_n, l_1,\dots,l_m) = 0 \qquad (k=1\cdots n-q)$$

in the space $x_1,\dots,x_n$ admits the transformation:

$$x_i' = f_i(x_1,\dots,x_n) \qquad (i=1\cdots n)$$

if and only if it is possible to add to this transformation in the x a corresponding transformation:

$$l_\mu' = \chi_\mu(l_1,\dots,l_m) \qquad (\mu=1\cdots m)$$

in the l in such a way that the system of equations: $\Omega_k(x,l) = 0$ in the $n+m$ variables $x_1,\dots,x_n$, $l_1,\dots,l_m$ allows the transformation:

$$x_i' = f_i(x_1,\dots,x_n), \qquad l_\mu' = \chi_\mu(l_1,\dots,l_m).$$

From the above considerations, it becomes clear that the transformation: $l_\mu' = \chi_\mu(l)$, when it actually exists, is the only one of its kind; indeed, it is completely determined when the transformation $x_i' = f_i(x)$ is known. The transformation: $l_\mu' = \chi_\mu(l)$ therefore contains no arbitrary parameters.

If the family of ∞^m manifolds: $\Omega_k(x,l) = 0$ allows two different transformations:

$$x_i' = f_i(x_1,\dots,x_n); \qquad x_i'' = F_i(x_1',\dots,x_n'),$$

then obviously, it also admits the transformation:

$$x_i'' = F_i\big(f_1(x),\,\dots,\,f_n(x)\big),$$

which comes into existence by executing each of the two transformations one after the other. From this, we conclude:

The totality of all transformations $x_i' = f_i(x_1,\dots,x_n)$ which leave invariant a family of ∞^m manifolds of the space $x_1,\dots,x_n$ forms a group.

Naturally, this group need neither be finite, nor continuous; in complete generality, we can only say: its transformations are ordered as inverses by pairs. Hence in particular, when it contains only a finite number of arbitrary parameters, then it belongs to the category of groups which was discussed in Chap. 18, and according to Theorem 56, p. 328, it contains a completely determined finite continuous subgroup generated by infinitesimal transformations. Evidently, this subgroup is the largest continuous subgroup by which the family $\Omega_k(x,l) = 0$ remains invariant.

We now turn ourselves to the consideration of finite continuous groups which leave invariant the family of ∞^m manifolds:

(1) $$\Omega_k(x_1,\dots,x_n,\,l_1,\dots,l_m) \qquad {\scriptstyle (k=1\cdots n-q)};$$

however, for reasons of simplicity we first restrict ourselves to the case of a one-term group having the concerned constitution.

Let the family of ∞^m manifolds (1) admit all transformations:

(1) $$x_i' = f_i(x_1,\dots,x_n,\,\varepsilon) \qquad {\scriptstyle (i=1\cdots n)}$$

of the one-term group:

$$Xf = \sum_{i=1}^{n} \xi_{ki}(x_1,\dots,x_n)\,\frac{\partial f}{\partial x_i}.$$

The transformation in the l which, according to p. 472, corresponds to the general transformation: $x_i' = f_i(x,a)$, can be read:

(4') $$l_\mu' = \chi_\mu(l_1,\dots,l_m,\,\varepsilon) \qquad {\scriptstyle (\mu=1\cdots m)}.$$

It is easy to see that the totality of all transformations of the form:

(4'') $$\begin{cases} x_i' = f_i(x_1,\dots,x_n,\,\varepsilon) & {\scriptstyle (i=1\cdots n)} \\ l_\mu' = \chi_\mu(l_1,\dots,l_m,\,\varepsilon) & {\scriptstyle (\mu=1\cdots m)} \end{cases}$$

forms in turn a one-term group in the $n+m$ variables: $x_1,\dots,x_n,\,l_1,\dots,l_m$.

In fact, according to p. 472, the transformations (4'') leave invariant the system of equations (1); hence if one first executes the transformation (4'') and then a transformation of the same form with the parameter ε_1, then one gets a transformation:

$$\begin{aligned} x_i'' &= f_i\big(f_1(x,\varepsilon),\,\dots,\,f_n(x,\varepsilon),\,\varepsilon_1\big) && {\scriptstyle (i=1\cdots n)} \\ l_\mu'' &= \chi_\mu\big(\chi_1(l,\varepsilon),\,\dots,\,\chi_m(l,\varepsilon),\,\varepsilon_1\big) && {\scriptstyle (\mu=1\cdots m)} \end{aligned}$$

which, likewise, leaves invariant the system of equations (1). Now, the transformation:

(5) $$x_i'' = f_i\big(f_1(x,\varepsilon),\,\dots,\,f_n(x,\varepsilon),\,\varepsilon_1\big)$$

belongs to the one-term group Xf and can hence be brought to the form:

$$x_i'' = f_i(x_1,\dots,x_n,\,\varepsilon_2) \qquad {\scriptstyle (i=1\cdots n)},$$

where ε_2 depends only on ε and on ε_1. Consequently, the transformation in the l which corresponds to the transformation (5) has the form:

$$l_\mu'' = \chi_\mu(l_1,\dots,l_m,\,\varepsilon_2) \qquad {\scriptstyle (\mu=1\cdots m)}$$

and we therefore deduce:

$$\chi_\mu(\chi_1(l,\varepsilon),\ldots,\chi_m(l,\varepsilon),\varepsilon_1)=\chi_\mu(l_1,\ldots,l_m,\varepsilon_2)$$
$$(\mu=1\cdots m).$$

As a result, it is proved that the equations (4”) effectively represent a group in the $n+m$ variables: $x_1,\ldots,x_n$, $l_1,\ldots,l_m$, and to be precise, a group which possesses the same parameter group as the given group: $x_i'=f_i(x,\varepsilon)$. At the same time, it is clear that the equations (4’) taken for themselves also represent a group in the variables: $l_1,\ldots,l_m$, however not necessarily a one-term group, because it is conceivable that the parameter ε in the equations (4’) is completely missing.

The transformations of the group (4) order as inverses by pairs, hence the same visibly holds true for the transformations of the group (4’). From this, we get (cf. Chap. 9, p. 184 above) that the group (4”) also contains the identity transformation, and in addition, an infinitesimal transformation:

$$\sum_{i=1}^{n}\xi_i(x_1,\ldots,x_n)\frac{\partial f}{\partial x_i}+\sum_{\mu=1}^{m}\lambda_\mu(l_1,\ldots,l_m)\frac{\partial f}{\partial \lambda_\mu}=Xf+Lf,$$

by which it is generated.

We recapitulate the obtained result in the:

Proposition 2. *If the family of ∞^m manifolds:*

$$\Omega_k(x_1,\ldots,x_n,l_1,\ldots,l_m)=0 \qquad (k=1\cdots n-q)$$

of the space $x_1,\ldots,x_n$ admits all transformations:

$$x_i'=f_i(x_1,\ldots,x_n,\varepsilon) \qquad (i=1\cdots n)$$

of the one-term group:

$$Xf=\sum_{i=1}^{n}\xi_i(x_1,\ldots,x_n)\frac{\partial f}{\partial x_i},$$

then the corresponding transformations:

$$x_i'=f_i(x_1,\ldots,x_n,\varepsilon) \qquad (i=1\cdots n)$$
$$l_\mu'=\chi_\mu(l_1,\ldots,l_m,\varepsilon) \qquad (\mu=1\cdots m)$$

which leave invariant the system of equations: $\Omega_1(x,l)=0,\ \ldots,\ \Omega_{n-q}(x,l)=0$ in the $n+m$ variables: $x_1,\ldots,x_n$, $l_1,\ldots,l_m$ form a one-term group with an infinitesimal transformation of the form:

$$\sum_{i=1}^{n}\xi_i(x_1,\ldots,x_n)\frac{\partial f}{\partial x_i}+\sum_{\mu=1}^{m}\lambda_\mu(l_1,\ldots,l_m)\frac{\partial f}{\partial \lambda_\mu}=Xf+Lf.$$

We set up the following definition:

Definition. *A family of ∞^m manifolds:*

$$\Omega_k(x_1,\ldots,x_n,\,l_1,\ldots,l_m)=0 \qquad {\scriptstyle (k=1\cdots n-q)}$$

of the space $x_1,\ldots,x_n$ admits the infinitesimal transformation:

$$Xf=\sum_{i=1}^{n}\xi_i(x_1,\ldots,x_n)\,\frac{\partial f}{\partial x_i}$$

when there is, in $l_1,\ldots,l_m$, an infinitesimal transformation:

$$Lf=\sum_{\mu=1}^{m}\lambda_\mu(l_1,\ldots,l_m)\,\frac{\partial f}{\partial l_\mu}$$

such that the system of equations: $\Omega_1(x,l)=0,\ \ldots,\ \Omega_{n-q}(x,l)=0$ in the $n+m$ variables: $x_1,\ldots,x_n,\,l_1,\ldots,l_m$ admits the infinitesimal transformation $Xf+Lf$.

It now remains to consider the question of whether the infinitesimal transformation Lf is completely determined by the transformation Xf. One easily sees that this is the case; indeed, if the system of equations: $\Omega_k(x,l)=0$ admits the two infinitesimal transformations: $Xl+Lf$ and $Xf+\mathfrak{L}f$, then it admits at the same time the transformation:

$$Xf+Lf-(Xf+\mathfrak{L}f)=Lf-\mathfrak{L}f;$$

but since the parameters of the family: $\Omega_k(x,l)=0$ are essential, the expression: $Lf-\mathfrak{L}f$ must vanish identically, hence the transformation $\mathfrak{L}f$ cannot be distinct from the transformation Lf.

By taking as a basis the above definition, we can obviously express the content of Proposition 2 as follows:

If the family of ∞^m manifolds:

$$\Omega_k(x_1,\ldots,x_n,\,l_1,\ldots,l_m)=0 \qquad {\scriptstyle (k=1\cdots n-q)}$$

admits the one-term group Xf, then at the same time, it admits the infinitesimal transformation Xf.

Conversely: when the family of ∞^m manifolds: $\Omega_k(x,l)=0$ admits the infinitesimal transformation Xf, then it also admits the one-term group Xf.

Indeed, under the assumption made, if the system of equations: $\Omega_k(x,l)=0$ in the $n+m$ variables x, l admits an infinitesimal transformation of the form:

$$Xf+Lf=\sum_{i=1}^{n}\xi_i(x_1,\ldots,x_n)\,\frac{\partial f}{\partial x_i}+\sum_{\mu=1}^{m}\lambda_\mu(l_1,\ldots,l_m)\,\frac{\partial f}{\partial l_\mu},$$

then[1] (Translator's note: See Chap. 6, Theorem 14, p. 127, for this general fact.) it admits at the same time all transformations of the one-term group: $Xf+Lf$, and therefore, the family of ∞^m manifolds: $\Omega_k(x,l)=0$ in the space $x_1,\ldots,x_n$ allows all transformations of the one-term group Xf.

With these words, we have proved the:

Proposition 3. *The family of* ∞^m *manifolds:*

$$\Omega_k(x_1,\dots,x_n,l_1,\dots,l_m)=0 \qquad (k=1\cdots n-q)$$

in the space $x_1,\dots,x_n$ *admits the one-term group* Xf *if and only if it admits the infinitesimal transformation* Xf.

If one wants to know whether the family of ∞^m manifolds: $\Omega_k(x,l)=0$ admits a given infinitesimal transformation Xf, then one will first solve the equations $\Omega_k(x,l)=0$ with respect to $n-q$ of the x:

$$\text{(2)} \qquad x_{q+k}=\psi_{q+k}(x_1,\dots,x_q,l_1,\dots,l_m) \qquad (k=1\cdots n-q),$$

and then, one will attempt to determine m functions: $\lambda_1(l),\dots,\lambda_m(l)$ of the l so that the system of equations (2) in the $n+m$ variables x, l admits the infinitesimal transformation:

$$Xf+Lf=\sum_{i=1}^{n}\xi_i(x_1,\dots,x_n)\frac{\partial f}{\partial x_i}+\sum_{\mu=1}^{m}\lambda_\mu(l_1,\dots,l_m)\frac{\partial f}{\partial l_\mu}.$$

If one denotes the substitution: $x_{q+1}=\psi_{q+1},\dots,x_n=\psi_n$ by the symbol [ ], then one obviously obtains for $\lambda_1(l),\dots,\lambda_m(l)$ the following equations:

$$\sum_{\mu=1}^{m}\lambda_\mu(l)\frac{\partial\psi_{q+k}}{\partial l_\mu}=[\xi_{q+k}]-\sum_{j=1}^{q}[\xi_j]\frac{\partial\psi_{q+k}}{\partial x_j}$$
$$(k=1\cdots n-q),$$

which must be satisfied independently of the values of the variables $x_1,\dots,x_q$. Theoretically, there is absolutely no difficulty in deciding whether this is possible. One finds either that there is no system of functions $\lambda_\mu(l)$ which has the constitution demanded, or one finds a system of this sort, but then also only one such system, for the infinitesimal transformation Lf is indeed, when it actually exists, completely determined by Xf.

Now we will prove that the following proposition holds:

Proposition 4. *If the family of* ∞^m *manifolds:*

$$\Omega_k(x_1,\dots,x_n,l_1,\dots,l_m)=0 \qquad (k=1\cdots n-q)$$

in the space $x_1,\dots,x_n$ *admits the two infinitesimal transformations:*

$$X_1f=\sum_{i=1}^{n}\xi_{1i}(x)\frac{\partial f}{\partial x_i}, \qquad X_2f=\sum_{i=1}^{n}\xi_{2i}(x)\frac{\partial f}{\partial x_i},$$

then it admits not only every infinitesimal transformation:

$$e_1X_1f+e_2X_2f$$

which can be linearly deduced from X_1f and X_2f, but also the transformation: $[X_1, X_2]$.

Under the assumptions that are made in the proposition, there are two infinitesimal transformations L_1f and L_2f in the l alone that are constituted in such a way that the system of equations: $\Omega_k(x,l) = 0$ in the $n+m$ variables x, l allows the two infinitesimal transformations $X_1f + L_1f, X_2f + L_2f$. According to Chap. 7, Proposition 5, p. 134, the system of equations $\Omega_k(x,l) = 0$ then also admits the infinitesimal transformation: $[X_1, X_2] + [L_1, L_2]$ coming into existence by combination; but this says precisely that the family of ∞^m manifolds $\Omega_k(x,l) = 0$ also admits $[X_1, X_2]$. As a result, our proposition is proved.

The considerations that we have applied in the proof just conducted also give the following:

If the family of ∞^m manifolds $\Omega_k(x,l) = 0$ admits the two infinitesimal transformations X_1f and X_2f, and if L_1f and L_2f, respectively, are the corresponding infinitesimal transformations in $l_1, \dots, l_m$ alone, then the infinitesimal transformation $[L_1, L_2]$ corresponds to the transformation $[X_1, X_2]$.

We assume that the family of ∞^m manifolds: $\Omega_k(x,l) = 0$ in the space: $x_1, \dots, x_n$ allows all transformations of an arbitrary r-term group:

(6) $$x_i' = f_i(x_1, \dots, x_n, a_1, \dots, a_r) \qquad (i=1\cdots n).$$

Let the group in question be generated by the r independent infinitesimal transformations:

$$X_kf = \sum_{i=1}^{n} \xi_{ki}(x_1, \dots, x_n)\frac{\partial f}{\partial x_i},$$

so that between: $X_1f, \dots, X_nf$, there are relations of the known form:

$$[X_i, X_k] = \sum_{s=1}^{r} c_{iks} X_s f \qquad (i,\, k=1\cdots r).$$

If:

(6') $$l_\mu' = \chi_\mu(l_1, \dots, l_m, a_1, \dots, a_r) \qquad (\mu=1\cdots m)$$

is the transformation in the l which corresponds to the general transformation: $x_i' = f_i(x,a)$ of the group: $X_1f, \dots, X_rf$, then the equations (6) and (6') taken together represent an r-term group in the $n+m$ variables: $x_1, \dots, x_n, l_1, \dots, l_m$.

In fact, if $T_{(a_1,\dots,a_r)}$, or shortly $T_{(a)}$, and $T_{(b_1,\dots,b_r)}$, or shortly $T_{(b)}$, are two transformations of the group: $x_i' = f_i(x,a)$, then when executed one after the other, they produce in the known way a transformation: $T_{(a)}\,T_{(b)} = T_{(c)}$ of the same group, where the parameters $c_1, \dots, c_r$ depend only on the a and on the b.

On the other hand, if $S_{(a)}$ and $S_{(b)}$ are the transformations (6') which correspond, respectively, to the transformations $T_{(a)}$ and $T_{(b)}$, then one obviously obtains the transformation in the l corresponding to the transformation $T_{(a)}\,T_{(b)}$ when one executes the two transformations $S_{(a)}$ and $S_{(b)}$ one after the other, that is to say: the transformation $S_{(a)}\,S_{(b)}$ corresponds to the transformation $T_{(a)}\,T_{(b)}$. But now, we have: $T_{(a)}\,T_{(b)} = T_{(c)}$ and the transformation $S_{(c)}$ in the l corresponds to the transformation $T_{(c)}$, whence we must have: $S_{(a)}\,S_{(b)} = S_{(c)}$.

With these words, we have proved that the equations (6) and (6') really represent an r-term group in the variables: $x_1,\dots,x_n$, $l_1,\dots,l_m$ and to be precise, a group which is holoedrically isomorphic with the group: $x_i' = f_i(x,a)$; indeed, both groups visibly have one and the same parameter group (cf. Chap. 21, p. 412 sq.).

At the same time, it is proved that the equations (6') in turn also represent a group in the variables $l_1,\dots,l_m$, and in fact, a group isomorphic with the group: $x_i' = f_i(x,a)$ (cf. Chap. 21, p. 430 sq.), as it results from the symbolic relations holding simultaneously:

$$T_{(a)}\,T_{(b)} = T_{(c)}$$

and:

$$S_{(a)}\,S_{(b)} = S_{(c)}.$$

If one associates to every transformation of the group (6) the transformation of the group (6') which is determined by it, then one obtains the two groups (6) and (6') related to each other in an isomorphic way.

The isomorphism of the two groups (6) and (6') need not be holoedric, and in certain circumstances, the group (6') can even reduce to the identity transformation, namely when the group $x' = f(x,a)$ leaves individually untouched each one of the ∞^m manifolds: $\Omega_k(x,l) = 0$.

We will show that the group which is represented by the joint equations (6) and (6') is generated by r independent infinitesimal transformations.

Since the family of ∞^m manifolds: $\Omega_k(x,l)$ allows all transformations of the group: $x_i' = f_i(x,a)$, it admits in particular each one of the r one-term groups: $X_1f,\dots,X_rf$, hence according to Proposition 3, p. 476, it also admits each one of the r infinitesimal transformations: $X_1f,\dots,X_rf$. From this, it results that to every infinitesimal transformation X_kf is associated a completely determined infinitesimal transformation:

$$L_kf = \sum_{\mu=1}^{m} \lambda_{k\mu}(l_1,\dots,l_m)\,\frac{\partial f}{\partial l_\mu}$$

of such a constitution that the system of equations: $\Omega_k(x,l) = 0$ in the $n+m$ variables: $x_1,\dots,x_n$, $l_1,\dots,l_m$ admits the infinitesimal transformation: $X_kf + L_kf$.

Naturally, the system of equations: $\Omega_k(x,l) = 0$ admits at the same time every infinitesimal transformation: $e_1\,(X_1f + L_1f) + \cdots + e_r\,(X_rf + L_rf)$, and in consequence of that, also every one-term group: $e_1\,(X_1f + L_1f) + \cdots + e_r\,(X_rf + L_rf)$. But since the group: $x_i' = f_i(x,a)$ consists of the totality of all one-term groups: $e_1\,X_1f + \cdots + e_rX_rf$, then the group represented by the equations (6) and (6') must obviously be identical to the totality of all one-term groups: $\sum e_k\,X_kf + \sum e_k\,L_kf$, hence it must be generated by the r independent infinitesimal transformations: $X_kf + L_kf$, which was to be shown.

Now, from this, it follows that two arbitrary infinitesimal transformations amongst the $X_k + L_k f$ must satisfy relations of the form:

$$[X_i + L_i f,\, X_k + L_k f] = \sum_{s=1}^{r} c'_{iks}\,(X_s f + L_s f).$$

By verifying this directly, we obtain a new proof of the fact that all finite transformations: $x'_i = f_i(x,a)$, $l'_\mu = \chi_\mu(l,a)$ form a group. But at the same time, we realize that $c'_{iks} = c_{iks}$, which is therefore coherent with the fact that our new group in the x and l possesses the same parameter group as the given group: $x'_i = f_i(x,a)$.

The system of equations: $\Omega_k(x,l) = 0$ admits, simultaneously with the infinitesimal transformations: $X_1 f + L_1 f, \dots, X_r f + L_r f$, the transformations:

$$[X_i,\, X_k] + [L_i,\, L_k] = \sum_{s=1}^{r} c_{iks}\, X_s f + [L_i,\, L_k]$$

$$(i,\, k = 1 \cdots r),$$

and therefore also the following:

$$[X_i,\, X_k] + [L_i,\, L_k] - \sum_{s=1}^{r} c_{iks}\,(X_s f + L_s f) = [L_i,\, L_k] - \sum_{s=1}^{r} c_{iks}\, L_s f$$

in the variables $l_1, \dots, l_m$ alone. But because of the constitution of the system of equations: $\Omega_k(x,l) = 0$, this is possible only when the infinitesimal transformations just written vanish identically, hence when the relations:

$$[L_i,\, L_k] = \sum_{s=1}^{r} c_{iks}\, L_s f$$

hold true. From this, it follows immediately that:

$$[L_i,\, L_k] + [X_i,\, X_k] = \sum_{s=1}^{r} c_{iks}\,(X_s f + L_s f),$$

whence the said property of the joint equations (6) and (6') is proved.

According to the above, it goes without saying that the group (6') is generated by the r infinitesimal transformations: $L_1 f, \dots, L_r f$. At the same time, in the preceding, we have a new proof of the fact that the equations (6') represent a group in the variables $l_1, \dots, l_m$ which is isomorphic with the group: $X_1 f, \dots, X_r f$.

At present, the isomorphic relationship between the two groups (6) and (6') mentioned on p. 478 can also be defined by saying that to every infinitesimal transformation: $e_1 X_1 f + \cdots + e_r X_r f$ is associated the infinitesimal transformation: $e_1 L_1 f + \cdots + e_r L_r f$ determined by it in the l.

We therefore have the:

Proposition 5. *If the family of* ∞^m *manifolds:*

$$\Omega_k(x_1,\ldots,x_n,l_1,\ldots,l_m)=0 \qquad (k=1\cdots n-q)$$

in the space: $x_1,\ldots,x_n$ admits the r independent infinitesimal transformations:

$$X_kf=\sum_{i=1}^{n}\xi_{ki}(x_1,\ldots,x_n)\frac{\partial f}{\partial x_i} \qquad (k=1\cdots r)$$

of an r-term group having the composition:

$$[X_i,\,X_k]=\sum_{s=1}^{r}c_{iks}X_sf \qquad (i,\,k=1\cdots r),$$

then to every X_kf there corresponds a completely determined infinitesimal transformation:

$$L_kf=\sum_{\mu=1}^{m}\lambda_{k\mu}(l_1,\ldots,l_m)\frac{\partial f}{\partial l_\mu}$$

of such a constitution that the system of equations: $\Omega_k(x,l)=0$ in the $n+m$ variables: $x_1,\ldots,x_n$, $l_1,\ldots,l_m$ admits the r infinitesimal transformations: X_kf+L_kf, so the r infinitesimal transformations L_kf stand pairwise in the relationships:

$$[L_i,\,L_k]=\sum_{s=1}^{r}c_{iks}L_sf \qquad (i,\,k=1\cdots r),$$

that is to say, they generate a group isomorphic with the group: $X_1f,\ldots,X_rf$.

Here, the following proposition may be expressly stated and proved independently:

Proposition 6. *If the r-term group: $X_1f,\ldots,X_rf$ of the space $x_1,\ldots,x_n$ contains exactly $p\leqslant r$ independent infinitesimal transformations which leave invariant the ∞^m manifolds:*

$$\Omega_k(x_1,\ldots,x_n,l_1,\ldots,l_m)=0 \qquad (k=1\cdots n-q),$$

then these p independent infinitesimal transformations generate a p-term subgroup of the group: $X_1f,\ldots,X_rf$.

The proof is very simple. Let:

$$\Xi_\pi f=\sum_{j=1}^{r}g_{\pi j}X_jf \qquad (\pi=1\cdots p)$$

be such independent infinitesimal transformations of the group: $X_1f,\ldots,X_rf$ which leave invariant the family: $\Omega_k(x,l)=0$, so that every other infinitesimal transformation: $e_1X_1f+\cdots+e_rX_rf$ which does the same can be linearly deduced from $\Xi_1f,\ldots,\Xi_pf$. Then according to Proposition 4, p. 476, the family: $\Omega_k(x,l)=0$ also admits every infinitesimal transformation: $[\Xi_\mu,\,\Xi_\nu]$ of the group: $X_1f,\ldots,X_rf$, and consequently, there are relations of the form:

$$[\Xi_\mu, \Xi_\nu] = \sum_{\pi=1}^{p} g_{\mu\nu\pi} \Xi_\pi f \qquad (\mu, \nu = 1 \cdots p)$$

in which the $g_{\mu\nu\pi}$ denote constants. From this, it follows that $\Xi_1 f, \dots, \Xi_p f$ effectively generate a p-term group. —

If the family of ∞^m manifolds:

$$\Omega_k(x_1, \dots, x_n, l_1, \dots, l_m) = 0 \qquad (k=1 \cdots n-q)$$

is presented, and if in addition an arbitrary r-term group: $X_1 f, \dots, X_r f$ of the space $x_1, \dots, x_n$ is also presented, one can ask how many independent infinitesimal transformations of the group: $X_1 f, \dots, X_r f$ the presented family admits. We now summarize how this question can be answered.

Indeed, if the family: $\Omega_k(x, l) = 0$ is supposed to admit an infinitesimal transformation of the form: $e_1 X_1 f + \cdots + e_r X_r f$, then it must be possible to indicate an infinitesimal transformation: Lf in the l alone such that the system of equations: $\Omega_k(x, l) = 0$ in the $n+m$ variables: $x_1, \dots, x_n, l_1, \dots, l_m$ admits the infinitesimal transformation: $\sum e_k X_k f + Lf$. Hence, if we imagine that the equations: $\Omega_k(x, l) = 0$ are resolved with respect to $n-q$ of the x:

$$x_{q+k} = \psi_{q+k}(x_1, \dots, x_q, l_1, \dots, l_m) \qquad (k=1 \cdots n-q),$$

and if, as on p. 476, we denote the substitution: $x_{q+1} = \psi_{q+1}, \dots, x_n = \psi_n$ by the symbol [ ], then we only have to determine the constants $e_1, \dots, e_r$ and the functions: $\lambda_1(l), \dots, \lambda_m(l)$ in the most general way so that the $n-q$ equations:

$$\sum_{\mu=1}^{m} \lambda_\mu(l) \frac{\partial \psi_{q+k}}{\partial l_\mu} = \sum_{\sigma=1}^{r} e_\sigma \left\{ [\xi_{\sigma, q+k}] - \sum_{j=1}^{q} [\xi_{\sigma j}] \frac{\partial \psi_{q+k}}{\partial x_j} \right\}$$

$$(k=1 \cdots n-q)$$

are satisfied identically, independently of the values of the variables $x_1, \dots, x_q$. In this way, we find the most general infinitesimal transformation: $e_1 X_1 f + \cdots + e_r X_r f$ which leaves invariant the family: $\Omega_k(x, l) = 0$.

§ 112. When the family of ∞^m manifolds:

(1) $$\Omega_k(x_1, \dots, x_n, l_1, \dots, l_m) = 0 \qquad (k=1 \cdots n-q)$$

admits the transformation: $x_i' = f_i(x_1, \dots, x_n)$, then there is, as we have seen on p. 471 sq., a completely determined transformation: $l_\mu' = \chi_\mu(l_1, \dots, l_m)$ of such a constitution that the system of equations: $\Omega_k(x, l) = 0$ in the $n+m$ variables x, l admits the transformation:

$$\begin{cases} x_i' = f_i(x_1, \dots, x_n) & (i=1 \cdots n) \\ l_\mu' = \chi_\mu(l_1, \dots, l_m) & (\mu=1 \cdots m). \end{cases}$$

Already on p. 471, we observed that the equations: $l'_\mu = \chi_\mu(l)$ determine the parameters of the manifold of the invariant family (1) into which the manifold with the parameters $l_1, \dots, l_m$ is transferred by the transformation: $x'_i = f_i(x)$. Hence, when we interpret $l_1, \dots, l_m$ virtually as coordinates of individual manifolds of the family (1), the equations: $l'_\mu = \chi_\mu(l)$ indicate the law according to which the manifolds of our invariant family are permuted with each other by the transformation: $x'_i = f_i(x)$.

For instance, if a special system of values: $l_1^0, \dots, l_m^0$ admits the transformation:

$$l'_\mu = \chi_\mu(l_1, \dots, l_m) \qquad (\mu = 1 \cdots m),$$

then at the same time, the manifold:

$$\Omega_k(x_1, \dots, x_n, l_1^0, \dots, l_m^0) = 0 \qquad (k = 1 \cdots n-q)$$

admits the transformation:

$$x'_i = f_i(x_1, \dots, x_n) \qquad (i = 1 \cdots n).$$

In fact, the manifold: $\Omega_k(x, l^0) = 0$ is transferred, by the execution of the transformation: $x'_i = f_i(x)$, to the new manifold:

$$\Omega_k\big(x'_1, \dots, x'_n, \chi_1(l^0), \dots, \chi_m(l^0)\big) = 0 \qquad (k = 1 \cdots n-q);$$

but under the assumption made, we have:

$$\chi_\mu(l_1^0, \dots, l_m^0) = l_\mu^0 \qquad (\mu = 1 \cdots m),$$

hence the new manifold coincides with the old one and the manifold: $\Omega_k(x, l^0) = 0$ really remains invariant.

However, by contrast, the converse does not always hold true. If an arbitrary special manifold:

$$\Omega_k(x_1, \dots, x_n, l_1^0, \dots, l_m^0) = 0 \qquad (k = 1 \cdots n-q)$$

of the family (1) admits the transformation: $x'_i = f_i(x)$, then it does not necessarily follow that the system of values: $l_1^0, \dots, l_m^0$ admits the transformation: $l'_\mu = \chi_\mu(l)$. Indeed, it is conceivable that, to the system of values: $l_1^0, \dots, l_m^0$, there is associated a continuous series of systems of values: $l_1, \dots, l_m$ which provide the same manifold as the system of values: $l_1^0, \dots, l_m^0$, when inserted in (1); in this case, from the invariance of the manifold: $\Omega_k(x, l^0) = 0$, it only follows that the individual systems of values: $l_1, \dots, l_m$ of the series just defined are permuted with each other by the transformation: $l'_\mu = \chi_\mu(l)$, but not that the system of values: $l_1^0, \dots, l_m^0$ remains invariant by the transformation in question. *Nevertheless, if the l_k do not have special values but general values, then the manifold: $\Omega_k(x, l) = 0$ admits the transformation: $x'_i = f_i(x, a)$ only when the system of values l_k allows the corresponding transformation: $l'_\mu = \chi_\mu(l)$, and also, always in this case too.*

Now, we consider the general case where the family of ∞^m manifolds:

(1) $$\Omega_k(x_1,\dots,x_n,\,l_1,\dots,l_m)=0 \qquad {\scriptstyle (k=1\cdots n-q)}$$

admits the r-term group:

$$x_i'=f_i(x_1,\dots,x_n,\,a_1,\dots,a_r) \qquad {\scriptstyle (i=1\cdots n)}$$

with the r independent infinitesimal transformations: $X_1f,\dots,X_rf$. Let the group in the l which corresponds to the group: $x_i'=f_i(x,a)$ have the form:

$$l_\mu'=\chi_\mu(l_1,\dots,l_m,\,a_1,\dots,a_r) \qquad {\scriptstyle (\mu=1\cdots m)},$$

and let it be generated by the r infinitesimal transformations: $L_1f,\dots,L_rf$ which in turn correspond naturally to the infinitesimal transformations: $X_1f,\dots,X_rf$, respectively.

At first, the question is how one decides whether the infinitesimal transformation: $e_1^0X_1f+\cdots+e_r^0X_rf$ leaves invariant a determined manifold contained in the family (1):

(7) $$\Omega_k(x_1,\dots,x_n,\,l_1^0,\dots,l_m^0)=0 \qquad {\scriptstyle (k=1\cdots n-q)}.$$

It is easy to see that the manifold (7) admits in any case the infinitesimal transformation: $e_1^0X_1f+\cdots+e_r^0X_rf$ when the system of values: $l_1^0,\dots,l_m^0$ admits the infinitesimal transformation: $e_1^0L_1f+\cdots+e_r^0L_rf$.

In fact, the system of equations (1) in the $m+n$ variables: $x_1,\dots,x_n,\,l_1,\dots,l_m$ admits, under the assumptions made, the infinitesimal transformation: $\sum e_j^0\,(X_jf+L_jf)$; the $n-q$ expressions:

$$\sum_{j=1}^r e_j^0\,(X_j\Omega_k+L_j\Omega_k)=\sum_{j=1}^r e_j^0\bigg\{\sum_{i=1}^n \xi_{ji}(x)\,\frac{\partial\Omega_k}{\partial x_i}+\sum_{\mu=1}^m \lambda_{j\mu}(l)\,\frac{\partial\Omega_k}{\partial l_\mu}\bigg\}$$

therefore all vanish by virtue of: $\Omega_1(x,l)=0,\ \dots,\ \Omega_{n-q}(x,l)=0$. This also holds true when we set: $l_1=l_1^0,\ \dots,\ l_m=l_m^0$; but now, the system of values $l_1^0,\dots,l_m^0$ admits the infinitesimal transformation: $\sum e_j^0L_jf$ and hence, the m expressions:

$$e_1^0\,\lambda_{1\mu}(l^0)+\cdots+e_r^0\,\lambda_{r\mu}(l^0) \qquad {\scriptstyle (\mu=1\cdots m)}$$

all vanish. Consequently, the $n-q$ expressions:

$$\sum_{i=1}^n\bigg\{\sum_{j=1}^r e_j^0\,\xi_{ji}(x)\bigg\}\frac{\partial\Omega_k(x,l^0)}{\partial x_i} \qquad {\scriptstyle (k=1\cdots n-q)}$$

all vanish by virtue of (7), that is to say, the manifold (7) really admits the infinitesimal transformation: $\sum e_j^0X_jf$.

Nevertheless, the sufficient criterion found above is not necessary.

Indeed, if our family of manifolds $\Omega_k(x,l) = 0$ remains invariant by the group: $X_1 f, \dots, X_r f$, and if at the same time, the special manifold: $\Omega_k(x,l^0) = 0$ is supposed to admit the infinitesimal transformation: $e_1^0 X_1 f + \cdots + e_r^0 X_r f$, then for that, it is only necessary that the $n-q$ expressions:

$$\sum_{\mu=1}^{m} \sum_{j=1}^{r} e_j^0 \lambda_{j\mu}(l^0) \frac{\partial \Omega_k(x,l^0)}{\partial l_\mu}$$

be equal to zero by virtue of the system of equations: $\Omega_k(x,l^0) = 0$; but it is not at all necessary that the r expressions: $\sum e_j^0 \lambda_{j\mu}(l^0)$ themselves vanish.

Thus, if one wants to know whether every manifold: $\Omega_k(x,l) = 0$ of the invariant family admits one or several infinitesimal transformations: $\sum e_k X_k f$, and if in addition, one wants for every manifold l_k to find the concerned infinitesimal transformation, then one has to determine the e_k in the most general way as functions of the l so that the equations:

$$\sum_{\mu=1}^{m} \sum_{j=1}^{r} e_j \lambda_{j\mu}(l) \frac{\partial \Omega(x,l)}{\partial l_\mu} = 0 \tag{8}$$

are a consequence of the system of equations: $\Omega_k(x,l) = 0$. But since, according to an assumption made earlier on, the l are essential parameters, the system of equations: $\Omega_k(x,l) = 0$ in the $n+m$ variables x, l can admit no non-identically vanishing transformation:

$$\sum_{\mu=1}^{m} \sum_{j=1}^{r} e_j(l) \lambda_{j\mu}(l) \frac{\partial f}{\partial l_\mu},$$

whence it necessarily follows that:

$$e_1 \lambda_{1\mu}(l) + \cdots + e_r \lambda_{r\mu}(l) = 0 \qquad (\mu = 1, 2 \cdots m).$$

We therefore get the following:

If: $\sum e_k^0 L_k f$ is the most general infinitesimal transformation contained in the group: $L_1 f, \dots, L_r f$ which leaves invariant the system of values: $l_1^0, \dots, l_m^0$ located in general position, then: $\sum e_k^0 X_k f$ is the most general infinitesimal transformation contained in the group: $X_1 f, \dots, X_r f$ which leaves invariant the manifold (7) *located in general position.*

We can assume that amongst the r infinitesimal transformations: $L_1 f, \dots, L_r f$, exactly $m-p$, say: $L_1 f, \dots, L_{m-p} f$, are linked together by no linear relation of the form:

$$\alpha_1(l_1, \dots, l_m) L_1 f + \cdots + \alpha_{m-p}(l_1, \dots, l_m) L_{m-p} f = 0,$$

while by contrast: $L_{m-p+1} f, \dots, L_r f$ can be expressed linearly in terms of $L_1 f, \dots, L_{m-p} f$:

$$L_{m-p+j}f \equiv \sum_{\mu=1}^{m-p} \alpha_{j\mu}(l_1,\dots,l_m)\,L_\mu f \qquad (j=1\cdots r-m+p).$$

We can always insure that this assumption holds, even when the infinitesimal transformations: $L_1f,\dots,L_rf$ are not mutually independent, which can very well occur.

It is now clear that the $r-m+p$ infinitesimal transformations:

$$\text{(9)}\qquad L_{m-p+j}f - \sum_{\mu=1}^{m-p} \alpha_{j\mu}(l_1^0,\dots,l_m^0)\,L_\mu f \qquad (j=1\cdots r-m+p)$$

leave invariant the system of values: $l_1^0,\dots,l_m^0$; moreover, because $l_1^0,\dots,l_m^0$ is a system of values in general position, one realizes immediately that every infinitesimal transformation: $\sum e_k L_k f$ by which the system of values remains invariant can be linearly deduced from the $r-m+p$ transformations (9). Consequently, the most general infinitesimal transformation: $\sum e_k X_k f$ which leaves invariant the manifold (7) located in general position can be linearly deduced from the $r-m+p$ transformations:

$$\text{(10)}\qquad X_{m-p+j}f - \sum_{\mu=1}^{m-p} \alpha_{j\mu}(l_1^0,\dots,l_m^0)\,X_\mu f \qquad (j=1\cdots r-m+p).$$

Of course, the infinitesimal transformations (10) are mutually independent and they generate an $(r-m+p)$-term group, namely the most general subgroup contained in the group: $X_1f,\dots,X_rf$ by which the manifold (7) remains invariant.

In particular, if the group: $L_1f,\dots,L_rf$ is transitive, then the whole number p has the value zero; so in this case, every manifold of the family (1) located in general position admits exactly $r-m$ independent infinitesimal transformations of the group: $X_1f,\dots,X_rf$.

We therefore have the

Proposition 7. *If the family of ∞^m manifolds:*

$$\Omega_k(x_1,\dots,x_n,\,l_1,\dots,l_m)=0 \qquad (k=1\cdots n-q)$$

admits the r-term group: $X_1f,\dots,X_rf$, if moreover $L_1f,\dots,L_rf$ are the infinitesimal transformations in the l alone which correspond to the X_jf and if lastly $L_1f,\dots,L_rf$ are linked together by exactly $r-m+p$ independent relations of the form: $\sum \beta_j(l)\,L_jf=0$, then the group: $X_1f,\dots,X_rf$ contains exactly $r-m+p$ independent infinitesimal transformations which leave invariant a manifold $l_1^0,\dots,l_m^0$ in general position. These transformations generate an $(r-m+p)$-term group. If the group: $L_1f,\dots,L_rf$ in the m variables $l_1,\dots,l_m$ is transitive, then every manifold of the family $\Omega_k(x,l)=0$ located in general position admits exactly $r-m$ independent infinitesimal transformations of the group: $X_1f,\dots,X_rf$ and these infinitesimal transformations generate an $(r-m)$-term subgroup.

We add here the obvious remark that the group: $L_1 f, \dots, L_r f$ in $l_1, \dots, l_m$ is transitive if and only if every manifold of the family $\Omega_k(x,l) = 0$ located in general position can be transferred, by means of at least one transformation of the group: $X_1 f, \dots, X_r f$, to every other manifold.

§ 113. In the n-times extended space $x_1, \dots, x_n$, let an r-term group: $X_1 f, \dots, X_r f$ be presented, and in addition, let an arbitrary manifold be given, which we denote by M. We assume that no arbitrary parameters of any kind appear in the equations of M.

If all ∞^r transformations of the group: $X_1 f, \dots, X_r f$ are executed on the manifold M, then this manifold is transferred to a series of new manifolds. We will prove that the totality of all these manifolds remains invariant by the group: $X_1 f, \dots, X_r f$, and that it forms a family invariant by the group.

Let M' be an arbitrary manifold which belongs to the totality just said, and let T_1 be a transformation of the group: $X_1 f, \dots, X_r f$ which transfers M to M', whence there is the symbolic equation:

$$(M)\,T_1 = (M').$$

Now, if T is an arbitrary transformation of the group, we have:

$$(M')\,T = (M')\,T_1\,T = (M)\,T_2,$$

where the transformation T_2 again belongs to the group; consequently, the manifold M' is transferred by the transformation T to another manifold of the totality in question. Since this holds for every manifold M' of the totality, we see that the manifolds belonging to the totality are permuted with each other by T, and hence by all transformations of the group: $X_1 f, \dots, X_r f$, hence we see that the totality of manifolds defined above effectively forms a family invariant by the group: $X_1 f, \dots, X_r f$.

It is easy to see that every manifold of this invariant family can be transferred, by means of at least one transformation of the group, to every other manifold of the family. Indeed, if:

$$(M') = (M)\,T_1, \qquad (M'') = (M)\,T_2,$$

then $(M) = (M')\,T_1^{-1}$, whence:

$$(M'') = (M')\,T_1^{-1}\,T_2,$$

whence the claim is proved. At the same time, it results from this that one also obtains the discussed family of manifolds when one executes all ∞^r transformations of the group on an arbitrary manifold of the family.

We therefore have the

Proposition 8. *If one executes all ∞^r transformations of an r-term group: $X_1 f, \dots, X_r f$ of R_n on a given manifold of this space, then the totality of all* *positions that the manifold takes on the occasion forms a family of manifolds invariant by the group. Every manifold of this family can be transferred, by means of at least one transformation of the group, to any other manifold of the family. The family can be derived from each one of its manifolds in the same way as it is deduced from the initially given manifold.*

The manifold M will admit a certain number, that we assume to be exactly equal to $r-m$, of infinitesimal transformations of the group: $X_1 f, \dots, X_r f$.[1] These transformations then generate an $(r-m)$-term subgroup (cf. Theorem 31, p. 219).

Now, let S be the general symbol of a transformation of this subgroup, so: $(M)\,S = (M)$; moreover, let T_1 be an arbitrary transformation of the group: $X_1 f, \dots, X_r f$ and let M be transferred to the new position M' by T_1: $(M') = (M)\,T_1$. Then it is easy to indicate *all* transformations T of the group $X_k f$ which transfer M to M'.

Indeed, one has:

$$(M)\,\mathsf{T} = (M') = (M)\,T_1,$$

whence:

$$(M)\,\mathsf{T}\,T_1^{-1} = (M),$$

hence $\mathsf{T}\,T_1^{-1}$ is a transformation S and $S\,T_1$ is the general symbol of all transformations of the group $X_k f$ which transfer M to M'. But there are as many transformations $S\,T_1$ as there are different transformations S, that is to say, ∞^{r-m}.

One finds in a similar way all transformations $\mathfrak{S}$ of our group which leave M' invariant. From $(M')\,\mathfrak{S} = (M')$ and $(M)\,T_1 = (M')$, one obtains:

$$(M)\,T_1\,\mathfrak{S}\,T_1^{-1} = (M),$$

whence:

$$T_1\,\mathfrak{S}\,T_1^{-1} = S, \qquad \mathfrak{S} = T_1^{-1}\,S\,T_1.$$

Likewise, there are ∞^{r-m} such different transformations and their totality forms an $(r-m)$-term subgroup which is conjugate to the group S inside the group $X_k f$.

We summarize this result as follows:

Proposition 9. *If a manifold M of R_n admits exactly $r-m$ independent infinitesimal transformations of the r-term group: $X_1 f, \dots, X_r f$, or briefly G_r, if moreover S is the general symbol of the ∞^{r-m} finite transformations of the $(r-m)$-term subgroup which is generated by these $r-m$ infinitesimal transformations, and lastly, if T is an arbitrary transformation of G_r: $X_1 f, \dots, X_r f$ and if M takes the new position M' after the execution of T, then G_r contains exactly ∞^{r-m} different transformations which likewise transfer M to M' and the general symbol of these transformations*

---

[1] The totality of all *finite* transformations of the group: $X_1 f, \dots, X_r f$ which leave invariant a manifold M always forms a subgroup (Theorem 32, p. 220), but of course, this group need not be a finite continuous group. Nevertheless, in the following developments of the text, when we make the implicit assumption that this subgroup is generated by infinitesimal transformations, it is not to be interpreted as an essential restriction, because we can indeed suitably narrow down the region $((a))$ introduced on p. 26.

is: ST; in addition, G_r contains exactly ∞^{r-m} transformations which leave M' invariant, these transformations have $T^{-1}ST$ for a general symbol and they form an $(r-m)$-term subgroup which is conjugate to the group of the S inside G_r.

If we imagine that the ∞^r transformations of the group: $x_i' = f_i(x,a)$ are executed on the equations of the manifold M, then we obtain the analytic expression of the discussed invariant family of manifolds. Formally, this expression contains the r parameters $a_1,\dots,a_r$, but these parameters need not all be essential. We now want to determine the number m' of the essential parameters amongst the parameters $a_1,\dots,a_r$.

Our invariant family consists of $\infty^{m'}$ different manifolds and each one of these manifolds can be transferred to every other manifold by means of a transformation of the group: $X_1f,\dots,X_rf$. According to Proposition 7, p. 485, each individual manifold amongst the $\infty^{m'}$ manifolds then admits exactly $r-m'$ independent infinitesimal transformations of G_r; but from the preceding, we know that under the assumptions made, each one of these manifolds admits exactly ∞^{r-m} finite transformations of G_r, and consequently, we have $m' = m$ and amongst the r parameters: $a_1,\dots,a_r$, there are exactly m that are essential.

We therefore have the

Proposition 10. *If a manifold of the n-times extended space R_n admits exactly $r-m$ independent infinitesimal transformations of an r-term group: $X_1f,\dots,X_rf$ of this space, then this manifold takes exactly ∞^m different positions by the ∞^r transformations of this group.*

§ 114. In the equations of our invariant family of manifolds, as said, the parameters need not all be essential, but we can always imagine that $m \leqslant r$ functions $l_1,\dots,l_m$ of the a are introduced as new parameters so that the equations of our family are given the form:

$$\Omega_k(x_1,\dots,x_n,\, l_1,\dots,l_m) = 0 \qquad (k=1\cdots n-q),$$

where now $l_1,\dots,l_m$ are essential parameters.

The family $\Omega_k = 0$ remains invariant by the group: $X_1f,\dots,X_rf$ while its individual manifolds are permuted with each other. The way the manifolds are permuted is indicated by the group: $L_1f,\dots,L_rf$ in the l alone which, as shown earlier on, is completely determined by the group: $X_1f,\dots,X_rf$.

The group L_kf in the variables $l_1,\dots,l_m$ is isomorphic with the group X_kf, hence it has at most r essential parameters; on the other hand it is certainly, under the assumption made, transitive (cf. p. 485 sq.), hence it has at least m essential parameters. For us, it only remains to indicate a simple criterion for determining how many essential parameters the group L_kf really has.

Let the group: $L_1f,\dots,L_rf$ be ρ-term, where $0 \leqslant \rho \leqslant r$. Because it is meroedrically isomorphic with G_r: $X_1f,\dots,X_rf$, then there must exist in G_r an $(r-\rho)$-term invariant subgroup which corresponds to the identity transformation in the group

$L_k f$ (cf. Theorem 54, p. 312). This $(r-\rho)$-term invariant subgroup of G_r then leaves individually fixed each one of the ∞^m manifolds: $\Omega_k(x,l)=0$, hence it is contained in the $(r-m)$-term subgroup g_{r-m} of G_r which leaves invariant the manifold M discussed above and at the same time, it is contained in all $(r-m)$-term subgroups contained in G_r which are conjugate, inside G_r, to the g_{r-m} just mentioned.

Conversely, if this g_{r-m} contains an $(r-\rho)$-term subgroup which is invariant in G_r, then this subgroup is at the same time contained in all subgroups of G_r that are conjugate to g_{r-m}, hence it leaves untouched every individual manifold: $\Omega_k(x,l) = 0$, and to it, there corresponds the identity transformation in the group: $L_1 f,\dots,L_r f$.

In order to decide how many parameters the group: $L_1 f,\dots,L_r f$ contains, we therefore only have to find the largest group contained in g_{r-m} which is invariant in G_r. When the group in question is exactly $(r-\rho)$-term ($\rho \geqslant m$), then the group: $L_1 f,\dots,L_r f$ is exactly ρ-term.

We do not want to state this result as a specific proposition, but instead, we want to recapitulate all the results of the previous two sections in one theorem.

Theorem 85. *If one has an r-term group: $X_1 f,\dots,X_r f$, or briefly G_r, of the space $x_1,\dots,x_n$ and if one has an arbitrary manifold M which allows exactly $r-m$ independent infinitesimal transformations of G_r and hence which also admits the $(r-m)$-term subgroup g_{r-m} generated by these infinitesimal transformations, then through the ∞^r transformations of G_r, M takes in total ∞^m different positions the totality of which remains invariant relative to the group G_r. If one marks the individual positions of M by means of m parameters: $l_1,\dots,l_m$, then one obtains a certain group in $l_1,\dots,l_m$:*

$$l'_\mu = \chi_\mu(l_1,\dots,l_m;\, a_1,\dots,a_r) \qquad (\mu=1\cdots m)$$

which indicates in which way the individual positions of M are permuted with each other by the group: $X_1 f,\dots,X_r f$. This group in the l is transitive and isomorphic with the group: $X_1 f,\dots,X_r f$. If the largest subgroup contained in g_{r-m} which is invariant in G_r is exactly $(r-\rho)$-term, then the group: $l'_\mu = \chi_\mu(l,a)$ has exactly ρ essential parameters. In particular, if G_r is simple, then the group in the l is always r-term and holoedrically isomorphic to G_r, with the only exception being the case $m=0$, for which the group: $l'_\mu = \chi_\mu(l,a)$ consists only of the identity transformation.[2]

The number $r-\rho$ mentioned in the theorem may have each one of the values: 0, 1, ..., $r-m$; if $r-\rho=0$, then the $(r-\rho)$-term subgroup of g_{r-m} consists of the identity transformation, hence the group: $l'_\mu = \chi_\mu(l,a)$ is holoedrically isomorphic to G_r; if $r-\rho=r-m$, then g_{r-m} itself is invariant in G_r and the group: $l'_\mu = \chi_\mu(l,a)$ is only m-term.

§ 115. In the last but one section we gave a method to find the families of manifolds which remain invariant by a given r-term group: $X_1 f,\dots,X_r f$. The invariant

[2] LIE, Archiv for Mathematik og Naturv., Vol. 10, Christiania 1885.

families that we obtained in this way were distinguished by the fact that every manifold of a family of this sort could be transferred, by means of at least one transformation of the group: $X_1 f, \dots, X_r f$, to every other manifold of the family.

Now we want to generalize the discussed method so that it produces *all* families of manifolds invariant by the group $X_1 f, \dots, X_r f$.

For this, we are led to the obvious observation that the considerations of pp. 486–488 also remain valid when the equations of the manifold M contain arbitrary parameters, that is to say: when in place of an individual manifold M we use directly a complete family of manifolds. Thus, we can also proceed in the following way in order to find invariant families of manifolds:

We take an arbitrary family:

$$(11) \qquad V_k(x_1, \dots, x_n, u_1, \dots, u_h) = 0 \qquad (k=1 \cdots n-q)$$

of ∞^h manifolds and we execute on it all ∞^r transformations of the group: $X_1 f, \dots, X_r f$; the totality of all manifolds that we obtain in this way always forms a family invariant by the group: $X_1 f, \dots, X_r f$.

It is clear that we obtain all families invariant by the group: $X_1 f, \dots, X_r f$ when we choose the family (11) *in all possible ways*. Indeed, if an arbitrary family invariant by the group is presented, say the family:

$$(1) \qquad \Omega_k(x_1, \dots, x_n, l_1, \dots, l_m) = 0 \qquad (k=1 \cdots n-q),$$

then this family can in any case be obtained when we just choose, as the family (11), the family (1). Besides, one can easily indicate infinitely many other families (11) out of which precisely the family (1) is obtained, but we do not want to spend time on this.

It remains to answer a question.

If the family (11) is presented and if all the transformations of the group: $X_1 f, \dots, X_r f$ are executed on it, then the equations of the invariant family which comes into existence in this way obviously have the form:

$$(12) \qquad W_k(x_1, \dots, x_n, u_1, \dots, u_h, a_1, \dots, a_r) = 0 \qquad (k=1 \cdots n-q),$$

and therefore, they formally contain $h+r$ arbitrary parameters, namely: $u_1, \dots, u_h$, $a_1, \dots, a_r$. How many parameters amongst these parameters are essential?

Every generally positioned manifold of the family (11) takes by the group: $X_1 f, \dots, X_r f$ a certain number, say ∞^p, of different positions; of these ∞^p positions, there is a certain number, say ∞^o, of different positions, which again belong to the family (11). In this manner, the complete family (11) is decomposed into ∞^{h-o} different subfamilies of ∞^o manifolds in such a way that every manifold of the family (11) can always be transferred to every manifold which belongs to one and the same subfamily by means of at least one transformation of the group: $X_1 f, \dots, X_r f$, and such that every manifold of the family (11), as soon as it remains inside the family (11) by a transformation of the group: $X_1 f, \dots, X_r f$, remains at the same time in the subfamily to which it belongs.

Now, if we imagine that all transformations of the group: $X_1 f, \dots, X_r f$ are executed on two arbitrary manifolds which belong to the *same* subfamily, we obviously obtain in the two cases the same family of ∞^p manifolds; on the other hand, if we imagine that all transformations of the group are executed on two manifolds of the family (11) which belong to *distinct* subfamilies, we obtain two distinct families of ∞^p manifolds which have absolutely no manifold in common.

Hence, when we choose on each one of the ∞^{h-o} subfamilies of the family (11) a manifold and when we execute all transformations of the group: $X_1 f, \dots, X_r f$ on the so obtained ∞^{h-o} manifolds, we receive ∞^{h-o} distinct families of ∞^p manifolds, in total ∞^{h-o+p} different manifolds. At the same time, it is clear that in this way, we obtain exactly the same family as when we execute all transformations of our group on the manifolds (11) themselves.

As a result, it is proved that the family (12) consists of ∞^{h-o+p} different manifolds, hence that amongst the $h+r$ parameters of the equations (12), exactly $h-o+p$ are essential.

We want to indicate how one has to proceed in order to determine the numbers p and o mentioned above.

The number $r-p$ is evidently the number of the infinitesimal transformations: $e_1 X_1 f + \cdots + e_r X_r f$ which leave invariant a generally positioned manifold of the family (11). Now, the most general infinitesimal transformation: $\sum e_j X_j f$ which leaves invariant the manifold:

$$V_k(x_1, \dots, x_n, u_1, \dots, u_h) = 0 \qquad (k=1 \cdots n-q) \tag{13}$$

necessarily has the form:

$$\sum_{j=1}^{r} e_j(u_1, \dots, u_h) X_j f, \tag{14}$$

where the $e_j(u_1, \dots, u_h)$ are functions of the u. So we need only to determine the functions $e_j(u)$ in the most general way so that the system of equations (13) in the variables: $x_1, \dots, x_n$ admits the infinitesimal transformation (14).

If we imagine that the equations (13) are resolved with respect to $n-q$ of the x:

$$x_{q+k} = \omega_{q+k}(x_1, \dots, x_q, u_1, \dots, u_h) \qquad (k=1 \cdots n-q),$$

and if we denote the substitution: $x_{q+1} = \omega_{q+1}, \dots, x_n = \omega_n$ by the symbol [ ], we visibly obtain for the functions $e_j(u)$ the equations:

$$\sum_{j=1}^{r} e_j(u_1, \dots, u_h) \left\{ [\xi_{j,q+k}] - \sum_{\pi=1}^{q} [\xi_{j\pi}] \frac{\partial \omega_{q+k}}{\partial x_\pi} \right\} = 0 \tag{15}$$

$$(k=1 \cdots n-q)$$

which must be satisfied independently of the values of $x_1, \dots, x_n$.

If we have determined from these equations the $e_j(u)$ in the most general way, we know the most general infinitesimal transformation $\sum e_j X_j f$ which leaves invariant the manifold (13) and from this, we can immediately deduce the number of independent infinitesimal transformations $\sum e_j X_j f$ of this sort.

For the determination of the number o, we proceed as follows:

First we seek the most general infinitesimal transformation: $\sum \varepsilon_j(u_1,\dots,u_h) X_j f$ that transfers the manifold (13) to an infinitely neighboring manifold, or, expressed differently: we seek the most general infinitesimal transformation:

$$\sum_{j=1}^{r} \varepsilon_j(u_1,\dots,u_h) X_j f + \sum_{\sigma=1}^{h} \Phi_\sigma(u_1,\dots,u_h) \frac{\partial f}{\partial u_\sigma}$$

which leaves invariant the system of equations (13) in the $n+h$ variables: $x_1,\dots,x_n$, $u_1,\dots,u_h$.

If we retain the notation chosen above, the functions: $\varepsilon_j(u)$ and $\Phi_\sigma(u)$ are obviously defined by the equations:

$$\text{(15')} \qquad \sum_{j=1}^{r} \varepsilon_j(u) \left\{ [\xi_{j,q+k}] - \sum_{\pi=1}^{q} [\xi_{j\pi}] \frac{\partial \omega_{q+k}}{\partial x_\pi} \right\} = \sum_{\sigma=1}^{h} \Phi_\sigma(u) \frac{\partial \omega_{q+k}}{\partial u_\sigma}$$
$$(k=1\cdots n-q)$$

which they must satisfy independently of the values of the variables $x_1,\dots,x_q$.

We imagine that the $\varepsilon_j(u)$ and the $\Phi_\sigma(u)$ are determined in the most general way from these equations and we form the expression:

$$\sum_{\sigma=1}^{h} \Phi_\sigma(u_1,\dots,u_h) \frac{\partial f}{\partial u_\sigma}.$$

Evidently, this expression can be deduced from a completely determined number, say $h' \leqslant h$, of expressions:

$$\sum_{\sigma=1}^{h} \Phi_{\tau\sigma}(u_1,\dots,u_h) \frac{\partial f}{\partial u_\sigma} \qquad (\tau=1\cdots h')$$

by means of additions and of multiplications by functions of the u. According to the nature of things, the h' equations:

$$\sum_{\sigma=1}^{h} \Phi_{\tau\sigma}(u_1,\dots,u_h) \frac{\partial f}{\partial u_\sigma} = 0 \qquad (\tau=1\cdots h')$$

form an h'-term complete system with $h-h'$ independent solutions:

$$w_1(u_1,\dots,u_h), \ \dots, \ w_{h-h'}(u_1,\dots,u_h),$$

and it is clear that the equations:

$$w_1 = \text{const.}, \; \dots, \; w_{h-h'} = \text{const.}$$

determine the ∞^{h-o} subfamilies in which the family (11) can be decomposed, as we saw above. Consequently, we have: $h - o = h - h'$, hence $o = h'$.

It should not remain unmentioned that the results of the present chapter can be generalized.

One can for instance, instead of starting from an individual manifold, start from a discrete number of manifolds or even more generally: instead of starting from an individual family of manifolds, one can start from several such families. We call a number of manifolds briefly a *figure* [FIGUR].

§ 116. Theorem 85, p. 489, contains a method for the determination of transitive groups which are isomorphic with a given r-term group; we have already given a method for the determination of *all* groups of this sort in Chap. 22, p. 443 sq. At present, we will show that our new method fundamentally amounts ultimately to the old one, and we will in this way come to state an important result found earlier in a new, much more general form.

Let:

$$V_k(x_1, \dots, x_n) = 0 \qquad (k=1\cdots n-q) \tag{16}$$

be an arbitrary manifold and let:

$$x_i' = f_i(x_1, \dots, x_n, a_1, \dots, a_r) \qquad (i=1\cdots n)$$

be an arbitrary r-term group generated by r independent infinitesimal transformations. By resolution with respect to the x, the equations: $x_i' = f_i(x,a)$ may give:

$$x_i = F_i(x_1', \dots, x_n', a_1, \dots, a_r) \qquad (i=1\cdots n).$$

Lastly, we want to assume that the manifold (16) admits exactly $r - m$ independent infinitesimal transformations of the group: $x_i' = f_i(x,a)$.

If we execute on the manifold (16) the general transformation: $x_i' = f_i(x,a)$ of our group, then according to Theorem 85, p. 489, we obtain a family of ∞^m manifolds invariant by the group. The equations of this family are:

$$V_k\big(F_1(x',a), \dots, F_n(x',a)\big) = W_k(x_1', \dots, x_n', a_1, \dots, a_r) = 0 \qquad (k=1\cdots n-q), \tag{17}$$

hence they formally contain r arbitrary parameters. But amongst these r parameters, only m are essential, hence it is possible to indicate m independent functions: $\omega_1(a), \dots, \omega_m(a)$ of the a so that the equations (17) take the form:

$$\Omega_k\big(x_1', \dots, x_n', \omega_1(a), \dots, \omega_m(a)\big) = 0 \qquad (k=1\cdots n-q). \tag{17'}$$

Here, we can lastly introduce: $l_1 = \omega_1(a), \dots, l_m = \omega_m(a)$ as new parameters in place of the a; then in the system of equations:

$$\Omega_k(x'_1,\dots,x'_n,l_1,\dots,l_m)=0 \qquad (k=1\cdots n-q) \tag{18}$$

which comes into existence in this way, the parameters $l_1,\dots,l_m$ are essential.

We find a new representation of our invariant family when we execute an arbitrary transformation: $x''_i = f_i(x',b)$ of our group on the system of equations (18). According to Theorem 85, p. 489, (18) receives on the occasion the form:

$$\Omega_k(x''_1,\dots,x''_n,l'_1,\dots,l'_m)=0 \qquad (k=1\cdots n-q), \tag{18'}$$

where the l' are completely determined functions of the l and of the b:

$$l'_\mu = \chi_\mu(l_1,\dots,l_m,b_1,\dots,b_r) \qquad (\mu=1\cdots m). \tag{19}$$

We must obtain this representation of our family when we execute the transformation:

$$x''_i = f_i\big(f_1(x,a),\dots,f_n(x,a),b_1,\dots,b_r\big) = f_i(x_1,\dots,x_n,a'_1,\dots,a'_r)$$

directly on the manifold (16), where the a' are completely determined functions of the a and of the b:

$$a'_k = \varphi_k(a_1,\dots,a_r,b_1,\dots,b_r) \qquad (k=1\cdots r). \tag{20}$$

If we do that, we first receive the equations of our family in the form:

$$V_k\big(F_1(x'',a'),\dots,F_n(x'',a')\big) = W_k(x''_1,\dots,x''_n,a'_1,\dots,a'_r)=0$$
$$(k=1\cdots n-q),$$

which we can also obviously write in the form:

$$\Omega_k\big(x''_1,\dots,x''_n,\omega_1(a'),\dots,\omega_m(a')\big)=0 \qquad (k=1\cdots n-q).$$

But since these equations must agree with the equations (18'), it follows that the parameters l' are linked to the a' by the relations:

$$l'_\mu = \omega_\mu(a'_1,\dots,a'_r) \qquad (\mu=1\cdots m).$$

Because of the equations (19), we therefore have:

$$\omega_\mu(a') = \chi_\mu(l_1,\dots,l_m,b_1,\dots,b_r) \qquad (\mu=1\cdots m),$$

or:

$$\omega_\mu(a') = \chi_\mu\big(\omega_1(a),\dots,\omega_m(a),b_1,\dots,b_r\big) \qquad (\mu=1\cdots m). \tag{21}$$

If we make here the substitution: $a_1' = \varphi_1(a,b)$, ..., $a_r' = \varphi_r(a,b)$, then we must obtain only identities, because the parameters $a_1,\dots,a_r$, $b_1,\dots,b_r$ are absolutely arbitrary, and hence, are linked together by no relation.

At present, if we remember that the equations (20) represent a simply transitive group in the variables $a_1,\dots,a_r$ equally composed with the group: $x_i' = f_i(x,a)$, namely the associated parameter group (Chap. 21, p. 413), and that the equations (19) represent a transitive group isomorphic with the group: $x_i' = f_i(x,a)$, then we realize immediately what follows:

The equations:

$$\omega_1(a_1,\dots,a_r) = \text{const.}, \quad \dots, \quad \omega_m(a_1,\dots,a_r) = \text{const.} \tag{22}$$

represent a decomposition of the r-times extended space $a_1,\dots,a_r$ invariant by the group (20), and to be precise, a decomposition into ∞^m $(r-m)$-times extended manifolds. But the group:

$$l_\mu' = \chi_\mu(l_1,\dots,l_m,b_1,\dots,b_r) \qquad (\mu=1\cdots m) \tag{19}$$

indicates in which way the ∞^m manifolds (22) are permuted with each other by the transformations of the simply transitive group (20).

We therefore see that the group (19) can be derived from the simply transitive group: $a_k' = \varphi_k(a,b)$ according to the rules of the preceding chapter (p. 444 sq.).

We can use this fact in order to decide under which conditions we obtain two groups (19) similar to each other when we start from two different manifolds (16). Thanks to similar considerations, we realize that the following statement holds, of which Theorem 80 in Chap. 22, p. 453, is only a special case, fundamentally:

Theorem 86. *If, in the space $x_1,\dots,x_n$, an r-term group: $X_1f,\dots,X_rf$ is presented, if in addition, two manifolds M and M' are given, and if one executes all ∞^r transformations of the group: $X_1f,\dots,X_rf$ on each one of these two manifolds, then the individual manifolds of the two invariant families that one obtains in this way are transformed by two transitive groups isomorphic with the group: $X_1f,\dots,X_rf$ which are similar to each other when and only when the following two conditions are satisfied:*

Firstly, the two manifolds M and M' must admit the same number of independent infinitesimal transformations of the form: $e_1X_1f+\cdots+e_rX_rf$, and:

Secondly, it must be possible to relate the group: $X_1f,\dots,X_rf$ to itself in a holoedrically isomorphic way so that, to every infinitesimal transformation which fixes the one manifold, there corresponds an infinitesimal transformation which leaves invariant the other manifold.

It goes without saying that this theorem also remains valid when one replaces the two manifolds by two figures.

§ 117. Especially worthy of note is the case where one has a manifold, or a figure, which admits absolutely no infinitesimal transformation of G_r: $X_1f,\dots,X_rf$. The group (19) isomorphic to G_r then has the form:

$$l'_k = \chi_k(l_1, \dots, l_r, a_1, \dots, a_r) \qquad (k=1\cdots r),$$

it is simply transitive and hence holoedrically isomorphic with G_r.[3]

We therefore have here a general method in order to set up the simply transitive groups that are equally composed with a given r-term group.

The method applied in the preceding chapter, Proposition 1, page 441, is only a special case of the present more general method. Indeed at that time — as we can now express it — we used as a figure the totality of r different points of R_n.

When no special assumption is made about the position of these points, then the figure consisting of them can allow none of the infinitesimal transformations of the group $X_k f$. Since the group $X_k f$ certainly leaves invariant no point in general position, there can be in it at most $r-1$ independent infinitesimal transformations which fix such a point; from these possible $r-1$ infinitesimal transformations, one can again linearly deduce at most $r-2$ independent infinitesimal transformations by which a second point in general position remains untouched, and so on; one realizes at the end that there is no infinitesimal transformation in the group by which r points in general position remain simultaneously invariant.

Hence if we take a figure which consists of r points of this sort, and if we execute on them the ∞^r transformations of the group: $X_1 f, \dots, X_r f$, then the figure takes ∞^r different positions, the totality of which remains invariant by the group. These ∞^r positions are transformed by a simply transitive group.

If we set up this simply transitive group, we obtain exactly the same group as the one obtained thanks to the method of the preceding chapter.

§ 118. It appears to be desirable to illustrate the general developments of §§ 114 and 116 by means of a specific example. However, we restrict ourselves here to giving indications, and we leave to the reader the effective execution of the concerned simple computations.

The most general projective group of the plane which leaves invariant the non-degenerate conic section: $x^2 - 2y = 0$ is three-term and contains the following three independent infinitesimal transformations:

$$X_1 f = \frac{\partial f}{\partial x} + x\frac{\partial f}{\partial y}, \qquad X_2 f = x\frac{\partial f}{\partial x} + 2y\frac{\partial f}{\partial y}$$
$$X_3 f = (x^2 - y)\frac{\partial f}{\partial x} + xy\frac{\partial f}{\partial y}.$$

One sees immediately that this group — which we call G_3 for short — is transitive and that its composition is determined through the relations:

$$[X_1, X_2] = X_1 f, \qquad [X_1, X_3] = X_2 f, \qquad [X_2, X_3] = X_3 f.$$

[3] LIE, Gesellsch. d. W. zu Christiania, 1884.

From this, it follows by taking account of Chap. 15, Proposition 8, p. 275, that G_3 contains no two-term invariant subgroup; one convinces oneself that it also does not contain any one-term invariant subgroup. Consequently, G_3 is *simple* (Chap. 15, p. 276).

Every tangent to the fixed conic section $x^2 - 2y = 0$ admits exactly two independent infinitesimal transformations of the group; furthermore, it can be proved that the tangents are the only curves of the plane which possess this property. In the same way, the conic sections which enter into contact [BERÜHREN] in two points with the fixed conic section are the only curves which admit one and only one infinitesimal transformation of G_3; as a conic section which enters twice into contact, one must also certainly count every line of the plane which cuts [SCHNEIDET] the conic section in two separate points. Lastly, it is clear that every point not lying on the conic section admits one and only one infinitesimal transformation of the group.

Now, if one chooses as a manifold M another arbitrary curve, hence a curve which admits absolutely no infinitesimal transformation of G_3, then by the group, this curve takes ∞^3 different positions and the totality of these positions is obviously transformed by means of a three-term group. One therefore finds a simply transitive group of the R_3 equally composed with the original G_3. All groups that one obtains in this way are similar to each other; one amongst them is for instance the three-term group of all projective transformations of the R_3 which leaves invariant a winding curve of third order.

If one introduces as manifold M a conic section (irreducible or decomposable) entering into contact in two separate points [with the fixed conic section], then one obtains a group holoedrically isomorphic to G_3 in a twice-extended manifold; all groups obtained in this way are similar to each other. By contrast, if one uses as a manifold M a conic section having four points entering into contact [with the fixed conic section], one obtains a completely different type of three-term group of a twice-extended manifold.

Lastly, if one introduces as manifold M a tangent to the fixed conic section, then one obtains a three-term group in a once-extended manifold which is similar to the general projective group of the straight line.

§ 119. Let a linear homogeneous group:

$$X_k f = \sum_{\mu,\,\nu}^{1\cdots n} a_{k\mu\nu}\, x_\mu \frac{\partial f}{\partial x_\nu} \qquad {\scriptstyle (k=1\cdots r)}$$

of R_n be presented. This group leaves invariant the family of ∞^n planes M_{n-1} of R_n:

$$u_1 x_1 + \cdots + u_n x_n + 1 = 0.$$

We want to set up the group corresponding to it in the parameters $u_1, \dots, u_n$.

According to § 111, the infinitesimal transformations:

$$U_k f = \sum_{\nu=1}^{n} v_{k\nu}(u_1, \dots, u_n) \frac{\partial f}{\partial u_\nu}$$

of the sought group are to be determined so that the equation $\sum u_\nu x_\nu + 1 = 0$ admits the infinitesimal transformation $X_k f + U_k f$. So, the r expressions:

$$\sum_{\mu,\,\nu}^{1\cdots n} a_{k\mu\nu}\, x_\mu\, u_\nu + \sum_{\nu}^{1\cdots n} x_\nu\, v_{k\nu}$$

must vanish by means of $\sum u_\nu x_\nu + 1 = 0$. This is possible only when they vanish identically, that is to say when $v_{k\nu} = -\sum_\mu a_{k\nu\mu}\, u_\mu$.

We therefore find:

$$U_k f = -\sum_{\mu,\,\nu}^{1\cdots n} a_{k\nu\mu}\, u_\mu \frac{\partial f}{\partial u_\nu}.$$

Consequently, the group $U_k f$ is linear homogeneous too; we know from the beginning that it is isomorphic with the group $X_k f$.

We call the group: $U_1 f, \dots, U_r f$ the group *dualistic* [DUALISTISCH] to the group: $X_1 f, \dots, X_r f$.

In order to give an example, we want to find the group dualistic to the adjoint group of an r-term group having the composition:

$$[Y_i, Y_k] = \sum_{s=1}^{r} c_{iks}\, Y_s f.$$

The adjoint group reads (cf. p. 287):

$$\sum_{i,\,s}^{1\cdots r} c_{iks}\, e_i \frac{\partial f}{\partial e_s} \qquad (k=1\cdots r),$$

whence the group dualistic to it reads:[4]

$$\sum_{i,\,s}^{1\cdots r} c_{kis}\, \varepsilon_s \frac{\partial f}{\partial \varepsilon_i} \qquad (k=1\cdots r),$$

where the relation: $c_{iks} = -c_{kis}$ is used.

Similar considerations can actually be made for all projective groups of R_n, since they all leave invariant the family of ∞^n straight, $(n-1)$-times extended manifolds of R_n. However, we do not want to enter into these considerations, and rather, we refer to the next volume in which the concept of duality [DUALITÄT] is considered under a more general point of view, namely as a special case of the general concept of contact transformation [BERÜHRUNGSTRANSFORMATION].

---

4 LIE, Math. Ann. Vol. XVI, p. 496, cf. also Archiv for Mathematik, Vol. 1, Christiania 1876.

§ 120. Finally, we want to consider an important example of a more general nature.

We assume that we know all q-term subgroups of G_r: $X_1 f, \dots, X_r f$. Then the question is still to decide what are the different *types* of such q-term subgroups.

In Chap. 16, p. 292, we have already explained the concept of "types of subgroups"; according to that, we include two q-term subgroups in the same type when they are conjugate to each other inside G_r; of all subgroups which belong to the same type, we therefore need only to indicate a single one, and in addition, all these subgroups are perfectly determined by this one.

We know that every subgroup of the group: $X_1 f, \dots, X_r f$ is represented by a series of linear homogeneous relations between the parameters: $e_1, \dots, e_r$ of the general infinitesimal transformation: $e_1 X_1 f + \cdots + e_r X_r f$ (cf. Chap. 12, p. 223). Moreover, we know that two subgroups are conjugate to each other inside the group: $X_1 f, \dots, X_r f$ if and only if the system of equations between the e which represents the one subgroup can be transferred, by means of a transformation of the adjoint group, to the system of equations which represents the other subgroup (Chap. 16, p. 292).

Now according to our assumption, we know all q-term subgroups of G_r, hence we know all systems of $r-q$ independent linear homogeneous equations between the e:

$$\sum_{j=1}^{r} h_{kj}\, e_j = 0 \qquad (k=1 \cdots r-q)$$

which represent q-term subgroups.

For reasons of simplicity, amongst all these systems of equations, we want to take those which can be resolved with respect to $e_{q+1}, \dots, e_r$, hence which can be brought to the form:

$$e_{q+k} = g_{q+k,1}\, e_1 + \cdots + g_{q+k,q}\, e_q \qquad (k=1 \cdots r-q); \tag{23}$$

we leave the remaining ones which cannot be brought to this form, because they could naturally be treated in exactly the same way as those of the form (23).

All systems of values $g_{q+k,j}$ which, when inserted in (23), provide q-term subgroups are defined by means of certain equations between the $g_{q+k,j}$; however, in general, it is not possible to represent all these systems of values by means of a single system of equations between the g, and rather, a discrete number of such systems of equations will be necessary if one wants to have all q-term subgroups which are contained in the form (23). Naturally, two different systems of equations of this sort then provide nothing but different types of q-term subgroups.

We restrict ourselves to an arbitrary system of equations amongst the concerned systems of equations, say the following one:

$$\Omega_\mu(g_{q+1,1}, g_{q+1,2}, \dots, g_{r,q}) = 0 \qquad (\mu=1,2\cdots), \tag{24}$$

and we now want to see what types of q-term subgroups this system determines. The equations (23) determine, when the $g_{q+k,j}$ are completely arbitrary, the family of all straight q-times extended manifolds of the space $e_1, \dots, e_r$ which pass through the point: $e_1 = 0, \dots, e_r = 0$. Of course, this family of manifolds remains invariant by the adjoint group:

$$e'_k = \sum_{j=1}^{r} \rho_{kj}(a_1, \dots, a_r)\, e_j \qquad (k=1 \cdots r) \tag{25}$$

of the group: $X_1 f, \dots, X_r f$. Hence, if we execute the transformation (25) on the system of equations (23), we obtain a system of equations in the e' of the corresponding form:

$$e'_{q+k} = \sum_{j=1}^{q} g'_{q+k,j}\, e'_j \qquad (k=1 \cdots r-q),$$

where the g' are linear homogeneous functions of the g with coefficients which depend upon the a:

$$g'_{q+k,j} = \sum_{\mu=1}^{r-q} \sum_{\nu=1}^{q} \alpha_{kj\mu\nu}(a_1, \dots, a_r)\, g_{q+\mu,\nu} \qquad (k=1 \cdots r-q;\; j=1 \cdots q). \tag{26}$$

According to p. 476 sq., the equations (26) determine a group in the variables g. The system of equations (24) remains invariant by this group, because every system of values $g_{q+k,j}$ which provides a subgroup is naturally transferred to a system of values $g'_{q+k,j}$ which determines a subgroup; but since the group (26) is continuous, it leaves individually invariant all discrete regions of systems of values $g_{q+k,j}$ of this sort, hence in particular also the system of equations:

$$\Omega_\mu(g_{q+1,1}, \dots, g_{r,q}) = 0 \qquad (\mu = 1, 2 \cdots). \tag{24}$$

Now, the question is how the systems of values (24) are transformed by the group (26), and whether every system of values can be transferred to every other, or not.

This question receives a graphical sense when we imagine that the $g_{q+k,j}$ are point coordinates in a space of $q(r-q)$ dimensions. Indeed, the equations (24) then represent a certain manifold in this space which remains invariant by the group (26). Each point of the manifold belongs to a certain smallest invariant subsidiary domain of the manifold and the points of such a subsidiary domain represent nothing but conjugate q-term subgroups of G_r, and to be precise, all the q-term subgroups of G_r which belong to one and the same type.

Besides, one must draw attention on the fact that only the points $g_{q+k,j}$ of such a smallest invariant subsidiary domain that belong in turn to no smaller invariant subsidiary domain are to be counted, because it is only when one delimits the subsidiary domain in this way that each one of its points can be transferred to any other point by means of a nondegenerate transformation of the group (26).

Hence, if one wants to determine all types of q-term subgroups which are contained in the system of equations (24), then one only has to find all smallest invariant subsidiary domains of the manifold $\Omega_\mu = 0$. Every such subsidiary domain determines all subgroups which belong to the same type, and an arbitrary point of the subsidiary domain provides a group which can be chosen as a representative of the concerned type.

Furthermore, in order to be able to determine the discussed invariant subsidiary domains, one does not need to assume that the finite equations (25) of the adjoint group are known; it suffices that one has the infinitesimal transformations of this group, because one can then immediately indicate the infinitesimal transformations of the group (26) and then, following the rules of Chap. 14, one can determine the desired invariant subsidiary domains; in the case present here, this determination requires only executable operations.

One observes that the preceding developments also remain applicable when one knows not an r-term *group*, but only a possible *composition* of such a group, hence a system of c_{iks} which satisfies the known relations:

$$\begin{cases} c_{iks} + c_{kis} = 0 \\ \sum_{\nu=1}^{r} \left\{ c_{ik\nu}\, c_{\nu js} + c_{kj\nu}\, c_{\nu is} + c_{ji\nu}\, c_{\nu ks} \right\} = 0 \\ (i,\, k,\, j,\, s = 1 \cdots r). \end{cases}$$

If G_r: $X_1 f, \dots, X_r f$ is invariant in a larger group $\mathfrak{G}$, then the question is often whether two subgroups of G_r are conjugate in this larger group $\mathfrak{G}_r$, or not. In this case, one can also define differently the concept of "type of subgroup of G_r", by reckoning two subgroups of G_r as being distinct only when they are not conjugate to each other in $\mathfrak{G}$.

If one wants to determine all types, in this sense, of subgroups of G_r, then this task presents no special difficulty. Indeed, the problem in question is obviously a part of the more general problem of determining all types of subgroups of $\mathfrak{G}$, when the word "type" is understood in the sense of Chap. 16, p. 292.

This study is particularly important when the group $\mathfrak{G}$ is actually the largest subgroup of R_n in which G_r: $X_1 f, \dots, X_r f$ is invariant.

Chapter 24
Systatic and Asystatic Transformation Groups

In the s-times extended space $x_1, \dots, x_s$, let an r-term group:

$$X_k f = \sum_{i=1}^{s} \xi_{ki}(x_1, \dots, x_n) \frac{\partial f}{\partial x_i} \qquad {\scriptstyle (k=1\cdots r)},$$

or shortly G_r, be presented. Of the r infinitesimal transformations $X_1 f, \dots, X_r f$, let there be precisely n, say: $X_1 f, \dots, X_n f$, which are linked together by no linear relation, while by contrast $X_{n+1} f, \dots, X_r f$ can be expressed as follows:

$$X_{n+j} f \equiv \sum_{\nu=1}^{n} \varphi_{j\nu}(x_1, \dots, x_n) X_\nu f \qquad {\scriptstyle (j=1\cdots r-n)}. \tag{1}$$

Now, if $x_1^0, \dots, x_s^0$ is a point for which not all $n \times n$ determinants of the matrix:

$$\begin{vmatrix} \xi_{11}(x) & \cdot & \cdot & \xi_{1s}(x) \\ \cdot & \cdot & \cdot & \cdot \\ \xi_{n1}(x) & \cdot & \cdot & \xi_{ns}(x) \end{vmatrix} \tag{2}$$

vanish, then according to Chap. 11, p. 215, there are in G_r exactly $r-n$ independent infinitesimal transformations whose power series expansions with respect to the $x_i - x_i^0$ contain no term of zeroth order, but only terms of first order or of higher order; the point $x_1^0, \dots, x_s^0$ therefore admits exactly $r-n$ independent infinitesimal transformations of G_r which generate an $(r-n)$-term subgroup G_{r-n} of G_r (cf. Chap. 12, p. 218). The infinitesimal transformations of this G_{r-n} can, according to p. 216, be linearly deduced from the $r-n$ independent transformations:

$$X_{n+j} f - \sum_{\nu=1}^{n} \varphi_{j\nu}(x_1^0, \dots, x_s^0) X_\nu f \qquad {\scriptstyle (j=1\cdots r-n)}.$$

In the sequel, we want to briefly call a point $x_1^0, \dots, x_s^0$ for which not all $n \times n$ determinants of the matrix (2) vanish, a point *in general position*.

S. Lie, *Theory of Transformation Groups I*,
DOI 10.1007/978-3-662-46211-9_24

§ 121. According to the above, to every point $x_1^0,\dots,x_s^0$ in general position is associated a completely determined $(r-n)$-term subgroup of G_r, namely the most general subgroup of G_r by which it remains invariant.

If we let the point x^0 change its position, we receive a new $(r-n)$-term subgroup of G_r, and since there are ∞^s different points, we receive in total ∞^s subgroups of this sort; however, it is not said that we obtain ∞^s different subgroups.

If for example $r=n$, then the discussed ∞^s subgroups all coincide, namely they all reduce to the identity transformation, since G_r contains absolutely no infinitesimal transformation which leaves at rest a point x^0 in general position.

Disregarding this special case, it can happen that to the ∞^s points of the space $x_1,\dots,x_s$, only ∞^{s-1}, or fewer, different groups of the said constitution are associated; evidently, this will always occur in any case when *there is a continuous family of individual points which simultaneously keep their positions by the* $(r-n)$*-term subgroup*.

Now, we want to find the analytic conditions under which such a phenomenon occurs. At first, we take up the question: when do two points in general position remain invariant by the same $(r-n)$-term subgroup of G_r?

The answer to this question has a great similarity with the considerations in Chap. 19, p. 369 sq.

Let the one point be $x_1^0,\dots,x_s^0$ and let us call G_{r-n} the associated $(r-n)$-term subgroup of G_r; then the general infinitesimal transformation of G_{r-n} reads:

$$\sum_{j=1}^{r-n}\varepsilon_j\left(X_{n+j}f-\sum_{\nu=1}^{n}\varphi_{j\nu}^0\,X_\nu f\right),$$

where it is understood that the ε are arbitrary parameters.

Let the other point be: $\overline{x}_1,\dots,\overline{x}_s$, and let the general infinitesimal transformation of the subgroup $\overline{G}_{r-n}$ associated to it then be:

$$\sum_{j=1}^{r-n}\overline{\varepsilon}_j\left(X_{n+j}f-\sum_{\nu=1}^{n}\overline{\varphi}_{j\nu}\,X_\nu f\right).$$

Now, if the two points are supposed to remain invariant by the same $(r-n)$-term subgroup, then G_{r-n} and $\overline{G}_{r-n}$ coincide; for this, it is necessary and sufficient that all infinitesimal transformations of the one belong to those of the other group, and conversely; when expressed analytically, one must be able to satisfy identically the equation:

$$\sum_{j=1}^{r-n}\varepsilon_j\left\{X_{n+j}f-\sum_{\nu=1}^{n}\varphi_{j\nu}^0\,X_\nu f\right\}=\sum_{j=1}^{r-n}\overline{\varepsilon}_j\left\{X_{n+j}f-\sum_{\nu=1}^{n}\overline{\varphi}_{j\nu}\,X_\nu f\right\}$$

for arbitrarily chosen ε thanks to suitable values of the $\overline{\varepsilon}$, and also for arbitrarily chosen $\overline{\varepsilon}$ thanks to suitable values of the ε.

The latter equation can also be written:

$$\sum_{j=1}^{r-n} (\varepsilon_j - \overline{\varepsilon}_j) X_{n+j} f - \sum_{\nu=1}^{n} \sum_{j=1}^{r-n} (\varepsilon_j \varphi_{j\nu}^0 - \overline{\varepsilon}_j \overline{\varphi}_{j\nu}) X_\nu f = 0,$$

hence, because of the independence of the infinitesimal transformations $X_1 f, \dots, X_r f$, it can hold only if $\overline{\varepsilon}_j = \varepsilon_j$; in addition, since the ε_j are absolutely arbitrary, we obtain:

$$\varphi_{j\nu}(x_1^0, \dots, x_s^0) = \varphi_{j\nu}(\overline{x}_1, \dots, \overline{x}_s) \qquad (j=1 \cdots r-n;\ \nu=1 \cdots n).$$

Consequently, the two $(r-n)$-term subgroups of G_r which are associated to two distinct points in general position are identical with each other if and only if each one of the $n(r-n)$ functions $\varphi_{j\nu}$ takes the same numerical values for the one point as for the other point.

Hence, if we want to know all points in general position which, under the group G_{r-n}, keep their positions simultaneously with the point $x_1^0, \dots, x_s^0$, then we only have to determine all systems of values x which satisfy the equations:

$$\text{(3)} \qquad \varphi_{j\nu}(x_1, \dots, x_s) = \varphi_{j\nu}^0 \qquad (j=1 \cdots r-n;\ \nu=1 \cdots n);$$

every such system of values provides a point having the constitution demanded.

Here, two cases have to be distinguished.

Firstly the number of mutually independent functions amongst the $n(r-n)$ functions $\varphi_{k\nu}(x)$ can be equal to s exactly. In this case, the $(r-n)$-term subgroup G_{r-n} which fixes the point: $x_1^0, \dots, x_s^0$ leaves untouched at most a discrete number of points in general position.

Secondly the number of independent functions amongst the $\varphi_{k\nu}(x)$ can be smaller than s. In this case, there is a continuous manifold of points in general position which all remain invariant by the group G_{r-n}; to every point of the manifold in question is then associated the same $(r-n)$-term subgroup of G_r as to the point $x_1^0, \dots, x_s^0$. At the same time, the point $x_1^0, \dots, x_s^0$ evidently lies inside the manifold, that is to say: there are, in the manifold, also points that are infinitely close to the point: $x_1^0, \dots, x_s^0$.

We assume that the second case happens, so that amongst the $n(r-n)$ functions $\varphi_{k\nu}(x)$, there are only $s-\rho < s$ that are mutually independent, and we may call them $\varphi_1(x), \dots, \varphi_{s-\rho}(x)$.

Under this assumption, all $\varphi_{k\nu}(x)$ can be expressed by means of $\varphi_1(x), \dots, \varphi_{s-\rho}(x)$ alone, and the equations (3) can be replaced by the $s-\rho$ mutually independent equations:

$$\text{(3')} \qquad \varphi_1(x_1, \dots, x_s) = \varphi_1(x_1^0, \dots, x_s^0), \ \dots, \ \varphi_{s-\rho}(x_1, \dots, x_s) = \varphi_{s-\rho}(x_1^0, \dots, x_s^0).$$

We therefore see that to every generally located point: $x_1^0, \dots, x_s^0$ of the space is associated a completely determined ρ-times extended manifold (3') which is formed of the totality of all points to which is associated the same $(r-n)$-term subgroup of the group: $X_1 f, \dots, X_r f$ as to the point: $x_1^0, \dots, x_s^0$.

It is possible that the equations:

$$\varphi_{j\nu}(x_1,\dots,x_s) = \varphi_{j\nu}(x_1^0,\dots,x_s^0) \qquad (j=1\cdots r-n;\ \nu=1\cdots n)$$

represent, for every system of values $x_1^0,\dots,x_s^0$, a manifold which decomposes into a discrete number of different manifolds.

An example is provided by the three-term group:

$$X_1 f = \frac{\partial f}{\partial x_1}, \qquad X_2 f = \frac{1}{\cos x_2}\frac{\partial f}{\partial x_2}, \qquad X_3 f = \tan x_2 \frac{\partial f}{\partial x_2}.$$

Here, we have:

$$X_3 f \equiv \sin x_2 \, X_2 f,$$

while $X_1 f$ and $X_2 f$ are linked by no linear relation. So, if one fixes the point x_1^0, x_2^0, all points whose coordinates x_1, x_2 satisfy the equation:

$$\sin x_2 = \sin x_2^0$$

also remain fixed, that is to say: simultaneously with the point x_1^0, x_2^0, every point which lies in one of the infinitely many lines:

$$x_2 = x_2^0 + 2k\pi$$

parallel to the x_1-axis keeps its position, where it is understood that k is an arbitrary, positive or negative, whole number.

Since, under the assumption made above, the group: $X_1 f,\dots,X_r f$ associates to every point: $x_1^0,\dots,x_s^0$ a ρ-times extended manifold passing through it, then the whole space $x_1,\dots,x_s$ obviously decomposes into a family of $\infty^{s-\rho}$ ρ-times extended manifolds:

$$\text{(4)} \qquad \varphi_1(x_1,\dots,x_s) = \text{const.}, \ \dots, \ \varphi_{s-\rho}(x_1,\dots,x_s) = \text{const.},$$

and to be precise, in such a way that all transformations of the group which fix an arbitrarily chosen point in general position leave at rest all points of the manifold (4) that passes through this point.

Certainly, it is to be remarked here that one can speak of a real decomposition of the space only when the number ρ is smaller than s; if $\rho = s$, no real decomposition of the space occurs, since every transformation of our group which fixes a point in general position actually leaves invariant all points of the space; in other words: the identity transformation is the only transformation of the group which leaves at rest a point in general position.

The preceding considerations give an occasion for an important division [EINTHEILUNG] of all r-term groups $X_1 f,\dots,X_r f$ of the space $x_1,\dots,x_s$, and to be precise, for a division into two different classes.

If a group of the space $x_1,\dots,x_s$ is constituted in such a way that all its transformations which leave invariant a point in general position simultaneously fix all *points of a continuous manifold passing through this point, then we include this group in the one class and we call them systatic* [SYSTATISCH]. *We include the remaining groups, hence those which are not systatic, in the other class, and we call them asystatic* [ASYSTATISCH].[1]

Using this terminology, we can express the obtained result in the following way:

Theorem 87. *If the independent infinitesimal transformations $X_1 f, \dots, X_r f$ of an r-term group in s variables $x_1, \dots, x_s$ are linked together by $r-n$ linear relations of the form:*

$$X_{n+k} f \equiv \sum_{\nu=1}^{n} \varphi_{k\nu}(x_1, \dots, x_s) X_\nu f \qquad (k=1 \cdots r-n),$$

while between $X_1 f, \dots, X_n f$ alone no relation of this sort holds, then the group is systatic when amongst the $n(r-n)$ functions $\varphi_{k\nu}(x_1, \dots, x_s)$ there are fewer than s that are mutually independent; by contrast, if amongst the functions $\varphi_{k\nu}(x_1, \dots, x_s)$ there are s functions that are mutually independent, then the group is asystatic.

§ 122. We want to maintain all the assumptions that we have made in the introduction of the chapter concerning the r-term group: $X_1 f, \dots, X_r f$ of the space $x_1, \dots, x_s$, and at present, we only want to add the assumption that the group is systatic. Thus, we assume that, amongst the $n(r-n)$ functions $\varphi_{k\nu}(x_1, \dots, x_s)$, only $0 \leqslant s-\rho < s$ are mutually independent, and as above, we may call them $\varphi_1(x), \dots, \varphi_{s-\rho}(x)$.

We consider at first the case where the number $s-\rho$ has the value zero.

If the number $s-\rho$ vanishes, then n is obviously equal to r, that is to say the r independent infinitesimal transformations: $X_1 f, \dots, X_r f$ are linked together by no linear relation of the form:

$$\chi_1(x_1, \dots, x_s) X_1 f + \cdots + \chi_r(x_1, \dots, x_s) X_r f = 0.$$

From this, it follows that the number r is in any case not larger than the number s of the variables x, hence that the group: $X_1 f, \dots, X_r f$ is either intransitive, or at most simply transitive. Hence in the two cases, the group: $X_1 f, \dots, X_r f$ is imprimitive (cf. Chap. 13, p. 233 and Chap. 20, Proposition 6, p. 394.

So we see that the systatic group: $X_1 f, \dots, X_r f$ is always imprimitive when the whole number $s-\rho$ has the value zero.

We now turn ourselves to the case $s-\rho > 0$.

In this case, the equations:

$$(4) \qquad \varphi_1(x_1, \dots, x_s) = \text{const.}, \; \dots, \; \varphi_{s-\rho}(x_1, \dots, x_s) = \text{const.}$$

[1] The concepts of systatic and asystatic groups, and the theory of these groups, stem from LIE (Ges. d. W. zu Christiania 1884, Archiv for Math. Vol. 10, Christiania 1885). The expressive terminologies: systatic "leaving fixed" [MITSTEHENDLASSEND] and asystatic "not leaving fixed" [NICHTMITSTEHENDLASSEND] are from ENGEL. As for the rest, the terminology characteristic of the present work was introduced by LIE.

provide a real decomposition of the space $x_1,\dots,x_s$ into $\infty^{s-\rho}$ ρ-times extended manifolds, and in fact, as we have seen above, a decomposition which is completely determined by the group: $X_1 f,\dots,X_r f$ and which stands in a completely characteristic relationship to this group. We are very close to presuming that this decomposition remains invariant by the group: $X_1 f,\dots,X_r f$; from this, it would then follow that the systatic group: $X_1 f,\dots,X_r f$ is also imprimitive in the case $s-\rho>0$ (Chap. 13, p. 232).

One can easily see that the above assumption is true. In fact, according to Chap 19, p. 356 sq., the $r(s-\rho)$ expressions: $X_k\,\varphi_1,\,\dots,\,X_k\,\varphi_{s-\rho}$ can be expressed as functions of $\varphi_1,\dots,\varphi_{s-\rho}$ alone:

$$X_k\,\varphi_j=\pi_{jk}(\varphi_1,\dots,\varphi_{s-\rho}) \qquad (k=1\cdots r;\ j=1\cdots s-\rho).$$

In this (cf. Chap. 8, p. 153 and 157) lies the reason why the decomposition (4) admits the r infinitesimal transformations $X_k f$, and therefore actually, the complete group: $X_1 f,\dots,X_r f$.

The developments just carried out prove that a systatic group of the space $x_1,\dots,x_s$ is always imprimitive. We therefore have the

Theorem 88. *Every systatic group is imprimitive.*

It is not superfluous to establish, also by means of conceptual considerations, that in the case $s-\rho>0$, the decomposition (4) remains invariant by the systatic group: $X_1 f,\dots,X_r f$.

Let us denote by M an arbitrary manifold amongst the $\infty^{s-\rho}$ ρ-times extended manifolds (4), let P be the general symbol of a point of the manifold M, let S be the general symbol of the ∞^{r-n} transformations of our group which leave untouched all the points of M, and lastly, let us understand by T an arbitrary transformation of our group.

If we execute the transformation T on M, we obtain a certain ρ-times extended manifold M', the ∞^{ρ} points P' of which are defined by the equation:

$$(P')=(P)\,T.$$

Now, since every point P remains invariant by all ∞^{r-n} transformations S of our group, it is clear that every point P' keeps its position by the ∞^{r-n} transformations $T^{-1}\,ST$, which also belong to our group. From this, it follows that M' also belongs to the $\infty^{s-\rho}$ manifolds (4); consequently, it is proved that the $\infty^{s-\rho}$ manifolds (4) are permuted with each other by every transformation of the group: $X_1 f,\dots,X_r f$, so that the decomposition (4) effectively remains invariant by our group.

Since according to Theorem 88, every systatic group is imprimitive, every primitive group must be asystatic, inversely. But there are also imprimitive groups which are asystatic, for instance the four-term group:

$$\frac{\partial f}{\partial x}, \qquad \frac{\partial f}{\partial y}, \qquad x\frac{\partial f}{\partial y}, \qquad y\frac{\partial f}{\partial y}$$

of the plane x, y. This group is imprimitive, since it leaves invariant the family of straight lines: $x = \text{const.}$, but it is at the same time asystatic, because from the two identities:

$$x\frac{\partial f}{\partial y} \equiv x\frac{\partial f}{\partial y}, \qquad y\frac{\partial f}{\partial y} \equiv y\frac{\partial f}{\partial y},$$

it becomes clear that the functions $\varphi_{k\nu}$ associated to the group are nothing but x and y themselves; but they are obviously independent of each other. The three-term intransitive group:

$$\frac{\partial f}{\partial y}, \qquad x\frac{\partial f}{\partial y}, \qquad y\frac{\partial f}{\partial y}$$

shows that there are even intransitive asystatic groups.

§ 123. If one knows r independent infinitesimal transformations: $X_1 f, \dots, X_r f$ of an r-term group of the space $x_1, \dots, x_s$, then according to Theorem 87, p. 507, one can easily decide whether the concerned group is systatic or not.

For this purpose, one forms the matrix:

$$\begin{vmatrix} \xi_{11}(x) & \cdot & \cdot & \xi_{1s}(x) \\ \cdot & \cdot & \cdot & \cdot \\ \xi_{r1}(x) & \cdot & \cdot & \xi_{rs}(x) \end{vmatrix}$$

and one studies its determinants. If all the $(n+1)\times(n+1)$ determinants vanish identically, but not all $n \times n$ determinants, and in particular, not all $n \times n$ determinants that can be formed with the topmost n rows of the matrix, then there are identities of the form:

$$X_{n+k} f \equiv \sum_{\nu=1}^{n} \varphi_{k\nu}(x_1, \dots, x_s)\, X_\nu f \qquad {\scriptstyle (k=1\cdots r-n)}, \tag{1}$$

while $X_1 f, \dots, X_n f$ are linked together by no linear relation. If one has set up the identities (1), then one determines the number of independent functions amongst the $n(r-n)$ functions $\varphi_{k\nu}(x_1, \dots, x_s)$.

However, in order to be able to decide whether a determined r-term group is systatic or not, one need not know the finite expressions for the infinitesimal transformations of the group, but rather, it suffices that one knows the *defining equations* of the group. This sufficiency is based on the fact that, as soon as the defining equations of the group are presented, one can always indicate an unrestricted integrable system of total differential equations, the only integral functions of which are just the $\varphi_{k\nu}(x)$ and the functions of them. Indeed, from this it clearly turns out that one can determine the number of independent functions amongst the $\varphi_{k\nu}(x)$ without knowing the $\varphi_{k\nu}(x)$ themselves.

Now, we will derive this important result.

The independent equations amongst the equations:

$$(5) \qquad d\varphi_{k\nu} = \sum_{i=1}^{s} \frac{\partial \varphi_{k\nu}(x)}{\partial x_i}\, \mathrm{d}x_i = 0 \qquad (k=1\cdots r-n;\ \nu=1\cdots n)$$

form an unrestricted integrable system of total differential equations, the only integral functions of which are the $\varphi_{k\nu}(x)$, and the functions of them (cf. Chap. 5, p. 106 sq.). We start from this system of total differential equations.

Because of (1), we have identically:

$$\xi_{n+j,i} - \sum_{\nu=1}^{n} \varphi_{j\nu}\, \xi_{\nu i} \equiv 0 \qquad (j=1\cdots r-n;\ i=1\cdots s),$$

whence we get by differentiation:

$$d\xi_{n+j,i} - \sum_{\nu=1}^{n} \varphi_{j\nu}\, d\xi_{\nu i} \equiv \sum_{\nu=1}^{n} \xi_{\nu i}\, d\varphi_{j\nu},$$

or, since according to the assumption, not all $n \times n$ determinants of the matrix:

$$\begin{vmatrix} \xi_{11}(x) & \cdot & \cdot & \xi_{1s}(x) \\ \cdot & \cdot & \cdot & \cdot \\ \xi_{r1}(x) & \cdot & \cdot & \xi_{rs}(x) \end{vmatrix}$$

vanish, we obtain:

$$d\varphi_{j\pi} \equiv \sum_{i=1}^{s} \chi_{\pi i}(x_1,\dots,x_s) \left\{ d\xi_{n+j,i} - \sum_{\nu=1}^{n} \varphi_{j\nu}\, d\xi_{\nu i} \right\}$$
$$(j=1\cdots r-n;\ \pi=1\cdots n).$$

From this, it follows that the system of the total differential equations (5) can be replaced by the following one:

$$(6) \qquad \sum_{\pi=1}^{s} \left\{ \frac{\partial \xi_{n+j,i}}{\partial x_\pi} - \sum_{\nu=1}^{n} \varphi_{j\nu} \frac{\partial \xi_{\nu i}}{\partial x_\pi} \right\} \mathrm{d}x_\pi = 0 \qquad (j=1\cdots r-n;\ i=1\cdots s).$$

Of course, the independent equations amongst the equations (6) form an unrestricted integrable system of total differential equations, the integral functions of which are just the $\varphi_{k\nu}(x)$, and the functions of them.

However, it is not possible now to set up the individual equations (6) when one only knows the defining equations of the group: $X_1 f,\dots,X_r f$; by contrast, it is possible to replace the system of equations (6) by another system which, aside from the $\mathrm{d}x_\pi$, contains only the coefficients of the defining equations. We arrive at this in the following way.

By $x_1^0,\dots,x_s^0$, we understand a point in general position. According to p. 215, the most general infinitesimal transformation: $e_1 X_1 f + \cdots + e_r X_r f$ whose power series expansion with respect to the $x_i - x_i^0$ contains only terms of first order or of higher order has the form:

$$\sum_{j=1}^{r-n} \mathfrak{e}_j \left(X_{n+j}f - \sum_{\nu=1}^{n} \varphi_{j\nu}(x_1^0, \dots, x_s^0) X_\nu f \right),$$

where it is understood that the $\mathfrak{e}_j$ are arbitrary parameters. However, when we execute the power series expansion with respect to the $x_i - x_i^0$ and when at the same time we take into account only the terms of first order, this expression receives the form:

$$(7) \quad \sum_{j=1}^{r-n} \mathfrak{e}_j \sum_{i,\,\pi}^{1\cdots s} \left\{ \left[\frac{\partial \xi_{n+j,i}}{\partial x_\pi} \right]_{x=x^0} - \sum_{\nu=1}^{n} \varphi_{j\nu}(x^0) \left[\frac{\partial \xi_{\nu i}}{\partial x_\pi} \right]_{x=x^0} \right\} (x_\pi - x_\pi^0) \frac{\partial f}{\partial x_i} + \cdots.$$

But we can compute the terms of first order in the expression (7) from the defining equations of the group, and to be precise, without integration. Indeed, according to Chap. 11, p. 203 sq., the most general infinitesimal transformation whose power series expansion with respect to the $x_i - x_i^0$ contains only terms of first order or of higher order has the form:

$$\sum_{i,\,\pi}^{1\cdots s} g'_{i\pi} (x_\pi - x_\pi^0) \frac{\partial f}{\partial x_\pi} + \cdots,$$

where a certain number of the s^2 quantities $g'_{\pi i}$ which we denoted by $\varepsilon_1 - \nu_1$ at that time were arbitrary, while the $s^2 - \varepsilon_1 + \nu_1$ remaining ones were linear homogeneous functions of these $\varepsilon_1 - \nu_1$ quantities with coefficients which could be computed immediately from the defining equations. We therefore obtain, when we start from the defining equations, the following representation for the expression (7):

$$(7') \qquad \sum_{j=1}^{\varepsilon_1 - \nu_1} \mathfrak{e}'_j \sum_{i,\,\pi}^{1\cdots s} \alpha_{j\pi i}(x_1^0, \dots, x_s^0) (x_\pi - x_\pi^0) \frac{\partial f}{\partial x_i} + \cdots,$$

where the $\mathfrak{e}'_j$ denote arbitrary parameters, while the $\alpha_{j\pi i}(x^0)$ are completely determined analytic functions of the x^0 which can, as said above, be computed from the defining equations.

Because (7) and (7') are only different representations of the same infinitesimal transformation, the factors of $(x_\pi - x_\pi^0)\,\partial f / \partial x_i$ in the two expressions must be equal to each other, that is to say, there are the following s^2 relations:

$$\sum_{j=1}^{r-n} \mathfrak{e}_j \left[\frac{\partial \xi_{n+j,i}}{\partial x_\pi} - \sum_{\nu=1}^{n} \varphi_{j\nu}(x) \frac{\partial \xi_{\nu i}}{\partial x_\pi} \right]_{x=x^0} = \sum_{j=1}^{\varepsilon_1 - \nu_1} \mathfrak{e}'_j \, \alpha_{j\pi i}(x^0) \qquad (i, \pi = 1 \cdots s).$$

For arbitrarily chosen $\mathfrak{e}_j$, one must always be able to satisfy these relations thanks to suitable choices of the values of the $\mathfrak{e}'_j$, and for arbitrarily chosen $\mathfrak{e}'_j$, always thanks to suitable choices of the values of the $\mathfrak{e}_j$.

All of this holds for every point $x_1^0, \dots, x_s^0$ in general position; hence this also holds true when we consider the x^0 as variables and when we substitute them by

$x_1,\dots,x_n$. Consequently, when we have chosen the $\mathfrak{e}$ in a completely arbitrary way as functions of the x, we can always determine the $\mathfrak{e}'$ as functions of the x in such a way that the equations:

$$(7'')\qquad \sum_{j=1}^{r-n} \mathfrak{e}_k \left\{ \frac{\partial \xi_{n+j,i}}{\partial x_\pi} - \sum_{\nu=1}^{n} \varphi_{j\nu}(x) \frac{\partial \xi_{\nu i}}{\partial x_\pi} \right\} = \sum_{j=1}^{\varepsilon_1-\nu_1} \mathfrak{e}'_j \, \alpha_{j\pi i}(x) \qquad (i,\, \pi = 1 \cdots s)$$

are identically satisfied, and when we have chosen the $\mathfrak{e}'$ in a completely arbitrary way as functions of the x, we can always satisfy (7") identically thanks to suitable functions $\mathfrak{e}_1,\dots,\mathfrak{e}_{r-n}$ of the x.

Now, one can obviously replace the equations (6) by the following s equations:

$$(6')\qquad \sum_{j=1}^{r-n} \mathfrak{e}_j \sum_{\pi=1}^{s} \left\{ \frac{\partial \xi_{n+j,i}}{\partial x_\pi} - \sum_{\nu=1}^{n} \varphi_{j\nu} \frac{\partial \xi_{\nu i}}{\partial x_\pi} \right\} \mathrm{d}x_\pi = 0 \qquad (i=1\cdots s),$$

provided only that one regards the $\mathfrak{e}$ as arbitrary functions of the x in them. Hence from what has been said above, it follows that the totality of all equations (6) is equivalent to the totality of all equations of the form:

$$\sum_{j=1}^{\varepsilon_1-\nu_1} \mathfrak{e}'_j \sum_{\pi=1}^{s} \alpha_{j\pi i}(x_1,\dots,x_s)\,\mathrm{d}x_\pi = 0 \qquad (i=1\cdots s),$$

in which the $\mathfrak{e}'$ are to be interpreted as arbitrary functions of the x. Lastly, the latter equations can evidently be replaced by the $(\varepsilon_1 - \nu_1)s$ equations:

$$(8)\qquad \sum_{\pi=1}^{s} \alpha_{j\pi i}(x_1,\dots,x_s)\,\mathrm{d}x_\pi = 0 \qquad (j=1\cdots\varepsilon_1-\nu_1;\; i=1\cdots s).$$

With these words, we have proved that the two systems of total differential equations: (6) and (8) are equivalent to each other; thus, it follows that the independent equations amongst the equations (8) form an unrestricted integrable system of total differential equations, and to be precise, a system, the only integral functions of which are the $\varphi_{k\nu}(x)$ and the functions of them.

We can therefore say:

Theorem 89. *If the defining equations of an r-term group of the space $x_1,\dots,x_s$ are presented, then one decides in the following way whether the concerned group is systatic or not:*

One understands by $x_1^0,\dots,x_s^0$ an arbitrary point in the neighborhood of which the coefficients of the resolved defining equations behave regularly and one determines the terms of zeroth order and of first order in the power series expansion of the general infinitesimal transformation of the group with respect to the powers of $x_1-x_1^0$, $\dots$, $x_s-x_s^0$. Then, one searches for the terms of first order in the most general infinitesimal transformation of the group which contains no term of zeroth *order. These terms will have the form:*

$$\sum_{j=1}^{\varepsilon_1-\nu_1} e'_j \sum_{i,\,\pi}^{1\cdots s} \alpha_{j\pi i}(x_1^0,\dots,x_s^0)\,(x_\pi - x_\pi^0)\,\frac{\partial f}{\partial x_i},$$

where the e'_j *denote arbitrary parameters, while the* $\alpha_{j\pi i}(x^0)$ *are completely determined analytic functions of the* x^0 *and can be computed without integration from the coefficients of the defining equations. Now, one forms the system of the total differential equations:*

$$(8)\qquad \sum_{\pi=1}^{s} \alpha_{j\pi i}(x_1,\dots,x_s)\,\mathrm{d}x_\pi = 0 \qquad (j=1\cdots\varepsilon_1-\nu_1\,;\; i=1\cdots s)$$

and one determines the number $s-\rho$ *of the independent equations amongst these equations. If* $s-\rho < s$*, then the group is systatic, but if* $s-\rho = s$*, the group is asystatic.*[2]

We can add:

Proposition 1. *The* $s-\rho$ *mutually independent equations amongst the differential equations* (8) *form an unrestricted integrable system with* $s-\rho$ *independent integral functions:* $\varphi_1(x),\dots,\varphi_{s-\rho}(x)$*. These integral functions stand in the following relationship to the group:* $X_1 f,\dots,X_r f$*:*

If, amongst the r independent infinitesimal transformations: $X_1 f,\dots,X_r f$ *there are exactly n, say* $X_1 f,\dots,X_n f$*, that are linked together by no linear relation, while* $X_{n+1} f,\dots,X_r f$ *can be expressed linearly in terms of* $X_1 f,\dots,X_n f$*:*

$$X_{n+k} f \equiv \sum_{\nu=1}^{n} \varphi_{k\nu}(x_1,\dots,x_s)\,X_\nu f \qquad (k=1\cdots r-n),$$

then all $n(r-n)$ *functions* $\varphi_{k\nu}(x)$ *can be expressed in terms of* $\varphi_1(x),\dots,\varphi_{s-\rho}(x)$ *alone.*

One does not even need to know the defining equations themselves in order to be able to decide whether a determined group is systatic or not. *For this, one only needs to know the initial terms in the power series expansions of the infinitesimal transformations of the group in the neighborhood of an individual point:* $x_1^0,\dots,x_s^0$ *in general position.* Indeed, if one knows these initial terms, one can obviously compute the numerical values $\alpha_{j\pi i}^0$ that the functions $\alpha_{j\pi i}(x)$ take for $x_1 = x_1^0$, ..., $x_s = x_s^0$. The number $s-\rho$ defined above is then nothing but the number of mutually independent equations amongst the linear equations in $\mathrm{d}x_1,\dots,\mathrm{d}x_s$:

$$\sum_{\pi=1}^{s} \alpha_{j\pi i}^0\,\mathrm{d}x_\pi = 0 \qquad (j=1\cdots\varepsilon_1-\nu_1\,;\; i=1\cdots s).$$

---

2 LIE, Archiv for Math., Vol. 10, Christiania 1885.

The system of total differential equations has a very simple conceptual interpretation. Indeed, as one easily realizes directly, it defines all the points: $x_1 + \mathrm{d}x_1, \dots, x_s + \mathrm{d}x_s$ infinitely close to the point $x_1, \dots, x_s$ that remain invariant by all transformations of our group which leave at rest the point $x_1, \dots, x_s$. Here lies the inner reason [INNERE GRUND] why the $\varphi_{k\nu}(x)$ and the functions of them are the only integral functions of the system (6) or (8), because indeed, the equations:

$$\varphi_{k\nu}(y_1, \dots, y_s) = \varphi_{k\nu}(x_1, \dots, x_s) \qquad (k=1 \cdots r-n;\ \nu=1 \cdots n)$$

define all points $y_1, \dots, y_s$ which remain invariant simultaneously with the point $x_1, \dots, x_s$.

§ 124. We have found that the r-term group: $X_1 f, \dots, X_r f$ of the space $x_1, \dots, x_s$ is asystatic or is systatic according to whether there are exactly s, or fewer than s, mutually independent functions amongst the functions $\varphi_{k\nu}(x_1, \dots, x_s)$ defined on p. 345 and 503. Now, according to Chap. 20, Theorem 67, p. 388, there is always an infinitesimal transformation Zf which is interchangeable with all $X_k f$ when, and only when, the number of independent functions amongst the $\varphi_{k\nu}(x)$ is smaller than s. Consequently, we can also say:

Proposition 2. *The r-term group: $X_1 f, \dots, X_r f$ of the space $x_1, \dots, x_s$ is systatic if and only if there is an infinitesimal transformation Zf which is interchangeable with all $X_k f$; if there is no such infinitesimal transformation, the group: $X_1 f, \dots, X_r f$ is asystatic.*[3]

Since the excellent infinitesimal transformations of a group are interchangeable with all the other infinitesimal transformations of the group, we have in addition:

Proposition 3. *Every group which contains one or several excellent infinitesimal transformations is systatic.*

Thus for such groups, one realizes already from the composition that they are systatic.

Finally, if we remember that the adjoint group of a group without excellent infinitesimal transformation contains r essential parameters (cf. Theorem 49, p. 289), we see that the following proposition holds:

Proposition 4. *The adjoint group of an asystatic group is always r-term.*

Proposition 2 and Proposition 4 are the generalizations of Propositions 1 and 2 of Chap. 16 (p. 290) announced at that time.

Already from the developments of Chap. 20 we could have taken a division of all groups into two classes; in the first class, we had included every r-term group: $X_1 f, \dots, X_r f$ for which there is at least one infinitesimal transformation interchangeable with all $X_k f$, and in the other class, all the remaining groups. According to what

---

[3] LIE, Archiv for Math., Vol. 10, p. 377, Christiania 1885.

was said above, it is clear that this division would coincide with our present division of the groups in systatic and asystatic groups; the first class would consist of all systatic group, and the second class, of all asystatic groups. In what follows, we want to explain this fact by means of conceptual considerations and at the same time, we want to derive new important results.

Let Θ be a transformation which is interchangeable with all transformations of a given r-term group: $X_1 f, \dots, X_r f$ of the space $x_1, \dots, x_s$, and moreover, let S be the general symbol of the transformations of the group: $X_1 f, \dots, X_r f$ which leave invariant an arbitrary, generally positioned point P of the space $x_1, \dots, x_s$.

If P is transferred to the point P_1 by the execution of the transformation Θ, then obviously, $\Theta^{-1} S \Theta$ is the general symbol of all transformations of the group: $X_1 f, \dots, X_r f$ which leave invariant the point P_1. But now, since Θ is interchangeable with all transformations of the group: $X_1 f, \dots, X_r f$, then the totality of all transformations $\Theta^{-1} S \Theta$ is identical to the totality of all transformations S, hence we see that all transformations of the group: $X_1 f, \dots, X_r f$ which fix the point P also leave at rest the point P_1.

We now apply this to the case where there is a continuous family of transformations Θ which are interchangeable with all transformations of the group: $X_1 f, \dots, X_r f$.

Since P is a point in general position, then by the execution of the transformations Θ, it takes a continuous series of different positions. But as we saw just now, each one of these positions remains invariant by all transformations S, and consequently, the group: $X_1 f, \dots, X_r f$ is systatic.

As a result, it is proved that the r-term group: $X_1 f, \dots, X_r f$ of the space $x_1, \dots, x_s$ is in any case systatic when there is an infinitesimal transformation Zf interchangeable with all $X_k f$. It remains to show that the converse also holds true, namely that for every systatic group: $X_1 f, \dots, X_r f$, one can indicate a continuous family of transformations which are interchangeable with all transformations $X_1 f, \dots, X_r f$.

Thus, we imagine that a systatic r-term group: $X_1 f, \dots, X_r f$ of the space $x_1, \dots, x_s$ is given. We will indicate a construction that provides infinitely many transformations which are interchangeable with all transformations of this group.

Every transformation Θ which is interchangeable with all transformations of the group: $X_1 f, \dots, X_r f$ transfers every point P of the space to a point P_1 which admits exactly the same transformations of the group as the point P; this is what we showed above. *Hence, we choose two arbitrary points P and P_1 which admit the same transformations of our group, and we attempt to determine a transformation Θ which is interchangeable with all transformations of our group and which in addition transfers P to P_1.*

Let P' be a point to which P can be transferred by means of a transformation T of the group: $X_1 f, \dots, X_r f$; furthermore, let, as earlier on, S be the general symbol of all transformations of this group which leave invariant P and hence P_1 too. Then (Chap. 14, Proposition 1, p. 238), ST is the general symbol of all transformations of our group which transfer P to P'. Obviously, P_1 takes the same position by all these transformations ST, for one indeed has:

$$(P_1)ST = (P_1)T.$$

At present, by assuming the existence of *one* transformation Θ having the constitution just demanded, we can easily see that every point $P' = (P)T$ receives a completely determined new position by all possible Θ; indeed, this follows immediately from the equations:

$$(P')\Theta = (P)T\Theta = (P)\Theta T = (P_1)T.$$

Hence if T is an arbitrary transformation of our group, then by every transformation Θ which actually exists, the point $(P)T$ takes the new position $(P_1)T$.

We add that we obtain in this way no overdetermination [ÜBERBESTIMMUNG] of the new position of the point P'. Indeed, if we replace in the latter equations the transformation T by another arbitrary transformation of the group: $X_1f,\dots,X_rf$ which transfers in the same way P to P', hence if write ST in place of T, then again we get:

$$(P')\Theta = (P)ST\Theta = (P)\Theta ST = (P_1)ST = (P_1)T.$$

We first consider the special case where the systatic group: $X_1f,\dots,X_rf$ is *transitive*.

When the group: $X_1f,\dots,X_rf$ is transitive, by a suitable choice of T, the point $(P)T$ can be brought to coincidence with every other point $(P_1)T$ of the space; hence, if we associate to every point $(P)T$ of the space the point $(P_1)T$, a completely determined transformation Θ' is defined in this way. If we succeed in proving that Θ' is interchangeable with all transformations of our group, then it is clear that Θ' possesses all properties which we have required of the transformation Θ, and that Θ' is the only transformation Θ which actually exists.

The fact that Θ' is interchangeable with all transformations of our group can easily be proved. Indeed, we have:

$$(P)T\Theta' = (P_1)T = (P)\Theta' T,$$

where T means a completely arbitrary transformation of our group. Hence if we understand in the same way by T a completely arbitrary transformation of our group, we obtain:

$$(P)T\mathsf{T}\Theta' = (P)\Theta' T\mathsf{T} = (P)T\Theta'\mathsf{T},$$

and therefore, the transformation: $\mathsf{T}\Theta'\mathsf{T}^{-1}\Theta'^{-1}$ leaves invariant the point $(P)T$, that is to say, every point of the space. From this, it follows that $\mathsf{T}\Theta'\mathsf{T}^{-1}\Theta'^{-1}$ is the identity transformation, that is to say: Θ' is really interchangeable with all transformations of our group.

Thus, when the systatic group: $X_1f,\dots,X_rf$ is transitive, to every pair of points P, P_1 having the constitution defined above there corresponds one and only one transformation interchangeable with all transformations of the group. If one chooses the pair of points in all possible ways, one obtains infinitely many such transformations, and it is easy to determine how many: In any case, when ∞^ρ different points remain untouched by all transformations of the group which leave invariant an arbitrarily chosen point, then there are exactly ∞^ρ different transformations which are interchangeable with all transformations of the group: $X_1 f, \dots, X_r f$. This is coherent with Theorem 67, p. 388, for because of the assumption made above, amongst the $n(r-n)$ functions $\varphi_{k\nu}(x)$, there are exactly $s-\rho$ that are mutually independent, hence there are exactly ρ independent infinitesimal transformations $Z_1 f, \dots, Z_\rho f$ which are interchangeable with all $X_k f$.

We can therefore state the following proposition:

Proposition 5. *If the r-term group: $X_1 f, \dots, X_r f$ of the space $x_1, \dots, x_s$ is transitive, and if P and P_1 are two points which admit exactly the same infinitesimal transformations of the group: $X_1 f, \dots, X_r f$, then there is one and only one transformation Θ which is interchangeable with all transformations of the group: $X_1 f, \dots, X_r f$ and which transfers P to P_1. If one understands by T the general symbol of a transformation of the group: $X_1 f, \dots, X_r f$, then Θ can be defined as the transformation which transfers every point $(P)T$ to the point $(P_1)T$. If each time exactly ∞^ρ different points admit precisely the same infinitesimal transformations of the group: $X_1 f, \dots, X_r f$, then there are exactly ∞^ρ different transformations which are interchangeable with all transformations of the group $X_1 f, \dots, X_r f$.*

The results concerning simply transitive groups that we gave in Chap. 20, p. 400–404, are obviously contained as a special case of the present results.

Now, we turn to the case where the r-term systatic group $X_1 f, \dots, X_r f$ is *intransitive*. However, we want to be brief here.

If the systatic group: $X_1 f, \dots, X_r f$ is intransitive, then there is not only a single transformation which transfers the point P to the point defined on p. 515 and which is interchangeable with all transformations of our group, and rather, there are infinitely many different transformations of this kind. We will indicate how one finds such transformations.

The intransitive systatic group: $X_1 f, \dots, X_r f$ determines several invariant decompositions of the space $x_1, \dots, x_s$.

A first decomposition is represented by the $s-\rho < s$ equations:

$$\text{(a)} \qquad \varphi_1(x_1, \dots, x_s) = \text{const.}, \ \dots, \ \varphi_{s-\rho}(x_1, \dots, x_s) = \text{const.}$$

A second decomposition is determined by $s-n > 0$ arbitrary independent solutions: $u_1(x), \dots, u_{s-n}(x)$ of the n-term complete system: $X_1 f = 0, \dots, X_n f = 0$; the analytic expression of this decomposition reads:

$$\text{(b)} \qquad u_1(x_1, \dots, x_s) = \text{const.}, \ \dots, \ u_{s-n}(x_1, \dots, x_s) = \text{const.}$$

In what follows, by M_ρ, we always understand one of the $\infty^{s-\rho}$ ρ-times extended manifolds (a), and by M_n, we understand one of the ∞^{s-n} n-times extended manifolds (b).

Amongst the solutions of the complete system: $X_1 f = 0, \dots, X_n f = 0$, there is a certain number, say $s-q \leqslant s-n$, which can be expressed in terms of $\varphi_1(x), \dots, \varphi_{s-\rho}(x)$ alone; we want to assume that $u_1(x), \dots, u_{s-q}(x)$ are such solutions, so that $s-q \leqslant s-\rho$ relations of the form:

$$u_1(x) = \mathfrak{U}_1\big(\varphi_1(x),\dots,\varphi_{s-\rho}(x)\big), \;\dots,\; u_{s-q}(x) = \mathfrak{U}_{s-q}\big(\varphi_1(x),\dots,\varphi_{s-\rho}(x)\big)$$

hold, and therefore, every solution of the complete system: $X_1 f = 0, \dots, X_n f = 0$ which can be expressed in terms of $\varphi_1(x),\dots,\varphi_{s-\rho}(x)$ alone is a function of $u_1(x),\dots,u_{s-q}(x)$ (cf. Chap. 19, p. 358). Then the $s-q$ equations:

$$\text{(c)} \qquad u_1(x_1,\dots,x_s) = \text{const.}, \;\dots,\; u_{s-q}(x_1,\dots,x_s) = \text{const.}$$

represent a third decomposition invariant by the group: $X_1 f,\dots,X_r f$. The individual manifolds of this decomposition are visibly the smallest manifolds which consist both of the M_ρ and of the M_n (cf. Chap. 8, p. 159). By $\mathfrak{M}_q$, we always understand in what follows one of the ∞^{s-q} q-times extended manifolds (c).

Lastly, a fourth invariant decomposition is determined by the manifold sections [SCHNITTMANNIGFALTIGKEITEN] of the M_ρ and the M_n (Chap. 8, p. 159), and this decomposition is obviously determined by the $s-\rho+q-n$ equations:

$$\text{(d)} \qquad \begin{cases} \varphi_1(x) = \text{const.}, \;\dots,\; \varphi_{s-\rho}(x) = \text{const.} \\ u_{s-q+1}(x) = \text{const.}, \;\dots,\; u_{s-n}(x) = \text{const.}, \end{cases}$$

which, according to Chap. 19, p. 358 sq., are independent of each other. In what follows, by $N_{\rho+n-q}$, we always understand one of the $\infty^{s-\rho+q-n}$ $(\rho+n-q)$-times extended manifolds (d).

Now, in order to find a transformation Θ which is interchangeable with all transformations of the group: $X_1 f,\dots,X_r f$, we proceed in the following way:

Inside every $\mathfrak{M}_q$, we associate to every M_n another M_n which we may call M_n', and to be precise, we make this association according to an arbitrary analytic law. Then, on each one of the ∞^{s-n} M_n, we choose an arbitrary point P, and to each one of the ∞^{s-n} chosen points, we associate an arbitrary point P_1 on the $N_{\rho+n-q}$ in which the M_ρ passing through the point cuts the M_n' which corresponds to the M_n passing through the point.

There is one and only one transformation Θ' which transfers the ∞^{s-n} chosen points P to the point P_1 corresponding to them. This transformation is defined by the symbolic equation:

$$(P)\,T\,\Theta = (P_1)\,T,$$

in which P is the general symbol of the ∞^{s-n} chosen points, while T is the general symbol of the ∞^r transformations of our group.

One easily convinces oneself that the transformation Θ just defined is interchangeable with all transformations of the group: $X_1 f,\dots,X_r f$ and that one obtains all transformations Θ of this constitution when one chooses in the most general way the arbitrary elements which are contained in the definition of Θ.

We do not spend time proving this; let it only be remarked that when one sets up the analytic expression of the transformation Θ, one sees immediately that the number of arbitrary functions appearing in this expression and the number of arguments appearing in these functions agree with Theorem 67, p. 388.

Finally, we make a remark which applies both to the intransitive and to the transitive systatic groups: When the manifold:

$$\varphi_{k\nu}(x_1,\dots,x_s) = \varphi_{k\nu}(x_1^0,\dots,x_s^0) \qquad (k=1\cdots r-n;\ \nu=1\cdots n)$$

decomposes for every system of values $x_1^0,\dots,x_s^0$ into several discrete manifolds, then the totality of all transformations which are interchangeable with the transformations of the group: $X_1 f,\dots,X_r f$ also decompose into several discrete families.

§ 125. The functions $\varphi_{k\nu}(x)$ which decide whether a group is systatic or not have already played a great rôle in the chapter on the similarity of r-term groups. At present, we want to return to the results of that chapter, and we want to complete them in a certain direction.

If, in the same number of variables, two r-term groups are presented:

$$X_k f = \sum_{i=1}^{s} \xi_{ki}(x_1,\dots,x_s)\,\frac{\partial f}{\partial x_i} \qquad (k=1\cdots r)$$

and:

$$Y_k f = \sum_{i=1}^{s} \eta_{ki}(y_1,\dots,y_s)\,\frac{\partial f}{\partial y_i} \qquad (k=1\cdots r),$$

and if at the same time the relations:

$$[X_i, X_k] = \sum_{\sigma=1}^{r} c_{iks}\,X_\sigma f \qquad \text{and} \qquad [Y_i, Y_k] = \sum_{\sigma=1}^{r} c_{iks}\,Y_\sigma f,$$

hold with the same constants c_{iks} in the two cases, then according to Chap. 19, p. 365 sq., there is a transformation:

$$y_\nu = \Phi_\nu(x_1,\dots,x_s) \qquad (\nu=1\cdots s)$$

which transfers $X_1 f,\dots,X_r f$ to $Y_1 f,\dots,Y_r f$, respectively, if and only if the following conditions are satisfied: if, between $X_1 f,\dots,X_r f$, there are relations of the form:

$$X_{n+k} f = \sum_{\nu=1}^{n} \varphi_{k\nu}(x_1,\dots,x_s)\,X_\nu f \qquad (k=1\cdots r-n),$$

while $X_1 f,\dots,X_r f$ are linked together by no linear relation, then between $Y_1 f,\dots,Y_r f$, there must exist analogous relations:

$$Y_{n+k} f = \sum_{\nu=1}^{n} \psi_{k\nu}(y_1,\dots,y_s)\,Y_\nu f \qquad (k=1\cdots r-n)$$

but $Y_1 f,\dots,Y_n f$ should not be linked together by linear relations; in addition, the $n(r-n)$ equations:

(e) $$\varphi_{k\nu}(x_1,\dots,x_s) = \psi_{k\nu}(y_1,\dots,y_s) \qquad (k=1\cdots r-n;\ \nu=1\cdots n)$$

should neither contradict each other, nor provide relations between the x alone or the y alone.

If, amongst the functions $\varphi_{k\nu}(x)$, there were present fewer than s that are mutually independent, then the determination of a transformation which transfers $X_1 f,\dots,X_r f$ to $Y_1 f,\dots,Y_r f$ would require certain integrations; by contrast, if the number of independent functions $\varphi_{k\nu}(x)$ were equal to s, the equations (e) would represent by themselves a transformation transferring $X_1 f,\dots,X_r f$ to $Y_1 f,\dots,Y_r f$, respectively, and in fact, the most general transformation of this nature. Hence, if we remember that in the latter case the group: $X_1 f,\dots,X_r f$ is asystatic, and naturally also the group: $Y_1 f,\dots,Y_r f$, then we obtain the:

Proposition 6. *If one knows that two r-term asystatic groups in s variables are similar and if one has already chosen, in each one of the two groups, r infinitesimal transformations:*

$$X_k f = \sum_{i=1}^{s} \xi_{ki}(x_1,\dots,x_s)\,\frac{\partial f}{\partial x_i} \qquad (k=1\cdots r)$$

and:

$$Y_k f = \sum_{i=1}^{r} \eta_{ki}(y_1,\dots,y_s)\,\frac{\partial f}{\partial y_i} \qquad (k=1\cdots r)$$

such that there exists a transformation: $y_i = \Phi_i(x_1,\dots,x_s)$ *which transfers* $X_1 f,\dots,X_r f$ *to* $Y_1 f,\dots,Y_r f$*, respectively, then one can set up without integration the most general transformation that achieves the concerned transfer; this most general transformation contains neither arbitrary functions nor arbitrary parameters.*

From this, it follows that one can find without integration the most general transformation which actually transfers the asystatic group $X_1 f,\dots,X_r f$ to the group: $Y_1 f,\dots,Y_r f$ similar to it. To this end, one has to proceed as follows:

One determines in the group: $Y_1 f,\dots,Y_r f$ in the most general way r independent infinitesimal transformations:

$$\mathsf{Y}_j f = \sum_{k=1}^{r} \overline{g}_{jk}\, Y_k f \qquad (j=1\cdots r)$$

such that firstly, the relations:

$$[\mathsf{Y}_i, \mathsf{Y}_k] = \sum_{\sigma=1}^{r} c_{iks}\, \mathsf{Y}_\sigma f$$

hold, and secondly such that there exists a transformation which transfers $X_1 f,\dots,X_r f$ to $Y_1 f,\dots,Y_r f$, respectively. Then according to what was said a short while ago, one can find without integration the most general transformation which achieves the transfer in question, and as a result, one obtains at the same time the most general transformation which converts the group: $X_1 f, \dots, X_r f$ into the group: $Y_1 f, \dots, Y_r f$. Obviously, this transformation contains only arbitrary parameters.

In particular, if the group: $Y_1 f, \dots, Y_r f$ coincides with the group: $X_1 f, \dots, X_r f$, then in the way indicated, one obtains all transformations which leave invariant the group: $X_1 f, \dots, X_r f$. According to Chap. 19, p. 372, the totality of all these transformations forms a group, and in fact in our case, visibly a finite group. Thus:

Theorem 90. *The largest subgroup in which an r-term asystatic group: $X_1 f, \dots, X_r f$ of the space $x_1, \dots, x_s$ is contained as an invariant subgroup contains only a finite number of parameters. One can find the finite equations of this group without integration as soon as the infinitesimal transformations of the group: $X_1 f, \dots, X_r f$ are given.*[4]

It is of importance that one can also find without integration the finite equations of the asystatic group: $X_1 f, \dots, X_r f$ itself, as soon as its infinitesimal transformations are given.

One simply sets up the finite equations:

$$e'_k = \sum_{j=1}^{r} \psi_{kj}(\varepsilon_1, \dots, \varepsilon_r)\, e_j \qquad (k=1\cdots r) \tag{f}$$

of the adjoint group associated to the group: $X_1 f, \dots, X_r f$; this demands only executable operations (cf. Chap. 16, p. 286). Next, if the sought finite equations of the group: $X_1 f, \dots, X_r f$ have the form: $x'_i = f_i(x_1, \dots, x_n, \varepsilon_1, \dots, \varepsilon_r)$, then according to Theorem 48, p. 287, after the introduction of the new variables: $x'_i = f_i(x, \varepsilon)$, the infinitesimal transformations $X_k f$ take the form:

$$X_k f = \sum_{j=1}^{r} \psi_{jk}(\varepsilon_1, \dots, \varepsilon_r)\, X'_j f \qquad (k=1\cdots r),$$

where, as usual, we have set:

$$\sum_{i=1}^{n} \xi_{ki}(x'_1, \dots, x'_s)\, \frac{\partial f}{\partial x'_i} = X'_k f.$$

Now, since the group: $X_1 f, \dots, X_r f$ is asystatic, there is a completely determined transformation between the x and the x' that transfers $X_1 f, \dots, X_r f$ to:

$$\sum_{j=1}^{r} \psi_{jk}(\varepsilon_1, \dots, \varepsilon_r)\, X'_j f \qquad (k=1\cdots r),$$

respectively. If one computes this transformation according to the former rules, one finds the sought equations $x'_i = f_i(x, \varepsilon)$.

---

4 LIE, Archiv for Math. Vol. 10, p. 378, Christiania 1885.

In particular, if the equations (f) are a canonical form of the adjoint group (cf. Chap. 9, p. 187), then evidently, one obtains the finite equations of the group: $X_1 f, \dots, X_r f$ also in canonical form. We therefore have the

Proposition 7. *If one knows the infinitesimal transformations of an asystatic group of the space $x_1, \dots, x_s$, then one can always find the finite equations of this group by means of executable operations and, to be precise, in canonical form.*

There exist yet more general cases for which the finite equations of an r-term group, the infinitesimal transformations of which one knows, can be determined without integration. However, we do not want to involve ourselves further in such questions, we only want to remark that the determination of the finite equations succeeds, amongst other circumstances, when there is no infinitesimal transformation interchangeable with all $X_k f$ which does not belongs to the group $X_1 f, \dots, X_r f$.

§ 126. Let $X_1 f, \dots, X_r f$, or shortly G_r, be an r-term *systatic* group of the space $x_1, \dots, x_s$, and let G_{r-n} be the $(r-n)$-term subgroup of G_r which is associated to a determined point $x_1^0, \dots, x_s^0$ in general position. The manifold:

$$\varphi_{k\nu}(x_1, \dots, x_s) = \varphi_{k\nu}(x_1^0, \dots, x_s^0) \qquad (k=1 \cdots r-n;\ \nu=1 \cdots n)$$

which consists of all points invariant by G_{r-n} will be denoted by M.

Since G_{r-n} fixes all points of M, it naturally leaves invariant M itself; but is it conceivable that G_r contains transformations which also leave invariant the manifold M without fixing all of its points. We want to assume that the largest subgroup of G_r which leaves M invariant contains exactly $r-l$ parameters and we want to call this subgroup G_{r-l}.

Of course, G_{r-n} is either identical to G_{r-l} or contained in it as a subgroup. The latter case always occurs when G_r is transitive; indeed, in this case, the point $x_1^0, \dots, x_s^0$ can be transferred to all points of M by means of suitable transformations of G_r, and since every transformation of G_r which transfers $x_1^0, \dots, x_s^0$ to another point of M visibly leaves invariant the manifold M, G_r contains a continuous family of transformations which leave invariant M without fixing all of its points; consequently, in the case of a transitive group G_r, the number $r-l$ is surely larger than $r-n$. But if G_r is intransitive, then it is quite possible that all transformations of G_r which leave M invariant also fix all points of M, so that $r-l = r-n$. This is shown for example by the three-term intransitive systatic group:

$$\frac{\partial f}{\partial x_2}, \qquad x_2 \frac{\partial f}{\partial x_2}, \qquad x_2^2 \frac{\partial f}{\partial x_2}$$

of the plane x_1, x_2.

It is easy to see that G_{r-n} is invariant in G_{r-l}. In fact, G_{r-n} is generated by all infinitesimal transformations of G_{r-l} which leave untouched all the points of M; but according to Chap. 17, Proposition 7, p. 319, these infinitesimal transformations generate an invariant subgroup of G_{r-n}. We therefore have the:

Proposition 8. *If: $X_1 f,\dots,X_r f$, or G_r, is an r-term systatic group of the space $x_1,\dots,x_s$, if moreover P is a point in general position, and lastly, if M is the manifold of all points that remain invariant by all transformations of G_r which fix P, then the largest subgroup of G_r which leaves P invariant is either identical to the largest subgroup which leaves M at rest, or is contained in this subgroup as an invariant subgroup. The first case can occur only when G_r is intransitive; when G_r is transitive, the second case always occurs.*

On the other hand, if one knows an arbitrary r-term group G_r of the space $x_1,\dots,x_s$ and if one knows that the largest subgroup G_{r-n} of G_r which leaves invariant an arbitrary point P in general position is invariant in a yet larger subgroup G_{r-h} with $r-h > r-n$ parameters, then one can conclude that G_r belongs to the class of systatic groups.

In fact, the point P admits exactly $r-n$ independent infinitesimal transformations of G_{r-h} and hence (Chap. 23, Theorem 85, p. 489), it takes, by all ∞^{r-h} transformations of G_{r-h}, exactly ∞^{n-h} different positions, where the number $n-h$, under the assumption made, is at least equal to 1. Now, since G_{r-n} is invariant in G_{r-h}, then to each of these ∞^{n-h} positions of P is associated exactly the same $(r-n)$-term subgroup of G_r as to the point P, that is to say: G_r is effectively systatic. Thus:

Proposition 9. *If the r-term group G_r of the space $x_1,\dots,x_s$ is constituted in such a way that its largest subgroup G_{r-n} which leaves invariant an arbitrary point in general position is invariant in a larger subgroup of G_r, or even in G_r itself, then G_r belongs to the class of systatic groups.*

From this proposition, it follows immediately that for an r-term asystatic group of the space $x_1,\dots,x_s$, the subgroup associated to a point in general position is invariant neither in G_r itself, nor in a larger subgroup of G_r. But conversely, if G_r is constituted in such a way that the subgroup which is associated to a point in general position is invariant in no larger subgroup, then it need not be asystatic for this reason, and it is necessarily so only when it is also transitive; this follows immediately from Proposition 8, p. 522. We can therefore say:

Proposition 10. *An r-term transitive group G_r of the space $x_1,\dots,x_s$ is asystatic if and only if the subgroup associated to a point in general position is invariant in no larger subgroup.*

In Chap. 22, Theorem 79, p. 451, we provided a method for the determination of all *transitive* groups which are equally composed with a given r-term group Γ. Now, we can specialize this method so that it provides in particular all *transitive asystatic* groups which are equally composed with the group Γ.

One determines all subgroups of Γ which neither contain a subgroup invariant in Γ nor are invariant in a larger subgroup of Γ. Each of these subgroups provides, when one proceeds according to the rules of Theorem 78, a transitive asystatic group which is equally composed with Γ, and in fact, one finds in this way all transitive asystatic groups having the concerned composition.

Taking Proposition 10 as a basis, one can answer the question of whether a given transitive group is systatic, or asystatic. One can set up a similar statement by means of which one can answer the question *whether a given transitive group is primitive, or not*.

Let the r-term *transitive* group G_r of the space $x_1,\dots,x_s$ be imprimitive, and let:

$$u_1(x_1,\dots,x_s) = \text{const.}, \quad \dots, \quad u_{s-m}(x_1,\dots,x_s) = \text{const.} \qquad (0<m<s)$$

be a decomposition of the space into ∞^{s-m} m-times extended manifolds which is invariant by the group.

Since G_r is transitive, all its transformations which leave invariant a point $x_1^0,\dots,x_s^0$ in general position form an $(r-s)$-term subgroup G_{r-s}, and on the other hand, all its transformations which leave invariant the m-times extended manifold:

$$u_1(x_1,\dots,x_s) = u_1(x^0), \quad \dots, \quad u_{s-m}(x_1,\dots,x_s) = u_{s-m}(x^0)$$

form an $(r-s+m)$-term subgroup G_{r-s+m} (cf. Chap. 23, p. 485). Here naturally, G_{r-s} is contained as subgroup in G_{r-s+m}, which in turn is obviously a subgroup of G_r.

Now we imagine conversely that an arbitrary r-term transitive group $\mathfrak{G}_r$ of the space $x_1,\dots,x_s$ is presented, and we assume that the $(r-s)$-term subgroup $\mathfrak{G}_{r-s}$ of $\mathfrak{G}_r$ which is associated to a point P in general position is contained in a larger subgroup:

$$\mathfrak{G}_{r-s+h} \qquad (r-s<r-s+h<r).$$

If all transformations of $\mathfrak{G}_{r-s+h}$ are executed on P, then the point takes ∞^h different positions. These ∞^h positions form an h-times extended manifold M which remains invariant by $\mathfrak{G}_{r-s+h}$ (cf. Chap. 23, p. 489). Lastly, if we execute on M all transformations of $\mathfrak{G}_r$, we obtain ∞^{s-h} different h-times extended manifolds which determine a decomposition of the space $x_1,\dots,x_s$.

In fact, since the manifold M remains invariant by $\mathfrak{G}_{r-s+h}$, then by the ∞^r transformations of $\mathfrak{G}_r$, it takes at most ∞^{s-h} different positions (cf. Chap. 23, p. 489), and on the other hand, thanks to the transitivity of $\mathfrak{G}_r$, it takes at least ∞^{s-h} positions, hence it receives by $\mathfrak{G}_r$ exactly ∞^{s-h} different positions which fill the space and hence determine a decomposition. It follows from Theorem 85, p. 489, that this decomposition remains invariant by $\mathfrak{G}_r$.

From this, it results that $\mathfrak{G}_r$, under the assumptions made, is imprimitive. If we combine this result with the one obtained above, we then first have:

Theorem 91. *An r-term transitive group G_r of the space $x_1,\dots,x_s$ is primitive if and only if the $(r-s)$-term subgroup G_{r-s} which is associated to a point in general position is contained in no larger subgroup of G_r.*

Furthermore, we obviously obtain at the same time a method for determining all possible decompositions of the space $x_1,\dots,x_s$ which remain invariant by a given transitive group of this space:

Theorem 92. *If an r-term transitive group G_r of the space $x_1, \dots, x_s$ is presented, then one finds all possible decompositions of the space invariant by the group in the following way:*

One first determines the $(r-s)$-term subgroup G_{r-s} of G_r by which an arbitrary point P in general position remains invariant, a point which does not lie on any manifold invariant by G_r. Then, one looks up at all subgroups of G_r which contain G_{r-s}. If G_{r-s+h} is one of these subgroups, then one executes all transformations of G_{r-s+h} on P; in the process, P takes ∞^h different positions which form an h-times extended manifold M; now, if one executes on M all transformations of G_r, then M takes ∞^{s-h} different positions which determine a decomposition of the space invariant by G_r. If one treats in this way all subgroups of G_r which comprise G_{r-s}, one obtains all decompositions invariant by G_r.

We can also indicate how one has to specialize the method explained in Chap. 22, Theorem 79, p. 451, in order to find all *primitive* groups which are equally composed with a given r-term group Γ.

Since all primitive groups are transitive, one has to proceed in the following way:

One determines all subgroups of Γ which contain no subgroup invariant in Γ and which are contained in no larger subgroup of Γ. Each of these subgroups provides, when one proceeds according to the rules of Theorem 78, a primitive group equally composed with Γ and in fact, one obtains in this way all primitive groups of this sort. —

Theorem 92 shows that one can find without integration all decompositions invariant by a *transitive* group when the finite equations of the group are known. Now, the same also holds true for intransitive groups, although in order to see this, one needs considerably longer considerations that might not be advisable to develop here.

Chapter 25
Differential Invariants

In n variables, we imagine that a transformation is presented:

$$(1) \qquad y_i = f_i(x_1,\dots,x_n) \qquad (i=1\cdots n),$$

and we imagine that this transformation is executed on a system of $n-q$ independent equations:

$$(2) \qquad \Omega_k(x_1,\dots,x_n) = 0 \qquad (k=1\cdots n-q);$$

we obtain in this way a new system of equations:

$$(2') \qquad \overline{\Omega}(y_1,\dots,y_n) = 0 \qquad (k=1\cdots n-q)$$

in the y.

Now, by virtue of the equations (2), $n-q$ of the variables $x_1,\dots,x_n$ can be represented as functions of the q remaining ones and these $n-q$ naturally possess certain differential quotients: p_1, p_2, ..., with respect to the q remaining. On the other hand, by virtue of the equations (2'), $n-q$ of the variables $y_1,\dots,y_n$ can be represented as functions of the q remaining ones and these $n-q$ possess in their turn certain differential quotients: p_1', p_2', ..., with respect to the q remaining.

Between the two series of differential quotients so defined, there exists a certain connection which is essentially independent of the form of the $n-q$ relations (2). In the course of the chapter, we will explain this known connection thoroughly, and we recall here that the p' from the first order up to the m-th order can be represented as functions of $x_1,\dots,x_n$ and of the p from the first order up to the m-th order, when the p are interpreted as functions of $x_1',\dots,x_n'$ and of the p'. From this, it results that one can derive from the transformation (1) a new transformation which, aside from the x, also transforms the differential quotients from the first order up to the m-th order. We say that *this new transformation is obtained by* extension [ERWEITERUNG] *of the transformation* (1).

Visibly, there are several very varied ways that one can extend the transformation (1), since one can choose how many and which ones of the variables $x_1,\dots,x_n$

S. Lie, *Theory of Transformation Groups I*,
DOI 10.1007/978-3-662-46211-9_25

will be functions of the others. One can even increase the number of possibilities by taking in addition auxiliary variables: $t_1, t_2, \dots$, that are not transformed by the transformation (1), so that one adds the identity transformation in the auxiliary variables: $t_1, t_2, \dots$, to the equations of the transformation (1).

If an r-term group is presented, we can extend [ERWEITERN] all its ∞^r transformations and obtain in this way ∞^r extended transformations; the latter transformations constitute in their turn, as we will see, an r-term group which is equally composed with the original group.

§ 127. We first consider a special type of extension.

In the transformation:

$$y_i = f_i(x_1, \dots, x_n) \qquad (i=1\cdots n), \tag{1}$$

we consider the variables $x_1, \dots, x_n$ as functions of an auxiliary variable t which is not transformed by the transformation (1). Obviously, $y_1, \dots, y_n$ are then also to be interpreted as functions of t; hence if we set:

$$\frac{\mathrm{d}x_i}{\mathrm{d}t} = x_i^{(1)}, \qquad \frac{\mathrm{d}y_i}{\mathrm{d}t} = y_i^{(1)},$$

then it follows from (1) by differentiation with respect to t that:

$$y_i^{(1)} = \sum_{\nu=1}^{n} \frac{\partial f_i}{\partial x_\nu}\, x_\nu^{(1)} \qquad (i=1\cdots n). \tag{3}$$

If we take (1) and (3) together, we obtain a transformation in the $2n$ variables x_i and $x_i^{(1)}$.

The transformation (1), (3) is in a certain sense the simplest transformation that one can derive from (1) by extension.

If we apply the above special extension to all ∞^r transformations of the r-term group: $X_1 f, \dots, X_r f$, or to:

$$y_i = f_i(x_1, \dots, x_n, a_1, \dots, a_r) \qquad (i=1\cdots n), \tag{4}$$

then we obtain ∞^r extended transformations which evidently have the form:

$$y_i = f_i(x_1, \dots, x_n, a_1, \dots, a_r), \qquad y_i^{(1)} = \sum_{\nu=1}^{n} \frac{\partial f_i(x,a)}{\partial x_\nu}\, x_\nu^{(1)} \qquad (i=1\cdots n). \tag{5}$$

It can be proved that the equations (5) represent an r-term group in the $2n$ variables x_i, $x_i^{(1)}$.

In fact, the equations:

$$y_i = f_i(x_1, \dots, x_n, a_1, \dots, a_r) \quad \text{and} \quad z_i = f_i(y_1, \dots, y_n, b_1, \dots, b_r)$$

have by assumption as a consequence:

$$z_i = f_i(x_1, \dots, x_n, c_1, \dots, c_r),$$

where the c depend only on the a and on the b. Therefore, from:

$$y_i^{(1)} = \sum_{\nu=1}^{n} \frac{\partial f_i(x,a)}{\partial x_\nu} x_\nu^{(1)} \quad \text{and} \quad z_j^{(1)} = \sum_{i=1}^{n} \frac{\partial f_j(y,b)}{\partial y_i} y_i^{(1)},$$

it follows by elimination of the $y_i^{(1)}$ that:

$$\begin{aligned} z_j^{(1)} &= \sum_{i,\,\nu}^{1\cdots n} \frac{\partial f_j(y,b)}{\partial y_i} \frac{\partial f_i(x,a)}{\partial x_\nu} x_\nu^{(1)} \\ &= \sum_{\nu=1}^{n} \frac{\partial f_j(y,b)}{\partial x_\nu} x_\nu^{(1)} = \sum_{\nu=1}^{n} \frac{\partial f_j(x,c)}{\partial x_\nu} x_\nu^{(1)}, \end{aligned}$$

whence the announced proof is complete.

In addition, from what has been said, it follows that the new group has the same parameter group as the original group:

$$y_i = f_i(x_1, \dots, x_n, a_1, \dots, a_r).$$

The new group:

$$\text{(5)} \qquad y_k = f_k(x,a), \qquad y_k^{(1)} = \sum_{i=1}^{n} \frac{\partial f_k}{\partial x_i} x_i^{(1)} \qquad {\scriptstyle (k=1\cdots n)}$$

which is a first example of an *extended group* has a very simple conceptual meaning.

Indeed, if one interprets $x_1, \dots, x_n$ as ordinary point coordinates of an n-times extended space, then one can interpret the $2n$ quantities: $x_1, \dots, x_n$, $x_1^{(1)}, \dots, x_n^{(1)}$ as the determination pieces [BESTIMMUNGSSTÜCKE] of a line element [LINIENELEMENT]; here $x_1^{(1)}, \dots, x_n^{(1)}$ are homogeneous coordinates in the domain of the ∞^{n-1} directions which pass through the point $x_1, \dots, x_n$.

The new group (5) in the x, $x^{(1)}$ indicates in which way the line elements of the space $x_1, \dots, x_n$ of the original group: $y_i = f_i(x,a)$ are permuted with each other.

Since the group: $y_i = f_i(x,a)$ is generated by the r infinitesimal transformations: $X_1 f, \dots, X_r f$, its finite equations in canonical form read:

$$y_k = x_k + \sum_{j=1}^{r} e_j \, \xi_{jk} + \cdots \qquad {\scriptstyle (i=1\cdots n)}.$$

If we set this form of the group: $y_i = f_i(x,a)$ as fundamental in order to make up the extended group (5), then the equations of this group take the form:

$$y_k = x_k + \sum_{j=1}^{r} e_j\,\xi_{jk} + \cdots, \qquad y_k^{(1)} = x_k^{(1)} + \sum_{j=1}^{r} e_j\,\xi_{jk}^{(1)} + \cdots \tag{6}$$

$$(k = 1 \cdots n),$$

where we have set for short:

$$\sum_{i=1}^{n} \frac{\partial \xi_{jk}}{\partial x_i}\, x_i^{(1)} = \xi_{jk}^{(1)}.$$

From this, we realize immediately that the extended group contains the identity transformation, and at the same time, we come to the presumption that it is generated by the r infinitesimal transformations:

$$X_k^{(1)} f = \sum_{i=1}^{n} \xi_{ki}\,\frac{\partial f}{\partial x_i} + \sum_{i=1}^{n} \xi_{ki}^{(1)}\,\frac{\partial f}{\partial x_i^{(1)}} \qquad (k = 1 \cdots r).$$

The correctness of this presumption can be established in the following way:

According to Theorem 3, p. 40, the n functions: $y_i = f_i(x,a)$ satisfy differential equations of the form:

$$\frac{\partial y_i}{\partial a_k} = \sum_{j=1}^{r} \psi_{kj}(a_1,\dots,a_r)\,\xi_{ji}(y_1,\dots,y_n) \qquad (i = 1 \cdots n;\; k = 1 \cdots r) \tag{7}$$

If we differentiate these equations with respect to t and if take into account that the a are independent of the t, then:

$$\frac{\partial y_i^{(1)}}{\partial a_k} = \sum_{j=1}^{r} \psi_{kj}(a) \sum_{\nu=1}^{n} \frac{\partial \xi_{ji}(y)}{\partial y_\nu}\, y_\nu^{(1)} \qquad (i = 1 \cdots n;\; k = 1 \cdots r). \tag{8}$$

Now, if $a_1^0,\dots,a_r^0$ are the parameters of the identity transformation in the group: $y_i = f_i(x,a)$, then the determinant: $\sum \pm \psi_{11}(a) \cdots \psi_{rr}(a)$ does not vanish for: $a_1 = a_1^0$, ..., $a_r = a_r^0$; but $a_1^0,\dots,a_r^0$ are also the parameters of the identity transformation in the extended group (5); consequently, from the equations (7) and (8), we can conclude in the known way (cf. Chap. 4, p. 82 sq.), that the extended group is generated by the r independent infinitesimal transformations $X_k^{(1)} f$. We call the $X_k^{(1)} f$ the *extended infinitesimal transformations*.

As the independent infinitesimal transformations of an r-term group, the $X_k^{(1)} f$ satisfy pairwise relations of the form:

$$X_k^{(1)} X_j^{(1)} f - X_j^{(1)} X_k^{(1)} f = \sum_{s=1}^{r} c'_{kjs}\, X_s^{(1)} f.$$

Here, the coefficients of the $\partial f / \partial x_i$ must coincide on the right and on the left, hence one must have: $X_k X_j f - X_j X_k f = \sum_s c'_{kjs} X_s f$, and consequently, one has:

$$\sum_{s=1}^{r}(c'_{kjs}-c_{kjs})X_s f=0,$$

or, because of the independence of $X_1 f,\dots,X_r f$:

$$c'_{kjs}=c_{kjs}\qquad (k,j,s=1\cdots r).$$

Thus, the extended group (5) is holoedrically isomorphic to the original group, which is consistent with the fact that, according to p. 529, both groups have the same parameter group. —

We give a second direct proof of the result just found.

If $X_k f$ and $X_j f$ are *arbitrary* infinitesimal transformations and if the associated two extended infinitesimal transformations are denoted by $X_k^{(1)}f$ and $X_j^{(1)}f$ as earlier on, it follows that:

$$\begin{aligned} X_k^{(1)}X_j^{(1)}f-X_j^{(1)}X_k^{(1)}f &= X_kX_jf-X_jX_kf \\ &+\sum_{\pi,\,i,\,\nu}^{1\cdots n}\left\{\xi_{ki}\frac{\partial^2\xi_{j\pi}}{\partial x_i\partial x_\nu}-\xi_{ji}\frac{\partial^2\xi_{k\pi}}{\partial x_i\partial x_\nu}\right\}x_\nu^{(1)}\frac{\partial f}{\partial x_\pi^{(1)}} \\ &+\sum_{\pi,\,i,\,\nu}^{1\cdots n}\left\{\frac{\partial\xi_{ki}}{\partial x_\nu}\frac{\partial\xi_{j\pi}}{\partial x_i}-\frac{\partial\xi_{ji}}{\partial x_\nu}\frac{\partial\xi_{k\pi}}{\partial x_i}\right\}x_\nu^{(1)}\frac{\partial f}{\partial x_\pi^{(1)}}, \end{aligned}$$

or that:

$$(9)\qquad \begin{cases} X_k^{(1)}X_j^{(1)}f-X_j^{(1)}X_k^{(1)}f=\displaystyle\sum_{\pi,\,i}^{1\cdots n}\left\{\xi_{ki}\frac{\partial\xi_{j\pi}}{\partial x_i}-\xi_{ji}\frac{\partial\xi_{k\pi}}{\partial x_i}\right\}\frac{\partial f}{\partial x_\pi} \\ \qquad\qquad\qquad\qquad +\displaystyle\sum_{\pi,\,i}^{1\cdots n}\frac{\mathrm{d}}{\mathrm{d}t}\left\{\xi_{ki}\frac{\partial\xi_{j\pi}}{\partial x_i}-\xi_{ji}\frac{\partial\xi_{k\pi}}{\partial x_i}\right\}\frac{\partial f}{\partial x_\pi^{(1)}}. \end{cases}$$

In particular, if we assume that the $X_k f$ are infinitesimal transformations of an r-term group, hence that:

$$X_kX_jf-X_jX_kf=\sum_{s=1}^{r}c_{kjs}X_sf,$$

then it follows that:

$$\begin{aligned} X_k^{(1)}X_j^{(1)}f-X_j^{(1)}X_k^{(1)}f &= \sum_{s=1}^{r}c_{kjs}\sum_{\pi=1}^{n}\xi_{s\pi}\frac{\partial f}{\partial x_\pi} \\ &+\sum_{s=1}^{r}c_{kjs}\sum_{\pi=1}^{n}\frac{\mathrm{d}}{\mathrm{d}t}\xi_{s\pi}\frac{\partial f}{\partial x_\pi}=\sum_{s=1}^{r}c_{kjs}X_s^{(1)}f, \end{aligned}$$

which is what was to be shown.[1]

The general formula (9) in the computations executed above is of special interest and it can be expressed in words as follows:

Proposition 1. *If, for two arbitrary infinitesimal transformations:*

$$X_1 f = \sum_{i=1}^{n} \xi_{1i} \frac{\partial f}{\partial x_i}, \qquad X_2 f = \sum_{i=1}^{n} \xi_{2i} \frac{\partial f}{\partial x_i},$$

one forms the extended infinitesimal transformations:

$$X_k^{(1)} f = \sum_{i=1}^{n} \xi_{ki} \frac{\partial f}{\partial x_i} + \sum_{i=1}^{n} \xi_{ki}^{(1)} \frac{\partial f}{\partial x_i^{(1)}} \qquad (k=1,2),$$

then $X_1^{(1)} X_2^{(1)} f - X_2^{(1)} X_1^{(1)} f$ is the extended infinitesimal transformation associated to $X_1 X_2 f - X_2 X_1 f$.

§ **128.** Now, we ask how one can decide whether the *equation*:

$$\sum_{\nu=1}^{n} U_\nu(x_1, \dots, x_n)\, x_\nu^{(1)} = 0$$

admits every finite transformation of the extended one-term group:

$$X^{(1)} f = \sum_{i=1}^{n} \xi_i \frac{\partial f}{\partial x_i} + \sum_{i,\,\nu}^{1\cdots n} \frac{\partial \xi_i}{\partial x_\nu} x_\nu^{(1)} \frac{\partial f}{\partial x_i^{(1)}}.$$

According to Chap. 7, p. 127, this holds true if and only if $X^{(1)}(\sum U_\nu x_\nu^{(1)})$ vanishes by virtue of $\sum U_\nu x_\nu^{(1)} = 0$. By computation, one finds:

$$X^{(1)}\left(\sum_{\nu=1}^{n} U_\nu x_\nu^{(1)}\right) = \sum_{\nu=1}^{n} X U_\nu x_\nu^{(1)} + \sum_{\nu=1}^{n} U_\nu \sum_{i=1}^{n} \frac{\partial \xi_\nu}{\partial x_i} x_i^{(1)},$$

hence an expression which is linear in the $x_i^{(1)}$. *Consequently, the equation: $\sum U_\nu x_\nu^{(1)} = 0$ admits the one-term group $X^{(1)} f$ when and only when a relation of the form:*

$$X^{(1)}\left(\sum_{\nu=1}^{n} U_\nu x_\nu^{(1)}\right) = \rho \sum_{\nu=1}^{n} U_\nu x_\nu^{(1)}$$

holds, where it is understood that ρ is a function of $x_1, \dots, x_n$ alone.

[1] The preceding analytic developments present great similarities with certain developments of Chap. 20 (cf. p. 382 sq.). The main reason of this connection lies in the fact that the quantities ζ on p. 381 are nothing but certain differential quotients: $\delta y/\delta t$ of the y with respect to t.

Incidentally, it may be observed that the *expression* $\sum U_\nu x_\nu^{(1)}$ always remains invariant by every finite transformation of the extended group $X^{(1)}f$ if and only if the expression: $X^{(1)}\left(\sum U_\nu x_\nu^{(1)}\right)$ vanishes identically.

Lastly, a system of m equations:

$$\sum_{i=1}^{n} U_{ki}(x_1,\dots,x_n)\,x_i^{(1)} = 0 \qquad (k=1\cdots m) \tag{10}$$

will admit all transformations of the extended group $X^{(1)}f$ if and only if every expression: $X^{(1)}\left(\sum U_{ki}x_i^{(1)}\right)$ vanishes by virtue of the system, hence when relations of the form:

$$X^{(1)}\left(\sum_{i=1}^{n} U_{ki}x_i^{(1)}\right) = \sum_{j=1}^{m} \rho_{kj} \sum_{i=1}^{n} U_{ji}x_i^{(1)} \qquad (k=1\cdots m)$$

hold, where the ρ_{kj} denote functions of the x alone.

If we combine Proposition 5 of Chap. 7, p. 134, with the proposition proved above (p. 532), we immediately get:

Proposition 2. *If a system of equations of the form:*

$$\sum_{i=1}^{n} U_{ki}(x_1,\dots,x_n)\,x_i^{(1)} = 0 \qquad (k=1\cdots m) \tag{10}$$

admits the two one-term groups: $X_1^{(1)}f$ and $X_2^{(1)}f$ which are derived from X_1f and X_2f, respectively, by means of the special extension defined above, then it admits at the same time the one-term group: $X_1^{(1)}X_2^{(1)}f - X_2^{(1)}X_1^{(1)}f$, which is derived by means of extension from the group: $X_1X_2f - X_2X_1f$.

We have obtained the preceding proposition by applying Proposition 5 of Chap. 7 to the special case of a system of equations of the form (10). But it must be mentioned that our present proposition can be proved in a way substantially easier than the general Proposition 5 of Chap. 7. Indeed, one can convince oneself without difficulty by a calculation that a system of equations (10) which admits $X_1^{(1)}f$ and $X_2^{(1)}f$ also allows $X_1^{(1)}X_2^{(1)}f - X_2^{(1)}X_1^{(1)}f$.

We call an equation of the form:

$$\sum_{i=1}^{n} U_i(x_1,\dots,x_n)\,\mathrm{d}x_i = 0$$

a *Pfaffian equation* [PFAFFSCHE GLEICHUNG], and we call its left-hand side a *Pfaffian expression* [PFAFFSCHE AUSDRUCK].

From the developments of the present chapter so far, we have important propositions concerning systems of Pfaffian equations:

$$\text{(10')} \qquad \sum_{i=1}^{n} U_{ki}(x_1,\dots,x_n)\,\mathrm{d}x_i = 0 \qquad {\scriptstyle (k=1\cdots m)}.$$

The execution of the transformation: $y_i = f_i(x_1,\dots,x_n)$ on the differential expression: $\sum_i U_{ki}(x_1,\dots,x_n)\,\mathrm{d}x_i$ obviously happens in such a way that one transforms the $2n$ quantities $x_1,\dots,x_n$, $\mathrm{d}x_1,\dots,\mathrm{d}x_n$ by means of the transformation:

$$y_i = f_i(x_1,\dots,x_n), \qquad y_i = \sum_{\nu=1}^{n} \frac{\partial f_i}{\partial x_\nu}\,\mathrm{d}x_\nu \qquad {\scriptstyle (i=1\cdots n)}.$$

From this, it follows that the system of Pfaffian equations:

$$\text{(10')} \qquad \sum_{i=1}^{n} U_{ki}(x_1,\dots,x_n)\,\mathrm{d}x_i = 0 \qquad {\scriptstyle (k=1\cdots m)}$$

remains invariant by the one-term group Xf if and only if the system of equations:

$$\sum_{i=1}^{n} U_{ki}(x_1,\dots,x_n)\,x_i^{(1)} = 0 \qquad {\scriptstyle (k=1\cdots m)}$$

admits the one-term extended group:

$$X^{(1)}f = \sum_{i=1}^{n} \xi_i \frac{\partial f}{\partial x_i} + \sum_{i=1}^{n} \xi_i^{(1)} \frac{\partial f}{\partial x_i^{(1)}},$$

hence when we have m relations of the form:

$$\sum_{i=1}^{n} X\,U_{ki}\,x_i^{(1)} + \sum_{i=1}^{n} U_{ki} \sum_{\nu=1}^{n} \frac{\partial \xi_i}{\partial x_\nu}\,x_\nu^{(1)} = \sum_{j=1}^{m} \rho_{kj}(x) \sum_{i=1}^{n} U_{ji}\,x_i^{(1)}$$

$$(k=1\cdots m),$$

or, what amounts to the same, m relations of the form:

$$\sum_{i=1}^{n} X\,U_{ki}\,\mathrm{d}x_i + \sum_{i=1}^{n} U_{ki}\,\mathrm{d}(X\,x_i) = \sum_{j=1}^{m} \rho_{kj}(x) \sum_{i=1}^{n} U_{ji}\,\mathrm{d}x_i$$

$$(k=1\cdots m).$$

Hence if we introduce the shorthand:

$$\sum_{i=1}^{n} X\,U_{ki}\,\mathrm{d}x_i + \sum_{i=1}^{n} U_{ki}\,\mathrm{d}(X\,x_i) = X\left(\sum_{i=1}^{n} U_{ki}\,\mathrm{d}x_i\right),$$

we obtain the

Proposition 3. *If the system of m Pfaffian equations:*

$$\text{(10')} \qquad \sum_{i=1}^{n} U_{ki}(x_1,\ldots,x_n)\,\mathrm{d}x_i = 0 \qquad (k=1\cdots m).$$

admits all transformations: $y_i = x_i + \varepsilon X x_i + \cdots$ *of the one-term group* Xf*, hence if for every value of* ε*, relations of the form:*

$$\sum_{i=1}^{n} U_{ki}(y_1,\ldots,y_n)\,\mathrm{d}y_i = \sum_{j=1}^{m} \omega_{kj}(x_1,\ldots,x_n,\varepsilon)\sum_{i=1}^{n} U_{ji}(x)\,\mathrm{d}x_i \qquad (k=1\cdots m)$$

hold, then for this, it is necessary and sufficient that the m expressions $X(\sum U_{ki}\,\mathrm{d}x_i)$ *can be represented in the form:*

$$X\left(\sum_{i=1}^{n} U_{ki}\,\mathrm{d}x_i\right) = \sum_{j=1}^{m} \rho_{kj}(x)\sum_{i=1}^{n} U_{ji}\,\mathrm{d}x_i \qquad (k=1\cdots m).$$

Lastly, if we introduce the terminology: *the system of Pfaffian equations* (10') *admits the infinitesimal transformation* Xf *when m relations of the form:*

$$X\left(\sum_{i=1}^{n} U_{ki}\,\mathrm{d}x_i\right) = \sum_{j=1}^{m} \rho_{kj}(x)\sum_{i=1}^{n} U_{ji}\,\mathrm{d}x_i$$

hold, then we can also state the latter proposition in the following way:

The system of m Pfaffian equations (10') *admits all transformations of the one-term group* Xf *if and only if it admits the infinitesimal transformation* Xf.

In addition, from Proposition 2, it follows easily:

Theorem 93. *If a system of m Pfaffian equations:*

$$\sum_{i=1}^{n} U_{ki}(x_1,\ldots,x_n)\,\mathrm{d}x_i = 0 \qquad (k=1\cdots m)$$

admits the two one-term groups: X_1f *and* X_2f*, then at the same time, it also admits the one-term group:* $X_1X_2f - X_2X_2f$.

We remark that the Pfaffian *expression*: $\sum_i U_i(x_1,\ldots,x_n)\,\mathrm{d}x_i$ remains invariant by every transformation of the one-term group Xf when the expression: $X(\sum U_i\,\mathrm{d}x_i)$ vanishes identically.

Earlier (p. 68 sq.), we agreed that in place of ξ_i, we will occasionally write $\delta x_i/\delta t$ and in place of Xf, $\delta f/\delta t$. In a similar way, in place of $X(\sum U_i\,\mathrm{d}x_i)$, we also want to write: $\frac{\delta}{\delta t}\sum U_i\,\mathrm{d}x_i$; then we have the equation:

$$\delta\left(\sum_{i=1}^{n} U_i\,\mathrm{d}x_i\right) = \sum_{i=1}^{n} \delta U_i\,\mathrm{d}x_i + \sum_{i=1}^{n} U_i\,\mathrm{d}\,\delta x_i,$$

an expression which appears with the same meaning as in the Calculus of Variations.

§ 129. The propositions of the preceding section perfectly suffice in order derive in complete generality the theory of the extension of a group by means of the addition of differential quotients. However, we first want to consider a special case, before we tackle the treatment of the general case. At the same time, we return to known geometrical concepts of thrice-extended space.

By x, y, z, we understand ordinary point coordinates of thrice-extended space and moreover, we let:

$$x' = \Xi(x,y,z), \qquad y' = \mathsf{H}, \qquad z' = \mathsf{Z} \tag{11}$$

be a transformation of this space.

By the transformation (11), all points x, y, z are transformed, that is to say, they are transferred to the new positions: x', y', z'. At the same time, all surfaces take new positions: every surface: $\chi(x,y,z) = 0$ is transferred to a new surface: $\psi(x',y',z') = 0$, the equation of which is obtained by means of elimination of x, y, z from the equations (11) of the transformation in combination with $\chi = 0$.

It is in the nature of things that the transformation (11) transfers surfaces which enter into contact to surfaces which stand in the same relationship, at least in general. Hence, if by the expression *surface element* [FLÄCHENELEMENT] we call *the totality of a point x, y, z located on the surface and of the tangential plane:*

$$z_1 - z = p\,(x_1 - x) + q\,(y_1 - y) \qquad \left(p = \frac{\partial z}{\partial x}, \quad q = \frac{\partial z}{\partial y}\right)$$

passing through it, then we can say that our transformation (11) converts every surface element x, y, z, p, q to a new surface element x', y', z', p', q'. In other words, there must exist certain equations:

$$p' = \Pi(x,y,z,p,q), \qquad q' = \mathsf{K}(x,y,z,p,q) \tag{12}$$

which we will set up below. The equations (11) and (12) taken together represent a transformation which comes into existence by extension of the transformation (11).

At present, we assume that the ∞^r transformations:

$$x' = \Xi(x,y,z,a_1,\dots,a_r), \qquad y' = \mathsf{H}, \qquad z' = \mathsf{Z} \tag{13}$$

are given which represent an r-term group of point transformations of the space, and we imagine that for each of these ∞^r transformations, the extended transformation just defined is set up. We claim that the ∞^r extended transformations:

$$\begin{cases} x' = \Xi(x,y,z,a_1,\dots,a_r), \qquad y' = \mathsf{H}, \qquad z' = \mathsf{Z} \\ p' = \Pi(x,y,z,p,q,a_1,\dots,a_r), \qquad q' = \mathsf{H} \end{cases} \tag{14}$$

that we obtain in this way form an r-term group in the variables x, y, z, p, q.

For the proof, we interpret the ∞^r transformations (13) as operations and we observe that these operations permute with each other both the points x,y,z and the surface elements x,y,z,p,q. When regarded as permutations of the points, these operations form a group, and consequently, when interpreted as permutations of the surface elements, they must also form a group, whence the fact that the equations (14) represent a group finds its analytic expression.

We will make more precise the considerations made up to now by developing them in a purely analytic way.

Let:

$$x' = \Xi(x,y,z), \qquad y' = \mathsf{H}(x,y,z), \qquad z' = \mathsf{Z}(x,y,z) \tag{11}$$

be a transformation between the variables x,y,z and x',y',z'. If we imagine z as an arbitrarily chosen function: $z = \varphi(x,y)$ of x and y, there exist partial differential quotients of first order of z with respect to x and y; these are defined by means of the equation:

$$\mathrm{d}z - p\,\mathrm{d}x - q\,\mathrm{d}y = 0.$$

But on the other hand, our transformation converts every dependence relationship [ABHÄNGIGSKEITSVERHÄLTNISS]: $z = \varphi(x,y)$ between x,y,z into just such a relationship between z',x',y' which can in general be given the form: $z' = \overline{\varphi}(x',y')$; hence z' also has two partial differential quotients p' and q' which, in their turn, satisfy the condition:

$$\mathrm{d}z' - p'\,\mathrm{d}x' - q'\,\mathrm{d}y' = 0.$$

If, in the equation just written, we substitute z',x',y' with their values Z, Ξ, H and if we organize the result with respect to $\mathrm{d}z$, $\mathrm{d}x$, $\mathrm{d}y$, we get:

$$\left(\frac{\partial \mathsf{Z}}{\partial z} - p'\frac{\partial \Xi}{\partial z} - q'\frac{\partial \mathsf{H}}{\partial z}\right)\mathrm{d}z + \left(\frac{\partial \mathsf{Z}}{\partial x} - p'\frac{\partial \Xi}{\partial x} - q'\frac{\partial \mathsf{H}}{\partial x}\right)\mathrm{d}x$$
$$+\left(\frac{\partial \mathsf{Z}}{\partial y} - p'\frac{\partial \Xi}{\partial y} - q'\frac{\partial \mathsf{H}}{\partial y}\right)\mathrm{d}y = 0,$$

or, because of $\mathrm{d}z = p\,\mathrm{d}x + q\,\mathrm{d}y$:

$$\begin{cases} \left\{\dfrac{\partial \mathsf{Z}}{\partial x} - p'\dfrac{\partial \Xi}{\partial x} - q'\dfrac{\partial \mathsf{H}}{\partial x} + p\left(\dfrac{\partial \mathsf{Z}}{\partial z} - p'\dfrac{\partial \Xi}{\partial z} - q'\dfrac{\partial \mathsf{H}}{\partial z}\right)\right\}\mathrm{d}x \\ + \left\{\dfrac{\partial \mathsf{Z}}{\partial y} - p'\dfrac{\partial \Xi}{\partial y} - q'\dfrac{\partial \mathsf{H}}{\partial y} + q\left(\dfrac{\partial \mathsf{Z}}{\partial z} - p'\dfrac{\partial \Xi}{\partial z} - q'\dfrac{\partial \mathsf{H}}{\partial z}\right)\right\}\mathrm{d}y = 0. \end{cases} \tag{15}$$

In the equation (15), the two factors of $\mathrm{d}x$ and of $\mathrm{d}y$ must vanish, because $\mathrm{d}x$ and $\mathrm{d}y$ are not linked together by a relation. We therefore obtain the two equations:

$$(16)\qquad \begin{cases} p'\left(\dfrac{\partial \Xi}{\partial x}+p\dfrac{\partial \Xi}{\partial z}\right)+q'\left(\dfrac{\partial \mathsf{H}}{\partial x}+p\dfrac{\partial \mathsf{H}}{\partial z}\right)=\dfrac{\partial \mathsf{Z}}{\partial x}+p\dfrac{\partial \mathsf{Z}}{\partial z} \\ p'\left(\dfrac{\partial \Xi}{\partial y}+q\dfrac{\partial \Xi}{\partial z}\right)+q'\left(\dfrac{\partial \mathsf{H}}{\partial y}+q\dfrac{\partial \mathsf{H}}{\partial z}\right)=\dfrac{\partial \mathsf{Z}}{\partial y}+q\dfrac{\partial \mathsf{Z}}{\partial z}. \end{cases}$$

These equations can be solved with respect to p' and q'. Indeed, if the determinant:

$$(17)\qquad \begin{vmatrix} \frac{\partial \Xi}{\partial x}+p\frac{\partial \Xi}{\partial z} & \frac{\partial \Xi}{\partial y}+q\frac{\partial \Xi}{\partial z} \\ \frac{\partial \mathsf{H}}{\partial x}+p\frac{\partial \mathsf{H}}{\partial z} & \frac{\partial \mathsf{H}}{\partial y}+q\frac{\partial \mathsf{H}}{\partial z} \end{vmatrix}$$

vanished for arbitrary functions z of x and y, then it would actually vanish identically, that is to say, for every value of the variables x,y,z,p,q. Evidently, this case can occur only when all 2×2 determinants of the matrix:

$$\begin{vmatrix} \frac{\partial \Xi}{\partial x} & \frac{\partial \Xi}{\partial y} & \frac{\partial \Xi}{\partial z} \\ \frac{\partial \mathsf{H}}{\partial x} & \frac{\partial \mathsf{H}}{\partial y} & \frac{\partial \mathsf{H}}{\partial z} \end{vmatrix}$$

vanish identically. But this is excluded from the beginning.

We therefore see that the equations (16) are in general solvable with respect to p' and q', whatever the function z of x and y is, and that the resolution is impossible only when the function $z=\varphi(x,y)$ satisfies the partial differential equation which arises by setting to zero the determinant (17). By executing the resolution of the equations (16), we obtain for p' and q' completely determined functions of x,y,z,p,q:

$$p'=\Pi(x,y,z,p,q),\qquad q'=\mathsf{K}(x,y,z,p,q).$$

This determination is generally valid, because we have made no special assumption on the function $z=\varphi(x,y)$.

Besides, all of this has long been known.

The equation (15) shows that it is possible to determine a quantity $\mathfrak{a}$ in such a way that the relation:

$$(18)\qquad \mathrm{d}\mathsf{Z}-p'\,\mathrm{d}\Xi-q'\,\mathrm{d}\mathsf{H}=\mathfrak{a}\,(\mathrm{d}z-p\,\mathrm{d}x-q\,\mathrm{d}y)$$

holds identically in $\mathrm{d}x$, $\mathrm{d}y$, $\mathrm{d}z$. Indeed, if one expands (18) with respect to $\mathrm{d}x$, $\mathrm{d}y$, $\mathrm{d}z$, one first obtains by comparing the factors of $\mathrm{d}z$ in both sides:

$$\mathfrak{a}=\frac{\partial \mathsf{Z}}{\partial z}-p'\frac{\partial \Xi}{\partial z}-q'\frac{\partial \mathsf{H}}{\partial z};$$

but if one sets this value in the equation (18), then this equation converts into (15).

In turn, equation (18) has a very simple meaning: *it expresses that the extended transformation:*

$$x'=\Xi,\qquad y'=\mathsf{H},\qquad z'=\mathsf{Z},\qquad p'=\Pi,\qquad q'=\mathsf{K}$$

leaves invariant the Pfaffian equation: $\mathrm{d}z - p\mathrm{d}x - q\mathrm{d}y = 0$.[2]

For what follows, it is very important to observe that the extended transformation: $x' = \Xi, \dots, q' = \mathsf{K}$ is perfectly determined by this property. In other words: if one knows Ξ, H, Z, then Π and K are uniquely determined by the requirement that the transformation: $x' = \Xi, \dots, q' = \mathsf{K}$ should leave invariant the Pfaffian equation: $\mathrm{d}z - p\mathrm{d}x - q\mathrm{d}y = 0$.

Now, we again pass to the consideration of an r-term group:

$$x' = \Xi(x,y,z,a_1,\dots,a_r), \qquad y' = \mathsf{H}, \qquad z' = \mathsf{Z}, \tag{19}$$

and we imagine that each of the ∞^r transformations of this group is extended in the way formerly indicated by adding the equations:

$$p' = \Pi(x,y,z,p,q,a_1,\dots,a_r), \qquad q' = \mathsf{K}. \tag{20}$$

Then we already know (cf. p. 536) that the equations (19) and (20) taken together represent an r-term group, but at present, we also want to prove this in an analytic way.

If the equations:

$$x'' = \Xi(x',y',z',b_1,\dots,b_r), \qquad y'' = \mathsf{H}, \qquad z'' = \mathsf{Z}$$

are combined with (19), then one obtains in the known way:

$$x'' = \Xi(x,y,z,c_1,\dots,c_r), \qquad y'' = \mathsf{H}, \qquad z'' = \mathsf{Z},$$

where c depends only on the a and on the b. But on the other hand, the two equations:

$$\mathrm{d}z' - p'\mathrm{d}x' - q'\mathrm{d}y' = \mathfrak{a}(\mathrm{d}z - p\mathrm{d}x - q\mathrm{d}y)$$
$$\mathrm{d}z'' - p''\mathrm{d}x'' - q''\mathrm{d}y'' = \mathfrak{a}'(\mathrm{d}z' - p'\mathrm{d}x' - q'\mathrm{d}y')$$

give the analogous equation:

$$\mathrm{d}z'' - p''\mathrm{d}x'' - q''\mathrm{d}y'' = \mathfrak{a}\mathfrak{a}'(\mathrm{d}z - p\mathrm{d}x - q\mathrm{d}y),$$

which, according to the observation made above, shows that p'' and q'' depend on x,y,z,p,q and on the c in exactly the same way as p' and q' depend on x,y,z,p,q and on the a. We therefore have the

Proposition 4. *If the equations:*

$$x' = \Xi(x,y,z,a_1,\dots,a_r), \qquad y' = \mathsf{H}, \qquad z' = \mathsf{Z} \tag{19}$$

represent an r-term group and if one determines the functions $\Pi(x,y,z,a_1,\dots,a_r)$, $\mathsf{K}(x,y,z,a_1,\dots,a_r)$ in such a way that the extended transformation equations:

---

[2] LIE, Göttinger Nachr. 1872, p. 480, Verhandl. d. G. d. W. zu Christiania 1873; Math. Ann. Vol. VIII.

$$(21)\qquad x' = \Xi, \qquad y' = \mathsf{H}, \qquad z' = \mathsf{Z}, \qquad p' = \Pi, \qquad q' = \mathsf{K}$$

leave invariant the Pfaffian equation: $\mathrm{d}z - p\,\mathrm{d}x - q\,\mathrm{d}y = 0$*, so that a relation of the form:*

$$\mathrm{d}z' - p'\,\mathrm{d}x' - q'\,\mathrm{d}y' = \mathfrak{a}(x,y,z,p,q,a_1,\dots,a_r)\,(\mathrm{d}z - p\,\mathrm{d}x - q\,\mathrm{d}y)$$

holds, then these transformation equations represent in the same way an r-term group and, to be precise, a group which has the same parameter group as the original group.

We now claim: If the r-term group (19) is generated by r independent infinitesimal transformations — and as always, we naturally also assume this here —, then at the same time, the same holds true of the extended group (21).

We will prove this claim at first for the simple case: $r = 1$. Therefore let $r = 1$ and let the group (19) be generated by the infinitesimal transformation:

$$Xf = \xi\,\frac{\partial f}{\partial x} + \eta\,\frac{\partial f}{\partial y} + \zeta\,\frac{\partial f}{\partial z}.$$

In order to prove that the extended group is generated by an infinitesimal transformation, we only need to show that from Xf, an extended infinitesimal transformation:

$$X^{(1)}f = \xi\,\frac{\partial f}{\partial x} + \eta\,\frac{\partial f}{\partial y} + \zeta\,\frac{\partial f}{\partial z} + \pi\,\frac{\partial f}{\partial p} + \chi\,\frac{\partial f}{\partial q}$$

can be derived which leaves invariant the Pfaffian equation:

$$(22)\qquad \mathrm{d}z - p\,\mathrm{d}x - q\,\mathrm{d}y = 0.$$

If this is proved, then it is clear that the one-term group in the variables x, y, z, p, q generated by $X^{(1)}f$ is identical to the one-term group (21), for this last group then leaves invariant the Pfaffian equation (22) and comes into existence by extension of the group Xf which, according to the assumption, is just the group (19).

According to p. 535, the infinitesimal transformation $X^{(1)}f$ leaves invariant the Pfaffian equation (22) if and only if π and χ satisfy an equation of the form:

$$\mathrm{d}\zeta - p\,\mathrm{d}\xi - q\,\mathrm{d}\eta - \pi\,\mathrm{d}x - \chi\,\mathrm{d}y = \mathfrak{b}\,(\mathrm{d}z - p\,\mathrm{d}x - q\,\mathrm{d}y),$$

where it is understood that $\mathfrak{b}$ is a function of x, y, z, p, q. This equation decomposes into three equations:

$$\begin{aligned}
\mathfrak{b} &= \frac{\partial \zeta}{\partial z} - p\,\frac{\partial \xi}{\partial z} - q\,\frac{\partial \eta}{\partial z}\\
\pi &= \frac{\partial \zeta}{\partial x} - p\,\frac{\partial \xi}{\partial x} - q\,\frac{\partial \eta}{\partial x} + \mathfrak{b}\,p\\
\chi &= \frac{\partial \zeta}{\partial y} - p\,\frac{\partial \xi}{\partial y} - q\,\frac{\partial \eta}{\partial y} + \mathfrak{b}\,q,
\end{aligned}$$

whence we obtain for π and χ the completely determined values:

$$\pi = \frac{\mathrm{d}\zeta}{\mathrm{d}x} - p\frac{\mathrm{d}\xi}{\mathrm{d}x} - q\frac{\mathrm{d}\eta}{\mathrm{d}x}, \qquad \chi = \frac{\mathrm{d}\zeta}{\mathrm{d}y} - p\frac{\mathrm{d}\xi}{\mathrm{d}y} - q\frac{\mathrm{d}\eta}{\mathrm{d}y},$$

where for brevity, we have set:

$$\frac{\partial\Phi(x,y,z)}{\partial x} + p\frac{\partial\Phi(x,y,z)}{\partial z} = \frac{\mathrm{d}\Phi}{\mathrm{d}x}, \qquad \frac{\partial\Phi(x,y,z)}{\partial y} + q\frac{\partial\Phi(x,y,z)}{\partial z} = \frac{\mathrm{d}\Phi}{\mathrm{d}y}.$$

With these words, the existence of the infinitesimal transformation $X^{(1)}f$ is proved, and also at the same time, the correctness for $r = 1$ of the claim stated above.

Now, let r be arbitrary. Next, if $e_1 X_1 f + \cdots + e_r X_r f$ is an arbitrary infinitesimal transformation of the group (19), then evidently, the one-term group which is generated by the extended infinitesimal transformation: $e_1 X_1^{(1)} f + \cdots + e_r X_r^{(1)} f$ belongs to the group (21), because the concerned one-term group indeed comes into existence by extension of the one-term group: $e_1 X_1 f + \cdots + e_r X_r f$. From this, it follows that the group (21) contains the ∞^{r-1} one-term groups: $e_1 X_1^{(1)} f + \cdots + e_r X_r^{(1)} f$, so that it is generated by the r independent infinitesimal transformations: $X_1^{(1)} f, \dots, X_r^{(1)} f$.

Of course, the $X_k^{(1)} f$ satisfy pairwise relations of the form:

$$X_i^{(1)} X_k^{(1)} f - X_k^{(1)} X_i^{(1)} f = \sum_{s=1}^{r} c'_{iks} X_s^{(1)} f;$$

we will verify this by means of a computation, and hence in addition, we will recognize that the c'_{iks} here have the same values as the c_{iks} in the relations:

$$X_i X_k f - X_k X_i f = \sum_{s=1}^{r} c_{iks} X_s f.$$

According to Theorem 93, p. 535, the Pfaffian equation: $\mathrm{d}z - p\,\mathrm{d}x - q\,\mathrm{d}y = 0$ admits, simultaneously with the two infinitesimal transformations: $X_i^{(1)} f$ and $X_k^{(1)} f$ and also the following: $X_i^{(1)} X_k^{(1)} f - X_k^{(1)} X_i^{(1)} f$, which visibly has the form:

$$X_i^{(1)} X_k^{(1)} f - X_k^{(1)} X_i^{(1)} f = X_i X_k f - X_k X_i f + \alpha\frac{\partial f}{\partial p} + \beta\frac{\partial f}{\partial q}.$$

From this, it follows that the infinitesimal transformation: $X_i^{(1)} X_k^{(1)} f - X_k^{(1)} X_i^{(1)} f$ is obtained by extension of:

$$X_i X_k f - X_k X_i f = \sum_{s=1}^{r} c_{iks} X_s f,$$

so that the relations:

$$X_i^{(1)} X_k^{(1)} f - X_k^{(1)} X_i^{(1)} f = \sum_{s=1}^{r} c_{iks} X_s^{(1)} f$$

hold.

From this, we see that the extended group (21) is holoedrically isomorphic to the group (19), which is coherent with the fact that the two groups have the same parameter group (Proposition 4, p. 539).

The group:

$$\text{(19)} \qquad x' = \Xi(x, y, z, a_1, \dots, a_r), \qquad y' = \mathsf{H}, \qquad z' = \mathsf{Z}$$

can still be extended further, namely by also taking differential quotients of order higher than the first order. Here, we only want to consider the extension by means of the addition of differential quotients of second order.

Since z is considered as a function of x and y, apart from p and q, we also have to take into account the three differential quotients:

$$\frac{\partial^2 z}{\partial x^2} = r, \qquad \frac{\partial^2 z}{\partial x \partial y} = s, \qquad \frac{\partial^2 z}{\partial y^2} = t.$$

Through the transformations of our group, x', y', z' depend in the known way upon x, y, z, and according to the preceding, p', q' depend likewise upon x, y, z, p, q. As should at present be shown, r', s', t' can also be represented as functions of x, y, z, p, q, r, s, t:

$$r' = \mathsf{P}(x, y, z, p, q, r, s, t; a_1, \dots, a_r), \qquad s' = \Sigma, \qquad t' = \mathsf{T}.$$

The quantities r', s', t' are defined by:

$$\mathrm{d}p' - r'\,\mathrm{d}x' - s'\,\mathrm{d}y' = 0, \qquad \mathrm{d}q' - s'\,\mathrm{d}x' - t'\,\mathrm{d}y' = 0.$$

Here, if the $\mathrm{d}x'$, $\mathrm{d}y'$, $\mathrm{d}p'$, $\mathrm{d}q'$ are expanded with respect to $\mathrm{d}x$, $\mathrm{d}y$, $\mathrm{d}z$, $\mathrm{d}p$, $\mathrm{d}q$, and then, if the values for $\mathrm{d}z$, $\mathrm{d}p$, $\mathrm{d}q$ are inserted from:

$$\mathrm{d}z - p\,\mathrm{d}x - q\,\mathrm{d}y = 0, \quad \mathrm{d}p - r\,\mathrm{d}x - s\,\mathrm{d}y = 0, \quad \mathrm{d}q - s\,\mathrm{d}x - t\,\mathrm{d}y = 0,$$

then one obtains two equations of the form: $A\,\mathrm{d}x + B\,\mathrm{d}y = 0$ whose coefficients, apart from x', y', z' and their differential quotients, also contain p', q', r', s', t', p, q, r, s, t. Since $\mathrm{d}x$ and $\mathrm{d}y$ are independent of each other, A and B must in the two cases vanish individually, and hence we find the four equations:

$$r'\left(\frac{\partial \Xi}{\partial x}+p\frac{\partial \Xi}{\partial z}\right)+s'\left(\frac{\partial \mathsf{H}}{\partial x}+p\frac{\partial \mathsf{H}}{\partial z}\right)=\frac{\partial \Pi}{\partial x}+p\frac{\partial \Pi}{\partial z}+r\frac{\partial \Pi}{\partial p}+s\frac{\partial \Pi}{\partial q}$$

$$r'\left(\frac{\partial \Xi}{\partial y}+q\frac{\partial \Xi}{\partial z}\right)+s'\left(\frac{\partial \mathsf{H}}{\partial y}+q\frac{\partial \mathsf{H}}{\partial z}\right)=\frac{\partial \Pi}{\partial y}+q\frac{\partial \Pi}{\partial z}+s\frac{\partial \Pi}{\partial p}+t\frac{\partial \Pi}{\partial q}$$

and so on.

Of these equations, the first two are solvable with respect to r' and s', the last two with respect to s' and t', exactly as the equations (16) were solvable with respect to p' and q'. Still, the question is only whether the two values which are found in this way for s' coincide with each other. But as is known, this is indeed the case, since otherwise, we would obtain a relation between x, y, z, p, q, r, s, t not holding identically which should be satisfied identically after the substitution: $z=\varphi(x,y)$, and this is impossible, because φ is submitted to absolutely no restriction.

Similarly as before, we also realize here that the quantities r', s', t' are defined uniquely as functions of x, y, z, p, q, r, s, t by the condition that equations of the form:

$$\begin{aligned} \mathrm{d}p'-r'\,\mathrm{d}x'-s'\,\mathrm{d}y' &= \alpha_1(\mathrm{d}z-p\,\mathrm{d}x-q\,\mathrm{d}y)+\beta_1(\mathrm{d}p-r\,\mathrm{d}x-s\,\mathrm{d}y) \\ &\quad +\gamma_1(\mathrm{d}q-s\,\mathrm{d}x-t\,\mathrm{d}y) \\ \mathrm{d}q'-s'\,\mathrm{d}x'-t'\,\mathrm{d}y' &= \alpha_2(\mathrm{d}z-p\,\mathrm{d}x-q\,\mathrm{d}y)+\beta_2(\mathrm{d}p-r\,\mathrm{d}x-s\,\mathrm{d}y) \\ &\quad +\gamma_2(\mathrm{d}q-s\,\mathrm{d}x-t\,\mathrm{d}y) \end{aligned}$$

hold identically in $\mathrm{d}p$, $\mathrm{d}q$, $\mathrm{d}z$, $\mathrm{d}y$, $\mathrm{d}x$. For this, if we remember the former identity:

$$\mathrm{d}z'-p'\,\mathrm{d}x'-q'\,\mathrm{d}y'=\mathfrak{a}(\mathrm{d}z-p\,\mathrm{d}x-q\,\mathrm{d}y)$$

which determine p' and q', we can say that for given Ξ, H, Z, the transformation equations:

(23) $\quad x'=\Xi(x,y,z,a_1,\dots,a_r),\ \dots,\ t'=\mathsf{T}(x,y,z,p,q,r,s,t,a_1,\dots,a_r)$

are completely and uniquely determined by the condition that they should leave invariant the system of Pfaffian equations:

(24) $\quad \mathrm{d}z-p\,\mathrm{d}x-q\,\mathrm{d}y=0, \qquad \mathrm{d}p-r\,\mathrm{d}x-s\,\mathrm{d}y=0, \qquad \mathrm{d}q-s\,\mathrm{d}x-t\,\mathrm{d}y=0.$

From this, it becomes clear that after the succession of two transformations (23), a transformation comes into existence with again leaves invariant the system (24). But by assumption, the equations:

$$\begin{aligned} x'&=\Xi(x,y,z,a_1,\dots,a_r), & y'&=\mathsf{H}, & z'&=\mathsf{Z} \\ x''&=\Xi(x',y',z',b_1,\dots,b_r), & y''&=\mathsf{H}, & z''&=\mathsf{Z} \end{aligned}$$

have as a consequence:

(19') $\quad x''=\Xi(x,y,z,c_1,\dots,c_r) \qquad y''=\mathsf{H}, \qquad z''=\mathsf{Z},$

where the c depend only on the a and on the b. Therefore, the two transformations:

$$x' = \Xi(x,y,z,a_1,\dots,a_r), \ \dots, \quad t' = \mathsf{T}(x,y,z,p,q,r,s,t,a_1,\dots,a_r)$$
$$x'' = \Xi(x',y',z',b_1,\dots,b_r), \ \dots, \ t'' = \mathsf{T}(x',y',z',p',q',r',s',t',b_1,\dots,b_r)$$

when executed one after the other, give a transformation which results from (19') by the same extension as the one by which (23) comes into existence from:

$$(19) \qquad x' = \Xi(x,y,z,a_1,\dots,a_r), \qquad y' = \mathsf{H}, \qquad z' = \mathsf{Z},$$

a transformation therefore which belongs in the same way to the family (23). Thus, the transformations (23) form an r-term group.

We will convince ourselves directly that the group (23) is generated by r independent infinitesimal transformations.

To begin with, we again assume that $r = 1$ and that the group (19) is generated by the infinitesimal transformation Xf. Then obviously, we need only to prove that there is an extended infinitesimal transformation:

$$X^{(2)}f = Xf + \pi\frac{\partial f}{\partial p} + \chi\frac{\partial f}{\partial q} + \rho\frac{\partial f}{\partial r} + \sigma\frac{\partial f}{\partial s} + \tau\frac{\partial f}{\partial t}$$

which leaves invariant the system of Pfaffian equations (24). But there is no difficulty in doing this; for π and χ, we find the same values as before, and for ρ, σ, τ, we obtain in a similar way expressions which depend linearly and homogeneously on the ξ, η, ζ and their differential quotients of first order and of second order.

Also, we find here four equations for ρ, σ, τ; but it can easily be proved that these equations are compatible with each other. We postpone the proof to the consideration of the general case, which appears to be clearer than the case at hand here.

From the existence of $X^{(2)}f$, it naturally follows that the group (23) is just generated by $X^{(2)}f$ in the case $r = 1$.

One realizes in the same way that for an arbitrary r, the group (23) is generated by the r infinitesimal transformations: $X_1^{(2)}f,\dots,X_r^{(2)}f$ which are obtained by extension of the infinitesimal transformations: $X_1f,\dots,X_rf$. As a result, it is proved at the same time that relations of the form:

$$X_i^{(2)}X_k^{(2)}f - X_k^{(2)}X_i^{(2)}f = \sum_{s=1}^{r} c''_{iks}X_s^{(2)}f$$

hold.

It can be seen easily that $c''_{iks} = c_{iks}$. In fact, together with $X_i^{(2)}f$ and $X_k^{(2)}f$, the system of Pfaffian equations (24) also admits the infinitesimal transformation: $X_i^{(2)}X_k^{(2)}f - X_k^{(2)}X_i^{(2)}f = \big[X_i^{(2)}, X_k^{(2)}\big]$, but this transformation obviously has the form:

$$\big[X_i^{(2)}, X_k^{(2)}\big] = [X_i, X_k] + \alpha\frac{\partial f}{\partial p} + \beta\frac{\partial f}{\partial q} + \lambda\frac{\partial f}{\partial r} + \mu\frac{\partial f}{\partial s} + \nu\frac{\partial f}{\partial t},$$

hence it comes into existence by extension of the transformation:

$$[X_i, X_k] = \sum_{s=1}^{r} c_{iks} X_s f,$$

and it can be represented in the following way:

$$\big[X_i^{(2)}, X_k^{(2)}\big] = \sum_{s=1}^{r} c_{iks} X_s^{(2)} f.$$

Here lies the reason why the extended group (23) is equally composed with the original group (19).

§ 130. After the realization of the special results of the preceding section, the general theory of the extension of a finite continuous group by addition of differential quotients will cause us no more difficulties.

To begin with, we consider an individual transformation in the $n+m$ variables: $x_1, \dots, x_n$, $z_1, \dots, z_m$, say the following:

$$\text{(25)} \qquad \begin{aligned} x_i' &= f_i(x_1, \dots, x_n, z_1, \dots, z_m) \qquad && (i=1\cdots n) \\ z_k' &= F_k(x_1, \dots, x_n, z_1, \dots, z_m) \qquad && (k=1\cdots m). \end{aligned}$$

If we want to extend this transformation by adding differential quotients, then we must at first agree on how many and which of the variables $x_1, \dots, x_n$, $z_1, \dots, z_m$ should be considered as independent, and which ones should be considered as dependent. This can occur in very diverse ways and to each such possible way there corresponds a completely determined extension of the transformation (25).

In the sequel, we will always consider $x_1, \dots, x_n$ *as independent variables and* $z_1, \dots, z_m$ *as functions of* $x_1, \dots, x_n$, *but which can be chosen arbitrarily. Under this assumption,* $x_1', \dots, x_n'$ *are in general mutually independent, while* $z_1', \dots, z_m'$ *are functions of* $x_1', \dots, x_n'$.

For the differential quotients of the z with respect to the x and of the z' with respect to the x', we introduce the following notation:

$$\frac{\partial z_\nu}{\partial x_k} = z_{\nu,k}, \qquad \frac{\partial^{\alpha_1+\cdots+\alpha_n} z_\mu}{\partial x_1^{\alpha_1} \cdots \partial x_n^{\alpha_n}} = z_{\mu,\alpha_1,\dots,\alpha_n}, \qquad \frac{\partial^{\alpha_1+\cdots+\alpha_n} z_\mu'}{\partial x_1'^{\alpha_1} \cdots \partial x_n'^{\alpha_n}} = z_{\mu,\alpha_1,\dots,\alpha_n}',$$

and we claim that $z_{\mu,\alpha_1,\dots,\alpha_n}'$ can be expressed by means of $x_1, \dots, x_n$, $z_1, \dots, z_m$, and by means of the differential quotients $z_{\nu,\beta_1,\dots,\beta_n}$ of first order up to the $(\alpha_1+\cdots+\alpha_n)$-th order:

$$\text{(26)} \qquad z_{\mu,\alpha_1,\dots,\alpha_n}' = F_{\mu,\alpha_1,\dots,\alpha_n}\big(x_1, \dots, x_n, z_1, \dots, z_m, z_{\nu,\beta_1,\dots,\beta_n}\big)$$
$$(\nu=1\cdots m;\ \beta_1+\cdots+\beta_n \leqslant \alpha_1+\cdots+\alpha_n).$$

If we set $\alpha_1 + \cdots + \alpha_n = N$, then the existence of equations of the form (26) is clear for $N = 0$; in order to establish this existence for an arbitrary N, we therefore need only to show that equations of the form (26) also hold for $\alpha_1 + \cdots + \alpha_n = N+1$ as soon as such equations hold for $\alpha_1 + \cdots + \alpha_n \leqslant N$.

Thus, let the functions $F_{\mu,\alpha_1,\dots,\alpha_n}$ $(\alpha_1 + \cdots + \alpha_n \leqslant N)$ be known; then the values of the $z'_{\mu,\alpha_1,\dots,\alpha_n}$ $(\alpha_1 + \cdots + \alpha_n = N+1)$ are to be determined from the equations:

$$\text{(27)} \qquad \mathrm{d}z'_{\mu,\alpha_1,\dots,\alpha_n} - \sum_{i=1}^{n} z'_{\mu,\alpha_1,\dots,\alpha_i+1,\dots,\alpha_n}\,\mathrm{d}x'_i = 0 \qquad (\alpha_1 + \cdots + \alpha_n = N),$$

and to be precise, (27) must hold identically by virtue of the system of equations:

$$\text{(28)} \qquad \begin{gathered} \mathrm{d}z_{\nu,\beta_1,\dots,\beta_n} - \sum_{i=1}^{n} z_{\nu,\beta_1,\dots,\beta_i+1,\dots,\beta_n}\,\mathrm{d}x_i = 0 \\ (\nu = 1\cdots m;\ 0 \leqslant \beta_1 + \cdots + \beta_n \leqslant N), \end{gathered}$$

while $\mathrm{d}x_1,\dots,\mathrm{d}x_n$ are fully independent of each other. For $z'_{\mu,\alpha_1,\dots,\alpha_i+1,\dots,\alpha_n}$, we therefore obtain the equations:

$$\text{(29)} \quad \sum_{i=1}^{n} z'_{\mu,\alpha_1,\dots,\alpha_i+1,\dots,\alpha_n} \left\{ \frac{\partial f_i}{\partial x_k} + \sum_{\nu=1}^{m} z_{\nu,k}\,\frac{\partial f_i}{\partial z_\nu} \right\} = \frac{\mathrm{d}F_{\mu,\alpha_1,\dots,\alpha_n}}{\mathrm{d}x_k} \qquad (k = 1\cdots n),$$

where the right-hand side means the complete differential quotient:

$$\begin{gathered} \frac{\partial F_{\mu,\alpha_1,\dots,\alpha_n}}{\partial x_k} + \sum_{\nu=1}^{m} z_{\nu,k}\,\frac{\partial F_{\mu,\alpha_1,\dots,\alpha_n}}{\partial z_\nu} + \sum_{\nu=1}^{m}\sum_{\beta} z_{\nu,\beta_1,\dots,\beta_k+1,\dots,\beta_n}\,\frac{\partial F_{\mu,\alpha_1,\dots,\alpha_n}}{\partial z_{\nu,\beta_1,\dots,\beta_n}} \\ (1 < \beta_1 + \cdots + \beta_n \leqslant \alpha_1 + \cdots + \alpha_n) \end{gathered}$$

of $F_{\mu,\alpha_1,\dots,\alpha_n}$ with respect to x_k.

If the equations (29) were not solvable with respect to the $z'_{\mu,\alpha_1,\dots,\alpha_i+1,\dots,\alpha_n}$ $(i = 1,\dots,n)$, then the determinant:

$$\begin{vmatrix} \frac{\partial f_1}{\partial x_1} + \sum_{\mu=1}^{m} z_{\mu,1}\,\frac{\partial f_1}{\partial z_\mu} & \cdot & \cdot & \frac{\partial f_1}{\partial x_n} + \sum_{\mu=1}^{m} z_{\mu,n}\,\frac{\partial f_1}{\partial z_\mu} \\ \cdot & \cdot & \cdot & \cdot \\ \cdot & \cdot & \cdot & \cdot \\ \frac{\partial f_n}{\partial x_1} + \sum_{\mu=1}^{m} z_{\mu,1}\,\frac{\partial f_n}{\partial z_\mu} & \cdot & \cdot & \frac{\partial f_n}{\partial x_n} + \sum_{\mu=1}^{m} z_{\mu,n}\,\frac{\partial f_n}{\partial z_\mu} \end{vmatrix}$$

would necessarily be zero, whichever functions of $x_1,\dots,x_n$ one could substitute for $z_1,\dots,z_m$, that is to say: this determinant should vanish identically for every value of the variables x_i, z_μ, $z_{\mu,i}$. One easily convinces oneself that this can happen only when all $m \times m$ determinants of the matrix:

$$\begin{vmatrix} \frac{\partial f_1}{\partial x_1} & \cdot & \cdot & \frac{\partial f_1}{\partial x_n} & \frac{\partial f_1}{\partial z_1} & \cdot & \cdot & \frac{\partial f_1}{\partial z_m} \\ \cdot & \cdot & \cdot & \cdot & \cdot & \cdot & \cdot & \cdot \\ \frac{\partial f_n}{\partial x_1} & \cdot & \cdot & \frac{\partial f_n}{\partial x_n} & \frac{\partial f_n}{\partial z_1} & \cdot & \cdot & \frac{\partial f_n}{\partial z_m} \end{vmatrix}$$

vanish identically, hence when the functions $f_1, \dots, f_n$ are not independent of each other. But since this is excluded, it follows that the equations (29) are solvable with respect to $z'_{\mu,\alpha_1,\dots,\alpha_i+1,\dots,\alpha_n}$ $(i = 1, \dots, n)$.

It still remains to eliminate an objection. Apparently, the equations (29) provide in general different values for the differential quotients $z'_{\mu,\alpha_1,\dots,\alpha_i+1,\dots,\alpha_n}$. But in reality, this is only fictitious, because otherwise this would give certain not identical relations between the x_i, the z_μ and their differential quotients which should hold for completely arbitrary functions $z_1, \dots, z_m$ of $x_1, \dots, x_n$, which is impossible.

We therefore see: the equations (29) determine all $z'_{\mu,\alpha_1,\dots,\alpha_i+1,\dots,\alpha_n}$ completely and uniquely, and consequently, there are also, under the assumptions made, equations of the form (26) for $\alpha_1 + \dots + \alpha_n = N + 1$, whence their general existence is established. But in addition, we see that the transformation:

(30)

$$x'_i = f_i(x_1, \dots, x_n, z_1, \dots, z_m) \qquad z'_\mu = F_\mu(x_1, \dots, x_n, z_1, \dots, z_m)$$

$$z'_{\mu,\alpha_1,\dots,\alpha_n} = F_{\mu\,\alpha_1,\dots,\alpha_n}(x_1, \dots, x_n, z_1, \dots, z_m, z_{\nu,\beta_1,\dots,\beta_n}) \qquad (\textstyle\sum_\chi \beta_\chi \leqslant \sum_\chi \alpha_\chi)$$

$$(i = 1 \cdots n;\ \mu = 1 \cdots m;\ 0 < \alpha_1 + \cdots + \alpha_n \leqslant N)$$

leaves invariant the system of equations:

$$\text{(31)} \qquad \mathrm{d}z_{\mu,\beta_1,\dots,\beta_n} - \sum_{i=1}^{n} z_{\mu,\beta_1,\dots,\beta_i+1,\dots,\beta_n}\,\mathrm{d}x_i = 0 \qquad (\mu = 1 \cdots m;\ 0 \leqslant \textstyle\sum_\chi \beta_\chi < N)$$

and that it is completely defined by this property. Lastly, it is also clear that the succession of two transformations (30) which leave invariant the Pfaffian system of equations (31) again provides a transformation of this form.

At present, we assume that the original transformation equations (25) contain a certain number, say r, of parameters:

$$\text{(32)} \qquad \begin{cases} x'_i = f_i(x_1, \dots, x_n, z_1, \dots, z_m, a_1, \dots, a_r) & (i = 1 \cdots n) \\ z'_\mu = F_\mu(x_1, \dots, x_n, z_1, \dots, z_m, a_1, \dots, a_r) & (\mu = 1 \cdots m) \end{cases}$$

and that they represent an r-term group generated by r independent infinitesimal transformations. From (32), we imagine that all equations of the form (26) are derived, in which $\alpha_1 + \dots + \alpha_n \leqslant N$; we claim that these equations taken together with (32):

$$\text{(33)} \qquad \begin{cases} x'_i = f_i(x, z, a), \qquad z'_\mu = F_\mu(x, z, a) \\ z'_{\mu,\alpha_1,\dots,\alpha_n} = F_{\mu,\alpha_1,\dots,\alpha_n}(x, z, z_{\nu,\beta_1,\dots,\beta_n}, a) \\ \qquad (i = 1 \cdots n;\ \mu = 1 \cdots m;\ 0 < \alpha_1 + \cdots + \alpha_n \leqslant N) \end{cases}$$

again represent an r-term group.

The proof is very simple.

The transformations:

$$\begin{aligned} x_i' &= f_i(x,z,a), & z_k' &= F_k(x,z,a) \\ x_i'' &= f_i(x',z',b), & z_k'' &= F_k(x',z',b) \end{aligned}$$

executed one after the other give a transformation:

$$x_i'' = f_i(x,z,c), \qquad z_k'' = F_k(x,z,c)$$

in which the c are certain functions of the a and of the b. According to what has been said above, the $z''_{\mu,\alpha_1,\dots,\alpha_n}$ can be expressed in terms of the x_i, z_k, $z_{\nu,\beta_1,\dots,\beta_n}$ and of the c_j in exactly the same way as the $z'_{\mu,\alpha_1,\dots,\alpha_n}$ can be expressed in terms of the x_i, z_k, $z_{\nu,\beta_1,\dots,\beta_n}$ and of the a. As a result, our claim is proved.

It still remains to prove that the extended group (33) is generated by r infinitesimal transformations, just as the original group (32). In order to be able to perform this proof, first make the following considerations:

We start from an arbitrary infinitesimal transformation:

$$Xf = \sum_{i=1}^{n} \xi_i(x,z)\frac{\partial f}{\partial x_i} + \sum_{\mu=1}^{m} \zeta_\mu(x,z)\frac{\partial f}{\partial z_\mu}$$

and we attempt to form an extended infinitesimal transformation from it:

$$X^{(N)}f = Xf + \sum_{\nu=1}^{m}\sum_{\alpha} \zeta_{\nu,\alpha_1,\dots,\alpha_n}\frac{\partial f}{\partial z_{\nu,\alpha_1,\dots,\alpha_n}}$$
$$(0 < \alpha_1 + \cdots + \alpha_n \leqslant N)$$

which leaves invariant the Pfaffian system of equations:

$$\text{(31)} \qquad \mathrm{d}z_{\nu,\beta_1,\dots,\beta_n} - \sum_{i=1}^{n} z_{\nu,\beta_1,\dots,\beta_i+1,\dots,\beta_n}\,\mathrm{d}x_i = 0$$
$$(\nu = 1\cdots m;\ 0 \leqslant \beta_1 + \cdots + \beta_n < N).$$

For $N = 0$ and $N = 1$, there certainly exists an infinitesimal transformation $X^{(N)}f$ of the form just described, and this does not require any justification. Hence we can carry out the general proof for the existence of $X^{(N)}f$ so we show that, as soon as $X^{(N-1)}f$ and $X^{(N)}f$ exist, then $X^{(N+1)}f$ also exists.

So the assumption is that $X^{(N-1)}f$ and $X^{(N)}f$ exist. Now, if there is an infinitesimal transformation $X^{(N+1)}f$ which leaves invariant the system of equations:

$$\text{(34)} \qquad \mathrm{d}z_{\nu,\beta_1,\dots,\beta_n} - \sum_{i=1}^{n} z_{\nu,\beta_1,\dots,\beta_i+1,\dots,\beta_n}\,\mathrm{d}x_i = 0$$
$$(\nu = 1\cdots m;\ 0 \leqslant \beta_1 + \cdots + \beta_n < N+1)$$

then the still unknown coefficients:

$$\frac{\delta z_{\mu,\gamma_1,\dots,\gamma_n}}{\delta t} = \zeta_{\mu,\gamma_1,\dots,\gamma_n} \qquad (\gamma_1+\cdots+\gamma_n=N+1)$$

must satisfy certain equations of the form:

$$\begin{aligned} &\mathrm{d}\,\zeta_{\mu,\alpha_1,\dots,\alpha_n} - \sum_{i=1}^{n} z_{\mu,\alpha_1,\dots,\alpha_i+1,\dots,\alpha_n}\,\mathrm{d}\,\xi_i - \sum_{i=1}^{n} \zeta_{\mu,\alpha_1,\dots,\alpha_i+1,\dots,\alpha_n}\,\mathrm{d}x_i \\ &= \sum_{\nu=1}^{m}\sum_{\beta} \mathsf{P}_{\nu,\beta_1,\dots,\beta_n}\left\{\mathrm{d}z_{\nu,\beta_1,\dots,\beta_n} - \sum_{i=1}^{n} z_{\nu,\beta_1,\dots,\beta_i+1,\dots,\beta_n}\,\mathrm{d}x_i\right\} \end{aligned}$$

$$(\alpha_1+\cdots+\alpha_n=N;\ 0\leqslant\beta_1+\cdots+\beta_n<N+1),$$

and to be precise, independently of the differentials: $\mathrm{d}x_i$, $\mathrm{d}z_{\nu,\beta_1,\dots,\beta_n}$.

From this, at first, the $\mathsf{P}_{\nu,\beta_1,\dots,\beta_n}$ determine themselves uniquely and there remain only equations between the mutually independent differentials: $\mathrm{d}x_1,\dots,\mathrm{d}x_n$. Hence if one inserts the values of the $\mathsf{P}_{\nu,\beta_1,\dots,\beta_n}$ and if one compares the coefficients of the $\mathrm{d}x_i$ in both sides, one obtains for $\zeta_{\mu,\alpha_1,\dots,\alpha_i+1,\dots,\alpha_n}$ the following expression:

$$\begin{aligned} &\frac{\delta}{\delta t} z_{\mu,\alpha_1,\dots,\alpha_i+1,\dots,\alpha_n} = \zeta_{\mu,\alpha_1,\dots,\alpha_i+1,\dots,\alpha_n} \\ &\frac{\mathrm{d}\,\zeta_{\mu,\alpha_1,\dots,\alpha_n}}{\mathrm{d}x_i} - \sum_{j=1}^{n} z_{\mu,\alpha_1,\dots,\alpha_j+1,\dots,\alpha_n}\frac{\mathrm{d}\,\xi_j}{\mathrm{d}x_i} \end{aligned} \tag{35}$$

$$(\alpha_1+\cdots+\alpha_n=N),$$

where $\mathrm{d}/\mathrm{d}x_i$ denotes a complete differential quotient with respect to x_i.[3]

But now, there remains a difficulty; indeed, in general, we obtain for each $\zeta_{\mu,\alpha_1,\dots,\alpha_i+1,\dots,\alpha_n}$ a series of apparently different expressions.

The expression on the right-hand side of (35) is the derivation of $\zeta_{\mu,\alpha_1,\dots,\alpha_i+1,\dots,\alpha_n}$ from the value of:

$$\frac{\delta}{\delta t}\frac{\partial z_{\mu,\alpha_1,\dots,\alpha_n}}{\partial x_i} = \frac{\delta}{\delta t} z_{\mu,\alpha_1,\dots,\alpha_i+1,\dots,\alpha_n}.$$

But on the other hand, we also have:

$$\begin{aligned} &\frac{\delta}{\delta t} z_{\mu,\alpha_1,\dots,\alpha_i+1,\dots,\alpha_n} = \frac{\delta}{\delta t}\frac{\partial z_{\mu,\alpha_1,\dots,\alpha_h-1,\dots,\alpha_i+1,\dots,\alpha_n}}{\partial x_h} \\ &= \frac{\mathrm{d}\,\zeta_{\mu,\alpha_1,\dots,\alpha_h-1,\dots,\alpha_i+1,\dots,\alpha_n}}{\mathrm{d}x_h} - \sum_{j=1}^{n} z_{\mu,\alpha_1,\dots,\alpha_j+1,\dots,\alpha_h-1,\dots,\alpha_i+1,\dots,\alpha_n}\frac{\mathrm{d}\,\xi_j}{\mathrm{d}x_h}, \end{aligned} \tag{36}$$

[3] The formula (35) is fundamentally identical to a formula due to POISSON in the Calculus of Variations.

where h denotes an arbitrary number amongst $1,2,\dots,n$ different from i. Thus, all possibilities are exhausted. Therefore, it only remains to prove that the latter value of $\frac{\delta}{\delta t} z_{\mu,\alpha_1,\dots,\alpha_i+1,\dots,\alpha_n}$ coincides with the value (35).

In order to prove this, we remember that the following equations hold:

$$\begin{aligned}\zeta_{\mu,\alpha_1,\dots,\alpha_n} &= \frac{\delta}{\delta t} \frac{\partial z_{\mu,\alpha_1,\dots,\alpha_h-1,\dots,\alpha_n}}{\partial x_h} \\ &= \frac{\mathrm{d}\,\zeta_{\mu,\alpha_1,\dots,\alpha_h-1,\dots,\alpha_n}}{\mathrm{d}x_h} - \sum_{j=1}^{n} z_{\mu,\alpha_1,\dots,\alpha_j+1,\dots,\alpha_h-1,\dots,\alpha_n} \frac{\mathrm{d}\xi_j}{\mathrm{d}x_h},\end{aligned}$$

and:

$$\begin{aligned}\zeta_{\mu,\alpha_1,\dots,\alpha_h-1,\dots,\alpha_i+1,\dots,\alpha_n} &= \frac{\delta}{\delta t} \frac{\partial z_{\mu,\alpha_1,\dots,\alpha_h-1,\dots,\alpha_n}}{\partial x_i} \\ &= \frac{\mathrm{d}\,\zeta_{\mu,\alpha_1,\dots,\alpha_h-1,\dots,\alpha_n}}{\mathrm{d}x_i} - \sum_{j=1}^{n} z_{\mu,\alpha_1,\dots,\alpha_j+1,\dots,\alpha_h-1,\dots,\alpha_n} \frac{\mathrm{d}\xi_j}{\mathrm{d}x_i}.\end{aligned}$$

If we insert the value of $\zeta_{\mu,\alpha_1,\dots,\alpha_n}$ in (35), we obtain:

$$\begin{aligned}\frac{\delta}{\delta t} \frac{\partial z_{\mu,\alpha_1,\dots,\alpha_n}}{\partial x_i} &= \frac{\mathrm{d}}{\mathrm{d}x_i} \frac{\mathrm{d}}{\mathrm{d}x_h} \zeta_{\mu,\alpha_1,\dots,\alpha_h-1,\dots,\alpha_n} - \sum_{j=1}^{n} z_{\mu,\alpha_1,\dots,\alpha_j+1,\dots,\alpha_h-1,\dots,\alpha_i+1,\dots,\alpha_n} \frac{\mathrm{d}\xi_j}{\mathrm{d}x_h} \\ &- \sum_{j=1}^{n} z_{\mu,\alpha_1,\dots,\alpha_j+1,\dots,\alpha_n} \frac{\mathrm{d}\xi_j}{\mathrm{d}x_i} - \sum_{j=1}^{n} z_{\mu,\alpha_1,\dots,\alpha_j+1,\dots,\alpha_h-1,\dots,\alpha_n} \frac{\mathrm{d}}{\mathrm{d}x_i} \frac{\mathrm{d}\xi_j}{\mathrm{d}x_h}.\end{aligned}$$

On the other hand, if we insert the value of $\zeta_{\mu,\alpha_1,\dots,\alpha_h-1,\dots,\alpha_i+1,\dots,\alpha_n}$ in (36) and if we take into account that $\frac{\mathrm{d}}{\mathrm{d}x_i} \frac{\mathrm{d}}{\mathrm{d}x_h} = \frac{\mathrm{d}}{\mathrm{d}x_h} \frac{\mathrm{d}}{\mathrm{d}x_i}$, we find:

$$\frac{\delta}{\delta t} \frac{\partial z_{\mu,\alpha_1,\dots,\alpha_h-1,\dots,\alpha_i+1,\dots,\alpha_n}}{\partial x_h} = \frac{\delta}{\delta t} \frac{z_{\mu,\alpha_1,\dots,\alpha_n}}{\partial x_i}.$$

But this is what was to be shown.

At present, it is shown that under the assumptions made, the $\zeta_{\mu,\gamma_1,\dots,\gamma_n}$ ($\gamma_1+\cdots+\gamma_n = N+1$) really exist and are uniquely determined; consequently, it is certain that, according to what was said above, to every Xf and for every value of N, there is associated a completely determined extended infinitesimal transformation $X^{(N)}f$.

The coefficients of $X^{(N)}f$ are obviously linear and homogeneous in the ξ_i, ζ_k and their partial differential quotients with respect to the x and the z. Hence if $X_i f$ and $X_j f$ are two infinitesimal transformations of the form Xf, and moreover, if $X_i^{(N)}f$ and $X_j^{(N)}f$ are the infinitesimal transformations extended in the way indicated, then:

$$c_i X_i^{(N)} f + c_j X_j^{(N)} f$$

results from $c_i X_i f + c_j X_j f$ by means of the extension in question. In addition, since $X_i^{(N)} X_j^{(N)} f - X_j^{(N)} X_i^{(N)} f$ leaves invariant the Pfaffian system of equations (31) simultaneously with $X_i^{(N)}f$ and $X_j^{(N)}f$, it follows that:

$$X_i^{(N)}X_j^{(N)}f - X_j^{(N)}X_i^{(N)}f$$

must come into existence from $X_iX_jf - X_jX_if$ by means of this extension.

Lastly, let $X_1f, \ldots, X_rf$ be independent infinitesimal transformations of the r-term group (32) so that they satisfy the known relations:

$$X_iX_jf - X_jX_if = \sum_{s=1}^{r} c_{ijs}X_sf.$$

If we form the extended infinitesimal transformations: $X_1^{(N)}f, \ldots, X_r^{(N)}f$, then $X_i^{(N)}X_j^{(N)}f - X_j^{(N)}X_i^{(N)}f$ also follows from $X_iX_jf - X_jX_if$ by means of this extension, and consequently, it follows from $\sum c_{iks}X_s$, which again means that we have:

$$X_i^{(N)}X_j^{(N)}f - X_j^{(N)}X_i^{(N)}f = \sum_{s=1}^{r} c_{ijs}X_s^{(N)}f.$$

Therefore, the r infinitesimal transformations $X_i^{(N)}f$ generate, for every value of N, a group equally composed with the group X_if; the former group is obviously identical to the group (33) discussed earlier which was obtained by extending the finite equations (32) of the group: $X_1f, \ldots, X_rf$.

Theorem 94. *If the ∞^r transformations:*

$$x_i' = f_i(x_1, \ldots, x_n, z_1, \ldots, z_m, a_1, \ldots, a_r) \qquad (i=1\cdots n)$$
$$z_k' = F_k(x_1, \ldots, x_n, z_1, \ldots, z_m, a_1, \ldots, a_r) \qquad (k=1\cdots m)$$

in the variables $x_1, \ldots, x_n$, $z_1, \ldots, z_m$ form an r-term group and if one considers the z_k as functions of the x_i which can be chosen arbitrarily, then the differential quotients of the z_k with respect to the x_i are also subjected to transformations. If one takes together all differential quotients from the first order up to, say, the N-th order, then one obtains certain equations:

$$z'_{\mu,\alpha_1,\ldots,\alpha_n} = F_{\mu,\alpha_1,\ldots,\alpha_n}(x_1, \ldots, x_n, z_1, \ldots, z_m, z_{\nu,\beta_1,\ldots,\beta_n}; a_1, \ldots, a_r)$$
$$(\beta_1 + \cdots + \beta_n \leqslant \alpha_1 + \cdots + \alpha_n \leqslant N)$$

which, when joined to the equations of the original group, represent an r-term group equally composed with the original group.[4]

Above, we already mentioned that every given group can be extended in very many different ways; indeed, it is left just as one likes which variables one wants to choose as the independent ones.

4 LIE, Math. Annalen Vol. XXIV, 1884; Archiv for Math., Christiania 1883.

In addition, one can, from the beginning, substitute the given group for a group equally composed with it by adding a certain number of variables: $t_1,\dots,t_\sigma$ which are absolutely not transformed by the group, or put differently, that are transformed only by the identity transformation:

$$t_1' = t_1, \ \dots, \ t_\sigma' = t_\sigma.$$

Now, if one regards as the independent variables an arbitrary number amongst the original variables and amongst the t_i, one can add differential quotients and extend; one always comes to an equally composed group.

As one sees, the number of possibilities is very large here.

§ 131. The theory of the invariants of an arbitrary group developed earlier in Chap. 13 can immediately be applied to our extended groups.

Since one can always choose the number N so large that the infinitesimal transformations $X_i^{(N)}f$ contain more than r independent variables, one can always arrange that the r equations $X_k^{(N)}f = 0$ form a complete system with one or more solutions. These solutions are functions of the x, of the z and of the differential quotients of the latter, they admit every finite transformation of the extended group $X_k^{(N)}f$ and hence are absolute invariants of this group; they shall be called the *differential invariants* [DIFFERENTIALINVARIANTEN] of the original group.

A function Ω of $x_1,\dots,x_n$, $z_1,\dots,z_m$ and of the differential quotients of the z with respect to the x is called a differential invariant *of the r-term group:*

$$x_i' = f_i(x_1,\dots,x_n,\, z_1,\dots,z_m\,;\; a_1,\dots,a_r)$$
$$z_k' = F_k(x_1,\dots,x_n,\, z_1,\dots,z_m\,;\; a_1,\dots,a_r)$$

when a relation of the form:

$$\Omega\left(x_1',\dots,x_n',\, z_1',\dots,z_m',\, z'_{\mu,\alpha_1,\dots,\alpha_n}\right) = \Omega(x_1,\dots,x_n,\, z_1,\dots,z_m,\, z_{\mu,\alpha_1,\dots,\alpha_n})$$

holds identically.

Since we can choose N arbitrarily large, we have instantly:

Theorem 95. *Every continuous transformation group: $X_1f,\dots,X_rf$ determines an infinite series of differential invariants which define themselves as solutions of complete systems.*[5]

If one knows the finite equations of the group: $X_1f,\dots,X_rf$, then in the way explained above, one finds the finite equations of the extended group and afterwards, under the guidance of Chap. 13, the differential invariants of any order without integration. But in general, this method for the determination of the differential invari-

[5] LIE, Gesellsch. d. W. zu Christiania 1882; Math. Ann. Vol. XXIV, 1884.

ants is not practically applicable. In most cases, direct integration of the complete system: $X_k^{(N)} f = 0$ is preferable.

We shall not enter into these considerations here. Still, we should observe that from sufficiently well known differential invariants, one can derive new differential invariants by differentiation and by formation of determinants.

In the variables $x_1, \dots, x_n, z_1, \dots, z_m$, if a group is represented by *several* systems of equations, each one with r parameters, then naturally, there are in the same way extended groups whose differential invariants are those of the original group. The former general developments not only show that each such group possesses differential invariants, but also at the same time, they show how these differential invariants can be found.

§ 132. One can also ask for possible *systems of differential equations* which remain invariant by a given group. The determination of a system of this sort can obviously be carried out by extending the concerned group in a suitable way and by applying the developments of Chap. 14, on the basis of which all systems of equations invariant by the group can be determined. Each system of equations found in this way then represents a system of differential equations invariant by the original group.

However, in each individual case, it must be specially studied whether the concerned system of differential equations satisfies the condition of integrability. —

Conversely, one can imagine that a system of differential equations is given — integrable or not integrable — and one can ask the question whether this system admits a given group. The answer to this question also presents no difficulty at present. One only has to extend the given group in the right way and then, to study whether the given system of equations admits the extended group; according to Chap. 7, this study can be conducted without integration.

§ 133. In order to give a simple application of the preceding theory, we want to seek the conditions under which a system of differential equations of the form:

$$A_k \varphi = \sum_{i=1}^{n} \alpha_{ki}(x_1, \dots, x_n) \frac{\partial \varphi}{\partial x_i} = 0 \qquad (k=1 \cdots q)$$

admits the r-term group:

$$X_j f = \sum_{i=1}^{n} \xi_{ji}(x_1, \dots, x_n) \frac{\partial f}{\partial x_i} \qquad (j=1 \cdots r)$$

in the variables $x_1, \dots, x_n$; here naturally, the q equations $A_k \varphi = 0$ are assumed to be mutually independent.

In order to be able to answer the question raised, to the variables $x_1, \dots, x_n$ of the group $X_k f$, we must add the variable φ which is not at all transformed by the group.

The x are to be considered as independent variables, φ as dependent and the group: $X_1 f, \dots, X_r f$ is thus to be extended by adding the n differential quotients:

$$\frac{\partial \varphi}{\partial x_i} = \varphi_i \qquad (i=1\cdots n).$$

Then, one has to examine whether the system of equations: $\sum_k \alpha_{ki}\,\varphi_i = 0$ allows the extended group.

To begin with, we compute the infinitely small increment $\delta\,\varphi_i$ that φ_i is given by the infinitesimal transformation $X_j f$. This increment is to be determined so that the expression:

$$\delta\left(\mathrm{d}\varphi - \sum_{\nu=1}^{n} \varphi_\nu\,\mathrm{d}x_\nu\right) = \mathrm{d}\,\delta\,\varphi - \sum_{\nu=1}^{n} \{\varphi_\nu\,\mathrm{d}\,\delta\,x_\nu + \delta\,\varphi_\nu\,\mathrm{d}x_\nu\}$$

vanishes by virtue of $\mathrm{d}\,\varphi = \sum \varphi_\nu\,\mathrm{d}x_\nu$. But since, as remarked above, $\delta\varphi$ is zero, we obtain for the $\delta\,\varphi_\nu$ the equation:

$$\sum_{\nu=1}^{n} \{\varphi_\nu\,\mathrm{d}\,\xi_{j\nu}\,\delta t + \delta\,\varphi_\nu\,\mathrm{d}x_\nu\} = 0$$

which must hold identically. Consequently, we have:

$$\delta\,\varphi_\nu = -\sum_{\mu=1}^{n} \frac{\partial \xi_{j\mu}}{\partial x_\nu}\,\varphi_\mu\,\delta t,$$

and the extended infinitesimal transformation $X_j f$ has the form:

$$X_j^{(1)} f = \sum_{i=1}^{n} \xi_{ji}\,\frac{\partial f}{\partial x_i} - \sum_{i=1}^{n} \left\{\sum_{\mu=1}^{n} \frac{\partial \xi_{j\mu}}{\partial x_i}\,\varphi_\mu\right\} \frac{\partial f}{\partial \varphi_i}.$$

Now, if the system of equations: $\sum_i \alpha_{ki}(x)\,\varphi_i = 0$ in the $2n$ variables $x_1, \dots, x_n$, $\varphi_1, \dots, \varphi_n$ is supposed to admit the infinitesimal transformation $X_j^{(1)} f$, then all the q expressions:

$$X_j^{(1)}\left(\sum_{i=1}^{n} \alpha_{ki}\,\varphi_i\right) = \sum_{i=1}^{n} X_j\,\alpha_{ki}\,\varphi_i - \sum_{\mu=1}^{n} \left\{\sum_{i=1}^{n} \alpha_{ki}\,\frac{\partial \xi_{j\mu}}{\partial x_i}\right\} \varphi_\mu \qquad (k=1\cdots q)$$

must vanish by virtue of the system of equations. This necessary, and at the same time sufficient, condition is then satisfied only when relations of the form:

$$X_j^{(1)}\left(\sum_{i=1}^{n} \alpha_{ki}\,\varphi_i\right) = \sum_{\sigma=1}^{q} \rho_{jk\sigma}(x) \sum_{i=1}^{n} \alpha_{\sigma i}\,\varphi_i$$

hold, where the $\rho_{jk\sigma}$ denote functions of $x_1, \dots, x_n$ only which do not depend on the φ_i.

With these words, we have found the desired conditions; they can be written:

$$\sum_{i=1}^{r} (X_j\,\alpha_{ki} - A_k\,\xi_{ji})\,\varphi_i = \sum_{\sigma=1}^{q} \rho_{jk\sigma} \sum_{i=1}^{n} \alpha_{\sigma i}\,\varphi_i,$$

or, if we again insert $\partial\varphi/\partial x_i$ in place of φ_i:

$$X_j\big(A_k(\varphi)\big) - A_k\big(X_j(\varphi)\big) = \sum_{\sigma=1}^{q} \rho_{jk\sigma}(x)\,A_\sigma\varphi.$$

Relations of this sort must be satisfied for every arbitrary function $\varphi(x_1,\dots,x_n)$. The system of q linear partial differential equations: $A_1\varphi = 0$, ..., $A_q\varphi = 0$ always remains invariant by every transformation of the group: $X_1 f,\dots,X_r f$ when these relations hold, and only when they hold.

In the case where the q equations: $A_k\varphi = 0$ form a q-term complete system, this result is not new for us. Indeed, in this special case, we have already indicated in Chap. 8, Theorem 20, p. 155, the necessary and sufficient condition just found. However, our present results accomplish more than the results found at that time, for we have now shown that the criterion in question holds generally, also when the equations: $A_k\varphi = 0$ do not form a q-term complete system.

The origins of the theory of differential invariants goes back a long time; indeed, mathematicians of the previous century already considered and integrated the differential invariants associated to several especially simple groups.

For instance, it has long been known that every differential equation of first order between x and y in which one variable, say y, does not explicitly appear, can be integrated by quadrature; but obviously:

$$f\left(x,\,\frac{\mathrm{d}y}{\mathrm{d}x}\right) = 0 = f(x,y')$$

is nothing but the most general differential equation of first order that admits the one-term group:

$$\mathfrak{x} = x, \qquad \mathfrak{y} = y + a \tag{37}$$

with the parameter a. The most general integral equation [INTEGRALGLEICHUNG] can be deduced from a particular integral equation, and in fact, as we can now say, in such a way that one executes the general transformation of the one-term group (37) on the particular solution in question; thanks to this, one indeed obtains the equation: $\mathfrak{y} = \varphi(\mathfrak{x}) + a$ with the arbitrary constant a.

Furthermore, the homogeneous differential equation:

$$f\left(\frac{y}{x},\,\frac{\mathrm{d}y}{\mathrm{d}x}\right) = 0 = f\left(\frac{y}{x},\,y'\right)$$

is the general form of a differential equation of first order which admits the one-term group: $\mathfrak{x} = ax$, $\mathfrak{y} = ay$; here, $f(y/x, y')$ is the most general first order differential invariant associated to this group.

It has long been observed that every particular integral equation: $F(x,y) = 0$ of a homogeneous differential equation: $f = 0$ can, by executing the general transformation: $\mathfrak{x} = ax$, $\mathfrak{y} = ay$, be transferred to the general integral equation:

$$F\left(\frac{\mathfrak{x}}{a}, \frac{\mathfrak{y}}{a}\right) = 0.$$

However, the equation $xy' - y = 0$ must be disregarded here.

A third example is provided by differential equations of the form:

$$f\left(\frac{\mathrm{d}^m y}{\mathrm{d}x^m}, \frac{\mathrm{d}^{m+1} y}{\mathrm{d}x^{m+1}}\right) = 0 = f\big(y^{(m)}, y^{(m+1)}\big);$$

nevertheless, it is not necessary here to write down the group of all equations of this form.

In the invariant theory of linear transformations, there often appear true differential invariants relative to all linear transformations. They were first considered by CAYLEY. Nevertheless, it is to be observed on the occasion *firstly* that the differential invariants of CAYLEY are not the simplest ones which are associated to the general linear homogeneous group, and *secondly*, that CAYLEY has not considered invariants of *differential* equations, and even less has integrated such equations.

In a prize-winning essay [PREISSCHRIFT] of 1867, published in 1871 (Determination of a special minimal surface, Akad. d. W. zu Berlin), H. SCHWARZ considered differential equations of the form:

$$J = \frac{y'y''' - \frac{3}{2}y''^2}{y'^2} - f(f) = 0$$

which, as he himself indicated, had already appeared occasionally in the writings of LAGRANGE. SCHWARZ observed that the most general solution can be deduced from every particular solution: $y = \varphi(x)$, namely the former has the form:

$$y = \frac{a + b\varphi}{c + d\varphi},$$

with the arbitrary constants: $a\colon b\colon c\colon d$. Thus, as we can say, the expression J is a differential invariant, and to be precise, the most general third order differential invariant by the group:

$$\mathfrak{x} = x, \qquad \mathfrak{y} = \frac{a + by}{c + dy}.$$

But now, although all these special theories are undoubtedly valuable, it is however still to be remarked that the central connection between them, the general principle from which they flow [FLIESSEN], has been missed by the mathematicians.

They have not observed that differential invariants are associated to *every* finite continuous group.

In the years 1869–1871, LIE occupied himself with differential equations which allow interchangeable infinitesimal transformations and in 1874, he published a work already announced in 1872 concerning a *general integration theory of ordinary differential equations which admit an arbitrary continuous group of transformations*.[6]

Afterwards, HALPHEN computed the simplest differential invariants relative to all projective transformations and he gave in addition nice applications to the theory of *linear* differential equations.[7]

After that, LIE developed in the years 1882–1884 a *general theory of the differential invariants* of finite and infinite continuous groups, where he specially occupied himself with finite groups in two variables.[8] As far as it relates to finite groups, this general theory is explained in the preceding.

Finally, after 1884, SYLVESTER and many other English and American mathematicians have published detailed but *special* studies on differential invariants.

---

[6] Verhandlungen der Gesellsch. d. W. zu Christiania 1870–1874; Math. Ann. Vol. V, XI; Gött. Nachr. 1874.

[7] Thèse sur les invariants différentiels 1878; Journal de l'École Pol. 1880. Cf. also Comptes Rendus Vol. 81, 1875, p. 1053; Journal de Liouville Novemb. 1876. Mémoire sur la réduction des équat. diff. lin. aux formes intégrables 1880–1883.

[8] Verh. d. Gesellsch. d. W. zu Christiania, 1882, 1883 and February 1875; Archiv for Math. 1882, 1883; Math. Ann. Vol. XXIV, 1884.

Chapter 26
The General Projective Group

The equations:

$$(1) \qquad x'_\nu = \frac{a_{1\nu}\,x_1 + \cdots + a_{n\nu}\,x_n + a_{n+1,\nu}}{a_{1,n+1}\,x_1 + \cdots + a_{n,n+1}\,x_n + a_{n+1,n+1}} \qquad (\nu = 1 \cdots n)$$

determine a group, as one easily convinces oneself, the so-called *general projective group* of the manifold $x_1,\dots,x_n$. In the present chapter, we want to study more closely this important group, which is also called the group of all *collineations* [COLLINEATIONEN] of the space $x_1,\dots,x_n$, by focusing our attention on its subgroups.

§ 134. The $(n+1)^2$ parameters a are not all essential: indeed only their ratios appear; one of the parameters, say $a_{n+1,n+1}$, can hence be set equal to 1. The values of the parameters are subjected to the restriction that the substitution determinant [SUBSTITUTIONSDETERMINANT] $\sum \pm a_{11} \cdots a_{n+1,n+1}$ should not be equal to zero, because simultaneously with it, the functional determinant [FUNCTIONALDETERMINANT]: $\sum \pm \frac{\partial x'_1}{\partial x_1} \cdots \frac{\partial x'_n}{\partial x_n}$ would also vanish.

The identity transformation is contained in our group, it corresponds to the parameter values:

$$a_{\nu\nu} = 1, \qquad a_{\mu\nu} = 0 \qquad (\mu, \nu = 1 \cdots n+1,\ \mu \neq \nu),$$

for which we indeed have $x'_i = x_i$. As a consequence of this, one obtains the infinitesimal transformations of the group by giving to the $a_{\mu\nu}$ the values:

$$a_{\nu\nu} = 1 + \omega_{\nu\nu}, \qquad a_{n+1,n+1} = 1, \qquad a_{\mu\nu} = \omega_{\mu\nu},$$

where the $\omega_{\mu\nu}$ mean infinitesimal quantities. Thus one finds:

S. Lie, *Theory of Transformation Groups I*,
DOI 10.1007/978-3-662-46211-9_26

$$x'_\nu = \left(x_\nu + \sum_{\mu=1}^{n} \omega_{\mu\nu}\, x_\mu + \omega_{n+1,\nu}\right)\left(1 - \sum_{\mu=1}^{n} \omega_{\mu,n+1}\, x_\mu + \cdots\right),$$

or by leaving out the quantities of second or higher order:

$$x'_\nu - x_\nu = \sum_{\mu=1}^{n} \omega_{\mu\nu}\, x_\mu + \omega_{n+1,\nu} - x_\nu \sum_{\mu=1}^{n} \omega_{\mu,n+1}\, x_\mu .$$

If one sets all the $\omega_{\mu\nu}$ equal to zero with the exception of a single one, then one gradually recognizes that our group contains the $n\,(n+2)$ independent infinitesimal transformations:

$$\frac{\partial f}{\partial x_i}, \qquad x_i\,\frac{\partial f}{\partial x_k}, \qquad x_i \sum_{j=1}^{n} x_j\,\frac{\partial f}{\partial x_j} \qquad (i,\, k=1 \cdots n). \tag{2}$$

The general projective group of the n-times extended space $x_1,\dots,x_n$ therefore contains $n\,(n+2)$ essential parameters and is generated by infinitesimal transformations. The analytic expressions of the latter behave regularly for every point of the space.

From now on, we will as a rule write p_i for $\partial f/\partial x_i$. In addition, for reasons of convenience, we want in this chapter to introduce the abbreviations:

$$x_i\, p_k = T_{ik}, \qquad x_i \sum_{k=1}^{n} x_k\, p_k = P_i.$$

Lastly, we still want to agree that ε_{ik} should mean zero whenever i and k are distinct from each other, whereas by contrast ε_{ii} is assumed to have the value 1; this is a fixing of notation that we have already adopted from time to time. On such a basis, we can write as follows the relations which emerge through a combination of the infinitesimal transformations p_i, T_{ik}, P_i:

$$\big[p_i,\, p_k\big] = 0, \qquad \big[P_i,\, P_k\big] = 0, \qquad \big[p_i,\, P_k\big] = T_{ki} + \varepsilon_{ik} \sum_{\nu=1}^{n} T_{\nu\nu},$$

$$\big[p_i,\, T_{k\nu}\big] = \varepsilon_{ik}\, p_\nu, \qquad \big[P_i,\, T_{k\nu}\big] = -\,\varepsilon_{i\nu}\, P_k,$$

$$\big[T_{ik},\, T_{\mu\nu}\big] = \varepsilon_{k\mu}\, T_{i\nu} - \varepsilon_{\nu i}\, T_{\mu k}.$$

One easily convinces oneself that these relations remain unchanged when one substitutes in them the p_i, T_{ik} and P_i by the respective expressions standing under them in:

$$\begin{matrix} p_i, & T_{ik}, & P_i \\ P_i, & -T_{ki}, & p_i. \end{matrix} \tag{3}$$

Thus in this way, the general projective group is related to itself in a holoedrically isomorphic way.

One could presume that there is a transformation: $x_i' = \Phi_i(x_1, \dots, x_n)$ which transfers the infinitesimal transformations:

$$p_i, \qquad x_i p_k, \qquad x_i \sum_{k=1}^{n} x_k p_k$$

to, respectively:

$$x_i' \sum_{k=1}^{n} x_k' p_k', \qquad -x_k' p_i', \qquad p_i'.$$

But there is no such transformation, and this, for simple reasons, because the n infinitesimal transformations $p_1, \dots, p_n$ generate an n-term *transitive* group, while: $x_1' \sum x_k' p_k', \dots, x_n' \sum x_k' p_k'$ generate an n-term *intransitive* group.

In the next volume we will learn to see the full significance of this important property of the projective group, when the concept of contact transformation [BERÜHRUNGSTRANSFORMATION] and especially the duality will be introduced.

The general infinitesimal transformation:

$$\sum_{i=1}^{n} a_i p_i + \sum_{i,\,k=1}^{n} b_{ik} T_{ik} + \sum_{i=1}^{n} c_i P_i$$

of our group is expanded in powers of $x_1, \dots, x_n$ and visibly contains only terms of zeroth, first and second order in the x. One easily realizes that the group contains n independent infinitesimal transformations of zeroth order in x, out of which no infinitesimal transformation of first or second order in the x can be deduced linearly. For instance, $p_1, \dots, p_n$ are n such infinitesimal transformations. From this it follows that *the general projective group is transitive*.

Besides, there are n^2 infinitesimal transformations of first order in the x_i, for instance all $x_i p_k = T_{ik}$, out of which no transformation of second order can be deduced linearly. Finally, there arise n transformations of second order in the x:

$$x_i \sum_{k=1}^{n} x_k p_k = P_i.$$

In agreement with Proposition 9 of Chap. 15 on p. 276, the P_i are pairwise interchangeable and in addition, the T_{ik} together with the P_i generate a subgroup in which the group of the P_i is contained as an invariant subgroup.

As one sees, and also as follows from our remark above concerning the relationship between the p_i and the P_i, the p_i are also interchangeable in pairs and they generate together with the T_{ik} a subgroup in which the group of the p_i is invariant.

§ 135. For the most important subgroups of the general projective group, it is advisable to employ special names. If, in the general expression (1) of a projective transformation, one lets the denominator reduce to 1, then one gets a *linear* transformation:

$$x'_\nu = a_{1\nu}x_1 + \cdots + a_{n\nu}x_n + a_{n+1,\nu} \qquad (\nu = 1 \cdots n);$$

all transformations of this form constitute the so-called *general linear* group. We have already indicated at the end of the previous section the infinitesimal transformations of this group; they are deduced by linear combination from the following $n(n+1)$ transformations:

$$p_i, \quad x_i\,p_k \qquad (i,k = 1 \cdots n).$$

If one interprets $x_1, \dots, x_n$ as coordinates of an n-times extended space R_n and if one translates the language of ordinary space, then one can say that the general linear group consists of all projective transformations which leave invariant the infinitely far $(n-1)$-times extended straight manifold, or briefly, the *plane at infinity* [UNENDLICH FERNE EBENE] M_{n-1}.

Next, if one recalls that by execution of two finite linear transformations one after the other, the substitution determinants: $\sum \pm a_{11} \cdots a_{nn}$ multiply with each other, then one realizes without difficulty that the totality of all linear transformations whose determinant equals 1 constitutes a subgroup, and in fact, an invariant subgroup, which we want to call the *special linear group*. One easily finds that, as the $n(n+1)-1$ independent infinitesimal transformations of this group, the following can be chosen:

$$p_i, \quad x_i p_k, \quad x_i p_i - x_k p_k \qquad (i \gtrless k).$$

If, amongst all linear transformations, one restricts oneself to those homogeneous in x, then one obtains the *general linear homogeneous group*:

$$x'_\nu = a_{1\nu}x_1 + \cdots + a_{n\nu}x_n \qquad (\nu = 1 \cdots n),$$

all infinitesimal transformations of which have the form: $\sum b_{ik}\,x_i p_k$ and hence can be linearly deduced from the n^2 transformations: $x_i p_k$. This group also visibly contains an invariant subgroup, the *special linear homogeneous group*, for which: $\sum \pm a_{11} \cdots a_{nn}$ has the value 1. The $n^2 - 1$ infinitesimal transformations of the latter group are:

$$x_i p_k, \quad x_i p_i - x_k p_k \qquad (i \gtrless k);$$

hence, the general infinitesimal transformation of the group in question has the form: $\sum_{i,k} \alpha_{ik}\,x_i p_k$, where the n^2 arbitrary constants α_{ik} are only subjected to the condition $\sum \alpha_{ii} = 0$.

Since the expression: $\left[x_i p_k, \sum_j x_j p_j\right]$ always vanishes, it is obvious that the last two named groups are *systatic* and consequently *imprimitive*. Indeed, if one sets:

$$\frac{x_i}{x_n} = y_i, \qquad \frac{x'_i}{x'_n} = y'_i \qquad (i=1\cdots n-1),$$

then one receives:

$$y'_\nu = \frac{a_{1\nu}y_1 + \cdots + a_{n-1,\nu}y_{n-1} + a_{n\nu}}{a_{1,n}y_1 + \cdots + a_{n-1,n}y_{n-1} + a_{n,n}} \qquad (\nu=1\cdots n-1).$$

From this, it results that in both cases the y are transformed by the (n^2-1)-term general projective group of the $(n-1)$-times extended manifold $y_1,\dots,y_{n-1}$. Consequently, this group is isomorphic with the general linear homogeneous group of an n-times extended manifold and with the special linear homogeneous group as well, though the isomorphism is holoedric only for the special linear homogeneous group, since this one contains n^2-1 parameters.

Theorem 96. *The special linear homogeneous group:*

$$x_ip_k, \qquad x_ip_i - x_kp_k \qquad (i \gtrless k=1\cdots n)$$

in the variables $x_1,\dots,x_n$ is imprimitive and holoedrically isomorphic with the general projective group of an $(n-1)$-times extended manifold.

The formally simplest infinitesimal transformations of the general projective group are $p_1,\dots,p_n$; these generate a group, as already observed: the group of all *translations*:

$$x'_i = x_i + a_i \qquad (i=1\cdots n),$$

which is obviously simply transitive.

In fact, m arbitrary infinitesimal translations, for instance $p_1,\dots,p_m$, always generate an m-term group. For all of these groups, the following holds:

Proposition 1. *All m-term groups of translations are conjugate to each other inside the general projective group, and even already inside the general linear group.*

Indeed, the m independent infinitesimal transformations of such a group always have the form:

$$\sum_{\nu=1}^{n} b_{\mu\nu}\,p_\nu \qquad (\mu=1\cdots m),$$

where not all $m\times m$ determinants of the $b_{\mu\nu}$ vanish.

But we can very easily show that by means of some linear transformation, new variables $x'_1,\dots,x'_n$ can be introduced for which one has:

$$p'_\mu = \sum_{\nu=1}^{n} b_{\mu\nu}\,p_\nu \qquad (\mu=1\cdots m).$$

In fact, let $p'_\mu = p_1\frac{\partial x_1}{\partial x'_\mu} + \cdots + p_n\frac{\partial x_n}{\partial x'_\mu}$; then we only need to set:

$$\frac{\partial x_\nu}{\partial x'_\mu} = b_{\mu\nu} \qquad (\nu=1\cdots n;\ \mu=1\cdots m),$$

while the $\frac{\partial x_\nu}{\partial x'_{m+1}}, \dots, \frac{\partial x_\nu}{\partial x'_n}$ remain arbitrary. We can give to these last ones some values such that the equations:

$$x_\nu = \sum_{\mu=1}^{n} b_{\mu\nu}\, x'_\mu + \sum_{\pi=m+1}^{n} c_{\pi\nu}\, x'_\pi \qquad (\nu=1\cdots n)$$

determine a transformation, and then, this transformation transfers the given group of translations to the group $p'_1, \dots, p'_m$. From this, our proposition follows immediately.

We want to at least indicate a second proof of the same proposition. As already observed, the general linear group leaves invariant the plane at infinity M_{n-1}, and in fact, it is even the most general projective group of this nature. Now, every infinitesimal translation is directed by a point at infinity and is completely determined by this point; *every m-term group of translations can therefore be represented by an m-times extended straight manifold M_m at infinity*. But two straight M_m at infinity can always be transferred to one another by a linear transformation which leaves invariant the plane at infinity. Consequently, all m-term groups of translations are conjugate to each other inside the general linear group, and in the same way, inside the general projective group.

The correspondence indicated earlier which takes place between the p_i and the P_i yields, as we prove instantly, the

Proposition 2. *All m-term groups, whose infinitesimal transformations are of the form $\sum e_i P_i$, are conjugate to each other inside the general projective group.*

For the proof, we start from the fact that two subgroups are conjugate inside a group G_r when one can be transferred to the other by means of a transformation of the adjoint group of G_r; here, we have to imagine the subgroup as a straight manifold in the space $e_1, \dots, e_r$ which is transformed by the adjoint group (cf. Chap. 16, p. 292). If we now write the transformations of the projective group firstly in the sequence p_i, T_{ik}, P_i and secondly in the sequence P_i, $-T_{ki}$, p_i, then in the two cases we get the same adjoint group. But since two m-term groups of translations can always be transferred to one another by the adjoint group, this must also always be the case with two m-term groups whose infinitesimal transformations can be deduced linearly from the P_i. Furthermore, it immediately follows that two m-term groups of this sort are already conjugate to each other inside the group P_i, T_{ik}. With that, our proposition is proved.

§ 136. We consider now one after the other the general projective group, the general linear group and the general linear homogeneous group, and to be precise, we examine whether there are invariant subgroups and which ones are contained in these three groups.

We begin with the general projective group. Let:

$$S=\sum_{i=1}^{n}\alpha_i\,p_i+\sum_{i=1}^{n}\sum_{k=1}^{n}\beta_{ik}\,x_ip_k+\sum_{i=1}^{n}\gamma_ix_i\sum_{k=1}^{n}x_kp_k$$

be an infinitesimal transformation of an invariant subgroup; then necessarily $[p_\nu, S]$ and $[p_\mu, [p_\nu, S]]$ are also transformations of the same subgroup. Consequently, in our invariant subgroup, there would certainly appear an infinitesimal translation $\sum \rho_i\,p_i$.

But because all infinitesimal translations are conjugate to each other inside the general projective group, they would all appear. Furthermore, since it is invariant, the subgroup would necessarily contain all transformations: $[p_i, x_i \sum_j x_jp_j]$, or after computation:

$$x_i\,p_k \qquad (i \gtrless k), \qquad\qquad x_i\,p_i+\sum_{j=1}^{n}x_jp_j.$$

Adding the n transformations: $x_ip_i+\sum_j x_jp_j$, one obtains: $(n+1)\sum x_jp_j$, hence x_ip_i and therefore all x_ip_k. Finally, the invariant subgroup would contain all transformations: $[x_ip_i, x_i\sum_k x_kp_k]$, hence all $x_i\sum_k x_kp_k$ and thus it would be identical to the general projective group itself. Thus, our first result is:

Theorem 97. *The general projective group in n variables is simple.*[1]

Correspondingly, the special linear homogeneous group:

$$x_ip_k, \qquad x_ip_i-x_kp_k \qquad\qquad (i \gtrless k) \tag{4}$$

is also simple.

The general linear homogeneous group with the n^2 infinitesimal transformations x_ip_k contains, as we saw above, an invariant subgroup with n^2-1 parameters, namely the group (4) just named.

If there is a second invariant subgroup, then this subgroup obviously cannot comprise the group (4), and in the same way, it cannot have infinitesimal transformations in common with the same group, since such transformations would constitute an invariant subgroup in the simple group (4) (cf. Proposition 10 of Chap. 15 on p. 276). Taking Proposition 7 of Chap. 12 on p. 223 into account, it hence follows that a possible second invariant subgroup can contain only one infinitesimal transformation, and to be precise, one of the form:

$$\sum_{i=1}^{n}x_ip_i+\sum_{i,\,k}^{1\cdots n}\alpha_{ik}\,x_ip_k \qquad\qquad \left(\textstyle\sum_{i=1}^{n}\alpha_{ii}=0\right).$$

In addition, according to Proposition 11 of Chap. 15 on p. 276, the same transformation must be interchangeable with every transformation of the group (4), from which it follows that the transformation:

[1] Lie, Math. Ann., Vol. XXV, p. 130.

$$\sum_{i,\,k}^{1\cdots r} \alpha_{ik}\, x_i p_k \qquad \left(\Sigma_{i=1}^{n}\ \alpha_{ii}=0\right)$$

must be excellent inside the group (4). But there is no such transformation, whence all the α_{ik} vanish and it is revealed that $x_1 p_1 + \cdots + x_n p_n$ and (4) are the only two invariant subgroups of the group $x_i p_k$.

Theorem 98. *The general linear homogeneous group $x_i p_k$ in n variables contains only two invariant subgroups, namely the special linear homogeneous group and the one-term group: $x_1 p_1 + \cdots + x_n p_n$.*

Now, one can easily set up all invariant groups of the general linear group. Let:

$$S = \sum_{i=1}^{n} \alpha_i\, p_i + \sum_{i=1}^{n} \sum_{k=1}^{n} \beta_{ik}\, x_i p_k$$

be a transformation of such a subgroup. Then together with S, $[p_j, S]$ also belongs to the invariant subgroup; hence this subgroup certainly contains a translation, and because of Proposition 1, p. 563, it contains all of them. The smallest invariant subgroup therefore consists of the translations themselves; every other subgroup must contain, aside from the translations, a series of infinitesimal transformations of the form: $\sum_i \sum_k \alpha_{ik}\, x_i p_k$. But these last transformations visibly generate an invariant subgroup, the linear homogeneous group $x_i p_k$. So we find:

Theorem 99. *The general linear group: p_i, $x_i p_k$ contains only three invariant subgroups,*[2] *namely the three subgroups:*

$$p_i \qquad p_i,\ x_1 p_1 + \cdots + x_n p_n \qquad p_i,\ x_i p_k,\ x_i p_i - x_k p_k \qquad (i \gtrless k),$$

with respectively n, $n+1$ and n^2+n-1 parameters.

If, as we have already done several times, we employ the terminology which is common for ordinary space, we can say: the three invariant subgroups of the general linear group are firstly the group of all translations, secondly the group of all similitudes [AEHNLICHKEITSTRANSFORMATIONEN]: $(x_1 - x_1^0)p_1 + \cdots + (x_n - x_n^0)p_n$, and lastly the most general linear group which leaves all volumes unchanged.

§ 137. Before we pass to determining the largest subgroups of the general projective group, make an observation which will find several applications in the sequel.

Let $X_1 f, \dots, X_r f$ be independent infinitesimal transformations of an r-term group G_r and let a family of ∞^{r-m-1} infinitesimal transformations of this group be determined by m independent equations of the form:

$$\text{(5)} \qquad \sum_{j=1}^{r} \alpha_{kj}\, e_j = 0 \qquad (k=1\cdots m).$$

[2] Lie, Math. Ann., Vol. XXV, p. 130.

Assume furthermore that one knows that amongst the infinitesimal transformations: $e_1 X_1 f + \cdots + e_r X_r f$, no infinitesimal transformation of the form: $e_1 X_1 f + \cdots + e_m X_m f$ is contained in this family. Then it can be deduced that the equations (5) are solvable with respect to $e_1, \dots, e_m$, for if one sets: $e_{m+1} = \cdots = e_r = 0$ in these equations, it must follow that: $e_1 = \cdots = e_m = 0$, which happens only when the determinant: $\sum \pm \alpha_{11} \cdots \alpha_{mm}$ does not vanish. Hence if one chooses $e_{m+1}, \dots, e_r$ arbitrary, though not all zero, it follows that $e_1, \dots, e_m$ receive determined values and therefore, the family contains $r-m$ mutually independent infinitesimal transformations of the form:

$$X_{m+j} f + e_{1j} X_1 f + \cdots + e_{mj} X_m f \qquad (j = 1 \cdots r-m).$$

Thus, the following holds:

Proposition 3. *If, amongst the infinitesimal transformations: $\sum e_k X_k f$ of the r-term group: $X_1 f, \dots, X_r f$, a family is sorted by means of m independent linear equations:*

$$\sum_{j=1}^{r} \alpha_{kj} e_j = 0 \qquad (k = 1 \cdots m)$$

which embraces no infinitesimal transformation of the form: $e_1 X_1 f + \cdots + e_m X_m f$, then this family contains r − m infinitesimal transformations of the form:

$$X_{m+j} + \sum_{\nu=1}^{m} e_{j\nu} X_\nu f \qquad (j = 1 \cdots r-m).$$

§ 138. After these preparations, we turn ourselves in particular to the general projective group. We denote its number $n(n+2)$ of parameters shortly by N and we first seek all subgroups with more than $N-n$ parameters, hence some G_{N-m} for which $m < n$. Naturally, this way of putting the question is meaningful only when the number n is larger than 1.

According to an observation made earlier (cf. Chap. 12, Proposition 7, p. 223), the sought G_{N-m} has at least $n-m$ independent infinitesimal transformations in common with the n-term group $p_1, \dots, p_n$. So G_{N-m} contains in any case $n-m$ independent infinitesimal translations. If it contains no more than $n-m$ translations, then thanks to Proposition 1, we can assume that $p_{m+1}, \dots, p_n$ are these translations, while no translation of the form: $e_1 p_1 + \cdots + e_m p_m$ is extant. From Proposition 3, it then follows that there appears a transformation:

$$x_{m+1} p_1 + e_1 p_1 + \cdots + e_m p_m,$$

but by combination with p_{m+1}, it would give p_1, which would be a contradiction. Hence in our G_{N-m} there are surely more than $n-m$, say $n-q$ $(q < m)$, infinitesimal translations, and we want to assume that these are: $p_{q+1}, \dots, p_m, \dots, p_n$; by contrast, when $q > 0$, there appear no translations of the form: $e_1\,p_1 + \cdots + e_q\,p_q$. Now in any case (cf. Chap. 12, Proposition 7, p. 223), there is in G_{N-m} *one* transformation:

$$\lambda_n\,x_n\,p_1 + \cdots + \lambda_{q+1}\,x_{q+1}\,p_1 + e_q\,p_q + \cdots + e_1\,p_1$$

in which, according to what precedes, the λ cannot all vanish. By combination with one of the translations $p_{q+1}, \ldots, p_n$, we therefore obtain p_1 at least once, but this was excluded. Hence the number q cannot be bigger than zero, so $q = 0$ and the sought G_{N-m} therefore comprises, when $m < n$, all translations.

Thanks to completely analogous considerations, one realizes that G_{N-m} ($m < n$) must contain all transformations P_i. These considerations coincide even literally with those undertaken just now, when one replaces the p_i, $x_i p_k$, P_i by P_i, $-x_k p_i$, p_i, respectively, and when one relates to Proposition 2, p. 564.

Thus, our G_{N-m} contains all p_i and all P_i simultaneously, but then as was already shown earlier (on page 565), it also contains all $x_i p_k$ and is hence identical to the general projective group itself. Consequently:

Theorem 100. *The general projective group of the manifold $x_1, \ldots, x_n$ contains no subgroup with more than $n\,(n+2) - n = n\,(n+1)$ parameters.*

§ 139. At present, the question is to determine all subgroups contained in the general projective group having $N - n = n\,(n+1)$ parameters. In order to be able to settle this problem completely, we must individually treat a series of various possibilities.

First, we seek all $n\,(n+1)$-term subgroups which contain no infinitesimal translation $\sum e_k\,p_k$. According to Proposition 3, p. 567, there surely exists a transformation:

$$U = \sum_{i=1}^{n} x_i\,p_i - \sum_{i=1}^{n} \alpha_i\,p_i = \sum_{i=1}^{n} (x_i - \alpha_i)\,p_i$$

and for every value of i and k, there is in the same way a transformation of the form:

$$T = (x_i - \alpha_i)\,p_k + \sum_{j=1}^{n} \beta_{ikj}\,p_j.$$

By combination of the two infinitesimal transformations U and T, we obtain the expression: $[U, T] = -\sum_j \beta_{ikj}\,p_j$, and because our group contains no infinitesimal transformation of this form, all β_{ikj} must vanish.

Lastly, there are n infinitesimal transformations:

$$P_i + \sum_{k=1}^{n} \gamma_{ik}\,p_k,$$

or, what amounts to the same:

$$P_i' = (x_i - \alpha_i) \sum_{j=1}^{n} (x_j - \alpha_j)\, p_j + \sum_{k=1}^{n} \delta_{ik}\, p_k.$$

If we form the combination of P_i' with $\sum (x_k - \alpha_k)\, p_k = U$, we obtain:

$$[U, P_i'] = (x_i - \alpha_i) \sum_{j=1}^{n} (x_j - \alpha_j)\, p_j - \sum_{k=1}^{n} \delta_{ik}\, p_k,$$

so that all δ_{ik} vanish.

Thus, in the sought $n\,(n+1)$-term subgroups, the following $n\,(n+1)$ independent infinitesimal transformations must appear:

$$(6) \qquad (x_i - \alpha_i)\, p_k, \qquad (x_i - \alpha_i) \sum_{j=1}^{n} (x_j - \alpha_j)\, p_j \qquad (i, k = 1 \cdots n).$$

Thanks to pairwise combinations, one easily convinces oneself that these infinitesimal transformations effectively generate an $n\,(n+1)$-term group. Besides, this also follows from the fact that all the infinitesimal transformations (6) leave invariant the point $x_i = \alpha_i$ lying in the domain of the finite. Indeed, they are mutually independent and their number equals $n\,(n+1)$, that is to say, exactly equal to the number of independent infinitesimal transformations there are in the $n\,(n+2)$-term projective group that leave invariant the point $x_i = \alpha_i$. According to Proposition 2, p. 218, the infinitesimal transformations (6) therefore generate an $n\,(n+1)$-term group.

As a result, every $n\,(n+1)$-term projective group of R_n in which no infinitesimal translation $\sum e_k\, p_k$ appears consists of all projective transformations that fix a point located in the domain of the finite.

If, in the above computations, we had written $P_i, -x_i p_k, p_i$ in place of $p_i, x_k\, p_i, P_i$, respectively, then we would have found all $n\,(n+1)$-term subgroups which contain no infinitesimal transformation $\sum e_k\, P_k$. We can therefore make exactly the same exchange in the expressions (6) and we obtain in this way:

$$x_k\, p_i + \alpha_i\, P_k, \qquad p_i + \alpha_i \sum_{j=1}^{n} x_j\, p_j + \sum_{j=1}^{n} \alpha_j\, (x_j\, p_i + \alpha_i\, P_j).$$

Here, we may clearly take away the term $\sum \alpha_j\, (x_j\, p_i + \alpha_i\, P_j)$ and we therefore obtain the general form of the $n\,(n+1)$-term subgroups which contain no transformation $\sum e_k\, P_k$ as follows:

$$(7) \qquad p_i + \alpha_i \sum_{j=1}^{n} x_j\, p_j, \qquad x_i\, p_k + \alpha_k\, P_i.$$

The fact that these infinitesimal transformation generate a group follows from their derivation, but naturally, one could also corroborate this directly.

If all α_i vanish, then we have the general linear group already discussed earlier, which leaves invariant the plane at infinity M_{n-1}. Hence we are very close to presuming that in the general case where not all α_i are equal to zero, there exists in the same way a plane $M_{n-1}:\ \lambda_1 x_1 + \cdots + \lambda_n x_n + \lambda = 0$ which admits all infinitesimal transformations (7).

By execution of the infinitesimal transformations: $p_i + \alpha_i \sum x_j p_j$, we obtain the following conditions for the λ_i:

$$\lambda_i + \alpha_i \sum_{j=1}^{n} \lambda_j x_j = 0 = \lambda_i - \alpha_i \lambda .$$

The quantity λ can hence in any case not vanish and may be set equal to 1; so if there actually is an invariant plane M_{n-1}, this plane can only have the form: $\alpha_1 x_1 + \cdots + \alpha_n x_n + 1 = 0$. In fact, this latter plane also admits the infinitesimal transformations: $x_i p_k + \alpha_k P_i$.

Every subgroup (7) therefore leaves invariant a plane M_{n-1} of R_n and is in addition the most general projective group of R_n which leaves at rest the plane M_{n-1} in question. Every further infinitesimal projective transformation which leaves invariant the plane: $M_{n-1}:\ \alpha_1 x_1 + \cdots + \alpha_n x_n + 1 = 0$ could indeed be given the form:

$$\sum_{k=1}^{n} e_k P_k = \sum_{k=1}^{n} e_k x_k \sum_{i=1}^{n} x_i p_i .$$

But if one executes this infinitesimal transformation on the M_{n-1}, one gets:

$$\sum_{k=1}^{n} e_k x_k \sum_{i=1}^{n} \alpha_i x_i = 0 = -\sum_{k=1}^{n} e_k x_k ,$$

whence $e_1 = \cdots = e_n = 0$.

Thus, if an $n\,(n+1)$-term projective group of R_n contains no infinitesimal transformation $\sum e_k P_k$, it consists of all projective transformations which leave invariant a plane M_{n-1} not passing through the origin of coordinates.

From the previous results, we can derive, thanks to a simple transformation, a few other results which will be useful to us in the future. Indeed, if we transfer to infinity the former origin of coordinates by means of the collineation:

$$x_1 = \frac{1}{x_1'}, \quad x_2 = \frac{x_2'}{x_1'}, \quad \ldots, \quad x_n = \frac{x_n'}{x_1'}, \tag{7'}$$

we then obtain:

$$p_1' = \sum_{i=1}^{n} p_i \frac{\partial x_i}{\partial x_1'} = -\frac{1}{x_1'^{\,2}} \left(p_1 + \sum_{i=2}^{n} x_i' p_i \right),$$

and hence:

$$p_1' = -x_1 \sum_{i=1}^{n} x_i p_i = -P_1 ,$$

$$x_k' p_1' = -x_k \sum_{i=1}^{n} x_i p_i = -P_k \qquad (k=2\cdots n).$$

In the same way, we obtain:

$$x_1' \sum_{i=1}^{n} x_i' p_i' = -p_1, \qquad x_1' p_k' = -p_k \qquad (k=2\cdots n),$$

as every infinitesimal projective transformation is actually transferred to a transformation of the same kind after the introduction of the x' (cf. Chap. 3, Proposition 4, p. 48).

From this, we see: our collineation (7') converts every $n(n+1)$-term projective group in which no infinitesimal transformation $\sum e_k P_k$ appears into a projective group which contains no transformation $e_1 p_1 + e_2 x_2 p_1 + \cdots + e_n x_n p_1$. In the same way, every projective group free of all $\sum e_k p_k$ is transferred to a projective group which contains no transformation: $e_1 P_1 + e_2 x_1 p_2 + \cdots + e_n x_1 p_n$.

Thus, if in an $n(n+1)$-term projective group, there are no infinitesimal transformations: $e_1 p_1 + e_2 x_2 p_1 + \cdots + e_n x_n p_1$, then this group consists of all projective transformations which leave invariant a certain plane M_{n-1}. By contrast, if in the group there are no infinitesimal transformations: $e_1 P_1 + e_2 x_1 p_2 + \cdots + e_n x_1 p_n$, then this group consists of all projective transformations which leave invariant a certain point.

The general projective group possesses the property of being able to transfer *every* point of R_n to *every* other point, and *every* plane M_{n-1} to *every* other plane. From this, it results that every $n(n+1)$-term projective group of R_n which leaves invariant a point is conjugate, inside the general projective group of R_n, to every other group of the same sort, and likewise, it follows that every $n(n+1)$-term projective group of R_n which leaves invariant a plane M_{n-1} is conjugate to every other projective group of this sort.

Finally, we can now find all $n(n+1)$-term groups of R_n.

Since we know all groups that contain no translation, it only remains to find the groups in which some infinitesimal translations appear. We assume that there are exactly q independent such infinitesimal translations, say $p_n, \ldots, p_{n-q+1}$, so that, when $n-q>0$, no translation of the form: $e_1 p_1 + \cdots + e_{n-q} p_{n-q}$ is extant.

Then in our group, there surely exists an infinitesimal transformation of the form:

$$\sum_{i=1}^{q} x_{n-q+i} \sum_{k=1}^{n-q} \lambda_{ik} p_k + e_1 p_1 + \cdots + e_{n-q} p_{n-q}, \tag{8}$$

only when the number $(n-q)(q+1)$ of terms contained in this infinitesimal transformation is larger than n.

This is the case only when $(n-q)(q+1)-n = q(n-q-1)$ is larger than zero, from which it follows that we must temporarily disregard the cases $q=n-1$ and $q=n$. But if we assume that q is smaller than $n-1$ and if we form a combination of the transformation (8) — in which obviously not all λ_{ik} vanish — with each one of the extant translations $p_{n-q+1}, \ldots, p_n$, we then obtain in all circumstances a non-identically vanishing transformation of the form $\mu_1 p_1 + \cdots + \mu_{n-q} p_{n-q}$, and this is a contradiction. Thus, the number q cannot be smaller than $n-1$.

Therefore, if an $n(n+1)$-term projective group contains *one* infinitesimal translation $\sum e_k p_k$, it contains at least $n-1$ independent translations.

In this result we can again, as so often before, replace the p_i by the P_i and find that there always exist $n-1$ independent transformations $\sum e_k P_k$ in every group of the said sort as soon as there exists only a single transformation of this form.

Now, we seek all $n(n+1)$-term projective groups with exactly $n-1$ independent infinitesimal translations $\sum e_k p_k$, say with $p_2, \dots, p_n$. No such group can contain a transformation of the form:

$$e_1 p_1 + e_2 x_2 p_1 + \cdots + e_n x_n p_1,$$

because by combination with $p_2, \dots, p_n$, we would obtain p_1, which is excluded. According to what we have shown above, all these groups belong to the category of $n(n+1)$-term projective groups which leave invariant a plane M_{n-1}. Correspondingly, it follows that every $n(n+1)$-term group which contains exactly $n-1$ independent transformations $\sum e_k P_k$ leaves invariant a point.

It still remains to determine the $n(n+1)$-term projective groups which contain all the n translations $p_1, \dots, p_n$. Besides, the P_i cannot all appear, because otherwise, the group would coincide with the general projective group itself.

Therefore, according to what has been said above, there are only two possibilities: either there is absolutely no transformation $\sum e_k P_k$, or there are $n-1$ independent such transformations. Both cases are already settled above.

As a result, our study is brought to a conclusion. The result is the following:

Theorem 101. *The largest subgroups of the general projective group of an n-times extended manifold contain $n(n+1)$ parameters. Each such subgroup consists of all projective transformations which leave invariant either a plane M_{n-1}, or a point. In the first case, the subgroup is conjugate inside the general projective group to the general linear group p_i, $x_i p_k$, and in the second case, to the group $x_k p_i$, P_k.*[3]

Because the groups of the one category come from the groups of the other category through the exchange of p_i, $x_i p_k$, P_i with P_i, $-x_k p_i$, p_i, all $n(n+1)$-term subgroups of the general projective group are holoedrically isomorphic to each other. From the preceding, we obtain the following

Proposition 4. *The general projective group of the space $x_1, \dots, x_n$ can be related to itself in a holoedrically isomorphic way so that the largest subgroups which leave invariant one point correspond each time to the largest subgroups which leave invariant a plane M_{n-1}.*

If $n = 1$, the difference between point and plane M_{n-1} disappears; the general three-term projective group of the once-extended manifold therefore contains only *one* category of two-term subgroups, and all of these are conjugate in the three-term projective group.

---

[3] LIE, Math. Ann. Vol. XXV, p. 130.

According to the previous results, the general projective group of an n-times extended space contains $n(n+1)$ independent infinitesimal transformations which leave at rest a given point, and to be precise, these infinitesimal transformations generate a subgroup which is contained in no larger subgroup. From this, it follows (Chap. 24, Theorem 91, p. 524) that *the general projective group is primitive and all the more asystatic.*

§ 140. We derive here a few general considerations concerning the determination of all subgroups of the general linear group $p_i, x_i p_k$ $(i, k=1 \cdots n)$ of R_n.

The general infinitesimal transformation of a linear group of R_n can be written:

$$\text{(9)} \qquad \begin{cases} \sum_{i,k}^{i \gtrless k} a_{ik} x_i p_k + \sum_{i=1}^{n-1} b_i (x_i p_i - x_n p_n) \\ + c \sum_{k=1}^{n} x_k p_k + \sum_{k=1}^{n} d_k p_k. \end{cases}$$

If one combines two infinitesimal transformations of this form, one obtains a transformation:

$$\sum_{i,k}^{i \gtrless k} A_{ik} x_i p_k + \sum_{i=1}^{n-1} B_i (x_i p_i - x_n p_n) + C \sum_{k=1}^{n} x_k p_k + c \sum_{k=1}^{n} D_k p_k,$$

in which the A_{ik}, B_i and C $(C=0)$ depend only on a_{ik}, b_i and c. Consequently, the reduced infinitesimal transformation:

$$\text{(10)} \qquad \sum_{i,k}^{i \gtrless k} a_{ik} x_i p_k + \sum_{i=1}^{n-1} b_i (x_i p_i - x_n p_n) + c \sum_{k=1}^{n} x_k p_k$$

is in turn the general infinitesimal transformation of a linear homogeneous group in $x_1, \dots, x_n$.

From this, we realize that the problem of determining all subgroups of the general linear group decomposes into two problems which must be settled one after the other. First, all subgroups of the general linear homogeneous group $x_i p_k$ $(i, k=1 \cdots n)$ have to be sought; then, to the infinitesimal transformations of each of the found groups, one must add terms $\sum \beta_k p_k$ in the most general way so that one again obtains a group. Thus, if $X_1 f, \dots, X_r f$ is one of the found linear homogeneous groups, one has to determine all groups of the form:

$$X_k f + \sum_{i=1}^{n} \alpha_{ki} p_i, \qquad \sum_{i=1}^{n} \beta_{\mu i} p_i$$

$$(k=1 \cdots r;\ \mu=1 \cdots m,\ m \leqslant n).$$

We do not want to say anything more about the further treatment of these two reduced problems here; rather, we refer to the third volume where a detailed study of the projective groups of the plane and of thrice-extended space will be given.

By contrast, we shouldn't neglect the geometric significance of the above mentioned decomposition.

To this end, we imagine that the group (9) is extended, by regarding, as in Chap. 25, p. 528 sq., the x as functions of an auxiliary variable t and by taking with them the differential quotients $\mathrm{d}x_i/\mathrm{d}t = x_i'$. In the process, we obtain the group:

$$\begin{aligned}&\sum a_{ik}\,x_i\,p_k + \sum b_i\,(x_i p_i - x_n p_n) + c\sum x_k\,p_k + \sum d_k\,p_k\\&+\sum a_{ik}\,x_i'\,p_k' + \sum b_i\,(x_i' p_i' - x_n' p_n') + c\sum x_k'\,p_k',\end{aligned}$$

in which the terms in the x_i' determine for themselves a group, namely precisely the group (10) just found. But now, as we have seen loc. cit., the x_i' can be interpreted as homogeneous coordinates of the directions through the point $x_1, \dots, x_n$ of R_n. The fact that the x_i' are transformed for themselves by the above group therefore means nothing else than the fact that parallel lines are transferred to parallel lines by every linear transformation of the x; the directions which a determined system of values x_i' associates to all points of R_n are indeed parallel to each other. But every bundle [BÜNDEL] of parallel lines provides a completely determined point on the plane M_{n-1} at infinity of R_n, hence *$x_1', \dots, x_n'$ can be virtually interpreted as homogeneous coordinates of the points on the plane at infinity M_{n-1}* and the group:

$$\sum a_{ik}\,x_i'\,p_k' + \sum b_i\,(x_i' p_i' - x_n' p_n') + c\sum x_k'\,p_k' \tag{11}$$

therefore indicates how the points at infinity of R_n are transformed by the group (9). At the same time, it is yet to be observed that the infinitesimal transformation $\sum x_k'\,p_k'$ leaves fixed all points at infinity, so that these points are transformed by the last group exactly as if c were zero.

Here, we have the central reason for the decomposition indicated above for the problem of determining all linear groups of R_n. The groups in question are simply thought to be distributed in classes, and in each class are included all the groups for which the group (11) is the same, so that the points at infinity of R_n are transformed in the same way (cf. for this purpose Theorem 40, p. 243).

§ 141. In order to give at least *one* application of the general considerations developed just above, we want to determine all linear groups of R_n which transform the points of infinity of R_n in the most general way. For all of these groups, the associated group (11) has the form:

$$x_i'\,p_k', \quad x_i'\,p_i' - x_k'\,p_k' \qquad {\scriptstyle (i,\,k=1\cdots n,\ i\gtrless k)},$$

where in certain circumstances, the transformation $x_1'\,p_1' + \cdots + x_n'\,p_n'$ can occur, which leaves individually fixed all points at infinity. The points at infinity are thus transformed by an (n^2-1)-term group and, to be precise, by the general projective group of an $(n-1)$-times extended space.

Each one of the sought groups must contain n^2-1 infinitesimal transformations out of which none can be linearly deduced which leaves invariant all points at infinity, hence none which are of the form: $\gamma \sum_j x_j p_j + \sum \gamma_k p_k$. Therefore, the group surely contains n^2-1 infinitesimal transformations of the form:

$$(12) \qquad \begin{cases} x_i p_k + \alpha_{ik} \sum_{j=1}^{n} x_j p_j + \sum_{\nu=1}^{n} \beta_{ik\nu} p_\nu & (i \gtrless k) \\ x_i p_i - x_n p_n + \alpha_i \sum_{j=1}^{n} x_j p_j + \sum_{\nu=1}^{n} \beta_{i\nu} p_\nu . \end{cases}$$

In addition, one or several infinitesimal transformations of the form: $\gamma \sum_j x_j p_j + \sum_\nu \gamma_\nu p_\nu$ can occur.

If a group of the demanded sort contains a translation, then it contains all translations. Indeed, if $p_1 + e_2 p_2 + \cdots + e_n p_n$ is the translation in question, we combine it with:

$$x_1 p_k + \alpha_{1k} \sum_{j=1}^{n} x_j p_j + \sum_{\nu=1}^{n} \beta_{1k\nu} p_\nu \qquad (k=2\cdots n)$$

and we obtain in this way $p_2, \dots, p_n$ and therefore all p_i.

Consequently, we first want to assume that all translations appear. Then if the transformation $\sum x_j p_j$ appears, we have the general linear group itself. By contrast, if the transformation $\sum x_j p_j$ is not extant, we obtain by combination of the transformations (12) in which we can set beforehand the $\beta_{ik\nu}$ and $\beta_{i\nu}$ equal to zero:

$$\left[x_i p_k + \alpha_{ik} \sum_{j=1}^{n} x_j p_j,\; x_k p_i + \alpha_{ki} \sum_{j=1}^{n} x_j p_j \right] = x_i p_i - x_k p_k,$$

so that all the α_i are zero. But in addition, we have:

$$\left[x_i p_i - x_k p_k,\; x_i p_k + \alpha_{ik} \sum_{j=1}^{n} x_j p_j \right] = 2 x_i p_k .$$

The concerned group is therefore the special linear group.

If in the sought group there is no translation, but by contrast a transformation:

$$(13) \qquad \sum_{j=1}^{n} x_j p_j + \sum_{\nu=1}^{n} \gamma_\nu p_\nu = \sum_{j=1}^{n} (x_j + \gamma_j) p_j ,$$

then all α_{ik} and all α_i can be set equal to zero. If we write the infinitesimal transformations (12) in the form:

$$(x_i + \gamma_i)\, p_k + \sum_{\nu=1}^{n} \beta'_{ik\nu}\, p_\nu \qquad (i \gtrless k)$$

$$(x_i + \gamma_i)\, p_i - (x_n + \gamma_n)\, p_n + \sum_{\nu=1}^{n} \beta'_{i\nu}\, p_\nu,$$

we then realize immediately by combination with $\sum (x_i + \gamma_i)\, p_i$ that all $\beta'_{ik\nu}$ and $\beta'_{i\nu}$ vanish and we thus find the group:

$$(x_i + \gamma_i)\, p_k \qquad (i, k = 1 \cdots n).$$

Lastly, if there also occurs no transformation of the form (13), then one first obtains by combination from (12) that all α_{ik} and α_i are equal to zero. Furthermore, we have:

$$\left[x_i\, p_k + \sum_{\nu=1}^{n} \beta_{ik\nu}\, p_\nu,\; x_k\, p_i + \sum_{\nu=1}^{n} \beta_{ki\nu}\, p_\nu \right] = x_i\, p_i - x_k\, p_k + \beta_{ikk}\, p_i - \beta_{kii}\, p_k,$$

and in addition:

$$\left[(x_i + \beta_{ikk})\, p_i - (x_k + \beta_{kii})\, p_k,\; x_i\, p_k + \sum_{\nu=1}^{n} \beta_{ik\nu}\, p_\nu \right]$$
$$= 2\, x_i\, p_k + 2\, \beta_{ikk}\, p_k - \beta_{iki}\, p_i,$$

so that all $\beta_{ik\nu}$ vanish with the exception of the β_{ikk} and β_{kii}, by means of which the $\beta_{i\nu}$ can also be expressed.

Now, if $n > 2$, the β_{ikk} could vary with k; however, this is not the case. Indeed, if we replace i and k firstly by k, j and secondly by j, i in:

$$(x_i + \beta_{ikk})\, p_i - (x_k + \beta_{kii})\, p_k$$

and if we add together the three obtained infinitesimal transformations, then the sum must vanish, since no translation should occur. So we obtain: $\beta_{ikk} = \beta_{ijj}$ and so on. Thus, if we write shortly β_i for β_{ikk}, we have the group:

$$(x_i + \beta_i)\, p_k, \qquad (x_i + \beta_i)\, p_i - (x_k + \beta_k)\, p_k \qquad (i \gtrless k).$$

With that, all cases are settled. If, in the two latter forms of groups [GRUPPENFORMEN], we introduce $x_i + \gamma_i$ and $x_i + \beta_i$ as new x_i, respectively, we may recapitulate our result as follows:

Theorem 102. *The general linear group in n variables contains only three different sorts of subgroups which transform the points of the plane at infinity in an $(n^2 - 1)$-term way, as does the general linear group itself: firstly, the special linear group and secondly all groups that are conjugate to the two homogeneous groups:*[4]

[4] LIE, Archiv for Math. og Nat. Vol. IX, p. 103 and 104, Christiania 1884.

$$x_i p_k; \qquad x_i p_k \quad x_i p_k - x_k p_k \qquad (i \gtrless k).$$

Here, we thus have a characteristic property which is common to all these groups already known to us. By contrast, the distinctive marks of the four groups in question are briefly the following: The general linear group leaves invariant the ratios of all volumes; the special linear group leaves invariant all volumes. The general and the special linear homogeneous groups differ from the general and the special linear groups, respectively, in that they leave invariant the point $x_i = 0$.

§ 142. At the beginning of the previous section, we saw that the determination of all linear groups of R_n is essentially produced as soon as all linear homogeneous projective groups of the R_n are determined. There is no special difficulty in settling this last problem when one knows all projective groups of R_{n-1}. We now want to show this.

As we know, the general linear homogeneous group of R_n: $x_i p_k$ $(i, k = 1 \cdots n)$ contains an invariant subgroup with $n^2 - 1$ parameters, namely the special linear homogeneous group:

$$x_i p_k, \qquad x_i p_i - x_k p_k \qquad (i, k = 1 \cdots n, \; i \gtrless k).$$

This last group is equally composed with the general projective group of R_{n-1}, so its subgroups can immediately be written down when all projective groups of R_{n-1} are known (cf. the next chapter). Then, one finds the subgroups of the group $x_i p_k$, thanks to the following considerations:

An r-term subgroup G_r of the group $x_i p_k$ is either contained at the same time in the group $x_i p_k$, $x_i p_i - x_k p_k$ $(i \gtrless k)$ or it is not. In the first case it would be already known, and in the second case, according to Proposition 7, p. 223, it would have a G_{r-1} in common with the special linear homogeneous group. Hence, in order to find all linear homogeneous groups G_r of this sort, we need only to add, to every G_{r-1}: $X_1 f, \dots, X_{r-1} f$ of the group $x_i p_k$, $x_i p_i - x_k p_k$ $(i \gtrless k)$, an infinitesimal transformation of the form:

$$\mathsf{Y}f = \sum_{i=1}^{n} x_i p_i + \sum_{k,\, j}^{1 \cdots n} \alpha_{kj} x_k p_j \qquad \left(\Sigma_k \, \alpha_{kk} = 0\right),$$

and to determine, by combination with $X_1 f, \dots, X_{r-1} f$, all values of the α_{kj} which produce a group. Here, it is to be observed that the $(r-1)$-term group $X_1 f, \dots, X_{r-1} f$ must obviously be invariant in the sought r-term group; we therefore find, so to speak, the most general values of the α_{kj} when we seek the most general linear homogeneous infinitesimal transformation $\mathsf{Y}f$ which leaves invariant the given $(r-1)$-term group. Besides, we always find *one* r-term group, namely when we choose all α_{kj} equal to zero.

By taking into consideration the infinitesimal transformations $X_1 f, \dots, X_{r-1} f$, one realizes that $r - 1$ of the α_{kj} can be made equal to zero. Hence the smaller the number r is, the more constants must be determined. But for small values of r, the following method is otherwise often more convenient:

Indeed, the r-term subgroups of the group $x_i\,p_k$ can in another way be distributed into two categories; firstly, in the category of subgroups which contain the transformation $\sum x_i\,p_i$ — they can, under the assumptions made, be written down instantly — and secondly, in the category of subgroups which do not contain the transformation $x_1\,p_1 + \cdots + x_n\,p_n$. The r infinitesimal transformations of a group from the latter category must have the form:

$$X_k f + \alpha_k \sum_{i=1}^{n} x_i\,p_i, \tag{14}$$

where the $X_k f$ represent infinitesimal transformations of the special linear homogeneous group. Because the transformation $\sum x_i\,p_i$ is interchangeable with all $X_k f$, for the execution of the bracket operation [KLAMMEROPERATION], it is completely indifferent whether the α_i vanish or not; hence $X_1 f, \dots, X_r f$ must themselves generate a group and, to be precise, an r-term subgroup of the group $x_i\,p_k$, $x_i\,p_i - x_k\,p_k$ $(i \gtrless k)$, one of those which we assume as known. Therefore, it only remains to determine the α_k in the most general way so that the infinitesimal transformations (14) generate a group. In certain circumstances, one can realize easily that certain of the α_k must vanish; indeed, if there is an equation of the form:

$$[X_i, X_k] = X_j f,$$

then α_j must necessarily be zero.

If all $[X_i, X_k]$ generate a ρ-term group (Chap. 15, Proposition 6, p. 274), then we can assume that the infinitesimal transformations are chosen so that all $[X_i, X_k]$ can be linearly deduced from $X_1 f, \dots, X_\rho f$. Then we have $\alpha_1 = \cdots = \alpha_\rho = 0$, while all other α_i can be different from zero. It must be specially studied in each individual case whether the different values of these parameters produce different *types* of linear homogeneous groups, or in other words, whether the concerned parameters are essential. The settlement of this problem for $n = 2$ and for $n = 3$ appears in the third volume.

§ 143. If an arbitrary linear homogeneous group G_r is presented which is not contained in the special linear homogeneous group, then as we have already observed above, this G_r comprises an *invariant* $(r-1)$-term subgroup whose infinitesimal transformations are characterized by the fact that they are of the form:

$$\sum_{i,\,k}^{i \gtrless k} a_{ik}\,x_i\,p_k + \sum_{i=1}^{n-1} a_i\,(x_i\,p_i - x_n\,p_n).$$

If we now apply this observation to the adjoint associated to an *arbitrary* r-term group: $X_1 f, \dots, X_r f$:

$$E_\nu f = \sum_{k,s}^{1\cdots r} c_{k\nu s}\, e_k \frac{\partial f}{\partial e_s} \qquad (\nu = 1 \cdots r),$$

then we recognize immediately that this adjoint group, when the r sums $\sum_k c_{k\nu k}$ do not all vanish, contains an invariant subgroup whose infinitesimal transformations: $\lambda_1 E_1 f + \cdots + \lambda_r E_r f$ are defined by means of the condition-equations:

$$\sum_{\nu=1}^{r} \lambda_\nu \sum_{k=1}^{r} c_{k\nu k} = 0.$$

Lastly, if we remember that every group is isomorphic with its adjoint group, we obtain the

Proposition 5. *If r independent infinitesimal transformations $X_1 f, \dots, X_r f$ of an r-term group stand pairwise in the relationships: $[X_i, X_k] = \sum_s c_{iks} X_s f$ and if at least one of the r sums $\sum_k c_{\nu kk}$ is different from zero, then all infinitesimal transformations $\lambda_1 X_1 f + \cdots + \lambda_r X_r f$ which satisfy the condition:*

$$\sum_{\nu=1}^{r} \lambda_\nu \sum_{k=1}^{r} c_{\nu kk} = 0$$

generate an invariant $(r-1)$-term subgroup.[5]

For a thorough study of the general projective group, one must naturally devote special attention to its adjoint group $\sum e_k E_k f$ and to the associated invariant systems of equations in the e_k as well. Here we make only two brief, but important observations.

If, as usual, one interprets the e_k as homogeneous point coordinates of a space with $n^2 + 2n - 1$ dimensions, then amongst all invariant manifolds of this space, there is a determined one whose dimension number has the smallest value. This important manifold consists of all points e_k which represent either translations or transformations conjugate to translations. This manifold is contained in no even [EBEN] manifold of the space in question. Besides, the known classification of all projective transformations naturally provides without effort *all* invariant manifolds of the space $e_1, \dots, e_r$.

For every infinitesimal projective transformation which is conjugate to a translation, all points of a plane M_{n-1} in the space x_k remain invariant and simultaneously, all planes M_{n-1} which pass through a certain point of this M_{n-1}. Every transformation of this sort is completely determined by means of the firstly said plane M_{n-1} and by the distinguished invariant point of this plane.

If $n = 2$, then as we can say, every projective transformation of the plane x_1, x_2 which is conjugate to an infinitesimal translation is completely represented by

---

[5] LIE, Archiv for Math., Vol. IX, p. 89, Christiania 1884; Fortschritte der Mathematik, Vol. XVI, p. 325.

means of a *line element*. Accordingly, in thrice-extended space x_1, x_2, x_3, every projective transformation conjugate to an infinitesimal translation is represented by a *surface element*, and so on. These observations will be exploited in the third volume.

Chapter 27
Linear Homogeneous Groups

In the previous chapter, p. 577, we characterized the general linear homogeneous group in n variables $x_1, \dots, x_n$ as the most general projective group of the n-times extended space $x_1, \dots, x_n$, or shortly R_n, that leaves invariant the plane at infinity M_{n-1} and simultaneously the point $x_1 = \cdots = x_n = 0$. This group receives another meaning when one interprets $x_1, \dots, x_n$ as homogeneous coordinates in an $(n-1)$-times extended space R_{n-1}. In the present chapter, this interpretation shall be followed at the foundation.

§ 144. We imagine that the transition from the ordinary Cartesian coordinates $y_1, \dots, y_{n-1}$ of the $(n-1)$-times extended space R_{n-1} to the homogeneous coordinates $x_1, \dots, x_n$ of the same space is procured by means of the equations:

$$y_k = \frac{x_k}{x_n} \qquad (k=1 \cdots n-1).$$

To the n^2-term general linear homogeneous group:

$$x_i' = \sum_{k=1}^{n} \alpha_{ik}\, x_k \qquad (i=1 \cdots n) \tag{1}$$

then corresponds, in the variables $y_1, \dots, y_{n-1}$, the meroedrically isomorphic group:

$$y_i' = \frac{\sum_{k=1}^{n-1} \alpha_{ik}\, y_k + \alpha_{in}}{\sum_{k=1}^{n-1} \alpha_{nk}\, y_k + \alpha_{nn}} \qquad (k=1 \cdots n-1), \tag{2}$$

the (n^2-1)-term general projective group of R_{n-1}. Thus, to every linear homogeneous transformation (1) corresponds a single projective transformation (2), hence a completely determined collineation of R_{n-1}, whereas conversely, to every projective transformation (2), there correspond in total ∞^1 different linear homogeneous transformations (1).

S. Lie, *Theory of Transformation Groups I*,
DOI 10.1007/978-3-662-46211-9_27

In addition, one can also establish a univalent invertible association between linear homogeneous transformations in $x_1,\dots,x_n$ and projective transformations in $y_1,\dots,y_{n-1}$ when one submits the constants α_{ik} to the condition $\sum \pm \alpha_{11}\cdots\alpha_{nn}=1$, hence when one considers instead of the general the special linear homogeneous group which is holoedrically isomorphic to the general projective group (2) (cf. Chap. 26, p. 563). We want to develop in detail this association for the infinitesimal transformations of the two groups.

The special linear homogeneous group in $x_1,\dots,x_n$ contains the following n^2-1 independent infinitesimal transformations:

$$x_i\,p_k, \qquad x_i\,p_i - x_k\,p_k \qquad (i \gtrless k). \tag{3}$$

In order to find the corresponding infinitesimal transformations in $y_1,\dots,y_{n-1}$, we only have to compute, for each of the individual infinitesimal transformations just written, the increment:

$$\delta y_i = \frac{x_n\,\delta x_i - x_i\,\delta x_n}{x_n^2} \qquad (i=1\cdots n-1).$$

In this way, we find the following table:

$$\begin{cases} x_n\,p_k \equiv q_k, \qquad x_k\,p_n \equiv -y_k\,(y_1\,q_1+\cdots+y_{n-1}\,q_{n-1}) \\ x_k\,p_k - x_n\,p_n \equiv y_k\,q_k + y_1\,q_1+\cdots+y_{n-1}\,q_{n-1}, \qquad x_i\,p_k \equiv y_i\,q_k \\ \qquad (i,\,k=1\cdots n-1;\ i \gtrless k), \end{cases} \tag{4}$$

where q_i is written for $\partial f/\partial y_i$. This table also provides inversely the infinitesimal transformations of the special linear homogeneous group (3) corresponding to every infinitesimal transformation of the projective group (2); indeed, from the equations (4), we obtain without effort:

$$\begin{aligned} n\,(y_1\,q_1+\cdots+y_{n-1}\,q_{n-1}) &\equiv x_1\,p_1+\cdots+x_n\,p_n - n\,x_n\,p_n \\ n\,y_k\,q_k &\equiv n\,x_k\,p_k - (x_1\,p_1+\cdots+x_n\,p_n). \end{aligned}$$

Now, it is also easy to indicate which ∞^1 infinitesimal transformations of the general linear homogeneous group (1) correspond to a given infinitesimal transformation of the projective group (2). Indeed, the infinitesimal transformation $x_1\,p_1 + \cdots + x_n\,p_n$ reduces in the variables $y_1,\dots,y_{n-1}$ to the identity, for all increases of the y_k vanish. Consequently, if the infinitesimal transformation Yf of the projective group (2) corresponds to the infinitesimal transformation Xf of the special homogeneous group (3), then all ∞^1 infinitesimal transformations of the general homogeneous group (1) which correspond to Yf are contained in the expression $Xf + c\,(x_1\,p_1+\cdots+x_n\,p_n)$, where c denotes an arbitrary constant.

If a system of equations:

$$\Omega_k(y_1,\dots,y_{n-1}) = 0 \qquad (k=1\cdots m)$$

in $y_1, \dots, y_{n-1}$ admits a finite or an infinitesimal projective transformation, then naturally, the corresponding system of equations in the x:

$$\Omega_k\left(\frac{x_1}{x_n}, \dots, \frac{x_{n-1}}{x_n}\right) = 0 \qquad (k=1\cdots m)$$

admits the corresponding finite or infinitesimal transformation of the group (3); but in addition, because of its homogeneity, it also admits the infinitesimal transformation: $x_1 p_1 + \cdots + x_n p_n$.

Conversely, every system of equations in $x_1, \dots, x_n$ which admits $x_1 p_1 + \cdots + x_n p_n$ is homogeneous. But now, when we write a projective group of R_{n-1} in the homogeneous variables $x_1, \dots, x_n$, we are concerned only with the ratios of the x, hence, we are concerned only with systems of equations that are homogeneous in the x. So, when we have written in a homogeneous way the infinitesimal transformations of a projective group of R_{n-1} with the help of the table (4) and when we want to find the associated invariant systems of equations, that is why we will always add the infinitesimal transformation $x_1 p_1 + \cdots + x_n p_n$. The group in $x_1, \dots, x_n$ obtained in this way is the true analytic representation in homogeneous variables of the concerned projective group of R_{n-1}.

For the study of a projective group, if one also wants to take the infinite into consideration, then one must write the group in homogeneous variables.

§ 145. In the preceding section, we have shown that the infinitesimal projective transformations of R_{n-1} can be replaced by infinitesimal linear homogeneous transformations in n variables. Now, we want to imagine that an arbitrary transformation of this sort in $x_1, \dots, x_n$ is presented, say:

$$Xf = \sum_{i,\,k}^{1\cdots n} a_{ki}\, x_i\, p_k,$$

and we want to submit it to a closer examination. Namely, we want to look for plane manifolds of R_{n-1} which admit the infinitesimal transformation in question. In this way, we will succeed in showing that Xf can always be given a certain canonical form after the introduction of appropriate new variables: $x_i' = \sum c_{ik}\, x_k$.

If a plane: M_{n-2}: $\sum c_i x_i = 0$ of R_{n-1} admits the infinitesimal transformation Xf, then according to Theorem 14, p. 127, it admits at the same time all finite transformations of the associated one-term group. Now, the M_{n-2} admits the infinitesimal transformation if and only if the expression $X(\sum c_i x_i)$ vanishes simultaneously with $\sum c_i x_i$. Since $X(\sum c_i x_i)$ is linear in the x_i, this condition amounts to the fact that a relation of the form:

$$X\left(\sum_{k=1}^{n} c_k\, x_k\right) = \rho \sum_{i=1}^{n} c_i\, x_i$$

holds identically, where ρ denotes a constant. The condition-equation following from this:

$$\sum_{i=1}^{n}\left(\sum_{k=1}^{n} a_{ki}\,c_k - \rho\,c_i\right)x_i = 0$$

decomposes into the n equations:

$$(5)\qquad a_{1i}\,c_1 + \cdots + (a_{ii} - \rho)\,c_i + \cdots + a_{ni}\,c_n = 0 \qquad (i=1\cdots n),$$

which can be satisfied only when the determinant:

$$(6)\qquad \begin{vmatrix} a_{11}-\rho & a_{21} & \cdot\;\cdot\;\cdot & a_{n1} \\ a_{12} & a_{22}-\rho & \cdot\;\cdot\;\cdot & a_{n2} \\ \cdot & \cdot & \cdot\;\cdot\;\cdot & \cdot \\ a_{1n} & a_{2n} & \cdot\;\cdot\;\cdot & a_{nn}-\rho \end{vmatrix} = \Delta(\rho)$$

vanishes. This produces for ρ an equation of degree n with n roots, amongst which some can be multiple. Therefore in all circumstances, there is at least one plane $M_{n-2}:\; c_1\,x_1 + \cdots + c_n\,x_n = 0$ which remains invariant by the one-term group Xf.

It is known that one sees in exactly the same way that every *finite* projective transformation, or, when written homogeneously, every transformation $x_i' = \sum b_{ik}\,x_k$ likewise leaves fixed at least one plane M_{n-2}. In addition, this follows from the fact that every transformation $x_i' = \sum b_{ik}\,x_k$ is associated to a one-term group Xf.

If the equation $\Delta(\rho) = 0$ found above has exactly n different roots ρ, then there are in total n separate planes M_{n-2} which remain invariant by the group Xf; indeed, two distinct roots ρ_1 and ρ_2 of the equation for ρ always provide, because of the form of the equations (5), two distinct systems of values $c_1 : c_2 : \cdots : c_n$. By contrast, if there are multiple roots ρ, then various cases can occur. If, for an m-fold root ρ, the determinant (6) vanishes itself, but not all of its $(n-1)\times(n-1)$ determinants, then for the concerned value of ρ, exactly $n-1$ of the equations remain independent of each other and the ratios of the c_i are then all determined. To the m-fold root is hence associated only a single invariant plane M_{n-2}, but this plane is counted m times. By contrast, if for an m-fold root not only the determinant (6) itself is equal to zero, but if all its $(n-1)\times(n-1)$, ..., $(n-q+1)\times(n-q+1)$ subdeterminants also vanish, whereas not all $(n-q)\times(n-q)$ subdeterminants do ($q \leqslant m$), then the equations (5) reduce to exactly $n-q$ independent equations and amongst the ratios of the c_i, there remain $q-1$ that can be chosen arbitrarily. Thus in this case, the m-fold root ρ gives a family of ∞^{q-1} planes M_{n-2} which remain individually invariant.

It is easy to see that a root of the equation $\Delta(\rho) = 0$ is at least q-fold when all $(n-q+1)\times(n-q+1)$ subdeterminants of $\Delta(\rho)$ vanish for this root. Indeed, the differential quotients of order $(q-1)$ of $\Delta(\rho)$ with respect to ρ express themselves as sums of the $(n-q+1)\times(n-q+1)$ subdeterminants of $\Delta(\rho)$.

If we assume that the duality theory is already known at this point, then we could immediately conclude that the infinitesimal transformation Xf leaves invariant at least one *point* in R_{n-1}. But we prefer to prove this directly, particularly because we gain on the occasion a deeper insight.

In the homogeneous variables $x_1, \dots, x_n$, a point is represented by $n-1$ equations of the form $x_i x_k^0 - x_k x_i^0$; so the point will admit all transformations of the one-term group Xf when $x_k^0 X x_i - x_i^0 X x_k$ vanishes by virtue of the equations $x_i x_k^0 - x_k x_i^0 = 0$, hence when n relations of the form:

$$X x_i = \sigma x_i \qquad (i=1\cdots n)$$

hold. For the x_i and for σ, we therefore obtain the n condition-equations:

$$a_{i1} x_1 + \cdots + (a_{ii} - \sigma) x_i + \cdots + a_{in} x_n = 0 \qquad (i=1\cdots n).$$

If we disregard the nonsensical solution $x_1 = \cdots = x_n = 0$, σ must be a root of the equation:

$$\Delta(\sigma) = \begin{vmatrix} a_{11}-\sigma & a_{12} & \cdots & a_{1n} \\ a_{21} & a_{22}-\sigma & \cdots & a_{2n} \\ \cdot & \cdot & \cdots & \cdot \\ a_{n1} & a_{n2} & \cdots & a_{nn}-\sigma \end{vmatrix}$$

and each such root produces an invariant point. The determination of the points which stay fixed by Xf therefore leads to the same algebraic equation as does the determination of the invariant plane: $M_{n-2}\colon\ c_1 x_1 + \cdots + c_n x_n = 0$.

Thus, if this equation of degree n has n different roots, then in $R_{n-1}\colon$ $x_1 : x_2 : \cdots : x_n$, not only n different planes $M_{n-2}\colon\ \sum c_k x_k = 0$ remain invariant, but also simultaneously, n separate points. In the process, these n points do not lie all in one and the same plane M_{n-2}, because if n points stay fixed in a plane M_{n-2}, then necessarily, infinitely many points of the M_{n-2} keep their positions, which is excluded under the assumption made. We can briefly express this property of the n invariant points by saying that a nondegenerate n-frame [n-FLACH] remains invariant by Xf.

If there appear multiple roots, then two cases must again be distinguished. If, for an m-fold root, not all $(n-1) \times (n-1)$ subdeterminants of the determinant (6) vanish, then this root gives an invariant plane M_{n-2} counting m times and an invariant point counting m times. By contrast, if, for the root in question, all $(n-1) \times (n-1)$, $\dots$, $(n-q+1) \times (n-q+1)$ subdeterminants are equal to zero ($q \leqslant m$) without all $(n-q) \times (n-q)$ subdeterminants vanishing, then to this root is associated a family of ∞^{q-1} individually invariant planes M_{n-2} and a plane manifold of ∞^{q-1} individually invariant points. Therefore, the following holds

Proposition 1. *Every infinitesimal transformation:*

$$\sum_{i,\,k}^{1\cdots n} a_{ki} x_i p_k$$

in the homogeneous variables $x_1, \dots, x_n$, or, what is the same, every infinitesimal projective transformation in $n-1$ variables:

$$\frac{x_1}{x_n}, \ \dots, \ \frac{x_{n-1}}{x_n},$$

leaves invariant a series of points $x_1 : x_2 : \cdots : x_n$ *and a series of planes* M_{n-2}: $c_1 x_1 + \cdots + c_n x_n = 0$. *The points which stay fixed fill a finite number of, and to be precise at most n, separate manifolds. Likewise, the invariant planes* M_{n-2} *form a finite number of, and to be precise at most n, separate linear pencils.*

If the infinitesimal transformation Xf has the form: $\sum x_k p_k$, then it actually leaves invariant all points $x_1 : x_2 : \cdots : x_n$ and naturally also, all planes M_{n-2}: $\sum c_i x_i = 0$.

From what has been found up to now, one can now draw further conclusions. First, we consider once more the special case where the equation of degree n discussed above has n different roots. Then there are n separate invariant planes M_{n-2}: $\sum_i c_{ki} x_i = 0$ which form a true n-frame according to the preceding. We can therefore introduce:

$$x_k' = \sum_{i=1}^{n} c_{ki} x_i \qquad (k = 1 \cdots n)$$

as new homogeneous variables and in the process, we must obtain an infinitesimal transformation in $x_1', \dots, x_n'$ which leaves invariant the n equations $x_k' = 0$, hence which possesses the form:

$$Xf = a_1' x_1' p_1' + \cdots + a_n' x_n' p_n'.$$

Under the assumptions made, Xf can be brought to this canonical form. Naturally, no two of the quantities $a_1', \dots, a_n'$ are equal to each other here, for the equation:

$$(a_1' - \rho) \cdots (a_n' - \rho) = 0$$

must obviously have n distinct roots.

Now, similar canonical forms of Xf also exist when the equation for ρ possesses multiple roots. However, we do not want to get involved in the consideration of them, and we only want to show that there is a canonical form to which *every* infinitesimal transformation:

$$Xf = \sum_{i,\,k}^{1 \cdots n} a_{ki} x_i p_k$$

can be brought thanks to an appropriate change of variables: $x_k' = \sum h_{ki} x_i$, that is to say hence, thanks to an appropriate collineation of R_{n-1}, completely without paying heed to the constitution of the equation $\Delta(\rho) = 0$.

Since Xf always leaves a point invariant, we can imagine that our coordinates are chosen so that the point: $x_1 = \cdots = x_{n-1} = 0$ stay fixed. On the occasion, we find:

$$Xf = \sum_{k=1}^{n-1} \sum_{i=1}^{n-1} a_{ik}' x_k p_i + \sum_{k=1}^{n} a_{nk}' x_k p_n = X'f + \sum_{k=1}^{n} a_{nk}' x_k p_n. \tag{7}$$

To the linear homogeneous infinitesimal transformation $X'f$ in the $n-1$ variables $x_1,\dots,x_{n-1}$ we can apply the same process as the one which provided us with the reduction of Xf to the form (7), and we obtain in this way:

$$Xf = \sum_{k=1}^{n-2}\sum_{i=1}^{n-2} a''_{ik}x_k\,p_i + \sum_{k=1}^{n-1} a''_{n-1,k}x_k\,p_{n-1} + \sum_{k=1}^{n} a'_{nk}x_k\,p_n.$$

Here, we can again treat in an analogous way the first term of the right-hand side. Finally, we obtain the

Theorem 103. *In every linear homogeneous infinitesimal transformation with n variables, one can introduce as new independent variables n linear homogeneous functions of these variables so that the concerned infinitesimal transformation takes the form:*

$$a_{11}x_1\,p_1 + (a_{21}x_1 + a_{22}x_2)\,p_2 + \cdots + (a_{n1}x_1 + \cdots + a_{nn}x_n)\,p_n.$$

It is of some interest to interpret the way in which this result has been obtained.

Since Xf leaves a point invariant in any case, if above we chose such a point as corner of coordinates [COORDINATENECKPUNKT]: $x_1 = \cdots = x_{n-1} = 0$, then the ratios $x_1 : x_2 : \cdots : x_{n-1}$ represent the straight lines through the chosen point; simultaneously with the points $x_1 : x_2 : \cdots : x_n$, these straight lines are permuted with each other by the transformation Xf, and to be precise, by means of the linear transformation $X'f$ in the $n-1$ variables $x_1,\dots,x_{n-1}$. But according to the preceding, a system of ratios $x_1 : \cdots : x_{n-1}$ must remain unchanged by $X'f$, that is to say, a straight line through the point just said. If we choose this line as an edge $x_1 = \cdots = x_{n-2} = 0$ of our coordinate system, then the ratios $x_1 : \cdots : x_{n-2}$ represent the planes M_2 passing through this edge. In the same way, these M_2 are permuted with each other by Xf; one amongst them surely remains invariant and again gives a closer determination of the coordinate system, and so on.

In this way, one realizes that the coordinate system can be chosen so that Xf receives the normal form indicated above. However, we want to recapitulate in a specific proposition the considerations just made, since they express a general property of the infinitesimal projective transformations of R_{n-1}:

Proposition 2. *By every infinitesimal projective transformation of R_{n-1} there remain invariant: at least one point; through every invariant point: at least one straight line; ...; and lastly, through every invariant plane M_{n-2}: at least one plane M_{n-2}.*

§ **146.** The results of the previous section have long been known since they essentially coincide in with the reduction to a normal form, due to CAUCHY, of a system of linear ordinary differential equations with constant coefficients.

In what follows, we will generalize in an essential way the results found up to now, but beforehand, we must make a few observations which, strictly speaking, are certainly subordinate to the general developments in Chap. 29, p. 486 sq.

Let an arbitrary r-term group G_r with an $(r-m)$-term *invariant* subgroup G_{r-m} be presented, and let T denote an arbitrary transformation of G_r, while S denotes an arbitrary transformation of G_{r-m}. Then according to the assumption, there are certain equations of the form:

$$T^{-1}ST = S_1, \qquad TS_1T^{-1} = S,$$

where S_1 is again a transformation of G_{r-m}, and in fact, a completely arbitrary one, provided only that one chooses the S appropriately.

Now, if every transformation S leaves invariant a certain point figure [PUNKTFIGUR] M, so that:

$$(M)S = (M),$$

then the following equation also holds:

$$(M)TT^{-1}ST = (M)T,$$

and it shows that the figure $(M)T$ admits every transformation $T^{-1}ST$, hence actually, every transformation S.

Proposition 3. *If G_r is an r-term group, if G_{r-m} is an invariant subgroup of it, and lastly, if M is a point figure which admits all transformations of G_{r-m}, then every position that M takes by means of a transformation G_r also remains invariant by all transformations of G_{r-m}.*

Now in particular, let the two groups G_r and G_{r-m} be projective groups of R_{n-1} and let them be written down in n homogeneous variables. Let the figure M be a point. Then every individual point invariant by all S is transferred, by the execution of all transformations T, only to points which again admit all transformations S. Consequently, the *totality of all* points invariant by the group G_{r-m} also remains invariant by the group G_r, though in general, the individual points of this totality are permuted with each other by G_r.

Above, we saw that all the points which remain invariant by a one-term projective group can be ordered in at most n mutually distinct plane manifolds. Naturally, this also holds true for the points which keep their position by our G_{r-m}. Hence, let $M_1, M_2, \ldots, M_\rho$ ($\rho \leqslant n$) be distinct plane manifolds all points of which remain fixed by G_{r-m}, whereas, out of these manifolds, there are no points invariant by G_{r-m}. Now, if the group G_r were discontinuous — the considerations made also remain valid for this case —, then the manifolds $M_1, \ldots, M_\rho$ could be permuted with each other by G_r. This is not so when G_r is continuous, hence when it is generated by infinitesimal transformations, because if for instance M_1 took new positions by means of the transformations contained in G_r, then these positions would form a continuous family. But now, we have just seen that M_1 can at most be given the finitely different positions $M_1, \ldots, M_\rho$. Consequently, no transformation of G_r can change the position of M_1. In the same way, every other of the ρ manifolds M_k naturally remains invariant by all transformations of G_r.

Now, we want to add the assumption that m has the value 1. Let $X_1 f, \dots, X_{r-1} f$ be independent infinitesimal transformations of G_{r-1} invariant in G_r, and let $X_1 f, \dots, X_{r-1} f$, Yf be the transformations of G_r; moreover, let $x_1 = 0, \dots, x_q = 0$ be one of the plane manifolds $M_1, \dots, M_\rho$ all points of which remain invariant by $X_1 f, \dots, X_{r-1} f$.

Since, as we know, Yf surely leaves invariant the manifold $x_1 = 0, \dots, x_q = 0$, it must have the form:

$$Yf = \sum_{i,\,k}^{1\cdots q} a_{ki}\, x_i\, p_k + \sum_{j=q+1}^{n} \sum_{\nu=1}^{n} a_{j\nu}\, x_\nu\, p_j.$$

The points of the manifold in question are transformed by Yf (cf. Theorem 40, p. 243) and, to be precise, by means of the transformation which comes from Yf after the substitution $x_1 = \cdots = x_q = 0$, namely:

$$\sum_{j=q+1}^{n} \sum_{\nu=q+1}^{n} a_{j\nu}\, x_\nu\, p_j; \tag{8}$$

here, the variables $x_{q+1}, \dots, x_n$ are to be considered as coordinates of the points of the manifold. Now, the transformation (8) leaves invariant at least one point $x_{q+1}^0 : \cdots : x_n^0$ *inside* the manifold $x_1 = \cdots = x_q = 0$, further, it leaves invariant a straight line passing through this point, and so on.

Saying this, we have the

Theorem 104. *If an $(r+1)$-term projective group $X_1 f, \dots, X_r f$, Yf of the R_{n-1} contains an r-term invariant subgroup $X_1 f, \dots, X_r f$ and if there are points of R_{n-1} which remain invariant by all $X_k f$, then every plane manifold consisting of such invariant points which is not contained in a larger manifold of this sort also admits all transformations of the $(r+1)$-term group. Furthermore, the points of each such manifold are permuted with each other, but so that at least one of these points keeps its position by all transformations of the $(r+1)$-term group.*

Now, we will apply this theorem to a special category of linear homogeneous groups.

Let $X_1 f, \dots, X_r f$ be the infinitesimal transformations of an r-term linear homogeneous group, and let, as is always possible, $X_1 f, \dots, X_\rho f$ generate a ρ-term subgroup which is invariant in the $(\rho+1)$-term subgroup $X_1 f, \dots, X_{\rho+1} f$, when ρ is smaller than r. Analytically, these assumptions find their expressions in certain relations of the form:

$$[X_i, X_{i+k}] = \sum_{s=1}^{i+k-1} c_{iks}\, X_s f \qquad (i = 1 \cdots r-1;\ k = 1 \cdots r-i).$$

Now, if we interpret as before $x_1, \dots, x_n$ as homogeneous coordinates of an R_{n-1}, we then see that there always is in this R_{n-1} a point invariant by $X_1 f$, moreover that there is always a point invariant by $X_1 f$ and by $X_2 f$ as well, and finally, a point which actually admits all transformations $X_1 f, \dots, X_r f$, hence for which relations of the form:

$$X_k x_i = \alpha_k x_i \qquad (k=1\cdots r;\, i=1\cdots n)$$

hold. If we choose the variables x_i so that $x_1 = \cdots = x_{n-1} = 0$ is this invariant point, then in each of the r expressions:

$$X_k f = \xi_{k1}\, p_1 + \cdots + \xi_{kn}\, p_n,$$

the $n-1$ first coefficients $\xi_{k1}, \dots, \xi_{k,n-1}$ depend only upon $x_1, \dots, x_{n-1}$. Consequently, the reduced expressions:

$$X_k' f = \xi_{k1}\, p_1 + \cdots + \xi_{k,n-1}\, p_{n-1}$$

stand again in the relationships:

$$[X_i', X_{i+k}'] = \sum_{s=1}^{i+k-1} c_{iks}\, X_s' f.$$

However, $X_1' f, \dots, X_r' f$ need not be mutually independent any more, but nevertheless, we can apply to them the same considerations as above for $X_1 f, \dots, X_r f$, because on the occasion, no use was made of the independence of the $X_k f$. Similarly as above, we can hence imagine that the variables $x_1, \dots, x_{n-1}$ are chosen in such a way that all $\xi_{k1}, \dots, \xi_{k,n-2}$ depend only on $x_1, \dots, x_{n-2}$. At present, the expressions:

$$X_k'' f = \xi_{k1}\, p_1 + \cdots + \xi_{k,n-2}\, p_{n-2}$$

can be treated in exactly the same way, and so on.

We therefore obtain the

Theorem 105. *If $X_1 f, \dots, X_r f$ are independent infinitesimal transformations of an r-term linear homogeneous group in the variables $x_1, \dots, x_n$ and if relations of the specific form:*

$$[X_i, X_{i+k}] = \sum_{s=1}^{i+k-1} c_{iks}\, X_s f \qquad (i=1\cdots r-1;\, k=1\cdots r-i)$$

hold, then one can always introduce linear homogeneous functions of $x_1, \dots, x_n$ as new independent variables so that all $X_k f$ simultaneously receive the canonical form:[1]

$$a_{k11}\, x_1\, p_1 + (a_{k21}\, x_1 + a_{k22}\, x_2)\, p_2 + \cdots + (a_{kn1}\, x_1 + \cdots + a_{knn}\, x_n)\, p_n.$$

On the other hand, if we remember that $X_1 f, \dots, X_r f$ is also a projective group of R_{n-1}, we can say:

---

[1] LIE, Archiv for Math., Vol. 3, p. 110 and p. 111, Theorem 3; Christiania 1878.

Proposition 4. *If $X_1f,\dots,X_rf$ is an r-term projective group of R_{n-1} having the specific composition:*

$$[X_i, X_{i+k}] = \sum_{s=1}^{i+k-1} c_{iks} X_s f \qquad (i=1\cdots r-1;\ k=1\cdots r-i),$$

then in R_{n-1}, there is at least one point M_0 invariant by the group; through every invariant point, there passes at least one invariant straight line; ...; through every invariant plane M_{n-3}, there passes at least one invariant plane M_{n-2}. In certain circumstances, several (infinitely many) such series of invariant manifolds M_0, M_1, ..., M_{n-2} are associated to the group.

If we maintain the assumptions of this last proposition and if we assume in addition that in the space $x_1 : x_2 : \cdots x_n$, several invariant plane manifolds $\mathsf{M}_{\rho_1}, \dots, \mathsf{M}_{\rho_q}$ are already known, about which each one is contained in the one following next, then amongst the series of invariant manifolds $M_0, M_1, \dots, M_{n-2}$ mentioned in Proposition 4, there is obviously at least one for which M_{ρ_1} coincides with M_{ρ_1}, at least one for which M_{ρ_2} coincides with M_{ρ_2}, ..., at least for which M_{ρ_q} coincides with M_{ρ_q}.

§ 147. Apparently, the results found above certainly have a considerably special character, but this special character is the very reason why they nevertheless possess a general significance for any finite continuous group, because for every such group, an isomorphic linear homogeneous group can be shown, namely the associated adjoint group (Chap. 16).

Amongst other things, we can employ our theory mentioned above in order to prove that every group with more than two parameters contains two-term subgroups, and that every group with more than three parameters contains three-term subgroups.

Let $X_1f,\dots,X_rf$ be an arbitrary r-term group and let:

$$E_k f = \sum_{i,\,j}^{1\cdots r} c_{jki}\, e_j \frac{\partial f}{\partial e_i} \qquad (k=1\cdots r)$$

be the associated adjoint group, so that the relations:

$$[X_i, X_k] = \sum_{s=1}^{r} c_{iks} X_s f, \qquad [E_i, E_k] = \sum_{s=1}^{r} c_{iks} E_s f$$

hold simultaneously.

We claim that in any case, X_1f belongs to *one* two-term group, hence that an infinitesimal transformation $\lambda_2 X_2 f + \cdots + \lambda_r X_r f$ exists for which one has:

$$[X_1, \lambda_2 X_2 + \cdots + \lambda_r X_r] = \rho X_1 f + \mu \sum_{k=2}^{r} \lambda_k X_k f.$$

This condition represents itself in the form:

$$\sum_{k=2}^{r} \lambda_k \sum_{s=1}^{r} c_{1ks} X_s f = \rho X_1 f + \mu \sum_{k=2}^{r} \lambda_k X_k f,$$

and it decomposes itself in the r equations:

$$(9) \qquad \begin{cases} \sum_{k=2}^{r} \lambda_k \, c_{1k1} = \rho \\ \sum_{k=2}^{r} \lambda_k \, c_{1ks} = \mu \, \lambda_s & (s=2\cdots r). \end{cases}$$

Here, this obviously amounts to satisfying only the last $r-1$ equations; but one sees directly in the simplest way that this is possible. Indeed, one obtains for μ the equation:

$$\begin{vmatrix} c_{122}-\mu & c_{132} & \cdot\;\cdot\;\cdot & c_{1r2} \\ c_{123} & c_{133}-\mu & \cdot\;\cdot\;\cdot & c_{1r3} \\ \cdot & \cdot & \cdot\;\cdot\;\cdot & \cdot \\ c_{12r} & c_{13r} & \cdot\;\cdot\;\cdot & c_{1rr}-\mu \end{vmatrix}$$

which always possesses roots; for this reason, there always exists a system of λ_k not all vanishing which satisfies the above conditions and which also determines ρ.

But the fact that the equations (9) may be satisfied is also an immediate consequence of our theory mentioned above; although this observation seems hardly necessary here, where the relationships are so simple, we nevertheless do not want to miss this, because for the more general cases that are to be treated next, we cannot do without this theory.

In the infinitesimal transformation:

$$E_1 f = \sum_{i=1}^{r} \sum_{k=1}^{r} c_{k1i} \, e_k \frac{\partial f}{\partial e_i} = \sum_{i=1}^{r} \varepsilon_i \frac{\partial f}{\partial e_i}$$

of the adjoint group, all ε_i are free of e_1, since c_{11i} is always zero. Now, the cut linear homogeneous infinitesimal transformation $\varepsilon_2 \, p_2 + \cdots + \varepsilon_r \, p_r$ in the $r-1$ variables $e_2, \dots, e_r$ surely leaves invariant a system $e_2 : \cdots : e_r$; so it is possible to satisfy the equations:

$$\sum_{k=2}^{r} c_{k1i} \, e_k = \sigma \, e_i \qquad (i=2\cdots r),$$

but these equations do not differ from the last $r-1$ equations (9), since one has $c_{k1i} = -c_{1ki}$.

Now, we can state more precisely our claim above that every group with more than two parameters contains two-term subgroups, in the following way:

Proposition 5. *Every infinitesimal transformation of a group with more than two parameters is contained in at least one two-term subgroup.*[2]

Now, we want to assume that $X_1 f$ and $X_2 f$ generate a two-term subgroup, so that a relation of the form:

$$[X_1, X_2] = c_{121} X_1 f + c_{122} X_2 f$$

holds; we claim that, as soon as r is larger than 3, there also always exists a three-term subgroup in which $X_1 f$ and $X_2 f$ are contained.

First, for the constants c_{121} and c_{122}, when the two are not already zero, we can always make one equal to zero. In fact, if, say, c_{121} is different from zero, then we introduce $X_1 f + \frac{c_{122}}{c_{121}} X_2 f$ as a new $X_1 f$ and we obtain:

$$[X_1, X_2] = c_{121} X_1 f.$$

Thus, every two-term group can be brought to this form.

If $\lambda_3 X_3 f + \cdots + \lambda_r X_r$ is assumed to generate a three-term group together with $X_1 f$ and $X_2 f$, then one must have:

$$\text{(10)} \quad \begin{cases} [X_1, \lambda_3 X_3 + \cdots + \lambda_r X_r] = \alpha_1 X_1 f + \alpha_2 X_2 f + \mu\,(\lambda_3 X_3 f + \cdots + \lambda_r X_r f) \\ [X_2, \lambda_3 X_3 + \cdots + \lambda_r X_r] = \beta_1 X_1 f + \beta_2 X_2 f + \nu\,(\lambda_3 X_3 f + \cdots + \lambda_r X_r f). \end{cases}$$

From this, we obtain the following condition-equations:

$$\text{(11)} \quad \begin{cases} \sum_{k=3}^{r} \lambda_k c_{1k1} = \alpha_1, & \sum_{k=3}^{r} \lambda_k c_{1k2} = \alpha_2, \\ \sum_{k=3}^{r} \lambda_k c_{2k1} = \beta_1, & \sum_{k=3}^{r} \lambda_k c_{2k2} = \beta_2, \\ \sum_{k=3}^{r} c_{1ks} \lambda_k = \mu \lambda_s, & \sum_{k=3}^{r} c_{2ks} \lambda_k = \nu \lambda_s \qquad {\scriptstyle (s=3\cdots r)}. \end{cases}$$

As we have observed, it is only necessary to prove that the $2(r-2)$ equations in the last row can be satisfied, since the remaining equations can always be satisfied afterwards.

In order to settle this question, we form the infinitesimal transformations:

$$E_1 f = \sum_{i=1}^{r} \sum_{k=1}^{r} c_{k1i} e_k \frac{\partial f}{\partial e_i} = \sum_{i=1}^{r} \varepsilon_{1i} \frac{\partial f}{\partial e_i}$$

$$E_2 f = \sum_{i=1}^{r} \sum_{k=1}^{r} c_{k2i} e_k \frac{\partial f}{\partial e_i} = \sum_{i=1}^{r} \varepsilon_{2i} \frac{\partial f}{\partial e_i},$$

which, as we know, stand in the relationship:

[2] LIE, Archiv for Math. Vol. 1, p. 192. Christiania 1876.

$$[E_1, E_2] = c_{121}\, E_1 f.$$

Since all c_{11i}, c_{22i} and the $c_{122}, \dots, c_{12r}$ are equal to zero, both e_1 and e_2 do not appear at all in $\varepsilon_{13}, \dots, \varepsilon_{1r}$, $\varepsilon_{23}, \dots, \varepsilon_{2r}$. The reduced expressions:

$$\overline{E}_1 f = \sum_{i=3}^{r} \varepsilon_{1i} \frac{\partial f}{\partial e_i}, \qquad \overline{E}_2 f = \sum_{i=3}^{r} \varepsilon_{2i} \frac{\partial f}{\partial e_i}$$

are therefore linear homogeneous infinitesimal transformations in $e_3, \dots, e_r$ and they satisfy the relation:

$$\big[\overline{E}_1, \overline{E}_2\big] = c_{121}\, \overline{E}_1 f.$$

According to the preceding, it follows from this that there exists at least one system $e_3 : \cdots : e_r$ which admits the two infinitesimal transformations $\overline{E}_1 f$ and $\overline{E}_2 f$, hence which satisfies conditions of the form:

$$\sum_{k=3}^{r} c_{k1i}\, e_k = \sigma\, e_i, \qquad \sum_{k=3}^{r} c_{k2i}\, e_k = \tau\, e_i \qquad {\scriptstyle (i=3\cdots r)}.$$

But this is exactly what was to be proved, for these last equations are nothing else than the last equations (11), about which the question was whether they could be satisfied. Naturally, the quantities α_1, α_2, β_1, β_2 are also determined together with the λ_i.

Since we can choose in all circumstances $\lambda_3, \dots, \lambda_r$ so that equations of the form (3) hold, we can say:

Theorem 106. *Every infinitesimal transformation and likewise every two-term subgroup of a group with more than three parameters is contained in at least one three-term subgroup.*[3]

One could be led to presume that every three-term subgroup is also contained in at least one four-term subgroup, and so on, but this presumption is not confirmed. Our process of proof could be employed further only when *every* three-term group $X_1 f, X_2 f, X_3 f$ could be brought to the form:

$$[X_1, X_2] = c_{121}\, X_1 f, \qquad [X_1, X_3] = c_{131}\, X_1 f + c_{132}\, X_2 f$$
$$[X_2, X_3] = c_{231}\, X_1 f + c_{232}\, X_2 f,$$

hence not only when $X_1 f$ would be invariant in the group $X_1 f$, $X_2 f$, but when this last group would also be invariant in the entire three-term group.

But for all three-term groups of the composition:

$$[X_1, X_2] = X_1 f, \qquad [X_1, X_3] = 2X_2 f, \qquad [X_2, X_3] = X_3 f,$$

this is *not* the case (Chap. 15, Proposition 12, p. 280).

---

[3] LIE, Archiv for Math. Vol. 1, p. 193, Vol. 3, pp. 114–116, Christiania 1876 and 1878.

Still, we want to briefly dwell on a special case in which an m-term subgroup is really contained in an $(m+1)$-term subgroup.

Let an arbitrary r-term group $X_1 f, \dots, X_r f$ be presented which contains an m-term subgroup $X_1 f, \dots, X_m f$ having the characteristic composition:

$$[X_i, X_{i+k}] = \sum_{s=1}^{i+k-1} c_{i,i+k,s} X_s f \qquad (i < m,\ i+k \leqslant m)$$

already mentioned. We claim that there always exists an $(m+1)$-term subgroup of the form:

$$X_1 f, \dots, X_m f, \qquad \lambda_{m+1} X_{m+1} f + \cdots + \lambda_r X_r f.$$

Our assertion amounts to the fact that certain relations of the form:

$$\big[X_j,\ \lambda_{m+1} X_{m+1} + \cdots + \lambda_r X_r\big] = \sum_{k=1}^{m} \alpha_{jk} X_k f + \mu_j \sum_{s=m+1}^{r} \lambda_s X_s f$$
$$(j = 1 \cdots m)$$

hold. By decomposition, we get:

$$(12) \qquad \begin{cases} \displaystyle\sum_{i=m+1}^{r} \lambda_i c_{jik} = \alpha_{jk} & (j, k = 1 \cdots m) \\ \displaystyle\sum_{i=m+1}^{r} \lambda_i c_{jis} = \mu_j \lambda_s & (j = 1 \cdots m;\ s = m+1 \cdots r). \end{cases}$$

Thus, the question is whether the last $m(r-m)$ equations can be satisfied; then always, the first m^2 equations can be satisfied.

In order to settle this question, we form the m infinitesimal transformations:

$$E_k f = \sum_{i=1}^{r} \sum_{j=1}^{r} c_{jki} e_j \frac{\partial f}{\partial e_i} = \sum_{i=1}^{r} \varepsilon_{ki} \frac{\partial f}{\partial e_i}$$
$$(k = 1 \cdots m)$$

which satisfy in pairs the relations:

$$[E_i, E_{i+k}] = \sum_{s=1}^{i+k-1} c_{i,i+k,s} E_s f \qquad (i+k \leqslant m).$$

We observe that all $c_{j,k,m+1}, \dots, c_{jkr}$ vanish for which j and k are smaller than $m+1$, and from this we deduce that in $E_1 f, \dots, E_m f$, all coefficients $\varepsilon_{k,m+1}, \dots, \varepsilon_{kr}$ are free of $e_1, \dots, e_m$. The reduced infinitesimal transformations:

$$\overline{E}_k f = \sum_{i=m+1}^{r} \varepsilon_{ki} \frac{\partial f}{\partial e_i} \qquad (k = 1 \cdots m)$$

in the variables $e_{m+1},\dots,e_r$ therefore stand pairwise in the relationships:

$$\big[\overline{E}_i,\overline{E}_k\big]=\sum_{s=1}^{i+k-1} c_{i,i+k,s}\,\overline{E}_s f \qquad (i+k\leqslant m)$$

and consequently, according to Proposition 4, p. 591, there is a system $e_{m+1}:\ \cdots : e_r$ which admits all infinitesimal transformations $\overline{E}_1 f,\dots,\overline{E}_m f$. But this says nothing but that it is possible to satisfy the equations:

$$\sum_{j=m+1}^{r} c_{jks}\,e_j=\sigma_k\,e_s \qquad (k=1\cdots m;\ s=m+1\cdots r).$$

But these equations are exactly the same as the equations (12) found above and thus, everything we wanted to show is effectively proved.

Theorem 107. *If, in an r-term group $X_1 f,\dots,X_r f$, there is an m-term subgroup $X_1 f,\dots,X_m f$ having the specific composition:*

$$[X_i,X_{i+k}]=\sum_{s=1}^{i+k-1} c_{i,i+k,s}\,X_s f \qquad (i<m,\ i+k\leqslant m),$$

then this m-term subgroup is always contained in at least one $(m+1)$-term subgroup.[4]

§ 148. We now want to derive the above stated theorem thanks to conceptual considerations, by interpreting, as in Chap. 16, the infinitesimal transformations $e_1 X_1 f+\cdots+e_r X_r f$ of our group as points of an $(r-1)$-times extended space with the homogeneous coordinates $e_1,\dots,e_r$.

Since the group $X_1 f,\dots,X_m f,\dots,X_r f$ is isomorphic with its adjoint group: $E_1 f,\dots,E_m f,\dots,E_r f$, then under the assumptions made, $E_1 f,\dots,E_m f$ satisfy relations of the form:

$$[E_i,E_{i+k}]=\sum_{s=1}^{i+k-1} c_{i,i+k,s}\,E_s f \qquad (i+k\leqslant m),$$

hence they generate a subgroup which, in the space e_k, is represented by the $(m-1)$-times extended plane manifold:

$$e_{m+1}=0,\ \dots,\ e_r=0.$$

This plane manifold naturally admits the subgroup $E_1 f,\dots,E_m f$, and the same obviously holds true of the *family* of all m-times extended plane manifolds:

---

4 LIE, Archiv for Math., Vol. 3, pp. 114–116, Christiania 1878.

$$\frac{e_{m+1}}{\mathfrak{e}_{m+1}} = \frac{e_{m+2}}{\mathfrak{e}_{m+2}} = \cdots = \frac{e_r}{\mathfrak{e}_r}$$

which pass through the manifold $e_{m+1} = 0, \dots, e_r = 0$. But since the parameters $\mathfrak{e}_{m+1} : \cdots : \mathfrak{e}_r$ are transformed by a linear homogeneous group which is isomorphic with the subgroup $E_1 f, \dots, E_m f$ (Chap. 23, Proposition 5, p. 479), then amongst the m-times extended manifolds of our invariant family, there is at least one which remains invariant by the subgroup $E_1 f, \dots, E_m f$. This m-times extended manifold is the image of an $(m+1)$-term subgroup of the group: $X_1 f, \dots, X_r f$ (cf. Proposition 5, p. 298).

The conceptual considerations just made, which have again led us to Theorem 107, are in essence identical to the analytical considerations developed earlier. However, the synthetical explanation is more transparent than the analytical one [DOCH IST DIE SYNTHETISCHE BEGRÜNDUNG DURCHSICHTIGER ALS DIE ANALYTISCHE].

For the results on the composition of transformation groups, it is actually advisable to set as fundamental the interpretation of all infinitesimal transformations $e_1 X_1 f + \cdots + e_r X_r f$ as the points of a space which is transformed by the linear homogeneous adjoint group. Thanks to an example, we will highlight the fruitfulness and the simplicity of this method which shall also find multiple applications in the third volume, and at the same time, we will derive a new remarkable statement.

We consider an r-term group $X_1 f, \dots, X_r f$, the infinitesimal transformations of which are linked together by relations of the form:

$$[X_i, X_{i+k}] = \sum_{s=1}^{i+k-1} c_{i,i+k,s} X_s f.$$

The infinitesimal transformations $E_1 f, \dots, E_r f$ of the adjoint group then satisfy the analogous equations:

$$[E_i, E_{i+k}] = \sum_{s=1}^{i+k-1} c_{i,i+k,s} E_s f.$$

Consequently (Proposition 4, p. 591), the space e_k contains at least one point invariant by the adjoint group. Through every point of this sort, there passes at least one invariant straight line M_1, through every such straight line, there passes at least one invariant plane M_2, and so on.

Now, if we interpret the points e_k as infinitesimal transformations, we see that our r-term group: $X_1 f, \dots, X_r f$ contains at least one invariant one-term subgroup; next, that every invariant one-term subgroup is contained in at least one invariant two-term subgroup; next, that every invariant two-term subgroup is contained in at least one invariant three-term subgroup; and so on (cf. p. 291).

Thus, the following holds true.

Theorem 108. *If an r-term group contains r independent infinitesimal transformations $Y_1 f, \dots, Y_r f$ which satisfy relations of the form:*

$$[Y_i, Y_{i+k}] = \sum_{s=1}^{i+k-1} c_{i,i+k,s} Y_s f,$$

then at the same time, it contains r independent infinitesimal transformations $Z_1 f, \ldots, Z_r f$ *between which relations of the form:*

$$[Z_i, Z_{i+k}] = \sum_{s=1}^{i+k-1} c_{i,i+k,s} Z_s f,$$

hold; then $Z_1 f, \ldots, Z_i f$ *generate for every* $i < r$ *an i-term subgroup* $\mathfrak{G}_i$ *and to be precise, every* $\mathfrak{G}_i$ *is invariant in every* $\mathfrak{G}_{i+k}$ *and also in the group* $Y_1 f, \ldots, Y_r f$ *itself.*[5]

If we maintain the assumptions of this theorem and if we assume in addition that by chance, several invariant subgroups, say G_{ρ_1}, G_{ρ_2}, ..., G_{ρ_q} are known, of which each one comprises the one following next, then it becomes evident (cf. the concluding remarks of § 146) that the subgroups $\mathfrak{G}_i$ discussed in Theorem 108 can be chosen in such a way that $\mathfrak{G}_{\rho_1}$ coincides with G_{ρ_1}, that $\mathfrak{G}_{\rho_2}$ coincides with G_{ρ_2}, ..., that $\mathfrak{G}_{\rho_q}$ coincides with G_{ρ_q}.

On the other hand, when an r-term group: $X_1 f, \ldots, X_r f$ of the specific composition considered here is presented, it is always possible to derive certain invariant subgroups by differentiation. Indeed, all $[X_i, X_k]$ generate a *firstly derived* invariant subgroup with say $r_1 < r$ independent infinitesimal transformations. If, as in Chap. 15, p. 278, we denote them by $X_1' f, \ldots, X_{r_1}' f$, then all $[X_i', X_k']$ generate a *secondly derived* invariant subgroup of the r-term group with, say, $r_2 < r_1$ independent infinitesimal transformations $X_1'' f, \ldots, X_{r_2}'' f$, and so on.

It is therefore possible to bring our group to a form $U_1 f, \ldots, U_i f, \ldots, U_r f$ such that *firstly* for every i, the infinitesimal transformations $U_1 f, \ldots, U_i f$ generate an invariant subgroup, such that *secondly* all $[U_i, U_k]$ can be linearly deduced from $U_1 f, \ldots, U_{r_1} f$, such that *thirdly* all $\big[[U_i, U_k], [U_\alpha, U_\beta]\big]$ can be linearly deduced from $U_1 f, \ldots, U_{r_2} f$, and so on.

---

[5] LIE, Archiv for Math., Vol. 3, p. 112 and p. 113, Christiania 1878; cf. also Vol. IX, pp. 79–82.

Chapter 28
Approach [Ansatz] towards the Determination of All Finite Continuous Groups of n-times Extended Space

It is unlikely that we are close to determining all finite continuous transformation groups; indeed, it is even uncertain whether this will ever succeed. Therefore, instead of the general problem of determining *all* finite continuous groups, one would do well to first tackle more special problems which concern the determination of certain categories of finite continuous groups. Namely, more special problems of this kind are the following three:

Firstly the determination of all r-term groups in n variables.
Secondly the determination of all r-term groups in general.
Thirdly the determination of all finite continuous groups in n variables.

In Chap. 22, p. 439 sq., we have shown that the settlement of the first of these problems, aside from executable operations, requires in any case only the integration of simultaneous systems of ordinary differential equations. Moreover, we found that the second of our three problems can be led back to the first one (Theorem 84, p. 465).

By contrast, the third problem is not settled by means of the developments of Chap. 22. Namely if $n > 1$, then for every value of r (how large can it be though?), there are always r-term groups in n variables. For instance, if one chooses r functions $F_1, \dots, F_r$ of $x_1, \dots, x_{n-1}$ so that between them, no linear relation of the form:

$$c_1 F_1 + \cdots + c_r F_r = 0$$

with constant coefficients holds, then the r infinitesimal transformations:

$$F_1(x_1, \dots, x_{n-1}) \frac{\partial f}{\partial x_n}, \quad \dots, \quad F_r(x_1, \dots, x_{n-1}) \frac{\partial f}{\partial x_n}$$

are mutually independent and pairwise exchangeable, hence they generate an r-term group in the n variables $x_1, \dots, x_n$.

S. Lie, *Theory of Transformation Groups I*,
DOI 10.1007/978-3-662-46211-9_28

In spite of the important results of Chap. 22, the third of the indicated problems therefore still awaits a solution. This is why we shall tackle this problem in the present chapter and at least provide an approach towards its solution.

In the sequel, we decompose the problem of determining all groups of an n-times extended space R_n into a series of individual problems which are independent of each other. This is possible thanks to a natural division of the groups of R_n into classes which are selected in such a way that the groups of some given class can be determined without it being necessary to know any of the groups of the remaining classes. Admittedly, we cannot indicate a general method which accomplishes in every case the determination of all groups of a class; nonetheless, we provide important statements about the groups belonging to a given class and on the other hand, we develop a series of considerations which facilitate the determination of all groups in a class; in the next chapter, this will be illustrated by special applications. For these general discussions, we essentially restrict ourselves to transitive groups, because our classification is of specific practical meaning precisely for the transitive groups.

This is justified by the fact that later (in the third volume), for the determination of all finite continuous groups in one, two or three variables, we will exploit only in part the results of the present chapter. For the determination of the *primitive* groups of a space, the process explained here is firmly effective; but not so for the determination of the imprimitive groups, where it is advisable to take a different route. Every imprimitive group of R_n decomposes the space into an invariant family of manifolds and transforms the manifolds of this family by means of an isomorphic group in fewer than n variables. From this, it results that one would do well to first tackle the determination of all imprimitive groups of R_n, when one already know all groups in fewer than n variables. Applying this fundamental principle, we shall undertake in Volume III the determination of all imprimitive groups of R_2 and of certain imprimitive groups of R_3.

§ 149. Let:

$$X_k f = \sum_{i=1}^{n} \xi_{ki}(x_1,\dots,x_n)\,\frac{\partial f}{\partial x_i} \qquad (k=1\cdots r)$$

be a completely arbitrary r-term group of n-times extended space $x_1,\dots,x_n$, or briefly of R_n.

Under the guidance of Chap. 25, p. 527, we extend this group by viewing $x_1,\dots,x_n$ as functions of an auxiliary variable t which is not transformed by our group and by considering that the differential quotients: $\mathrm{d}x_i/\mathrm{d}t = x_i'$ are transformed by the group. In the $2n$ variables: $x_1,\dots,x_n$, $x_1',\dots,x_n'$, we then obtain the extended group:

$$\overline{X}_k f = \sum_{i=1}^{n} \xi_{ki}(x)\,\frac{\partial f}{\partial x_i} + \sum_{i=1}^{n}\bigg(\sum_{\nu=1}^{n} \frac{\partial \xi_{ki}}{\partial x_\nu}\,x_\nu'\bigg)\frac{\partial f}{\partial x_i'} \qquad (k=1\cdots r).$$

As we know, this extended group indicates in which way the ∞^{2n-1} line elements: $x_1,\dots,x_n$, $x_1' : x_2' : \cdots : x_n'$ of the space $x_1,\dots,x_n$ are permuted with each other by the group: $X_1 f,\dots,X_r f$ (cf. p. 529).

Every point: $x_1^0,\dots,x_n^0$ in general position remains invariant by a completely determined number of independent infinitesimal transformations: $e_1 X_1 f + \cdots + e_r X_r f$ and to be precise, at least by $r-n$ such transformations, and at most by $r-1$. We want to denote by $r-q$ the number of these independent infinitesimal transformations and to suppose that: $X_1^0 f,\dots,X_{r-q}^0 f$ are such independent infinitesimal transformations; then they certainly generate an $(r-q)$-term subgroup of the group: $X_1 f,\dots,X_r f$ (cf. Chap. 12, p. 218 sq.).

In the series expansions of $X_1^0 f,\dots,X_{r-q}^0 f$ with respect to the powers of the $x_i - x_i^0$, all terms of order zero are naturally lacking and there appear only terms of first or higher order:

$$X_k^0 f = \sum_{\nu=1}^{n} \left\{ \sum_{i=1}^{n} \alpha_{ki\nu}(x_1^0,\dots,x_n^0)\,(x_i - x_i^0) + \cdots \right\} \frac{\partial f}{\partial x_\nu} \qquad (k=1\cdots r-q).$$

The group: $X_1^0 f,\dots,X_{r-q}^0 f$ leaves the point $x_1^0,\dots,x_n^0$ invariant, but it permutes the line elements which pass through this point; how? this is what the associated extended group shows:

$$\begin{aligned}\overline{X}_k^0 f = &\sum_{\nu=1}^{n} \left\{ \sum_{i=1}^{n} \alpha_{ki\nu}(x_1^0,\dots,x_n^0)\,(x_i - x_i^0) + \cdots \right\} \frac{\partial f}{\partial x_\nu} \\ &+ \sum_{\nu=1}^{n} \left\{ \sum_{i=1}^{n} \alpha_{ki\nu}(x_1^0,\dots,x_n^0)\,x_n' + \cdots \right\} \frac{\partial f}{\partial x_\nu'} \qquad (k=1\cdots r-q),\end{aligned}$$

and it transforms the ∞^{2n-1} line elements of the space $x_1,\dots,x_n$ in exactly the same way as the group: $X_1^0 f,\dots,X_{r-q}^0 f$. Hence, if one wants to restrict oneself to the line elements which pass through the point $x_1^0,\dots,x_n^0$, and to disregard the remaining ones, then under the guidance of Chap. 14, p. 244 sq., one has to leave out, in $\overline{X}_k^0 f$, all terms with the differential quotients of f with respect to $x_1,\dots,x_n$ and to make the substitution: $x_1 = x_1^0$, $\dots$, $x_n = x_n^0$ in the terms remaining. The so obtained reduced infinitesimal transformations:

$$(1) \qquad L_k f = \sum_{i,\,\nu}^{1\cdots n} \alpha_{ki\nu}(x_1^0,\dots,x_n^0)\,x_i' \frac{\partial f}{\partial x_\nu'} \qquad (k=1\cdots r-q)$$

generate *a linear homogeneous group in the n variables* $x_1',\dots,x_n'$; this group, which is isomorphic with the group: $\overline{X}_1^0 f,\dots,\overline{X}_{r-q}^0 f$, and naturally also, with the group: $X_1^0 f,\dots,X_{r-q}^0 f$, indicates in which way the two groups just mentioned transform the ∞^{n-1} line elements: $x_1' : x_2' : \cdots x_n'$ through the point $x_1^0,\dots,x_n^0$.

Visibly, the linear homogeneous group: $L_1 f,\dots,L_{r-q} f$ is perfectly determined by the terms of first order in the power series expansions of $X_1^0 f,\dots,X_{r-q}^0 f$, and therefore, it contains as many essential parameters as the group: $X_1^0 f,\dots,X_{r-q}^0 f$ contains independent infinitesimal transformations of first order out of which no transformation of second order, or of higher order, can be linearly deduced. From this, it follows that one can set up the group: $L_1 f, \dots, L_{r-q} f$ as soon as one knows how many independent infinitesimal transformations of first order the group: $X_1, \dots, X_r$ contains in the neighborhood of $x_1^0, \dots, x_n^0$ out of which no transformation of second or higher order can be linearly deduced, and in addition, as soon as one knows the terms of first order in the power series expansions of these infinitesimal transformations. Conversely, as soon as one knows the group: $L_1 f, \dots, L_r f$, one can indicate the number of and the terms of those independent infinitesimal transformations of first order in the group: $X_1 f, \dots, X_r f$ out of which no transformation of second or higher order in the $x_i - x_i^0$ can be linearly deduced.

One can also derive the linear homogeneous group: $L_1 f, \dots, L_{r-q} f$ in the following way:

Since the question is about the way in which the directions through the fixed point: $x_1^0, \dots, x_n^0$ are transformed, one may substitute the group: $X_1^0 f, \dots, X_{r-q}^0 f$ for the following:

$$\sum_{i,\,\nu}^{1\cdots n} \alpha_{ki\nu}(x_1^0, \dots, x_n^0)\,(x_i - x_i^0)\,\frac{\partial f}{\partial x_\nu} \qquad (k = 1 \cdots r-q)$$

which is obtained by leaving out all terms of second and higher order. Now, $x_1 - x_1^0$, $\dots$, $x_n - x_n^0$ can here be directly conceived as homogeneous coordinates of the line elements through the point $x_1^0, \dots, x_n^0$, whence the group:

$$\sum_{i,\,\nu}^{1\cdots n} \alpha_{ki\nu}(x_1^0, \dots, x_n^0)\,(x_i - x_i^0)\,\frac{\partial f}{\partial (x_\nu - x_\nu^0)} \qquad (k = 1 \cdots r-q)$$

indicates how these line elements are transformed. This is consistent with the above.

We have seen that the r-term group: $X_1 f, \dots, X_r f$ associates to every point $x_1^0, \dots, x_n^0$ in general position a completely determined linear homogeneous group (1) which, however, turns out to be different for different points. We now study how this linear homogeneous group behaves after the introduction of some new variables.

In place of $x_1, \dots, x_n$, we introduce the new variables:

$$(2) \qquad y_i = y_i^0 + \sum_{\nu=1}^{n} a_{i\nu}(x_\nu - x_\nu^0) + \cdots \qquad (i = 1 \cdots n)$$

which are ordinary power series of $x_1 - x_1^0$, $\dots$, $x_n - x_n^0$; as always, we assume on the occasion that the determinant:

$$\sum \pm a_{11} \cdots a_{nn}$$

is different from zero, so that conversely, the x_i are ordinary power series of $y_1 - y_1^0$, $\dots$, $y_n - y_n^0$.

In $y_1,\dots,y_n$, let the X_k^0 be of the form:

$$Y_k^0 f = \sum_{\nu=1}^{n}\left\{\sum_{i=1}^{n}\beta_{ki\nu}\,(y_i-y_i^0)+\cdots\right\}\frac{\partial f}{\partial y_\nu} \qquad {\scriptstyle (k=1\cdots r-q).}$$

Then it is clear that the group which comes into existence from $X_1f,\dots,X_rf$ after the introduction of the new variables $y_1,\dots,y_n$ associates to the point $y_1^0,\dots,y_n^0$ the linear homogeneous group:

$$\text{(1')} \qquad \mathfrak{L}_k f = \sum_{i,\,\nu}^{1\cdots n}\beta_{ki\nu}\,y_i'\,\frac{\partial f}{\partial y_\nu'} \qquad {\scriptstyle (k=1\cdots r-q).}$$

But on the other hand, it is clear (cf. Chap. 11, p. 209 sq.) that the terms of first order:

$$\sum_{i,\,\nu}^{1\cdots r}\beta_{ki\nu}\,(y_i-y_i^0)\,\frac{\partial f}{\partial y_\nu}$$

of Y_k^0 can be obtained from the terms of first order:

$$\sum_{i,\,\nu}^{1\cdots n}\alpha_{ki\nu}\,(x_i-x_i^0)\,\frac{\partial f}{\partial x_\nu}$$

of $X_k^0 f$ after the introduction of the new variables:

$$y_i = y_i^0 + \sum_{\nu=1}^{n} a_{i\nu}\,(x_\nu - x_\nu^0) \qquad {\scriptstyle (i=1\cdots n).}$$

Consequently, it follows that the linear homogeneous group (1') comes into existence from the linear homogeneous group (1) when one introduces in (1), by means of the linear homogeneous transformation:

$$y_i' = \sum_{\nu=1}^{n} a_{i\nu}\,x_\nu' \qquad {\scriptstyle (i=1\cdots n),}$$

the new variables: $y_1',\dots,y_n'$ in place of the x'.

In this lies the reason why the linear homogeneous group (1) is essentially independent from the analytic representation of the group: $X_1f,\dots,X_rf$, that is to say, from the choice of the variables; indeed, if, by means of a transformation (2), one introduces new variables in the group: $X_1f,\dots,X_rf$, then the linear homogeneous group (1) converts into another linear homogeneous group (1') which is conjugate to (1) inside the general linear homogeneous group of R_n (cf. Chap. 16, p. 292). Thanks to an appropriate choice of the constants $a_{i\nu}$ in the transformation (2), one can obviously insure that the group (1') associated to the point y_k^0 becomes an arbitrary group conjugate to (1).

Now, we specially assume that the transformation (2) belongs to the group: $X_1 f, \dots, X_r f$ itself. In this case, (1') is visibly the linear homogeneous group that the group: $X_1 f, \dots, X_r f$ associates to the point: $y_1^0, \dots, y_n^0$, and consequently, (1') comes from (1) when one replaces the x^0 in the $\alpha_{kiv}(x_1^0, \dots, x_n^0)$ by the y^0. But since the two groups (1) and (1') are conjugate inside the general linear homogeneous group of R_n, we have the following

Theorem 109. *Every r-term group: $X_1 f, \dots, X_r f$ of the space $x_1, \dots, x_n$ associates to every point $x_1^0, \dots, x_n^0$ in general position a completely determined linear homogeneous group of R_n which indicates in which way the line elements through this point are transformed, as soon as the point is fixed. To those points which can be transferred to one another by means of transformations of the group: $X_1 f, \dots, X_r f$ are associated linear homogeneous groups which are conjugate inside the general linear homogeneous group of R_n. In particular, if the group: $X_1 f, \dots, X_r f$ is transitive, then to all points which lie in no invariant manifold, it associates linear homogeneous groups that are conjugate to each other inside the general linear homogeneous group.*

The above theorem which recapitulates the most important result up to now, provides the classification of all groups of R_n announced in the introduction.

We first consider the *transitive* groups.

We include two transitive groups G and Γ in the same class when the linear homogeneous group that G associates to an arbitrary point in general position is conjugate to the linear homogeneous group that Γ associates to such an arbitrary point. In the opposite case, we include G and Γ in different classes.

Thus, we differentiate as many classes of transitive groups of R_n as there are *types* of subgroups of the general linear homogeneous group of R_n (cf. p. 292). Later, we will see that to each one of such classes, there belongs in any case a transitive group of R_n.

If two transitive groups G and Γ of the space $x_1, \dots, x_n$ belong to the same class, then in the neighborhood of any point $x_1^0, \dots, x_n^0$ in general position, they obviously contain the same number of independent infinitesimal transformations of first order in the $x_i - x_i^0$ out of which no transformation of second or of higher order can be linearly deduced. In addition, since one can always, by introducing new variables, reshape Γ so that it associates to the point $x_1^0, \dots, x_n^0$ exactly the same linear homogeneous group as does G, then in all cases, one can insure that the terms of first order in the infinitesimal transformations of first order in question are the same for the two groups. In addition, because they are transitive, G and Γ contain, in the neighborhood of $x_1^0, \dots, x_n^0$, n independent infinitesimal transformations of zeroth order out of which no transformation of first or higher order can be linearly deduced; by contrast, the numbers of terms and the initial terms of second, third, ... orders can very well be different for Γ and for G. *Here lies the reason why two transitive groups of the space $x_1, \dots, x_n$ which belong to the same class need not have the same number of parameters.*

We now turn ourselves to the *intransitive* groups.

To every transitive group of R_n was associated a completely determined type of linear homogeneous group of R_n; for the intransitive groups, this is in general not the case. Every intransitive group G of R_n decomposes this space into a family of ∞^{n-q} $(0 < q < n)$ individually invariant q-times extended manifolds M_q, but so that the points of every individual M_q are transformed transitively (cf. Chap. 13, p. 228). From this, it follows that G always associates to all points of one and the same M_q conjugate linear homogeneous groups, but not necessarily to the points of different M_q.

In general, our ∞^{n-q} M_q are gathered in continuous families so that conjugate linear homogeneous groups are associated only to the points which belong to the M_q in the same family. If each such family consists of exactly ∞^m M_q, then the whole R_n decomposes into ∞^{n-q+m} $(q+m)$-times extended manifolds M_{q+m} and to every M_{q+m} is associated a completely determined type of linear homogeneous group, while to different M_{q+m} are associated different types too. So to our group is associated a family of ∞^{n-q-m} different types; the totality of all these types can naturally be represented by certain analytic expressions with $n-q-m$ essential arbitrary parameters. We can also express this as follows: all the concerned types belong to the same kind of types [TYPENGATTUNG] (cf. Chap. 22, p. 455).

Now, we include two intransitive groups of R_n in the same class when the same kind of type of linear homogeneous group of R_n is associated to both of them.

§ 105. We can use the classification of all groups of R_n just described in order to provide an approach [ANSATZ] to the determination of these groups. But we will only undertake this for the *transitive* groups.

If we imagine the variables chosen so that the origin of coordinates: $x_1 = 0, \dots, x_n = 0$ is a point in general position, then every transitive group of R_n contains, in the neighborhood of the origin of coordinates, n independent infinitesimal transformations of zeroth order in the x_i:

$$T_i^{(0)} = \frac{\partial f}{\partial x_i} + \cdots \qquad (i=1\cdots n)$$

out of which no transformation of first or higher order can be linearly deduced.

Furthermore, every transitive group of R_n also contains in general certain infinitesimal transformations of first order in the x_i that depend, according to the preceding, on the class to which the group belongs. Now, since a completely determined class of transitive groups of R_n is associated to every type of linear homogeneous group of this space, we want to choose any such type and to restrict ourselves to the consideration of those transitive groups which belong to the corresponding class.

Let the m_1-term group:

$$\text{(3)} \qquad \sum_{i,\,\nu}^{1\cdots n} \alpha_{ji\nu}\, x_i' \frac{\partial f}{\partial x_\nu'} \qquad (j=1\cdots m_1\,;\; 0 \leqslant m_1 \leqslant n^2)$$

be a representative of the chosen type of linear homogeneous groups. Then according to the above, every transitive group of R_n which belongs to the corresponding class can, by means of an appropriate choice of variables, be brought to a form such that in the neighborhood of: $x_1 = 0, \dots, x_n = 0$, it contains the following m_1 independent infinitesimal transformations of first order:

$$T_j^{(1)} = \sum_{\nu=1}^{n} \left\{ \sum_{i=1}^{n} \alpha_{ji\nu}\, x_i + \cdots \right\} \frac{\partial f}{\partial x_\nu} \qquad (j=1\cdots m_1).$$

These m_1 infinitesimal transformations $T_j^{(1)}$ are constituted in such a way that out of them, no transformation of second or higher order can be linearly deduced and on the other hand, such that in every first order infinitesimal transformation of the group, the terms of first order can be linearly deduced from the terms of first order in $T_1^{(1)}, \dots, T_{m_1}^{(1)}$.

Besides, we already see now that in all circumstances, there is at least one transitive group which belongs to the class chosen by us; indeed, *one* such group can be immediately indicated, namely the $(n+m_1)$-term group:

$$\frac{\partial f}{\partial x_1}, \dots, \frac{\partial f}{\partial x_n}, \qquad \sum_{i,\,\nu}^{1\cdots n} \alpha_{ji\nu}\, x_i \frac{\partial f}{\partial x_\nu} \qquad (j=1\cdots m_1),$$

and it is obtained by leaving out all terms of first, second, and higher orders in the P_i and in the $T_j^{(1)}$.

Aside from the infinitesimal transformations of zeroth and first order already indicated, every group which belongs to our class can contain, in the neighborhood of: $x_1 = 0, \dots, x_n = 0$, a certain number of independent infinitesimal transformations of second order, out of which no transformation of third or higher order can be linearly deduced, and moreover, a certain number of infinitesimal transformations of third, fourth, ... orders; but according to Theorem 29, p. 206, there is always a whole number $s \geqslant 1$ characteristic to the group of such a nature that the group contains infinitesimal transformations of second, third, ..., s-th orders, while by contrast, it contains no transformations of orders $(s+1)$ or higher.[1] From this, it follows that the totality of all groups of our class decomposes into a series of subclasses: to each value of s there corresponds a subclass.

Since there are infinitely many whole numbers s which are $\geqslant 1$, the number of subclasses just defined is infinitely large, though every subclass need not be represented effectively by a group. We will show how it can be decided for every individual value of s whether some groups belong to the concerned subclass; for this, the question is fundamentally about values of s larger than 1 only, for we already know that the subclass: $s = 1$ contains some groups.

Let $s_0 \geqslant 1$ be an arbitrarily chosen but completely determined whole number; we ask whether there are groups in our class which belong to the subclass: $s = s_0$.

---

[1] What we denoted by s in Theorem 29 is called here $s+1$.

If there are groups of this sort, let for instance G be one of them. If, for $k = 0,1,2,\dots,s_0$, we omit from the infinitesimal transformations of order k of G all terms of orders $(k+1)$ and higher, then we obviously obtain independent infinitesimal transformations which generate a certain group Γ and, to be precise, a group which belongs to the subclass $s = s_0$ just as G.

From this, it follows that the subclass: $s = s_0$, as soon as it actually comprises groups, contains at least one group Γ having the following specific constitution: Γ is generated by the $n+m_1$ infinitesimal transformations of zeroth and of first orders:

$$\mathsf{T}_i^{(0)} = \frac{\partial f}{\partial x_i}, \qquad \mathsf{T}_j^{(1)} = \sum_{i,\,\nu}^{1\cdots n} \alpha_{ji\nu}\, x_i\, \frac{\partial f}{\partial x_\nu}$$
$$(i=1\cdots n;\; j=1\cdots m_1),$$

and in addition, by m_2 independent infinitesimal transformations of second order, by m_3 independent ones of third order, ..., by m_{s_0} independent ones of s_0-th order; the general form of these transformations is:

$$\mathsf{T}_{i_k}^{(k)} = \sum_{\nu=1}^{n} \xi_{i_k,\,\nu}^{(k)}(x_1,\dots,x_n)\,\frac{\partial f}{\partial x_\nu}$$
$$(k=2,3\cdots s_0;\; i_k=1,2\cdots m_k),$$

where the $\xi^{(k)}$ are entire k-th order homogeneous functions of their arguments. At the same time, it follows that to every group G which belongs to the subclass $s = s_0$ is associated a completely determined group Γ having the constitution just described.

By executable operations, it can be decided whether there is a group Γ which possesses the properties just described; by means of executable operations, one can even determine all possibly existing groups Γ.

In fact, the number of all possible systems: m_2, m_3, ..., m_{s_0} is at first finite. Furthermore, if one has chosen such a system, one can always determine, in the most general way and by means of algebraic operations, $m_2 + \cdots + m_{s_0}$ independent infinitesimal transformations:

$$\mathsf{T}_{i_k}^{(k)} \qquad (k=2,3\cdots s_0;\; i_k=1,2\cdots m_k)$$

which, together with the $\mathsf{T}_i^{(0)}$, $\mathsf{T}_j^{(1)}$ generate an $(n+m_1+\cdots+m_{s_0})$-term group. Indeed, for that, one only needs to determine in the most general way the coefficients in the functions $\xi^{(k)}$ so that every transformation:

$$\big[\mathsf{T}_{i_k}^{(k)},\, \mathsf{T}_{j_\mu}^{(\mu)}\big] \qquad (k,\,\mu=2,3\cdots s_0)$$

may be linearly deduced from:

$$\mathsf{T}_{\pi}^{(k+\mu-1)} \qquad (\pi=1\cdots m_{k+\mu-1}),$$

as soon as $k+\mu-1 \leqslant s_0$, and vanishes identically as soon as $k+\mu-1 > s_0$. It is clear on the occasion that one obtains only algebraic equations for the unknown coefficients.

In the preceding, it is shown that for every *individual* whole number $s_0 \geqslant 1$, one can determine by means of executable operations whether there are groups of our class which belong to the subclass: $s = s_0$. But we possess no general method which accomplishes this for *all* whole numbers $s > 1$ *in one stroke*. Only in special cases, for special constitutions of the linear homogeneous group (3) did we succeed in recognizing how many and which ones of the infinitely many subclasses are represented by some groups. On the occasion, it happens that a maximum exists for the number s, so that only the classes whose number s does not exceed a certain maximum really contain some groups (cf. Chap. 29); however, for the existence of such a maximum, we do not have a general criterion. Nevertheless, we believe that it is possible to set up such a criterion.

In consequence of that, we will restrict ourselves to explaining how one can find all groups of our class which belong to a determined subclass, say in the subclass: $s = s_0 \geqslant 1$.

According to p. 607, to every group of the subclass: $s = s_0$ is associated a completely determined group Γ of the same subclass with infinitesimal transformations which have a specific form described above. Now, since all groups Γ of this sort which belong to the subclass: $s = s_0$ can be determined, as we know, by means of executable operations, we need only to show in which way one can find the groups of the subclass: $s = s_0$ to which are associated an arbitrarily chosen concerned group Γ.

Let:

$$(4)\qquad \begin{cases} \mathsf{T}_i^{(0)} = \dfrac{\partial f}{\partial x_i}, \qquad \mathsf{T}_j^{(1)} = \sum\limits_{\nu,\,\pi}^{1\cdots n} \alpha_{j\nu\pi}\, x_\nu \dfrac{\partial f}{\partial x_\pi} = \sum\limits_{\pi=1}^{n} \xi_{j\pi}^{(1)}(x_1,\dots,x_n)\dfrac{\partial f}{\partial x_\pi} \\ \qquad\qquad \mathsf{T}_{i_k}^{(k)} = \sum\limits_{\nu=1}^{n} \xi_{i_k,\nu}^{(k)}(x_1,\dots,x_n)\dfrac{\partial f}{\partial x_\nu} \\ \qquad\qquad (i=1\cdots n;\ j=1\cdots m_1;\ i_k=1\cdots m_k;\ k=2\cdots s_0) \end{cases}$$

be the $n+m_1+\cdots+m_{s_0}$ independent infinitesimal transformations of an arbitrary group amongst the discussed groups Γ, and let the composition of this group be determined by the relations:

$$(5)\qquad \big[\mathsf{T}_{i_k}^{(k)},\,\mathsf{T}_{j_\mu}^{(\mu)}\big] = \sum_{\pi=1}^{m_{k+\mu-1}} c_{i_k\,j_\mu\,\pi}\,\mathsf{T}_\pi^{(k+\mu-1)}$$
$$(k,\,\mu=0,\,1\cdots s_0;\ i_k=1\cdots m_k;\ j_\mu=1\cdots m_\mu;\ m_0=n),$$

in which the c of the right-hand side are to be considered as known and in particular, vanish as soon as $k+\mu-1$ exceeds the number s_0.

Every group $\mathfrak{G}$ belonging to the subclass: $s = s_0$ to which is associated the group (4) contains $n+m_1+\cdots+m_{s_0}$ parameters and is generated by the same number of independent infinitesimal transformations; these transformations have the form:

$$T_i^{(0)} = \frac{\partial f}{\partial x_i} + \cdots, \qquad T_{i_k}^{(k)} = \sum_{\nu=1}^{n} \xi_{i_k,\nu}^{(k)}(x_1,\dots,x_n)\,\frac{\partial f}{\partial x_\nu} + \cdots \tag{6}$$

$$(i=1\cdots n;\; i_k=1\cdots m_k;\; k=1\cdots s_0),$$

where, generally, the omitted terms are of higher order than those written. *The question is nothing but to determine all groups, the infinitesimal transformations of which have the form* (6).

Evidently, the composition of a group with the infinitesimal transformations (6) is represented by relations of the form:

$$\begin{cases} \big[T_{i_k}^{(k)},\, T_{j_\mu}^{(\mu)}\big] = \displaystyle\sum_{\pi=1}^{m_{k+\mu-1}} c_{i_k\, j_\mu\, \pi}\, T_\pi^{(k+\mu-1)} \\ \qquad\qquad + \displaystyle\sum_{\tau=k+\mu}^{s_0} \sum_{\pi_\tau}^{1\cdots m_\tau} C_{i_k\, j_\mu\, \pi_\tau}\, T_{\pi_\tau}^{(\tau)} \end{cases} \tag{7}$$

$$(k,\mu=0,1\cdots s_0;\; i_k=1\cdots m_k;\; j_\mu=1\cdots m_\mu;\; m_0=n).$$

Here, the C are certain constants which, as is known, satisfy relations derived from the Jacobi identity (cf. Chap. 9, p. 184 sq.).

We imagine that the concerned relations between the C are set up and that the most general system of C which satisfies them is computed. Since all the relations are algebraic, this computation requires only executable operations. In addition, the form of the relations (7) can be simplified by replacing every infinitesimal transformation of k-th order: $T_{i_k}^{(k)}$ by another transformation of order k:

$$\mathfrak{T}_{i_k}^{(k)} = T_{i_k}^{(k)} + \sum_{\tau=k+1}^{s_0} \sum_{\pi_\tau}^{1\cdots m_\tau} \mathsf{P}_{i_k\, \pi_\tau}\, T_{\pi_\tau}^{(\tau)} \tag{8}$$

$$(i_k=1\cdots m_k;\; k=0,1\cdots s_0),$$

where it is understood that the P are arbitrary numerical quantities. Indeed, if one introduces the $\mathfrak{T}$ in place of the T, one obtains in place of (7) relations of the form:

$$\begin{cases} \big[\mathfrak{T}_{i_k}^{(k)},\, \mathfrak{T}_{j_\mu}^{(\mu)}\big] = \displaystyle\sum_{\pi=1}^{m_{k+\mu-1}} c_{i_k\, j_\mu\, \pi}\, \mathfrak{T}_\pi^{(k+\mu-1)} \\ \qquad\qquad + \displaystyle\sum_{\tau=k+\mu}^{s_0} \sum_{\pi_\tau}^{1\cdots m_\tau} \mathfrak{C}_{i_k\, j_\mu\, \pi_\tau}\, \mathfrak{T}_{\pi_\tau}^{(\tau)}, \end{cases} \tag{7'}$$

where, between the $\mathfrak{C}$ and the C, a connection holds which can be easily indicated. Now, one will provide the P, which are perfectly arbitrary in order that the coefficients $\mathfrak{C}$ receive the simplest possible numerical values; thanks to this, one achieves a certain simplification.

If, in the equations (8), all P are chosen fixed, then we say that the infinitesimal transformation of k-th order: $\mathfrak{T}_{i_k}^{(k)}$ is *normalized*.

If one knows all systems of $\bar{C}$ which satisfy the relations mentioned a short while ago, then one therefore knows at the same time all compositions that a group of the form (6) can possibly have. Still, the question is only whether, for each of the so defined compositions, there are groups of the form (6) which have precisely the composition in question, and how these groups can be found, in case they exist.

Thus, we imagine that an arbitrary system of values C is given in the equations (7) which satisfies the relations discussed, so that the equations (7) represent a possible composition of an $(n+m_1+\cdots+m_{s_0})$-term group (cf. p. 309).

First, we easily see that if there are groups of the form (6) which have the composition (7), then they are all similar to each other, whence they all belong to the same type of transitive groups of the space $x_1,\dots,x_n$ (Chap. 22, p. 443).

In fact, if we have two groups $\mathfrak{G}$ and $\mathfrak{G}'$ which both possess the form (6) and the composition (7), then these two groups can obviously be related to each other in a holoedrically isomorphic way thanks to a choice of their infinitesimal transformations so that the largest subgroup of $\mathfrak{G}$ which leaves invariant the point: $x_1=0$, $\dots$, $x_n=0$ corresponds to the largest subgroup of $\mathfrak{G}'$ which fixes this point. Now, since under the assumptions made, the point: $x_1=0,\dots,x_n=0$ is a point in general position, then according to Theorem 76, p. 434, these two *transitive* groups $\mathfrak{G}$ and $\mathfrak{G}'$ are similar to each other. But this is what was to be proved.

Moreover, we will show that there are always groups of the form (6) which possess the composition (7).

At the outset we determine in a space R_N of $N=n+m_1+\cdots+m_{s_0}$ dimensions a simply transitive group:

$$(9) \qquad W_i f, \quad W_{i_k}^{(k)} f \qquad (i=1\cdots n;\ k=1\cdots s_0;\ i_k=1\cdots m_k)$$

of the composition (7); according to Chap. 22, pp. 439–442, this is always possible and this requires at most the integration of ordinary differential equations. Then, we choose any manifold $\mathfrak{M}$ of R_N which admits the $(N-n)$-term subgroup:

$$(10) \qquad W_{i_k}^{(k)} f \qquad (k=1\cdots s_0;\ i_k=1\cdots m_k),$$

of the group (9), but no larger subgroup; this property is possessed by every characteristic manifold (p. 116) of the $(N-n)$-term complete system:

$$W_{i_k}^{(k)} f=0 \qquad (k=1\cdots s_0;\ i_k=1\cdots m_k).$$

Through the ∞^N transformations of the group (9), $\mathfrak{M}$ takes precisely ∞^n different positions whose totality forms an invariant family. If we characterize the individual manifolds of this family by means of n coordinates $x_1,\dots,x_n$, we obtain a transitive group in $x_1,\dots,x_n$:

$$\text{(11)} \qquad X_i^{(0)} f, \quad X_{i_k}^{(k)} f \qquad (i = 1 \cdots n;\ k = 1 \cdots s_0;\ i_k = 1 \cdots m_k)$$

that indicates in which way the manifolds of our invariant family are permuted with each other by the group (9). This new group is isomorphic with the group (9) so that relations of the form:

$$\text{(7'')} \qquad \big[X_{i_k}^{(k)}, X_{j_\mu}^{(\mu)}\big] = \sum_{\pi=1}^{m_{k+\mu-1}} c_{i_k j_\mu \pi} X_\pi^{(k+\mu-1)} f + \sum_{\tau=k+\mu}^{s_0} \sum_{\pi_\tau}^{1 \cdots m_\tau} C_{i_k j_\mu \pi_\tau} X_{\pi_\tau}^{(\tau)} f$$
$$(k, \mu = 0, 1 \cdots s_0;\ i_k = 1 \cdots m_k;\ j_\mu = 1 \cdots m_\mu;\ m_0 = n)$$

hold (cf. Theorem 85, p. 489).

It can easily be proved that the infinitesimal transformations of the group (11) receive the form (6) after an appropriate choice of the variables $x_1, \dots, x_n$.

In order to conduct the concerned proof, we imagine above all that the variables are chosen so that $\mathfrak{M}$ receives the coordinates: $x_1 = 0, \dots, x_n = 0$. Then obviously, all infinitesimal transformations:

$$X_{i_k}^{(k)} f \qquad (k = 1 \cdots s_0;\ i_k = 1 \cdots m_k)$$

in the x_ν are of first or higher order, while by contrast: $X_1^0 f, \dots, X_n^0 f$ are independent infinitesimal transformations of zeroth order, out of which no transformation of first or higher order can be linearly deduced; this last fact follows from the transitivity of the group (11). It is therefore clear that, notwithstanding our assumption just made, we can choose the variables $x_1, \dots, x_n$ so that $X_1^0 f, \dots, X_n^0 f$ receive the form:

$$X_i^{(0)} f = \frac{\partial f}{\partial x_i} + \cdots \qquad (i = 1 \cdots n).$$

After these preparations, we want to determine the initial terms in the power series expansions of the m_1 infinitesimal transformations $X_j^{(1)} f$.

We have:

$$X_j^{(1)} f = \sum_{\nu=1}^{n} \zeta_{j\nu}^{(1)}(x_1, \dots, x_n) \frac{\partial f}{\partial x_\nu} + \cdots \qquad (j = 1 \cdots m_1),$$

where it is understood that the $\zeta^{(1)}$ are linear homogeneous functions of their arguments. If we insert this expression in the n relations:

$$\big[X_i^{(0)}, X_j^{(1)}\big] = \sum_{\pi=1}^{n} c_{ij\pi} X_\pi^{(0)} f + \sum_{\tau=1}^{s_0} \sum_{\pi_\tau}^{1 \cdots m_\tau} C_{ij\pi_\tau} X_{\pi_\tau}^{(\tau)} f \qquad (i = 1 \cdots n)$$

and if we compare the terms of zeroth order in the two sides, we find:

$$\sum_{\nu=1}^{n} \frac{\partial \zeta_{j\nu}^{(1)}}{\partial x_i} \frac{\partial f}{\partial x_\nu} = \sum_{\pi=1}^{n} c_{ij\pi} \frac{\partial f}{\partial x_\pi} \qquad (i = 1 \cdots n),$$

hence all first order differential quotients of $\zeta_{j\nu}^{(1)}$ are completely determined, and because of the known equation:

$$\sum_{i=1}^{n} x_i \frac{\partial \zeta_{j\nu}^{(1)}}{\partial x_i} = \zeta_{j\nu}^{(1)},$$

$\zeta_{\nu}^{(1)}$ itself is also completely determined. Now, since $\xi_{j\nu}^{(1)}$ obviously satisfies all differential equations which we have just found, it follows that: $\zeta_{j\nu}^{(1)} = \xi_{j\nu}^{(1)}$, and consequently:

$$X_j^{(1)} f = \sum_{\nu=1}^{n} \xi_{j\nu}^{(1)}(x_1,\dots,x_n) \frac{\partial f}{\partial x_\nu} + \cdots \qquad (j=1\cdots m_1).$$

In the same way, we find generally:

$$X_{i_k}^{(k)} f = \sum_{\nu=1}^{n} \xi_{i_k \nu}^{(k)}(x_1,\dots,x_n) \frac{\partial f}{\partial x_\nu} + \cdots \qquad (k=1\cdots s_0\,;\; i_k=1\cdots m_k),$$

or in other words: we find that the infinitesimal transformations of the group (11) have the form (6). Evidently, here lies the reason why the N infinitesimal transformations (11) are mutually independent, so that the group (11) is N-term and has the same composition as the group (9).

Since, as we have seen just now, the group (11) is N-term, the group:

(9) $$W_i^{(0)} f, \quad W_{i_k}^{(k)} f \qquad (i=1\cdots n;\; k=1\cdots s_0\,;\; i_k=1\cdots m_k)$$

can contain no invariant subgroup which belongs to the group:

(10) $$W_{i_k}^{(k)} f \qquad (k=1\cdots s_0\,;\; i_k=1\cdots m_k)$$

(Theorem 85, p. 489). But one can also see directly that there is no such invariant subgroup of the group (9); in this way, by taking Theorem 85 into consideration, one obtains a new proof of the fact that the group (11) has N essential parameters.

Every invariant subgroup of (9) which belongs to the group (10) obviously contains an infinitesimal transformation:

$$\mathfrak{W} f = \sum_{\tau=1}^{m_p} \rho_\tau W_\tau^{(p)} f + \sum_{k=p+1}^{s_0} \sum_{\tau_k}^{1\cdots m_k} \sigma_{\tau_k} W_{\tau_k}^{(k)} f \qquad (p \geqslant 1),$$

in which not all the m_p quantities ρ_τ vanish. Simultaneously with $\mathfrak{W} f$, there also appear in $\mathfrak{g}$ the n infinitesimal transformations:

$$[W_1^0, \mathfrak{W}], \;\cdots,\; [W_n^0, \mathfrak{W}]\,;$$

consequently, as one sees from the composition (7) of the group (9), there surely appears in $\mathfrak{g}$ a transformation of the form:

$$\mathfrak{W}'f = \sum_{\tau=1}^{m_p-1} \rho'_\tau W_\tau^{(p-1)} f + \sum_{k=p}^{s_0} \sum_{\tau_k}^{1\cdots m_k} \sigma'_{\tau_k} W_{\tau_k}^{(k)} f,$$

in which not all the m_{p-1} quantities ρ'_τ are zero. In this way, one realizes finally that $\mathfrak{g}$ must contain an infinitesimal transformation which does not belong to the group (10); but this contradicts the assumption that $\mathfrak{g}$ should be contained in the group (10), hence there is no group $\mathfrak{g}$ having the supposed constitution.

The results found up to now provide the

Theorem 110. *In the variables $x_1, \dots, x_n$, let a transitive group be presented which, in the neighborhood of the point: $x_1 = 0, \dots, x_n = 0$ in general position contains the following infinitesimal transformations: firstly, n zeroth order independent transformations of the form:*

$$\mathsf{T}_1^{(0)} = \frac{\partial f}{\partial x_1}, \dots, \mathsf{T}_n^{(0)} = \frac{\partial f}{\partial x_n},$$

and secondly, for every $k = 1, 2, \dots, s$, $m_k > 0$ independent k-th order transformations of the form:

$$\mathsf{T}_{i_k}^{(k)} = \sum_{\nu=1}^{n} \xi_{i_k\,\nu}^{(k)}(x_1, \dots, x_n) \frac{\partial f}{\partial x_\nu} \qquad (k=1\cdots s;\ i_k=1\cdots m_k),$$

where the $\xi^{(k)}$ denote completely homogeneous functions of order k. Also, let the composition of the group be determined by the relations:

$$\big[\mathsf{T}_{i_k}^{(k)}, \mathsf{T}_{j_\mu}^{(\mu)}\big] = \sum_{\pi=1}^{m_{k+\mu-1}} c_{i_k\,j_\mu\,\pi}\, \mathsf{T}_\pi^{(k+\mu-1)}$$
$$(k, \mu=0, 1\cdots s;\ i_k=1\cdots m_k;\ j_\mu=1\cdots m_\mu;\ m_0=n),$$

where the c in the right-hand side all vanish as soon as $k+\mu-1$ is larger than s. Then one finds in the following way all $(n+m_1+\cdots+m_s)$-term groups of the space $x_1, \dots, x_n$, the infinitesimal transformations of which are of the form:

$$(6)\qquad T_i^{(0)} = \frac{\partial f}{\partial x_i} + \cdots, \qquad T_{i_k}^{(k)} = \sum_{\nu=1}^{n} \xi_{i_k,\nu}^{(k)}(x_1, \dots, x_n) \frac{\partial f}{\partial x_\nu} + \cdots$$
$$(i=1\cdots n;\ i_k=1\cdots m_k;\ k=1\cdots s),$$

in the neighborhood of: $x_1 = 0, \dots, x_n = 0$.

One determines the constants C in the equations:

$$(7)\quad \begin{cases} \left[T_{i_k}^{(k)},\, T_{j_\mu}^{(\mu)}\right] = \sum\limits_{\pi=1}^{m_{k+\mu-1}} c_{i_k\, j_\mu\, \pi}\, T_\pi^{(k+\mu-1)} \\ \qquad + \sum\limits_{\tau=k+\mu}^{s_0} \sum\limits_{\pi_\tau}^{1\cdots m_\tau} c_{i_k\, j_\mu\, \pi_\tau}\, T_{\pi_\tau}^{(\tau)} \end{cases}$$

$$(k,\, \mu = 0,\, 1 \cdots s;\; i_k = 1 \cdots m_k;\; j_\mu = 1 \cdots m_\mu;\; m_0 = n)$$

in the most general way so that they satisfy the relations following from the Jacobi identity. If this takes place, then the equations (7) *represent all compositions that the sought groups may have. To every individual composition amongst these compositions correspond groups of the form* (6) *which are all similar to each other and which can be found in any case by integrating ordinary differential equations.*[2]

§ 151. In the preceding sections, we have reduced the problem of determining all transitive groups of R_n to the following four problems:

𝔄. *To find all types of linear homogeneous groups in n variables.*

If one has solved this problem, then in the neighborhood of any point in general position, one knows the possible forms of the initial terms in all infinitesimal transformations of first order which can appear in a transitive group of R_n.

𝔅. *Assuming that, in the neighborhood of a point in general position, the initial terms of the first order infinitesimal transformations of a transitive group of* R_n *are given, to determine all possible forms of the initial terms in the infinitesimal transformations of second and higher order.*

ℭ. *To determine all compositions that a transitive group of* R_n *can have, a group which, in the neighborhood of a point in general position contains certain infinitesimal transformations of first, second, ..., s-th order with given initial terms, and which by contrast, contains no transformation of* $(s+1)$*-th or higher order.*

𝔇. *Assuming that one of the compositions found in the preceding problem is given, to set up a transitive group which has this composition and whose infinitesimal transformations of first, ..., s-th order have the form given in the preceding problem.*

If one knows *one* group having the constitution demanded in the last problem, then one knows *all* such groups, since these are, according to Theorem 110, similar to each other.

The settlement of the first amongst these four problems requires only executable, or put more precisely: only algebraic operations (cf. Chap. 12, p. 222 and Chap. 23, p. 499 sq.). By contrast, we have not succeeded in reducing problem 𝔅 to a finite number of executable operations. Problem ℭ again requires only algebraic operations. Lastly, as we have shown in the previous section, problem 𝔇 can in any case be settled by integrating ordinary differential equations.

[2] LIE, Archiv for Math. Vol. X, pp. 381–389. 1885.

At present, we want to consider a special case which presents a characteristic simplification and we want to carry it out in detail.

Assume that the $(n+m_1+\cdots+m_s)$-term group:

$$(12)\qquad \mathsf{T}_i^{(0)} = \frac{\partial f}{\partial x_i}, \qquad \mathsf{T}_{i_k}^{(k)} = \sum_{\nu=1}^{n} \xi_{i_k\,\nu}^{(k)}(x_1,\dots,x_n)\,\frac{\partial f}{\partial x_\nu}$$

$$(i=1\cdots n;\ k=1\cdots s;\ i_k=1\cdots m_k)$$

which we imagined as given in Theorem 110, p. 613, contains amongst its first-order infinitesimal transformations:

$$e_1\,\mathsf{T}_1^{(1)}+\cdots+e_{m_1}\,\mathsf{T}_{m_1}^{(1)}$$

in particular one of the form:

$$x_1\,\frac{\partial f}{\partial x_1}+\cdots+x_n\,\frac{\partial f}{\partial x_n}$$

and let, say, $\mathsf{T}_{m_1}^{(1)}$ have this form. We will show that under this assumption, it is always possible to determine all transitive $(n+m_1+\cdots+m_s)$-term groups of the space whose transformations, in the neighborhood of the point: $x_1=0,\ \dots,\ x_n=0$ in general position, have the form:

$$(12')\qquad T_i^{(0)} = \frac{\partial f}{\partial x_i}+\cdots, \qquad T_{i_k}^{(k)} = \sum_{\nu=1}^{n} \xi_{i_k\,\nu}^{(k)}(x_1,\dots,x_n)\,\frac{\partial f}{\partial x_\nu}+\cdots$$

$$(i=1\cdots n;\ k=1\cdots s;\ i_k=1\cdots m_k).$$

For:

$$T_{m_1}^{(1)} = \sum_{\nu=1}^{n} x_\nu\,\frac{\partial f}{\partial x_\nu}+\cdots,$$

we write U, a notation which we shall also employ on later occasions.

First, it is clear that between U and the infinitesimal transformations of order s of a group of the form (12'), the following relations hold:

$$\big[T_\pi^{(s)},\,U\big] = (1-s)\,T_\pi^{(s)} \qquad (\pi=1\cdots m_s).$$

In the same way, between U and the infinitesimal transformations of order $(s-1)$, there are relations of the form:

$$\big[T_j^{(s-1)},\,U\big] = (2-s)\,T_j^{(s-1)}+\sum_{\pi=1}^{m_s} K_{j\pi}\,T_\pi^{(s)} \qquad (j=1\cdots m_{s-1}),$$

where the K are unknown constants. In order to simplify these relations, we set (cf. p. 609):

$$\mathfrak{T}_j^{(s-1)} = T_j^{(s-1)} + \sum_{\pi=1}^{m_s} \mathsf{P}_{j\pi}\, T_\pi^{(s)} \qquad (j=1\cdots m_{s-1}),$$

and we find:

$$\left[\mathfrak{T}_j^{(s-1)},\, U\right] = (2-s)\,\mathfrak{T}_j^{(s-1)} + \sum_{\pi=1}^{m_s} \left(K_{j\pi} + (1-s-2+s)\,\mathsf{P}_{j\pi}\right) T_{j\pi}^{(s)},$$

hence when we choose the P in an appropriate way:

$$\left[\mathfrak{T}_j^{(s-1)},\, U\right] = (2-s)\,\mathfrak{T}_j^{(s-1)} \qquad (j=1\cdots m_{s-1}).$$

As a result, the infinitesimal transformations of order $(s-1)$ of the group (12') are *normalized*.

In exactly the same way, we can normalize the infinitesimal transformations of order $(s-2)$ by setting:

$$\mathfrak{T}_j^{(s-2)} = T_j^{(s-2)} + \sum_{\pi=1}^{m_{s-1}} \mathsf{P}'_{j\pi}\, \mathfrak{T}_\pi^{(s-1)} + \sum_{\pi=1}^{m_s} \mathsf{P}''_{j\pi}\, T_\pi^{(s)}$$
$$(j=1\cdots m_{s-2}),$$

and dispose appropriately of the P' and of the P''; in this way, we get:

$$\left[\mathfrak{T}_j^{(s-2)},\, U\right] = (3-s)\,\mathfrak{T}_j^{(s-2)} \qquad (j=1\cdots m_{s-2}).$$

By proceeding similarly, we obtain finally, when we generally write T for $\mathfrak{T}$:

$$\text{(13)} \qquad \left[T_{i_k}^{(k)},\, U\right] = (1-k)\, T_{i_k}^{(k)} \qquad (k=0,\,1\cdots s;\; i_k=1\cdots m_k;\; m_0=n).$$

As a result, the infinitesimal transformations of zeroth, first, up to s-th order are all normalized, *except for U itself*.

Now, we recall that because of the composition of the group (12'), between the T, there are relations of the form:

$$\text{(14)} \qquad \begin{cases} \left[T_{i_k}^{(k)},\, T_{j_\mu}^{(\mu)}\right] = \displaystyle\sum_{\pi=1}^{m_{k+\mu-1}} c_{i_k\, j_\mu\, \pi}\, T_\pi^{(k+\mu-1)} \\ \qquad\qquad + \displaystyle\sum_{\tau=k+\mu}^{s} \sum_{\pi_\tau}^{1\cdots m_\tau} C_{i_k\, j_\mu\, \pi_\tau}\, T_{\pi_\tau}^{(\tau)} \end{cases}$$
$$(k,\,\mu=0,\,1\cdots s;\; i_k=1\cdots m_k;\; j_\mu=1\cdots m_\mu;\; m_0=n),$$

in which the c actually vanish as soon as $k+\mu-1 > s$. In order to determine the unknown constants C, we form the Jacobi identity:

$$\left[\left[T_{i_k}^{(k)},\, T_{j_\mu}^{(\mu)}\right],\, U\right] + \left[\left[T_{j_\mu}^{(\mu)},\, U\right],\, T_{i_k}^{(k)}\right] + \left[\left[U,\, T_{i_k}^{(k)}\right],\, T_{j_\mu}^{(\mu)}\right] = 0,$$

which, because of (13), obviously takes the form:

$$\big[[T_{i_k}^{(k)},\,T_{j_\mu}^{(\mu)}],\,U\big]=(2-k-\mu)\,\big[T_{i_k}^{(k)},\,T_{j_\mu}^{(\mu)}\big].$$

Here, if we insert the expression (14) and if we use once more the equations (13), we get:

$$\sum_{\pi=1}^{m_{k+\mu-1}} c_{i_k\,j_\mu\,\pi}\,(2-k-\mu)\,T_\pi^{(k+\mu-1)}+\sum_{\tau=k+\mu}^{s}\;\sum_{\pi_\tau}^{1\cdots m_\tau} C_{i_k\,j_\mu\,\pi_\tau}\,(1-\tau)\,T_{\pi_\tau}^{(\tau)}$$
$$=(2-k-\mu)\,\big[T_{i_k}^{(k)},\,T_{j_\mu}^{(\mu)}\big],$$

or, because the T are independent infinitesimal transformations:

$$C_{i_k\,j_\mu\,\pi_\tau}\,(\tau+1-k-\mu)=0 \qquad (k+\mu\leqslant\tau\leqslant s).$$

From this, it follows that all the C vanish.

Thus, under the assumption made, all groups of the form (12') have the same composition as the group (12), hence according to Theorem 110, p. 613, they all are similar to each other and similar to the group (12).

In consequence of this, we can say:

Theorem 111. *If a transitive $(n+m_1+\cdots+m_s)$-term group in the variables $x_1,\dots,x_n$ contains, in the neighborhood of the point in general position: $x_1=0$, $\dots$, $x_n=0$, aside from the n independent infinitesimal transformations of zeroth order in the x:*

$$T_i=\frac{\partial f}{\partial x_i}+\cdots \qquad (i=1\cdots n),$$

m_k, independent infinitesimal transformations of order k, for $k=1,2,\dots,s$, out of which no transformation of order $k+1$ or higher can be linearly deduced, and if it in particular contains a first order infinitesimal transformation of the form:

$$\sum_{\nu=1}^{n} x_\nu\,\frac{\partial f}{\partial x_\nu}+\cdots,$$

then thanks to the introduction of new variables $x_1',\dots,x_n'$, it can be brought to the form:

$$T_i=\mathsf{T}_i=\frac{\partial f}{\partial x_i'},\qquad T_{i_k}^{(k)}=\mathsf{T}_{i_k}^{(k)}=\sum_{\nu=1}^{n}\xi_{i_k\,\nu}^{(k)}(x_1',\dots,x_n')\,\frac{\partial f}{\partial x_\nu'}$$
$$(i=1\cdots n;\;k=1\cdots s;\;i_k=1\cdots m_k);$$

here, the $\xi^{(k)}$ are the entire homogeneous functions of order k which determine the terms of order k in the infinitesimal transformations of order k:

$$T_{i_k}^{(k)}=\sum_{\nu=1}^{n}\xi_{i_k\,\nu}^{(k)}(x_1,\dots,x_n)\,\frac{\partial f}{\partial x_\nu}+\cdots \qquad (k=1\cdots s;\;i_k=1\cdots m_k)$$

of the group.

Chapter 29
Characteristic Properties of the Groups Which are Equivalent to Certain Projective Groups

In the preceding chapter, we gave a classification of all *transitive* groups: $X_1 f, \dots, X_r f$ of the n-fold extended space $x_1, \dots, x_n$. We chose a point $x_1^0, \dots, x_n^0$ in general position and considered all infinitesimal transformations of the group which leave this point at rest, hence all transformations whose power series expansion with respect to the $x_i - x_i^0$ have the form:

$$\sum_{i,\nu=1}^{n} \alpha_{ji\nu}(x_1^0, \dots, x_n^0) \cdot (x_i - x_i^0)\, \frac{\partial f}{\partial x_\nu} + \cdots \qquad (j=1,2,\dots),$$

where the omitted terms are of second or of higher order.[1] (Translator's note: Such terms are systematically written "$+\cdots$" by Engel and Lie.) Then the linear homogeneous group:

$$L_j f = \sum_{i,\nu=1}^{n} \alpha_{ji\nu}(x_1^0, \dots, x_n^0)\, x_i'\, \frac{\partial f}{\partial x_\nu'} \qquad (j=1,2,\dots)$$

showed in which way the ∞^{n-1} line-elements $x_1' : x_2' : \cdots : x_n'$ through the point $x_1^0, \dots, x_n^0$ are transformed by those transformations of the group: $X_1 f, \dots, X_r f$, which leave this point invariant.

In the present chapter, to begin with, we solve the problem of determining all *transitive* groups: $X_1 f, \dots, X_r f$ of the space $x_1, \dots, x_n$ for which the linear homogeneous group: $L_1 f, L_2 f, \dots$ assigned to a point in general position coincides either with the general linear homogeneous group or with the special linear homogeneous group.[1] Then we obtain the curious result that every such group: $X_1 f, \dots, X_r f$ is

[1] It is easy to see that a group of R_n which assigns the general or the special linear homogeneous group to a point $x_1^0, \dots, x_n^0$ in general position is always transitive. Indeed, in the neighborhood of $x_1^0, \dots, x_n^0$, the group certainly comprises an infinitesimal transformation of zeroth order in the $x_i - x_i^0$, hence a transformation: $\sum \alpha_i p_i + \cdots$, in which not all α_i are equally null. In addition, the group surely comprises $n(n-1)$ first order transformations of the form:

$$(x_i - x_i^0)\, p_k + \cdots \qquad (i,k=1\cdots n;\ i \neq k).$$

S. Lie, *Theory of Transformation Groups I*,
DOI 10.1007/978-3-662-46211-9_29

equivalent either to the general projective group, or to the general linear group, or to the special linear group of the space $x_1,\dots,x_n$.

In addition, we show in the last section of the chapter that in the space $x_1,\dots,x_n$ there is no finite continuous group which can transfer $m > n+2$ arbitrarily chosen points in general position to just the same kind of m points; at the same time, we show that the general projective group and the groups that are equivalent to it are the only groups of the space $x_1,\dots,x_n$ which can transfer $n+2$ arbitrarily chosen points in general position to just the same kind of $n+2$ points.

§ 152. Every transitive group of R_n comprises, in the neighborhood of the point: $x_1^0,\dots,x_n^0$ in general position, n independent infinitesimal transformations of zeroth order in the $x_i - x_i^0$:

$$p_1 + \cdots, \quad p_2 + \cdots, \quad \dots, \quad p_n + \cdots,$$

where, according to the notation fixed on p. 555, p_i is written in place of $\frac{\partial f}{\partial x_i}$.

Now, if G assigns to the point $x_1^0,\dots,x_n^0$ the general linear homogeneous group as group: $L_1 f, L_2 f,\dots$, then it comprises in the neighborhood of $x_1^0,\dots,x_n^0$ the largest possible number, namely n^2, of such infinitesimal transformations of first order in $x_1 - x_1^0,\dots,x_n - x_n^0$, out of which no transformation of second or of higher order can be deduced by linear combination. These n^2 first order transformations have the form:

$$(x_i - x_i^0)\, p_k + \cdots \qquad (i,k = 1 \cdots n).$$

If, on the other hand, G assigns to the point $x_1^0,\dots,x_n^0$ the special linear homogeneous group, then it comprises, in the neighborhood of the point, only $n^2 - 1$ independent first order infinitesimal transformations out of which no transformation of second or of higher order can be deduced by linear combination; the same transformations have the form:

$$(x_i - x_i^0)\, p_k + \cdots, \quad (x_i - x_i^0)\, p_i - (x_k - x_k^0)\, p_k + \cdots$$
$$(i,k = 1 \cdots n;\ i \gtrless k).$$

Therefore, when we choose the point $x_1^0,\dots,x_n^0$ as the origin of coordinates, we can enunciate as follows the problem indicated in the introduction to this chapter:

In the variables $x_1,\dots,x_n$, to seek all groups $X_1 f,\dots,X_r f$, or shortly G, which, in the neighborhood of the point in general position: $x_1 = 0,\dots,x_n = 0$, comprise the following infinitesimal transformations of zeroth and of first order: either the $n+n^2$:

$$p_i + \cdots, \quad x_i\, p_k + \cdots \qquad (i,k = 1 \cdots n);$$

or the $n+n^2-1$:

If one forms a Combination [bracketing] of the latter with $\sum \alpha_i\, p_i + \cdots$, then one recognizes that n transformations of the form: $p_1 + \cdots, \dots, p_n + \cdots$ appear, whence the group is actually transitive.

$$p_i + \cdots, \quad x_i p_k + \cdots, \quad x_i p_i - x_k p_k + \cdots$$
$$(i,k=1\cdots n;\, i \gtrless k).$$

To begin with, the two cases can be treated simultaneously; one must only, as long as possible, disregard the fact that in the first case, aside from the transformations which occur in the second case, one transformation appears: $\sum^i x_i p_i$.

The group $X_1 f, \dots, X_r f$ comprises infinitesimal transformations whose expansion in power series begins with terms of second, or of relatively higher order in the x, shortly, infinitesimal transformations of second or higher order, respectively. We search for the highest order number[2] (Translator's note: Actually, s will be shown to be equal to 2, not more, and the corresponding transformations of second order to be necessarily of the form $x_i(x_1 p_1 + \cdots + x_n p_n) + \cdots$, $i = 1, \dots, n$.) s of the existing transformations and we even look for the determination of such transformations.

We can assume this number s to be bigger than 1, since we already know all the first order infinitesimal transformations that appear. Let

$$K = \vartheta_1 p_1 + \cdots + \vartheta_n p_n + \cdots$$

be an s-th order infinitesimal transformation of the group; at the same time, the ϑ's denote completely homogeneous functions of order s in $x_1, \dots, x_n$, while the omitted terms are of higher order. Obviously, $\vartheta_1, \dots, \vartheta_n$ do not all vanish, because there should not be given any infinitesimal transformation $\sum c_k X_k f$ whose power series expansion starts with terms of $(s+1)$-th or of higher order; so we can assume in any case that ϑ_1 is not identically null.

Let the lowest power of x_1 which appears in ϑ_1 be the α_1-th and let

$$x_1^{\alpha_1} x_2^{\alpha_2} \cdots x_n^{\alpha_n} \qquad (\alpha_1 + \cdots + \alpha_n = s)$$

be a term of ϑ_1 with nonvanishing coefficient. Now, we form a Combination of the transformation $x_1 p_2 + \cdots$ with K, and we form a Combination of the result once more with $x_1 p_2 + \cdots$, and so on — in total α_2 times. We form a Combination of the s-th order infinitesimal transformation obtained this way with $x_1 p_3 + \cdots$, and we then proceed α_3 times, one after the other, etc., and finally, we apply α_n times $x_1 p_n + \cdots$ one after the other. In this way, we finally see that a transformation of the form:

$$K' = x_1^s p_1 + \vartheta_2' p_2 + \cdots + \vartheta_n' p_n + \cdots$$

belongs to our group.

All remaining terms of ϑ_1 are canceled; indeed, the same terms also contain either the power $x_1^{\alpha_1}$ or a higher one, and in both cases, the power of a variable x_i ($i > 1$) is certainly not the α_i-th, but a lowest one; consequently, the corresponding term vanishes by forming a Combination α_i times with $x_1 p_i + \cdots$. As always, the terms of $(s+1)$-th and of higher order are also left out of consideration here[3] (Translator's note: Here is an equivalent reformulation. Adapting the notation slightly, let $\vartheta_1 = \sum_{|\alpha|=s} A_\alpha x^\alpha$ and choose $\beta \in \mathbb{N}^n$ with $|\beta| = s$ so that $\beta_1 = \inf\{\alpha_1 : A_\alpha \neq 0\}$. For any other monomial $A_\alpha x^\alpha$ with $A_\alpha \neq 0$ we have either $\alpha_1 > \beta_1$ or $\alpha_1 = \beta_1$. In the first case, since $\alpha_1 + \alpha_2 + \cdots + \alpha_i + \cdots + \alpha_n = \beta_1 + \beta_2 + \cdots + \beta_i + \cdots + \beta_n = s$, there must exist an i with $2 \leqslant i \leqslant n$ such that $\alpha_i < \beta_i$. In the second case, namely if $\alpha_1 = \beta_1$, then $\alpha_2 + \cdots + \alpha_i + \cdots + \alpha_n = \beta_2 + \cdots + \beta_i + \cdots + \beta_n$, and again, there must exist an i with $2 \leqslant i \leqslant n$ such that $\alpha_i < \beta_i$, because otherwise, $\alpha_2 \geqslant \beta_2, \ldots, \alpha_n \geqslant \beta_n$ together with $|\alpha| = |\beta|$ implies $\alpha_2 = \beta_2, \ldots, \alpha_n = \beta_n$, hence $\alpha = \beta$, a contradiction.) (cf. Chap. 11, Theorem 30, p. 207).

In order to determine the form of the transformation K' more precisely, we make use of an auxiliary proposition which can be exploited several times with benefit.

Proposition 1. *If the infinitesimal transformations Cf and $B_1 f + \cdots + B_m f$ belong to a group and if, furthermore, there are m relations of the form $[C, B_k] = \varepsilon_k B_k f$, where the constants ε_k are all distinct from one another, then the group comprises all m infinitesimal transformations $B_1 f, \ldots, B_m f$.*

The proof of this auxiliary proposition is very simple. Aside from $B_1 f + \cdots + B_m f$, the group obviously comprises the following infinitesimal transformations:

$$\begin{gathered}
\big[C,\, B_1 + \cdots + B_m\big] = \varepsilon_1 B_1 f + \cdots + \varepsilon_m B_m f \\
\big[C,\, \varepsilon_1 B_1 + \cdots + \varepsilon_m B_m\big] = \varepsilon_1^2 B_1 f + \cdots + \varepsilon_m^2 B_m f \\
\cdots\cdots\cdots\cdots\cdots\cdots\cdots\cdots\cdots\cdots \\
\big[C,\, \varepsilon_1^{m-2} B_1 + \cdots + \varepsilon_m^{m-2} B_m\big] = \varepsilon_1^{m-1} B_1 f + \cdots + \varepsilon_m^{m-1} B_m f.
\end{gathered}$$

But since the determinant:

$$\begin{vmatrix} 1 & \cdots & 1 \\ \varepsilon_1 & \cdots & \varepsilon_m \\ \vdots & \ddots & \vdots \\ \varepsilon_1^{m-1} & \cdots & \varepsilon_m^{m-1} \end{vmatrix} = \prod_{i>k} (\varepsilon_i - \varepsilon_k)$$

does not vanish by assumption, each individual transformation $B_k f$ can be deduced, by multiplication with appropriate constants and by subsequent addition, from the just found transformations. Thus the proposition is proved.

In order to apply the same proposition, we now form a Combination of the transformation K', written in detail:

$$x_1^s\, p_1 + \big\{\Sigma A_\beta\, x_1^{\beta_1} \cdots x_n^{\beta_n}\big\}\, p_2 + \cdots + \big\{\Sigma N_\nu\, x_1^{\nu_1} \cdots x_n^{\nu_n}\big\}\, p_n + \cdots$$
$$(\beta_1 + \cdots + \beta_n = \cdots = \nu_1 + \cdots + \nu_n = s)$$

with the transformation:

$$x_1\, p_1 - x_i\, p_i + \cdots,$$

on the understanding that i is any of the numbers $2, \ldots, n$. Then one obtains:

$$(s-1)\,x_1^s\,p_1+\big\{\Sigma A_\beta\,(\beta_1-\beta_i+\varepsilon_{2i})\,x_1^{\beta_1}\cdots x_n^{\beta_n}\big\}\,p_2+\cdots+$$
$$+\big\{\Sigma N_\nu\,(\nu_1-\nu_i+\varepsilon_{ni})\,x_1^{\nu_1}\cdots x_n^{\nu_n}\big\}\,p_n+\cdots,$$

where ε_{ii} has the value 1, while all ε_{ki} $(i\neq k)$ vanish. Therefore the s-th order terms of the infinitesimal transformation K' have the form $B_1f+\cdots+B_mf$ discussed above, where the B_kf are reproduced through Combination with $x_1\,p_1-x_i\,p_i$, but with distinct factors.

Now, since from the expansion in power series of our group a new group Γ can be derived in such a way that in each power series expansion, only the terms of lowest order are kept (Chap. 28, p. 607), then our Proposition 1 shows that each individual B_kf belongs to the group Γ. Next, there is evidently a B_kf which embraces the term $x_1^s\,p_1$ and hence is reproduced with the factor $s-1$. The remaining terms of this B_kf are defined through the equations:[4] (Translator's note: Here is an enlightenment. The derived homogeneous group $\widehat{G}\equiv\Gamma$ contains the infinitesimal transformations $\widehat{D}_i=x_1\,p_1-x_i\,p_i$, for $i=1,\dots,n$, and also:

$$\widehat{K}'=x_1^s\,p_1+\big\{\Sigma A_\beta\,x^\beta\big\}\,p_2+\cdots+\big\{\Sigma N_\gamma\,x^\gamma\big\}\,p_n,$$

all the remainders being suppressed (one must have a group for Proposition 1 to apply). At first, for *fixed* i, one looks at the Lie bracket, joint with $\widehat{D}_i$, of each *monomial* homogeneous infinitesimal transformation $\big\{B_\gamma x^\gamma\big\}\,p_k$, $k\geqslant 2$, $|\gamma|=s$, which appears in $\widehat{K}'$ and which is distinct from $x_1^s\,p_1$:

$$\big[\widehat{D}_i,\,\big\{B_\gamma x^\gamma\big\}\,p_k\big]=(\gamma_1-\gamma_i+\varepsilon_{ki})\,\big\{B_\gamma x^\gamma\big\}\,p_k.$$

One then collects all the monomials $\big\{B_\gamma x^\gamma\big\}\,p_k$ of $\widehat{K}'$ having the *same* reproducing factor $\gamma_1-\gamma_i+\varepsilon_{ki}=s-1$ *as* the infinitesimal transformation $x_1^s\,p_1$, and one calls $\widehat{B}_1^i$ the corresponding sum (it depends on i), which is a part of $\widehat{K}'$. On the other hand, there are finitely many *other* values of the integers $\gamma_1-\gamma_i+\varepsilon_{ki}$, say $m(i)$, and one decomposes accordingly $\widehat{K}'=\widehat{B}_1^i+\widehat{B}_2^i+\cdots+\widehat{B}_{m(i)}^i$. Then Proposition 1 yields that $\widehat{B}_1^i$ belongs to $\widehat{G}$. Other $\widehat{B}_1^k$ are omitted.

Let $i'\neq i$ be another integer and consider bracketing with $\widehat{D}_{i'}$. The same reasoning applied to $\widehat{B}_1^i$ (instead of $\widehat{K}'$) yields a decomposition $\widehat{B}_1^i=\widehat{B}_{1,1}^{i,i'}+\widehat{B}_{1,2}^{i,i'}+\cdots+\widehat{B}_{1,m(i')}^{i,i'}$ with the first term $\widehat{B}_{1,1}^{i,i'}$ collecting monomials of $\widehat{B}_1^i$ that are reproduced with the factor $s-1$; clearly, $x_1^s\,p_1$ still belongs to $\widehat{B}_{1,1}^{i,i'}$. Proposition 1 yields again that $\widehat{B}_{1,1}^{i,i'}$ belongs to $\widehat{G}$. By induction, letting $i=1$, $i'=2$, $\dots$, $i^{(n)}=n$, one gets an infinitesimal transformation $\widehat{B}_{1,1,\dots,1}^{1,2,\dots,n}$ of $\widehat{G}$ — denoted by B_kf in the translated text — such that $\big[\widehat{D}_i,\,\widehat{B}_{1,1,\dots,1}^{1,2,\dots,n}\big]=(s-1)\,\widehat{B}_{1,1,\dots,1}^{1,2,\dots,n}$, *for all* $i=1,2,\dots,n$.)

$$\beta_1-\beta_i+\varepsilon_{2i}=\cdots=\nu_1-\nu_i+\varepsilon_{ni}=s-1.$$

Since $\varepsilon_{23},\dots,\varepsilon_{2n}$ vanish, from the same equations, we obtain immediately $\beta_3=\cdots=\beta_n$ and consequently:

$$\beta_1+\beta_2+\cdots+\beta_n=\beta_1+\beta_2+(n-2)\,\beta_3=s,$$

whence we get:

$$\beta_2=\beta_1-s+2, \qquad\qquad \beta_3=\beta_1-s+1.$$

By elimination of β_2 and β_3, it follows that:

$$(\beta_1-s+1)\,n=0$$

$$\text{hence}\quad \beta_1=s-1, \qquad \beta_2=1, \qquad \beta_3=\cdots=\beta_n=0.$$

In the same way, the v_k determine themselves, and so on. In brief, we realize that our group: $X_1 f,\dots,X_r f$ comprises an infinitesimal transformation of the form:

$$K''=x_1^s\,p_1+A_2\,x_1^{s-1}\,x_2\,p_2+\cdots+A_n\,x_1^{s-1}\,x_n\,p_n+\cdots.$$

By forming a Combination of K'' with $p_1+\cdots$, we get:

$$\begin{aligned}&s\,x_1^{s-1}\,p_1+(s-1)\,A_2\,x_1^{s-2}\,x_2\,p_2+\\&\qquad+\cdots+(s-1)\,A_n\,x_1^{s-2}\,x_n\,p_n+\cdots=L,\end{aligned}$$

and hence our group comprises an infinitesimal transformation, namely $\big[L,\,K''\big]$, which is of the form:

$$s\,x_1^{2s-2}\,p_1+\eta_2\,p_2+\cdots+\eta_n\,p_n+\cdots.$$

But now, since $2s-2$ should not be larger than s and since, on the other hand, s is larger than 1, it follows that:

$$s=2,$$

so that we have:

$$K''=x_1^2\,p_1+A_2\,x_1\,x_2\,p_2+\cdots+A_n\,x_1\,x_n\,p_n+\cdots.$$

Furthermore, we get:

$$\big[x_1\,p_i+\cdots,\,K''\big]=(A_i-1)\,x_1^2\,p_i+\cdots.$$

If now A_i were different from 1, then we would obtain, one after the other, both transformations:

$$\begin{aligned}&\big[x_1^2\,p_i+\cdots,\,x_i\,p_1+\cdots\big]=x_1^2\,p_1-2x_1x_i\,p_i+\cdots,\\&\big[x_1^2\,p_1-2x_1x_i\,p_i+\cdots,\,x_1^2\,p_i+\cdots\big]=4x_1^3\,p_i+\cdots.\end{aligned}$$

But since no infinitesimal transformation of third order should be found, all the A_i are equal to 1. Therefore K'' has the form:

$$x_1\,\big(x_1\,p_1+x_2\,p_2+\cdots+x_n\,p_n\big)+\cdots$$

and thereupon lastly, by forming a Combination with $x_i\, p_1 + \cdots$, we find generally:

$$x_i\,\big(x_1\, p_1 + \cdots + x_n\, p_n\big) + \cdots.$$

Consequently, if a group $X_1 f, \dots, X_r f$ of the nature required on p. 620 comprises infinitesimal transformations of order higher than the first, then it comprises only such transformations that are of order two, and in fact in any case, n of the form:

$$x_i\,\big(x_1\, p_1 + \cdots + x_n\, p_n\big) + \cdots \qquad (i = 1 \cdots n).$$

If we add up all the n transformations:

$$\Big[p_i + \cdots,\, x_i \textstyle\sum_{k=1}^n x_k\, p_k + \cdots\Big] = x_i\, p_i + \textstyle\sum_{k=1}^n x_k\, p_k + \cdots,$$

together, then we obtain the transformation $x_1\, p_1 + \cdots + x_n\, p_n + \cdots$. Hence if the group $X_1 f, \dots, X_r f$ contains infinitesimal transformations of second order, then the associated linear homogeneous group $L_1 f, L_2 f, \dots$ is the general linear homogeneous group. Or conversely:

The group $X_1 f, \dots, X_r f$ never contains infinitesimal transformations of order higher than the first, when the associated group $L_1 f, L_2 f, \dots$ is the special linear homogeneous group.

For short, we write the infinitesimal transformation $x_i\,\big(x_1\, p_1 + \cdots + x_n\, p_n\big) + \cdots$ in the form $H_i + \cdots$, with the understanding that H_i denotes the terms of second order. Now, it is conceivable that in one group $X_1 f, \dots, X_r f$, except the n transformations $H_k + \cdots$, there may appear others of second order. Let such a transformation be, for instance:

$$\tau_1\, p_1 + \cdots + \tau_n\, p_n + \cdots = \mathsf{T} + \cdots,$$

where the τ_i mean homogeneous functions of second order in $x_1, \dots, x_n$ and T the sum $\sum \tau_k\, p_k$. Then the expression $\big[H_i,\, \mathsf{T}\big]$ must obviously vanish, since the lowest term of this bracket represents an infinitesimal transformation of third order. Consequently, $\tau_1, \dots, \tau_n$ satisfy the equation:

$$\left[x_i \sum_{k=1}^n x_k\, p_k,\, \sum_{j=1}^n \tau_j\, p_j\right] = 0;$$

this equation decomposes into the following n equations:

$$x_i \left(\sum_{k=1}^n x_k\, \frac{\partial \tau_j}{\partial x_k} - \tau_j\right) - x_j\, \tau_i = 0$$

and from this it follows, aside from the natural equation:

$$\sum_{k=1}^n x_k\, \frac{\partial \tau_j}{\partial x_k} = 2\, \tau_j$$

also $x_i\,\tau_j - x_j\,\tau_i = 0$. Therefore the τ_i and T have the form:

$$\tau_i = \sum_{k=1}^{n} \alpha_k\,x_k x_i, \qquad\qquad \mathsf{T} = \sum_{k=1}^{n} \alpha_k\,H_k.$$

Thus we have proved that a group: $X_1 f, \dots, X_r f$ of the indicated constitution can comprise no second order infinitesimal transformation apart from the n transformations $H_k + \cdots$.

In total, we therefore have the following cases:

If the linear homogeneous group $L_1 f, L_2 f, \dots$ is the general linear homogeneous group, then the concerned group $X_1 f, \dots, X_r f$ comprises exactly n infinitesimal transformations of zeroth order:

$$P_i = p_i + \cdots \qquad\qquad {\scriptstyle (i=1\cdots n)}$$

and n^2 of first order:

$$T_{ik} = x_i\,p_k + \cdots \qquad\qquad {\scriptstyle (i,k=1\cdots n)}.$$

Either transformations of higher order do not occur at all, or there are extant the following n:

$$S_i = x_i\,\big(x_1\,p_1 + \cdots + x_n\,p_n\big) + \cdots \qquad\qquad {\scriptstyle (i=1\cdots n)}.$$

If the group: $L_1 f, L_2 f, \dots$ is the special linear homogeneous group, then the group: $X_1 f, \dots, X_r f$ comprises exactly n infinitesimal transformations of zeroth order:

$$P_i = p_i + \cdots \qquad\qquad {\scriptstyle (i=1\cdots n)}$$

and in addition, $n^2 - 1$ of first order:

$$T_{ik} = x_i\,p_k + \cdots, \qquad T_{ii} - T_{kk} = x_i\,p_i - x_k\,p_k + \cdots \qquad\qquad {\scriptstyle (i \gtrless k=1\cdots n)}$$

but none of higher order.

We will treat these three cases one after the other. First, we bring the Relations [brackets] between the infinitesimal transformations, and then, these transformations themselves, to a form which is as simple as possible. On the occasion, we remark that among the three cases, the first two are already completed by the results of section 151 on p. 616 sq. Nonetheless, we maintain that it is advisable to treat these two cases in detail.

§ **153**. The first case, where n transformations of zeroth order and n^2 of first order appear, is the simplest one.

We can indicate without effort the Relations between the n^2 infinitesimal transformations. They are:

$$\big[T_{ik},\, T_{\nu\pi}\big] = \varepsilon_{k\nu}\,T_{i\pi} - \varepsilon_{\pi i}\,T_{\nu k},$$

where ε_{ik} vanishes as soon as i and k are distinct, whereas ε_{ii} has the value 1. In particular, it is of importance that each expression:

$$\left[T_{ik},\,\sum_{\nu=1}^{n} T_{\nu\nu}\right]$$

vanishes. Further, if, for reasons of abbreviation, we introduce the symbol:

$$U = \sum_{\nu=1}^{n} T_{\nu\nu},$$

there exist Relations of the form:

$$\left[P_i,\,U\right] = P_i + \sum_{\nu=1}^{n}\sum_{\pi=1}^{n} \alpha_{\nu\pi}\,T_{\nu\pi},$$

or, when we introduce as new P_i the right-hand side: $\left[P_i,\,U\right] = P_i$.

Lastly, there are Relations of the form:

$$\left[P_i,\,T_{kj}\right] = \varepsilon_{ik}\,P_j + \sum\sum \beta_{\nu\pi}\,T_{\nu\pi},$$
$$\left[P_i,\,P_k\right] = \sum \gamma_\nu\,P_\nu + \sum\sum \delta_{\nu\pi}\,T_{\nu\pi}.$$

But the Jacobian identities:

$$\left[\left[P_i,\,T_{kj}\right],U\right] - \left[P_i,\,T_{kj}\right] = 0,$$
$$\left[\left[P_i,\,P_k\right],U\right] - 2\left[P_i,\,P_k\right] = 0$$

show immediately that all constants β, γ, δ vanish, whence we have:

$$\left[P_i,\,T_{kj}\right] = \varepsilon_{ik}\,P_j, \qquad \left[P_i,\,P_k\right] = 0.$$

As a result, all the Relations between the infinitesimal transformations of our group are known.

The r infinitesimal transformations $P_i = p_i + \cdots$ generate a simply transitive group which has *the same composition*[5] (Translator's note: Namely here: they both have the same, in fact vanishing, Lie brackets.) [GLEICHZUSAMMENGESETZ IST] as the group $p_1', \dots, p_n'$ and is hence also equivalent to it (Chap. 19, p. 339, Prop. 1).[6] (Translator's note: The so-called *Frobenius theorem* is needed in the proposition below. Classically (cf. [1]), one performs a preliminary reduction to a commuting system of vector fields, reducing the proof to the following concrete

Proposition. *If the $r \leqslant s$ independent infinitesimal transformations:*

$$X_k f = \sum_{i=1}^{s} \xi_{ki}(x_1,\dots,x_s)\,\frac{\partial f}{\partial x_i} \qquad {\scriptstyle (k=1\cdots r)}$$

stand in the Relationships:

$$\left[X_i,\,X_k\right] = 0 \qquad {\scriptstyle (i,k=1\cdots r)}$$

without being tied together by a linear relation of the form:

$$\sum_{k=1}^{n} \chi_k(x_1,\dots,x_s)\,X_s f = 0,$$

then they generate an r-term group which is equivalent to the group of translations:

$$Y_1 f = \frac{\partial f}{\partial y_1},\dots,\, Y_r f = \frac{\partial f}{\partial y_r}.$$

(Namely, in a neighborhood of a generic point, there is a local diffeomorphism $x \mapsto y = y(x)$ straightening each X_k to $Y_k = \frac{\partial}{\partial y_k}$.)) So we can introduce such new variables $x_1',\dots,x_n'$ achieving that:

$$P_i = p_i' \qquad {\scriptstyle (i=1\cdots n)}.$$

The form $\xi_1\, p_1' + \cdots + \xi_n\, p_n'$ which U takes in the new variables determines itself from the Relations $\big[P_i,\, U\big] = P_i$; the same Relations yield:

$$U = \sum_{k=1}^{n} (x_k' + \alpha_k)\, p_k'$$

and, when $x_k' + \alpha_k$ is introduced as new x_k, which does not change the form of the P_i, one achieves $U = \sum x_k\, p_k$. From the Relations:

$$\big[P_i,\, T_{kj}\big] = \varepsilon_{ik}\, P_j$$

one finds in the same way:

$$T_{kj} = x_k\, p_j + \sum_{\nu=1}^{n} \alpha_{kj\nu}\, p_\nu;$$

but since $\big[T_{kj},\, U\big]$ must vanish, all the $\alpha_{kj\nu}$ are equally null. Therefore we have the group:

$$P_i = p_i, \qquad T_{ik} = x_i\, p_k \qquad {\scriptstyle (i,\,k=1\cdots n)},$$

that is to say, *all groups which belong to the first case are equivalent to the general linear group of the manifold* $x_1,\dots,x_n$.

§ 154. We now come to the second case, where, apart from the n transformations of zeroth order $P_i = p_i + \cdots$ and the n^2 of first order $T_{ik} = x_i\, p_k + \cdots$, there still appear the n transformations of second order:

$$S_i = x_i\,\big(x_1\, p_1 + \cdots + x_n\, p_n\big) + \cdots.$$

The following Relations are obtained without any effort:

$$[S_i, S_k] = 0, \qquad [T_{ik}, S_j] = \varepsilon_{kj} S_i, \qquad [U, S_j] = S_j.$$

Moreover, there are equations of the form:

$$[T_{ik}, U] = \sum_{j=1}^{n} \alpha_{ikj} S_j.$$

If at first we suppose that i and k are not all both equal to n, then we can introduce $T_{ik} + \sum \alpha_{ikj} S_j$ as new T_{ik} and obtain correspondingly the Relation: $[T_{ik}, U] = 0$ for the concerned values of i and k. Since in addition $[U, U] = 0$ we have generally:

$$[T_{ik}, U] = 0 \qquad (i, k = 1 \cdots n).$$

Further, one has:

$$[T_{ik}, T_{\nu\pi}] = \varepsilon_{k\nu} T_{i\pi} - \varepsilon_{\pi i} T_{\nu k} + \sum_{j=1}^{n} \beta_j S_j;$$

but the identity:

$$[[T_{ik}, T_{\nu\pi}], U] = 0$$

enables us to see that all β_j vanish, so we have:

$$[T_{ik}, T_{\nu\pi}] = \varepsilon_{k\nu} T_{i\pi} - \varepsilon_{\pi i} T_{\nu k}.$$

From the relation:

$$[P_i, U] = P_i + \sum_{\nu=1}^{n} \sum_{\pi=1}^{n} \gamma_{\nu\pi} T_{\nu\pi} + \sum_{\nu=1}^{n} \delta_\nu S_\nu$$

when:

$$P_i + \sum_{\nu=1}^{n} \sum_{\pi=1}^{n} \gamma_{\nu\pi} T_{\nu\pi} + \frac{1}{2} \sum_{\nu=1}^{n} \delta_\nu S_\nu$$

is introduced as a new P_i, we get:

$$[P_i, U] = P_i.$$

Furthermore, we have:

$$[P_i, S_k] = \varepsilon_{ik} U + T_{ki} + \sum_{\nu=1}^{n} \lambda_\nu S_\nu;$$

but if we form the identity:

$$[[P_i, S_k], U] + [P_i, S_k] - [P_i, S_k] = 0,$$

we then find that the λ_ν are null; consequently we have:

$$\left[P_i,\, S_k\right] = \varepsilon_{ik}\, U + T_{ki}.$$

Finally, there are relations of the form:

$$\begin{aligned}\left[P_i,\, T_{kj}\right] &= \varepsilon_{ik}\, P_j + \sum\sum \alpha_{\nu\pi}\, T_{\nu\pi} + \sum \lambda_\nu\, S_\nu,\\ \left[P_i,\, P_k\right] &= \sum g_\nu\, P_\nu + \sum\sum h_{\nu\pi}\, T_{\nu\pi} + \sum l_\nu\, S_\nu;\end{aligned}$$

but also here all constants vanish because of the identities:

$$\begin{aligned}\left[\left[P_i,\, T_{kj}\right],\, U\right] - \left[P_i,\, T_{kj}\right] &= 0,\\ \left[\left[P_i,\, P_k\right],\, U\right] - 2\left[P_i,\, P_k\right] &= 0.\end{aligned}$$

Therefore we have:

$$\left[P_i,\, T_{kj}\right] = \varepsilon_{ik}\, P_j, \qquad \left[P_i,\, P_k\right] = 0.$$

All the relations between the infinitesimal transformations of our group are now determined. One will observe that the infinitesimal transformations P_i, T_{ik} generate a subgroup which is of the form considered in the first case and hence takes the form:

$$P_i = p_i, \qquad T_{ik} = x_i\, p_k \qquad {\scriptstyle (i,k=1\cdots n)}$$

by an appropriate choice of variables. In the new variables, the infinitesimal transformations S_i are, say, equal to $\sum \xi_{ik}\, p_k$, where the ξ_{ik} satisfy the relations:

$$\left[P_\nu,\, S_i\right] = \varepsilon_{\nu i}\, U + T_{i\nu}, \qquad \left[U,\, S_i\right] = S_i.$$

We find from this:

$$\frac{\partial \xi_{ik}}{\partial x_\nu} = \varepsilon_{\nu i}\, x_k + \varepsilon_{\nu k}\, x_i, \qquad \sum_{\nu=1}^{n} x_\nu\, \frac{\partial \xi_{ik}}{\partial x_\nu} = 2\, \xi_{ik},$$

therefore $\xi_{ik} = x_i x_k$ and:

$$S_i = x_i\big(x_1\, p_1 + \cdots + x_n\, p_n\big).$$

Consequently we have the group:

$$P_i = p_i, \quad T_{ik} = x_i\, p_k, \quad S_i = x_i\big(x_1\, p_1 + \cdots + x_n\, p_n\big) \qquad {\scriptstyle (i,k=1\cdots n)};$$

that is to say, *all groups which belong to the second case are equivalent to the general projective group of the manifold* $x_1, \dots, x_n$.

§ 155.

There remains the third and last case with n infinitesimal transformations of zeroth order: $P_i = p_i + \cdots$ and $n^2 - 1$ of first order:

$$T_{ik} = x_i p_k + \cdots, \quad T_{ii} - T_{kk} = x_i p_i - x_k p_k + \cdots \qquad (i, \gtrless k = 1 \cdots n),$$

while transformations of higher order do not occur.

It is convenient to replace the infinitesimal transformations $T_{ii} - T_{kk}$ by one of the form: $\alpha_1 T_{11} + \cdots + \alpha_n T_{nn}$, where the α_i are arbitrary constants, though subjected to the condition $\sum_i \alpha_i = 0$. Then to begin with, the following Relations hold:

$$\left[T_{ik}, T_{\nu\pi}\right] = \varepsilon_{k\nu} T_{i\pi} - \varepsilon_{\pi i} T_{\nu k} \qquad (i \gtrless k, \nu \gtrless \pi),$$

$$\left[T_{ik}, \sum_{\nu=1}^{n} \alpha_\nu T_{\nu\nu}\right] = (\alpha_k - \alpha_i) T_{ik}, \qquad \left[T_{ii} - T_{kk}, \sum_{\nu=1}^{n} \alpha_\nu T_{\nu\nu}\right] = 0.$$

Further, there is an equation of the form:

$$\left[P_i, \sum_{\nu=1}^{n} \alpha_\nu T_{\nu\nu}\right] = \alpha_i P_i + \sum_{k=1}^{n} \sum_{j=1}^{n} \lambda_{ikj} T_{kj} \qquad (\textstyle\sum_{k=1}^{n} \lambda_{ikk} = 0).$$

Therefore if we set:

$$P_i' = P_i + \sum_{k=1}^{n} \sum_{j=1}^{n} l_{ikj} T_{kj} \qquad (\textstyle\sum_{k=1}^{n} l_{ikk} = 0),$$

we get:

$$\left[P_i', \sum_{\nu=1}^{n} \alpha_\nu T_{\nu\nu}\right] = \alpha_i P_i' + \sum_{k=1}^{n} \sum_{j=1}^{n} \left\{\lambda_{ikj} - (\alpha_i + \alpha_k - \alpha_j)\right\} T_{kj}.$$

Now, we imagine that a completely determined system of values is chosen for $\alpha_1, \dots, \alpha_n$ which satisfies the condition $\sum \alpha_i = 0$ and has in addition the property that no expression $\alpha_i + \alpha_k - \alpha_j$ vanishes, a demand which can always be satisfied. Under these assumptions, we can choose the l_{ikj} so that the equations:

$$\lambda_{ikj} - (\alpha_i + \alpha_k - \alpha_j) l_{ikj} = 0$$

are satisfied, where the condition $\sum_k l_{ikk} = 0$ is automatically fulfilled. Consequently, we obtain after utilizing again the initial designation:

$$\left[P_i, \sum_{\nu=1}^{n} \alpha_\nu T_{\nu\nu}\right] = \alpha_i P_i \qquad (\textstyle\sum_{\nu=1}^{n} \alpha_\nu = 0).$$

If the expression:

$$\left[P_i, \sum_{k=1}^{n} \beta_k T_{kk}\right] = \beta_i P_i + \sum_{\nu=1}^{n} \sum_{\pi=1}^{n} g_{\nu\pi} T_{\nu\pi}$$

$$(\textstyle\sum \beta_k = \sum g_{\nu\nu} = 0)$$

is inserted into the identity:

$$\left[\left[P_i,\, \sum_{k=1}^n \beta_k T_{kk}\right],\, \sum_{k=1}^n \alpha_k T_{kk}\right] - \alpha_i \left[P_i,\, \sum_{k=1}^n \beta_k T_{kk}\right] = 0,$$

then we get:

$$\sum_{\nu=1}^n \sum_{\pi=1}^n (\alpha_\pi - \alpha_\nu - \alpha_i)\, g_{\nu\pi}\, T_{\nu\pi} = 0.$$

Because of the nature of the α, it follows from this that all $g_{\nu\pi}$ are equally null, and therefore the equation:

$$\left[P_i,\, \sum_{k=1}^n \beta_k T_{kk}\right] = \beta_i T_i$$

holds for all systems of values $\beta_1, \dots, \beta_n$ which satisfy the condition $\sum \beta_k = 0$. Furthermore, there is a relation of the form:

$$\left[P_i,\, T_{kj}\right] = \varepsilon_{ik} P_j + \sum_{\nu=1}^n \sum_{\pi=1}^n h_{\nu\pi}\, T_{\nu\pi}$$

$$(k \gtrless j,\, \textstyle\sum_\nu h_{\nu\nu} = 0).$$

The identity:

$$\left[\left[P_i,\, T_{kj}\right],\, \sum_{\tau=1}^n \beta_\tau T_{\tau\tau}\right] - (\beta_i + \beta_j - \beta_k)\left[P_i,\, T_{kj}\right] = 0$$

therefore takes the form:

$$\varepsilon_{ik} (\beta_k - \beta_i) P_j + \sum_{\nu=1}^n \sum_{\pi=1}^n (\beta_\pi - \beta_\nu - \beta_i + \beta_k - \beta_j)\, h_{\nu\pi}\, T_{\nu\pi} = 0;$$

consequently the $h_{\nu\pi}$ must vanish, since the β_ν, while disregarding the condition $\sum_\nu \beta_\nu = 0$, are completely arbitrary:

$$\left[P_i,\, T_{kj}\right] = \varepsilon_{ik} P_j \qquad (k \gtrless j).$$

Finally, the relations:

$$\left[P_i,\, P_k\right] = \sum_{\nu=1}^n m_\nu P_\nu + \sum_{\nu=1}^n \sum_{\pi=1}^n m_{\nu\pi}\, T_{\nu\pi} \qquad \left(\textstyle\sum_{\nu=1}^n m_{\nu\nu} = 0\right)$$

are still to be examined. By calculating the identity:

$$\left[\left[P_i,\, P_k\right],\, \sum_{\tau=1}^n \beta_\tau T_{\tau\tau}\right] - (\beta_i + \beta_k)\left[P_i,\, P_k\right] = 0,$$

we find:

$$\sum_{\nu=1}^n (\beta_\nu - \beta_i - \beta_k)\, m_\nu P_\nu + \sum_{\nu=1}^n \sum_{\pi=1}^n (\beta_\pi - \beta_\nu - \beta_i - \beta_k)\, m_{\nu\pi}\, T_{\nu\pi} = 0;$$

because of the arbitrariness of the β_ν, all the m_ν and $m_{\nu\pi}$ must be equally null. Consequently, we have:

$$[P_i, P_k] = 0$$

and we therefore know all Relations between the infinitesimal transformations of our group.

In the same way as in the first case, we bring $P_1, \dots, P_n$ by means of an appropriate choice of variables to the form:

$$P_i = p_i \qquad (i=1\cdots n).$$

From this, by proceeding as in the end of section 153, we conclude that *all groups which belong to the third case are equivalent to the special linear group of the manifold* $x_1, \dots, x_n$.

By unifying the results found above we therefore obtain the

Theorem 112. *If a transitive group G in n variables is constituted so that all of its transformations which leave invariant a point in general position transform the line-elements passing through the point by means of the general or of the special linear homogeneous group* $L_1 f, L_2 f, \dots$, *then G is equivalent either to the general projective group, or to the general linear group, or to the special linear group, in n variables*[2].

If we call an r-term group of the space $x_1, \dots, x_n$ *m-fold transitive* when it contains at least one transformation which transfers any m given points in mutually general position to m other arbitrary given points in general position, then we can now easily prove *firstly*, that m is always $\leqslant n+2$ and *secondly*, that every group for which $m = n+2$ is equivalent to the general projective group of n-fold extended space.

Indeed, if the infinitesimal transformations: $X_1 f, \dots, X_r f$ in the variables $x_1, \dots, x_n$ generate an m-fold transitive group, then it immediately stands to reason that the linear homogeneous group $L_1 f, L_2 f, \dots$ assigned to a point x_k^0 in general position transforms the ∞^{n-1} line-elements passing through this point by means of an $(m-1)$-fold transitive projective group. But now, since the general projective group of an $(n-1)$-fold extended space is known to be $(n+1)$-transitive, we then realize that:

$$m - 1 \leqslant n + 1 \qquad \text{and hence:} \qquad m \leqslant n+2.$$

Therefore, the following holds.

Theorem 113. *A finite continuous group in n variables is at most* $(n+2)$*-fold transitive.*

If an r-term group: $X_1 f, \dots, X_r f$ in n variables is exactly $(n+2)$-transitive, then as observed earlier, the group $L_1 f, L_2 f, \dots$ assigned to a point x_k^0 in general position

[2] Lie, Archiv for Math., Vol. 3, Christiania 1878; cf. also Math. Ann. Vol. XVI, Vol. XXV and Götting. Nachr. 1874, p. 539.

is the general or the special linear homogeneous group in n variables; consequently, the group: $X_1 f, \dots, X_r f$ is equivalent either to the general projective group, or to the general linear group, or to the special linear group in n variables. But now, because only the first-mentioned among these three groups is $(n+2)$-transitive, we obtain the

Theorem 114. *If an r-term group in n variables is* $(n+2)$*-transitive, then it is equivalent to the general projective group of n-fold extended space.*

In the second and third volumes, among other things, a series of studies which are analogous to those conducted in this chapter shall be undertaken.

References

1. Stormark, O.: *Lie's structural approach to PDE systems*, Encyclopædia of mathematics and its applications, vol. 80, Cambridge University Press, Cambridge, 2000, xv+572 pp.

Chapter 30
Glossary of significantly used words

Symbols

n-FACH, n-fold
n-FLACH, n-frame
q-GLIEDRIG VOLLSTÄNDIG SYSTEM, q-term complete system

A

ABBILDEN, to represent
ABHÄNGIGSKEITSVERHÄLTNISS, dependence relationship
ABKÜRZEN, to abbreviate
ABSCHNITT, section
ABSCISSE, abscissa
ADJUNGIRTE GRUPPE, adjoint group
ÄHNLICH, similar
ÄHNLICHKEIT, similarity
ÄHNLICHKEITSTRANSFORMATION, similitude
ALGEBRAISCHE DISCUSSION, algebraic discussion
ANALYTISCH, analytic
ANALYTISCHE BEGRÜNDUNG, analytic explanation
ANFANGSWERTH, initial value
ANGEBEN, to indicate
ANGEHÖREN, to belong to
ANNAHME, assumption
ANORDNEN, to order
ANSATZ, approach
ANSCHAULICH, graphical
ANSCHAULICH AUFFASUNG, graphical interpretation
ANSCHAULICHKEIT, graphical nature
ANZEIGEN, to indicate
ASSOCIATIVE GESETZ, associative rule
ASYSTATISCH, asystatic
AUFEINANDERFOLGE, succession
AUFFASSEN, to interpret
AUFFINDUNG, finding
AUFFÜHREN, to list
AUFGABE, problem
AUFLÖSUNG, solution
AUFSTELLEN, to set up
AUFSTELLUNG, list
AUFZÄHLUNG, enumeration
AUSDEHNEN AUF, to extend to, generalize to
AUSDEHNUNG, (complete) extension
AUSFÜRLICH, in detail
AUSGEARTET, degenerate
AUSGEZEICHNET, excellent
AUSRECHNUNG, calculation
ÄUSSERST WICHTIG, utmost importance
AUSSPRECHEN, to enunciate
AUSSTELLEN, to show
AUSZEICHNEN, to distinguish

B

BAHNCURVE, integral curve
BEDENKEN, to consider
BEDENKEN, thought
BEDEUTEN, to mean
BEDEUTUNG, meaning, sense, signification
BEDIENEN SICH, to make use of
BEDINGEN, to give rise to
BEDINGUNGSGLEICHUNGEN, condition-equations
BEFRIEDIGEN, to satisfy
BEGRIFF, concept

S. Lie, *Theory of Transformation Groups I*,
DOI 10.1007/978-3-662-46211-9

BEGRIFFLICH SINN, conceptual sense
BEGRIFFLICHE INHALT, conceptual content
BEGRIFFSBILDUNG, forming of concepts
BEGRÜNDEN, to give reasons for
BEGRÜNDUNG, reason
BEHALTEN, to keep
BEHAUPTUNG, assertion
BEHUFE, for this purpose
BEKANNTLICH, as everybody knows
BEKOMMEN, to receive
BELIEBIG, arbitrary
BEMERKEN, to remark
BEMERKENSWERTH IST, it is noteworthy that
BENACHBARCHT, neighboring
BENENNUNG, naming, nomenclature
BENUTZEN, to use
BEQUEM, convenient
BERECHTIGT, legitimate
BEREICH, region
BERÜCKSICHTIGEN, to take into account
BERÜCKSICHTIGUNG, consideration
BERUHEN, to be based on
BERÜHREN, to come into contact with
BERÜHRUNGSTRANSFORMATION, contact transformation
BESCHAFFEN, to procure
BESCHAFFENHEIT, constitution
BESCHÄFTIGEN SICH, to occupy oneself in (with) doing something
BESCHRÄNKUNG, restriction
BESCHREIBEN, to describe
BESITZEN, to possess
BESONDER, special, particular, specific
BESONDERS, particularly, especially, notably,
BESPROCHENE (OBEN), discussed (above)
BESTÄTIGEN, to confirm
BESTEHEN, existence
BESTEHEN AUS, to be made of (up)
BESTIMMEN, to determine
BESTIMMT, determinate, definite
BESTIMMUNGSSTÜCKE, determination pieces
BETONEN, to stress, emphasize, underline
BETRACHTEN, to consider
BETRACHTUNG, consideration
BETREFFEND, concerned, in question
BEWEGUNG, movement
BEWEIS, proof
BEWEISEN, to prove, to demonstrate
BEWUSSTEN, said, in question
BEZEICHNUNG, notation, naming
BEZIEHUNG, relationship
BILDEN, to form
BILDPUNKT, image-point
BÜNDEL, bundle
BÜSCHEL, bundle, pencil

C

COLLINEATIONEN, collineation
COMBINATION, combination
CONSEQUENTERWEISE, consequently
COORDINATENECKPUNKT, corner point of coordinates

D

DARSTELLEN, to represent
DECKEN SICH, to coincide
DENKBAR, conceivable
DETERMINANT, determinant
DEUTEN, to interpret
DEUTLICH, clearly, distinctly
DIFFERENTIALINVARIANT, differential invariant
DREIFACH AUSGEDEHNTEN, thrice-extended
DUALISTISCH, dualistic
DUALITÄT, duality
DURCH ANALYTISCHE METHODEN, by means of the analytic method
DURCH DETERMINANTENBILDUNG, by forming determinants
DURCH UNENDLICHMALIGE WIEDERHOLUNG, by repeating infinitely many times
DURCHFÜHRUNG, realization
DURCHSICHTIG, transparent

E

EBEN, straight, even
EIGENSCHAFT, property
EIGENTHÜMLICH, characteristic
EINDEUTIG, univalent
EINFACH, simple
EINFACH AUSGEDEHNTEN, once-extended
EINFACH TRANSITIV, simply transitive
EINFÜHREN, to introduce
EINGEHEN AUF, to deal with
EINGEKLAMMERT, in brackets
EINSETZEN, to insert
EINTHEILEN, to distribute
EINTHEILUNG, division, classification
EINTHEILUNGSGRUNDE, principle of classification
EINTRETEN, to occur
EINZELN, individual
ENDERGEBNISS, final result
ENTHALTEN, to contain
ENTSPRECHEN, to correspond to
ENTSTANDEN, to be generated, to be produced
ENTSTEHEN, to come into being

ENTSTEHUNG, generation, production
ENTWICKELN, to develop
ENTWICKLUNG, development
ENTWICKELUNGSCOEFFICIENT, expansion coefficient
ERBRINGEN, to produce
ERFORDERN, to require
ERFÜLLEN, to fulfill, to satisfy
ERGEBEN, to yield, to produce, to result in
ERGEBNISS, to result
ERHALTEN, to get, to receive, to obtain
ERHELLEN, to become clear, to be evident
ERINNERN SICH, to remember
ERKENNEN, to recognize
ERKLÄREN, to explain
ERLEDIGEN, to settle, to carry out
ERLEICHTERUNG, simplification
ERREICHEN, to insure
ERSCHEINEN, to appear, to become visible
ERSCHÖPFEN, to exhaust
ERSETZEN DURCH, to replace by, to substitute for
ERSICHTLICH, clear, obvious, evident
ERWÄHNEN, to mention
ERWÄHNUNG, mention of
ERWEISEN, to show, to demonstrate, to establish
ERWEISEN SICH, to prove to be
ERWEITERN, to extend
ERWEITERUNG, extension
ERZHIELEN, to achieve
ERZEUGEN, to produce, to generate
ETWA, for instance, for example

F

FASSUNG, version, interpretation
FESTSETZEN, to fix
FESTSETZUNG, fixing of terminology
FESTSTELLEN, to find out
FIGUR, figure (number of manifolds), diagram
FLÄCHENELEMENT, surface element
FLIESSEN, to flow
FLÜSSIGKEITSTHEILCHENS, velocity of the fluid particle
FOLGENDERMASSEN, as follows
FOLGLICH, so, consequently, therefore
FÖRDERN, to produce
FORDERUNG (ERFÜLLEN), requirement (meet, satisfy)
FORMELL, formally
FORTSCHREITUNGSRICHTUNG, direction of progress
FRAGE, question
FRAGEN NACH, to ask for
FRAGESTELLUNG, way of putting the question
FUNCTIONALDETERMINANT, functional determinant

G

GANZE RATIONALE FUNCTION, entire rational function (polynomial)
GATTUNG, kind
GEBIETE, domain
GEGENSEITIG, mutually
GELANGEN, to reach, to get to, to arrive at
GENAU GENOMMEN, strictly speaking
GERADE, straight
GERADE, straight line
GESICHTSPUNKT, point of view
GESONDERT, separate
GESTALT, shape, form
GESTALTEN SICH, to take shape
GESTATTEN, to admit
GETRENNT, separate
GEWINNEN, to gain
GEWISS, certain
GEWÖHNLICH (WIE), as usual
GLEICHBERECHTIGT, conjugate
GLEICH NULL, identically null
GLEICHWERTIG, equally good
GLEICHZEITIG, simultaneously, at the same time
GLEICHZUSAMMENGESETZT, equally composed, identically compound as
GLEICHZUSAMMENGESETZTSEINS, property of being equally composed
GRUND, foundation, reason, background
GRUNDSATZ, principle
GRUNDZÜGE, foundations
GRUPPENEIGENSCHAFT, group property
GRUPPENFORM, form of group

H

HAUPTSACHE, main thing
HERLEITEN, to derive
HERLEITUNG, derivation
HERRÜHREN VON, to come from
HERVORGEHEN AUS, to result from, to come from
HERVORHEBEN, to emphasize, to underline, to stress
HERVORRAGEND, outstanding
HINAUSKOMMEN, to amount to the same, to come down to
HINEINFALLEN, to fall into
HINREICHEND, sufficient
HINSCHREIBEN, to write down
HINSICHTLICH, concerning, regarding

HINWEISEN, point out to
HINZUFÜGEN, to add
HINZUNEHMEN, to add

I

IM EINZELNEN, in(to) details
IM ENDLICHEN, in the domain of the finite
IM GANZEM, in sum, in total, in all
IM GRUNDE, fundamentally
IM WESENTLICHEN, essentially
IM WIDERSPRUCH STEHEN, to be inconsistent with, to contradict
IM ZUSAMMENHANGE WIEDERHOLEN, to recapitulate in cohesion
IMPRIMITIV, imprimitive
IN ALLER KÜRZE, very briefly
IN ANGRIFF NEHMEN, to tackle
IN COGREDIENTER WEISE, in cogredient way
IN DER PRAXIS, in practice
IN DER THAT, in fact, actually
IN EIN NEUES LICHT SETZEN, to place in a new light
IN EINZELNEN, in detail
IN FOLGE, as a consequence
INBEGRIFF, (complete) totality
INCREMENT, increment
INDIVIDUEN, individuals
INFOLGEDESSEN, as a result (of this), consequently
INHALT, content(s)
INNERE GRUND, central reason
INS AUGE FASSEN, to consider
INSBESONDERE, particularly, especially, notably
INTEGRALGLEICHUNG, integral equation
IRREDUCIBEL, irreducible
ISOMORPH, isomorphic
ISOMORPHISMUS, isomorphism
IN VERBINDUNG BRINGEN, to associate with

J

JETZT, now

K

KATEGORIE, category
KEGEL, cone
KEGELSCHNITT, conic section
KENNEN, to know
KENNEN LERNEN, to get to know
KENNZEICHNEN, mark, identify
KLAMMER, bracket
KLAMMEROPERATION, bracketing $[\cdot,\cdot]$
KLARLEGEN, to explain, to clarify
KLARSTELLEN, to get straight, to clarify
KLEIDEN, to express, to put into words
KREIS, circle
KRITERIUM, criterion
KUGEL, sphere
KÜNFTIG, from now on
KURZ, shortly
KURZWEG, briefly

L

LEHRE, theory
LEICHT, easily
LEISTEN, to accomplish, to achieve
LIEFERN, to deliver
LIEGEN, to lie on
LINIENELEMENT, line-element

M

MAL, time
MANNIGFALTIGKEITSBETRACHTUNGEN, manifold considerations
MANNIGFALTIGKEITSLEHRE, theory of manifolds
MASSREGEL, rule
MATERIELL, materially
MENGE, amount
MERKWÜRDIG, curious
MIT AUSNAHME, with the exception of
MITBEKOMMEN, to catch, to understand
MITSCHREIBEN, to write down
MITSTEHENDLASSEND, 'leaving fixed'

N

NACHSTEHEND, following, below
NÄCHSTFOLGEND, following next
NACHWEIS, proof, evidence, certificate
NACHWEISEN, to prove, to show
NÄHER, closer, in more detail, more precisely
NAMENTLICH, especially, particularly, notably
NÄMLICH, namely, that is to say
NATURE DER SACHE, nature of things
NATURGEMÄSS, natural
NEBENBEI BEMERKT, incidentally
NENNER, denominator
NICHTMITSTEHENDLASSEND, 'without leaving fixed'
NOTHWENDIG, nessary
NUNMEHR, now, at present

O

OBENSTEHEND (DAS), the above
OBIG, above, above-mentioned
OFFENBAR, obviously, evidently

ÖFTERS, frequently
OHNE WEITERES, without any effort
OPERATIONSGRUPPE, group operation
ORDNEN SICH, to organize, to arrange

P

PAARWEISE, pairwise
PARAMETERGRUPPE, parameter group
PFAFFSCHE GLEICHUNG, Pfaffian equation
PREISSCHRIFT, prize-winning essay
PRIMITIV, primitive
PUNKTCOORDINATEN, point-coordinates
PUNKTFIGUR, point figure
PUNKTTRANSFORMATION, point transformation

R

RECHTWINKLING, right-angled
RECIPROCITÄTSVERHÄLTNISS, relationship of reciprocity
RECIPROKE, reciprocal
REDEWEISE, language
REGULÄR VERHALTEN, to behave regularly, to be smooth, to be regular
REIHE, row, line
REIHENFOLGE, order (of execution)
REPRÄSENTANT, representative
RICHTUNG, direction
RUHE, rest

S

SCHAAR, family
SCHLUSS ZIEHEN, to draw a conclusion
SCHNEIDEN, to cut
SCHNITT, cutting
SCHNITTMANNIGFALTIGKEIT, manifold section
SELBSTSTÄNDIG, independently
SELBSTVERSTÄNDLICH, naturally, of course, obviously
SOGENNANTE, so-called
STANDPUNKT, point of view, standpoint
STEHEN LASSEN, leave untouched
STELLE, position, point, place
STELLEN, to put, to set
STELLUNG, position
STIMMEN, to be right, to be correct
STRAHLBÜSCHEL, bundle of rays
STRECKE, (infinitely small) line
SUBSTITUTIONENTHEORIE, theory of substitutions
SUBSTITUTIONSDETERMINANT, substitution determinant
SUCHEN, to look for, to search for, to seek
SYMBOLIK, symbolism
SYNTHETISCHE BEGRÜNDUNG, synthetic explanation
SYSTATISCH, systatic

T

TABELLE, table
TERMINOLOGIE, terminology
THATSACHE, fact
THEILBAR, divisible
THEILEN, to distribute
THEILGEBIETE, subsidiary domain
THEILWEISE, partially
THEORETISCH, theoretical
TRANSFORMATION, transformation
TYPE, type
TYPENGATTUNG, kind of type

U

ÜBERALL, everywhere
ÜBEREINSTIMMEN, to correspond
ÜBEREINSTIMMUNG, correspondence
ÜBERGANG, transition
ÜBERGEHEN, to transfer to
ÜBERHAUPT, generally, altogether, actually
ÜBERLEGUNG, consideration, reflection
ÜBERSEHEN, to overview, to have a view of
ÜBERSICHTLICH, clear,
ÜBERTRAGEN AUF, to translate into
ÜBERZEUGEN, to convince
UMFASSEN, to embrace, to comprise
UMGEBUNG, neighborhood
UMGEKEHRT, inversely, conversely
UMKEHREN, reverse
UMSTAND, circumstance
UNABHÄNGIG VON, independent of
UNBESCHRÄNKT INTEGRABEL, unrestrictedly integrable
UNENDLICH FERNE EBENE, plane at infinity
UNTERSCHIED, difference, distinction
UNTERSCHEIDEN, to distinguish
UNTERSUCHEN, to study
UNTERSUCHUNG, research on (into), study
URSPRÜNGLICH, initial(ly), original(ly)

V

VERBINDEN, to connect
VERBINDUNG, connection
VERBINDUNGSLINIE, connection line
VERDEUTLICHEN, to make clear, to elucidate
VEREINFACHUNG, simplification
VERFAHREN, to proceed

VERFAHREN, method, process
VERFOLGEN, to follow, to pursue
VERGLEICHUNG, comparison
VERHÄLTNIS, ratio
VERHANDLUNG, essay, memoir
VERKÜRZTE, reduced
VERMÖGE, by virtue of
VERMUTEN, to presume
VERMUTHUNG, presumption
VERNÜKPFT, linked with
VERSCHIEDENE, different, various
VERSUCHEN, to try, to attempt
VERTAUSCHBAR, interchangeable
VERTAUSCHEN, to permute
VERTAUSCHUNG, permutation
VERVOLLSTÄNDIGEN, to complete
VERWEILEN AUF, to dwell on
VOLLKOMMEN, perfectly, absolutely
VOLLSTÄNDIG, completely, fully, totally
VON JETZ AB, from now on
VON VORNERHEIN, from the beginning
VOR ALLEN DINGEN, above all
VORANGEHENDEN, previous, above
VORAUSGESETZT, provided that
VORAUSSCHICKEN, to begin by mentioning that
VORAUSSETZEN, to assume
VORGANG, process
VORHABEN, to have in mind
VORHANDEN, extant, in existence, present
VORHANDENSEIN, existence
VORHER, beforehand
VORHERGEHEND, preceding
VORHIN, earlier on, a short while ago
VORIG, previous, former
VORKOMMEN, to appear, to happen, to occur
VORLÄUFIG, provisionally
VORLESUNG, lecture
VORNEHMEN, to carry out
VORSCHREIBEN, to prescribe, to stipulate
VORSTEHENDEN, preceding, above
VORSTELLEN SICH, to imagine, to represent
VORSTELLUNG, idea

W

WAHRHEIT, truth
WEG EINSCHLAGEN, to take a route
WEGEN, because of, on account of, due to
WESENTLICH, essentially, substantially
WICHTIG, important
WIE DER AUGENSCHEIN LEHRT, as an examination reveals
WIEDERHOLEN, to recapitulate
WILLKÜRLICH, arbitrary
WILLKÜRLICHKEIT, arbitrariness
WIRKLICH, really
WÜNSCHENWERTH, desirable

Z

ZAHLENWERTH, numerical value
ZEICHEN, sign, symbol
ZEIGEN, to show
ZEIGEN SICH, to come out
ZEITABSCHNITT, (infinitely small) time interval
ZERFALLEN, to decompose
ZERLEGEN, to decompose
ZERLEGUNG, decomposition (of space)
ZIEHL, goal, objective
ZU DIESEM ZWECKE, to this end
ZUERST, first of all,
ZUFÄLLIGKEIT, incidental character
ZUGEORDNET, attached
ZUGLEICH, at the same time
ZUGRUNDELEGUNG, taking as a basis
ZULASSEN, to allow
ZULÄSSIG, admissible
ZUNÄCHST, at first, to begin with
ZUORDNEN, to associate, to attach
ZUORDNUNG, association, classification
ZUR ABKÜRZUNG, for abbreviation
ZURÜCKFÜHREN, to reduce to
ZURÜCKKEHREN, to return
ZUSAMMENFASSEN, to sum up, to summarize, to recapitulate
ZUSAMMENFASSEND, in summary, to sum up
ZUSAMMENGESETZ, made up of, compound
ZUSAMMENHANG, coherence
ZUSAMMENHANGEN, to be linked, to be connected
ZUSAMMENNEHMEN, to take together
ZUSAMMENSETZUNG, composition
ZUSAMMENSTELLEN, to put together
ZU TAGE TRETEN, to come to light
ZUWACHS, increase
ZWEIFACH AUSGEDEHNTEN, twice-extended

Index

S. Lie, *Theory of Transformation Groups I*,
DOI 10.1007/978-3-662-46211-9

MIX
Papier aus verantwortungsvollen Quellen
Paper from responsible sources
FSC® C105338

If you have any concerns about our products,
you can contact us on
ProductSafety@springernature.com

In case Publisher is established outside the EU,
the EU authorized representative is:
Springer Nature Customer Service Center GmbH
Europaplatz 3, 69115 Heidelberg, Germany

Printed by Libri Plureos GmbH
in Hamburg, Germany